P9-COO-539

DESIGN OF STEEL STRUCTURES

DESIGN OF STEEL STRUCTURES

SECOND EDITION

Edwin H. Gaylord, Jr.
Professor Emeritus of Civil Engineering
University of Illinois
at Urbana

Charles N. Gaylord
Professor of Civil Engineering
University of Virginia

McGRAW-HILL BOOK COMPANY

New York St. Louis San Francisco Düsseldorf Johannesburg
Kuala Lumpur London Mexico Montreal New Delhi
Panama Rio de Janeiro Singapore Sydney Toronto

Library of Congress Cataloging in Publication Data

Gaylord, Edwin Henry.
Design of steel structures.

1. Building, Iron and steel. 2. Aluminum construction. I. Gaylord, Charles N., joint author.
II. Title.
TA684.G3 1972 624'.1821 71-39801
ISBN 0-07-023110-9

DESIGN OF STEEL STRUCTURES

2 3 4 5 6 7 8 9 0 K P K P 7 9 8 7 6 5 4 3

This book was set in Times New Roman. The editors were B. J. Clark and Andrea Stryker-Rodda; the designer was Richard Paul Kluga; and the production supervisor was Sally Ellyson. The drawings were done by John Cordes, J & R Technical Services, Inc.
The printer and binder was Kingsport Press, Inc.

CONTENTS

PREFACE

In the preparation of the second edition the authors have been guided by the philosophy that was the basis for the first edition. The arrangement of material is the same, except that beam columns are now treated in Chapter 4 on columns. Since the behavior of structural components cannot be discussed without consideration of post-yielding phenomena, elements of plastic analysis and design are introduced at an earlier stage in the treatment of columns, beams, beam columns and connections. In addition, the chapter on plastic design has been expanded to include a treatment of analysis by virtual displacements and moment-balancing procedures. There is a considerable amount of new material covering the new structural steels and the manner in which structures respond to load.

Since design specifications are revised more often than in the past, some of the illustrative examples are intentionally based on strength predictions and factors of safety which differ from those of standard specifications. As in the first edition, the authors have presented and discussed design calculations for several structures which have been built.

Calculations for the highway bridge which was a feature of the first edition have been revised in accordance with the 1969 edition of the AASHO Specifications. Discussion and analysis of the wind-bracing system of the South Central Bell Telephone Company building in Birmingham are presented as an example of the high-rise building. The authors wish to express their appreciation to Messrs. M. H. Eligator and A. J. Grasso of Weiskopf and Pickworth, New York City, and especially to Mr. J. P. Rutigliano of W & P Engineers, San Francisco, who furnished the necessary drawings and calculations for this structure and who also reviewed the chapter on multistory buildings and made several valuable suggestions.

The authors are grateful to Dr. Eugene Chesson, Jr., who read the entire manuscript and made a number of valuable comments and suggestions, and to Dr. Peter C. Birkemoe and Dr. James B. Radziminski, who read portions of the manuscript. They are also indebted to Dr. T. R. Higgins and Miss Mary Anne Donohue of the American Institute of Steel Construction, and to various other organizations and individuals who furnished photographs and other material.

Edwin H. Gaylord, Jr.
Charles N. Gaylord

NOTATION

ABBREVIATIONS

kcf	kips per cubic foot
klf	kips per lineal foot
kli	kips per lineal inch
ksf	kips per square foot
ksi	kips per square inch
mph	miles per hour
pcf	pounds per cubic foot
plf	pounds per lineal foot
pli	pounds per lineal inch
psf	pounds per square foot
psi	pounds per square inch

AASHO	American Association of State Highway Officials
AISC	American Institute of Steel Construction
AISI	American Iron and Steel Institute
ANSI	American National Standards Institute

AREA	American Railway Engineering Association
ASCE	American Society of Civil Engineers
ASTM	American Society for Testing and Materials
AWS	American Welding Society
BBC	Basic Building Code
CRC	Column Research Council
NBC	National Building Code
NBS	National Bureau of Standards
RCRBSJ	Research Council on Riveted and Bolted Structural Joints
SSBC	Southern Standard Building Code
UBC	Uniform Building Code
WRC	Welding Research Council

NOMENCLATURE

A	area
A_b	area of bolt
A_f	area of beam flange
A_g	gross area
A_n	net area of tension member
A_s	area of steel section in composite beam; area of longitudinal stiffener; shear area of bolts in connection
A_{st}	area of transverse stiffener
A_w	area of beam or plate-girder web
a	length of plate; distance between vertical stiffeners
B	bending factor (area of cross section divided by section modulus)
b	spacing of stringers in bridge floor; width of flange; width of plate; length of bearing on beam web; effective width of concrete flange of composite beam
b_e	effective width of plate or web for determining postbuckling strength
C	compressive force
C_1, C_2	lateral-torsional buckling coefficients
C_b	moment-gradient coefficient for lateral-torsional buckling of beams
C_c	column slenderness ratio dividing elastic and inelastic buckling
C_m	coefficient of bending term in beam-column interaction formula
C_s	shape factor determining effective wind pressure
C_v	ratio of critical shear stress to yield shear stress
C_w	warping constant of cross section
c	distance from neutral axis to extreme fiber of beam
D, DL	dead load

d diameter of hole; diameter of rocker; depth of cross section; diameter of bolt
d_c column-web depth between fillets
E modulus of elasticity (Young's modulus)
E_c modulus of elasticity of concrete
E_r double modulus (also called reduced modulus)
E_s strain-hardening modulus
E_t tangent modulus
e eccentricity of load
F_a allowable axial compressive stress
F_b allowable bending stress
$F_{b,cr}$ critical bending stress
F_c allowable web-crippling stress
F_{cr} critical compressive stress
F_E Euler column stress
F_E' allowable Euler column stress
F_p allowable bearing stress; proportional-limit stress
F_r allowable fatigue stress; residual stress
F_s axial force in stiffener
F_{sr} range of stress for fatigue loading
F_t allowable tensile stress
F_u ultimate stress
F_v allowable shear stress
$F_{v,cr}$ elastic critical shear stress
$F_{v(cr)i}$ inelastic critical shear stress
F_{vp} proportional-limit shear stress
F_{vu} ultimate shear stress
F_{vy} shear yield stress
F_y yield stress
f calculated stress
f_a calculated axial compressive stress
f_b calculated bending stress
f_c calculated compressive stress
f_c' specified 28-day compressive strength of concrete
f_e postbuckling edge stress in plate
f_p calculated bearing stress
f_t calculated tensile stress
f_v calculated shear stress
G modulus of elasticity in shear (modulus of rigidity); center of gravity; ratio of sum of column stiffnesses to sum of beam stiffnesses at joint in frame
G_t tangent modulus in shear
g gage (transverse spacing of fastener lines); distance from shear center to point of application of load

h	depth of web of beam or plate girder; depth of section
I	impact ratio; moment of inertia
I_{eff}	effective moment of inertia
I_p	polar moment of inertia
I_{pG}	polar moment of inertia with respect to centroid
I_{pS}	polar moment of inertia with respect to shear center
I_s	moment of inertia of stiffener
I_{xy}	product of inertia
J	torsion constant of cross section
K	spring constant; effective-length coefficient
K'	effective-length coefficient of laced or battened column
K_1, K_2, K_3, K_4	net-section efficiency coefficients
k	distance from flange outside face to toe of web fillet; coefficient in plate-buckling formula; $\sqrt{P/EI}$
L	length
L, LL	basic live load
L_0	center-to-center spacing of battens
l	length
M	moment
M_1	smaller of end moments on beam column
M_2	larger of end moments on beam column
M_{cr}	critical moment
M_m	critical moment in plastic design
M_0	moment due to transverse loads on beam column
M_p	plastic moment
M_s	simple-beam moment
M_u	ultimate moment
M_y	yield moment
N	normal force
n	factor of safety; number of fasteners; shape factor for cross section in shear; E/E_c
P	load
P_{cr}	column critical load
P_E	Euler column load
P_n	probability that an event will be exceeded at least once in n years
P_r	column double-modulus load
P_T	column twist-buckling load
P_t	column tangent-modulus load
P_u	ultimate load
P_y	plastic axial load
p	pressure per unit of area; allowable bearing pressure of rocker on bearing plate
Q	shear; prying force; first moment of area in shear-stress formula
Q, Q_a, Q_s	coefficients used in evaluating postbuckling allowable loads

q velocity pressure (also called dynamic pressure or stagnation pressure); allowable load on shear connector; shear per lineal inch

q_u static strength of shear connector

R reaction; shear force on fastener; return period (also called recurrence interval) of snow, wind, etc.; reduction in live load on building floors, percent; ratio of maximum stress to minimum stress (in fatigue formula)

r radius of gyration

r_{eq} equivalent radius of gyration for lateral-torsional buckling

r_0 radius of gyration of chord of battened column

r_T radius of gyration of section consisting of compression flange and one-third the compression web area

r_t equivalent radius of gyration for torsional buckling

r_{tb} equivalent radius of gyration for twist-bend buckling

S elastic section modulus; shear center; effective span of bridge floor slab

s pitch (distance, in the direction of the gage lines, between two successive holes)

T tensile force; torsional moment

T_b specified pretension of high-strength bolt

T_e effective throat of fillet weld or partial-penetration groove weld

T_v St. Venant torsional resistance

T_w torsional resistance due to nonuniform warping

t thickness

V shear; velocity

V_{cr} critical shear in plate-girder web

V_t shear component of plate-girder web tension field

V_u shear strength of beam or plate-girder web

v velocity

W load on beam, kips

W_e external work during virtual displacement

W_i internal work during virtual displacement

WL wind load

w distributed load on beam, kips per foot

w_g gross width of part in tension

w_n net width of part in tension

y_0 deflection due to transverse loads on beam columns; distance from shear center to centroid

Z plastic section modulus

β angle of twist

Δ beam deflection

δ displacement; deflection of column at midlength

δ_0 midlength displacement due to initial crookedness

ε unit strain

ε_r residual strain

ε_s strain at onset of strain hardening

ε_y yield strain

θ rotation of joint in frame or of beam at support; angle of twist per unit length; angle of plate-girder panel diagonal with horizontal; mechanism angle

μ Poisson's ratio

ρ radius of curvature; mass density of air

τ ratio of tangent modulus to Young's modulus

ϕ inclination of tension field in plate-girder web; rotation at end of beam

1

LOADS ON STRUCTURES

1-1 ENGINEERING STRUCTURES

Engineering structures are of such variety that they defy any attempt to enumerate them except in a general way. The countless problems which arise in their design have prompted engineers to specialize in the design of particular structures or groups of related structures, and it is profitable to study design somewhat according to customary areas of specialization. Although the complete design of many structures is the result of the coordinated efforts of several branches of engineering, we refer sometimes to the design of a structure having in mind only that part of the design which comes within the province of one of the branches.

Among the structures that are designed by civil engineers are bridges, buildings, transmission towers, storage vessels, dams, retaining walls, docks, wharves, highway pavements, and aircraft landing strips. Even this group of structures is too large for convenient study as a unit. In this book we shall limit ourselves to a study of structural members in metal and the methods by which they are usually connected, together with applications to the design of bridges and buildings.

1-2 THE DESIGN PROCEDURE

The first and often the most difficult problem in design is the development of a plan that will enable the structure to fulfill effectively the purpose for which it is to be built. If the structure is a building, for example, the designer must create a plan that is adapted to the site; that provides a suitable arrangement of rooms, corridors, stairways, elevators, etc.; that will be aesthetically acceptable; and that can be built at a price the client is prepared to pay. This phase of design, sometimes called *functional planning*, calls for a designer with a high order of skill and imagination.

Although the structural scheme is never independent of the functional plan, it is convenient for the purposes of this discussion to think of its development as a second major step in the design procedure. The extent to which the scheme must be developed during the functional-planning stage depends upon the structure. For example, the location of the columns in a building usually must be worked out with the functional plan, and sufficient space must be anticipated between finished ceiling and finished floor of adjacent stories to accommodate the floor construction. The functional plan and structural scheme of a highway bridge usually are not so interdependent. The roadway grade and alignment of a highway bridge are influenced principally by clearance requirements with respect to whatever is to be bridged and the necessity of providing adequate approaches to collect and disperse traffic, while the width depends upon the number of lanes required to accommodate the expected traffic. Many different types of bridge structure can be adapted to a given functional plan.

It is usually necessary to make tentative cost estimates for several preliminary structural layouts. Sometimes this may have to be done while the functional plan is being developed; sometimes it can be done later. Selection of structural materials must be based upon consideration of availability of specific materials and the corresponding skilled labor, relative costs and wage scales, and the suitability of the materials for the structure. Successful development of an efficient structural scheme hinges on the engineer's familiarity with the many types of supporting structure which have been developed in the past. On the other hand, however, the designer who leans too heavily on tradition may fail to see the possibilities in new and better solutions.

The third stage of the design is a structural analysis. Although design specifications and building codes usually prescribe the nature and magnitude of the loads to which it is assumed the structure may be subjected, at times the engineer must make the decision. Once the loads are defined, a structural analysis must be made to determine the internal forces which will be produced in the various members of the framework. Although this is a fairly routine procedure, simplifying assumptions must invariably be made before the principles of mechanics can be applied.

In the fourth phase of the design the engineer proportions the members of the structural system. They must be chosen so that they will be able to resist, with an appropriate factor of safety, the forces which the structural analysis has disclosed. Familiarity with the methods and processes of fabrication (and their limitations) and with the techniques of construction (and their limitations) is indispensable.

The four steps in structural design discussed above are seldom, if ever, distinct, and in many cases they must be carried along more or less simultaneously. Furthermore, they assume varying degrees of importance relative to one another. For example, in the design of a house the architect will rarely, if ever, need a structural analysis and design. On the other hand, both the analysis and the design are fully as important as the functional planning in the case of a suspension bridge.

1-3 LOADS

The weight of a structure is called *dead load.* It can be determined with a high degree of precision, although not until after the structure has been designed. For this reason it is necessary to estimate dead load before a structural analysis is made, in order that the part of the internal forces due to weight can be taken into account. Experienced designers can often guess the weight of a structure or its parts with good accuracy. The actual weight should be determined and compared with the estimated weight and correction made if the difference is significant. The designers of the ill-fated first Quebec bridge across the St. Lawrence River neglected this point. After the collapse of the bridge during erection, with the resulting death of more than 100 workmen, investigation disclosed dead loads 20 to 30 percent larger than those assumed for the design. Although this discrepancy was not the cause of the failure, it would have handicapped the bridge had it been possible to complete it as originally designed.

All loads other than dead load are called *live loads.** Live loads may be steady or unsteady; they may be fixed, movable, or moving; they may be applied slowly or suddenly; and they may vary considerably in magnitude. The history of some live loads—notably the weights of the heaviest highway trucks—is one of more or less continual increase in magnitude. The live loads which usually must be considered are:

1 The weight of people, furniture, machinery, and goods in a building
2 The weight of traffic on a bridge
3 The weight of snow
4 Dynamic forces resulting from moving loads

* In some specifications (ANSI A58.1, for example) live load is defined as the weight due to occupancy, excluding such loads as wind, snow, etc.

5 Dynamic forces induced by wind and earthquakes
6 The pressure of liquids in storage vessels
7 Forces resulting from temperature change if expansion and contraction are impeded
8 The pressure of earth, as on retaining walls and column footings

The primary effect of gravity loads on structures is calculated from their weight; i.e., they are considered to be static loads. However, live loads in motion may produce forces that are considerably greater than those resulting from the same loads at rest. These are the dynamic forces mentioned in category 4 above. Dynamic force caused by motion is called *impact* if the effect is equivalent to additional gravity load and *lateral* or *longitudinal force* (depending upon its direction relative to the path of the vehicle) when the result is equivalent to load in the horizontal plane. Lateral force may result from motion in a curved path (centrifugal force) or from the swaying motion of a train on a straight track. Longitudinal forces are caused by acceleration and deceleration of moving vehicles.

The determination of the loads for which a given structure or class of structure should be proportioned is one of the most difficult problems in design. Several questions must be answered: What loads may the structure be called upon to support during its lifetime? In what combinations may these loads occur? To what extent should a possible but highly improbable load or combination of loads be allowed to dictate the design? The probability that a specific live load will be exceeded at some time during the life of the structure usually depends on the period of exposure (life) of the structure and the magnitude of the design load. For example, if the roof of a building is designed for a snow load of 30 psf, the chance that this load might be exceeded at some time during the life of the building is greater if the building stands for 50 years than if the building stands for only 10 years. This is because the maximum snowfall varies from year to year, and a given seasonal maximum can be expected to occur at a given locality only once in so many years in the long run. This period is called a *mean return period* or *mean recurrence interval.* Such return periods can be determined by statistical analysis of snowfall records. Of course, extremes of other natural phenomena such as wind, flood, etc., also occur infrequently, and return periods for specific extremes can be determined similarly.

The reciprocal of the recurrence interval of an extreme snowfall, wind velocity, etc., is the probability that the extreme value will be exceeded in any one year. Thus if the return period R of a wind speed of, say, 90 mph at a certain locality is 100 years, the probability that there will be a wind speed greater than 90 mph in any one year is $1/R = 1/100 = 0.01$. However, the probability which is of interest in choosing a wind velocity for design purposes is not the probability that the design speed will be exceeded in any one year but rather the probability that it will be exceeded during the life of the structure. This probability can be determined as follows.[1] Since $1/R$ is the probability that the specified velocity will be exceeded in any year, $1 - 1/R$ is the probability that it will *not* be exceeded. Then the probability that it will *not* be exceeded during n years, where n is the life of the structure, is $(1 - 1/R)^n$. Therefore, the probability P_n that it *will* be exceeded at least once in the n years is

$$P_n = 1 - \left(1 - \frac{1}{R}\right)^n \tag{1-1}$$

As an example, suppose a structure which is expected to have a life of 50 years is to be built in the locality mentioned above, where the mean recurrence interval of a wind speed of 90 mph is 100 years. The probability that the structure will encounter a wind speed exceeding 90 mph during its life is

$$P_{50} = 1 - (1 - 0.01)^{50} = 1 - 0.60 = 0.40$$

That is, there is a 40 percent chance that the structure will be exposed to a wind exceeding 90 mph. If this is an acceptable risk, it is sufficient to design the building to resist the pressures from a 90-mph wind. There will be a factor of safety in the design, of course, and so the structure would not be expected to collapse under a wind of this velocity.

1-4 LIVE LOADS ON BUILDING FLOORS

Buildings serve such diverse purposes and present such random arrangements of physical equipment and persons as to make it extremely difficult to estimate suitable design loads. Although a number of systematic surveys have been made, there is still a lack of adequate data. Many municipalities assume responsibility for the public safety by controlling the design of buildings through building codes which specify live-load requirements as well as other factors pertaining to design, and several organizations have issued codes for national or regional use. Among the latter are the following:

National Building Code, sponsored since 1905 by the American Insurance Association, New York. This code has been adopted by various communities throughout the United States.

Uniform Building Code, sponsored since 1927 by the International Conference of Building Officials, Pasadena, Calif. This code is used widely in the Western states.

Southern Standard Building Code, sponsored since 1945 by the Southern Building Code Congress, Birmingham, Ala. This code is used in the Southern and Southeastern states.

Basic Building Code, sponsored since 1950 by the Building Officials and Code Administrators International, Chicago. This code is used in the Eastern and North Central states.

Live-load recommendations are also published by the American National Standards Institute in American National Standard Building Code Requirements for Minimum Design Loads in Buildings and Other Structures, ANSI A58.1, New York.

Buildings may be classified according to occupancy as follows:

1 Residential (including hotels)
2 Institutional (hospitals, sanatoriums, jails)
3 Assembly (theaters, auditoriums, churches, schools)
4 Business (office-type buildings)
5 Mercantile (stores, shops, salesrooms)
6 Industrial (manufacturing, fabrication, assembly)
7 Storage (warehouses)

Except for studies of combustible contents made in connection with fire-resistance classifications, apparently there are no published reports of residential live loads which are the results of an actual weighing of contents. Building Materials and Structures Report 92, Fire-resistance Classifications of Building Constructions (U.S. Department of Commerce), reports average combustible contents of about 4 psf of floor area, with a maximum of 7.3 psf except in one portion which served as a library. These figures are probably good approximations of total contents excluding persons. A residential room containing 1 person to every 6 ft^2—surely adequate allowance for a crowd—would average, say, 25 psf. At the most, then, we might expect residential live loads to be 35 psf over relatively small areas, with an average of about 10 psf over an entire building. The 40-psf design requirement found in most building codes is ample.

Live loads for institutional occupancy may be expected to be about the same as those for residential occupancy. Several surveys of crowded hospital wards have been made, and even those wards which contained 1 bed for every 30 ft^2 supported average live loads of only 9 psf. Building codes are in substantial agreement on 40 psf as a minimum live load for institutional private rooms.

There have been a number of investigations of live loads due to crowds of people. At the University of Iowa, students packed for the purpose of testing dynamic loads on balcony construction resulted in a load of 116 psf. Observations of normal loading conditions on the elevators at Grand Central Terminal in New York City showed live loads of about 100 psf. In a test by the Milwaukee Board of Education in 1920, a room normally intended for 48 pupils was crowded with 258 pupils, filling all seats double and all aisles and open space. The resulting live load, including furniture, was 41.7 psf. The range of minimum-live-load requirements for assembly occupancy, as specified by various building codes, is given in Table 1-1. The table is not intended to be complete but rather to show representative loads.

The U.S. Public Buildings Administration sponsored live-load studies of the Internal Revenue Building and the Veterans Administration Building in Washington.[2] In the former building, 20 psf or less of actual average live load was found on 70 percent of its area, 40 psf or less on 88 percent, and 60 psf or less on 96.5 percent. The maximum average live load of 106 psf occupied 825 ft^2 (0.5 percent of the total area). In the latter building, 95 percent of the floor area supported an average load of 20 psf or less, 97.8 percent supported 40 psf or less, and 99.5 percent supported 60 psf or less. The maximum average live load of 90 psf was found on 1,176 ft^2 of the tenth floor (0.5 percent of the total area). Somewhat similar prior studies of the Equitable Building in New York City disclosed average loads of 11.6 psf over three selected floors. The maximum load was 78.3 psf, the minimum 0.87. The National Bureau of Standards surveyed their Administration Building and the U.S. Civil Service Commission Building.[3] The largest load intensity found was 72.5 psf. Building-code requirements for office-space buildings range from 50 to 80 psf.

The Office of Technical Services of the U.S. Department of Commerce undertook detailed studies, similar to those mentioned above, of mercantile, industrial, and storage occupancies.[4] The studies include two department stores, two mattress factories, one men's clothing factory, one dress factory, two furniture factories, one newspaper plant, one printing plant, and two warehouses. As might be expected, wide variations in live load were found. For example, the maximum live load in one of the mattress factories was only 41 psf, on 0.3 percent of the total floor area, while in the other it was 101 psf on 11.1 percent of the area. The latter load is somewhat misleading, because it was in a cotton storage area. Maximums for the furniture factories were 98 and 120 psf. The largest load in the printing plant was 168 psf on 1 percent of its area. The range of loads in one of the department stores is shown in Fig. 1-1. The ordinates of this diagram give the percent of area occupied by loads varying in increments of 5 psf. The lightest load, 21 psf, was found in the playground-

Table 1-1 LIVE LOADS FOR TYPICAL ASSEMBLY OCCUPANCIES

Range in values required by building codes

Type of space	Live load, psf
School classrooms (fixed seats)	40–60
School classrooms (movable seats)	40–100
Assembly halls (fixed seats)	50–60
Assembly halls (movable seats)	100
Theaters (not necessarily balconies)	50–60
Dance halls	100–120

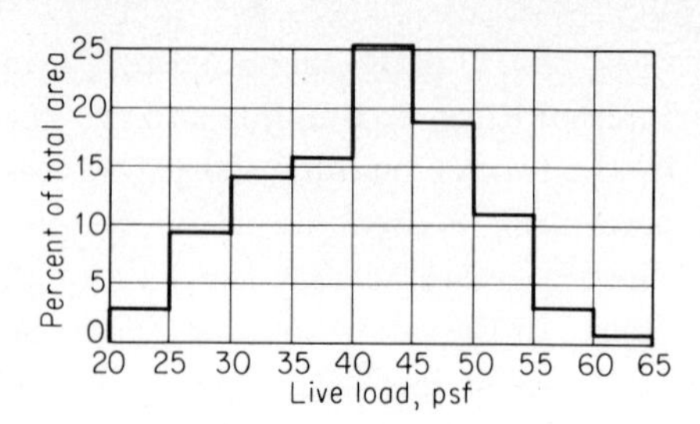

FIGURE 1-1

equipment department, the heaviest, 61 psf, in the phonograph-record department. In one warehouse, 0.4 percent of its area carried a load of 257 psf; in the other, 0.7 percent of the area supported 304 psf. The usual range of building-code requirements for the occupancies discussed in this paragraph is given in Table 1-2, which is not intended to be complete.

The load surveys mentioned above involve observations of load intensity at an instant of time, that is, at the time of the observation. However, floor loads in a particular structure vary randomly with time, and it is important to know what the peak lifetime loads may be. Very little information of this kind is available.

The tendency of average building live loads to decrease with an increase in the floor area considered poses a question with respect to the design of various kinds of supporting members. For example, it would be unrealistic to assume all the floor areas supported by a column to have loads of the same intensity as that supported by a joist. Similarly, because the maximum expected load is not likely to be realized on all floors simultaneously (except perhaps for some types of warehouse), it would be unrealistic to assume the same uniform load on all the floors supported by a lower column as on the floors supported by a column higher in the building. Building codes make varying provisions for this contingency. The American Standard Building Code Requirements, the Uniform Building Code, and the National Building Code permit a

Table 1-2 LIVE LOADS FOR VARIOUS OCCUPANCIES

Range in values required by building codes

Occupancy	Live load, psf
Mercantile	75–125
Light manufacturing	75–125
Heavy manufacturing	125–150
Light storage	120–125
Heavy storage	250 minimum

reduction in the basic design load at the rate of 0.08 percent for each square foot of area supported, provided the area exceeds 150 ft^2 and the live load is less than 100 psf. The reduction cannot exceed 60 percent, nor the value given by

$$R = 100\frac{D+L}{4.33L} = 23\left(1 + \frac{D}{L}\right) \tag{1-2}$$

where R = reduction, percent
D = dead load, psf
L = basic live load, psf

Equation (1-2) limits the overload to 30 percent in case the full basic live load should occur for a member which has a tributary area large enough to permit the maximum reduction. No reduction is allowed for roof loads or for live loads in excess of 100 psf, except that in the latter case column loads may be reduced by 20 percent.

The possibility of failure because of overload is a hazard which is often overlooked. The Department of Commerce recommends that the design live load be conspicuously posted in commercial and industrial buildings and that the occupant be held responsible for keeping the actual loads within the specified limit. This suggestion has not been widely adopted, and failures due to overload are not uncommon. Almost all these failures are the result of the conversion of buildings or portions of them to purposes for which they were not designed.

1-5 LIVE LOADS ON BRIDGE FLOORS

Highway traffic is made up of four principal kinds of vehicles—the truck-tractor with semitrailer, the truck, the bus, and the passenger car. These units vary widely in weight, and their average weights are considerably less than the weights of the heaviest units. The heaviest truck which is used to any considerable extent at present weighs about 50,000 lb when fully loaded, and since it is about 40 ft long, it represents an average load of, say, 1,250 plf of traffic lane. An average passenger car weighs perhaps 4,000 lb fully loaded, while the heaviest car may weigh something less than 10,000 lb, with corresponding lane intensities of 200 to 500 plf. The distance between vehicles in the same traffic lane is obviously important in its effect on load intensity. This distance may range from 25 to 50 ft center to center at speeds of 10 mph to upward of 200 ft at 60 mph.

Because of the widely variable character and distribution of highway traffic, it is expedient to adopt a conventionalized loading. Nearly all highway bridges in the United States are designed for one of the five classes of load recommended by the American Association of State Highway Officials (AASHO). These loads consist of a system of concentrated loads to represent a truck or of a load distributed uniformly along the traffic lane, together with a concentrated load, to represent a long line of medium-weight traffic with a heavy vehicle somewhere in the line. The two systems are necessary because of the differences in length of traffic responsible for maximum forces in the various parts of a bridge. A stringer will usually be of such length that it can support only one or two axles of a truck, and since the effect of one or two concentrated loads on a beam is quite different from the effect of an equal amount of load distributed uniformly, the actual concentrations must be considered. On the other hand, the force in a chord member of a simple truss span will be largest when the full length of the bridge is loaded. A relatively small error results from the substitution of a uniform load for a large number of concentrated loads; hence the lane load may be used, with considerable simplification in calculations. Since exceptionally heavy trucks usually operate at respectable distances from one another, a line of traffic is considered to have only one such vehicle (two in the case of continuous spans), represented by the concentrated load mentioned above. The heaviest loading of the AASHO specifications is the HS20-44, pictured in Fig. 1-2.

Although it is not an accurate representation of modern locomotives, the system of loads used in the design of most railroad bridges is one whose makeup was proposed a few years prior to 1890 by Cooper. The Cooper load (Fig. 1-3*a*) is intended to represent two locomotives followed by a uniformly loaded train. Although from time to time load systems intended to be closer approximations to the steam locomotives typical of the era preceding the advent of the diesel have been proposed, they have never been adopted. For comparison with the Cooper load, Fig. 1-3*b* shows the axle loads and spacing for two units of one of the heavier diesel locomotives.

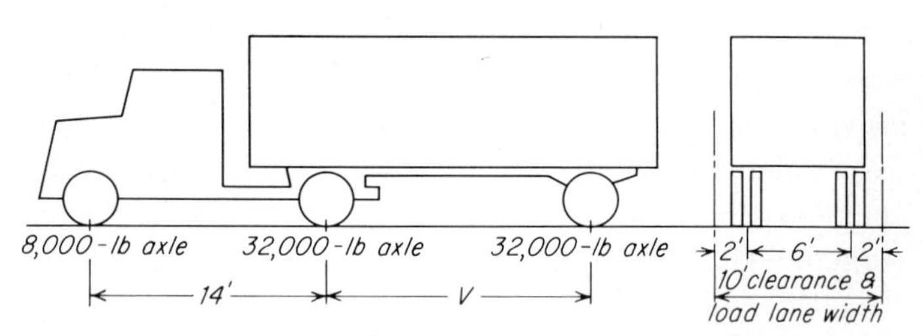

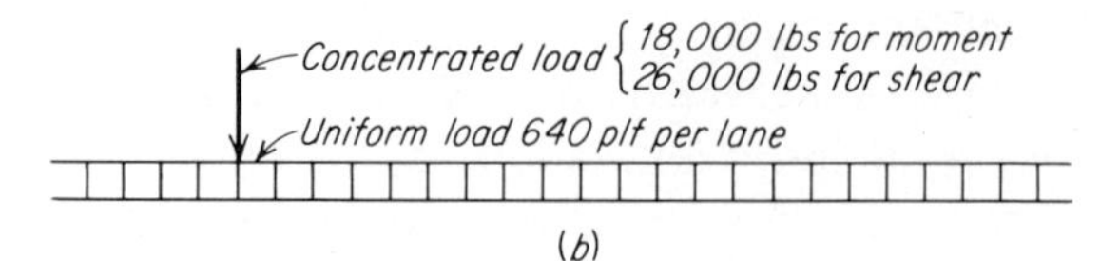

FIGURE 1-2
HS20-44 load.

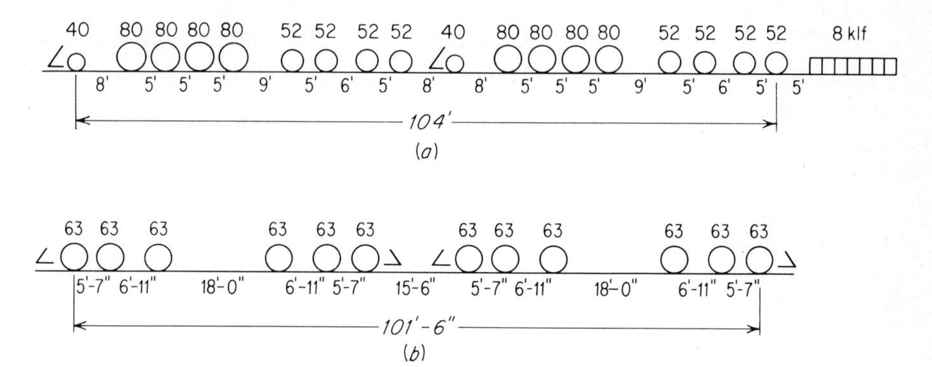

FIGURE 1-3
(*a*) Cooper E-80 train load. Axle load in kips. (*b*) Typical diesel locomotive (two units). Axle load in kips.

Bridge specifications of the American Railway Engineering Association (AREA) recommend that main-line bridges be designed for the Cooper E-80 train shown in Fig. 1-3*a*. In his original specifications, Cooper recommended three classes of load, which he called A, B, and C. The locomotive driving-axle loads for these three classes were 24, 22, and 25 kips, respectively, as compared with the 80-kip axles of the E-80 load.

1-6 IMPACT

The meaning of the word impact as it is used in structural design may be illustrated by describing two different ways in which a spring may be loaded. If one attaches a 20-lb weight to a suspended spring and supports it while it is being lowered to the position where it is supported entirely by the spring, the maximum force in the spring is 20 lb. On the other hand, if the weight is released after it is fastened to the spring, the maximum elongation of the spring will be almost double that required to support the weight in its position of static equilibrium and the corresponding force in the spring will be almost 40 lb. The 20 lb of force in excess of the 20-lb static force is called impact. It is customary to express impact as a percentage of the static force, and in this case, therefore, the impact factor is almost 100 percent.

The impact of moving live loads is a much more complex phenomenon than the one just described. The speed of a moving vehicle, its mass relative to the mass of a bridge, and irregularities in the track or floor and in the wheels of the vehicle are significant factors. Pulsating loads are particularly critical if the frequency of the pulses happens to be coincident, or nearly so, with the period of a fundamental mode of vibration of a structure.

Design-specification provisions for impact are frankly empirical and do not attempt to account for all the variables. The AASHO specifications require impact allowance given by the equation

$$I = \frac{50}{L + 125} \tag{1-3}$$

but not to exceed 0.3. In this formula, I is the ratio of impact to static load and L the length in feet of the part of the span which is loaded. The AREA specifications make similar but more severe provision for impact. Impact allowance for moving loads such as elevators, traveling cranes, reciprocating machinery, and the like, is usually specified as a fixed percentage of the load. Typical values may be found in the American Institute of Steel Construction's (AISC) Specification for the Design, Fabrication and Erection of Structural Steel for Buildings.

Since most building-floor live loads are essentially static, the values discussed in Art. 1-4 are considered to be sufficiently on the safe side to cover any impact likely to occur. One exception to this statement is an AISC stipulation that the live load on hangers supporting floor and balcony construction be increased by one-third for impact.

1-7 SNOW LOADS

Freshly fallen dry snow weighs 5 to 6 pcf, packed snow about 10 pcf. A number of studies of records of the National Weather Service have been made in an attempt to set reasonable snow loads for various sections of the country. In 1939 the Service published data from 166 weather stations on the greatest depth of snow in sheltered areas such as clearings in forests. The results disclosed snow loads ranging from zero in the southernmost parts of the southern tier of states, through about 5 psf for the remainder of the South, 10 psf for the central tier of states, 15 psf through southern New York, the southern Great Lakes region, and the Northern Plains states, up to 20 to 25 psf for the Lake Superior and New England areas. Exceptions to these values occur in mountainous regions, where the values may run about 5 psf higher than in the lower altitudes of the same section, and along the Pacific Coast to Seattle, where snow rarely falls.

A map published by the National Weather Service (formerly known as the United States Weather Bureau) gives ground snow loads having a 50-year mean recurrence interval (Fig. 1-4). These loads were based on the maximum annual water equivalents of snow on the ground. Except for the southern states, these loads tend to be higher than those discussed in the preceding paragraph. They range from 5 psf in the southern coastal states to 40 and 50 psf in the northern Great Lakes region, and to 70 and 80 psf in northeastern Maine. Data on the Rocky Mountain states are not given.

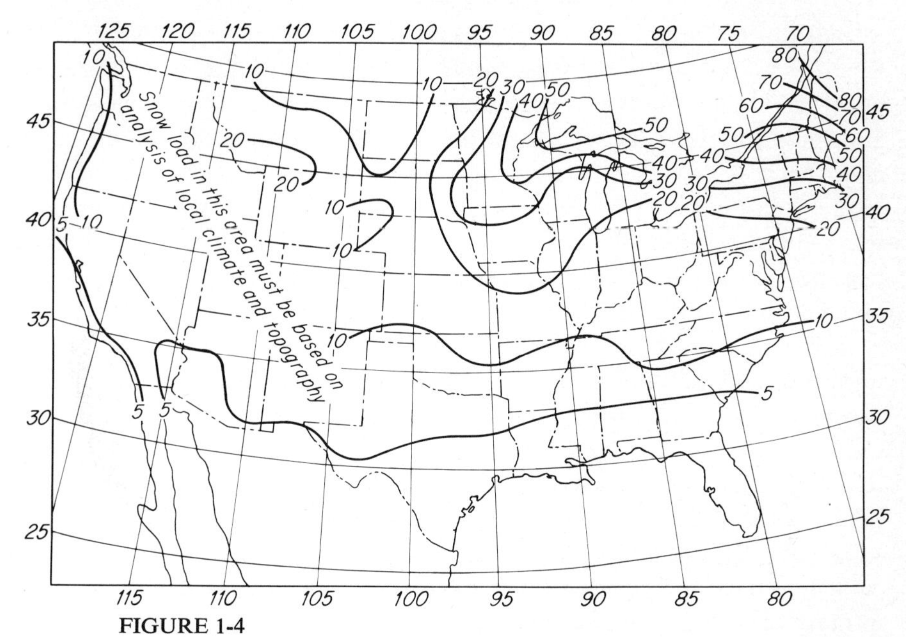

FIGURE 1-4
Snow load on the ground, 50-year mean recurrence interval. (*United States Weather Bureau Map 12158, January, 1969*).

The snow load on a roof is usually less than that on the ground. Based on a comprehensive study of roof loads relative to ground loads, the National Building Code of Canada specifies for flat roofs a basic snow load of 80 percent of the sum of the weights of the 30-year ground snowpack and the maximum one-day rainfall in late winter or early spring. The weight of snowpack plus rain runs to as much as 120 psf in the Canadian Rocky Mountains.[7] ANSI A58.1 specifies a basic snow load of 80 percent of the 50-year return ground snow. A further reduction (to 60 percent) is allowed in both codes for roofs that have a clear exposure to winds of sufficient intensity to remove snow. Still further reductions are allowed for sloping roofs, since they accumulate less snow than flat roofs. On the other hand, valleys of adjoining gabled roofs of multibay buildings may accumulate more than normal amounts of snow. Projections, such as penthouses on flat roofs, may also cause drifting. The Canadian Building Code recommends that snow loads near projections from a roof be increased by 50 percent for a width three times the height of the projection, but not to exceed 15 ft.

Based on a study of the weights of seasonal snowpacks having a mean recurrence interval of 10 years, the loads in Table 1-3 were suggested for the design of roofs.[5] It should be noted, however, that the mountain areas defined in this table were not included in the study because of extreme local variations in depths of snow. Therefore, the suggested 40 psf for flat roofs will be excessive for many localities in these areas and too low in others. For example, the 10-year-return snowpacks in Reno and Salt Lake City were 20 and 30 psf, respectively, but these cities are in an excluded area. On the other hand, measurements of the snowpack on the roof of a building at 10,000 ft in the Rocky Mountains in Colorado, where the winter-long accumulation was 7 ft deep at the valley and 5 ft deep at the eaves of the V-shaped roof, disclosed densities of 8, 16, and 25 pcf at depths of 3, 5, and 7 ft. Thus, the average density was about 16 pcf and the roof load at the 7-ft depth about 110 psf.[6]

The live load suggested in Table 1-3 for flat roofs in the southern states (20 psf) exceeds the 10-psf (or less) 10-year-return snowpack. This load was chosen because it conformed with the value commonly specified for this region by various building codes. It is intended to provide for live loads other than snow, such as those incidental to construction, maintenance, etc. In many cases, of course, the minimum live load prescribed by a governing building code may differ from the value in Table 1-3.

Snow load need not be considered in the design of bridges, since a fall heavy enough to be of consequence would make the bridge impassable or else compel traffic to move at such a pace as to reduce the dynamic effect.

1-8 WIND LOADS

The evaluation of the effects of wind on an object in its path is a complex problem in aerodynamics. If we consider air to be nonviscous and incompressible, which is a reasonable assumption for velocities of the magnitudes for which civil-engineering

Table 1-3 SUGGESTED MINIMUM SNOW OR OTHER LIVE LOAD FOR ROOFS*

On horizontal projection, psf

	Slope of roof			
Region	3 in 12 or less	6 in 12	9 in 12	12 in 12 or more
Southern states	20	15	12	10
Central states	25	20	15	10
Northern states	30	25	17	10
Great Lakes, New England, and mountain areas†	40	30	20	10

* From Ref. 5. These loads are based in part on weights of seasonal snowpacks, 10-year mean recurrence interval; see Art. 1-7.

† Great Lakes and New England areas include northern portions of Minnesota, Wisconsin, Michigan, New York, and Massachusetts and the states of Vermont, New Hampshire, and Maine. Mountain areas include Appalachians above 2,000 ft elevation, Pacific coastal ranges above 1,000 ft elevation, and Rocky Mountains above 4,000 ft elevation; however, see Art. 1-7.

structures are designed, Bernoulli's equation for streamline flow can be used to determine the local pressure at the stagnation point, as a column of air strikes (at 90°) an immovable body. Thus,

$$q = \tfrac{1}{2}\,\rho v^2 \qquad (a)$$

where q = pressure
ρ = mass density of air
v = velocity of air

This pressure is called *velocity* pressure, *dynamic* pressure, or *stagnation* pressure. It is important to note that this equation is based on steady flow and does not account for the dynamic effects of gusts or the dynamic response of the body.

The resultant wind pressure on a body depends upon the pattern of flow around it. Pressures vary from point to point on the surface, depending upon the local changes in velocity, which depend in turn upon the shape and size of the body. The resultant pressure P is expressed in terms of the drag component P_D and the lift component P_L

$$P_D = C_D A\,\frac{\rho v^2}{2} \qquad P_L = C_L A\,\frac{\rho v^2}{2} \qquad (b)$$

The drag coefficient C_D and the lift coefficient C_L depend on the shape of the body and its orientation with respect to the wind. A is a characteristic area of the body, usually the projection of the body's surface on a plane.

The terms drag and lift are not ordinarily used in describing wind pressures on buildings, bridges, and the like. Instead, the pressure p per square foot, normal to the surface, is expressed in terms of a shape factor C_s (also called pressure coefficient):

$$p = C_s q = C_s\,\rho\,\frac{v^2}{2} \qquad (c)$$

Air at a temperature of 15°C (59°F) at sea level weighs 0.0765 pcf. Substituting the corresponding mass density 0.0765/32.2 into Eq. (*c*) gives $p = 0.00119 C_s v^2$. With the symbol V to denote velocity in miles per hour, this gives

$$p = 0.00256 C_s V^2 \qquad (1\text{-}4)$$

It will be noted that a velocity of 100 mph will induce a pressure of 26 psf when $C_s = 1$. Shape factors are discussed later in this article.

Measured wind velocities are necessarily averages of the fluctuating velocities which are encountered during a finite interval of time. The usual reported value in the United States is the average of the velocities which are recorded during the time it takes a horizontal column of air 1 mile long to pass a fixed point (the measuring anemometer). For example, if a 1-mile column of air is moving at an average velocity of 60 mph, it passes a fixed point in 88 sec; the reported velocity is the average of the

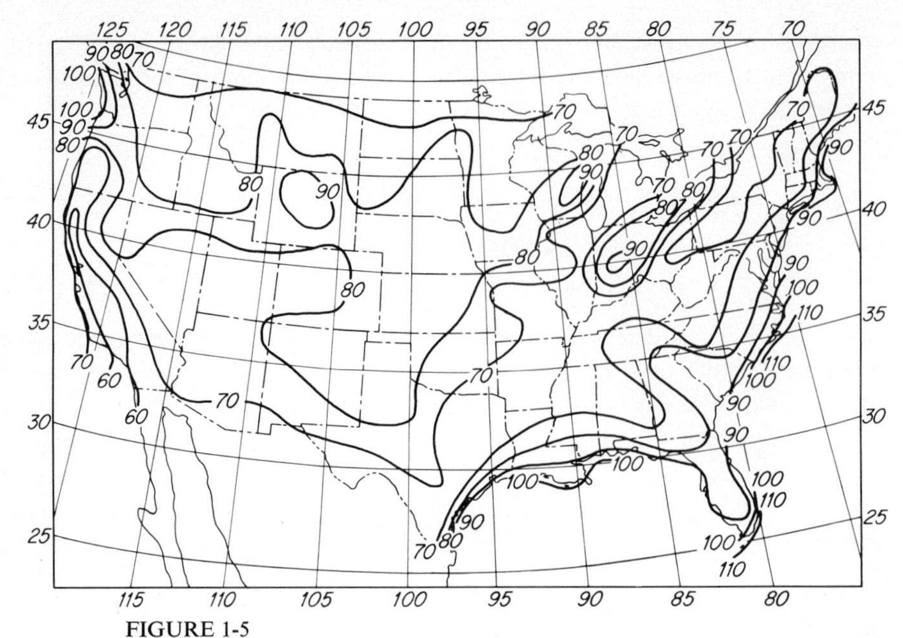

FIGURE 1-5
Annual extreme fastest-mile wind speed 30 ft above ground, 50-year mean recurrence interval. (*From H. C. S. Thom, J. Struct. Div. ASCE, July 1968.*)

velocities recorded during these 88 sec. The *fastest mile* is the highest velocity in 1 day. The *annual extreme mile* is the largest of the daily maximums. Furthermore, since the annual extreme mile varies from year to year, wind pressures to be used in design should be based on a wind velocity having a specific mean recurrence interval. Charts of annual extreme-mile velocities with mean recurrence intervals of 2, 10, 25, 50, and 100 years have been published.[8] These are based on statistical analysis of records from 138 stations, the average record covering a period of 21 years. The 50-year map is shown in Fig. 1-5. This recurrence interval has been suggested for all "permanent" structures except those that might have a high degree of sensitivity to wind and an unusually high loss of life and property in case of failure.[9] For the latter, a 100-year recurrence interval is suggested.* Furthermore, in recognition of the smaller risk with short periods of exposure, velocities of only 75 percent of the 50-year velocities are suggested for temporary structures, such as those used during construction.[9] The 50-year recurrence interval is also prescribed (with exceptions subject to the judgment of the engineer or authority having jurisdiction) by ANS1 A58.1, 1972.

The velocities shown in Fig. 1-5 are "open-country" velocities, i.e., they obtain

* On the average, the 100-year velocity in the United States is about 8 percent higher than the 50-year velocity.

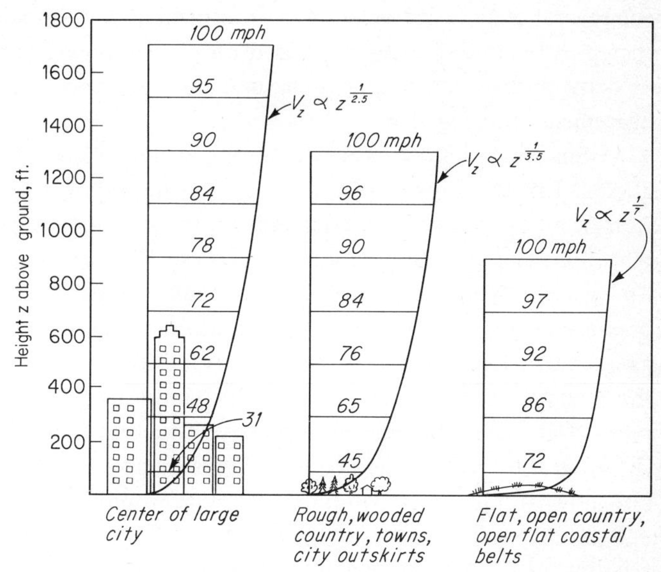

FIGURE 1-6
Velocity profiles over terrain with three different roughness characteristics for uniform-gradient wind velocity of 100 mph. (*From A. G. Davenport, Wind Loads on Structures, Nat. Res. Counc. Canada, Ottawa, Div. Building Res. Tech. Pap. 88, 1960.*)

where surface friction is relatively uniform for a fetch of about 25 miles. If the exposure is elevated, subject to channeling of the wind, etc., the map values must be adjusted accordingly. Locations with unobstructed exposure to large bodies of water may experience extreme winds 30 mph (or more) greater than for locations a short distance inland. However, this effect has been taken into account in Fig. 1-5.[8]

The National Building Code of Canada recommends pressures based on the 30-year *hourly* wind speed (average of the velocities measured during 1 hr). The hourly speed is the only speed measured at most weather stations in Canada.

Variation of wind velocity with height must be considered in the design of tall structures. The flow of air close to the ground is slowed by surface roughness, which is dependent on the density, size, and height of buildings, trees, vegetation, etc., on the ground. Fig. 1-6 shows velocity profiles and the corresponding exponential variation with height, according to Davenport.[10] Various other exponents have been suggested. However, the 1/7-power law of Fig. 1-6 is generally accepted for flat open country. Velocity at 30 ft above ground is used as the basic value for design purposes, and increases with height are provided for by specifying velocities (or wind pressures) for various height zones. Table 1-4 gives suggested velocities for various height zones for several basic wind velocities. Intermediate values can be interpolated. Velocity increases for inland areas are based on the 1/7-power law. It will be noted that the velocity profiles for inland areas and coastal areas differ considerably. This is in agreement with Fig. 1-6.

Since measured wind velocities are average values, it is necessary to consider the effects of fluctuations in velocity (gusts). The response of a structure to such fluctuations is a dynamic one, and depends on the size of the structure, its natural period of vibration, and its damping characteristics. The dynamic effect is usually accounted for by multiplying the wind velocity by a *gust factor* and computing the corresponding pressure by Eq. (1-4), which is to say that the response is evaluated as a static one. The gust factor depends on the wind velocity and the size of the structure. This is because the wind pressures are not fully developed until the structure is enveloped in the moving mass of air. For this reason, a massive structure is relatively insensitive to gusts of short duration while a sign is not. A gust factor of 1.3 will account for a one-second gust in a 90-mph basic wind. Such a gust would have a downwind length of 130 ft, and would be adequate for signs and small structures.[11] A gust factor of 1.1 will account for a 10-second gust in a 90-mph basic wind, which would have a downwind length of about 1,300 ft. This gust factor has been suggested for structures on the order of 125 ft wide transverse to the wind.[11] A method for evaluating the gust factor as a function of wind velocity and the characteristics of the structure has been developed.[12] It should be noted that the gust factors discussed in this paragraph do not provide for dynamic effects such as flutter, vortex shedding, etc. These effects are discussed in Ref. 13.

Design wind pressures can be determined for the velocities of Table 1-4 by Eq. (1-4). The shape factor C_s varies considerably with the proportions of the structure and the horizontal angle of incidence of the wind. The shape factor for the windward face of a flat-roofed rectangular building is about 0.9, regardless of the proportions of the building. There is negative pressure (suction) on the rear face, for which the shape factor varies from about -0.3 to -0.6, depending on the proportions of the building. Thus, the resultant pressure on such a building can be determined by using a shape factor ranging from 1.2 to 1.5 in Eq. (1-4). The value 1.3 is commonly used. It is not necessary to divide this force into pressure and suction when dealing with wind bracing of buildings. Sidewalls experience suction, for which C_s ranges from about -0.4 to -0.8. The roof also experiences suction, for which C_s ranges from about -0.5 to -0.8 for the average over the roof. However, the suction is larger on the windward side, and the average coefficient for the windward half may be as much as twice that on the leeward half.

The preceding discussion applies to an airtight building. Air leakage through small openings around doors, windows, etc., gives rise to internal pressures, with C_s as large as 0.25, if the openings are chiefly on the windward face and to internal suctions, with C_s as large as -0.35, if the openings are predominantly leeward. The following internal wind pressures were suggested in Ref. 14:

1 For buildings which are nominally airtight, a pressure or suction of 4.5 psf normal to the walls and roof
2 For buildings which have 30 percent or more of the wall surfaces open or subject to being opened or broken open, a pressure of 12 psf or a suction of 9 psf
3 For buildings which have wall openings between 0 and 30 percent of the wall area, pressures or suctions varying linearly between the values recommended in 1 and 2

The pressures in 1, 2, and 3 above are for a velocity which produces a resultant pressure (sum of windward-wall pressure and leeward-wall suction) of 20 psf, so that it would be consistent to increase (or decrease) them proportionately for larger (or smaller) basic pressures. When openings are large, as in hangars, internal wind pressures may be quite large.

Wind pressures on sloping roofs depend on the exposure, the slope, and the proportions of the building. For wind normal to a side parallel to the ridge, the leeward roof surface is always subjected to suction. There is suction on the windward surface for slopes less than about 30° and pressure for larger slopes. These pressures are not uniform but have maximum values at the eaves. The following pressures for single-ridged roofs are given in Ref. 9:

For the windward surface:
$$p = \begin{cases} -0.7q & 0 \le \alpha \le 20^\circ \\ (0.07\alpha - 2.1)q & 20^\circ \le \alpha \le 30^\circ \\ (0.03\alpha - 0.9)q & 30^\circ \le \alpha \le 60^\circ \\ 0.9q & 60^\circ \le \alpha \end{cases}$$

For the leeward surface:
$$p = -0.7q \qquad 0 \le \alpha \le 90^\circ$$

Table 1-4 FASTEST MILE OF WIND FOR VARIOUS ZONES ABOVE GROUND*

Zone height, ft	Design velocity, inland areas, mph			Zone height, ft	Design velocity, coastal areas,† mph			
0–50	60‡	80‡	100‡	0–50	60‡	80‡	100‡	130‡
50–150	70	95	120	50–150	85	105	125	150
150–400	80	110	140	150–400	115	135	155	180
400–700	90	120	150	400–600	140	165	185	195
700–1,000	100	130	160	600–1,500	150	170	190	200
1,000–1,500	105	135	165					

* From Ref. 9.
† Area to 30 miles inland from well-defined coast line along oceans or other large bodies of water.
‡ Basic velocity, i.e., velocity at 30 ft, from Fig. 1-5 or other source.

Negative values of p in these formulas denote suction; q is the velocity pressure from Eq. (a), and α is the angle of inclination with the horizontal.

Wind forces on trussed structures, such as bridges, transmission towers, and the like, and on beam bridges, girder bridges, etc., are at least as difficult to assess as those on enclosed structures. A complicating factor in this evaluation is the shielding of leeward parts of the structure. The amount of shielding depends principally on the distance between trusses or girders and on the angle of incidence of the wind. Shielding is discussed in Ref. 9.

A comprehensive tabulation of pressure coefficients for a wide variety of structures is given in Ref. 15 (reprinted in part in Refs. 9 and 16). Recommended coefficients for walls of buildings, gabled roofs, arched roofs, roofs over unenclosed structures (such as stadiums), chimneys, tanks, signs, transmission towers, etc., are also given in ANSI A58.1, 1972.

It is important to note that wind pressures specified by building codes include allowances for gust factors and shape factors, except where these factors are specified separately. For example, the National Building Code wind pressures on vertical surfaces (Table 1-5) include allowances for a gust factor of 1.3 and a shape factor of 1.3 (rectangular buildings). This code also specifies wind pressures on signs which differ from those for vertical surfaces because of the differences in shape factor. On the other hand, ANSI A58.1, 1972 specifies wind pressures in which only allowance for gusts is made, and specifies shape factors separately for various structures.

Until 1957 AASHO specified wind pressures on trusses at 50 psf on $1\frac{1}{2}$ times the exposed area of one truss. The exposed area of one truss is the area seen in elevation normal to the length of the bridge. Thus, the leeward truss was considered to be shielded to the extent that only half its area was effective. This was changed in 1957

Table 1-5 NATIONAL BUILDING CODE WIND REQUIREMENTS

Horizontal wind pressure on vertical surfaces, psf

Height zone, ft	Windstorm area		
	Minimum	Moderate	Severe*
Less than 30	15	25	35
30–49	20	30	45
50–99	25	40	55
100–499	30	45	70
500–1,199	35	55	80
1,200 and up	40	60	90

* Recommended for use within 50 miles of the Gulf Coast, and the Atlantic Coast from the southernmost part of Florida to Chesapeake Bay.

to 75 psf on the exposed area of one truss, which gives the same result. Since the shape factor of the H cross section commonly used for truss members is about 2, a pressure of 50 psf corresponds to a wind velocity of about 100 mph, including the effect of gusts [Eq. (1-4)]. The corresponding AREA requirement is 50 psf on the exposed area of the windward truss plus all the exposed area of all parts of the leeward truss not shielded by the floor. In both AASHO and AREA specifications these are pressures on an unloaded bridge, and smaller intensities are prescribed when wind pressure on both the structure and the live load are considered.

The dynamic response of long-span bridges to wind forces is a significant consideration. The failure in 1940 of the Tacoma suspension bridge only 6 months after it was opened to traffic stimulated research on the problems of aerodynamic instability. There have been several spectacular failures of bridges due to wind. In addition to the collapse of the Tacoma Bridge, it is of interest to recall the failure in 1879 of a railroad bridge across Scotland's Firth of Tay. Two years after its completion, 13 of the 84 truss spans of this bridge were blown from their piers, carrying with them a train and its seventy-odd passengers. The shock of the disaster was so great that its designer and builder, Sir Thomas Bouch, died within the year.

1-9 EARTHQUAKE LOADS

Earthquakes may happen in any part of the world, but they are more frequent and generally more violent in two great belts of the earth, of which one almost encircles the Pacific Ocean and the other stretches across southern Asia into the Mediterranean region. Although earthquakes of destructive or near-destructive proportions have occurred in almost every one of the United States, they have been far more frequent and disastrous on the Pacific Coast, particularly in California. Seismologists distinguish three types of waves set in motion by earthquakes. The type that seems to be most destructive travels over the earth's surface much like the waves generated by a stone dropped into water. Major wave movements may last from a few seconds to several minutes. Their period, amplitude, and acceleration change rapidly. Periods may vary from 0.1 to 0.2 sec, accelerations from very small values to more than that of gravitation, and amplitudes from fractions of an inch to 9 in. or more. The San Fernando, Calif. earthquake of February 9, 1971, had ground intensities larger than those for any earthquake ever recorded. A ground acceleration of 1.25 g was recorded at Pacoima Dam. Interpolation and examination of damage close to the center indicated ground accelerations of 0.25 g to 0.50 g. In the El Centro, Calif. earthquake of May 18, 1940, which covered a much larger area, the recorded accelerogram showed a maximum ground acceleration of 0.32 g. The maximum ground velocity, determined by integrating the accelerogram, was 13.7 in. per sec, and the maximum ground displacement, determined by integrating the velocity diagram, was 8.3 in.

A structure's response to an earthquake primarily depends upon its location in the affected region, its orientation relative to the direction of the most violent motion of the earth, its natural periods of vibration, its damping characteristics, the physical properties of the structural material, and the nature of the foundation material which supports it. Computer programs have been developed which enable these factors to be considered in an analysis of the complete response of the structure to ground motion. However, it is generally unnecessary to make these calculations in the design of a multistory building. The Uniform Building Code recommendations are, in general, consistent with forces and displacements determined by more elaborate procedures. A structure designed according to these recommendations will remain elastic, or nearly so, under moderate earthquakes of frequent occurrence but must be able to yield locally without serious consequences to resist an El Centro type earthquake. Thus, design for the required ductility is an important consideration.[17]

The ductility of the material itself is not a direct indication of the ductility of the structure. Laboratory and field tests and data from operational use of nuclear weapons indicate that structures of practical configurations having frames of ductile materials, or a combination of ductile materials, exhibit ductility factors μ ranging from a minimum of 3 to a maximum of 8. (The ductility factor is the ratio of maximum displacement to the yield displacement.) A minimum ductility factor of about 4 to 6 is a reasonable criterion for ordinary structures designed to UBC earthquake requirements.[17]

The UBC recommends that the minimum total lateral seismic force V, assumed to act nonconcurrently in the direction of each of the main axes of the building, be determined by

$$V = ZKCW \tag{1-5}$$

where

Z = zone factor, which depends on expected severity of earthquake in various regions of the United States
K = coefficient from Table 1-6
W = total dead load (for storage and warehouse occupancies total dead load plus 25 percent of floor live load)

The coefficient C is given by

$$C = \frac{0.05}{T^{1/3}} \tag{1-6}$$

except that for one- and two-story buildings $C = 0.10$. Except for parts and attached elements of buildings, which are discussed later, C need not exceed 0.10. In Eq. (1-6), T is the fundamental period of vibration, in seconds, in the direction considered. In the absence of properly substantiated technical data for the structure, T is determined from

$$T = \frac{0.05h_n}{\sqrt{D}} \tag{1-7}$$

where h_n is the height in feet above the base of the uppermost level in the main portion of the structure and D is the dimension of the building parallel to the applied forces, also in feet.

Equation (1-7) does not apply if the lateral resisting system is a moment-resisting space frame which resists all the lateral forces and which is not enclosed by or adjoined by more rigid elements that tend to prevent it from resisting lateral forces. In this case

$$T = 0.10N \tag{1-8}$$

where N is the number of stories above exterior grade.

The lateral force V is to be distributed over the height of the structure in the following manner.

Table 1-6 HORIZONTAL FORCE FACTOR K FOR BUILDINGS OR OTHER STRUCTURES*

Type or arrangement of resisting elements	K
All building framing systems except as hereinafter classified	1.00
Buildings with a box system†	1.33
Buildings with a dual bracing system consisting of a ductile moment-resisting space frame and shear walls, for which (*a*) the frame and the shear walls are designed to share the lateral force according to their relative rigidities, (*b*) the shear walls, acting independently, are designed to resist the entire lateral force, and (*c*) the space frame is designed to resist not less than 25 percent of the lateral force	0.80
Buildings with a ductile moment-resisting space frame designed to resist the entire lateral force	0.67
Elevated tanks plus full contents, on four or more cross-braced legs, not supported by a building	3.00‡
Structures other than buildings and other than those covered by Eq. (1-10)	2.00

* From Uniform Building Code, International Conference of Building Officials, Los Angeles, Calif., 1970.

† Defined as a structural system without a complete vertical load-carrying space frame, the lateral forces being resisted by shear walls.

‡ KC in Eq. (1-5) shall be at least 0.12 but need not exceed 0.25. Use $J = 1$ in Eqs. (1-11). For tanks not supported in the manner described and for tanks supported by buildings use Eq. (1-10) with $C_p = 0.2$.

A force F_t at the uppermost level n given by

$$F_t = 0.004\left(\frac{h_n}{D_s}\right)^2 \tag{1-9a}$$

A force F_x at each level x, including the uppermost level n, given by

$$F_x = (V - F_t)\frac{w_x h_x}{\sum_{i=1}^{n} w_i h_i} \tag{1-9b}$$

where

w_x, w_i = portion of W located at or assigned to level x, i
h_x, h_i = height of level x, i above the base, ft
D_s = plan dimension of vertical lateral-force resisting system, ft

The force F_t need not exceed $0.15V$ and may be taken zero if $h_n/D_s \lesssim 3$. Equations (1-9) do not apply to one- and two-story buildings, for which V is to be distributed uniformly.

Codes usually specify much larger coefficients for parts and appendages than for the structure itself. This is because such parts may experience accelerations much larger than those of the earthquake motion. The cantilevered parapet wall is an example. For such elements, the UBC specifies the force

$$V = ZC_p W_p \tag{1-10}$$

where W_p is the weight of the part. Values of C_p range from 0.20 for penthouses, chimneys, tanks, and towers attached to buildings, through 1.0 for exterior and interior ornamentations and appendages and for cantilever parapet walls, to 2 for connections of exterior panels attached to or enclosing the exterior.

Overturning moments for buildings of low or medium height can be determined with fair accuracy by assuming the building to be a cantilever beam loaded with the lateral earthquake forces acting simultaneously in the same direction. However, this procedure generally overestimates the moments for tall or slender buildings. Therefore, the UBC specifies the following overturning moment M at the base of the building:

$$M = J\left(F_t h_n + \sum_{i=1}^{n} F_i h_i\right) \tag{1-11a}$$

where $J = 0.6/T^{1/3}$ but need not exceed 1.0. The overturning moment M_x at any level x is given by

$$M_x = J_x\left[F_t(h_n - h_x) + \sum_{i=x}^{n} F_i(h_i - h_x)\right] \tag{1-11b}$$

where $J_x = J + (1 - J)(h_x/h_n)^3$.

1-10 SAFETY OF STRUCTURES

Although it goes without saying that the strength of any member of a structure must be greater than the expected force in it, it is not easy to decide what the reserve strength should be. Reserve strength can be expressed in terms of the *factor of safety*, which is defined as the ratio of the resistance of a structure or structural member to the induced force. We may determine factors of safety with respect to strength, first yielding, acceptable deflection, etc.

The magnitude of the factor of safety must be based upon consideration of the following factors:

1 Variability of the material with respect to strength and other pertinent physical properties
2 Uncertainty in the expected loads in regard to possible future change as well as with respect to present magnitude
3 Precision with which the internal forces in the various parts of a structure are determined
4 Possibility of deterioration due to corrosion and other causes
5 The extent of damage and loss of life which might result from failure
6 Quality of workmanship

Since no two specimens of a given structural material will have identical strengths, it is important to know the pattern of variation in strength with respect to the average of a large number of specimens. Even steel, the manufacture of which is carefully controlled, exhibits an occasional relatively large deviation in strength from the average. Concrete is more variable than steel, and carefully selected samples of the same wood may vary widely in strength and other properties. Strength is therefore a statistical quantity, and we can determine only the probability that it will not be less than a specified value. As an illustration, results of 73 tensile tests on structural silicon-steel shapes for the towers of the Golden Gate Bridge are presented in Fig. 1-7. The ordinates of this figure show the relative frequency of the various tensile strengths (shown as abscissas) revealed by the tests. Strengths ranged from a low of 80.3 ksi to a high of 104.5 ksi, with an average of 88.7 ($\bar{f}_t$ on the figure). The points marked "observed frequencies" represent the number of test specimens having tensile strengths between the values of the abscissas of the vertical boundaries of the corresponding rectangles. The graph of a theoretical probability function chosen to fit the experimental data is labeled "theoretical frequencies." One of the properties of a probability curve is that the area lying under the curve and between any two ordinates gives the probability that the event will lie between the corresponding abscissas. For example, the crosshatched portion marked $P_0 = 3$ percent is an indication that there are only 3 chances in 100 that any specimen of this particular steel will have a tensile strength

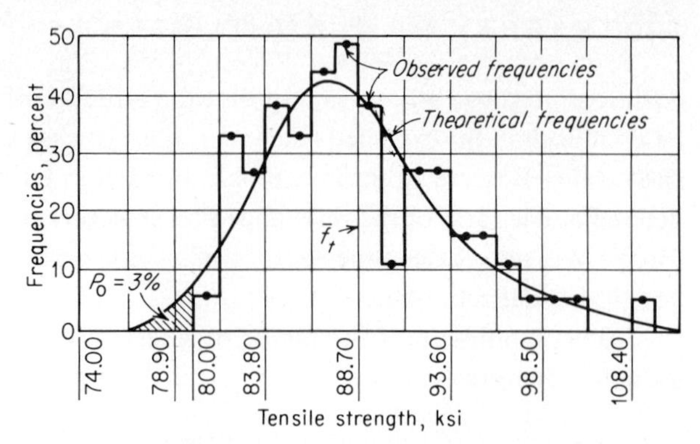

FIGURE 1-7
(*From A. M. Freudenthal, Trans. ASCE, vol. 112, 1947.*)

less than 80 ksi. Other physical properties, such as yield point, modulus of elasticity, and weight, show similar distributions of values about a mean.

The difficulties encountered in predicting live loads have been discussed in preceding articles. Most live loads are amenable to statistical analysis. For example, the similarity of Fig. 1-1, depicting variations in department-store live loads, to Fig. 1-7 suggests that if adequate data were available, we could predict the probability of occurrence of a department-store live load of any prescribed intensity. Statistical analysis of records of wind and snowfall to determine mean recurrence intervals was discussed in Arts. 1-3, 1-7, and 1-8. Because extreme intensities of such live loads can be assessed only in terms of their probabilities of occurrence, it is clear that the factor of safety must bear some relationship to the live-load intensity. Thus, assuming that the probability of failure of structures of a given use or occupancy should be about the same, it would be inconsistent to use the same factors of safety for, say, snow loads based on return periods of 10 years and those based on return periods of 100 years.

Occasionally certain live loads may be more or less rigidly restricted. For example, the amount of water in a storage tank cannot exceed the capacity of the tank. Airplane passenger, baggage, and freight loadings are also controlled, although in a different sense. The railroad can control the loads on its bridges, since they are used exclusively by the railroad. Many live loads are not subject to control, however, and there are obstacles to the control of some which might be restricted in theory. Uncertainty with respect to future increases makes the problem of predicting highway-bridge loadings even more difficult than it might otherwise be.

The internal forces in most structures can be determined with varying degrees of precision, and in general greater precision is attained only as the result of more detailed (and therefore usually more costly) analysis. Other things being equal, smaller factors of safety may be used with more precise knowledge of the forces in a structure. However, it is no advantage to the owner to pay for refined design procedures that do not produce at least an equal reduction in the cost of the structure or, as in the aircraft industry, greater revenue as a result of increased payloads or improved performance made possible by reduced dead load. In any case, there is a limit to the reduction in factor of safety which can accompany more precise evaluation of internal forces, unless it is feasible to verify the predicted structural bchavior by tests. The aeronautical structural engineer designs many airplane components for a factor of safety of only 1.5 with respect to failure but also substantiates many of the designs by tests and subjects the completed ship to extensive flight tests before it is put into service. A comparable program would serve no useful purpose in the case of most civil-engineering structures—in the interests of economy it is cheaper to use larger factors of safety. To be sure, these arguments dodge the larger question of what might be in the best interests of the nation and the world with respect to conservation of our natural resources.

Loss of strength as a result of corrosion sometimes must be considered. This is particularly true with steel. The designer must evaluate the hazards, not only with respect to exposure, but also with respect to preventive maintenance. Thin parts are more susceptible to severe damage from corrosion than thick parts. Corrosion hazards are greater for structures located near bodies of salt water. Exposure to chemical wastes, e.g., the sulfurous products of the combustion of coal, increases the danger of corrosion.

Probability of failure should not necessarily be the same for all parts of a structure or for all structures of a given class. For example, the failure of one of the floor-beam hangers of a suspension bridge would be less serious than the failure of one of the suspension cables. Localized damage would result from failure of the hanger, but failure of the cable would almost certainly precipitate a collapse of the entire floor. Similarly, the failure of a small highway bridge would not ordinarily be so disastrous as failure of a Golden Gate Bridge. Furthermore, abandonment of the smaller bridge at some later time as a result of its having insufficient reserve to support loads which may have turned out to be heavier than predicted would not be so economically significant as abandonment of a Verrazano-Narrows Bridge.

1-11 PROBABILISTIC CONSIDERATIONS OF SAFETY

The factor of safety was defined in Art. 1-10 as the ratio of resistance to induced force. However, it was also shown that there is an element of uncertainty in resistance, in the sense that there is a range of values which cluster around a mean (Fig. 1-7). Furthermore, it is clear that no absolute minimum tensile strength of any particular

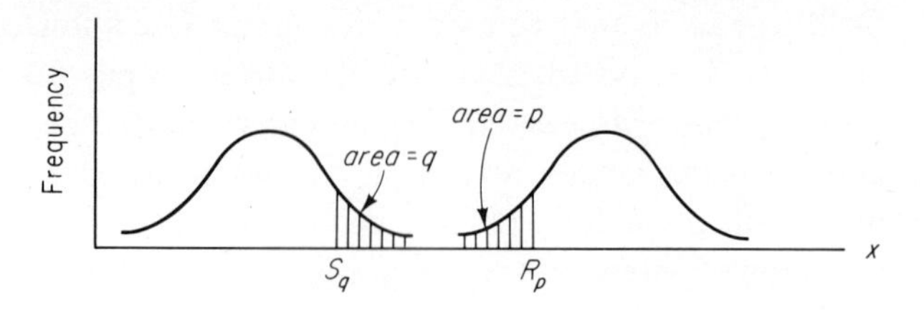

FIGURE 1-8

grade of steel can ever be known. Similarly, it was shown in preceding articles that most of the live loads for which structures are designed are also uncertain (Fig. 1-1). Again, no absolute maximum value can be determined for loads of this type. Because of these uncertainties, the reliability of a structure (conversely, the probability that it will fail) is difficult to determine. Probabilistic evaluation of structural safety will be discussed briefly in this article.

Figure 1-8 shows the frequency distributions of load S (or of the load effect, such as an induced force in a structural member) and of member resistance R (such as yield strength, tensile strength, etc.). Suppose S_q is the load effect caused by, say, the annual extreme wind speed for which the mean return period is 50 years. Points on x to the right of S_q represent effects of winds of higher velocity (return period greater than 50 years), and the shaded area q is the probability that S_q will be exceeded in any one year. Similarly, let R_p be the member resistance which is chosen for purposes of design. The area p gives the probability that the resistance will be less than R_p. By definition, the factor of safety n is R_p/S_q. This factor of safety must be determined so that the probability of failure is acceptably small.

The probability of failure p_f can be determined as follows.[19] The probability that the load effect S will be between the values x and $x + dx$ (Fig. 1-9) is $f_S(x)\,dx$, where $f_S(x)$ is the ordinate at x of the frequency distribution. The probability that the resistance R will be less than x is the shaded area $p(x)$ of the frequency distribution of R. Since these two events are independent, the probability that they occur simultaneously is the product of each separately. Therefore, the probability $R < S$ for $S = x$ is given by $p(x)f_S(x)\,dx$. Summing this for all values of S gives

$$p_f = \int_0^\infty p(x)f_S(x)\,dx \qquad (a)$$

If the frequency distributions are known, p_f can be evaluated. A variety of formulas is available for this purpose.

The probability p_f given by Eq. (a) is the probability of failure under a single application of load and is not a direct measure of the safety of a structure which must withstand repeated applications of load. Therefore, the probability of failure under a application of load is of interest largely to the extent that it can be used to

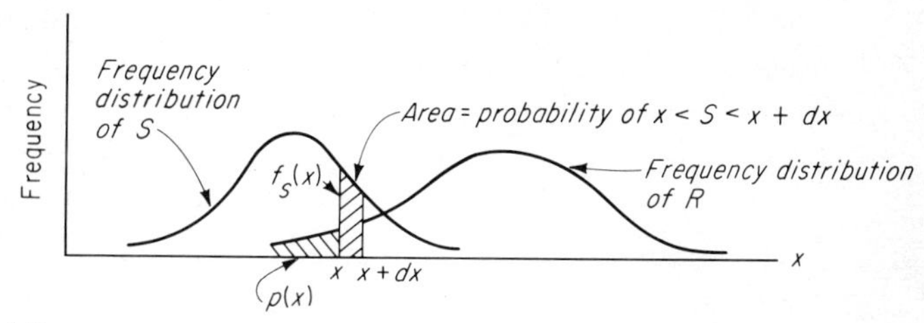

FIGURE 1-9

determine the probability that a structure can survive a random sequence of loads during its life. This probability is called the *probability of survival*. Procedures for evaluating it have been developed.[19, 20]

An additional complicating factor in evaluating the safety of structures has to do with a significant difference in the failure modes of statically determinate and statically indeterminate structures. A statically determinate structure can be expected to collapse if one of its members fails. On the other hand, a statically indeterminate structure does not necessarily collapse if one or more of its members fail. This problem is discussed briefly in Ref. 20.

Evaluations of probabilities of failure and of survival are very difficult to implement. Not only is there a lack of statistical data, but calculated probabilities are extremely sensitive to variations in the extreme values, i.e., the rare large values of S and the rare small values of R. Furthermore, there are other uncertainties and unknowns, some of which are not statistical or probabilistic. These difficulties have been summarized by Freudenthal, as follows:[21]

1. The existence of nonrandom phenomena affecting structural safety which cannot be included in a probabilistic approach
2. The impossibility of observing the relevant random phenomena within the ranges that are significant for safety analysis and the resulting necessity of extrapolation far beyond the range of actual observation
3. The assessment and justification of a numerical value for the "acceptable risk" of failure
4. The codification of the results of the rather complex probabilistic safety analysis in a simple enough form to be usable in actual design

1-12 FACTOR OF SAFETY AND LOAD FACTOR

The concept of factor of safety may be applied in either of two ways:

1. The expected resistance of the structural member, or other component, usually expressed as a tensile stress, compressive stress, etc., is divided by a

factor of safety to obtain an *allowable* or *working* stress, and the part is then chosen so that the stress induced by the expected service load is equal to or less than the allowable value. This procedure is called *allowable-stress design*, but because the force analysis is usually by methods based on Hooke's law, it is more often (and less accurately) called *elastic design*.

2 The structural member or other component is chosen so that its expected resistance equals or exceeds the expected service load times a factor of safety. With this procedure it is a simple matter to account for differing reliabilities in the prediction of loads. For example, dead load, which can usually be determined with good precision, can be multiplied by a smaller number than those used for live loads. Thus, the factors by which the various loads are multiplied are generally not equal, so that they differ somewhat in concept from the factor of safety. Therefore, they are called *load factors*, and the corresponding design procedure is called *load-factor design*.

Allowable-stress design and load-factor design are both used in the AISC and AASHO specifications. In the AISC specification, the allowable-stress procedure is used if the force analysis of the structure is based on elastic behavior, while the load-factor procedure is used if the force analysis takes the effects of inelastic behavior into account. The latter procedure is called *plastic design*; it is discussed in Chapter 8. The AASHO specifications permit either allowable-stress or load-factor design but require an elastic analysis of the structure in either case.

1-13 CODES AND SPECIFICATIONS

Standard design specifications and building codes have been developed as a means of supplying the engineer with a digest of the collective knowledge, judgment, and experience of his profession. They are intended to convey the pertinent information relative to service loads and member resistance. In addition, they attempt to cover in a general way the questions of form and proportion of a structure and its members, together with their connections, and acceptable methods of analysis, fabrication, erection, and construction. Specification writers evaluate the implications of the various clauses with a view to public safety, utility, and economy. Finally, codes provide the building official with an enforceable document.

The time lag between publication of research data and their incorporation into specifications is sometimes long, especially at the municipal level. Thus, in order to take advantage of the latest developments, the designer may sometimes have to file an appeal with the building department.

REFERENCES

1 Thom, H. C. S.: Distributions of Extreme Winds in the United States, *J. Struct. Div. ASCE*, April 1960.

2 Dunham, J. W.: Design Live Loads in Buildings, *Trans. ASCE*, vol. 112, 1947.

3 Bryson, J. O., and D. Gross: Techniques for the Survey and Evaluation of Live Floor Loads and Fire Loads in Modern Office Buildings, *Nat. Bur. Stand. Build. Sci. Ser.* 16, 1967.

4 Dunham, J. W., G. N. Brekke, and G. N. Thompson: Live Loads on Floors in Buildings, *NBS Build. Mater. Struct. Rep.* BMS133, December 1952.

5 Snow Load Studies, *Housing Home Finance Agency Washington, Housing Res. Pap.* 19, May 1952.

6 Andersen, A. E.: Snow Loads for Roofs, *Civ. Eng.*, July 1965.

7 Climatic Information for Building Design in Canada, *Nat. Res. Counc. Ottawa, Suppl.* 1 to *Nat. Building Code of Canada*, 1961.

8 Thom, H. C. S.: New Distributions of Extreme Winds in the United States, *J. Struct. Div. ASCE*, July 1968.

9 Wind Force on Structures, Final Report, Task Committee on Wind Forces, Committee on Loads and Stresses, Structural Division, ASCE, *Trans. ASCE*, vol. 126, 1961, pt. II.

10 Davenport, A. G.: Wind Loads on Structures, *Nat. Res. Counc. Canada, Ottawa, Div. Building Res.* Tech. Pap. 88, 1960.

11 Sherlock, R. H.: Gust Factors for the Design of Buildings, *Int. Assoc. Bridge Struct. Eng.*, Zurich, *Publ.* vol. 8, 1947.

12 Vellozzi, J., and E. Cohen: Gust Response Factors, *J. Struct. Div. ASCE*, June 1960.

13 Scruton, C.: Aerodynamics of Structures, Paper No. 4 in "Wind Effects on Buildings and Structures," Univ. of Toronto Press, 1968.

14 Wind Bracing in Steel Buildings, Final Report of Subcommittee 31, Committee on Steel of the ASCE Structural Division, *Trans. ASCE*, vol. 105, 1940.

15 Standards of the Swiss Association of Engineers and Architects on Load Assumptions, Acceptance and Supervision of Buildings, *Schweiz. Ing. Architek. Ver., Tech. Normen* 160, 1956.

16 McGuire, W.: "Steel Structures," Prentice-Hall, Englewood Cliffs, N.J., 1968.

17 Newmark, N. M.: Earthquake-resistant Building Design, sec. 3 in E. H. Gaylord and C. N. Gaylord (eds.), "Structural Engineering Handbook," McGraw-Hill, New York, 1968.

18 Freudenthal, A. M.: The Safety of Structures, *Trans. ASCE*, vol. 112, p. 125, 1947.

19 Freudenthal, A. M., J. M. Garrelts, and M. Shinozuka: The Analysis of Structural Safety, *J. Struct. Div. ASCE*, February 1966.

20 Ang, A. H.-S., and M. Amin: Reliability of Structures and Structural Systems, *J. Eng. Mech. Div. ASCE*, April 1968.

21 Freudenthal, A. M.: Critical Appraisal of Safety Criteria and Their Basic Concepts, *Prelim. Publ., 8th Congr. IABSE*, New York, 1968.

2

STRUCTURES, METALS, AND FASTENERS

The supporting systems used in buildings, bridges, transmission towers, radio towers, observation towers, airplanes, ships, storage vessels, and various other structures may be classified broadly into three categories: (1) framed systems, (2) suspension systems, and (3) shell systems. Examples of these systems are discussed in following articles. In many cases, structures are supported by combinations of supporting systems.

2-1 BUILDINGS

In addition to their function of partitioning space, walls in *bearing-wall* construction furnish reactions for the beams and joists which support the floors and roof. *Frame construction* uses a stable system of structural members to support *curtain walls* as well as the floor and roof construction. Frames are classified as *rigid, semirigid,* and *simple,* depending upon the stiffness of the beam-to-column connections. Framed buildings are also called *tier* buildings.

Simple-frame construction is illustrated in Fig. 2-1. Floors and roof in such buildings are supported on joists which are supported by beams framing into the

FIGURE 2-1
Simple frame with open-web steel joists.

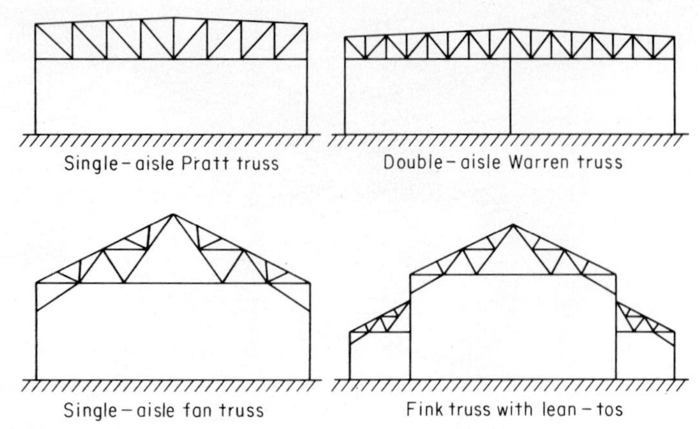

FIGURE 2-2
Types of industrial-building bents.

columns. Beam-to-column connections in simple frames are essentially rotationally unrestrained.

Because of the varied uses to which they are put, commercial and industrial buildings are framed in a number of ways. Frequently they have only a ground floor with the roof supported on beams or trusses. The latter may be supported by masonry walls or by columns. Sheet siding may be used instead of masonry. Industrial buildings often have two lines of columns forming a single-aisle structure. However, they may have one or more duplicate aisles, or additional aisles in the form of lean-tos.

Trusses which support flat roofs are usually of the Warren or Pratt type. Pitched roofs may be supported on Fink or fan trusses. A truss and its supporting columns is called a *bent*, and the space between successive bents is called a *bay*. The longitudinal beams which support the roof covering are called *purlins*. Some types of building bents are shown in Fig. 2-2.

Rigid frames may be single-span or multispan and single-story or multistory. The transverse member in the single-story frame may be straight, V-shaped, or arched. Single-story rigid frames are used in assembly halls, churches, gymnasiums, field houses, hangars, industrial buildings, etc., and have been built with spans of over 200 ft (Fig. 2-3). Steel arches are usually used for longer spans and, in general, for the same types of building. Arches may be trussed or of solid-webbed I or box cross section. They have been built with spans exceeding 350 ft.

Steel-framed domes may take a number of forms. The ribbed dome has radial members which frame into a compression ring at the crown and a tension ring or foundation at the perimeter (Fig. 2-4). The ribs may be solid-webbed or trussed. Domed structures may also be latticed. In the lamella dome secondary framing

FIGURE 2-3

FIGURE 2-4
Athletic and Convocation Center, University of Notre Dame. (*American Institute of Steel Construction.*)

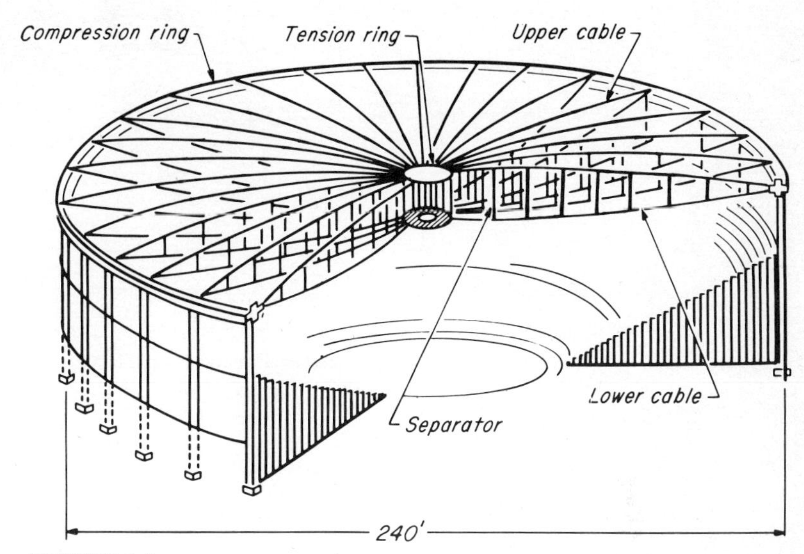

FIGURE 2-5
Utica Municipal Auditorium. Gehron & Seltzer, Architects. Lev Zetlin & Associates, Structural Engineers.

members parallel to the main ribs extend from the perimeter to their intersections with the ribs, forming a diamond-shaped pattern. Domed roofs have been built with spans exceeding 400 ft.

Cable-supported roofs may be suspended from a series of cables supported by towers and abutments, as in a suspension bridge. Another type consists of radial cables stretched between a central tension ring and an exterior compression ring, with a precast or cast-in-place concrete roof deck supported on the cables. These roofs may develop oscillatory motions, usually called *flutter,* from the effect of dynamic forces. Flutter may be controlled by tying the roof to the ground with cables or by using a heavy covering such as concrete, or it may be eliminated by using a system of interconnected cables which are internally self-damping. The Municipal Auditorium in Utica, N.Y. (Fig. 2-5) is an example of the internally self-damping system. Suspension roofs have been built with diameters of over 350 ft. Cables may be also used to support cantilever roofs (Fig. 2-6).

Shells may be built with standard cold-formed steel roof-deck panels. The *folded plate* and the *hyperbolic paraboloid* are suitable for this type of structure. In the folded-plate roof, flat or inclined light-gage steel panels span between ridge and valley members. The panels and ridge and valley members act together as longitudinal beams or girders. In addition, the panels themselves act as transverse beams to support

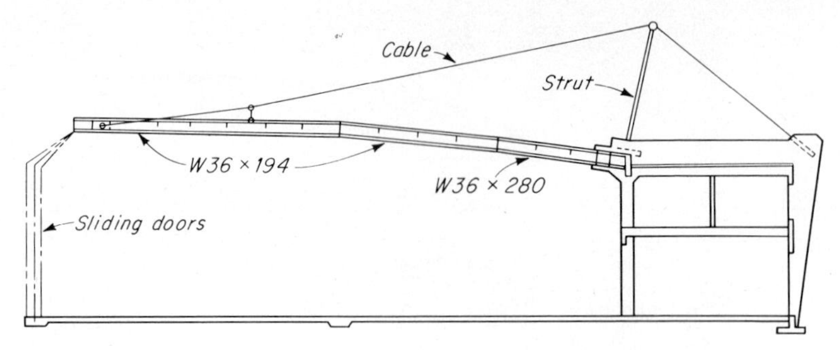

FIGURE 2-6
Framing system for TWA hangar, Philadelphia International Airport.

FIGURE 2-7
Folded-plate roof. Marple Township Library, Pennsylvania. (*American Institute of Steel Construction.*)

(a)

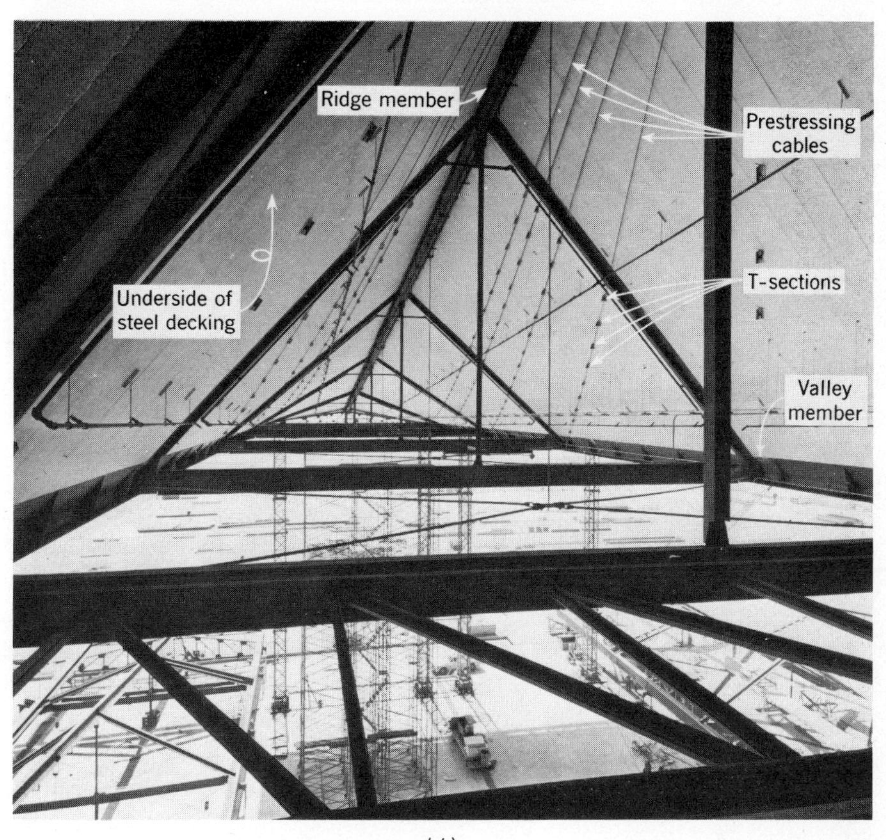

(b)

FIGURE 2-8
American Airlines superbay hangar, San Francisco. Lev Zetlin Associates, Inc., Consulting Engineers, New York.

snow or other live load (Fig. 2-7). Hyperbolic paraboloids are doubly curved shells made up of two families of straight generators intersecting at right angles. The steel panels are laid in the direction of one set of generators, which requires a slight warping of the individual panel (Fig. 2-8).

2-2 FIXED BRIDGES

Most bridges are built for the transportation of highway or railway traffic across natural or artificial obstacles. A *deck bridge* supports the roadway on its top chords or flanges, while a *through bridge* supports the floor system at or near the lower chords or flanges, so that traffic passes through the supporting structure.

The rolled-beam bridge supports its roadway directly on the top flanges of a series of rolled beams placed parallel to the direction of traffic and extending from abutment to abutment. It is simple and economical. It may also be used for multiple spans where piers or intermediate bents can be built economically. Beam bridges may be economical for spans up to about 60 ft. A typical beam bridge for highway traffic is illustrated in Fig. 2-9.

For crossings greater than those which can be spanned economically by a rolled-beam bridge, deck or through plate-girder bridges may be used. In its simplest form, a plate girder consists of three plates welded together in the form of an I. Ties and rails for railway bridges rest directly on the top flanges of the deck plate-girder bridge. When clearance below the structure is limited, a through girder span is used. The floor system may consist of a single line of stringers under each rail, supported by floor beams framing into the girders just above their lower flanges. If an open floor is objectionable, ballast may be laid on concrete or steel-plate decking supported by closely spaced floor beams without stringers. Knee braces are used to support the top flanges of through bridges, as illustrated in Fig. 2-10. Highway plate-girder bridges are usually of the deck type. The floor slab is usually supported directly on the girders, as in the beam bridge of Fig. 2-9.

In orthotropic steel-deck plate construction the floor consists of a steel deck plate stiffened in two mutually perpendicular directions by a system of longitudinal and transverse ribs welded to it (Fig. 2-11). The deck structure functions as the top flange of the main girders and floor beams. This system makes efficient and economical use of materials, particularly for long-span construction.

When the crossing is too long to be spanned economically by plate girders, a through or deck truss bridge may be used. Deck bridges are more economical than through bridges because the trusses can be placed closer together, so that the span of the floor beam is shortened. For multiple spans there is also a saving in the height of the piers.

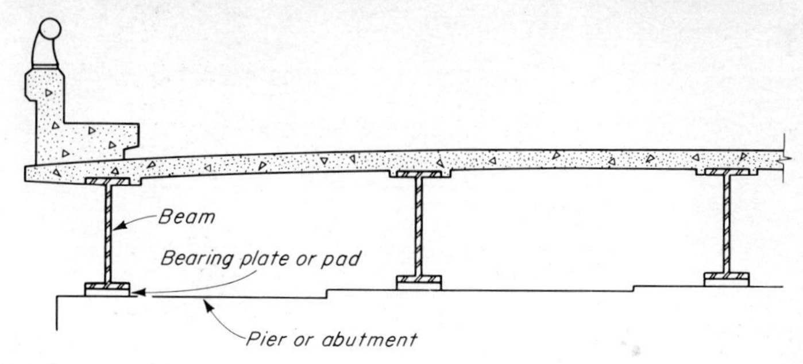

FIGURE 2-9
Beam bridge.

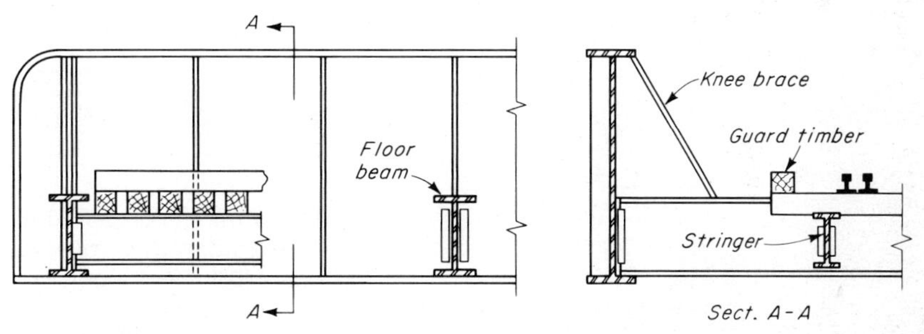

FIGURE 2-10
Plate-girder railroad bridge.

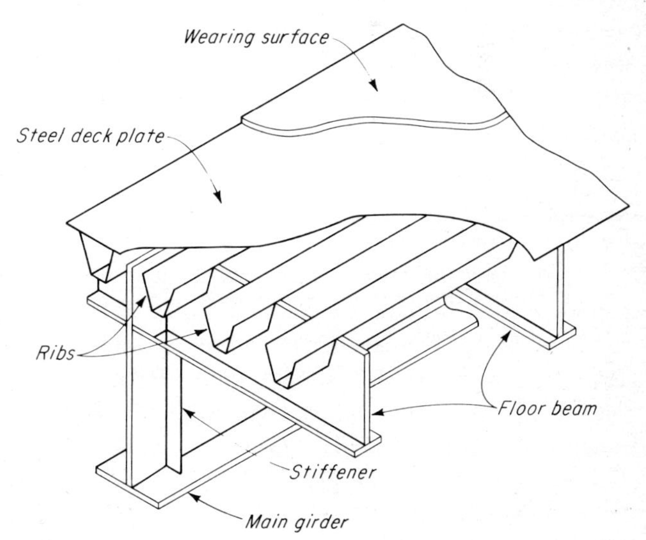

FIGURE 2-11
Elements of steel-deck bridges.

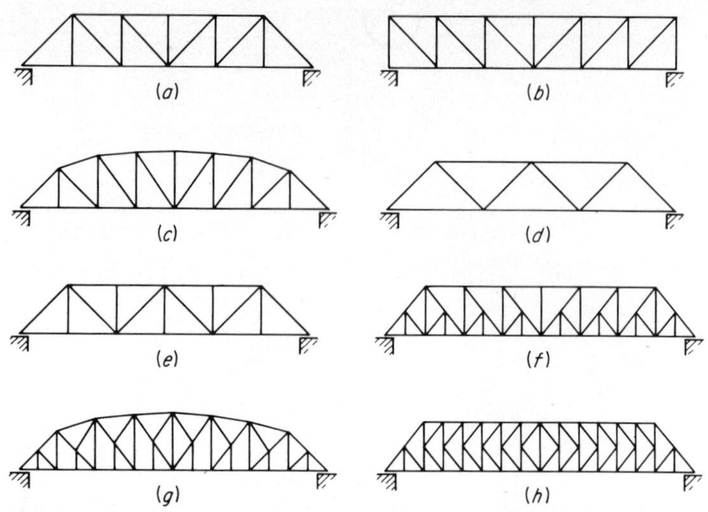

FIGURE 2-12
Common types of bridge trusses: (*a*) Pratt truss through bridge; (*b*) Pratt truss deck bridge; (*c*) curved-chord Pratt truss; (*d*) Warren truss; (*e*) Warren truss with verticals; (*f*) subdivided Pratt truss (Baltimore truss); (*g*) Petit or Pennsylvania truss; (*h*) K truss.

Figure 2-12 shows the common types of simple-span bridge trusses. By varying the depth of a truss throughout its length (Fig. 2-12*c*) forces in the chord members can be more nearly equalized and the forces in the web reduced. Trusses of economical proportions usually result if the angle between diagonals and verticals ranges from 45 to 60°. However, if long-span trusses are made deep enough for adequate rigidity as well as for economy, a suitable slope of the diagonals may produce panels too long for an economical floor system. The subdivided panels of the Baltimore and Petit trusses (Fig. 2-12*f* and *g*) solve this problem. Certain objections to subdivided panels were overcome with the invention of the K truss (Fig. 2-12*h*).

Cantilever bridges, continuous bridges (Fig. 2-13), arch bridges (Fig. 2-14), and suspension bridges are common types of structures suitable for long spans. A cantilever bridge consists of two shore, or anchor, spans flanked by cantilever arms supporting a suspended simple span. Positive bending moments are decreased because of the shorter simple span, while the cantilever and anchor arms are subjected to negative moments. Positive bending moments in continuous bridges are reduced because of the negative moments at the piers. Arch bridges may be fixed, single-hinged, two-hinged, or three-hinged. The principal supporting elements of the suspension-bridge superstructure are the cables which pass over the towers to be anchored in foundations at each end.

FIGURE 2-13
Rio Grande Gorge Bridge, Taos County, New Mexico. New Mexico State Highway Commission. (*American Institute of Steel Construction.*)

FIGURE 2-14
Satsop River Bridges, Satsop, Wash. Washington State Highway Commission.
(*American Institute of Steel Construction.*)

2-3 TOWERS

Towers are used to support transmission lines, radio and television antennas, radar and microwave equipment, tanks, bridges, etc. Freestanding towers (such as transmission towers and some radio and television towers) are usually rectangular in plan. The tall towers used in the electronic industry are usually guyed. The plan is usually an equilateral triangle, with the legs at the vertices. Most steel-tower members are hot-dipped galvanized for weather protection.

Aluminum towers are usually made of alloy 6061-T6, which is corrosion-resistant in all climates without surface protection. It can be extruded to produce optimum shapes for resisting stress and simplifying joints. Because of their lighter weight, aluminum towers can be flown by helicopter to relatively inaccessible sites.

2-4 MECHANICAL PROPERTIES OF STRUCTURAL METALS

The design of structures which support calculated loads at specified stresses is based on the assumption that certain mechanical properties of structural materials can be depended upon to meet definite requirements. The most widely used standards for structural materials are those of the American Society for Testing and Materials (ASTM).

Two typical stress-strain curves for structural-steel coupons tested in tension are shown in Fig. 2-15*a*. The tensile strength is the highest stress, based on the original cross-sectional area. After reaching this maximum stress, a localized reduction in area, called *necking,* begins, and elongation continues with diminishing load until the specimen breaks.

The *yield point* is the stress at which there is a marked increase in strain with no increase in load. The increase in strain may be as large as 1.5 or 2 percent; this is sometimes called *plastic strain*. The subsequent increase in stress, which continues until the tensile strength is reached, is called *strain hardening*. Yielding is sometimes accompanied by an abrupt decrease in load, as shown in the lower curves of Fig. 2-15, which results in *upper* and *lower* yield points. The upper yield point is influenced considerably by the shape of the test specimen and by the testing machine itself, and is sometimes completely suppressed. The lower yield point is much less sensitive and is considered to be more representative. Stress-strain curves of this type are typical of low-carbon (mild) steels. Both upper yield and lower yield tend to increase with increase in speed of loading (strain rate). The lower yield stress at zero strain rate is called the static yield level. It may be as much as 10 to 15 percent lower than the yield stress reported in rolling-mill acceptance tests.

High-carbon steels do not usually have a pronounced yield point. Instead,

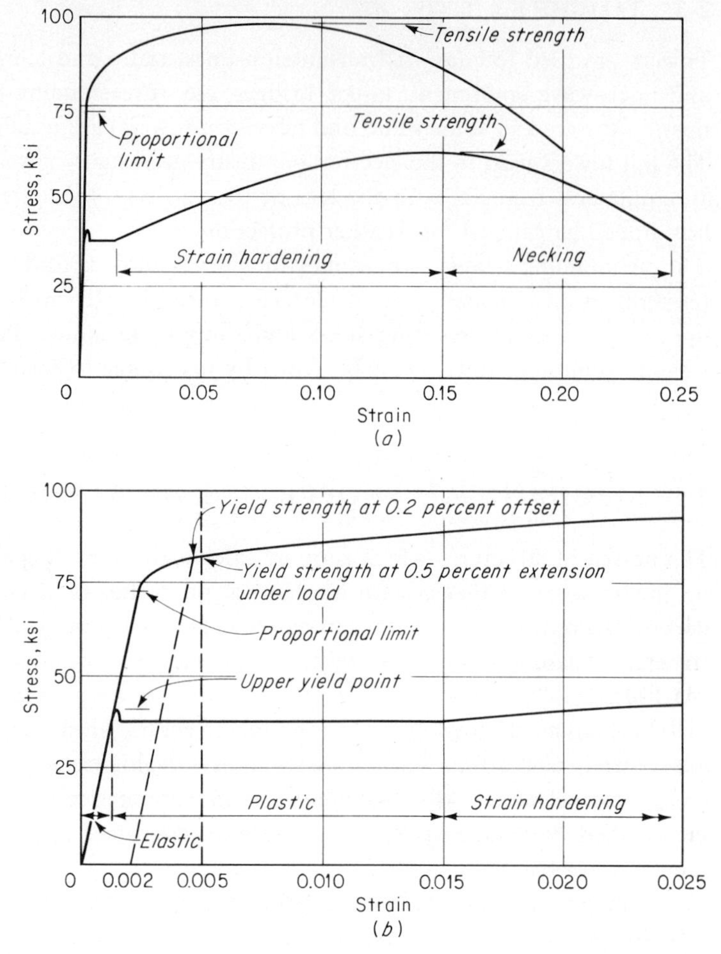

FIGURE 2-15

after a range of linearly elastic behavior which ends at the *proportional limit,* the rate of increase in stress begins to drop and continues to fall until the tensile strength is reached (upper curve of Fig. 2-15*a*). In this case, yielding is defined arbitrarily by a *yield strength,* which is usually taken to be that stress which leaves the specimen with a permanent set (plastic elongation) of 0.2 percent when the specimen is unloaded (Fig. 2-15*b*). However, yield strength may also be defined (ASTM Specification A370) as the stress corresponding to a 0.5 percent elongation under load (Fig. 2-15*b*).

The term *yield stress* is commonly used to mean either yield point or yield strength when it is not necessary to make the distinction.

Steel in compression has the same modulus of elasticity as in tension. The lower yield stress is also the same, and there is about the same length of level yielding (contraction).

The horizontal portion of the lower curve in Fig. 2-15*b* is not strictly a stress-strain curve for the material. This is because the unrestricted yielding which it represents is not a continuous process in the sense that the elongation is distributed uniformly over the length of the specimen. Instead, yielding is a discontinuous phenomenon. In a tension test, it generally begins with the sudden appearance in the specimen of one or more narrow slip bands, which are also called *Lüder's lines, flow lines,* or *yield lines.* These lines are usually inclined at about 45° to the direction of the tension. They result from sliding along the inclined planes, which are planes of maximum shearing stress. The strains in these planes of sliding increase suddenly from the yield strain to values of the order of 0.02 to 0.04. Flow lines are easily detected on a specimen whose surfacc has been polished. They are also revealed by the flaking of mill scale on members of hot-rolled steel. They can be detected more readily if the specimen is whitewashed, in which case they appear as dark lines or bands (Fig. 6-5).

Slip bands are plastic regions which are separated from one another by completely elastic regions. Thus, the elongation which is measured over a region yielded in tension is actually the sum of a sequence of alternating layers of elastic and plastic strains. Once they start, slip bands spread and increase in thickness until the yielded region is completely strain-hardened, at which stage stress begins to increase again.

Aluminum alloys do not have a pronounced yield point, so that the stress-strain curve is similar to the upper curve of Fig. 2-15*a*. The elongations of the aluminum alloys commonly used for structural purposes are roughly only half that of the structural steels (compare Tables 2-2 and 2-5). The modulus of elasticity is about 10,000 ksi, compared to 30,000 ksi for steel. Thus, the elastic deformation of an aluminum structure will be three times that of an identically loaded steel structure of the same dimensions.

Ductility implies a large capacity for inelastic deformation without rupture (opposite of *brittleness*). It is usually measured by the elongation, in percent, of a specific length, called the *gage length,* or by the reduction, in percent, of the cross-sectional area. It is also measured by a cold-bend test, which consists in bending a specimen 180° around a pin. The diameter of the pin is related to the thickness of the specimen and also varies with the steel. Ductility is an extremely important property, but there is no generally accepted minimum which is required of steels for structures.

Other important properties of structural metals are fatigue strength, resistance to brittle fracture, and toughness. *Fatigue* is a progressive, localized permanent damage under fluctuating stress that usually results in cracks which may eventually

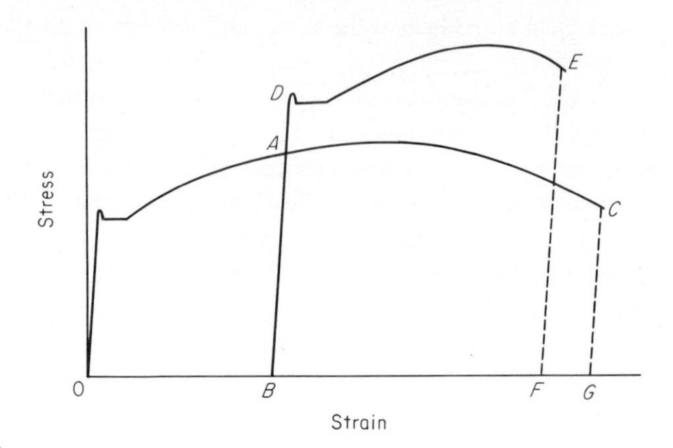

FIGURE 2-16

lead to complete fracture (Art. 2-15). *Brittle fractures* are usually catastrophic, low-ductility fractures that propagate rapidly at relatively low stresses. They often initiate at nominal stresses equal to or less than the yield strength and propagate at stresses as low as 20 percent of the yield strength (Art. 2-16). *Toughness* denotes the capacity of a material to resist fracture under impact loading. The area under the stress-strain curve is a measure of toughness, so that both strength and ductility contribute to toughness. Thus, because its modulus of elasticity is only one-third that of steel, an aluminum structural member can absorb three times as much energy, for the same stress, as a steel member of the same dimensions, provided the stress does not exceed the proportional limit.

Strain rate and strain aging are sometimes important in assessing the suitability of metals for structural purposes. Both yield stress and tensile strength increase with increase in strain rate. These increases can be substantial under tension impact. Strain aging has to do with a change in properties during a rest period following unloading of a specimen which has been stressed into the inelastic range. Such an unloading path is parallel to the elastic loading path of the material (*AB* of Fig. 2-16). If the specimen is immediately reloaded, it retraces the unloading path to the level of stress at which unloading began and, if strain continues, follows the path it would have taken had no unloading occurred (*BAC* of Fig. 2-16). However, if the specimen is allowed to "age" at room temperature for a few days after it is unloaded, it may reload along *BADE.* Aging for 1 or 2 weeks may result in marked increases in the proportional limit, the yield strength, and the ultimate strength.[12] It will be noted that these increases are accompanied by a loss in ductility (*BF,* compared with *BG,* in Fig. 2-16).

2-5 STRUCTURAL METALS

A wide variety of steels is produced for the construction of bridges, buildings, towers, tanks, and other structures. Those used regularly in bridges and buildings are described in Table 2-1. Tensile properties are given in Tables 2-2 to 2-4. Producers

Table 2-1 STEEL FOR STRUCTURAL PURPOSES

ASTM designation	Product	Use
A36	Carbon-steel shapes, plates, and bars	Welded, riveted, and bolted construction; bridges, buildings, towers, and general structural purposes
A53	Welded or seamless pipe, black or galvanized	Welded, riveted, and bolted construction; primary use in buildings, particularly columns and truss members
A242	High-strength, low-alloy shapes, plates, and bars	Welded, riveted, and bolted construction; bridges, buildings, and general structural purposes; atmospheric-corrosion resistance about four times that of carbon steel; a weathering steel
A245	Carbon-steel sheets, cold- or hot-rolled	Cold-formed structural members for buildings, especially standardized buildings; welded, cold-riveted, bolted, and metal-screw construction
A374	High-strength, low-alloy, cold-rolled sheets and strip	Cold-formed structural members for buildings, especially standardized buildings; welded, cold-riveted, bolted, and metal-screw construction
A440	High-strength shapes, plates, and bars	Riveted or bolted construction; bridges, buildings, towers, and other structures; atmospheric-corrosion resistance double that of carbon steel
A441	High-strength low-alloy manganese-vanadium steel shapes, plates, and bars	Welded, riveted, or bolted construction but intended primarily for welded construction; bridges, buildings, and other structures; atmospheric-corrosion resistance double that of carbon steel
A446	Zinc-coated (galvanized) sheets in coils or cut lengths	Cold-formed structural members for buildings, especially standardized buildings; welded, cold-riveted, bolted, and metal-screw construction
A500	Cold-formed welded or seamless tubing in round, square, rectangular, or special shapes	Welded, riveted, or bolted construction; bridges, buildings, and general structural purposes
A501	Hot-formed welded or seamless tubing in round, square, rectangular, or special shapes	Welded, riveted, or bolted construction; bridges, buildings, and general structural purposes

Table 2-1 (continued)

ASTM designation	Product	Use
A514	Quenched and tempered plates of high yield strength	Intended primarily for welded bridges and other structures; welding technique must not affect properties of the plate, especially in heat-affected zone
A529	Carbon-steel plates and bars to $\frac{1}{2}$ in. thick	Buildings, especially standardized buildings; welded, riveted, or bolted construction
A570	Hot-rolled carbon-steel sheets and strip in coils or cut lengths	Cold-formed structural members for buildings, especially standardized buildings; welded, cold-riveted, bolted, and metal-screw construction
A572	High-strength low-alloy columbian-vanadium steel shapes, plates, sheet piling, and bars	Welded, riveted, or bolted construction of buildings in all grades; welded bridges in grades 42, 45, and 50 only
A588	High-strength low-alloy steel shapes, plates, and bars	Intended primarily for welded bridges and buildings; atmospheric-corrosion resistance about four times that of carbon steel; a weathering steel
A606	High-strength, low-alloy hot- and cold-rolled sheet and strip	Intended for structural and miscellaneous purposes where savings in weight or added durability are important

Table 2-2 MINIMUM TENSILE PROPERTIES OF STRUCTURAL STEELS

ASTM designation	Yield, ksi	Strength, ksi	Elongation, % (in 8 in. unless noted)
Carbon steels:			
A36	36	58–80	20
A529	42	60–85	19
High-strength steels:			
A242, A440, A441:			
To $\frac{3}{4}$ in. thick	50	70	18
Over $\frac{3}{4}$ in. to $1\frac{1}{2}$ in.	46	67	19
Over $1\frac{1}{2}$ in. to 4 in.	42	63	16
A572:			
Grade 42, to 4 in. incl.	42	60	20
Grade 45, to $1\frac{1}{2}$ in. incl.	45	60	19
Grade 50, to $1\frac{1}{2}$ in. incl.	50	65	18
Grade 55, to $1\frac{1}{2}$ in. incl.	55	70	17
Grade 60, to 1 in. incl.	60	75	16
Grade 65, to $\frac{1}{2}$ in. incl.	65	80	15
A588:			
To 4 in. thick	50	70	19–21*
Over 4 in. to 5 in.	46	67	19–21*
Over 5 in. to 8 in.	42	63	19–21*
Quenched and tempered steels:			
A514:			
To $2\frac{1}{2}$ in. thick	100	115–135	18*
Over $2\frac{1}{2}$ in. to 4 in.	90	105–135	17*

* In 2 in.

Table 2-3 MINIMUM TENSILE PROPERTIES OF STRUCTURAL PIPE AND TUBING

ASTM designation	Yield, ksi	Strength, ksi	Elongation, % (in 8 in. unless noted)
Welded or seamless pipe:			
A53:			
Grade A	30	48	*
Grade B	35	60	*
Welded or seamless tubing:			
A500 (cold-formed):			
Round, grade A	33	45	25†
Round, grade B	42	58	23†
Shaped, grade A	39	45	25†
Shaped, grade B	46	58	23†
A501 (hot-formed), shaped	36	58	20

* Varies, see specification.

† In 2 in.

Table 2-4 MINIMUM TENSILE PROPERTIES OF SHEET AND STRIP STEELS

ASTM designation	Yield, ksi	Strength, ksi	Elongation, % (in 8 in. unless noted)
Carbon steels:			
A245:			
Grade A	25	45	18–20
Grade B	30	49	17–19
Grade C	33	52	16–18
Grade D	40	55	14–16
A570:			
Grade A	25	45	18–20
Grade B	30	49	17–19
Grade C	33	52	16–18
Grade D	40	55	14–16
Grade E	42	58	12–14
Low-alloy steels:			
A374	45	65	20–22*
Zinc-coated (galvanized):			
A446:			
Grade A	33	45	20*
Grade B	37	52	18*
Grade C	40	55	16*
Grade D	50	65	12*
Grade E†	80	82	‡

* In 2 in.

† This grade is a full-hard product for roofing and similar applications. Properties are obtained by cold working.

‡ Not specified.

have given many of these steels brand names, some of which are not covered by ASTM specifications. Only ASTM steels are discussed here.

Steels for structural purposes can be classified in various ways, but are commonly called carbon steels, high-strength steels, high-strength low-alloy steels, and quenched and tempered steels. This terminology is somewhat confusing and inconsistent, however. All steels contain carbon, which is the most important element except for the ferrite itself, but carbon steels are generally understood to mean steels whose properties are controlled largely by controlling the carbon content. Most of the high-strength steels are really of intermediate strength compared to the quenched and tempered steels. It will be noted that the yield stresses in Table 2-2 range from 36 to 42 ksi for carbon steel, 42 to 65 ksi for the high-strength steels, and 90 to 100 ksi for quenched and tempered steels.

Structural steel is produced in the form of shapes and flat-rolled products. Flat-rolled steel is called *bar, plate, sheet,* or *strip,* depending on its width and thickness, and may be either cold-rolled or hot-rolled. Tensile properties of the steels furnished in shapes, plates, and bars are given in Table 2-2; properties of structural pipe and tubing are given in Table 2-3 and of sheet and strip steels in Table 2-4. All the steels in these three tables are listed in Table 2-1. It will be noted that both yield stress and tensile strength are specified as "minimum values." Since the specification requires that test specimens conform to these values, rolling mills control the manufacture of the product so that the average yield stress and average tensile strength are well above the specified minimums. Figure 2-17 shows the yield-stress distribution of 3,974 mill-test specimens taken from 33,000 tons of A7 steel* on nine projects erected between 1938 and 1951 (Ref. 1). The figure shows that about 5 percent of the specimens tested less than 35 ksi, with possibly 2 percent below the specified minimum of 33 ksi. More than one-third had yield points in excess of 40 ksi.

The most important factor affecting mechanical properties of the steels is the chemical composition. Some of the other factors are the total reduction from ingot to finished product, finishing temperatures, and rate of cooling. Because thin plates involve a larger reduction in the ingot than thick plates, which requires more passes through the rolls, they have higher yield stresses. However, yield stress may be kept independent of thickness by varying the chemical composition with the thickness of shape to be rolled. This is done with A36 steel for thicknesses to 8 in. On the other hand, the chemical composition of A242, A440, and A441 is held constant, so that the yield stress of these steels is smaller for thicker material (Table 2-2).

Steel which will be concealed (as in buildings) usually needs only one coat of paint, but steel which will be exposed requires additional paint after erection.

* A carbon steel with a minimum yield stress of 33 ksi which was used extensively from 1936 to about 1965. It was discontinued in 1967.

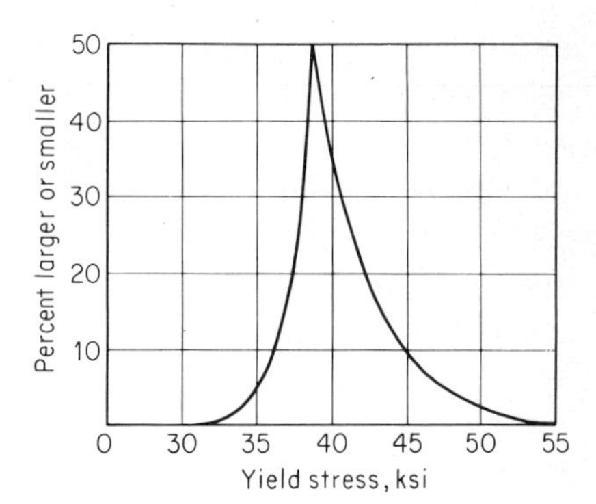

FIGURE 2-17
Yield-stress frequency distribution, A7 steel.

However, certain corrosion-resistant steels that produce a tight, dense, hard, protective skin during oxidation have been developed. Only about 0.002 in. of metal is lost, through erosion by wind and rain, in the process. The film is deep russet in color and pleasing in texture. It does not crack or flake, and it retards further corrosion so that painting is unnecessary except where the steel may be exposed to concentrated industrial fumes or to salt spray and fog. The protective coating is darker in color and forms faster (18 to 36 months) in industrial atmospheres. These steels are called *weathering steels*. They have been used in bridges and in buildings with exposed frames. A242 and A588 are weathering steels. They have about four times the atmospheric-corrosion resistance of A36 steel.

ASTM specification A36 covers weldable structural carbon steel with a minimum yield point of 36 ksi. It is available in all the standard rolled shapes and in plates to 8 in. in thickness. It is also furnished in plates over 8 in. thick, up to 15 in., but only with a yield point of 32 ksi. Copper can be specified as an alloying element in an amount which doubles the resistance to atmospheric corrosion. A529 is also a weldable structural carbon steel. It has a minimum yield stress of 42 ksi. It is available in plates and bars to $\frac{1}{2}$ in. thick and in the lighter standard shapes. It has the same atmospheric-corrosion resistance as copper-bearing A36 steel. It is used principally in standardized steel buildings.

A242, A440, A572, and A588 are called high-strength steels. A242, A440, and A441 have the same minimum yield points (Table 2-2). A242 steel is suitable for riveted, bolted, and welded construction. It is not as economical as A441 and is not used unless corrosion is a factor. A440 steel is intended for riveted and bolted construction. It is cheaper than A242 and A441 because its high yield point is obtained by increased percentages of carbon and manganese. Vanadium and silicon are the principal alloying elements of A441 steel, which is produced for welded construction. A572 covers steels available in six different yield stresses. Each is classified as a "grade," with the grade number denoting the yield stress. Grades 42, 45, and 50 are intended for riveted, bolted, or welded construction of bridges, buildings, and other structures. Grades 55, 60, and 65 are also intended for riveted, bolted, and welded construction of buildings and other structures but only for riveted and bolted construction in bridges. A588 is a high-strength low-alloy steel with 50 ksi minimum yield point in thicknesses to 4 in. It is also available in larger thicknesses at smaller yield stresses (Table 2-2). It is suitable for welded, riveted, and bolted construction but is intended primarily for welded bridges and buildings. There are seven grades, all with the same yield stress. Each grade is a proprietary steel, and each is produced by a different company.

Quenched and tempered alloy steel plate is covered by A514. This steel is furnished in 10 types, each under a proprietary brand name. It has the highest yield stress of the structural steels. It has outstanding toughness and is intended primarily for use in welded bridges and other structures.

Although all the steels in Table 2-2 except A440 are classified as weldable, this does not mean that identical welding procedures can be used for all. The ASTM specification usually cautions the user that welding procedures must be suitable for the steel and the intended service. Weldability of many of these steels is discussed in Ref. 5.

Steel pipe and tubing is often used in building construction (Table 2-3). A53 covers welded and seamless pipe. However, only Grade B of this specification, which is electric-resistance welded, is permitted by the AISC specification. It has about the same tensile properties as A36 steel (Table 2-2). This pipe is furnished in standard sizes ranging from $\frac{1}{2}$ to 24 in. in outside diameter. Carbon-steel structural tubing in round, square, rectangular, or special shapes may be cold-formed, welded or seamless (A500), or hot-formed, welded or seamless (A501). Properties of these steels are given in Table 2-3. Common sizes of the square tube range from 3 × 3 in. to 12 × 12 in. The rectangular tube ranges from 3 × 2 in. to 12 × 6 in.

Table 2-5 MINIMUM TENSILE PROPERTIES OF ALUMINUM ALLOYS

Alloy and temper	Yield, ksi	Strength, ksi	Elongation, % in 2 in.
6061-T6*	35	38–42	10
6062-T6*	35	38–42	10
6063-T5	16	22	8
6063-T6	25	30	8
2014-T6	53	60	7

* May be used interchangeably.

The sheet and strip steels (Table 2-4) are used for cold-formed structural members and find wide application in standardized steel buildings. Yield stresses range from 5 to 80 ksi. A374 is intended for service requiring greater strength and atmospheric-corrosion resistance equal to or greater than that of plain copper-bearing steel. A446 is produced with eight classes of hot-dip zinc coatings so that sheets with coating consistent with forming hazards and expected service life will be available.

The aluminum alloys usually used for structural work are known commercially as 2014-T6, 6061-T6, 6062-T6, 6063-T5, and 6063-T6 (Table 2-5). The number in these designations identifies the composition of the alloy, the T means that the metal has been heat-treated, and the final numeral indicates the type of heat treatment. Although these alloys weigh only about 36 percent as much as steel, they can compete only when their higher initial cost is offset by such advantages as light weight, resistance to corrosion, reduced maintenance, and appearance.

2-6 STRUCTURAL SHAPES

A wide variety of structural-steel shapes is manufactured. Round and square bars are extruded. Flat steel is rolled from the ingot and is classified according to width and thickness as bars, plates, and bearing plates. The common rolled shapes are the angle, the tube, the channel, and the I. The I is available in two classifications. The most widely used is the W shape (formerly WF, meaning "wide flange"). The other, once called the "American Standard Beam," is now called the S shape. Miscellaneous column and beam shapes used for lightweight construction are rolled by a few mills. One producer manufactures a complete range of wide-flange shapes by passing an assembly of two flange plates and a web plate through submerged-arc welders which simultaneously weld both flanges to one side of the web. The section is turned over to weld the flanges to the other side. The structural tee is obtained by splitting the web of an I, generally by the use of rotary shears. There is a more or less constant demand for all these shapes, and therefore they are readily procurable. A group of special shapes such as subway columns, special channels, bearing piles, tees, and zees are rolled only by arrangement with the mills and should not be used unless the quantity needed is sufficient to warrant a rolling. All these shapes are manufactured to certain tolerances with respect to dimensional variations such as camber, cross section, diameter, squareness, flatness, length, straightness, sweep, thickness, weight, and width. The specific limitations are contained in ASTM A6.

Aluminum structural shapes are produced by rolling or by extrusion. Standard angles, channels, I's, tees, and zees are available in a wide range of sizes. Limited selections of wide-flange shapes are also available. Hollow tubular shapes, interlocking shapes, sections with integral backup strips and bevels for welding, structural shapes with stiffening lips or bulbs to support outstanding flanges, and stiffened sheet panels are a few of the sections readily extruded.

Cold-formed steel shapes are formed in rolls or brakes from sheet or strip steel. Because of the great variety which can be produced, shapes of this type, unlike hot-rolled shapes, have not been standardized. Although a number of fabricators have developed their lines of members, the designer may devise special shapes for particular jobs. While shapes up to thicknesses of $\frac{1}{2}$ and even $\frac{3}{4}$ in. can be formed, cold-formed steel construction is usually restricted to thicknesses ranging from 30 gage (0.012 in.) to 4 (0.224 in.). Shapes such as channels, zees, and angles in the thinner gages usually require lips or other edge stiffeners. I shapes are made by spot-welding two or more shapes, e.g., two channels or a channel and two angles.

Wire rope is made of a number of strands laid helically around a core, which may be a fiber rope, another steel strand, or a small wire rope. A strand is made of a number of wires laid helically around a center wire. Strand itself may be used as an individual load-carrying tension member where flexibility or bending is not a major requirement. Wire rope provides increased flexibility. For structural purposes it usually consists of six strands plus the central core.

Bridge strand is approximately four times as strong as A36 steel but costs only about twice as much per pound. A large portion of the cost of suspension members is in the fittings, connections, and anchorage members.

2-7 RIVETS

The components which make up the completed metal member or structure are fastened together by means of rivets, bolts, or welds. Rivets are made from rivet bar stock in a machine which forms one head and shears the rivet to the desired length. Rivet heads are usually of a rounded shape called a buttonhead. The head may be flattened when clearance is limited. When very little clearance is available, countersunk heads are used. Countersunk heads are chipped flush if no clearance is available. Information concerning conventional signs for riveting, sizes of heads, weights, lengths, and other data are given in the AISC Steel Construction Manual.

Steel rivets are almost always heated before driving. In the shop the rivet is heated to a minimum temperature identified by a light cherry-red color. Most shop rivets are driven by pressure-type riveters which complete the riveting operation in one stroke. Riveting guns are portable hand tools, operated by compressed air, which drive the rivet by a rapid succession of blows.

Rivets are made from steel conforming to the specifications for rivet steel, ASTM A502. This specification covers two grades of steel. Grade 1 is a carbon steel for general purposes. Grade 2 is a carbon-manganese steel for riveting high-strength alloy structural steels. Rivet heads are marked to identify the manufacturer and with a numeral 1 or 2 to identify the grade; the manufacturer may omit the numeral 1.

Rivets used in fabricating structures of aluminum alloys may be driven either hot or cold. Cold-driven rivets for structures of alloys 6061-T6 and 6062-T6 are made from alloy 6061-T6 and for structures of 6063-T5 and 6063-T6 from alloy 6053-T61. Cold-driven rivets for structures of alloy 2014-T6 are made from alloy 2117-T3. Hot-driven rivets for the five alloys are made from alloy 6061-T43.

2-8 STRUCTURAL BOLTS

The two commonly used types of bolts for steel structures are the *unfinished bolt* (A307) and the *high-strength bolt* (A325, A449, and A490).

The A307 bolt is known by a variety of names—unfinished, rough, common, ordinary, and machine. It is furnished in two grades, A and B, the former for general purposes and the latter for joints in pipe systems. They are made of low-carbon steel with a minimum tensile strength of 60 ksi. They are tightened by using long-handled manual wrenches, so that the induced tension is relatively small and unpredictable. They are satisfactory for use in building frames not subject to shock or vibration, and are used in both hot-rolled and cold-formed steel construction. Castellated nuts with cotter pins, jam nuts, and various types of locknuts can be used to prevent loosening where shock and vibration are a consideration.

The A325 bolt (Fig. 2-18*a*) is made of medium-carbon steel. It is also used in both hot-rolled and cold-formed steel construction. The tensile strength of this bolt decreases with increase in diameter of the bolt, so that two ranges of diameter are specified (Table 2-6). The A449 bolt, also of medium-carbon steel, is furnished in three ranges of diameter. The A490 bolt is made of alloy steel in one tensile-strength grade. It should be noted that the tensile properties are based on the "stress area." This is larger than the section at the root of the thread but smaller than the unthreaded area. The tensile strength of the bolt, based on this area, is about the same as the coupon strength.

High-strength bolts can be tightened to large tensions, which produce high clamping forces between the connected parts. Specifications for A325 and A490 bolts, written by the Research Council on Riveted and Bolted Structural Joints, require that these bolts be installed with an initial tension equal to about 70 percent of the specified minimum tensile strength of the bolt. AISC specifies the same initial tension and also includes the A449 bolt (Table 2-7). This tension produces enough frictional resistance to prevent nuts from working loose under service loads. The bolts are installed by one of two tightening procedures which control the tension. Hand wrenches or power impact wrenches can be used with either method.

FIGURE 2-18
Structural bolts: (*a*) A325 bolt, (*b*) A490 bolt, (*c*) interference-body bolt, (*d*) ribbed bolt.

Table 2-6 PROPERTIES OF STRUCTURAL BOLTS

ASTM designation	Bolt diameter, in.	Tensile strength, ksi*		Minimum yield strength, ksi,* 0.2% offset	Proof load, length measurement, ksi*
		Minimum	Maximum		
Low-carbon steel:					
A307:					
Grade A	All	60			
Grade B	All	60	100		
High-strength structural bolts:					
Medium-carbon steel:					
A325 and A449	$\frac{1}{2}$–1	120	...	92	85
A325 and A449	$1\frac{1}{8}$–$1\frac{1}{2}$	105	...	81	74
A449	$1\frac{3}{4}$–3	90	...	58	55
Alloy steel:					
A490	$\frac{1}{2}$–$1\frac{1}{2}$	150	180	130	120

* On the area $\frac{\pi}{4}\left(D - \frac{0.9743}{n}\right)^2$, where D = nominal size and n = threads per inch.

Manual torque wrenches or adjustable power impact wrenches are used in the *calibrated-wrench method* of tightening. Wrenches are calibrated by tightening, in a hydraulic tension-measuring device, a minimum of three bolts of the same diameter. Impact wrenches are set to stall when the prescribed bolt tension is reached. Manual torque wrenches have a torque-indicating device, so that the torque required to produce the initial tension is measured. The calibration should be checked at least once a day or when the wrench is to be used on a bolt of different size. A hardened washer must be used under the nut or the head, whichever is turned in tightening. This is because friction between the nut or head and the connected material is likely to vary considerably from one connection to another, or even from one bolt to another.

Ordinary spud wrenches or standard power impact wrenches can be used in the *turn-of-the-nut method* of tightening. Bolt elongation, and thus the tension, is controlled in this method. After the bolts in a connection have been tightened to the *snug-tight* condition, they are given an additional half turn if the bolt length is less than eight diameters, or an additional two-thirds turn if the length exceeds eight diameters or 8 in. The snug-tight condition is reached when the turned element ceases to rotate freely, at which stage the impact wrench begins to impact. If the spud wrench is used, the snug-tight condition corresponds to the full effort of a man. One full turn from a *finger-tight* condition corresponds approximately to the one-half turn from the snug-tight condition.

The AISC specification does not require washers if the turn-of-the-nut method is used with A325 bolts or with A490 bolts connecting steel whose specified yield is 40 ksi or more. The AASHO specifications require a hardened washer (under the element to be turned) with either method of tightening. The AREA specifications allow only the turn-of-the-nut method and require a washer under the turned element.

Table 2-7 MINIMUM INSTALLATION TENSION FOR HIGH-STRENGTH BOLTS

Bolt size, in.	ASTM designation		
	A325	A449	A490
$\frac{1}{2}$	12	12	15
$\frac{5}{8}$	19	19	24
$\frac{3}{4}$	28	28	35
$\frac{7}{8}$	39	39	49
1	51	51	64
$1\frac{1}{8}$	56	56	80
$1\frac{1}{4}$	71	71	102
$1\frac{3}{8}$	85	85	121
$1\frac{1}{2}$	103	103	148
Over $1\frac{1}{2}$	...	*	

* 0.7 × tensile strength.

Washers are required under both head and nut of A490 bolts used in steel whose specified yield stress is less than 40 ksi, no matter which method of tightening is used. This is to prevent galling of the connected materials. They are also required under the heads of A449 bolts with either method of tightening. If the surface of a part to be connected by any high-strength bolt has a slope exceeding 1 in 20, a beveled washer must be used. Since this is to compensate for the angle between the axis of the bolt and the sloping surface, the washer must be used with either method of tightening.

Other fasteners are used occasionally. The high-strength, interference-body, bearing bolt (Fig. 2-18*c*) has relatively hard, rolled, serrated ribs which produce a solid bearing for the full thickness of the connected parts. This bolt gives the strength and clamping force of the A325 bolt and, because it fills the hole, prevents slip should shear loads exceed the frictional resistance of the joint. Another type of interference fastener is the ribbed bolt (Fig. 2-18*d*). It is made of carbon steel and has strength equal to or greater than that of the A502 Grade 1 rivet of the same diameter. It has a standard rivet head, a fluted shank with triangular-shaped ribs, and a self-locking nut. These fasteners are especially useful where it is impractical to use power tools.

A394 galvanized steel bolts and nuts are the common fasteners for towers and similar structures. These bolts are available with hexagonal or square heads and nuts. A nonloosening connection is provided by lock washers, jam nuts, or locknuts.

Aluminum bolts and nuts are recommended for aluminum structures, although steel fasteners can be used. Steel bolts should be aluminized, hot-dip galvanized, electrogalvanized, or of stainless steel to prevent galvanic corrosion that might result from direct contact between steel and aluminum. Alloy 2024-T4 is the common alloy for aluminum bolts, while the nuts are usually of alloy 6061-T6 or 6062-T6.

2-9 BOLTED AND RIVETED CONNECTIONS

Figure 2-19 illustrates the different types of failure that may occur in bolted or riveted joints with the fasteners in shear. Failure in which the fastener is sheared along the plane of slip is illustrated in Fig. 2-19*a* and *b*. The area subjected to shear is the cross-sectional area of the fastener. For the lap joint *a* failure is on one plane, and the fastener is in single shear. For the butt joint *b* failure occurs on two planes, and the fastener is in double shear.

Failure of the plate at a hole, caused by compression between the cylindrical surface of the plate hole and the fastener, is shown in Fig. 2-19*c*. This is called a *bearing failure*. The variation of the compressive stresses around the perimeter of the hole is unknown. For design purposes the stress distribution is assumed to be

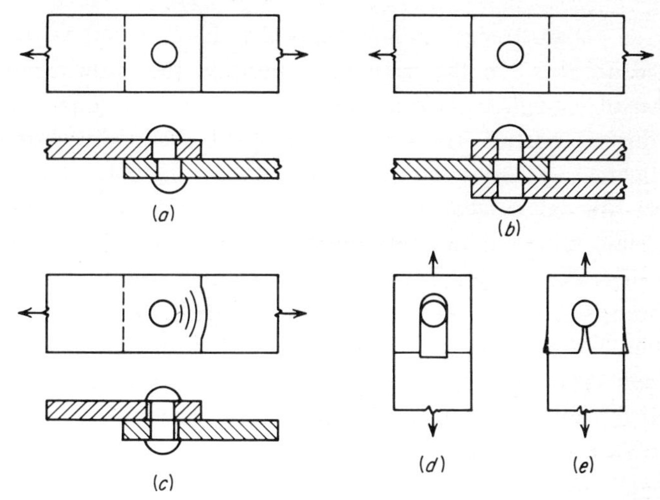

FIGURE 2-19

uniform over the rectangular area equal to the thickness of the plate times the nominal diameter of the fastener.

Failure may also occur in the plate between the hole and the end of the plate by shearing in the direction of the load, as illustrated in Fig. 2-19*d*, or by transverse tension, as in Fig. 2-19*e*. These failures are unlikely to occur if the fastener is placed at a distance from the end sufficient to make the shearing strength of the plate equal to the shearing strength of the fastener. For structural metals this requires a distance from the center of the hole to the end of the plate of 1.25 to 2 times the diameter of the fastener.

Owing to the eccentricity of the applied forces, the plates of a lap joint tend to bend as shown in Fig. 2-20*a*. This bending produces some tension in the fastener and nonuniform bearing of the fastener on the plate, as shown in Fig. 2-20*b*. However, tests indicate that the bearing strength of the fastener is about the same for single-shear bearing and double-shear bearing. Nevertheless, some specifications allow a larger unit stress for bearing in double shear than for bearing in single shear.

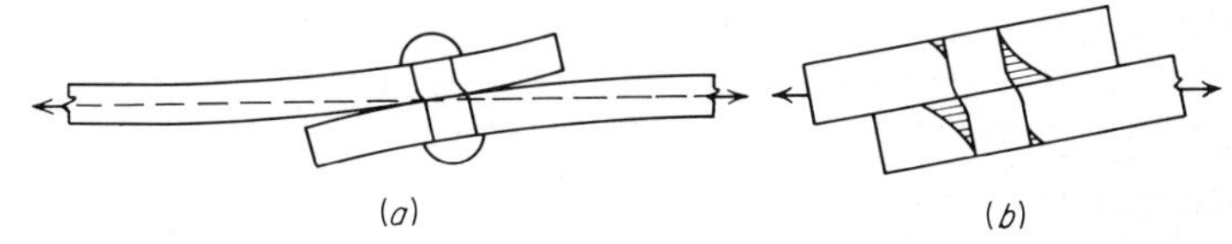

FIGURE 2-20

High-strength bolted joints may be designed as *friction connections* or *bearing connections*. In the friction connection, the allowable shear stress on the bolt is small enough to give a margin of safety with respect to slip of the joint. Thus, this connection transmits service-load shears by friction produced between the faying surfaces by the clamping action of the bolts. The faying surfaces must be free of oil, paint, lacquer, or galvanizing. Friction connections are recommended for joints subjected to stress reversal, severe stress fluctuation, impact, or vibration or where slip is objectionable. Maximum service-load shears on bearing connections are transferred by shear or bearing on the bolt. The shear strength depends on whether the shear plane intersects the body of the bolt or the threaded portion. Tests of bearing connections show that the strength of single-shear joints is increased by 40 to 45 percent if threads are excluded from the shear plane. Of course, this is a consequence of the difference in area.

Allowable stresses for high-strength bolts according to the Research Council on Riveted and Bolted Structural Joints are given in Table 2-8. Values for bridges are about 10 percent lower than those for buildings, which is consistent with allowable-stress provisions of the AISC, AASHO, and AREA specifications. It will be noted that allowable tension is independent of the size of the bolt, even though tensile properties vary with the size (Table 2-6). In assessing the factors of safety which these allowable stresses provide, it will be noted that the ASTM tensile properties are based on the stress area, while the allowable stresses are based for convenience on the unthreaded ("nominal") area of the bolt. Thus, for a 1-in. bolt the stress area is

Table 2-8 ALLOWABLE STRESSES* (ksi) FOR HIGH-STRENGTH STRUCTURAL BOLTS†

Condition	A325		A490	
	Bridges	Buildings	Bridges	Buildings
Tension	36	40	48‡	54‡
Shear, friction-type connection	13.5	15	18	20
Shear, bearing-type connection:				
With threads in shear plane	13.5	15	20	22.5
With no threads in shear plane	20	22	29	32
Bearing§	$1.22F_y$	$1.35F_y$	$1.22F_y$	$1.35F_y$

* On unthreaded-body area, also called nominal area.
† From Ref. 2.
‡ Static load only.
§ F_y = yield stress of lowest-strength connected material. Bearing stress not to exceed minimum tensile strength of connected material.

0.606 in.^{2} while the nominal area is 0.785 in.^{2} The recommended allowable tensile stress for A325 bolts in buildings gives a factor of safety for 1-in. bolts of

$$n = \frac{120 \times 0.606}{40 \times 0.785} = 2.32$$

Factors of safety for the sizes listed in Table 2-6 range from 1.76 to 2.35 for buildings.

Results of numerous tests show that the shear strength of a single A325 bolt is 65 to 70 ksi. Thus, the factor of safety for a single bolt at the allowable stress for a bearing-type connection, 22 ksi, is about 3. However, if there are a number of bolts in the line of stress, as in a long connection, this factor of safety is reduced if the shear is assumed to be divided equally among the bolts, as is customary (Art. 3-11).

The allowable bearing stresses in Table 2-8 would appear to offer a very small factor of safety. However, it must be remembered that this is not a true stress on the surface in bearing; instead, it is a fictitious stress on the projected area of the surface. Tests have shown that the strength of the net section of a tension member is not affected if the bearing stress is not larger than 2.25 times the net-section stress.[3] This was the basis for allowable bearing stress. Thus, with the customary allowable net-section tension $0.6F_y$ for buildings, we get for the allowable bearing $F_p = 2.25 \times 0.6F_y = 1.35F_y$.

Standard specification allowable-stress provisions for high-strength bolts are given in Table 2-9. The AISC values are identical with those for buildings in Table 2-8, but there are some differences in the values adopted by AASHO and AREA.

Table 2-9 ALLOWABLE STRESSES* (ksi) FOR HIGH-STRENGTH STRUCTURAL BOLTS

	A325			A490	
Condition	AISC†	AASHO	AREA	AISC	AREA
Tension	40	36	36	54	36
Shear, friction-type connection	15	13.5	...	20	
Shear, bearing-type connection:					
With threads in shear plane	15	‡	...	22.5	
With no threads in shear plane	22	20	20	32	27
Bearing§	$1.35F_y$¶	40	...	$1.35F_y$¶	

* On unthreaded-body area, also called nominal area.
† These allowable stresses also apply to the A449 bolt.
‡ This type of connection not permitted.
§ Not a consideration in friction-type connections.
¶ Yield stress of lowest-strength connected material.

Table 2-10 gives allowable stresses for A502 rivets according to AISC, AASHO, and AREA. Although there are no tensile requirements for A502 rivets (hardness is specified instead), Grade 1 corresponds to a former ASTM designation, A141, for which the specified tensile properties were 28 ksi yield and 52 to 62 ksi tensile strength. Thus, it would appear that the allowable tension for Grade 1 rivets, 20 ksi, gives a low factor of safety relative to yield. However, the yield stress of the in-place rivet exceeds that of the undriven rivet because of the work hardening caused by the driving. Furthermore, allowable stresses are based on the area of the undriven rivet, which is smaller than that of the driven rivet because rivet holes are $\frac{1}{16}$ in. larger than the nominal diameter of the rivet. The combination of these two effects gives an effective yield stress 20 to 25 percent more than that of the rivet stock. Thus, the allowable tension of 20 ksi gives a factor of safety with respect to yield on the order of 1.7. The factor of safety with respect to tensile ultimate of the undriven rivet is at least 2.6. The Grade 2 rivet also corresponds to a former ASTM designation, A195, for which the tensile properties were 38 ksi yield and 68 to 82 ksi tensile strength. The ratio of the allowable tensions for the two grades, 27/20, is identical with that of the yield stresses, 38/28.

Allowable stresses for aluminum rivets, suggested by a task committee of ASCE,[4] are given in Table 2-11. It will be noted that tension is not mentioned, which means, of course, that connections should be designed to avoid tension on the rivets as much as possible. Contrary to the practice of evaluating steel rivets in terms of their undriven diameters, tests on aluminum rivets are interpreted in terms of the diameter of the hole. Diameters and areas of rivets and the recommended diameters of the corresponding holes are given in Tables 2-12 and 2-13. The allowable shear stresses in Table 2-11 are based on factors of safety, with respect to shear strength, of about 2.25 for buildings and 2.65 for bridges. The allowable bearing stresses are based on factors of safety, with respect to bearing yield strength, of about 1.65 for buildings and 1.85 for bridges.

Table 2-10 ALLOWABLE STRESSES (ksi) FOR A502 RIVETS

	Grade 1			Grade 2		
Condition	AISC	AASHO	AREA	AISC	AASHO	AREA
Tension	20	...	...	27		
Shear	15	13.5	13.5	20	20	20
Bearing	$1.35F_y$*	40	27†	$1.35F_y$*	40	$0.75F_y$*†
			36‡			F_y*‡

* Yield stress of connected part.
† Rivet in single shear.
‡ Rivet in double shear.

Table 2-11 ALLOWABLE STRESSES FOR ALUMINUM RIVETS*

These values apply for a ratio of edge distance to rivet diameter of 2 or greater. For smaller ratios, multiply the allowable stress by the ratio (edge distance)/(2 × rivet diameter).

Designation after driving	Driving procedure	Designation of alloy fabricated	Shear, ksi		Bearing, ksi	
			Bridges	Buildings	Bridges	Buildings
6061-T6	Cold, as received	6061-T6	10	11	30	34
		6062-T6	10	11	30	34
6053-T61	Cold, as received	6063-T5	7.5	8.5	13.5	15
		6063-T6	7.5	8.5	21.5	24
6061-T43	Hot, 990–1,050°F	6061-T6	8	9	30	34
		6062-T6	8	9	30	34
		6063-T5	8	9	13.5	15
		6063-T6	8	9	21.5	24

* From Ref. 4. These allowable stresses do not apply to material within 1 in. of a weld. See Ref. 4 for reductions in heat-affected zone of weld.

Table 2-12 HOT-DRIVEN ALUMINUM RIVETS

Rivet diameter, in.	Hole diameter, in.	Rivet area, in.^{2}
$\frac{3}{8}$	0.397	0.124
$\frac{7}{16}$	0.469	0.173
$\frac{1}{2}$	0.531	0.222
$\frac{9}{16}$	0.594	0.277
$\frac{5}{8}$	0.656	0.338
$\frac{3}{4}$	0.781	0.479
$\frac{7}{8}$	0.922	0.668
1	1.063	0.888

Table 2-13 COLD-DRIVEN ALUMINUM RIVETS

Rivet diameter, in.	Hole diameter, in.	Rivet area, in.^{2}
$\frac{3}{8}$	0.386	0.117
$\frac{7}{16}$	0.453	0.161
$\frac{1}{2}$	0.516	0.209
$\frac{9}{16}$	0.578	0.262
$\frac{5}{8}$	0.641	0.323
$\frac{3}{4}$	0.766	0.461
$\frac{7}{8}$	0.891	0.624
1	1.016	0.811

Allowable shear stresses for the aluminum 2024-T4 bolt, according to Ref. 4 are 14 ksi (bridges) and 16 ksi (buildings) on the effective area in shear. Allowable tensions on the area at the root of the thread (not the stress area) are 23 ksi (bridges) and 26 ksi (buildings).

2-10 WELDING PROCESSES

Welding is a process of joining metal parts by means of heat and pressure, which causes fusion of the parts (resistance welding), or by heating the metal to the fusion temperature, with or without the addition of weld metal (fusion welding). Fusion welding usually employs either an electric arc or an oxyacetylene flame to heat the metal to the fusion temperature. The electric arc is used for most structural welding.

Welds are classified according to their type as groove, fillet, plug, and slot. A groove weld is made in the opening (called a groove) between two parts being joined, while the fillet weld, which is triangular in shape, joins surfaces which are at an angle with one another (Fig. 2-21). A plug weld is made by depositing weld metal in a circular hole in one of two lapped pieces. The hole must be filled completely. A slot weld is similar, the only difference being that the hole is elongated. Holes and slots can also be fillet-welded around the circumference, but these are not plug or slot welds.

Welds are classified according to the position of the weld during welding as flat (also called downhand), horizontal, vertical, and overhead. Welding in the flat position is executed from above, the weld face being approximately horizontal (Fig. 2-21*a*). The horizontal position is similar, but the weld is harder to make (Fig. 2-21*b*). Work in the overhead position is from the underside of the joint; this is the most difficult weld to make (Fig. 2-21*c*). The longitudinal axis of the weld is vertical in vertical-position welding (Fig. 2-21*d*).

In metal-arc welding, the arc is a sustained spark between a metallic electrode and the work to be welded. At the instant the arc is formed, the temperature of the work and the tip of the electrode are brought to the melting point. Only that portion of the work at the arc is melted. As the tip of the electrode melts, tiny globules of molten metal form and are forced across the arc to be deposited in the molten base metal. It is because these globules are actually impelled across the arc that the process can be used in overhead welding. The molten metal, when exposed to the air, combines chemically with oxygen and nitrogen to form oxides and nitrides which tend to embrittle it and make it less resistant to corrosion. Tough, ductile welds which are more resistant to corrosion are produced if the arc is shielded by an inert gas which completely envelops the molten metal and the tip of the electrode. Shielding is obtained in the manual process by the use of electrodes heavily coated with a material

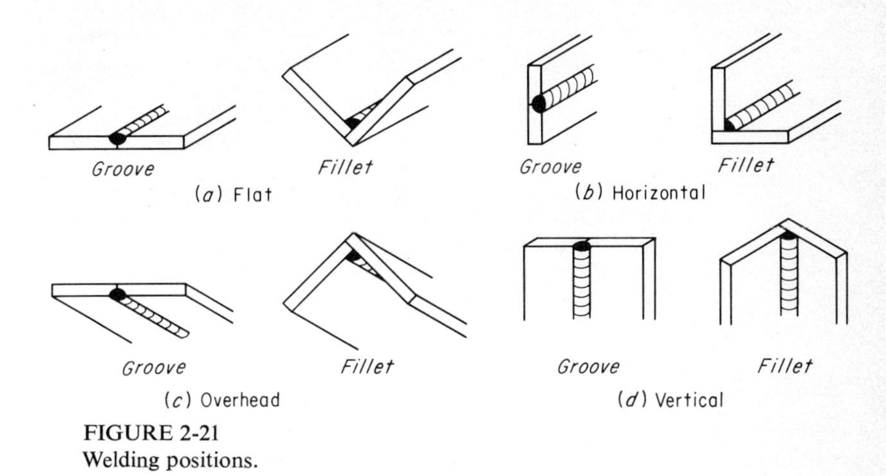

FIGURE 2-21
Welding positions.

of such composition that large quantities of gas are produced in the heat of the arc. The coating burns at a slower rate than the metal core does, thus directing and concentrating the arc stream as it protects it from the atmosphere. The coating also forms a slag which floats on top of the molten metal and protects it from the atmosphere while cooling. The slag is easily removed after the weld has cooled. Figure 2-22 illustrates the shielding of the arc and the slag protection of the weld metal. Shielded metal-arc welding, sometimes called *stick welding,* is used primarily in hand welding.

Automatic arc-welding processes produce high-quality welds at very high welding speeds. In the submerged or hidden-arc process a bare weld wire is fed automatically through the welding head at a rate to maintain a constant arc length. The welding is shielded by a blanket of granular, fusible material which is fed onto the work area by gravity in an amount sufficient to submerge the arc completely (Fig. 2-23). Some of the granular material fuses to form a covering over the weld. In addition to protecting the weld from the atmosphere, the covering aids in controlling the rate of cooling of the weld. Multiple-electrode welding uses two or more small weld wires, instead of the single wire, for increased welding speed at reduced cost. Submerged-arc welding must be performed downhand. Welding speeds for one-pass groove welds range from 30 in. per min in $\frac{1}{4}$-in. plate to 8 in. per min in $1\frac{1}{2}$-in. plate. The high currents used cause considerable melting of the base metal, so that less filler metal is required and the joint opening may be less than that necessary for other types of welding.

Inert-gas shielded-arc welding is usually performed without flux. The arc and the weld region are shielded from the atmosphere by a gas or gas mixture or by a combination of a gas and a flux. In addition to the inert gases argon and helium, carbon dioxide, which is heavier than air, may also be used. Carbon dioxide is

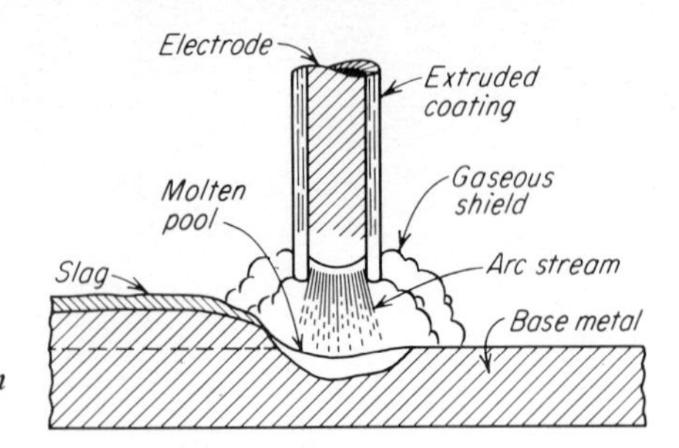

FIGURE 2-22
Shielded metal-arc welding. (*Lincoln Electric Company.*)

popular because of its low cost. Since they are completely protected from atmospheric elements, the welds are stronger, more ductile, and more corrosion-resistant. High welding speeds with thorough penetration of the weld and little distortion are obtained because of the intense arc heat. In gas-shielded metal-arc welding (called *Mig* for metal-arc, inert-gas) a bare-wire consumable electrode is fed automatically into the arc and deposited as weld metal. In the tungsten-arc process (*Tig* for tungsten-arc, inert-gas), an arc is struck between a virtually nonconsumable tungsten electrode and the base metal. Filler metal, if required, is added by feeding a welding rod into the weld pool.

The basic elements of the electroslag welding process are shown in Fig. 2-24. Heat is generated by passing an electric current through molten flux which melts the electrode and the edges of the base metal. Welding is usually done in a vertical position. The cavity formed between the water-cooled molding shoes and the edges of the base metal contains the molten flux pool, the molten weld metal, and the solidified weld metal. Joint preparation is simple since only square, oxygen-cut edges are necessary. Plates 18 in. thick can be welded in a single pass. A starting tab is needed to build up the proper depth of flux to ensure complete fusion of the base metal. A runoff tab is required at the top because the slag depth must be carried beyond the base metal. The slag bath is $1\frac{1}{2}$ to 2 in. deep.

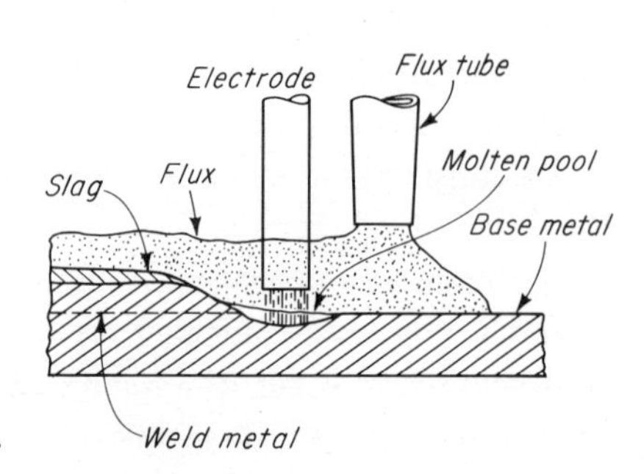

FIGURE 2-23
Submerged metal-arc welding.

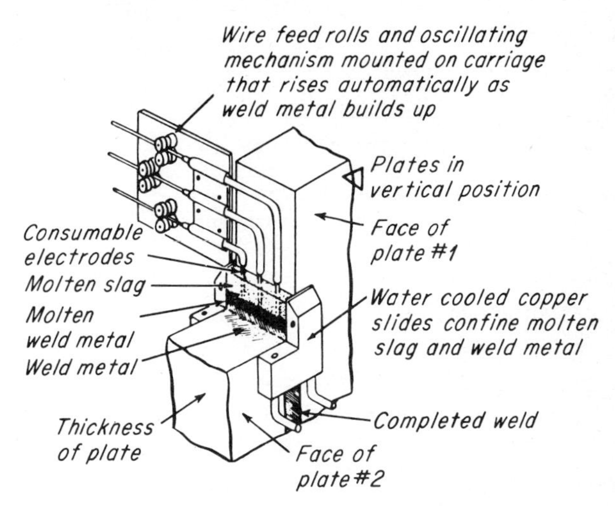

FIGURE 2-24
Electroslag welding.

Although similar in its basic elements to the electroslag process, the electrogas welding process uses an open electric arc instead of a molten flux to melt the base metal and electrode. Since the arc causes instantaneous fusion and no flux bath is required, starting and runoff tabs are not necessary. Joint thicknesses of $\frac{1}{2}$ to 3 in. can be produced with a single pass.

Electroslag or electrogas welding of ASTM A514 steel is not permitted unless the weldment is quenched and tempered after welding.

Aluminum alloys can be joined by arc welding, resistance welding, gas welding, or brazing. The most commonly used processes are Mig and Tig welding. They can be used on metal $\frac{1}{16}$ in. or more thick. Tig welding is generally used for material thicknesses from 0.05 to 0.25 in. The commonly used structural aluminum alloys are all readily weldable. Butt joints made in aluminum alloys which are in the annealed condition are usually 100 percent efficient. With alloys in the strain-hardened or heat-treated tempers, however, the heat of welding affects the metal on each side of the weld so that it is not as strong as the parent metal.

2-11 WELDED JOINTS

Welded joints are classified as butt, lap, tee, corner, and edge (Fig. 2-25). The butt joint is groove-welded while the lap joint is fillet-welded. The tee joint can be groove-welded, as shown, or it can be fillet-welded with one fillet on each side. Groove-welded joints can be *complete-penetration* joints or *partial-penetration* joints. In some

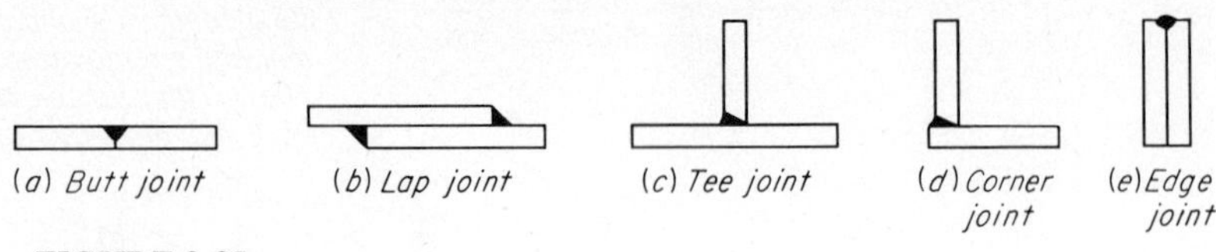

FIGURE 2-25
Types of welded joints.

cases the penetration is intentionally partial, in that the weld is less in depth than the thickness of the part joined, while in other cases it is partial because the welding procedure does not produce effective penetration in what might appear otherwise to be a complete-penetration joint.

There are a number of joints which can be used without qualification, which means that no tests are required to demonstrate their adequacy. Some of the manual shielded metal-arc joints for buildings, prequalified by the American Welding Society, are shown in Fig. 2-26. Four of them, *a* to *d*, are complete-penetration joints but are limited in use to the thicknesses shown. Another four, *e* to *h*, are complete-penetration joints of unlimited thickness. The last four, *i* to *l*, are partial-penetration joints. The square-groove joints, *a, b,* and *i,* require no preparation of the edges of the parts to be joined. The type of groove for other than square-butt joints depends in part upon the thickness of the material and the position of the weld and whether one side or both sides are accessible for welding. Single grooves, such as the bevel and the V, J, and U, are cheaper to form but require more weld metal than the double-grooved joints. For example, the single-V joint requires approximately twice as much weld metal as the double-V. The choice between single and double grooves is usually a question of whether the higher cost of preparation is offset by the saving in weld metal. Bevel or V joints are usually preferred for horizontal welds because it is difficult to make a good U or J joint in this position.

The part of a weld which is assumed to be effective in transferring stress is called the *throat*. The throat thickness of each of the complete-penetration welds in Fig. 2-26 is the dimension T. The effective throats of the partial-penetration welds, denoted by T_e, are also shown in the figure. Five of the joints in the figure (*a, g, i, j,* and *k*) are welded from only one side of the joint. However, *a* and *g* are welded against backing strips, which accounts for their being complete-penetration joints. Weld *i* is partial-penetration because it is welded from one side without backup. In the remaining complete-penetration joints, the root of the weld deposited first must be "gouged" before welding from the other side, which means that it must be cleaned.

The fillet weld is quite common in structural connections. The faces of the weld which are in contact with the parts joined are called its legs. The size of an equal-

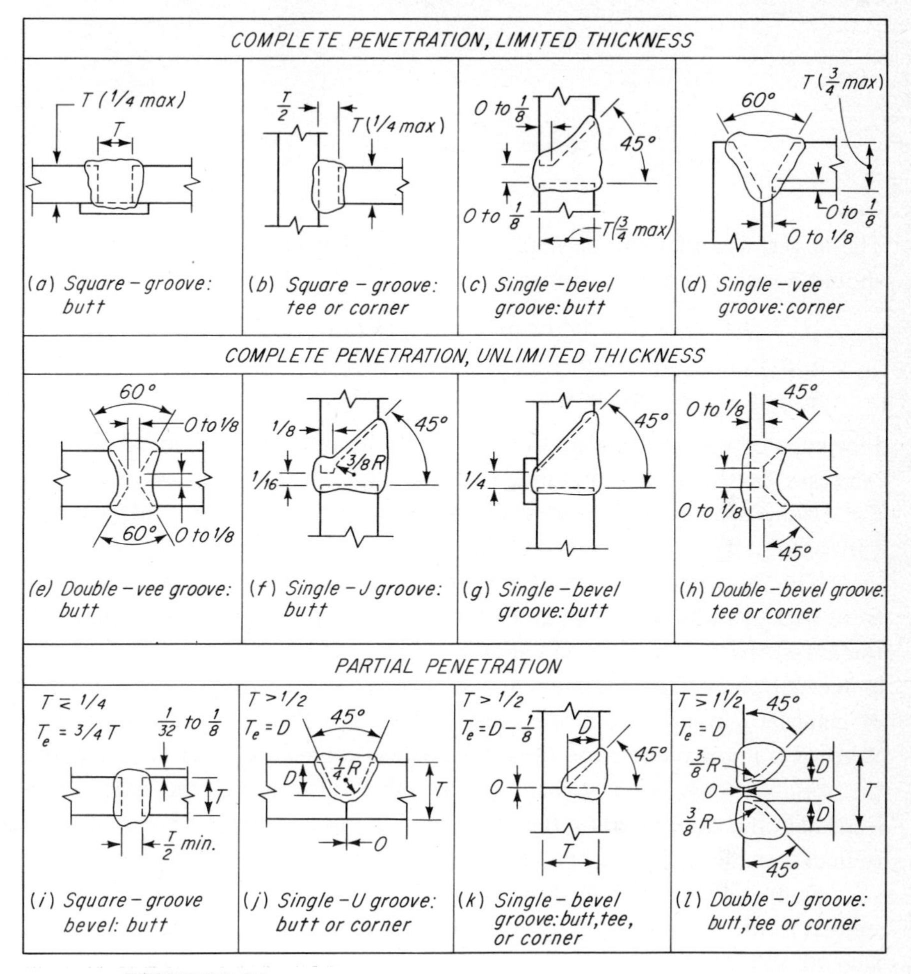

FIGURE 2-26
Typical prequalified manual shielded metal-arc joints for buildings. (*From "Welding Handbook," 6th ed., 1968.*)

legged fillet weld is given by the length of the side of the largest isosceles right triangle that can be inscribed within the weld cross section. The throat is the shortest distance from the root of the weld to the hypotenuse. (In the case of the submerged-arc weld, there is an exception to this definition in the AISC specification, discussed in Art. 2-13.) The legs are usually equal, but conditions sometimes require unequal legs. Large fillet welds made manually require two or more passes, as indicated in Fig. 2-27. Each pass must cool, and the slag must be removed, before the next pass is made. Therefore, the most efficient fillet welds are those which can be made in one pass.

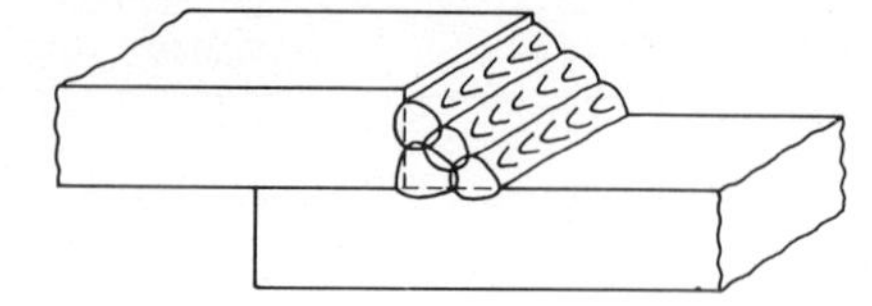

FIGURE 2-27
Multiple-pass fillet weld.

The largest size that can be made in one pass depends on the welding position, and should not exceed the following:

$\frac{5}{16}$ in. in the horizontal or overhead position
$\frac{3}{8}$ in. in the flat position
$\frac{1}{2}$ in. in the vertical position

Efficiency of welding is also affected by the amount of filler metal required, which increases by the square of the leg size as the throat dimension increases linearly. For example, a $\frac{5}{16}$-in. fillet weld has a throat only 25 percent greater than that of a $\frac{1}{4}$-in. weld, but its volume is 56 percent greater.

The most commonly used fillet welds increase in size by sixteenths of an inch from $\frac{1}{8}$ to $\frac{1}{2}$ and by eighths of an inch for sizes greater than $\frac{1}{2}$ in. To ensure full throat thickness, the size of the weld should be less than the thickness of the edge of the connected part. The maximum size of fillet permitted by the AWS along the edges of material $\frac{1}{4}$ in. or more in thickness is $\frac{1}{16}$ in. less than the thickness of the material. When it is impracticable to obtain a sufficient connection with these maximum sizes, the specifications permit a weld of the same thickness as the edge, provided this information is designated on the drawings. Along the edges of material less than $\frac{1}{4}$ in. in thickness, the weld size may be equal to the thickness of the material.

A fillet weld that is too small compared with the thickness of the material being welded is affected adversely during cooling. The amount of heat required to deposit a small weld is not sufficient to produce appreciable expansion of the thick material, and as the hotter weld contracts during cooling it is restrained by being attached to the cooler material. Thus, a lengthwise tensile stress is produced in the weld. Furthermore, conduction of heat by the relatively cold material accelerates the rate of cooling of the weld, and this tends to cause brittleness. The weld may actually crack because of the combination of the two effects. To help control this situation, specifications limit the thickness of the part, as given in Table 2-14.

Groove-welded joints are more efficient than fillet-welded lap joints, and because of their greater resistance to repeated stress and impact are to be preferred for dynamically loaded members. Groove welds not only require less weld metal than fillet welds of equal strength, but they also frequently eliminate the need for extra metal in the form of connecting plates or other structural shapes. Nevertheless, fillet welds are often used in structural work, partly because of the fact that many connections are more easily made with fillet welds and partly because groove welds require the members of a structure to be cut to rather close tolerances.

Cold-formed members are usually shop fabricated by spot welding (resistance welding). In this process the parts to be welded are clamped between two electrodes. Resistance of the metal to a strong current passed through the electrodes generates sufficient heat to melt a small area of the metal. Fusion of the parts between the electrodes is effected by pressure. Fusion welds are used for on-site welding to connect cold-formed members to other cold-formed members or to hot-rolled framing members. Shapes of welds and welding techniques are often different from those in ordinary structural welding. For example, the puddle weld, which is the standard way of connecting floor or roof deck to structural framing, is made by burning through the deck and then filling the hole with weld metal. This is analogous to the plug weld.

Standard welding symbols are shown in Fig. 2-28. The fillet weld is shown by a triangle and the groove weld by a symbol denoting the type of groove. The symbol for the type of weld is drawn on a reference line which has an arrow pointing to the joint. If the symbol is on the near side of the reference line, the weld is to be deposited at the side of the joint to which the arrow points. This is shown in Fig. 2-28*a*. Dimensions of the weld are to be specified as shown. In this case, the instruction is to deposit a $\frac{3}{8}$-in. fillet weld, 6 in. long, at the arrow side of the joint. If the symbol is on the far side of the reference line, the weld is to be made on the other side of the joint, as shown in Fig. 2-28*b*. Similarly, a weld symbol on both sides of the reference line is an instruction to weld both sides of the joint (Fig. 2-28*c*).

The arrow for a bevel or J-groove welding symbol points with a definite break toward the member which is to be chamfered (Fig. 2-28*d*). The root opening and the groove angle are specified as shown. The horizontal lines above and below the weld symbol in Fig. 2-28*g* mean that the weld is to be flush. If it is to be finished, a letter denoting the type of finish (G = grind, M = machine, etc.) is added. A weld which is to be made all around the joint is denoted by an open circle (Fig. 2-28*h*). The blacked-in circle in *i* signifies a weld to be made in the field.

Table 2-14 FILLET-WELD SIZES

Size of fillet weld, in.	Maximum thickness of part, in.
$\frac{1}{8}$	To $\frac{1}{4}$
$\frac{3}{16}$	Over $\frac{1}{4}$ to $\frac{1}{2}$
$\frac{1}{4}$	Over $\frac{1}{2}$ to $\frac{3}{4}$
$\frac{5}{16}$	Over $\frac{3}{4}$ to $1\frac{1}{2}$
$\frac{3}{8}$	Over $1\frac{1}{2}$ to $2\frac{1}{4}$
$\frac{1}{2}$	Over $2\frac{1}{4}$ to 6
$\frac{5}{8}$	Over 6

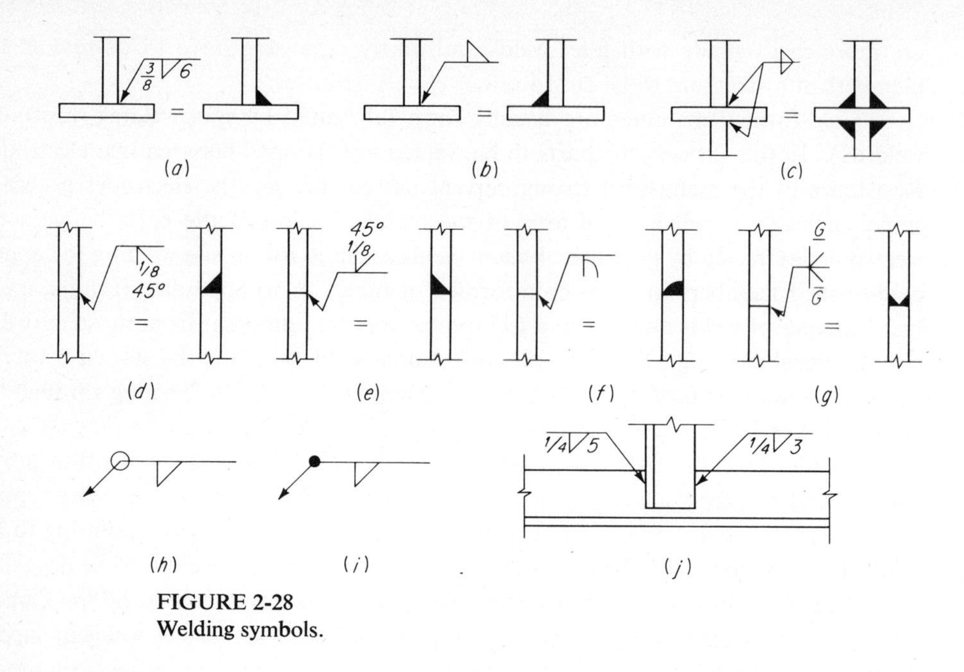

FIGURE 2-28
Welding symbols.

A situation which is common in structural drawings is shown in Fig. 2-28*j*, which represents two angles (vertical) welded to a tee, as in a truss. The weld symbols are shown for the angle which is seen in the drawing, in this case, the angle on the near side of the tee. It is understood that the duplicate angle on the far side is to be welded identically, and it would be incorrect to try to indicate this by using triangles on both sides of the weld reference line. In other words, the weld symbol can point to only one joint. On the other hand, if duplicate parts on opposite sides of a piece appear on the drawing, as in Fig. 2-28*c*, the required welding for both parts is shown.

These and other standard weld symbols and notations are described in detail in Ref. 6.

2-12 STRESSES IN WELDS

The groove weld may be stressed in tension, compression, shear, or combinations of tension, compression, and shear, depending upon the direction and position of the load relative to the weld. For example, the groove weld in the tee joint shown in Fig. 2-29*a* is in tension, while those in Fig. 2-29*b* are in shear. In either case, the stress is assumed to be distributed uniformly over the area of the throat. Thus for the weld in Fig. 2-29*a*, the tensile stress f_t is given by

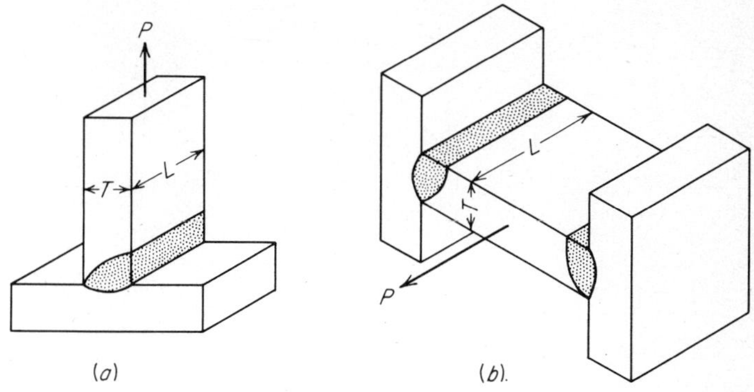

FIGURE 2-29

$$f_t = \frac{P}{LT_e} \tag{a}$$

where L is the length of the weld and T_e is the effective throat thickness. If the weld is a full-penetration one, $T_e = T$. In the connection of Fig. 2-29*b*, the shearing stress f_v for each weld is

$$f_v = \frac{P/2}{LT_e} \tag{b}$$

where $T_e = T$ for full-penetration welds.

The load P in Fig. 2-30*a* is resisted by a shearing force $P/2$ on the throat of each fillet weld. Therefore, the shearing stress is

$$f_v = \frac{P/2}{LT_e} \tag{c}$$

where T_e is the thickness of the throat. In the usual case of the 45° fillet, $T_e = 0.707T$, where T is the width of the leg, unless the weld is made by the submerged-arc process, as noted in Art. 2-13. It is customary to take the force on a fillet weld as a shear on the throat irrespective of the direction of the load relative to the throat. Thus, in Fig. 2-30*b*, the load P produces on the throat of each weld both a shearing component and a tensile component, each equal to $P\sqrt{2}/4$. In spite of this fact the stress is determined from Eq. (*c*).

Tests have shown that a fillet weld transverse to the load, as in Fig. 2-30*b*, is much stronger than a fillet weld of the same size parallel to the load, as in Fig. 2-30*a* (Art. 2-13). The explanation probably lies in the fact that the force on the throat of the transverse weld has both a shearing component and a normal component, as was shown above. Therefore, it is reasonable to expect the strength of the transverse weld

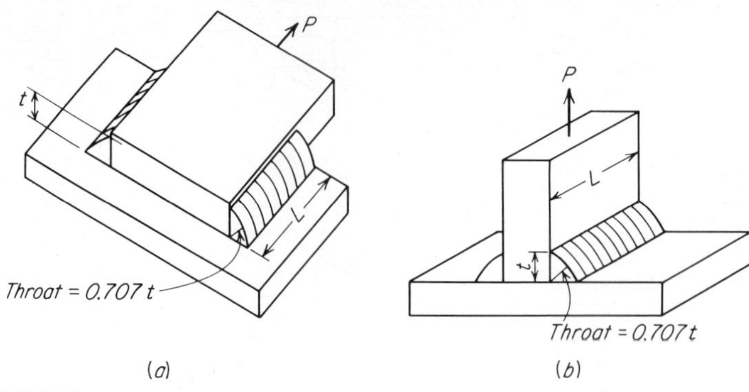

FIGURE 2-30

to be intermediate between the shearing and the tensile strengths of the weld metal, which is actually the case. It is also well known that at working loads the shearing stress is not uniform over the length of a longitudinal weld and that for long welds it may be considerably larger at the ends than at the center of the weld. However, since the more highly stressed portions yield first under increasing load, it is probable that the stress approaches a uniform distribution at failure under static loading.

Figure 2-31 shows a lap joint with two longitudinal fillet welds and one transverse fillet weld. A portion of the load P will be transmitted by each weld, but it is impossible to determine by statics alone how the load will be divided among the three welds. However, it is customary to assume that the stress is distributed uniformly along each weld and is of the same intensity for all the welds, with the result that the load is assumed to be distributed among the welds in proportion to their lengths.

Any abrupt discontinuity or change in the section of a member, such as a notch or a sharp reentrant corner, interrupts the transmission of stress along smooth lines. The magnitude of the stress concentration increases as the sharpness of the notch

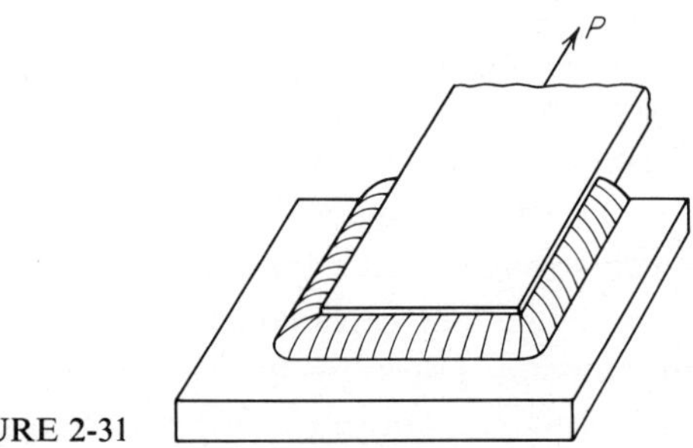

FIGURE 2-31

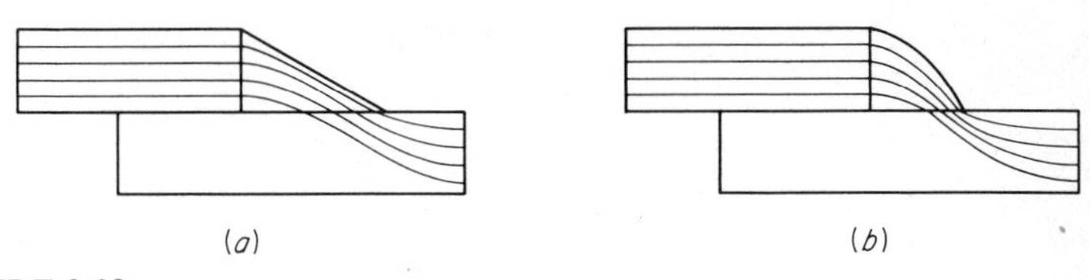

FIGURE 2-32

or the abruptness of the discontinuity increases. These concentrations, including those at the ends of longitudinal welds discussed above, are of no consequence for static loads or for cases where only a few thousand repetitions of maximum stress are likely to occur. However, they are significant where fatigue is involved (Art. 2-15). Figure 2-32*a* illustrates a transverse joint in which the weld is elongated in the direction of the load to produce a more uniform transfer of stress than in the conventional weld of Fig. 2-32*b*. Specifications do not permit an increase in the allowable unit stress for such a weld.

2-13 SPECIFICATIONS FOR WELDED CONNECTIONS

Welding electrodes are classified on the basis of the mechanical properties of the weld metal, the welding position, the type of coating, and the type of current required. Electrodes for shielded metal-arc welding are covered by ASTM A233 and A316. Each electrode is identified by a code number EXXXXX, where E stands for electrode and each X represents a number. The first two (or three) numbers indicate the tensile strength (kips per square inch) of the weld metal. The next number denotes the positions in which the electrode can be used, the number 1 meaning all positions, the number 2 flat and horizontal fillet welds, and the number 3 flat welding only. The last number denotes the type of covering, the type of current (ac or dc), and the polarity (straight or reversed). Straight polarity means that the electrode is negative. For example, the E7018 electrode has a tensile strength of 70 ksi, the number 1 means that it can be used in all positions, and the number 8 means that it is an iron-powder, low-hydrogen electrode which can be used with either alternating or direct current but only in reverse polarity.

Combinations of flux and electrodes for submerged-arc welding are covered by ASTM A558. A flux is designated by the letter F followed by two digits denoting the tensile strength and Charpy V-notch impact strength of the test weld. This is followed by a set of letters and numbers denoting the electrode used to classify the flux. For example, the letters EL in EL8 signify a low-manganese electrode, while the number 8 denotes the chemical composition in percent of carbon, manganese, and silicon.

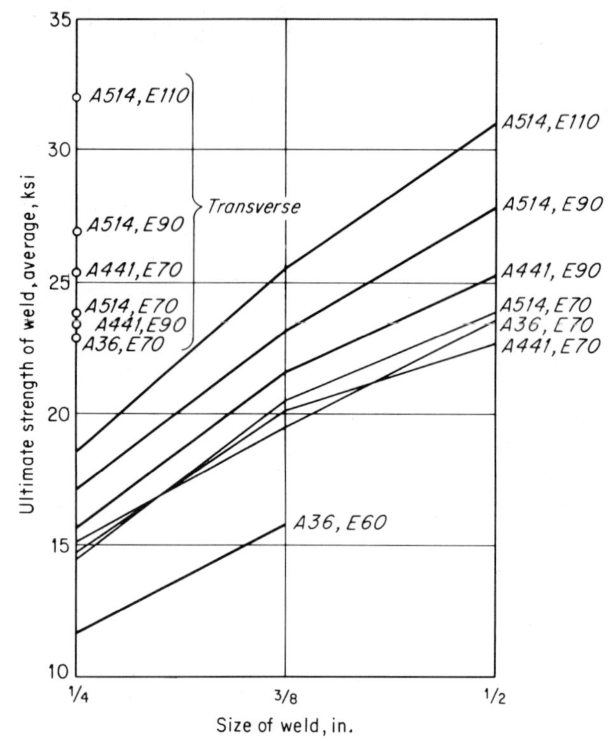

FIGURE 2-33
Strength of fillet welds. (*From AWS-AISC Fillet Weld Study, Report to AISC by Testing Engineers, Inc., Oakland, Calif., 1968.*)

Electrodes for gas metal-arc welding are covered by A559 and are identified by similar notation. Thus, E70S-X and E70-U1 are solid and emissive electrodes, respectively, where the last number identifies the chemical composition. Only one emissive electrode is specified in A559.

The mechanical properties of electrodes are given in Table 2-15.

For the full-penetration groove weld, most specifications prescribe tensile, compressive, and shearing stresses equal to those for the base metal, provided welds of the strength specified in Table 2-16 are used. Thus, since the tensile properties of the weld metal exceed those of the base metals on which they are used, factors of safety for full-penetration groove welds are greater than those for the base metal.

Partial-penetration groove welds and fillet welds may be made with filler metals whose tensile properties are less than those of the base metal. Allowable stresses for this situation are based on test results evaluated according to the assumptions as to stress distribution which were discussed in Art. 2-12. Figure 2-33 shows the results

Table 2-15 MECHANICAL PROPERTIES OF ELECTRODES

Welding process				Mechanical properties			
Shielded metal-arc	Submerged arc	Gas metal-arc	Flux-cored arc	Minimum tensile strength, ksi	Minimum yield point, ksi	Minimum elongation in 2 in., %	Minimum impact strength, ft-lb at 0°F
E60XX				62–67	50–55	17–25	Required for certain electrodes
	F60-XXXXX			60–80	45	25	Not required
		E60S-X		62	50	22	
E70XX				72	60	17–22	
	F70-XXXXX			70–95	50	22	
		E70S-X		70	60	20–22	
E80XX				80	67	16–19	
	F80-XXXXX	E80S-X	E80T	80	65	18	20
E90XX				90	77	14–17	
	F90-XXXXX	E90S-X	E90T	90	78	17	20
E100XX				100	87	13–16	
	F100-XXXXX	E100S-X	E100T	100	90	16	20
E110XX				110	97	15	
	F110-XXXXX	F110S-X	F110T	110	98	15	20

Table 2-16 FILLER-METAL REQUIREMENTS FOR COMPLETE-PENETRATION GROOVE WELDS*

Lower specified minimum yield point of base metals being joined	Welding process			
	Shielded metal-arc	Submerged arc	Gas metal-arc	Flux-cored arc
A36, A53 grade B, A375, A500, A501, A529, A570 grades D and E	E60XX or E70XX*†	F6X or F7X-EXXX	E70S-X or E70U-1	E60T-X or E70T-X
A242, A441, A572 Grades 42 to 60, and A588‡	E70XX§	F7X-EXXX	E70S-X or E70U-1	E70T-X
A572 Grade 65	E80XX§	Grade F80	Grade E80S	Grade E80T
A514	E110XX§	Grade F110	Grade E110S	Grade E110T

* Use of same type filler metal having next higher mechanical properties is permitted.

† Low-hydrogen electrodes must be used for welding A36 steel more than 1 in. thick.

‡ For architectural exposed bare unpainted applications, the deposited weld metal and the base metal must have similar atmospheric-corrosion resistance and coloring characteristics. Follow steel manufacturer's recommendation.

§ Low-hydrogen classifications.

of 168 tests on fillet-welded joints.[7] Weld sizes were $\frac{1}{4}$, $\frac{3}{8}$, and $\frac{1}{2}$ in., base metals A36, A441, and A514, and electrode classifications E60, 70, 90, and 110. The combinations of base metal and electrode which were tested are given in Table 2-17. Combinations of strong weld metal with weaker base metal, and vice versa, showed that the effect of dilution upon weld strength is quite small. For example, the strength of $\frac{1}{4}$-in. fillets made with E110 electrodes on A441 steel, whose average tensile strength was 79 ksi, was only 8 percent less than that of the same electrode on A514 steel with an average tensile strength of 118 ksi. Conversely, the strength of $\frac{1}{4}$-in. welds of E70 electrodes on A514 steel was only 2 percent greater than that of the same electrodes on A441 steel.

The 36 transverse-shear tests were made only on joints with $\frac{1}{4}$-in. welds. It will be noted (Fig. 2-33) that weld strength is considerably greater in transverse shear. For example, the E110 electrode on A514 steel produced a weld which was 32/18.6 = 1.72 times as strong when tested transverse to the load as it was in the direction of the load. Similarly, the E70 weld on A36 steel was 22.8/15 = 1.52 times stronger transversely. The AISC allowable stresses for fillet welds and partial-penetration groove welds are based on the test results for the longitudinal welds. The allowable stress for each type of weld is 0.3 times the minimum tensile strength of the electrode. Thus, the allowable stress for the E70 electrode is $0.3 \times 70 = 21$ ksi. The factors of safety for these allowable stresses are given in Table 2-18. Factors of safety are about half again as large for welds transverse to the direction of load.

The superior penetration of the submerged-arc fillet weld justifies a more liberal definition of the throat. Only the AISC specification makes such an allowance,

Table 2-17 BASE METAL AND ELECTRODES FOR TESTS OF FIG. 2-33

	Electrode classification										
	E60XX Weld size		E70XX Weld size			E90XX Weld size			E110XX Weld size		
Base metal	$\frac{1}{4}$	$\frac{3}{8}$	$\frac{1}{4}$	$\frac{3}{8}$	$\frac{1}{2}$	$\frac{1}{4}$	$\frac{3}{8}$	$\frac{1}{2}$	$\frac{1}{4}$	$\frac{3}{8}$	$\frac{1}{2}$
	Longitudinal fillet welds										
A36	X	X	X	X	X	X					
A441			X	X	X	X	X	X	X		
A514			X	X	X	X	X	X	X	X	X
	Transverse fillet welds										
A36			X								
A441			X			X					
A514			X			X			X		

however. It defines the throat thickness T_e as the leg size for $\frac{3}{8}$-in. and smaller fillets, and 0.11 in. more than the theoretical throat for larger welds.

Allowable weld stresses in the AASHO and AREA specifications are much more conservative than those of the AISC specification. AASHO requires the yield stress of the weld metal to be equal to or greater than that of the base metal and prescribes the following allowable stresses:

Groove welds	Same as that of the lower-yield base metal joined
	Only full-penetration welds permitted
Fillet welds	12.4 ksi on base metal with $F_y = 36$ ksi
	14.7 ksi on base metal with $40 \leqq F_y \leqq 50$ ksi
	25 ksi on base metal with $90 \leqq F_y \leqq 100$ ksi

AREA specifies $0.55F_y$ for tension or compression and $0.35F_y$ for shear in groove welds, where F_y is the yield stress of the base metal or weld metal, whichever is smaller. The allowable shear in fillet welds is 14.7 ksi. In this specification the highest-strength weldable steel is A572 Grade 50 and the highest-strength electrode the E80.

Allowable shears for resistance-welding spot welds according to the AISI specification for cold-formed members are given in Table 2-19. The factor of safety is about 2.5. Values for other thicknesses can be obtained by interpolation.

Table 2-18 AISC FACTORS OF SAFETY FOR FILLET AND PARTIAL-PENETRATION GROOVE WELDS

Base metal	Electrode	Longitudinal weld		Transverse weld	
		Average	Minimum	Average	Minimum
A36	E60	2.88	2.67		
A441	E70	2.95	2.67	4.62	4.06
A514	E110	2.41	2.21	3.48	3.30

Table 2-19 AISI ALLOWABLE SHEAR FORCE FOR SPOT WELDS

Sheet thickness, in.	Allowable shear per spot, lb
0.010	50
0.020	125
0.040	350
0.060	725
0.080	1,075
0.125	2,000
0.188	4,000

2-14 WELDING QUALITY CONTROL

The production of sound welds is governed by many factors. The type of joint, its preparation and fit-up, the root opening, etc., are important, as are the welding position, the welding current and voltage, the arc length, and the rate of travel. Accessibility for the welding operation is also important, as the quality of a weld is determined to a considerable degree by the position of the electrode. For a fillet weld, the electrode ordinarily should bisect the angle between the two legs of the weld. Furthermore, it must lean about 20° in the direction of travel. The joints in Fig. 2-34*a* and *b* emphasize the significance of inclination in the direction of travel. The welding procedure was identical for these two joints except that the electrode leaned about 45° in the direction of travel in Fig. 2-34*a* and 20° in Fig. 2-34*b*. The defective weld in Fig. 2-34*a* shows incomplete penetration, slag inclusion at the root, and a slight undercutting at the upper edge. This weld is also likely to be brittle.

The defective weld in Fig. 2-34*c* is shown to emphasize the importance of current. Too much current relative to the rate of travel of the electrode was used in producing these welds. The excessive penetration results in a weld with a large inner portion that cools more slowly than the rest. Shrinkage of the outer portion may cause cracking, as in the weld at the left. The pear-shaped weld at the right in Fig. 2-34*d* shows a similar crack.

The American Welding Society publishes weld qualification procedures. Procedure qualification deals with properties of the metals, type of groove and position of welding, electrode type and size, current and voltage, and requirements for preheating the base metal. The operator must also be qualified by welding prescribed test specimens, which must demonstrate the required strength and ductility. However, qualification of the procedure and the operator are not enough to guarantee satisfactory welds, and inspection is important. In addition to visual inspection, nondestructive tests can be used to determine the types and distribution of weld defects.[8]

Magnetic-particle inspection (magnaflux) is based on inducing a strong magnetic field in a short section of ferromagnetic material. Poles develop where there are leakages in the field, which occur at discontinuities in the weld, and iron powder sprinkled on the test area migrates to the poles. The resulting pattern tends to outline the discontinuities.

In dye-penetrant inspection a dye is brushed or sprayed on the surface of the weld. It seeps into surface irregularities. A developer which is sprayed on is stained red by the dye, which rises from the surface defects by capillary action. A similar procedure is based on using a fluorescent liquid which detects surface imperfections upon exposure to black light.

Radiographic inspection uses shortwave radiations, such as x-rays or gamma

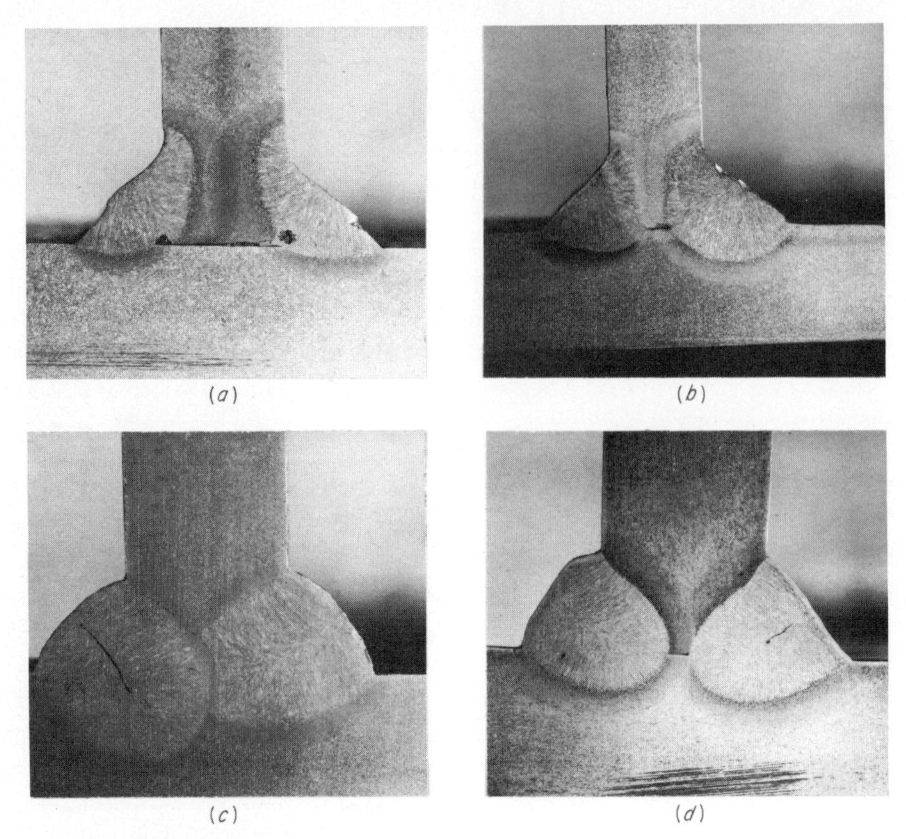

FIGURE 2-34
Effect of welding technique on quality of weld. (*State of California, Department of Public Works, Division of Highways.*)

rays, to discover surface and subsurface flaws in the weld. The beam encounters less resistance at a defect, and when the radiation is recorded on film, the defect is disclosed.

Ultrasonic inspection is also effective in locating subsurface weld defects as well as those on the surface. High-frequency sound waves sent through the area to be inspected are reflected by discontinuities and density differences. The reflected sound waves are monitored by a receiver, converted to electric energy, and displayed as visual patterns on an oscilloscope screen.

In interpreting the results of inspection of a weld, it is important to assess the severity of a defect as it relates to service requirements. Some defects may be relatively unimportant, while others may be critical in specific service situations.

2-15 FATIGUE

Fracture of metals is not always preceded by yielding and the subsequent elongation described in Art. 2-4. Instead, it may occur at stresses even less than the yield stress if a load is repeated a large number of times. This kind of failure is called *fatigue.* It is progressive in nature and is believed to begin with a dislocation or slip in the crystalline structure of the metal, followed by the development of a crack which gradually increases in size. Crack initiation is brittle, rather than ductile, and almost always occurs at a point of stress concentration such as a hole, weld, notch, or even a scratch. These stress raisers are usually on the surface, but they can also be internal, as at a defect in a weld. Because there is no plastic deformation of the material, fatigue cracks are hard to detect even on the surface of a member. They usually propagate very slowly and intermittently.

Fatigue failure of a laboratory specimen consisting of two plates joined by a groove weld is shown in Fig. 2-35. The specimen was tested in repeated tension in the longitudinal direction of the weld. Cracking began at an irregularity at the surface of the weld and propagated radially into the plate. This progressive cracking usually results in the "oystershell" or "beach" markings which are visible in the photograph.

Behavior under repeated load is evaluated in rotating-beam tests, flexure tests, and axial-load tests. In the rotating-beam test a round, polished specimen supported as a simple beam is rotated at constant speed while being subjected to a bending moment, so that every fiber of the specimen alternates sinusoidally between tension and compression. A polished specimen is also used in the flexure test, but it is tested by bending it in one plane. In the axial-load test, the specimen is subjected to alternating axial stress. Test data are plotted with the maximum stress S (*fatigue strength*) as ordinate and the number of cycles N to failure (*fatigue life*) as abscissa. The result is called an S-N curve. Figure 2-36 shows S-N curves for three different ranges of axial stress on polished specimens of T1 (A514) steel. The range of stress is denoted by R, the ratio of maximum stress to minimum stress. Thus, $R = 0$ (upper curve) denotes stress ranging from zero to tension, $R = -\frac{1}{2}$ (middle curve) denotes stress alternating between tension and compression equal to half the tension, etc. A positive value of R signifies a stress alternating between a maximum tension and a smaller tension. S-N curves tend to become horizontal at large values of N; the corresponding fatigue strength is called the *fatigue limit*. The fatigue limit is generally considered to correspond to a fatigue life of about 2 million cycles.

It will be noted that the fatigue limit for $R = -1$ in Fig. 2-36 is about one-half the tensile strength of the steel. Although this is a good approximation for polished specimens, severe stress concentrations and exposure to corrosion may produce drastic reductions, as in Fig. 2-37. This figure shows that these reductions can be so large as to make high-strength steels little better in fatigue than low-strength steels.

FIGURE 2-35
Typical fatigue fracture in a welded plate. (*Department of Civil Engineering, University of Illinois.*)

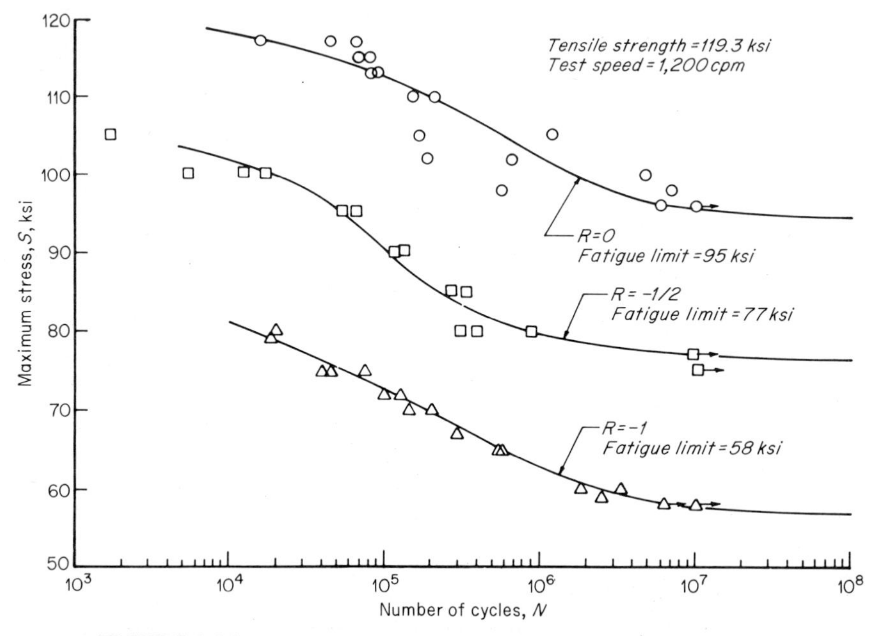

FIGURE 2-36
S-N diagrams for polished specimens of T1 steel in axial-load fatigue. (*From "USS Steel Design Manual," U.S. Steel Corporation, November 1968.*)

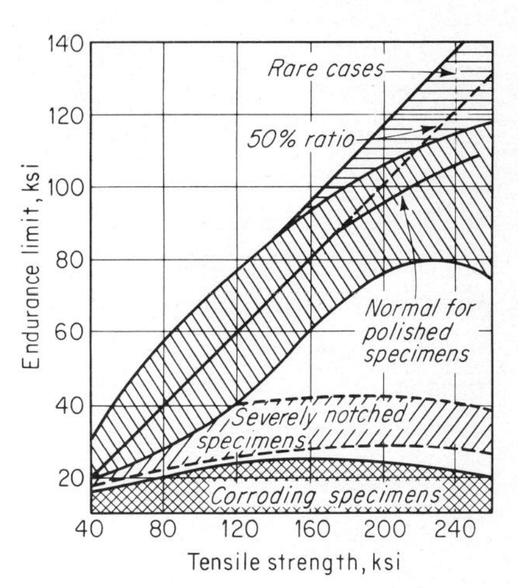

FIGURE 2-37
Relationship between fatigue limit and ultimate tensile strength of various steels. (*From Battelle Memorial Institute, "Prevention of Failure of Metals under Repeated Stress," Wiley, New York, 1946.*)

The data in Fig. 2-36 are presented in Fig. 2-38 in a form which is better adapted to design. This figure, called a *modified Goodman diagram,* is a variation of a diagram published by Goodman in 1899. Each curve is the locus of all the points corresponding to a given fatigue life, and the diagram covers the full range of stress ratios $-1 \leq R \leq 1$. Minimum stresses are plotted as abscissas and maximum stress as ordinate. Radial lines from the origin correspond to the various stress ratios. Since

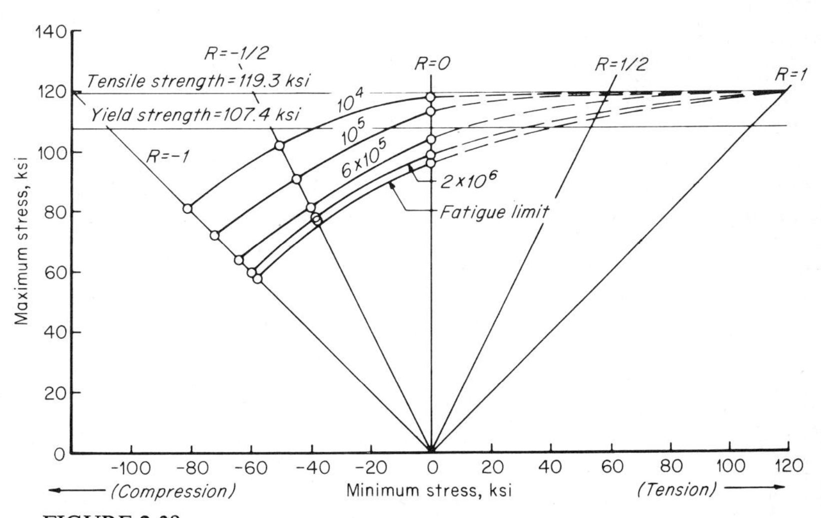

FIGURE 2-38
Modified Goodman diagram for specimens of Fig. 2-36. (*From "USS Steel Design Manual," U.S. Steel Corporation, November 1968.*)

$R = 1$ denotes no reversal of stress, the ray $R = 1$ corresponds to static tension and all the curves for a given steel join at the intersection of $R = 1$ with the ordinate corresponding to the tensile strength of the steel.

Taking the yield stress as the limiting useful strength, it will be seen that the curves in Fig. 2-38 can be represented quite closely by the straight lines *AB* and *BC* of Fig. 2-39. This enables fatigue-design criteria to be put in the form of simple equations. Allowable-stress formulas are obtained by applying a factor of safety to *ABC* to get *DEF*. This factor of safety can be smaller than the factor of safety for static load because of the smaller probability of occurrence of the much larger number of cycles of service-load magnitude needed to cause a fatigue failure.

The equation for *DE* in Fig. 2-39 is

$$f_{max} = f_0 + af_{min} \tag{a}$$

where f_0 is the value of f_{max} at $R = 0$. This equation can be written in terms of the stress ratio as

$$f_{max} = \frac{f_0}{1 - af_{min}/f_{max}} = \frac{f_0}{1 - aR} \tag{b}$$

Such formulas are needed for the various fasteners and, for each grade of steel, for the member (base metal) itself and for the base metal adjacent to connections and splices. The following AWS formulas for the allowable fatigue stress F_r in tension on transverse groove-welded butt joints not ground smooth, in bridges of A36 steel, are typical:

$$F_r = \begin{cases} \dfrac{20{,}500}{1 - 0.55R} \not> F_t & \text{for 100,000 cycles or less} \quad (c) \\[2ex] \dfrac{17{,}200}{1 - 0.62R} \not> F_t & \text{for 100,000 to 500,000 cycles} \quad (d) \\[2ex] \dfrac{15{,}000}{1 - R} \not> F_t & \text{for 500,000 cycles or more} \quad (e) \end{cases}$$

F_t in these formulas is the allowable tensile stress for static loading. The cycles of loading are repetitions of the maximum service load. Formulas of this type are used in the AASHO and AREA specifications.

In order to simplify design where fatigue is a factor, the AISC specification prescribes an allowable *range of stress*, $F_{sr} = f_{max} - f_{min}$, rather than an allowable maximum stress F_r, for various loading conditions. This is equivalent to taking $a = 1$ in Eqs. (*a*) and (*b*), which puts the fatigue-strength line *DE* in Fig. 2-39 at 45° and results in the stress range being independent of f_{max}. This is easily proved. Thus,

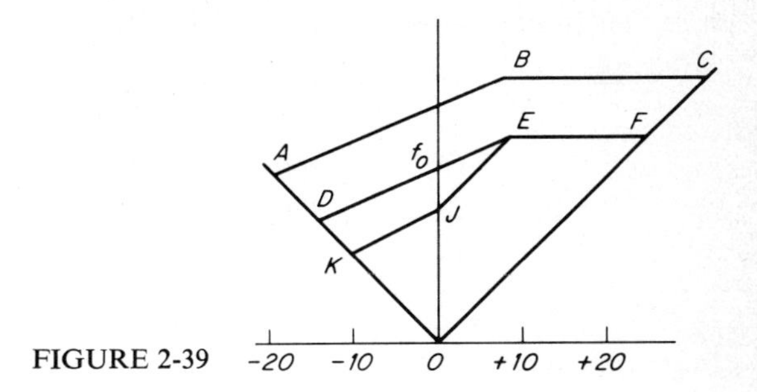

FIGURE 2-39

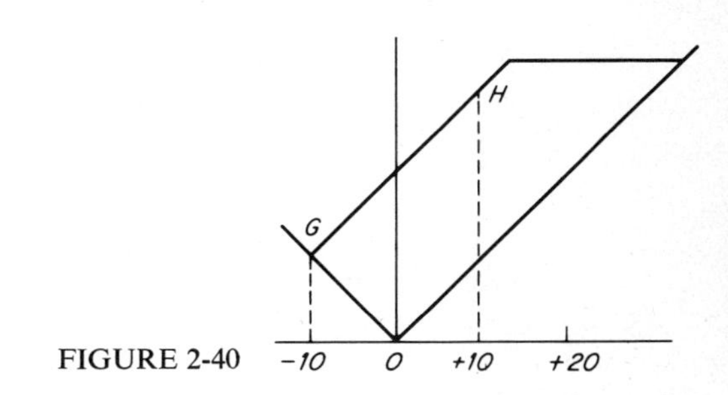

FIGURE 2-40

if the coordinates of G and H in Fig. 2-40 are $(-10, +10)$ and $(+10, +30)$, respectively, the stress range is 20 ksi in each case but $f_{max} = 10$ ksi for point G and 30 ksi for point H. The AISC allowable ranges of stress are obtained by dividing the fatigue strength ABC of Fig. 2-39 by 1.4 to obtain the allowable-stress diagram DEF and then drawing a line at 45° from a point in the near neighborhood of E to its intersection J with the ray $R = 0$. Then $F_{sr} = OJ$. These allowable ranges are based on fatigue data for A36 steel but are to be used for all grades of steel except for one loading condition in A514 steel. Therefore, but for this one exception, they do not recognize the increased fatigue strengths of higher-strength steels.

Since the fatigue line DE in Fig. 2-39 tends to be considerably flatter than EJ in some situations, the AISC provision may become too conservative if stress reversal is involved. For these cases, the allowable range of stress may be increased by multiplying the prescribed value of F_{sr} by the factor $(f_t + |f_c|)/(f_t + 0.6|f_c|)$. This is equivalent to

$$F_r = F_{sr} + 0.6f_c = F_{sr} - 0.6|f_c|$$

where F_r is the allowable fatigue stress in tension. The corresponding increased allowable stress is represented by JK in Fig. 2-39.

The AISC allowable values of F_{sr} for the situation in which Eqs. (*c*), (*d*), and (*e*) apply are

$$F_{sr} = \begin{cases} 28 \text{ ksi} & \text{for 20,000 to 100,000 cycles} \\ 21 \text{ ksi} & \text{for 100,000 to 500,000 cycles} \\ 14 \text{ ksi} & \text{for 500,000 to 2,000,000 cycles} \\ 12 \text{ ksi} & \text{for more than 2,000,000 cycles} \end{cases}$$

It should be noted that fatigue is not often a factor in the design of buildings, except for crane runways and the like. For example, 100,000 cycles of load correspond to 10 applications of maximum service load daily for 25 years, and buildings ordinarily do not experience such a pattern of load. On the other hand, it is obvious that bridges may well experience many more daily cycles of maximum service load.

The importance of details, such as splices, connections, etc., must be appreciated in situations where fatigue is a factor. For example, the fatigue strength of a complete-penetration, groove-welded butt splice in a tension member will usually be increased if the weld is ground flush with the surface of the connecting parts. This eliminates the stress concentrations that would arise in the as-welded condition.

The following suggestions should be kept in mind when designing structures subject to fatigue:[9]

1. Avoid details of design that produce severe stress concentrations or poor stress distribution.
2. Provide gradual changes in the section and avoid reentrant, notchlike corners.
3. Avoid abrupt changes of section or stiffness in members or components.
4. Align parts so as to eliminate eccentricities or reduce them to a minimum.
5. Avoid making attachments on parts subjected to severe fatigue loadings.
6. Use continuous welds rather than intermittent welds.
7. Avoid details that introduce high, localized constraint.
8. Provide suitable inspection to guarantee proper riveting, adequate clamping in high-strength bolts, and the deposition of sound welds.
9. Provide for suitable inspection during the fabrication and erection of structures.
10. When fatigue cracks are discovered, take immediate steps to prevent their propagation.

2-16 BRITTLE FRACTURE

Steel structures sometimes fail suddenly and without warning in the form of excessive deformation. These failures are often catastrophic, in the sense that the structure is more or less completely destroyed. In contrast to fatigue failures, which result from cyclic or repetitive load, brittle fractures usually occur under static load. For example,

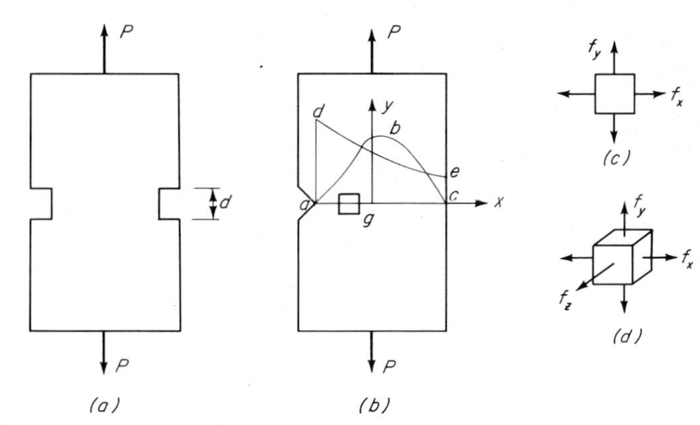

FIGURE 2-41

a riveted steel tank 90 ft in diameter and 50 ft high, which contained 2 million gallons of molasses, fractured in an explosive manner on Jan. 15, 1919, in Boston. Similar sudden failures of steel water tanks, oil tanks, transmission lines, ships, plate-girder bridges, etc., have occurred. Most of these failures occurred under normal service conditions, rather than overload, and originated at a point of concentration of stress such as a defect or a geometrical stress raiser. Furthermore, most of the structures were welded and failed at low temperatures.

There is little or no plastic deformation in advance of brittle fracture. In effect, this is another way of saying that the sliding on planes of maximum shear stress described in Art. 2-4, which is the source of the plastic deformation that precedes fracture, is somehow inhibited. This may happen because the material's resistance to sliding exceeds its resistance to separation (cleavage) or because sliding is prevented.

The relation between resistance to sliding and resistance to cleavage is not constant for a given material but depends on the speed of deformation and on the temperature. Resistance to sliding increases with increase in speed of deformation, while resistance to separation is affected to a smaller degree. Asphalt is a good example of sensitivity to strain rate; it may flow under its own weight over a long period of time, but it is brittle under suddenly applied forces. Again, both kinds of resistance increase with decrease in temperature, but in steel the difference between the two tends to become smaller and may finally disappear. Therefore, a material that may fail in a ductile manner at one temperature may fail in a brittle manner at a lower temperature.

Sliding in a ductile material may be inhibited by local geometry of the member, such as a hole or notch. For example, in a tensile test of a notched specimen, such as that shown in Fig. 2-41*a*, sliding that would ordinarily develop in the notched portion is restrained by the larger, lower-stressed portions on either side of the notch. The smaller the length d of the notch in relation to its depth and to the size of the specimen, the more the sliding is inhibited. This may increase resistance to sliding to the point where it exceeds resistance to separation, so that the specimen fails in a brittle manner on the reduced area. An analogous situation may develop at a small notch in a plate, as in Fig. 2-41*b*. In this case, there are tensile stresses f_x in the region of the notch, distributed somewhat as shown by the curve *abc*, in addition to the tensile stresses f_y, which are distributed as shown by the curve *de*. Therefore, an element such as the one at g is in the biaxial state of stress shown in Fig. 2-41*c*. When the load P reaches the value at which yielding would normally begin at a, the necessary contraction, which must be largely in the direction of the thickness of the plate (z axis), is restrained by the adjacent less highly stressed portions, in a manner similar to the restraint in the specimen of Fig. 2-41*a*. As a consequence, sliding is inhibited, and a brittle fracture may initiate at point a. Once it begins, such a fracture propagates rapidly, and in mild steel it may travel at a speed of 4,000 to 5,000 fps at stress levels as low as 6 to 8 ksi.[10] The resulting fracture surface is granular in appearance, in contrast to the silky or fibrous appearance of a shear-failure surface, and usually has a herringbone or chevron appearance, with the apexes of the chevrons pointing toward the point of fracture initiation (Fig. 2-42).

It will be noted that the restraint of contraction in Fig. 2-41*b* is accompanied by a tensile stress in the direction of the thickness of the plate, so that an element in the interior of the plate near the notch is in the triaxial state of stress shown in Fig. 2-41*d*. Brittle fracture can also be explained in terms of this condition. Assuming $f_z < f_x < f_y$, the maximum shear stress is $(f_y - f_z)/2$, which acts on the section through the x axis

FIGURE 2-42
Brittle fracture of structural steel. (*Department of Civil Engineering, University of Illinois.*)

that bisects the angle between the y and z axes. Therefore, if f_z approaches f_y in magnitude the shear stress becomes small, so that fracture may be by cleavage rather than by sliding.

The Charpy V-notch test (ASTM E23) is commonly used to evaluate the behavior of a metal as it is affected by an abrupt change in cross section. In this test, a rectangular bar with a V notch at midlength is simply supported as a beam and struck by a pendulum released from a fixed height. The energy absorbed in fracturing the specimen is determined from the difference in the height of the pendulum before release and the height to which it returns after impact. In the case of steel, it turns out that energy-absorbing capacity in the notched specimen is not uniformly related to properties determined by the standard tension test. In other words, a steel that is ductile in the tension test might break in a brittle manner at a notch. Furthermore, the energy-absorbing capacity of a notched specimen may be drastically reduced at low temperatures. On the other hand, aluminum and many other nonferrous materials show consistent behavior; i.e., if they are ductile (or brittle) in the standard tension test they are ductile (or brittle) in their notch behavior independently of temperature.

Variation in energy-absorbing capacity with temperature is determined from notched-specimen tests covering a range of temperatures. Figure 2-43 shows a typical plot of absorbed energy vs. temperature. There is a range of temperature in which fracture is largely by cleavage (separation), another in which it is largely by shear, and an indeterminate zone of transition from one type of fracture to the other. The lower-temperature boundary of the transition zone is called the *ductility transition* and the higher-temperature boundary the *plastic-fracture transition*. The corresponding temperatures are called *transition temperatures*. A value of 15 ft-lb generally is used to define the lower transition temperature, and 45 ft-lb has sometimes been specified for the higher one.[10] The difference between the two is generally about 120°F. If the temperature is above the ductility-transition temperature, there will be appreciable plastic flow at the root of a notch before cracking begins.

Brittle-fracture behavior is affected by the chemistry of the steel. Small increases in carbon lower the energy-absorbing capacity and raise the transition temperature.[10] The size of the piece is also a factor. Thick plates have higher transition temperatures than thin plates. This is because they require less rolling than thin plates. They also cool more slowly. Residual stress (Art. 3-1) is another important factor. Most low-temperature, low-stress brittle fractures have been in situations where there were large residual stresses, usually because of welding, in addition to notches. Cold work is also detrimental because of the resultant lowering of ductility.

Charpy V-notch 15 ft-lb transition temperatures are on the order of +30°F for A36 and A441 steels and 40°F for A440 steel. On the other hand, A514 steel has a 15 ft-lb transition temperature of about −50°F. It is important to note that A36 and A441 steels have been used successfully and extensively in bridges, transmission towers,

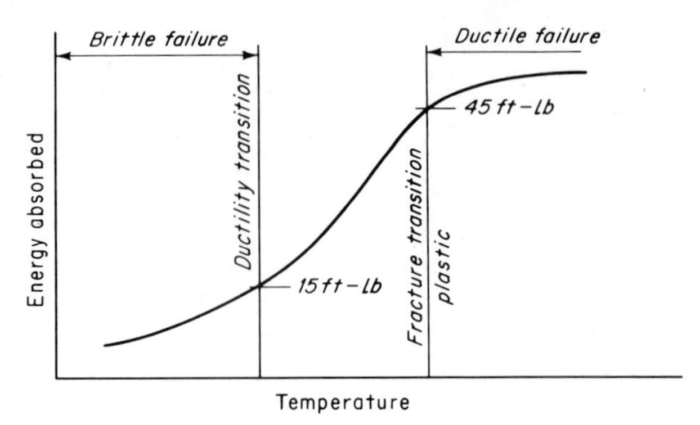

FIGURE 2-43
Transition from ductile to brittle behavior of mild structural steel.

and other exposed structures, in riveted, bolted, and welded construction throughout the continental United States with little or no history of brittle-fracture problems. Therefore, it is clear that a low 15 ft-lb transition temperature is not a necessary prerequisite to successful, low-temperature applications of these steels in these structures. Thus, the best guide to avoidance of brittle fracture is experience with structures and steels of various types. It must be kept in mind, however, that the lower the expected service temperatures the more important it becomes to avoid conditions which are conducive to brittle fracture. In other words, an exposed structure is much more likely to tolerate poor detail geometry (notches, abrupt changes in section, etc.), weld flaws (inclusions, undercut, incomplete fusion, etc.), residual stresses, cold working during fabrication, etc., in a warm climate than in a cold one. Guides to protecting against brittle fracture in structures for which there is no adequate specification or experience are discussed in Ref. 10.

REFERENCES

1 Julian, O. G.: Synopsis of First Progress Report of Committee on Factors of Safety, *J. Struct. Div. ASCE*, July 1957.
2 Research Council on Riveted and Bolted Structural Joints: Specification for Structural Joints Using A325 or A490 Bolts, 1970.
3 Jones, J.: Bearing Ratio Effect on Static Strength of Riveted Joints, *Trans. ASCE*, vol. 123, 1958.
4 Task Committee on Lightweight Alloys: Suggested Specifications for Structures of Aluminum Alloys 6061-T6 and 6062-T6 and Suggested Specifications for Structures of Aluminum Alloys 6063-T5 and 6063-T6, *J. Struct. Div. ASCE*, December 1962.

5 Doty, W. D.: Weldability of Construction Steels: USA Viewpoint, *Weld. J.*, February 1971.

6 "Standard Welding Symbols," American Welding Society, New York, 1968.

7 Higgins, T. R., and F. R. Preece: Proposed Working Stresses for Fillet Welds in Building Construction, *Weld. J.*, October 1968.

8 "Welding Handbook," 6th ed., sec. 1, American Welding Society, New York, 1968.

9 Designing and Making Welded Structural Steel Members for Cyclic Loading, Welding Research Council Committee on Fatigue of Welded Joints, *Weld. J.*, vol. 38, August 1954.

10 Munse, W. H.: Fatigue and Brittle Fracture, sec. 4 in E. H. Gaylord and C. N. Gaylord (eds.), "Structural Engineering Handbook," McGraw-Hill, New York, 1968.

11 Battelle Memorial Institute: "Prevention of Failure of Metals under Repeated Stress," Wiley, New York, 1946.

12 Chajes, A., S. J. Britvec, and G. Winter: Effects of Cold-straining on Structural Steel Sheets, *J. Struct. Div. ASCE*, April 1963.

3

TENSION MEMBERS

3-1 EFFECT OF RESIDUAL STRESSES

Tension members are efficient carriers of load and are used in many types of structures. In general, their response to load is much the same as that of the tensile-test coupon, but it is not identical. The stress-strain curves shown in Fig. 2-15 are typical of coupon results. Member behavior may differ from coupon behavior for various reasons, among which are slip in bolted and riveted connections, nonlinear behavior of the connections, and residual stresses in the member. Residual stresses result principally from nonuniform cooling of hot-rolled or welded shapes and from cold straightening of bent members.

The rolled I or H shape will be used to explain the manner in which stresses arise from nonuniform cooling after rolling (Fig. 3-1*a*). Because they have more surface exposure per unit of volume, the flange tips tend to cool faster than the flange-to-web junctures. Similarly, the central portion of the web tends to cool faster than the junctures. Therefore, the metal at the junctures continues to contract after the flange tips and web interior have cooled to the temperature of the surrounding atmosphere. This contraction is partially restrained by the cooler metal, so that tensile

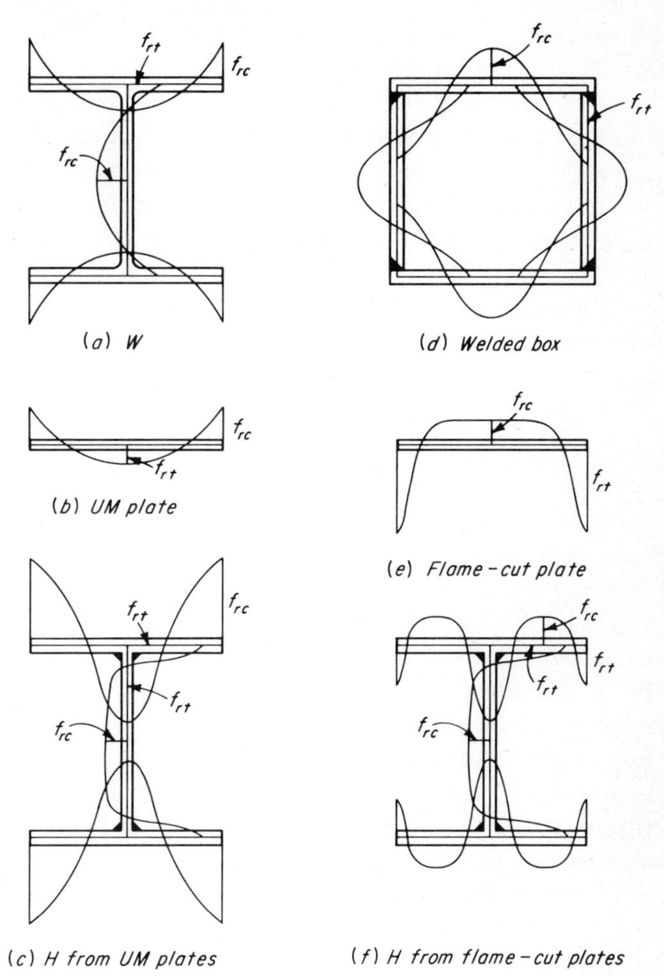

FIGURE 3-1
Thermal residual stresses.

stresses develop in the regions of the junctures and compressive stresses in the remainder of the cross section. These stresses are called *residual stresses*. Figure 3-1*a* shows a typical distribution of residual stress in a standard W shape. These stresses also vary across the thickness, so that the pattern shown represents averages of the across-thickness values. Variation across the thickness is discussed in Art. 4-7 (see Fig. 4-20).

Both magnitude and distribution of thermal residual stress are influenced to a considerable degree by the geometry of the cross section. Thus, in one investigation of

W sections of A7 steel, the flange-tip stress f_{rc} of Fig. 3-1*a* varied from 4.1 to 18.7 ksi, the average being 12.8 ksi.[1] The residual stress at the center of the web varied even more, ranging from 41 ksi compression to 18.2 ksi tension. This means that some W's developed residual tension over the entire web, instead of the pattern shown in Fig. 3-1*a*. Only one of the 20 cross sections in this investigation was thicker than 1 in., and since residual stresses tend to increase in magnitude with increase in thickness, these values are not representative of W's with thick flanges and webs.

Because of the high concentration of heat, tensile residual stresses at the weld in welded members usually equal the yield strength of the weld metal itself, which may be as much as 50 percent higher than that of the parent metal. Residual stresses in welded shapes are determined by the section geometry and the method of preparation of the components. Thus, a welded H may be fabricated from universal-mill (rolled-edge) plates or from plates flame-cut to width. The residual stress in the universal-mill plate is distributed as shown in Fig. 3-1*b*, where the magnitudes depend on both width and thickness and where f_{rc} may vary from as little as 4 or 5 ksi (for relatively thin plates) to as much as F_y for very thick plates and f_{rt} from 2 or 3 to 15 or 16 ksi. The residual-stress distribution in a welded H composed of such plates will be about as shown in Fig. 3-1*c*. Cooling after welding increases both the residual tension and the residual compression in the flange plate. In the case of the flame-cut plate, cooling after cutting leaves a residual-stress distribution such as that shown in Fig. 3-1*e*, where f_{rt} may equal F_y. The residual-stress distribution in a welded H formed from such plates will approximate that shown in Fig. 3-1*f*.

Large residual tensions develop at the corners of the welded box (Fig. 3-1*d*). On the other hand, residual stresses in the hot-rolled square box are very low and in one investigation averaged less than 5 ksi.

Because they are quenched and tempered, A514 rolled steel shapes are partially stress-relieved, so that residual stresses are small. In one investigation of A514 W shapes the maximum residual stress was about 7 ksi, except for one flange tip where it reached about 10 ksi.

Thermal residual stresses extend almost the full length of a member. Of course, they must vanish at the ends. However, they build up quite rapidly, so that they attain the values indicated in Fig. 3-1 at relatively short distances from the ends.

Fabricating operations such as cambering, straightening by cold bending, etc., also induce residual stresses. These stresses are superimposed on the thermal residual stresses. They are of about the same magnitude as thermal residual stresses but differ in distribution. If the member is straightened by rotorizing, which is a continuous straightening procedure, the residual-stress distribution will be changed along the entire length of the member. If it is straightened by gagging, which concentrates the straightening at a few points, thermal residual stresses may remain essentially unchanged over much of the length.

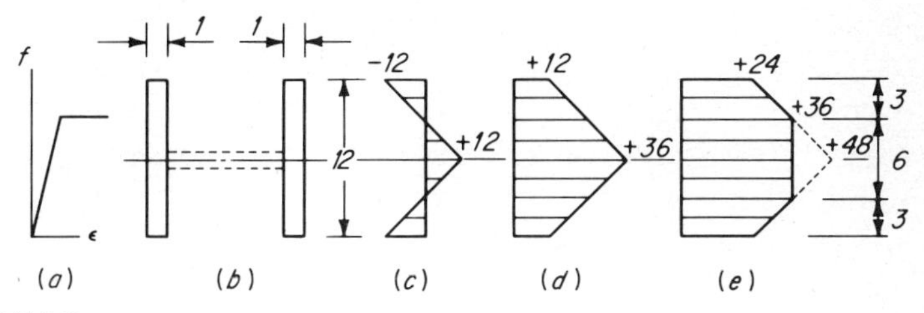

FIGURE 3-2

The effect of residual stresses on the behavior of a tension member can be demonstrated by considering the idealized (webless) H shown in Fig. 3-2*b*. The stress-strain curve for the steel is shown in *a* and the assumed residual-stress pattern in *c*. The area of the cross section is 24 in.^{2} If the member is subjected to a gradually increasing load P, the stress when $P = 576$ kips is $576/24 = 24$ ksi. This is superimposed on the residual stresses, with the result shown in *d*. The corresponding strain is $24/30{,}000 = 0.0008$. This gives point A on the stress-strain curve for the member (Fig. 3-3). If the load P is now increased until the stress at the flange tips is 24 ksi, the increase in strain is $12/30{,}000 = 0.0004$. The stress distribution is as shown in *e*, where each flange is yielded over a length of 6 in. The load is $P = 2(36 \times 6 \times 1 + 30 \times 6 \times 1) = 792$ kips, and the *average* stress on the cross section is $f = 792/24 = 33$ ksi. This gives point B in Fig. 3-3. If the load is next increased to the point where the stress

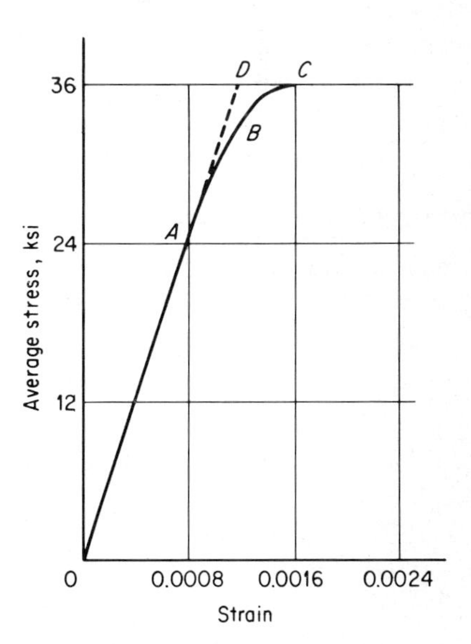

FIGURE 3-3

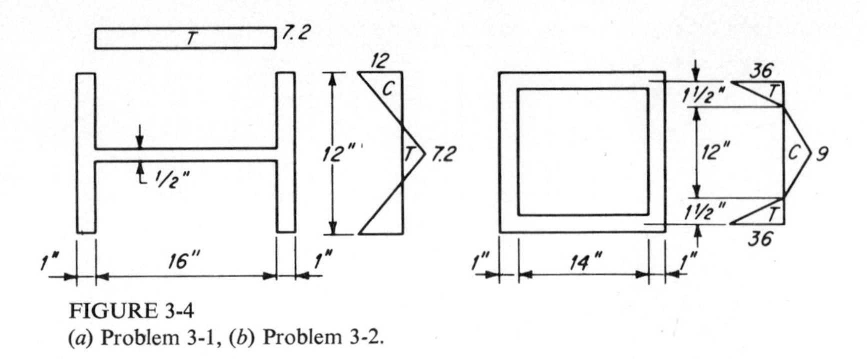

FIGURE 3-4
(*a*) Problem 3-1, (*b*) Problem 3-2.

at the flange tips is 36 ksi, the increase in strain is again 0.0004. This gives point C in Fig. 3-3. Each flange is now fully yielded, so that further straining produces no increase in stress until the metal at the juncture of the flange and web begins to strain harden.

If this member were free of residual stress, the stress-strain curve in the elastic region would be OAD. Thus, although the residual stresses do not affect the yield strength of the member, they do lower the proportional limit (point A) and increase the strain at initiation of overall yielding. However, they are of no consequence in regard to the static strength of the member. (They can be important if fatigue is involved.) On the other hand, residual stresses have a pronounced effect on the strength of columns (Arts. 4-6 and 4-7).

PROBLEMS

3-1 Construct the stress-strain curve to the beginning of unrestrained yield for a tension test of the member whose cross section is shown in Fig. 3-4*a*. The residual-stress distribution is given on the figure, with stresses in ksi. The coupon yield point is 40 ksi.

3-2 Same as Prob. 3-1, but for the square box consisting of four plates welded at the corners (Fig. 3-4*b*). The residual-stress distribution is the same for all four plates. The coupon yield point is 36 ksi.

3-2 ALLOWABLE STRESSES

Tension members are simple in principle, and their design is usually based on the assumption that stress is distributed uniformly over the cross section. Except for a few simple forms of member, however, this condition can be realized only if the member has no abrupt changes in cross section and is fully groove-welded to rigid connecting parts at each end.

Two hazards must be considered in establishing factors of safety: (1) excessive elongation at service loads and (2) fracture. Except for unusual situations, elongations less than those at the beginning of yield are tolerable. For example, an A36-steel tension member 10 ft long which is stressed just short of the yield point elongates $120 \times 36/30{,}000 = 0.144$ in. However, the elongation *after* yielding may be as much as 12 times the initial-yield value (1.8 in., say) before strain hardening begins. Obviously, then, there must be a margin of safety with respect to yielding at service loads. On the other hand, this margin of safety need not be as large as the margin with respect to fracture of the member, because of the possible differences in the consequences.

According to the AISC specification, the allowable stress F_t for tension members is the smaller of $0.6F_y$ or $0.5F_u$, where F_y and F_u are the minimum yield stress and minimum tensile strength, respectively. In the case of a member with welded connections, for which no reduction in area because of holes is involved, the corresponding factors of safety are 1.67 with respect to elongation and 2 with respect to strength. For A36 steel, $F_y = 36$ ksi and $F_u = 58$ ksi (Table 2-2), so that F_t is the smaller of $0.6 \times 36 = 21.6$ and $0.5 \times 58 = 29$ ksi, or 21.6 ksi. For A514 steel less than $2\frac{1}{2}$ in. thick, $F_y = 100$ ksi and $F_u = 115$ ksi (Table 2-2), which give $F_t = 60$ and 57.5 ksi, respectively. Thus, serviceability controls for A36 steel and strength for A514 steel.

Allowable stresses according to the AASHO specifications are 0.55 F_y for members without holes and the smaller of $0.55F_y$ and $0.46F_u$ for members with holes. AREA specifies only $0.55F_y$. However, the difference between F_y and F_u for the steels permitted by this specification is such that 0.55 F_y is always less than $0.46F_u$. The AISI specification for cold-formed members gives $F_t = 0.6F_y$. The steels recognized by this specification are also such that $0.6F_y$ is less than $0.5F_u$, which would be the comparable value according to the AISC specification.

Factors of safety for structural members of aluminum are suggested in Ref. 2. For structures that would be designed in steel in accordance with AASHO, AREA, or similar specifications, factors of safety of 1.85 on yield stress and 2.2 on tensile strength are recommended. The corresponding values for structures that would be designed in steel in accordance with AISC or similar specifications are 1.65 and 1.95. It will be noted that these factors of safety are the same as those for steel members.

3-3 TYPES OF TENSION MEMBER

The form of a tension member is governed to a large extent by the type of structure of which it is a part and by the method of joining it to connecting portions of the structure. Some of the more common types are described in the following paragraphs.

The simplest tension members are made of wire rope or cable, round and square bars, and rectangular bars or plate. Wire rope is used for guy wires, floor suspenders in suspension bridges, hoisting lines, etc. Bridge strand (Art. 2-6) is used for small suspension bridges, suspended roofs, and similar structures. Rope and cable are attached to other members or to anchorages by various kinds of sockets. Main cables of large suspension bridges consist of parallel wires which are strung individually, squeezed together, and wrapped.

Rods and bars are used principally in bracing systems, as in towers, sag rods for purlins in sloping roofs, etc. Round bars may be threaded at the ends and held in place by nuts. Standard clevises are available to fit threaded ends so that the bar can be pin-connected. The end of an unthreaded bar can be bent back along itself and welded to form a loop for a pin connection. Loop rods and rods with clevises are usually made in two sections joined with a turnbuckle which is adjusted to tighten the member in place. Standard dimensions and details of plain and upset rods, loop rods, and clevises are tabulated in the AISC Manual.

Eyebars are members of rectangular cross section with enlarged heads at each end which are bored for pin connections. They were used extensively in early bridge trusses and are still used occasionally as hangers.

Single shapes, such as the angle, the plate, the W and S shapes, and the tee, may be used as tension members. However, two or more shapes are often combined to form a "built-up" member, of which some of the more common are shown in Fig. 3-5. In this figure, shapes which extend full length are shown in solid lines, while intermittent connections whose function is to hold the shapes in line are shown in dashed lines. In welded structures, connections to adjoining members can usually be made directly by butting or lapping, while in riveted or bolted structures the connection must usually be made to a plate called a *gusset plate*. A single plate at each joint is sufficient for the lighter roof trusses (single-plane truss), but two parallel gusset plates are required in bridge trusses and large roof trusses (double-plane truss).

Single-angle members are used extensively in towers. Single-angle and double-angle members are common in roof trusses. Single angles, and double angles as in Fig. 3-5*a*, may be riveted or bolted to a single gusset plate at each end, or they may be welded directly to the webs or flanges of tee or I chord members. Two-angle members as in Fig. 3-5*b* usually require two gusset plates—one at each outstanding leg—although occasionally the other two legs may be connected to a single gusset plate. The four-angle section *c* usually connects to two gusset plates. Two channels may be arranged as in Fig. 3-5*d* to connect to a single gusset plate which fits between the webs or as in Fig. 3-5*e* to connect to a gusset plate at each web. Two channels may also have their flanges turned inward, as in Fig. 3-5*f*. Various shapes may be made up of plates welded together. For example, three plates welded in the form of an I are often used in trusses, and two plates in the form of an angle have been used for transmission towers.

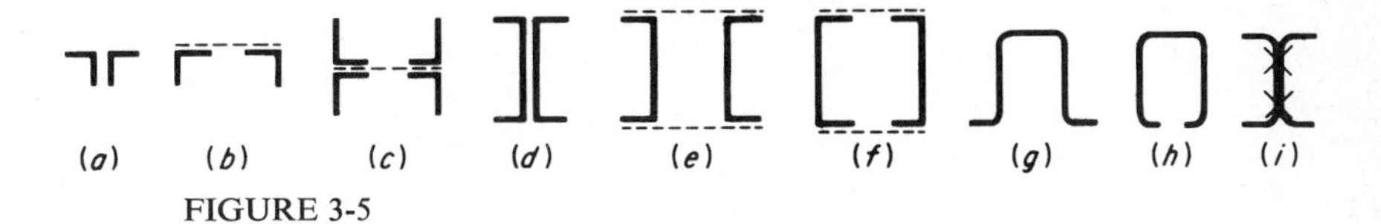

FIGURE 3-5

Aluminum shapes are produced in the same forms as steel and, by extrusion, in some different forms (Art. 2-6). Cold-formed steel tension members may be of angle, channel, or other shape. The cross sections shown in Fig. 3-5*g* and *h* are used as chord members in trusses. Cold-formed shapes can also be built-up, as in *i*, which shows two channels spot-welded together to form an I. Other cold-formed shapes, most of which can be used as tension members, are shown in Fig. 4-22.

3-4 TENSION MEMBERS WITH WELDED CONNECTIONS

The proportioning of tension members with welded connections is relatively simple. The principal problems are in deciding what form or forms of member are best suited and how the connections can be arranged to assure good welding position. Welded joints were discussed in Art. 2-11 and stresses in welds in Art. 2-12. In general, welds should be distributed so that there is no eccentricity of their force resultant with the centroid of the member cross section. Computations to determine the required welds and their arrangement for concentricity are usually simple. For example, in the case of the I-shaped member connected to gusset plates or to another member by welds on the flanges, it is only necessary to divide the required length of weld equally among the four flange tips. For the case of the double-angle member connected to a gusset plate as in Fig. 3-6*a*, distribution of the welds for concentricity can be determined as follows. Using $\frac{5}{16}$-in. E70 electrodes, the allowable stress according to the AISC specification is $0.3 \times 70 \times 0.707 \times \frac{5}{16} = 4.64$ kli. The required length of weld for each angle is $75/4.64 = 16.2$ in. If each angle is welded only at the toe and heel, the required length of weld along the toe is determined by taking moments about the heel:

$$l_2 = \frac{16.2 \times 1.63}{5} = 5.3 \text{ in.}$$

Subtraction gives the length of weld along the heel:

$$l_1 = 16.2 - 5.3 = 10.9 \text{ in.}$$

If each angle is also welded across the end, each of these lengths may be reduced by 2.5 in., which is half the length of the end weld. This is verified by taking moments about the heel:

$$5l_2 + 2.5 \times 5 = 1.63 \times 16.2$$

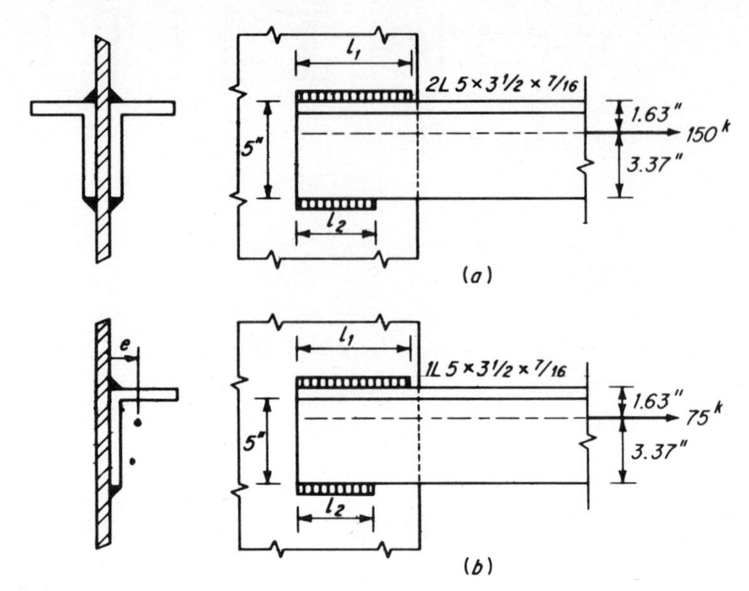

FIGURE 3-6

from which

$$l_2 = 5.3 - 2.5 = 2.8 \text{ in.}$$

Thus, the connection can be made concentric with $l_1 = 8.4$ in., $l_2 = 2.8$ in., and 5 in. of weld at the end, or with $l_1 = 10.9$ in., $l_2 = 5.3$ in., and no weld at the end.

In the case of a single-angle member connected by one leg to a gusset plate (Fig. 3-6*b*), it is impossible to arrange the welds so that the centroid of the weld forces coincides with that of the member. Eccentricity in the plane of the welds can be avoided by arranging the welds as described in the preceding paragraph, but there remains the eccentricity *e* shown in the figure. Consequently, stress in the member will not be uniformly distributed, so that the member cross section is not fully effective. Both AASHO and AREA provide for this condition by allowing only one-half the unconnected leg to be counted in computing the cross-sectional area. There is no such provision in the AISC specification.

With certain exceptions to be discussed in this paragraph, AASHO, AISC, and AREA all require that the centroid of the forces in the connecting welds on an axially loaded member be coincident with the centroid of the member cross section, and if this condition cannot be realized, the eccentricity must be taken into account. There is no exception to this requirement in the AASHO specifications. The exception in the AISC specification concerns double-angle members, single-angle members, and the like, for which disposition of the fillet welds at the connections to avoid the eccentricity discussed in the preceding paragraph is not required, provided there is no question of fatigue. For example, in the connections of Fig. 3-6, this specification would permit $l_1 = l_2$, so that the shortest connection with E70 $\frac{5}{16}$-in. welds would be $l_1 = l_2 = 6$ in. with 5 in. of weld at the end. The exception in the AREA specifications concerns a single-angle member such as that of Fig. 3-6*b*, in which case the eccentricity can be ignored if the welds are proportioned so that their force resultant lies between the centroid of the angle and the middle of the connected leg.

It will be shown in Art. 3-8 that an axially loaded member may not be 100 percent efficient even if the centroid of the weld-force resultant coincides with the centroid of the member cross section. However, standard specifications do not require that the resulting nonuniform distribution of stress be considered.

3-5 DP3-1 : TENSION MEMBERS FOR WELDED TRUSS

The design of the tension members of a roof truss for an industrial building is discussed in this example. There are three trolley beams attached to the bottom chord, one each at L_2 and L'_2 with a reaction of 6.4 kips, and one at L_5 with a reaction of 12.3 kips. The snow load is 40 psf, which is the recommended value for the Great Lakes and New England areas (Table 1-3). The wind load is 20 psf on vertical surfaces, which is the average of the National Building Code requirements for minimum and moderate windstorm areas (Table 1-5). The wind force on the roof is suction and is taken at $0.7q$, where q is the velocity pressure, as suggested in Art. 1-8. Since the 20-psf pressure on vertical surfaces is based on a shape factor of 1.3 (Art. 1-8), the velocity pressure (including the gust factor) is $q = 20/1.3 = 15.4$ psf. Therefore, the roof suction is $0.7 \times 15.4 = 10.8$, or 11 psf.

The dead load consists of the weights of covering, purlins, trusses, and bracing. The covering consists of built-up roofing weighing 5.5 psf, which is laid on a steel roof deck welded to the purlins and weighing 2.5 psf. The weight of purlins will vary from about 1.5 to 5 psf and is estimated at 3 psf for this building. The weight of a truss and its bracing seldom exceeds 10 percent of the load supported by the roof, so that a slight error in estimating its weight will have a negligible effect. For this building, the weight of the truss and bracing is estimated at 3 psf.

Since the wind force on the roof is suction, the trusses are designed for gravity loads alone, which total 54 psf. The corresponding member forces are shown on the truss outline on Sheet 2. The only member of the truss which may experience a significant wind force is considered in DP4-1, Art. 4-11.

Chord members for welded trusses are usually tees, W or S shapes, or two angles arranged as in Fig. 3-5*b*. Angles, channels, or W or S shapes are generally used for web

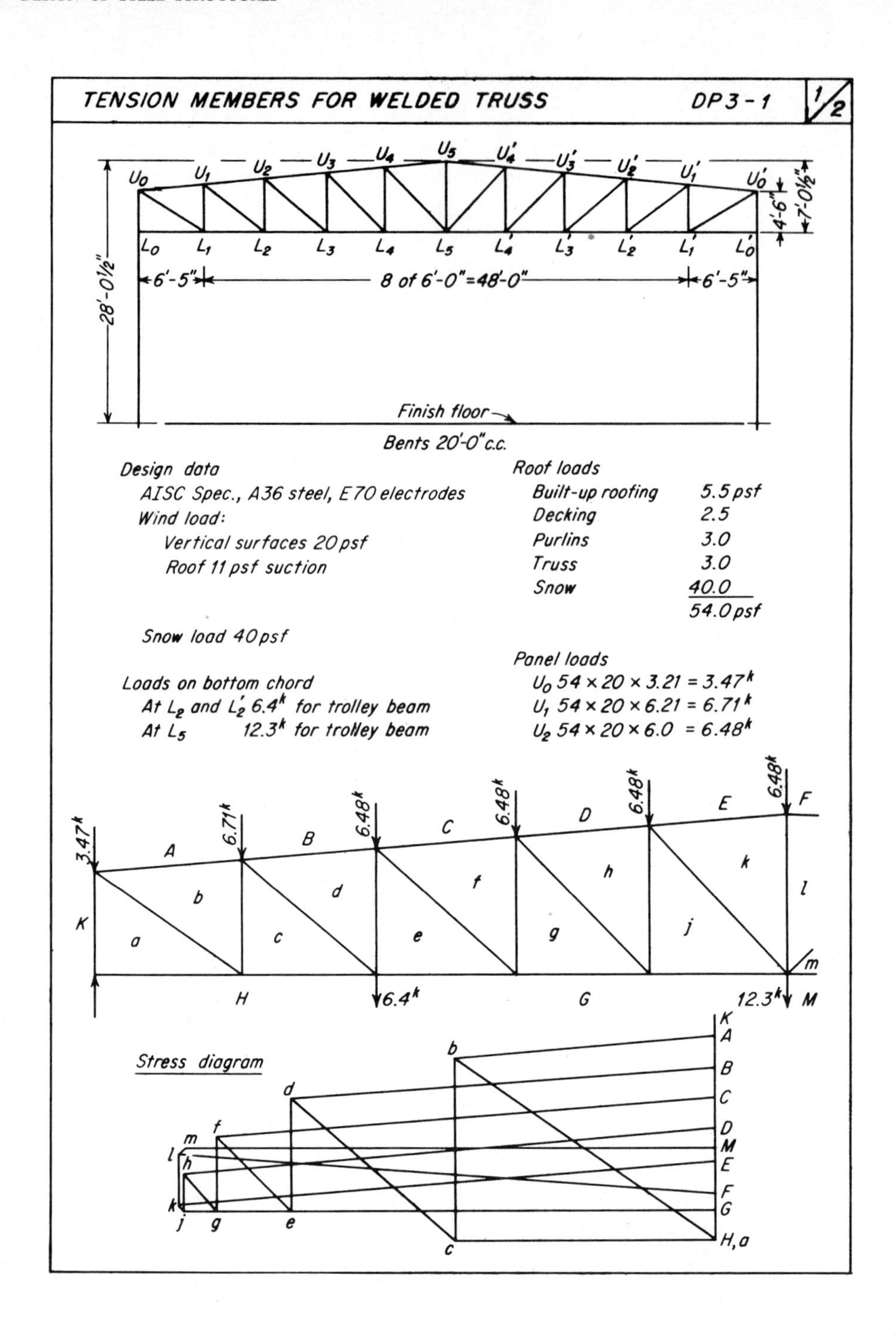

TENSION MEMBERS FOR WELDED TRUSS
DP3-1
1/2
U0 U1 U2 U3 U4 U5 U4' U3' U2' U1' U0'
L0 L1 L2 L3 L4 L5 L4' L3' L2' L1' L0'
6'-5"
8 of 6'-0"=48'-0"
6'-5"
28'-0½"
4'-6"
7'-0½"
Finish floor
Bents 20'-0" c.c.
Design data
AISC Spec., A36 steel, E70 electrodes
Wind load:
Vertical surfaces 20 psf
Roof 11 psf suction
Snow load 40 psf
Loads on bottom chord
At L2 and L2' 6.4k for trolley beam
At L5 12.3k for trolley beam
Roof loads
Built-up roofing 5.5 psf
Decking 2.5
Purlins 3.0
Truss 3.0
Snow 40.0
54.0 psf
Panel loads
U0 54 × 20 × 3.21 = 3.47k
U1 54 × 20 × 6.21 = 6.71k
U2 54 × 20 × 6.0 = 6.48k
3.47k 6.71k 6.48k 6.48k 6.48k 6.48k
A B C D E F
K a b c d e f g h j k l m
H 6.4k G 12.3k M
Stress diagram

TENSION MEMBERS FOR WELDED TRUSS — DP3-1 — 2/2

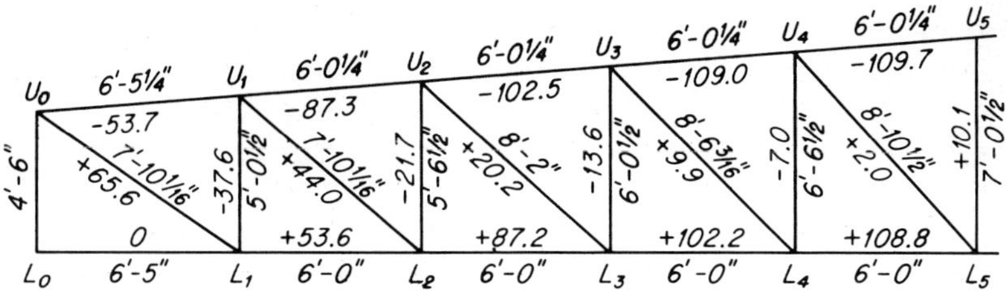

Design of tension members

Member	Force	Reqd. area	Section	Area in.^{2}	L/r
*$L_0L_1L_2$	53.6 @ 22 =	2.44 in.^{2}	WT7×17	5.01	77/1.52 = 51
L_2L_3	87.2 @ 22 =	3.96	do	5.01	72/1.52 = 47
L_3L_4	102.2 @ 22 =	4.65	do	5.01	do
L_4L_5	108.8 @ 22 =	4.94	do	5.01	do
U_0L_1	65.6 @ 22 =	2.98	2L 3½×3×¼	3.12	94/1.11 = 85
U_1L_2	44.0 @ 22 =	2.00	2L 2½×2×¼	2.12	94/0.78 = 121
U_2L_3	20.2 @ 22 =	0.92	2L 1½×1½×¼	1.38	98/0.45 = 218
U_3L_4	9.9 @ 22 =	0.45	do	1.38	102/0.45 = 225
U_4L_5	2.0 @ 22 =	0.09	do	1.38	106/0.45 = 235
U_5L_5	10.1 @ 22 =	0.46	do	1.38	84/0.45 = 187

*This member is checked in DP4-1 for compression due to wind load.

Joint L_1

Max. fillet weld 3/16": shear per in. = $0.3\times70\times\frac{3}{16}\times0.707 = 2.78$

Member	U_1L_1	U_0L_1
Reqd. weld	$\frac{37.6}{2\times2.78}=6.8$	$\frac{65.6}{2\times2.78}=11.8$
Toe	$\frac{6.8\times0.91}{3}=2.1$	$\frac{11.8\times1.04}{3.5}=3.5$
Heel	4.7	8.3

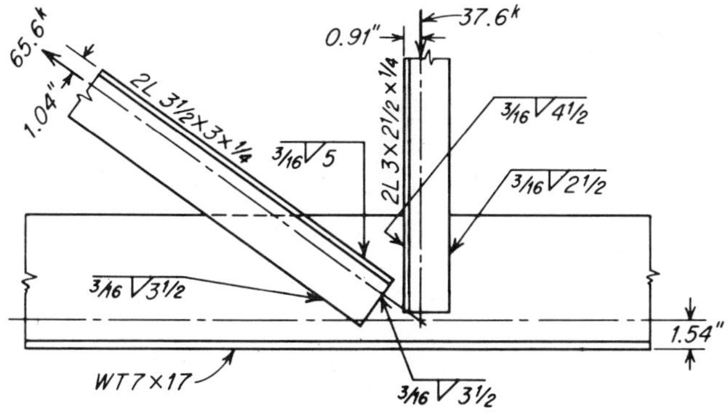

members. The WT7 × 17 has been chosen for the entire lower chord in this example, although a WT6 × 13.5 has enough cross-sectional area (3.97 in.^{2}) for L_2L_3 and L_0L_2. Although the 6-in. tee would save about 129 lb of steel per truss, the saving would probably be canceled by the cost of the two splices.

Some saving would result from the use of single angles for members U_2L_3, U_3L_4, U_4L_5, and U_5L_5. However, this results in an eccentricity which some designers object to. Although angles smaller than $1\frac{1}{2} \times 1\frac{1}{2} \times \frac{1}{4}$ would furnish enough area for a few of the members, they could not be used for U_2L_3, U_3L_4, and U_4L_5 without exceeding the AISC recommended maximum slenderness ratio of 240. This maximum slenderness ratio is intended to give the member a stiffness that reduces the chances of noticeable vibration.

The joint details of a truss of this kind are usually worked out by the fabricator. However, they are submitted to the engineer for his approval, and since responsibility for the adequacy of the structure rests with him, he must know how to design them. The design of the joint at L_1 is discussed here. Fastening is by fillet welds from U_0L_1 and U_1L_1 to the web of the bottom chord. The largest fillet weld that should be used is $\frac{1}{16}$ in. less than the thickness of the angle (Art. 2-11). For the $\frac{1}{4}$-in. angles in this example, a $\frac{3}{16}$-in. weld satisfies this limitation. The length of weld for each member is determined by dividing the load by the allowable shear on a $\frac{3}{16}$-in. weld, which for an E70 electrode is $0.3 \times 70 \times 0.707 \times \frac{3}{16} =$ 2.78 kli. The distribution of this length of weld is determined as explained in Art. 3-4. There is room enough for the required lengths of weld at the toe and the heel of U_1L_1. For U_0L_1, however, even with $3\frac{1}{2}$ in. of weld across the end, the weld at the heel would be $8.3 - 1\frac{3}{4} =$ $6\frac{1}{2}$ in. long, and there is room for no more than about 5 in. Therefore, the length of weld at the toe is increased to make up the difference. The resulting eccentricity is small and not in violation of the AISC specification.

It was mentioned in Art. 2-14 that ordinarily the electrode should bisect (approximately) the angle between the legs of a fillet weld. The outstanding legs of U_1L_1 would make this impossible at the corner of U_0L_1 nearest U_1L_1, while the flange of the tee would interfere to some extent at the opposite corner and at the end of U_1L_1. However, this inclination of the electrode is not so important as the inclination in the direction of travel, and, as a general rule, a good operator can produce a sound weld so long as he can see the tip of the arcing electrode with his hood in place and without using a mirror. Therefore, the disposition of the welds at this joint is not one to be concerned about.

Notation for designating welds on structural drawings was explained in Art. 2-11.

PROBLEMS

3-3 The end of a tension member consisting of a $\frac{1}{2} \times 10$ in. A36 plate is lap-welded to a member which allows a lap of 8 in. The load is 100 kips. Determine the required welds. AISC specification.

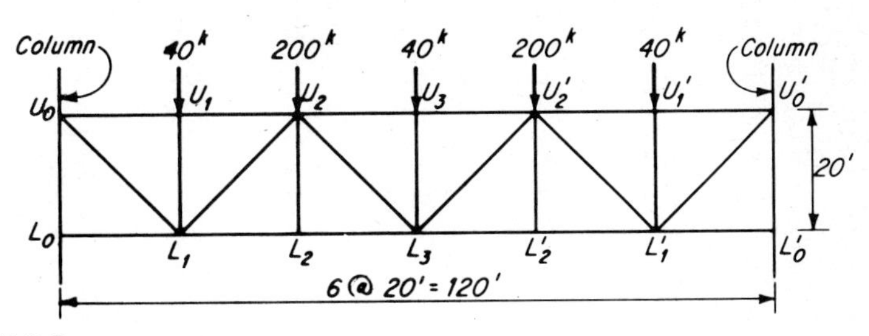

FIGURE 3-7
Problem 3-10.

3-4 The 6-in. leg of an A36 6 × 4 × ½ angle is fillet-welded to a ⅜-in. plate. The load is 95 kips. Proportion the welds to eliminate eccentricity of the connection in the plane of the welds. The connection cannot be longer than 12 in. AISC specification.

3-5 Design and detail a welded joint at L_3 for the truss of DP3-1. Member U_3L_3 consists of two angles 2 × 1¼ × ¼ with the long legs back to back (DP4-1). AISC specification.

3-6 Design the bottom-chord and tension web members for the truss of DP3-1 using W or S shapes. Position the bottom chord so that its web is in the vertical plane with the web members butting against its top flange. Design and detail the connection at L_1 if member U_1L_1 is an S5 × 10 with its web in the plane of the truss.

3-7 Design a W or S bottom-chord and single-angle web tension members for the truss of DP3-1. Design and detail the connection at L_1 if member U_1L_1 is a 3½ × 3½ × $\frac{7}{16}$ angle.

3-8 Suggest several types of web members that would connect satisfactorily if the bottom chord of DP3-1 were a W or S with its web in the horizontal plane.

3-9 A balcony in a building is 16 ft wide. It is supported at the outer edge by hangers welded to roof trusses spaced 20 ft on centers. The bottom chord of each truss is a WT7 × 15. Each hanger is welded to the end of a W16 × 40 beam which spans the width of the balcony and is connected at the opposite end to a column at the wall of the building. The live load is 100 psf and the dead load 60 psf. Design the hanger and its connections to the truss and the beam. AISC specification, A36 steel.

3-10 The truss shown in Fig. 3-7 is to be used in a steel-framed building to span a 120-ft column-free space. The truss supports floor loads of 40 kips at each upper-chord panel point and additional loads of 160 kips at U_2 and U_2' from columns supporting floors above. Design the tension members for welded construction. AISC specification.

3-6 THE NET SECTION

Holes for rivets or bolts in tension members affect the member in two ways: (1) they reduce the area of the cross section, and (2) they result in nonuniform strain on cross sections in the neighborhood of the hole (Fig. 3-10*b*).

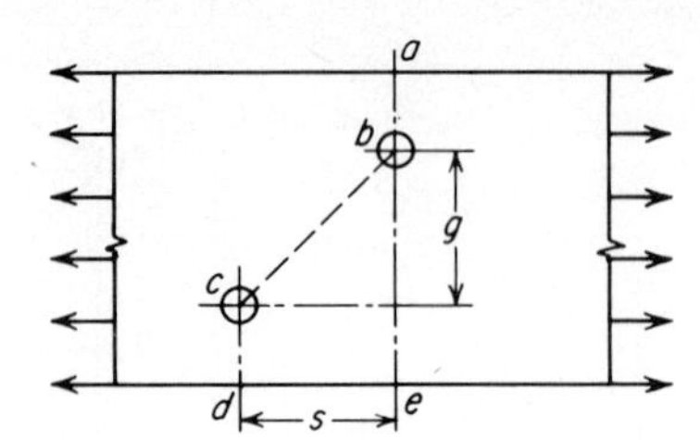

FIGURE 3-8

The area of the gross cross section minus the area which is lost because of holes is called the *net section*. The member may break on a net section normal to its axis, but it may also break on a zigzag section if the fasteners are staggered. Thus, for the case shown in Fig. 3-8, failure may be on the net section *abe* or on the net section *abcd*, depending on the relative values of the gage g, the pitch s, and the hole diameter d. (Note that gage refers to the distance between longitudinal fastener lines, while pitch refers to the distance between transverse rows.) If g is large relative to s, failure would be expected along *abcd*, while if it is small, failure is more likely along *abe*. On the other hand, for fixed values of g and s we see that failure along the zigzag section becomes more likely as d increases.

Practical design procedure is based on simple empirical formulas which have been proposed from time to time. The method prescribed by most specifications is based on the assumption that the effect of the hole c in Fig. 3-8 can be accounted for by deducting from the area of section *abe*, in addition to the area lost because of the hole at b, the normal deduction for the hole at c multiplied by the quantity $1 - s^2/4gd$. For section *abcd* this procedure gives

$$A_n = A_g - dt - dt\left(1 - \frac{s^2}{4gd}\right) \tag{3-1}$$

where A_n = net area
A_g = gross area on *abe*
t = thickness of plate

If the plate thickness is uniform, we can divide each term of Eq. (3-1) by t to get

$$w_n = w_g - d - d\left(1 - \frac{s^2}{4gd}\right) = w_g - 2d + \frac{s^2}{4g} \tag{3-2}$$

where w_n is the net width and w_g the gross width along *abe*. The formula shows that the net width is found by deducting from the gross width the sum of the diameters of both holes in the chain and adding the quantity $s^2/4g$ for the staggered hole.

To account for additional holes in the chain, the procedure just described is continued from one hole to the next, so that for any zigzag section the net width is determined by deducting from the gross width the sum of the diameters of all the holes in the chain and adding the quantity $s^2/4g$ for each gage space in the chain. Values of $s^2/4g$ must be computed in sequence from one hole to the next. This method of determining the net section, which is called the *$s^2/4g$ rule*, was proposed in 1922.[3] A limit analysis published in 1955 suggests that the rule is a reasonable one.[4] It is used in all standard specifications.

AISC, AASHO, and AREA specify that the net section be calculated on the basis of holes $\frac{1}{8}$ in. larger than the diameter of the fastener. Since the hole is only $\frac{1}{16}$ in. larger than the fastener, this procedure makes some allowance for damage to the material adjacent to the hole caused by the punching or drilling operation. However, suggested specifications for aluminum structures recognize the lesser damage by drilling by requiring deduction of only the actual hole diameter for a drilled or subpunched-and-reamed hole.[2]

EXAMPLE 3-1 A $7 \times 4 \times \frac{3}{4}$ angle is connected by two rows of $\frac{3}{4}$-in. bolts in the 7-in. leg and one row in the 4-in. leg (Fig. 3-9).

(*a*) Determine the pitch s so that only two holes for $\frac{3}{4}$-in. fasteners need be deducted in computing the net area.

The fastener gages shown in Fig. 3-9 are the usual values for a 7×4 angle. For the purpose of computing its cross-sectional area, the width of the angle and the distances between gage lines must be measured along the centerline of the cross section. Thus the gross width of the angle is the sum of the widths of the two legs less their thickness, or $7 + 4 - 0.75 = 10.25$ in. The distance between the gage line in the 4-in. leg and the adjacent line in the 7-in. leg is $2.5 + 2.5 - 0.75 = 4.25$ in. The net width for the section *abcde* must be not less than the gross width less the sum of the diameters of the two holes which are allowed. Therefore

$$10.25 - 3 \times 0.875 + \frac{s^2}{4 \times 3} + \frac{s^2}{4 \times 4.25} = 10.25 - 2 \times 0.875$$

$$s = 2.48 \text{ in.}$$

Use 2.50 in.

$$A_n = (10.25 - 2 \times 0.875)0.75 = 6.38 \text{ in.}^2$$

(*b*) Determine the net area of the 7×4 angle of Fig. 3-9 if the pitch s is 2 in.

$$A_n = \left(10.25 - 3 \times 0.875 + \frac{2^2}{4 \times 3} + \frac{2^2}{4 \times 4.25}\right)0.75 = 6.15 \text{ in.}^2$$

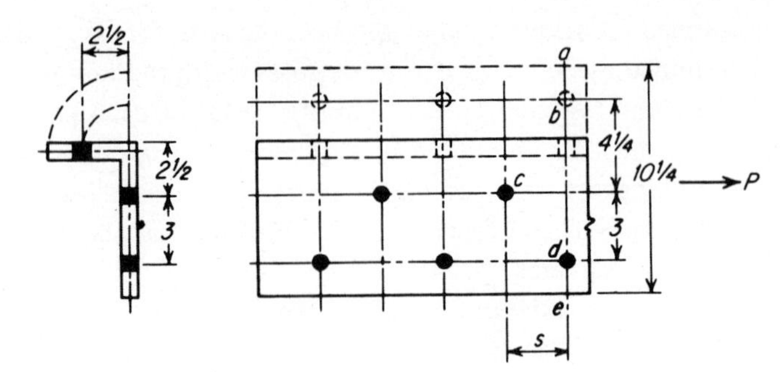

FIGURE 3-9

3-7 DISTRIBUTION OF STRESS ON NET SECTION

Nonuniform strain in the vicinity of a hole in a uniformly stretched sheet of rubber is shown in Fig. 3-10. The unloaded sheet, upon which an orthogonal grid was drawn, is shown in Fig. 3-10*a*, while the stretched sheet is shown in Fig. 3-10*b*. It will be noted that the strains at the edge of the elongated hole (at the ends of the minor diameter) are much larger than those elsewhere in the sheet. The disturbance is highly localized, however. Whether the corresponding stresses are nonuniform or not depends on the stress-strain relationship of the material. According to the theory of elasticity, the distribution of stress on the net section of an infinitely wide plate containing a hole at its centerline is given by[5]

$$f = f_1\left[1 + \frac{1}{2}\left(\frac{r}{x}\right)^2 + \frac{3}{2}\left(\frac{r}{x}\right)^4\right] \tag{3-3}$$

where f_1 = stress that would exist if there were no hole
r = radius of hole
x = distance from center of hole to any point on the transverse section

Because of the highly localized disturbance in stress, this equation can be applied with good accuracy to a plate of finite width. The results for a plate of the dimensions of that in Fig. 3-10*a*, subjected to a uniformly distributed tension of 12 ksi, are shown in Fig. 3-10*c*. It will be noted that the stress at the edge of the hole is equal to the yield stress of A36 steel.

Equation (3-3) is valid only if the stress at the edge of the hole does not exceed the proportional limit. If the load continues to increase after the proportional limit is reached, the stress distribution will depend on the nature of the stress-strain curve of the metal. The metal immediately adjacent to a hole in a plate of a flat-yielding steel will deform without increase in stress, but if the plate is of a steel with no definite yield the stress at the edge of the hole will increase, although at a slower rate than at

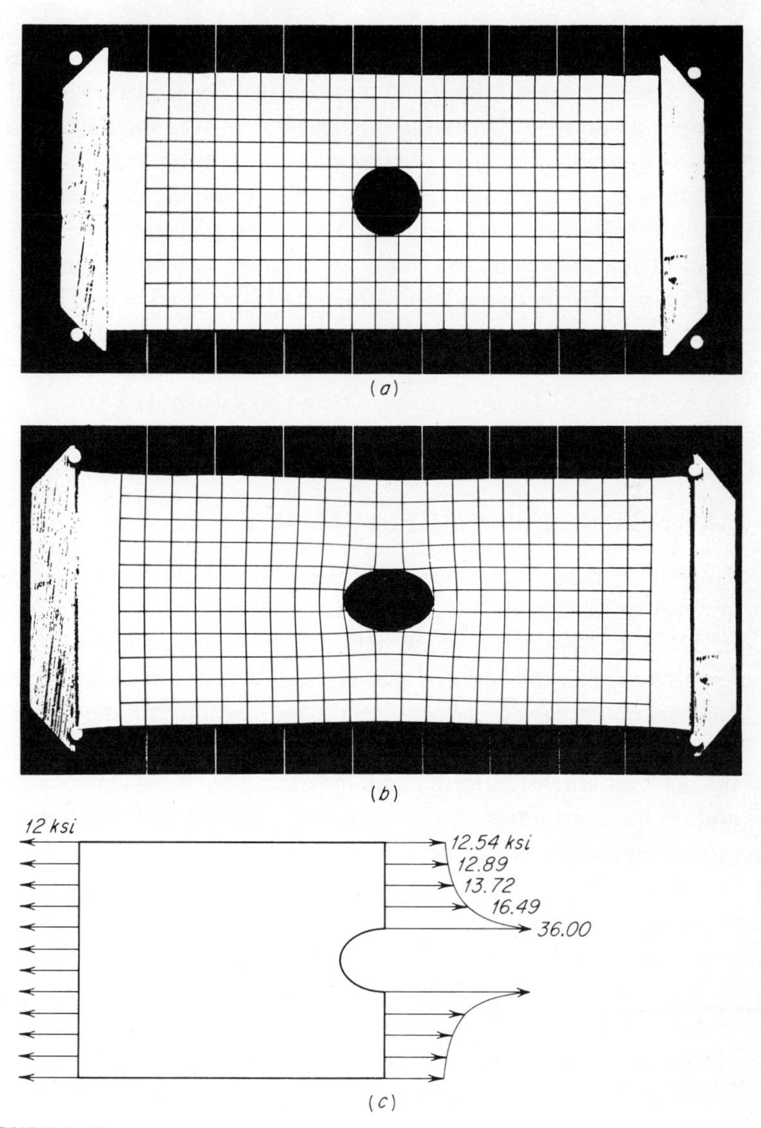

FIGURE 3-10

points on the cross section farther from the hole. In either case, stresses at points away from the edge of the hole will be greater than the value given by Eq. (3-3).

Stress concentrations at holes are usually neglected in structural design, and stress is assumed to be uniformly distributed over the net area of the cross section. This assumption is justified because structural metals are sufficiently ductile to equalize the stress over the area in most cases. For example, in a test of an A440 steel hanger 2.25 × 0.263 in. in cross section connected to a gusset plate with $\frac{3}{4}$-in. bolts in $\frac{13}{16}$-in. holes (actual diameter 0.817 in.), the member broke on the net section at a load of 29.75 kips. The net area was $(2.25 - 0.817) \times 0.263 = 0.377$ in.^{2}. Thus, the average stress at fracture was $29.75/0.377 = 78.9$ ksi. A test of a standard coupon cut from the hanger showed a tensile strength of 77.8 ksi and a yield stress of 55.6 ksi. Corresponding ASTM values for this steel are 70 and 50 ksi (Table 2-2).

3-8 NET-SECTION EFFICIENCY

Tests have shown that tension members do not always develop a net-section average stress equal to the tensile strength, as was the case for the $\frac{1}{4} \times 2\frac{1}{4}$ in. hanger described in Art. 3-7. Reductions in strength can be expressed in terms of the *efficiency* of the net section, i.e., the ratio of the average stress at fracture to the coupon strength. Net-section efficiency depends on (1) the ductility of the metal, (2) the method of making the holes, (3) the ratio of the gage g to the fastener diameter d, (4) the ratio of the net area in tension to the area in bearing on the fastener (called the *bearing ratio*), and (5) the distribution of the cross-sectional material relative to the gusset plates or other elements to which the member is connected.

Because of the nonuniform stress distribution which was discussed in Art. 3-7, it is not surprising that the efficiency of the net section is dependent on the ductility of the metal. Investigation of a large number of tests showed that the net section in a highly ductile material may be 15 to 20 percent stronger than the same section in a material with relatively low ductility.[6] This effect can be expressed by a net-section efficiency coefficient K_1, which is a function of the percent reduction R in the area of a standard test coupon (2-in. gage length), as follows:

$$K_1 = 0.82 + 0.0032R \not> 1 \tag{3-4}$$

Since values of R are not prescribed by ASTM specifications, data for evaluating K_1 are not readily available. However, R is 50 percent or more for A36 steel, which makes this steel 100 percent efficient as regards ductility. Also, it will be noted that efficiency is still high ($K_1 = 0.90$) even if R is only 25 percent.

Punched holes may reduce the efficiency of the net section by as much as 15 percent compared with drilled holes.[6] This effect can be expressed by an efficiency

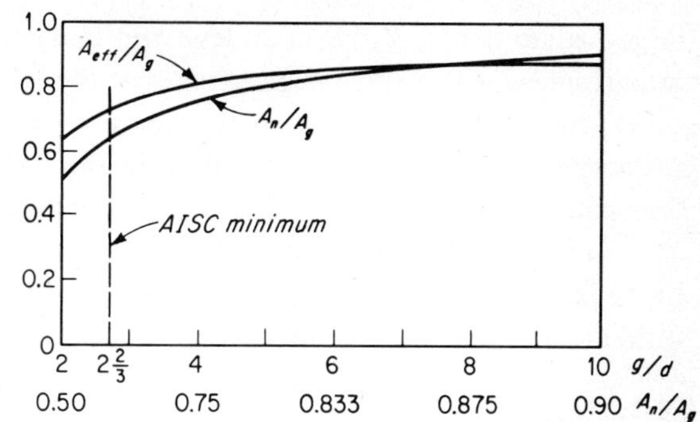

FIGURE 3-11
Effect of g/d ratio on net-section efficiency.

coefficient K_2, using $K_2 = 0.85$ for punched holes and $K_2 = 1$ for holes that are drilled or subpunched and reamed.

Tests show that the net section is more efficient if the ratio of the gage g to the diameter d is small than it is if the ratio is large. The increase for close transverse spacing is explained by the suppression of contraction, which was discussed in Art. 2-16 (Fig. 2-41*a*). Reduction in area at yielding of the metal on the net section between holes is restrained by the metal on the adjacent parallel gross cross sections. Therefore, a condition of biaxial stress develops, and there is an increase in tensile strength. On the other hand, a wide spacing of fasteners (large value of g/d) reduces the efficiency. This is because the concentrated forces applied to the member by the widely spaced fasteners produce nonuniform straining of such a nature that fracture begins before stress can be equalized over the net section. The following efficiency coefficient K_3 has been proposed to evaluate this effect:[7]

$$K_3 = 1.6 - 0.7\frac{A_n}{A_g} \tag{3-5}$$

where A_n and A_g are the net area and gross area, respectively. This equation was derived by fitting a curve to test results. Figure 3-11 shows a comparison of the net area of a plate, whose fastener lines are at a uniform gage g, with the effective area obtained by multiplying the net area by K_3. The horizontal scale in this figure is given in terms of both A_n/A_g and g/d. There is a direct relationship between these parameters for a plate with equally spaced fastener lines:

$$\frac{A_n}{A_g} = \frac{(g-d)t}{gt} = 1 - \frac{d}{g}$$

The dashed line at $g/d = 2\frac{2}{3}$ corresponds to the minimum distance between fasteners according to the AISC specification. At this value of g/d, $K_3 = 1.16$. At $A_n/A_g = \frac{6}{7}$, which corresponds to $g/d = 7$ (a fairly wide spacing of fasteners), $K_3 = 1$. For still larger values of g/d, K_3 becomes less than unity and approaches 0.9 as a limit. This is in agreement with tests, which show that the *effective* net section will not exceed 85 to 90 percent of the gross area no matter how widely fasteners are spaced in an attempt to increase it. This fact is recognized in the AISC and AREA specifications, which limit the value of A_n to be used in the computations to not more than $0.85A_g$.

It is difficult to isolate the effect of bearing pressure from the effect of the spacing of the fastener gage lines. In the case of a plate with uniform gage spacing, the bearing ratio and the fastener spacing are directly related, as is shown by

$$\frac{A_n}{A_b} = \frac{(g-d)t}{dt} = \frac{g}{d} - 1$$

Test data suggest that the strength of a connection is not impaired by bearing pressure so long as the bearing ratio is less than 2.25 (Art. 2-9). Since specification allowable bearing pressures are based on this fact, bearing pressure can be ignored in determining net-section efficiency.

Net-section efficiency is also influenced by the position of the shear planes of the fasteners relative to the cross section of the member. This is shown by a series of tests on A7-steel truss-type members reported in Ref. 7. The specimens consisted of four $5 \times 3 \times \frac{3}{8}$ in. angles riveted to a $\frac{1}{2} \times 16$ in. web plate and connected at each end to two gusset plates (Fig. 3-12). Holes were drilled. Eight of these members were tested (Table 3-1). The two specimens with riveted connections, 1 and 2, were among a number of truss-type members of various types of cross section which had been tested in an earlier investigation of riveted connections. With the intent of verifying the greater shear strength of A325 bolts as compared with rivets, two specimens fabricated to receive seven bolts per line (3 and 4 in Table 3-1) were tested. The first

Table 3-1 TESTS ON TRUSS-TYPE TENSION MEMBERS

Specimen	Fasteners per line	P_u, kips	$f_n = P/A_n$, ksi	$f_g = P/A_g$, ksi	Member properties F_u, ksi	F_y, ksi	$\frac{f_n}{F_u}$	$\frac{f_g}{F_y}$
1	10*	872	55.8	44.8	62.1	38.4	0.90	1.17
2	10*	902	57.6	46.2	62.5	39.4	0.92	1.17
3	7	866	55.4	44.5	64.8	39.4	0.86	1.13
4	6†	870	55.7	44.7	66.4	40.2	0.84	1.11
5	5	706	45.2	36.4	63.0	38.7	0.72	0.94
6	5	722	46.2	37.1	62.8	38.6	0.74	0.96
7	7	815	52.1	41.8				
8	5‡	796	51.0	40.9				

* $\frac{7}{8}$-in. rivets, all others $\frac{7}{8}$-in. high-strength bolts. Drilled holes.
† Last hole (at end of member) in each line empty.
‡ Last five holes (at end of member) in each line empty.

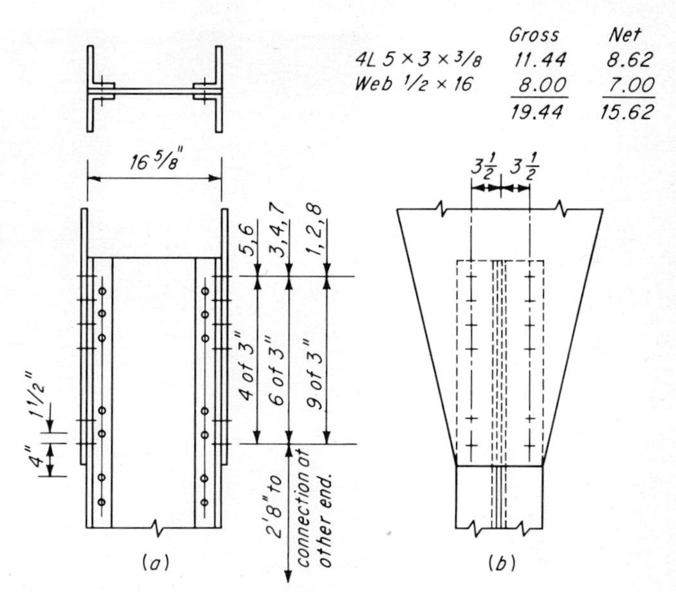

FIGURE 3-12
Test member reported in Table 3-1.

of these failed, at 866 kips, on the same net section as specimens 1 and 2. The fracture was through the net section of the angles in the legs connected to the gusset plates, at the first row of bolts (the bottom row in Fig. 3-12*b*), and thence through the net section of the other legs (Fig. 3-13). Specimen 4 was then tested with the last bolt in each of the four rows omitted (the bolts at the top in Fig. 3-12*b*). This specimen failed at 870 kips and on the same net section as specimen 3.

In an attempt to force a shear failure of the bolts, two specimens which had been fabricated for ten fasteners per line were shortened, by flame-cutting, so that there were only five bolts per line (specimens 5 and 6 in Table 3-1). However, instead of failing by shearing the bolts, both members broke on the same net section as the others had, one at 706 kips and one at 722 kips. It is of interest to note that this represented a reduction in the strength of the net section of about 18 percent. Thus, it is clear that the strength of the net section is influenced by the makeup of the connection. In this case, the evidence points to the length of the connection as a determinant. Strain measurements showed that the webs of specimens 3 (seven bolts per line), 4 (six bolts per line), and 6 (five bolts per line) were only 54, 52, and 38 percent effective, respectively, at the AISC allowable load of 312 kips for this member, and 88, 82, and 70 percent effective at twice this load. Effectiveness in this case is expressed as the ratio of the load corresponding to the measured strains to that corresponding to uniform stress over the entire net section.

FIGURE 3-13
Net-section failure of truss-type member. (*From W. H. Munse and E. Chesson, Jr., J. Struct. Div. ASCE, February 1963.*)

Specimens 7 and 8 were tested with high-strength bearing bolts (Fig. 2-18*c*). Specimen 8 had been fabricated for ten fasteners per line and was tested with only five per line without removing the excess length of member. It failed at 796 kips. This increase in strength in comparison with specimens 5 and 6 is due to the stiffening effect of the excess length, which results in more nearly uniform strain in the web at the section where the first transfer of stress from the gusset plates to the member occurs. Figure 3-14 shows the local buckling which developed in the outstanding legs of this specimen, demonstrating that tension was being carried into the portion of the web beyond the last row of bolts and was being reacted by the compression in the angles.

Average stresses at failure of the eight specimens in this series of tests are given in Table 3-1. The average net-section stress f_n is based on the net area by the $s^2/4g$ rule, which is 15.62 in.^{2} (Fig. 3-12). Also shown are the weighted averages of the coupon properties F_u and F_y of the angles and web plate. The decreasing efficiency of the net section with decrease in length of the connection is shown in the next to last

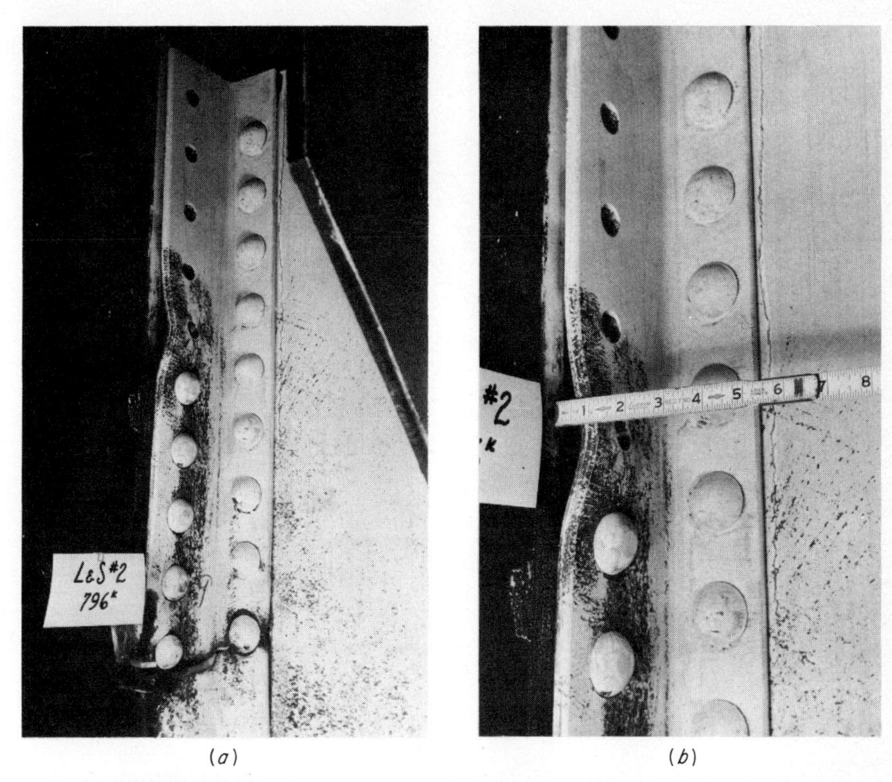

(a) (b)

FIGURE 3-14
Net-section failure of truss-type member. (*From W. H. Munse and E. Chesson, Jr., J. Struct. Div. ASCE, February 1963.*)

column. It will be noted (last column) that the first four specimens exceeded the coupon yield on the gross section before breaking while the last two did not.

Figure 3-15 shows the variation of load with overall deformation for the members of Table 3-1. Overall deformation is measured in the length of the specimen, including both joints. Calculated values based on $\Delta = PL/AE$ are also shown, one based on net area, the other on gross. It is of interest to note that nonlinear behavior of the specimens began at a load somewhat less than the AISC allowable load for members 5 and 6. It should also be noted that these plots stop short of the ultimate load. Thus, the plot for specimen 5 stops at about 33 ksi, but failure was at 36.4 ksi.

The phenomenon of nonuniform straining of the web in the test specimens discussed in this article is called *shear lag*. Figure 3-16, which shows the web of the test specimens in the unloaded and the loaded state, helps to explain the significance of this name. The four forces shown in Fig. 3-16*b* are the resultants of the bolt shears

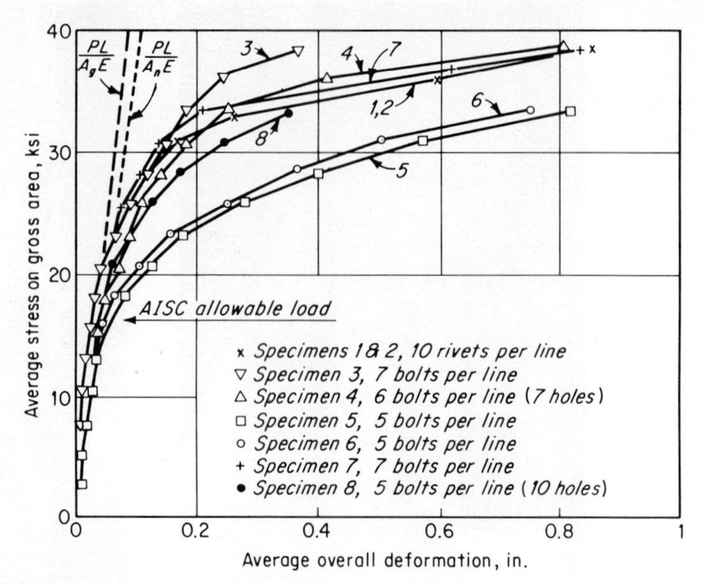

FIGURE 3-15
Load-deformation curves for member of Fig. 3-12. (*From W. H. Munse and E. Chesson, Jr., J. Struct. Div. ASCE, February 1963.*)

in the connections. Since the ends of the web are free, the distortion will be as shown. Therefore, an element such as that at A in the unloaded web will be deformed as shown in Fig. 3-16b when the member is loaded. This is a shear deformation, and the stress in the web is said to lag because of it. The shear-lag phenomenon can be analyzed by considering this deformation.[8]

Since shear lag reduces the effectiveness of tension-member components that are not connected directly to a gusset plate or other anchorage, the efficiency of a member can be increased by reducing the areas of such components relative to the

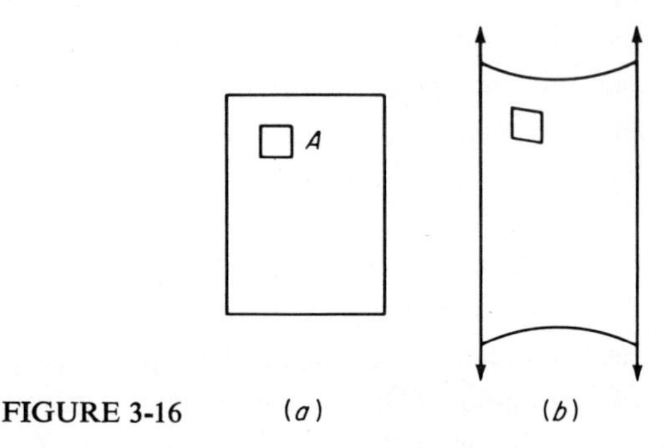

FIGURE 3-16

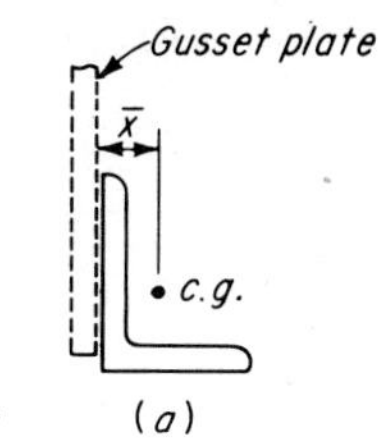

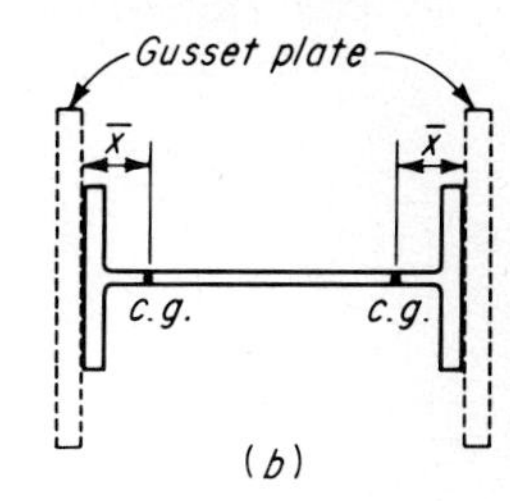

FIGURE 3-17

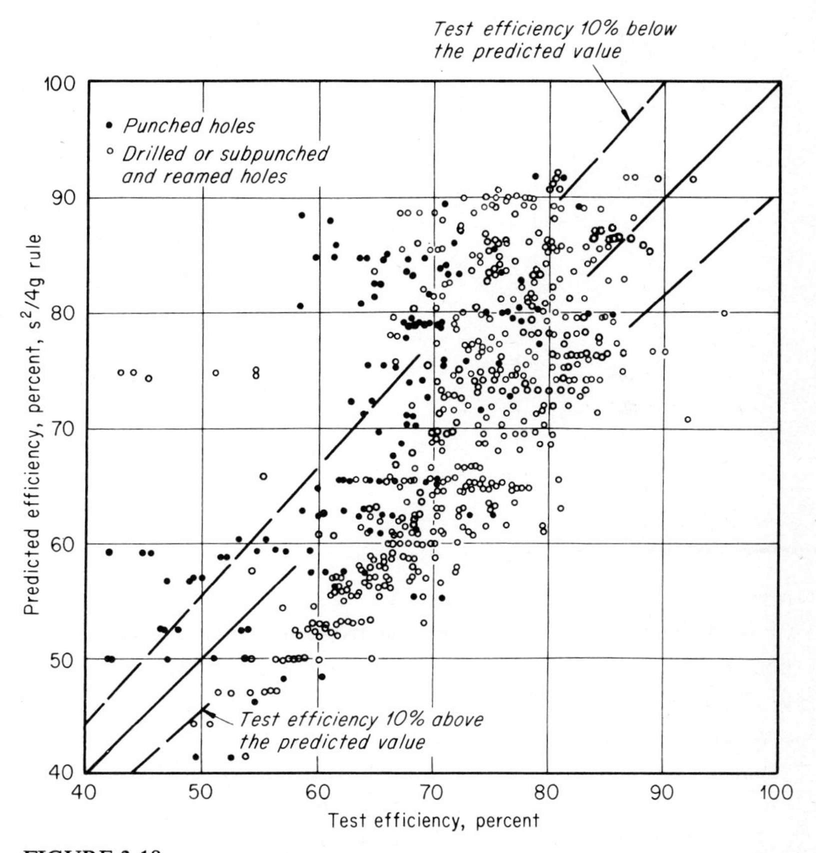

FIGURE 3-18
Comparison of tests with predicted net-section strength. (*From W. H. Munse and E. Chesson, Jr., J. Struct. Div. ASCE, February 1963.*)

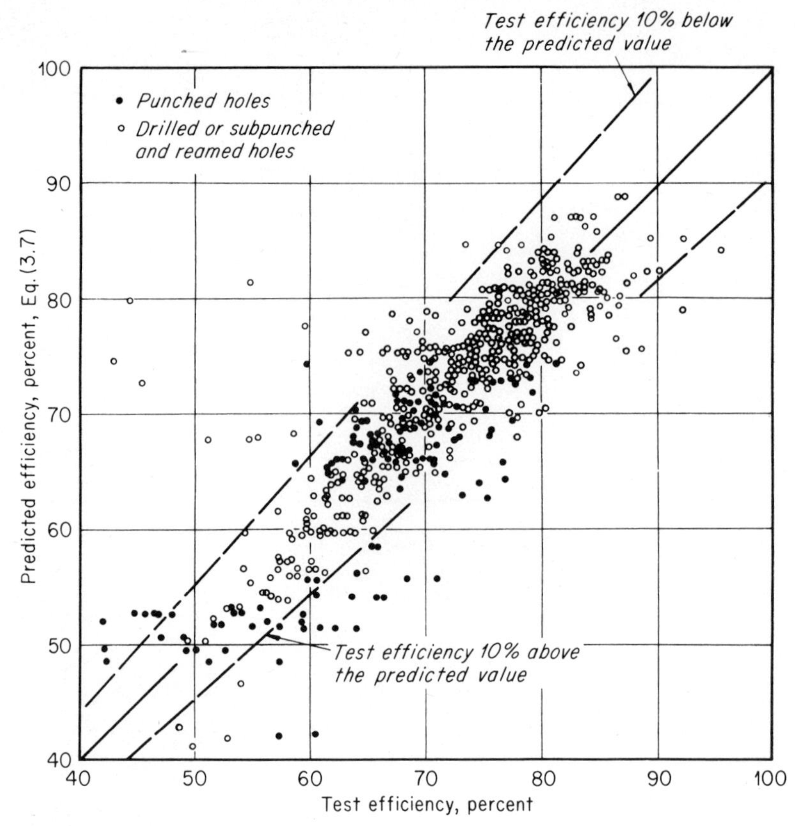

FIGURE 3-19
Comparison of tests with predicted net-section strength. (*From W. H. Munse and E. Chesson, Jr., J. Struct. Div. ASCE, February 1963.*)

area of the member as a whole. The distance from a fastener plane (gusset plate) to the center of gravity of the area tributary to it is a convenient measure of the distribution of the cross-sectional area of a member. For example, the tributary area of the single-angle member of Fig. 3-17*a* is the entire area of the angle, and the coordinate $\bar{x}$ from the fastener plane to the centroid of the area is a measure of the efficiency of the cross section. Similarly, in the double-plane member of Fig. 3-17*b*, the area tributary to each fastener plane is the area of half the cross section, and the coordinate $\bar{x}$ from each fastener plane to the centroid of each half cross section measures the relative importance of the unconnected web. Shear lag is also influenced by the length of the connection. This was seen to be the case for the members of

Table 3-1, where net-section effectiveness decreased with a decrease in the length of the fastener lines. The effect of these two parameters can be expressed as an efficiency coefficient given by[7]

$$K_4 = 1 - \frac{\bar{x}}{L} \tag{3-6}$$

where L is the length of the connection (distance from the first fastener to the last one).

The results of more than 1,000 tests on tension members are compared in Fig. 3-18 with values predicted by multiplying the net area according to the $s^2/4g$ rule by the tensile strength of the metal. The same tests are compared in Fig. 3-19 with predicted values based on corrections according to Eqs. (3-4) to (3-6) and adjustments for bearing ratio and method of fabrication of holes. Relatively few tests fall outside the 10 percent scatter bands in Fig. 3-19, despite the fact that the test data were from a large number of sources and involved various fastener sizes, many different joint configurations (including single angles with and without lug angles), and a variety of materials.

3-9 NET-SECTION DESIGN

It was shown in Art. 3-8 that the strength of a tension member with bolted or riveted connections can be predicted with good accuracy by taking into account the various factors affecting the strength of the net section. This suggests the following procedure for the design of such members. To provide the necessary margin of safety against fracture, the allowable load should be determined by multiplying the effective cross-sectional area by an allowable stress obtained by dividing the specified minimum tensile strength F_u by the desired factor of safety. The effective area A_{eff} is defined by

$$A_{eff} = K_1 K_2 K_3 K_4 A_n \tag{3-7}$$

where A_n is the net area by the $s^2/4g$ rule and K_1, K_2, K_3, and K_4 are the efficiency coefficients defined in Art. 3-8. Furthermore, to protect the member against unserviceability because of excessive elongation, the allowable load should also be computed by multiplying the gross area by an allowable stress obtained by dividing the specified minimum yield stress F_y by a factor of safety. The gross area should be used to investigate elongation, since it is the member itself, rather than the connections, which can be expected to contribute most of the elongation after the yield stress is reached. Furthermore, for reasons discussed in Art. 3-2, the factor of safety with respect to fracture would ordinarily be the larger.

The procedure suggested above is *not* used in standard specifications. Instead, only the *net* section as determined by the $s^2/4g$ rule is investigated. No adjustments of the net area are required, that is, $K_1 = K_2 = K_3 = K_4 = 1$ in Eq. (3-7), except that certain shear-lag reductions are specified by AASHO and AREA. These reductions are for single-angle members connected by only one leg and similar members (discussed in Art. 3-4) for which only half the area of the unconnected leg can be counted. Allowable stresses on the net section are those discussed in Art. 3-2, namely, the smaller of $0.6F_y$ and $0.5F_u$ for AISC, the smaller of $0.55F_y$ and $0.46F_u$ for AASHO, etc. This procedure gives factors of safety with respect to fracture that vary over a considerable range. Thus, for the specimens of Table 3-1, the AISC allowable loads give factors of safety on the strengths of the members ranging from 2.26 to 2.88, as shown in Table 3-2. Specimens 4 and 8 are omitted in this table since their connections were not typical because of the bolts that were omitted.

In a report which includes specimens 3, 4, 5, and 6 and ten additional specimens of several other types of cross section,[9] all of which failed on the net section, factors of safety relative to the AISC allowable load ranged from 2.26 to 3.53.

The results of the tests reported in Table 3-1 raise an additional question regarding criteria for design of tension members. It will be noted that members 5 and 6 broke before they had reached yield stress on the gross section. In general, yielding before fracture is a desirable attribute of a structural member, not only because of the warning it can give in the form of visible evidence of distress, but also because of the potentially greater capacity to absorb energy. For this reason, some engineers believe that a tension member should be designed to yield on the gross section at a load not greater than the strength of the net section. However, this criterion is difficult to meet with a material whose tensile strength is not much larger than its yield stress, as is the case with A514 steel, unless end stubs of larger cross section are used (Art. 3-13).

Table 3-2 FACTORS OF SAFETY FOR MEMBERS OF TABLE 3-1

Specimen	AISC allowable load, kips	P_u, kips	Factor of safety
1	312	872	2.79
2	312	902	2.88
3	312	866	2.77
5	312	706	2.26
6	312	722	2.31
7	312	815	2.60

TENSION MEMBER SPLICE *DP3-2*

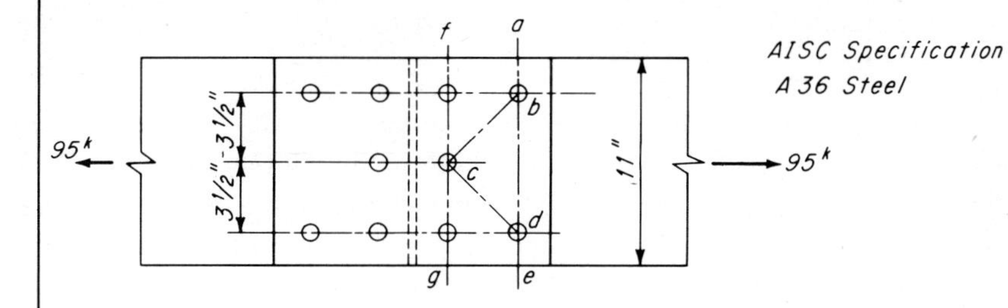

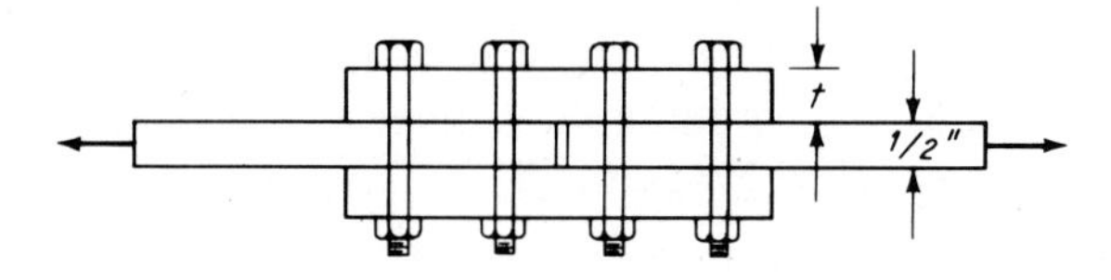

Bolts 7/8-in. A325, bearing-type connection with threads excluded from shear planes

$$\text{Shear} = 2 \times 22 \times 0.6 = 26.4^k \qquad (a)$$

$$\text{Brg} = 1.35 \times 36 \times 1/2 \times 7/8 = 21.3^k$$

$$n = 95/21.3 = 4.46$$

Use 5 bolts

$$-3 + \frac{2s^2}{4 \times 3.5} = -2 \qquad (b)$$

$$s = 2.65 \quad \text{Use } 3'' \qquad (c)$$

$$\text{Section } fcg \quad f_t = \frac{0.6 \times 95}{1/2\,(11-3)} = 14.3 \text{ ksi} < 22 \qquad (d)$$

Splice plates

$$t = \frac{95}{22 \times 2\,(11-3)} = 0.27 \text{ in. use } \frac{5}{16} \text{ in.} \qquad (e)$$

3-10 DP3-2 : TENSION-MEMBER SPLICE

The member is a $\frac{1}{2} \times 11$ in. plate, and two butting sections are to be spliced as shown. The comments which follow are intended to clarify the correspondingly lettered computations.

a The AISC allowable shear for bearing-type connections with no threads in the shear planes is 22 ksi. The bolts are in double shear. AISC allowable bearing is $1.35F_y$.

b The second row of bolts is to be placed far enough from the first row to let section *abde* be the critical net section. In computing the net width for section *abcde*, the diameters of the three holes are subtracted and the quantity $s^2/4g$ added twice, once for each gage space in the chain.

c The calculations show that the pitch need be no more than 2.65 in. However, the preferred minimum distance between bolt holes is three diameters of the bolt. The common pitches are $2\frac{1}{2}$, 3, and $3\frac{1}{2}$ in. for $\frac{3}{4}$-, $\frac{7}{8}$-, and 1-in. fasteners, respectively.

d If the five bolts share the load equally, the force on section *fcg* is only $0.6P$. The net width of this section is 7 in. The allowable tension is $0.6F_y = 0.6 \times 36 = 21.6$, which is rounded to 22 in the AISC specification.

e The critical section for the splice plates is *fg*.

PROBLEMS

3-11 A tension member consisting of two $6 \times 3\frac{1}{2} \times \frac{1}{2}$ angles with the long legs back to back (Fig. 3-5*a*) is connected to a $\frac{5}{8}$-in. gusset plate between the legs. There are two rows of bolts in the 6-in. legs. A gusset plate for diagonal bracing connects with one row of bolts to each of the $3\frac{1}{2}$-in. legs. Determine a bolt pattern that will require deduction for only two holes in each angle, and compute the allowable load. AISC specification, A36 steel, $\frac{3}{4}$-in. A325 bolts.

3-12 Design and detail a butt splice to join two tension members, each of which is a $\frac{3}{4} \times 12$ A36 plate. The axial tension is 165 kips. AISC specification, $\frac{7}{8}$-in. A502 grade 1 rivets.

3-13 Design a single-angle tension member and its connection by one leg to a $\frac{5}{16}$-in. gusset plate. The axial tension is 80 kips. AISC specification, A36 steel, A325 bolts with no threads in the shear plane.

3-14 Compute the factors of safety for the member of Prob. 3-13 (*a*) with respect to gross yielding and (*b*) with respect to fracture. Take the various efficiencies in Eq. (3-7) into account. Assume the ductility factor K_1 to be unity and assume punched holes.

3-15 Same as Prob. 3-13 but with AASHO specifications. AASHO allows only half the area of the unconnected leg to be counted.

3-16 Compute the factors of safety for the member of Prob. 3-15. Follow the instructions of Prob. 3-14.

3-17 Design a tension member of the type shown in Fig. 3-5*d* and its connection to a $\frac{3}{8}$-in. gusset plate with A325 bolts. The tension is 140 kips. A36 steel, AISC specification.

3-18 A $\frac{3}{4} \times 9$ in. A514 plate carries an axial tension of 300 kips. It connects to a $\frac{1}{2}$-in. gusset plate with A490 bolts in a bearing-type connection, with no threads in the shear plane. Design the connection. Use a bolt layout which will allow the member to yield on the gross section before breaking on the net section if it is feasible. AISC specification.

3-19 Design double-angle sections for the bottom chord and tension web members of the truss of DP3-1. AISC specification, A36 steel, A307 bolts.

3-20 Design the hanger for the balcony described in Prob. 3-9, using A325-bolted connections at both the balcony beam and the truss.

3-11 LONG JOINTS

In the design of riveted and bolted joints for tension members, it is generally assumed that the fasteners share the load equally if it is not eccentric. That this assumption may be in error when there are more than two fasteners in the same gage line can be shown by considering the connection shown in Fig. 3-20*a*. Assuming that each bolt carries a shear $P/3$ and that stress is proportional to strain, the bolts must deform equally and the elongations of the two plate segments between adjacent bolts must be equal. However, if the shear on bolt 1 is $P/3$, the force in the upper plate at section *a-a* is $2P/3$, while in the lower plate it is $P/3$. Therefore, the upper plate elongates twice as much as the lower, and the shear deformations of bolts 1 and 2 cannot be equal (Fig. 3-20*b*). Analysis of the joint between bolts 2 and 3 shows that here the lower plate elongates twice as much as the upper one, so that the shear deformation of bolt 3 equals that of bolt 1. Thus, the end bolts carry larger shears than the middle bolt. Of course, this means that the end bolts reach the proportional limit before the middle bolt does, after which the latter begins to pick up a larger share of the load. Thus, at failure of the joint the bolt forces may be equal, or nearly so. However, the extent to which this redistribution of load can develop depends on the ability of the bolts to tolerate large shear deformations.

Figure 3-21 shows a sawed section of a bolted butt joint of A7 steel plates with $\frac{7}{8}$-in. A325 bolts after it had been tested to its ultimate load in tension.[10] The test was stopped before rupture of the joint, but the large elongation of the holes in the splice plates at the left end shows that the stress on the net section was large, and the large shear deformation of the bolts at the ends suggests that fracture is imminent. The same effect is seen in the riveted joint of Fig. 3-22, in which the end rivets have sheared. Failure of the fasteners in a long joint is usually sequential, beginning with those at the ends and progressing toward the center, and is sometimes called *unbuttoning*. This phenomenon has been known for many years. For example, it was observed in a series of tests of large riveted joints in connection with the design of the San Francisco–Oakland Bay Bridge.[11]

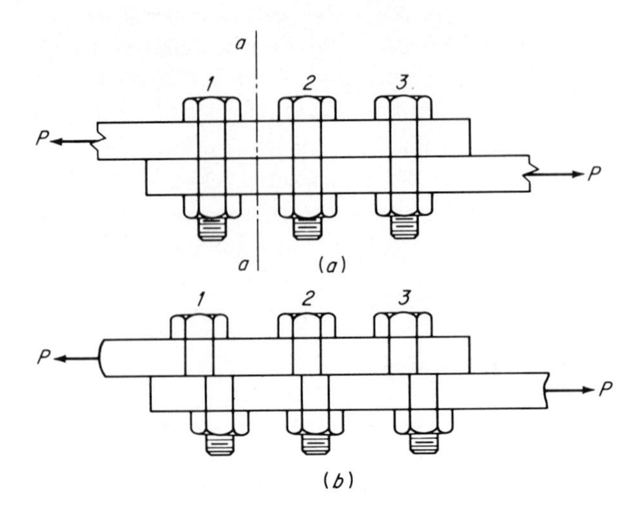

FIGURE 3-20

The joint of Fig. 3-21 was one of 16 joints that were tested to ultimate load (Fig. 3-23). Each specimen was designed so that the sum of the shear areas of the bolts (two for each bolt) was 10 percent more than the net area of the member; i.e., the tension-shear ratio A_n/A_s was 1 : 1.10. This is the ratio for balanced design of a compact (short) A7 steel joint with A325 bolts, which means a design for which the shear strength of the fasteners equals the tensile strength of the member. The ratio 1 : 1.10 had been determined in an earlier series of tests.[12] In eight of the specimens the tension-shear ratio was held constant by using a different width of plate for each joint, with the number of bolts in each line ranging from three to ten and the widths of the member from 5.84 to 15.10 in. The grip (sum of the thicknesses of the connected

FIGURE 3-21
Sawed section of bolted joint after testing. (*Fritz Engineering Laboratory, Lehigh University.*)

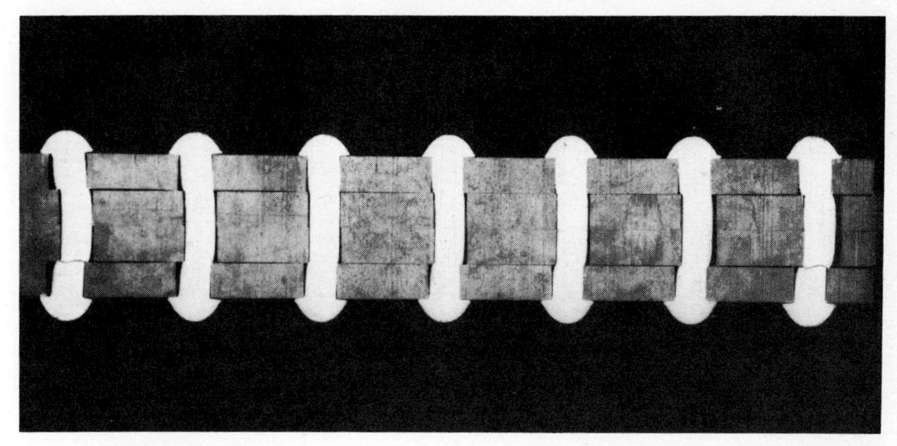

FIGURE 3-22
Sawed section of riveted joint after testing. (*Fritz Engineering Laboratory, Lehigh University.*)

material) was 4 in. Another four specimens had grips of 8 in., with the number of fasteners per line ranging from 10 to 16 and the specimen widths from 8.48 to 12.45 in. The remaining four specimens had widths of 9.88 in., with grips ranging from 4.75 to 6.75 in. to maintain the tension-shear ratio. Figure 3-24 shows the results of the 16 tests. The ordinate in the figure is bolt efficiency, which is the ratio of the average shear per bolt, at ultimate load, to the shear strength of a single bolt of the same grip. It will be noted that four short specimens failed by fracture of the net section, so that they are not representative of bolt efficiency. However, failure of the three short specimens from Ref. 12, which are also shown on the figure, was in the bolts.

The pitch of the bolts was $3\frac{1}{2}$ in. in all the specimens reported in Fig. 3-24 except one, in which a pitch of $2\frac{5}{8}$ in. was used. Specimen D13A had 13 bolts at $2\frac{5}{8}$ in., for

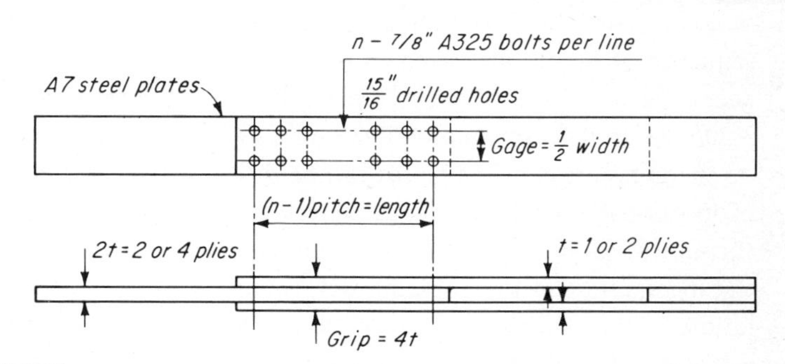

FIGURE 3-23
Test joint reported in Fig. 3-24.

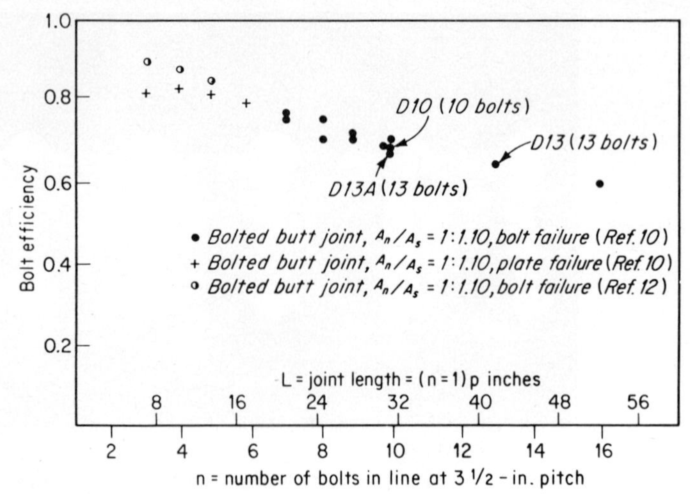

FIGURE 3-24
Efficiency of A325 bolts in A7 butt joints. (*From R. A. Bendigo, R. M. Hansen, and J. L. Rumpf, J. Struct. Div. ASCE, December 1963.*)

which the distance between the end bolts was $31\frac{1}{2}$ in., while specimen D10 had 10 bolts at $3\frac{1}{2}$ in., which gave the same length of connection, $31\frac{1}{2}$ in. The efficiencies of the bolts in these two joints were almost identical (Fig. 3-24). On the other hand, specimen D13 had 13 bolts at $3\frac{1}{2}$ in., which made the connection 42 in. long. The bolt efficiency of this connection was 0.65, compared to the 0.70 efficiency for D13A (Fig. 3-24). This suggests that fastener efficiency is a function of joint length, rather than of the number of fasteners in line.

It will be noted that bolt efficiency falls to about 60 percent for the longest joint in this series of tests. However, since the differences in the shearing deformations of the fasteners depend on the differential strains in the connected plate, as shown in Fig. 3-20*b*, the variation of fastener efficiency with joint length depends on the sequence of events in regard to nonproportional behavior of the plate and the fasteners. That this must be the case can be seen by observing that the three bolts in Fig. 3-20*b* would deform equally if the plates did not deform at all. Therefore, if yielding on the gross areas of the plates can be postponed until after there has been some nonproportional behavior of the fasteners, a better redistribution of the unequal fastener shears will result. On the other hand, of course, early yielding of the plates increases the inequalities in the shearing deformations of the fasteners and causes premature failure of those at the ends of the joint. For this reason, A325 bolts joining plates of a high-yield steel are more efficient than the A325 bolts joining plates of a low-yield steel. Similarly, A490 bolts joining A36 steel plates are less efficient than A325 bolts

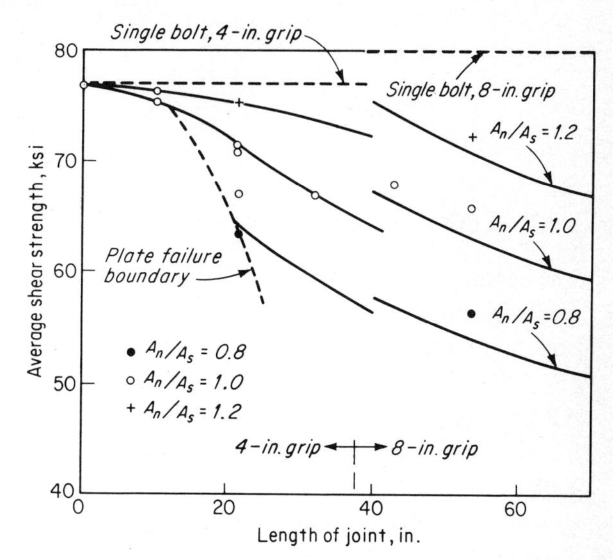

FIGURE 3-25
Efficiency of A325 bolts in A440 butt joints. (*From J. W. Fisher and J. L. Rumpf, J. Struct. Div. ASCE, October 1965.*)

joining A36 steel, given the same geometry. Finally, fastener efficiency for a given combination of materials can be improved by increasing the tension-shear ratio.

Results of tests on A440 steel joints with A325 bolts are compared in Fig. 3-25 with predicted values, for tension-shear ratios A_n/A_s of 0.8, 1.0, and 1.2. The predicted behavior is based on a detailed analysis using the measured properties of the plate material and the bolts.[13] The discontinuities in the curves are due in part to the fact that the longer bolts in the joints with 8-in. grip were somewhat stronger than the shorter bolts in the specimens with 4-in. grip, and in part to the changed geometry itself. The plate-failure boundary separates the zone in which net-section failures occur (to the left) and the zone in which bolt failures occur. These results suggest that the allowable shear stress on fasteners in connections of the type discussed in this article should vary with the length of the joint and the tension-shear ratio. Furthermore, since increasing the length of a joint decreases the efficiency of the fasteners but increases the efficiency of the net section of the member where shear lag is involved, there is a trade-off here in the design of a tension member and its connections. Although standard specifications do not require that fastener efficiency and the shear-lag effect on net-section efficiency be considered, there may be circumstances in which it would be wise to take these effects into account.

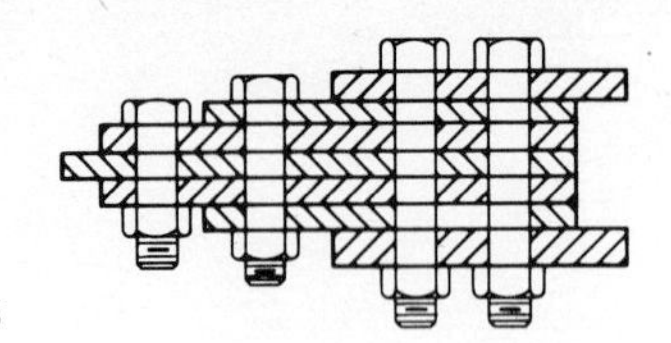

FIGURE 3-26

3-12 LONG-GRIP FASTENERS

The bending of the fasteners in the joints shown in Figs. 3-21 and 3-22 raises a question about the interaction of bending and shear in the case of long fasteners, since there is a possibility that this might further reduce the efficiencies of the fasteners. Some allowance for this has been, and is, customary for long rivets. Both AASHO and AREA require that the number of rivets in a connection be increased by 1 percent for each additional $\frac{1}{16}$ in. of grip in excess of $4\frac{1}{2}$ diameters of the rivet. AISC specifies a like increase if the grip exceeds 5 diameters. However, tests have shown that the shearing strength of high-strength bolts having grips as much as 9 diameters is not less than that of bolts with short grips.[10] Therefore, no penalty attaches to long-grip high-strength bolts.

Sometimes the plates or members forming a structural joint are separated by one or more filler plates, as shown in Fig. 3-26. Usually the plates are extended as shown to improve the transfer of stress through the connection. Specifications require that the filler extension be attached by fasteners sufficient in number to transmit a force equal to the area of the filler times the stress computed by dividing the load on the joint by the combined area of the fillers and the part connected. However, fillers thinner than $\frac{1}{4}$ in. need not be extended. Furthermore, fillers in friction-type connections with high-strength bolts need not be extended.

3-13 COMPENSATING FOR REDUCTION IN CROSS-SECTIONAL AREA

Ends of tension members may be strengthened to offset reduction in cross-sectional area from threads, fastener holes, etc. Round or square bars threaded on the normal end have a net area equal to the area of the cross section at the root of the thread. To compensate for this loss, the end of the bar may be upset to form an increased cross section into which the threads are cut. Upset bars have an area at the root of the thread at least 15 percent greater than the area of the bar.

Tension members which are field-riveted or bolted to gusset plates may be strengthened at the ends to make up for the loss of section caused by the holes. In the case of H-shaped truss members, either rolled (W shapes) or built-up, this can be

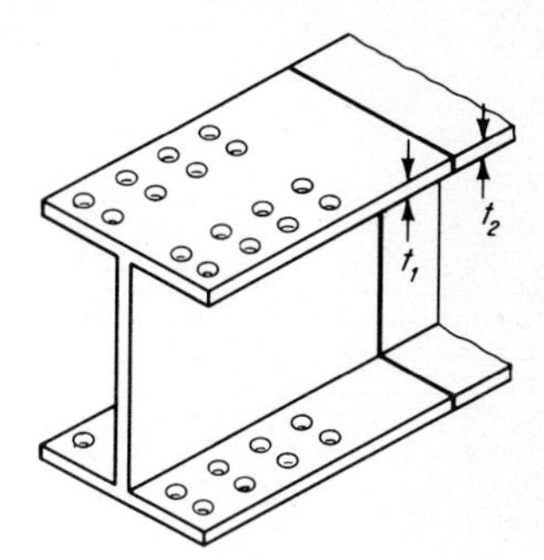

FIGURE 3-27

done by welding to the flanges thicker plates of the same steel or plates of the same thickness but of higher-strength steel (Fig. 3-27). The ends of certain tension members of the Carquinez Straits bridge in California were increased in area to offset bolt-hole losses by groove-welding thicker plates to the ends of the flange plates. For example, one of the diagonal members in the suspended span consists of 1 × 18 flange plates and a $\frac{3}{8} \times 20$ web. The cross-sectional area is

$$\begin{aligned} &\text{Flanges:} \quad 2 \times 1 \times 18 = 36 \\ &\text{Web:} \quad \tfrac{3}{8} \times 20 = 7.5 \\ &\qquad\qquad\qquad \overline{43.5} \text{ in.}^2 \end{aligned}$$

The end stubs are made up of $1\frac{3}{8} \times 18$ flanges and a $\frac{3}{8} \times 19\frac{1}{4}$ web, using the same steel. There are four lines of 1-in. A325 bolts in each flange. The resulting net area is

$$\begin{aligned} &\text{Flanges:} \quad 2 \times 1\tfrac{3}{8} \times (18 - 4 \times 1\tfrac{1}{8}) = 37.12 \\ &\text{Web:} \quad \tfrac{3}{8} \times 19\tfrac{1}{4} = 7.22 \\ &\qquad\qquad\qquad \overline{44.34} \text{ in.}^2 \end{aligned}$$

It will be noted that stubs and the member are of the same depth, 22 in.

In the Benicia-Martinez bridge, also in California, the end stubs were made of higher-strength steel, so that the thicknesses of the flanges and the web were the same in the stubs as in the member. For example, one of the main members which is of A242 steel has A514-steel stubs.

3-14 LUG ANGLES

Lug angles are sometimes used to reduce the length of the connection of a single-angle or double-angle member to a gusset plate (Fig. 3-28). According to Ref. 9, this device reduces shear lag to the extent that the net section of an angle connected in this way

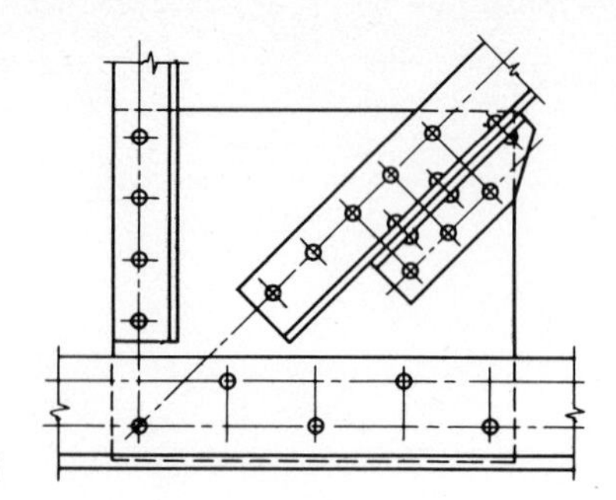

FIGURE 3-28

is fully effective, even though the connection is shorter than it would be without the lug. This is equivalent to saying that the lug angle reduces $\bar{x}$ in Eq. (3-6) to zero. However, the location of the lug angle is of some importance; it should be at the beginning of the joint, as in Fig. 3-28, rather than at the end.

3-15 DP3-3: TENSION MEMBERS FOR BRIDGE TRUSS

The truss of this example is one of the two trusses of the highway bridge discussed in detail in Chap. 10 (DP10-1). Figures 10-6 and 10-7 are photographs of the bridge. The forces shown on the design sheet are computed in DP10-1 and are the totals for dead load and the HS20-44 live load, including impact, of the AASHO specifications. Members of trusses of this size are usually W's or built-up H shapes, which require two gusset plates at each panel point. Therefore, all the members must be of the same nominal depth in order to avoid excessively thick fill plates at the joints. The following comments are clarifications of the correspondingly lettered computations on the design sheet.

a Net section of the chord members is based on two holes in the web and two in each flange. These are needed for the splice of L_0L_2 to L_2L_4, the detail of which is shown in Fig. 3-29.

b Net section of the web members is based on two holes in each flange for connections to the gusset plates. Typical joints are shown in Figs. 10-15 and 10-16.

c The W14 $\times$ 61 gives almost double the net area required for U_3L_4. No smaller W14 can be used without exceeding the specified slenderness limit ($L/r \lesssim 200$) for main tension members (AASHO 1.7.12).

d L_0L_2 is spliced to L_2L_4 just to the left of the gusset plates at panel point 2. Tension chords are usually spliced this way, rather than by the gusset plates, to facilitate erection.

e AASHO requires that the splice be designed for not less than the average of the actual stress in the member and the allowable stress.

TENSION MEMBERS FOR HIGHWAY-BRIDGE TRUSS — DP3-3

AASHO specs
7/8-in. A325 bolts

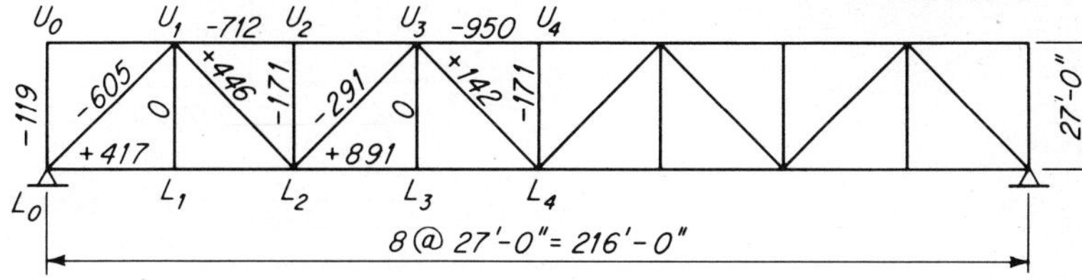

Member	Load	A_n	Section	A_g	Deduct	A_n	L/r	
L_0L_2	417 @ 20 =	20.9	W14 × 87	25.6	4 × 1 × 0.688 = 2.75 2 × 1 × 0.420 = 0.84	22.0	88	(a)
L_2L_4	891 @ 20 =	44.6	W14 × 176	51.7	4 × 1 × 1.313 = 5.25 2 × 1 × 0.820 = 1.64	44.8	81	
U_1L_2	446 @ 20 =	22.3	W14 × 87	25.6	4 × 1 × 0.688 = 2.75	22.8	124	(b)
U_3L_4	142 @ 20 =	7.1	W14 × 61	17.9	4 × 1 × 0.643 = 2.57	15.3	187	(c)

<u>Splice L_0L_2 to L_2L_4</u> (d)

L_0L_2: $P/A_n = 417/22 = 19$ ksi $(19 + 20)/2 = 19.5$ design stress for splice (e)

7/8-in. A 325 bolts, friction-type connection: $0.6 \times 13.5 = 8.1^k$ s.s. (f)
16.2^k d.s.

Web $A_g = 0.420\,(14 - 2 \times 0.688) = 5.3$ (g)
2 holes $= 2 \times 1 \times 0.420 = 0.8$
$A_n = 4.5$ in.^{2}

Use two 3/8 × 10 spl. pl. 3 holes out $A_n = 2 \times 7 \times 3/8 = 5.25$ in.^{2}

Bolts $= 4.5 \times 19.5/16.2 = 5.4$

Flanges W14 × 87 $A_n = 22.0$
Web $= 4.5$
two flgs. $= 17.5$ in.^{2}

Use 15" plates: $t = 17.5/(15 - 4) = 1.6$ (two plates) (h)

Use two 13/16 × 15 spl. pl.

Bolts $= 17.5 \times 19.5/8.1 = 42$ (two flanges)

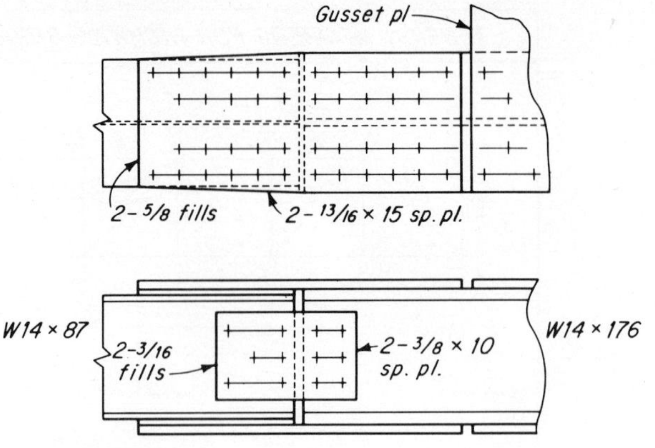

FIGURE 3-29
Splice of L_0L_2 to L_2L_4 of DP3-3.

f Friction-type connections are mandatory except in compression members and secondary members. Bearing stress is not a factor in friction-type connections.
g The web is spliced for its portion of the force in the member.
h There are four bolts at the critical section of the flange splice plates.

Because of the differences in section depth and web thickness, fill plates are needed on L_0L_2. Since this is a friction-type connection, it is not necessary to extend these fill plates beyond the splice plates (Art. 3-12).

PROBLEMS

3-21 Redesign the members of the truss of DP3-3 in A440 steel.

3-22 Design the tension members of the truss of Prob. 3-10 for bolted construction. Design a splice for L_0L_2 to $L_2L_{2'}$.

3-23 A bridge-truss tension member consists of two $1\frac{1}{4} \times 24$ flange plates and a $\frac{5}{8} \times 20$ web plate welded in the form of an I. The load is 1,800 kips. The member is designed on the basis of its gross cross-sectional area, so that a stub is required at each end for bolted connections (Art. 3-13). Design the stub, using six parallel rows of bolts in each flange (three on each side of the web). The member is of A441 steel, and the bolts are 1-in. A325. Use A441 steel for the stub. AASHO specifications.

3-24 Design stubs for the member of Prob. 3-23 so that the stub flanges and web are the same size as those of the member.

3-25 Design a single-angle tension member 12 ft long and its connection at each end to a $\frac{5}{16}$-in. gusset plate. The tension is 70 kips. The angle cannot extend more than 15 in. on the gusset. AISC specification, A36 steel, $\frac{7}{8}$-in. A502 grade 1 rivets.

RIVETED ALUMINUM TRUSS — DP3-4

Truss of DP3-1
Material: aluminum 6061-T6
3/4 rivets: aluminum 6061-T6
Specifications: ASCE committee (Ref. 2)

Tension members

Member	Force	Reqd. area	Section	Net area	L/r
$L_0L_1L_2$	53.6 @ 19 =	2.82 in.^{2}	2∠ 3½ × 3½ × ¼	3.00	77/1.07 = 72
L_2L_3	87.2 @ 19 =	4.58	2∠ 4 × 4 × 7/16	5.95	72/1.21 = 60
L_3L_4	102.2 @ 19 =	5.38	do	5.95	do
L_4L_5	108.8 @ 19 =	5.71	do	5.95	do
U_0L_1	65.6 @ 19 =	3.45	2∠ 3½ × 3½ × 5/16	3.70	94/1.06 = 89
U_1L_2	44.0 @ 19 =	2.32	2∠ 3 × 3 × ¼	2.48	94/0.91 = 103
U_2L_3	20.2 @ 19 =	1.06	2∠ 2½ × 1½ × 3/16	1.15	98/0.63 = 156
U_3L_4	9.9 @ 19 =	0.52	do	1.15	102/0.63 = 162
U_4L_5	2.0 @ 19 =	0.11	do	1.15	106/0.63 = 168
U_5L_5	10.1 @ 19 =	0.53	do	1.15	84/0.63 = 133

Joint L_3 (5/16" gusset plate)

Rivets: shear = 11 × 2 × 0.461 = 10.1^k

brg. = 34 × 5/16 × 0.766 = 8.14^k

U_2L_3 20.2/8.14 = 3, U_3L_3 13.6/8.14 = 2, $L_2L_3L_4$ 15/8.14 = 2

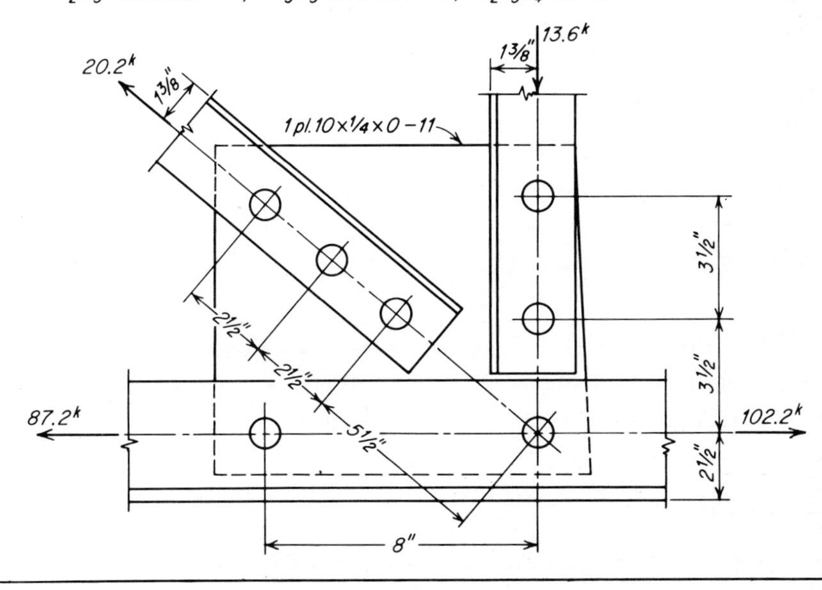

3-16 DP3-4: RIVETED ALUMINUM TRUSS

In this example the roof truss of DP3-1 is designed for riveted construction in aluminum 6061-T6, using the specifications of the ASCE Committee on Design in Lightweight Structural Alloys.[2] Because of its higher price per pound, the ratio of material cost to fabrication costs is greater for aluminum than it is for steel, so that it is more important to design for minimum weight. Nevertheless, availability of shapes must be taken into account, and any saving in material cost must be compared with the cost of splices needed to substitute two members of different size for a single member.

Cold-driven 6061-T6 rivets are usually used with members of alloy 6061-T6 (Table 2-11). Holes may be punched or drilled. Holes for connections in this truss will be drilled, in which case deduction for net section is based on the actual size of the hole (Art. 3-6). The hole for a cold-driven 3/4-in. rivet is 0.766 in. in diameter (Table 2-13). The bottom chord is designed in three sections—two members L_0L_2 and one member L_2L_2'. The design loads for L_2L_3, L_3L_4, and L_4L_5 are not significantly different, so that any saving in material cost that would result if a lighter member were used for L_2L_3 would almost certainly be less than the cost of the splice.

The $2\frac{1}{2} \times 1\frac{1}{2} \times \frac{3}{16}$ angles furnish more net area than is needed for U_3L_4, U_4L_5, and U_5L_5. However, material thinner than about one-third to one-fourth the diameter of the rivet is considered to be a minimum, and the leg of an angle should not be smaller than about three times the rivet diameter. Therefore, these angles are minimum for $\frac{3}{4}$-in. rivets.

Joint L_3 Allowable stresses for 6061-T6 rivets are given in Table 2-11. Since the chord is continuous at L_3, the rivets connecting it to the gusset plate resist only the difference between the tensions in L_2L_3 and L_3L_4. The gusset plate is dimensioned to provide the required distance of two diameters of the rivet from the edge toward which the rivet pressure is directed.

Although this truss weighs only slightly more than half as much as a comparable design for riveted steel construction, it would cost two to three or more times as much. Thus, unless the additional cost were justified by the corrosion hazard, or for some other reason, the aluminum truss could not possibly compete with the steel structure. On the other hand, we have patterned the aluminum truss after a form that has been found to be suited to steel, and it is important to remember that the effective use of new materials often awaits the development of new forms and new ideas. Aluminum can be extruded easily and inexpensively, and a great variety of shapes—shapes that could not possibly be rolled—can be produced by the extrusion process. Extruded shapes that can be interlocked (so as to eliminate fasteners either wholly or in part) are not outside the realm of possibility.

PROBLEMS

3-26 Design and detail joint L_1 for the truss of DP3-4. Member U_1L_1 consists of two angles $3 \times 2\frac{1}{2} \times \frac{5}{16}$ with long legs back to back (DP4-5).

3-27 Design and detail joint L_2 for the truss of DP3-4. Member U_2L_2 consists of two angles $3 \times 2 \times \frac{1}{4}$ with long legs back to back (DP4-5).

3-17 GUSSET PLATES

The riveted joints of the truss considered in DP3-4 are simple enough and usually require no more attention than was given to L_3. The lateral dimensions of a gusset plate are determined principally by the fastener requirements of the members, which leaves only the thickness to be based on other considerations. According to the AASHO specifications, "Gusset plates shall be of ample thickness to resist shear, direct stress, and flexure, acting on the weakest or critical section of maximum stress." This is all very good, but the only practicable method of estimating these stresses is based on the assumption that the elementary formulas for beams apply, and these formulas are valid only for beams whose span is more than twice the depth and at cross sections not closer to concentrated loads than about half the depth. The ordinary gusset plate falls considerably short of these requirements, so that the results obtained by the application of beam formulas are of questionable value, and may be misleading.

Experimental data from which one can judge the accuracy of the ordinary beam formulas when they are applied to gusset plates are contained in the report of an investigation of stress distributions in the gusset plates of a typical bottom-chord joint of a Warren truss.[14] The tests were made on a model. Strains in the plates were measured at service loads, but the joint was not tested to failure. As might be expected, the tests demonstrated that bending stresses are not distributed linearly and that the neutral axis of a cross section does not coincide with the centroidal axis. The maximum bending stress was found to occur at an interior point rather than on the extreme fiber, and the shearing stresses were not distributed according to the parabolic law which corresponds to a linear distribution of bending stresses.

We shall use the joint L_3 of DP3-4 to illustrate the conventional analysis of a gusset plate. A plate of this size would not be analyzed in practice, but the investigation is no different in principle for gusset plates of larger trusses. The stresses are determined on sections which, for convenience, are cut parallel to and normal to the chord member. It is impossible to tell by inspection what section will experience the largest stress.

The detail of the joint is shown in Fig. 3-30*a*. The portion of the gusset plate which lies above a horizontal section through the upper rivets of the gusset plate is shown in Fig. 3-30*b*. It is assumed that the forces on these rivets act on the part of the plate lying above the section. Therefore, one-half the force in the vertical member and one-third the force in the diagonal member must be resisted by an internal force on the section. The components of this force are the normal force N, the shearing force V, and the couple M. The 6.7-kip force from the diagonal member is resolved into two components at its intersection with the line of action of the vertical force. Then

$$N = 6.8 - 4.5 = 2.3 \text{ kips}$$
$$V = 5.0 \text{ kips}$$
$$M = 5.0 \times 7 + 2.3 \times 4 = 44.2 \text{ in.-kips}$$

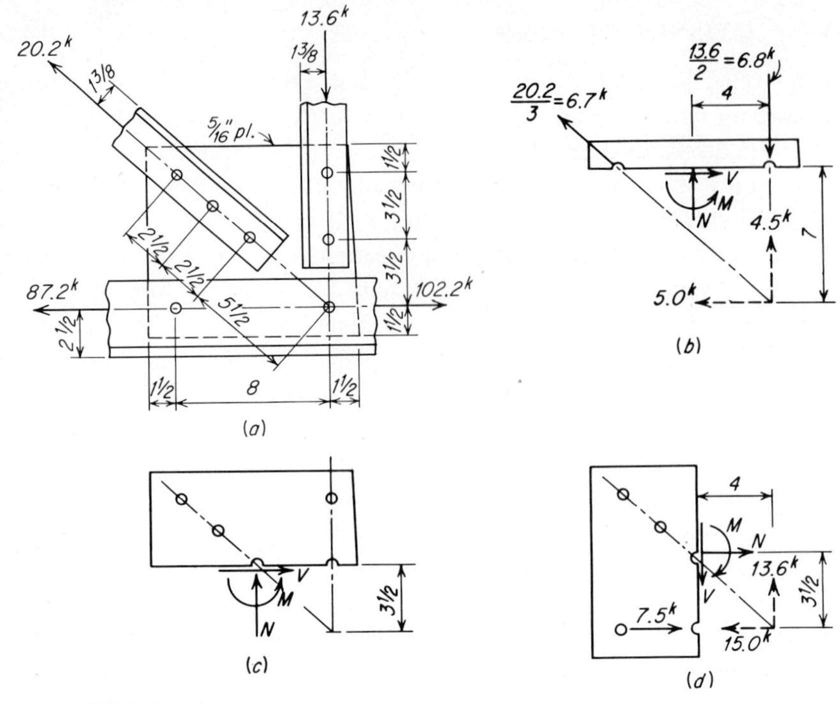

FIGURE 3-30

The corresponding stresses are

$$f_v = \frac{3}{2}\frac{V}{A} = \frac{1.5 \times 5.0}{11 \times 0.31} = 2.2 \text{ ksi}$$

$$f_b = \frac{N}{A} \pm \frac{Mc}{I} = \frac{-2.3}{11 \times 0.31} \pm \frac{44.2 \times 6}{0.31 \times 11^2} = -0.7 \pm 7.1 = +6.4, -7.8 \text{ ksi}$$

The portion of the plate above a horizontal section through the lower rivet hole in U_3L_3 is shown in Fig. 3-30*c*. At this section, the internal force is the resultant of the forces in L_2L_3 and L_3L_4. Therefore, $N = 0$, $V = 15.0$ kips, and $M = 15.0 \times 3.5 = 52.5$ in.-kips. Then

$$f_v = \frac{1.5 \times 15.0}{11 \times 0.31} = 6.6 \text{ ksi}$$

$$f_b = \frac{52.5 \times 6}{0.31 \times 11^2} = \pm 8.4 \text{ ksi}$$

A sketch of the part of the plate to the left of a vertical section through the lower rivet of U_2L_3 is shown in Fig. 3-30*d*. This portion of the plate contains all the rivets in U_2L_3 and one of the two rivets in the chord member. Therefore, the forces acting on it are the two components of the force in U_2L_3, shown at the common intersection of the three members at the joint, and one-half the difference between the forces in L_2L_3 and L_3L_4. The corresponding internal forces on the section are

$$N = 15.0 - 7.5 = 7.5 \text{ kips}$$
$$V = 13.6 \text{ kips}$$
$$M = 13.6 \times 4 - 7.5 \times 3.5 = 28.2 \text{ in.-kips}$$

Then

$$f_v = \frac{1.5 \times 13.6}{10 \times 0.31} = 6.6 \text{ ksi}$$

$$f_b = +\frac{7.5}{10 \times 0.31} \pm \frac{28.2 \times 6}{0.31 \times 10^2} = +2.4 \pm 5.5 = +7.9, \ -3.1 \text{ ksi}$$

The calculated stresses all are substantially less than allowable values. It should be reemphasized that no designer would analyze a gusset plate for a truss of this size. The analysis here is intended merely for demonstration. And, in any case, the results are of questionable value, for reasons that were discussed earlier in this article.

Compressive stresses may develop parallel to and at the edge of gusset plates. This is because deflection of a truss tends to change the angles between its members. Thus, the upper part of the gusset plate in Fig. 3-30*a* will be compressed if the angle between the adjacent members decreases. Therefore, the width of the top edge must not be too large, compared with the thickness, or the plate may buckle. Bending of this kind, which is called *local buckling*, is discussed in Art. 4-21. The AASHO specifications require that an unsupported edge of a gusset plate be stiffened if it is longer than $11{,}000/\sqrt{F_{y,\text{psi}}}$ times its thickness. This gives, for example, 58 for A36 steel.

An additional example of the analysis of gusset plates is given on Sheet 15 of DP10-1, and typical joints of bridge trusses are shown in Figs. 10-15 and 16.

3-18 SECONDARY STRESSES IN TRUSS MEMBERS

The principles of design which have been discussed in this chapter are based on the assumption that tension members resist only axial forces. Such a condition is rarely, if ever, realized. If the members of the truss *ABC* shown in Fig. 3-31*a* are connected with frictionless pins, the truss will be deformed by the force *P* to the shape *AB′C′*. If the members are bolted, riveted, or welded at the joints, so that the angles formed by the members resist change, the truss must assume the shape *AB′C′* of Fig. 3-31*b*. In the first case the tension member *AC* suffers only elongation. In the second case

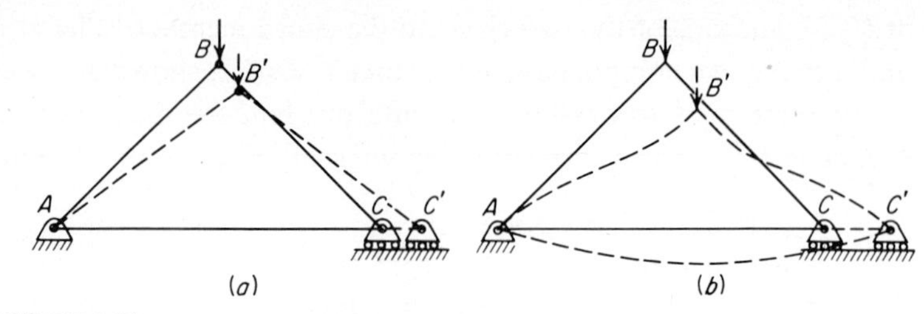

FIGURE 3-31

it undergoes both elongation and bending, and the maximum stress is no longer correctly given by $f = P/A$ but instead by $f = P/A + Mc/I$. The bending stress is called a *secondary stress*.

Joint rigidity may induce one-directional curvature, as in the tension member AC, or reversed curvature, as in the compression member AB. Although values of the corresponding end moments can be determined, secondary stresses are usually neglected in design. This is not to say that they are necessarily small but rather that there is sufficient evidence from both analytical studies and experience to indicate that they can safely be neglected, provided that the engineer is aware of certain precautions which should be observed in order to guard against their becoming excessive. Secondary stresses are inversely proportional to the ratio of the length of the member to its width in the plane of the truss. If this ratio is 10 or more, secondary stresses are not likely to exceed 20 to 30 percent of the primary stresses.

REFERENCES

1 Beedle, L. S., and L. Tall: Basic Column Strength, *J. Struct. Div. ASCE*, July 1960.
2 Suggested Specifications for Structures of Aluminum Alloys 6061-T6 and 6062-T6, *J. Struct. Div. ASCE*, December 1962.
3 Cochrane, V. H.: Rules for Rivet Hole Deduction in Tension Members, *Eng. News-Rec.* November 16, 1922.
4 Brady, W. G., and D. C. Drucker: Investigation and Limit Analysis of Net Area in Tension, *Trans. ASCE*, vol. 120, 1955.
5 Timoshenko, S. P., and J. N. Goodier: "Theory of Elasticity," 3d ed., McGraw-Hill, New York, 1970.
6 Schutz, F. W., and N. M. Newmark: The Efficiency of Riveted Structural Joints, *Univ. Ill. Struct. Res. Ser.* 30, 1952.
7 Munse, W. H., and E. Chesson, Jr.: Riveted and Bolted Joints: Net Section Design, *J. Struct. Div. ASCE*, February 1963.

8 Kuhn, P.: "Stresses in Aircraft and Shell Structures," McGraw-Hill, New York, 1956.

9 Chesson, E., and W. H. Munse: Riveted and Bolted Joints: Truss-type Tensile Connections, *J. Struct. Div. ASCE,* February 1963.

10 Bendigo, R. A., R. M. Hansen, and J. L. Rumpf: Long Bolted Joints, *J. Struct. Div. ASCE,* December 1963.

11 Davis, R. E., G. B. Woodruff, and H. E. Davis: Tension Tests of Large Riveted Joints, *Trans. ASCE,* vol. 105, 1940.

12 Foreman, R. T., and J. L. Rumpf: Static Tests of Compact Bolted Joints, *Trans. ASCE,* vol. 126, 1961, pt. II.

13 Fisher, J. W., and J. L. Rumpf: Analysis of Bolted Butt Joints, *J. Struct. Div. ASCE,* October 1965.

14 Whitmore, R. E.: Experimental Investigation of Stresses in Gusset Plates, *Univ. Tenn. Eng. Exp. Stn. Bull.* 16, 1952.

4

COMPRESSION MEMBERS

4-1 INTRODUCTION

Compression members are usually given names which identify them as particular members in a structure. The vertical compression members in building frames are called *columns* in the United States and *stanchions* in England. Compression members are sometimes called *posts*, and the diagonal members at the ends of through-bridge trusses are usually called *end posts*. Other compression members in trusses are known according to their position as chord members and web members. The principal compression member in a crane is called a *boom*. Some types of compression member are called *struts*. Members which connect adjacent frames at the eaves of some types of industrial buildings are called *eave struts*.

4-2 ELASTIC BUCKLING OF COLUMNS

There is a specific magnitude of load at which a straight, homogeneous, centrally loaded column becomes unstable. By this is meant that at this load the column may begin to bend, even though there is no apparent moment to initiate bending. It is

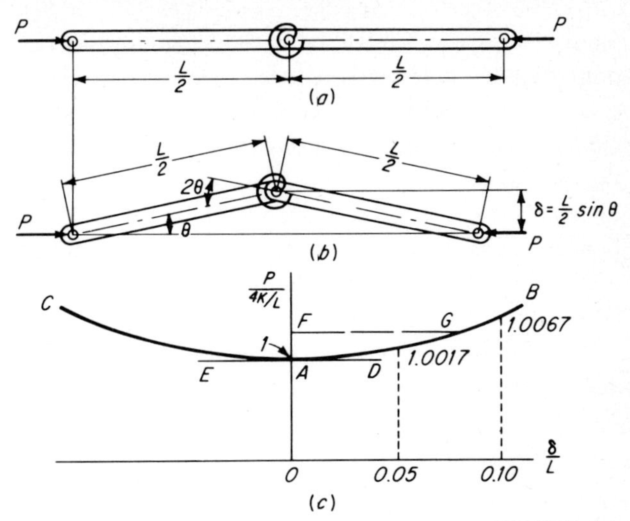

FIGURE 4-1

instructive to study this phenomenon for the pin-ended member of Fig. 4-1*a*, which consists of two rigid bars connected to one another by a linearly elastic torsion spring and which is loaded with axial compressive forces P. If the member assumes the deflected shape shown in Fig. 4-1*b*, equilibrium requires that

$$P\delta = P\frac{L}{2}\sin\theta = 2K\theta \qquad (a)$$

where δ is the deflection at midlength and K is the spring constant. Equation (*a*) gives

$$P = \frac{4K}{L}\frac{\theta}{\sin\theta} \qquad (b)$$

The curve *CAB* in Fig. 4-1*c* is a plot of Eq. (*b*). If $\delta = 0$, $\theta = 0$ and $P = 4K/L$. For any load less than this, $\delta = 0$ and the column is straight. This condition is represented by a point between O and A on OA in Fig. 4-1*c*. Thus, if the column is displaced laterally (by an accidental lateral force, say) while it is supporting a load $P \lessapprox 4K/L$, it will straighten when the accidental force disappears. This is because the moment in the spring exceeds the moment $P\delta$, so that it is a restoring moment. Therefore, any point between O and A represents *stable* equilibrium in the straight configuration of Fig. 4-1*a*. On the other hand, if the column is straight while it is supporting a load $P > 4K/L$, which corresponds to a point such as F on OA extended, it is in *unstable* equilibrium and the slightest disturbance will cause it to deflect the amount δ for which it will be in stable equilibrium (to G on AB). Furthermore, it will not straighten unless P is reduced or it is pushed back.

The value of P at which a straight column becomes unstable is called the *critical load.* When the column bends at the critical load, it is said to have buckled. Therefore, the critical load is also called the *buckling load.* At the critical load the column is extremely sensitive to increase in load, in the sense that a very slight increase is accompanied by a large lateral deflection. Thus, if the critical load $P = 4K/L$ for the column of Fig. 4-1a is increased by only $\frac{2}{3}$ percent, δ is 10 percent of the column length L (Fig. 4-1c).

If we assume that the deflection δ in Fig. 4-1b is small, we can equate it to $\theta L/2$, rather than $(L/2) \sin \theta$, so that

$$P\delta = P\frac{L}{2}\theta = 2K\,\theta \tag{c}$$

from which

$$P = \frac{4K}{L} \tag{d}$$

Thus, we obtain the critical value of P, but we get no information beyond this; that is, δ is indeterminate if we solve the problem in this fashion. This solution to the stability problem is depicted by OA and EAD in Fig. 4-1c, which show that δ is zero for any P less than the critical load but may have any small value, including zero, at the critical load. In this sense, there are two equilibrium configurations at the critical load, the straight one and one that is slightly bent. Because of this, the critical load is said to correspond to a *bifurcation* of the equilibrium configuration.

Figure 4-2 shows a straight, homogeneous, pin-ended, centrally loaded column which has buckled. At any point of the deflected centerline the bending moment $M = Py$, and equilibrium requires that

$$\frac{EI}{\rho} = Py \tag{e}$$

where ρ is the radius of curvature. Using the known expression for curvature $1/\rho$, we get

$$EI\frac{-d^2y/dx^2}{[1 + (dy/dx)^2]^{3/2}} = Py \tag{f}$$

where the negative sign is needed because d^2y/dx^2 itself is negative.

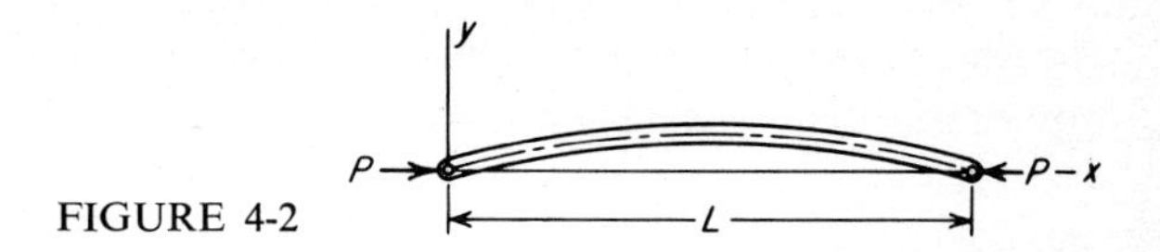

FIGURE 4-2

Equation (f) is not easy to solve. However, if we make the usual assumption that deflections are small, we can write

$$-EI\frac{d^2y}{dx^2} = Py \tag{g}$$

With the substitution

$$k^2 = \frac{P}{EI} \tag{h}$$

Eq. (g) gives

$$\frac{d^2y}{dx^2} + k^2y = 0 \tag{i}$$

the solution of which is

$$y = A \sin kx + B \cos kx \tag{j}$$

From the boundary condition $y = 0$ at $x = 0$, we find $B = 0$, and

$$y = A \sin kx \tag{k}$$

To satisfy the boundary condition $y = 0$ at $x = L$, we may have either $A = 0$ or $\sin kL = 0$. The first solution gives $y = 0$, in which case the column is straight, while the second gives

$$kL = \pi, 2\pi, \ldots, n\pi \tag{l}$$

From Eqs. (h) and (l)

$$P = \frac{\pi^2 EI}{L^2}, \frac{4\pi^2 EI}{L^2}, \ldots, \frac{n^2\pi^2 EI}{L^2} \tag{m}$$

If we substitute the several values from Eq. (l) into Eq. (k), we see that the column-deflection curve is a half sine wave for the first value of P, two half waves for the second, and so on. Therefore, since the higher values of P exist only if there are intermediate lateral supports, we take as the fundamental case

$$P_E = \frac{\pi^2 EI}{L^2} \tag{4-1}$$

where P_E denotes the Euler load, named for Leonhard Euler, who derived it in 1759. Since the formula is invalid if the stress exceeds the proportional limit, it is more convenient to write it in terms of the Euler (critical) stress F_E:

$$F_E = \frac{P_E}{A} = \frac{\pi^2 EI}{AL^2} = \frac{\pi^2 E}{(L/r)^2} \tag{4-2}$$

where A is the area and r the radius of gyration of the column cross section. Buckling at stresses exceeding the proportional limit is discussed in Art. 4-4.

Using δ for the deflection A at midlength, Eqs. (k) and (l) give the deflected shape for the Euler load,

$$y = \delta \sin \frac{\pi x}{L} \tag{4-3}$$

The deflection is indeterminate. This is because we used Eq. (g) as an approximation to Eq. (f). This is analogous to the situation for the two-bar column of Fig. 4-1 when we assume θ to be small, which gives us the critical load [Eq. (d)] but leaves θ undetermined. Thus, OA and EAD in Fig. 4-1c also depict the behavior of the Euler column, provided we change the axis of ordinates to P/P_E.

Exact solutions of Eq. (f) can be obtained in terms of elliptic integrals.[1] An approximate solution for the column with hinged ends is given by[2]

$$\frac{P}{P_E} = 1 + \frac{\pi^2}{8}\frac{\delta^2}{L^2} \tag{4-4}$$

This equation gives values of P accurate to within 1 percent up to $\delta/L = 0.25$. Except for the numerical values of the ordinates, CAB of Fig. 4-1c is a plot of Eq. (4-4), provided the axis of coordinates is labeled P/P_E. The value of P/P_E corresponding to $\delta/L = 0.1$ is 1.0123, which is of the same order of magnitude as that shown (1.0067) for the two-bar column. Thus, a column 10 ft long would be deflected 1 ft by a load only $1\frac{1}{4}$ percent larger than the Euler load, provided it is slender enough to accept such a deflection without being stressed beyond the proportional limit.

To determine the load at which Eq. (4-4) ceases to hold, we compute

$$f = \frac{P}{A} + \frac{Mc}{I} = \frac{P}{A}\left(1 + \frac{\delta c}{r^2}\right) = \frac{P}{A}\left(1 + \frac{\delta}{L}\frac{L}{r}\frac{c}{r}\right) \tag{n}$$

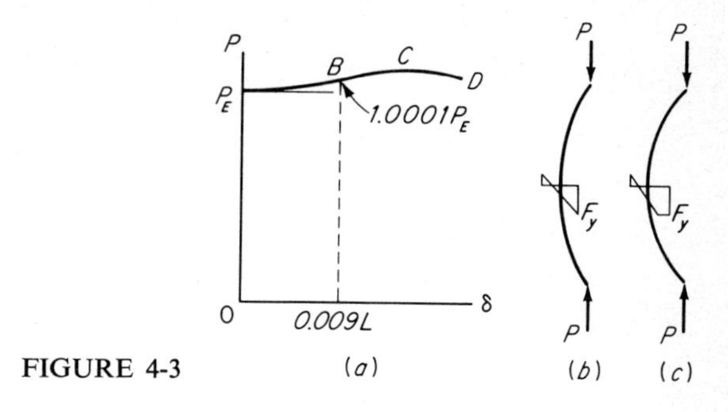

FIGURE 4-3

Assume the column to be 10 ft long with a slenderness ratio $L/r = 150$. For the load $P = 1.0001P_E$ Eq. (4-4) gives $\delta/L = 0.009$, so that δ is about 1 in. Also,

$$\frac{P}{A} = \frac{1.0001P_E}{A} = \frac{1.0001\pi^2 EI}{AL^2} = \frac{1.0001\pi^2 E}{(L/r)^2} = \frac{30{,}000\pi^2}{150^2} = 13.2 \text{ ksi}$$

Furthermore, $c \approx 1.25r$ if the cross section is an I. Substituting these values into Eq. (n) gives

$$f = 13.2(1 + 0.009 \times 150 \times 1.25) = 36 \text{ ksi}$$

Thus, if the column is of A36 steel, it reaches yield stress on the extreme fiber at midlength (Fig. 4-3b) when the axial load is only 0.01 percent more than the critical load (point B in Fig. 4-3a). Further increase in load produces strains in the region of maximum moment which exceed the yield strain ε_y. Therefore, with plane cross sections remaining plane, stress is no longer distributed linearly over a cross section where $\varepsilon > \varepsilon_y$; instead, it varies according to the f-ε curve for the material. Thus, for the A36 steel assumed in this example, which has a yield plateau, there will be uniform stress F_y in the yielded zones. For this case, the maximum load P is reached with a distribution of stress at midlength such as that in Fig. 4-3c.* Equilibrium at still larger deflections is possible only at reduced value of P (CD of Fig. 4-3a). This suggests that the difference between the postbuckling strength (point C) and the buckling strength P_E is so small that the Euler load is a practical measure of the ultimate strength of the perfectly straight axially loaded column, provided it buckles while the stress $f = P/A$ does not exceed the proportional limit.

Since $OP_E BCD$ depicts the behavior of the column, any lateral deflection δ, however small, is necessarily accompanied by some increase in the load P_E which initiates

* In some cases there will be penetration of yield stress on the convex side as well as on the concave side when the maximum load is reached.

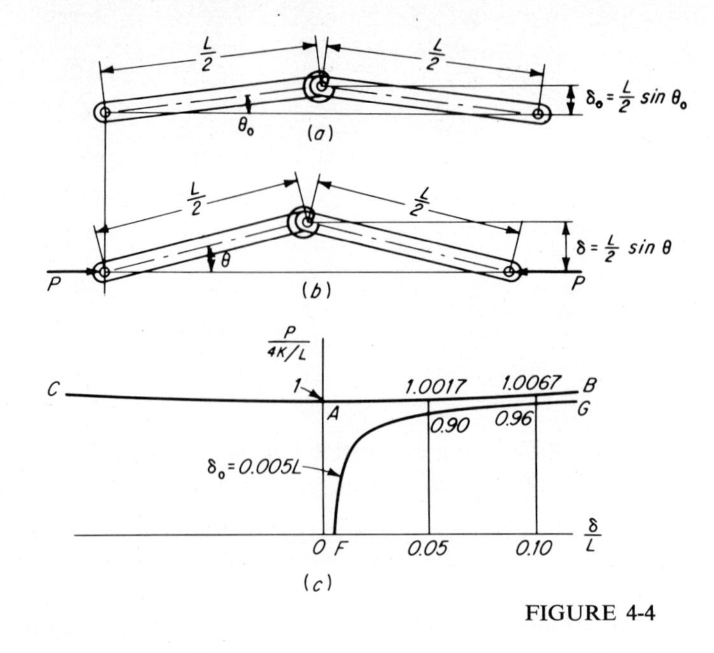

FIGURE 4-4

buckling. However, the plot of Eq. (4-4), $P_E B$ of Fig. 4-3*a*, is tangent to the horizontal at P_E, so that this increase in load for a small deflection at the beginning of buckling is a small quantity of the second order. This is why the deflection is indeterminate when we neglect second-order quantities to obtain Eqs. (*c*) and (*g*).

4-3 EFFECT OF INITIAL CROOKEDNESS

If the two-bar column of Fig. 4-1*a* is not initially straight, so that $\theta = \theta_0$ when $P = 0$ (Fig. 4-4*a*), equilibrium under the load P (Fig. 4-4*b*) requires that

$$P\delta = P\frac{L}{2}\sin\theta = 2K(\theta - \theta_0) \qquad (a)$$

from which

$$P = \frac{4K}{L}\frac{\theta - \theta_0}{\sin\theta} \qquad (b)$$

Curve *FG* in Fig. 4-4*c* shows the corresponding variation of P with δ for $\delta_0 = 0.005L$. Curve *CAB* from Fig. 4-1*c* for the initially straight column is also shown for com-

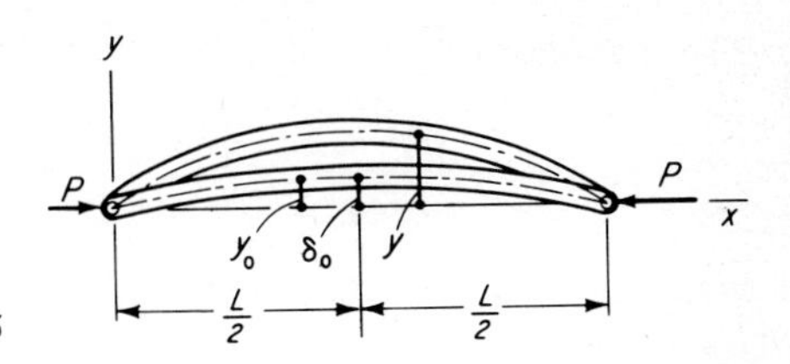

FIGURE 4-5

parison. Evidently, *FG* approaches *OAB* with diminishing δ_0. However, we see that the behavior represented by *OAB* cannot be realized in fact, since it requires the member to be absolutely straight in the unloaded condition.

Using the notation P_{cr} for the critical load $4K\theta/L \sin \theta$, from Eq. (*b*) of Art. 4-2, θ in Eq. (*b*) above is given by

$$\theta = \frac{\theta_0}{1 - P/P_{cr}} \tag{c}$$

This equation shows that θ increases more and more rapidly as P approaches the critical load.

Consider next the column in Fig. 4-5, which has the initial (unloaded) shape

$$y_0 = \delta_0 \sin \frac{\pi x}{L} \tag{d}$$

The equation of equilibrium for this column acted upon by the load P is

$$EI\left(\frac{d^2y}{dx^2} - \frac{d^2y_0}{dx^2}\right) = -Py \tag{e}$$

Substituting y_0 from Eq. (*d*) into Eq. (*e*) and using the abbreviation $k^2 = P/EI$ gives

$$\frac{d^2y}{dx^2} + k^2y = -\delta_0 \frac{\pi^2}{L^2} \sin \frac{\pi x}{L} \tag{f}$$

The solution of this equation is

$$y = A \sin kx + B \cos kx + \frac{\delta_0}{1 - k^2L^2/\pi^2} \sin \frac{\pi x}{L} \tag{g}$$

The boundary conditions $y = 0$ at $x = 0$ and $x = L$ give

$$y = \frac{\delta_0}{1 - k^2L^2/\pi^2} \sin \frac{\pi x}{L} = \frac{\delta_0}{1 - P/P_E} \sin \frac{\pi x}{L} \tag{h}$$

The maximum deflection is at $x = L/2$:

$$y_{max} = \frac{\delta_0}{1 - P/P_E} \tag{4-5}$$

Note that this equation is identical in form to Eq. (*c*) for the two-bar column of Fig. 4-4*a*.

The maximum stress in the column is

$$f = \frac{P}{A} + \frac{Mc}{I} = \frac{P}{A} + \frac{Py_{max}c}{I} = \frac{P}{A}\left(1 + \frac{\delta_0 c}{r^2}\frac{1}{1 - P/P_E}\right) \tag{4-6}$$

According to ASTM A6 (AISC 1.23.8.1) compression members must not deviate from straightness by more than $L/1{,}000$. Substituting this value into Eq. (4-6) gives

$$f = \frac{P}{A}\left(1 + \frac{Lc}{1{,}000r^2}\frac{1}{1 - P/P_E}\right) \tag{4-7}$$

This equation is valid only for stresses below the proportional limit. Assuming that the proportional limit equals the yield stress F_y and taking a rectangular cross section, for which $r = 2c/\sqrt{12}$, we get

$$F_y = f\left(1 + \frac{1.73}{1{,}000}\frac{L}{r}\frac{1}{1 - f/F_E}\right) \tag{4-8}$$

where $f = P/A$ and $F_E = P_E/A$. The graph of this equation for $F_y = 34$ ksi is shown in Fig. 4-6*a*. The variation of load with lateral deflection δ at midlength for a column with $L/r = 100$ is $\delta_0 CDE$ in Fig. 4-6*b*. Point C represents the load according to Eq. (4-8). The corresponding distribution of stress at midlength is plotted on the vertical line through C. This is not the ultimate strength. As in the postbuckling behavior of the straight column (Fig. 4-3), the ultimate load, point D, is attained at a stress distribution such as that plotted on the vertical line through D. Following this the load begins to decrease along the path DE.

The plot of F_u in Fig. 4-6*a* is based on numerical values from an approximate solution by Jezek for the eccentrically loaded column of rectangular cross section (Ref. 3, p. 40). Also shown is the Euler stress F_E from Eq. (4-2) and, in Fig. 4-6*b*, the load-deflection curve AB for a straight column with $L/r = 100$. The plot of Euler stress ends at $F_y = 34$ ksi at $L/r = 93$. This means that a perfect column made of steel with a yield plateau at 34 ksi will reach yield stress without buckling if L/r is less than 93. Exceptions to this type of behavior are discussed in Arts. 4-4 and 4-6.

Figure 4-6*c* shows the variation with L/r of the ratio of the strength F_u of the crooked column to that of the perfect column. It will be noted that the maximum effect of crookedness is in the range $80 < L/r < 120$, where reductions in strength of as much as 30 percent can be expected.

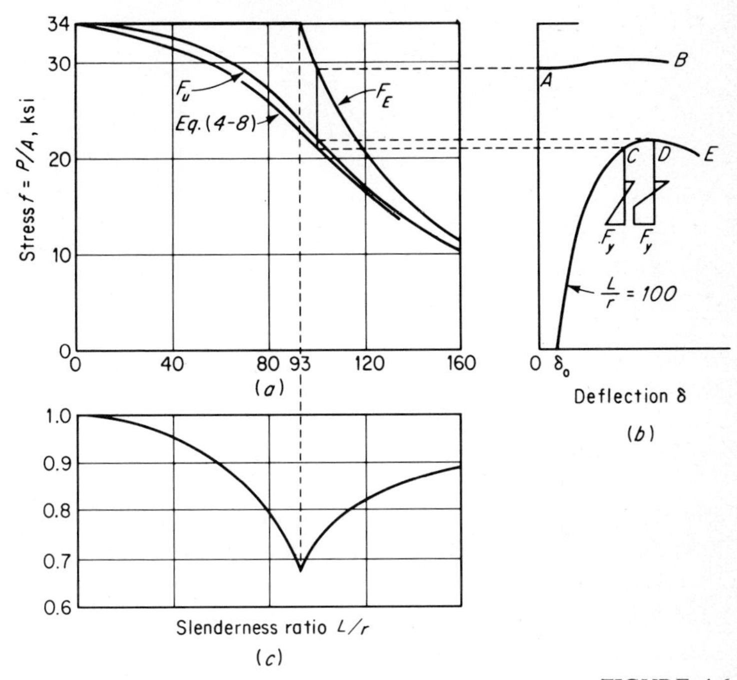

FIGURE 4-6

There is an important difference between the critical load for the perfect column (point A of Fig. 4-6b) and the ultimate load (point D) of the crooked column. When the straight column buckles, it assumes a *stable*, bent equilibrium configuration, but with a slightly larger load. On the other hand, the crooked column does not buckle in this sense at all. Instead, deflection increases from the beginning of loading, and the column is in an *unstable* condition when it reaches its maximum load. This is because deflection increases with decrease in load after we pass point D. Thus, we distinguish between two types of instability, that corresponding to a bifurcation of equilibrium (point A) and that corresponding to a peak load (point D). Of course, $\delta_0 CDE$ represents the behavior of the real (in contrast to perfect) column.

4-4 INELASTIC BUCKLING OF COLUMNS

The preceding discussions have been based on the assumption that the column is made of a metal whose stress-strain curve is linear until a yield plateau is reached. In this article we consider the behavior of columns made of metals which yield gradually

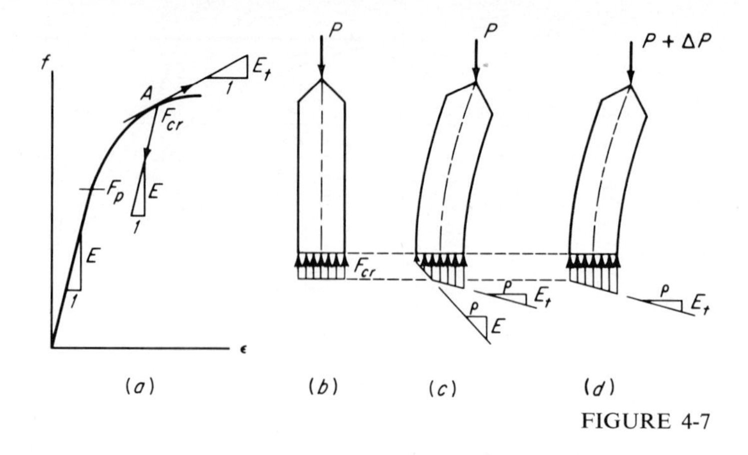

FIGURE 4-7

after a proportional limit F_p is reached (Fig. 4-7*a*). Assume that such a column is at its critical load at a stress F_{cr} exceeding the proportional limit (point A in the figure), so that the stress distribution at a typical cross section is that shown in Fig. 4-7*b*. Engesser suggested in 1889 that the critical load for this case could be found by replacing Young's modulus E in Eq. (4-2) with the tangent modulus E_t given by the slope of the tangent to the stress-strain curve at A. Thus,

$$F_{cr} = \frac{\pi^2 E_t}{(L/r)^2} \tag{4-9}$$

To determine a point on the tangent-modulus curve for inelastic buckling, a value of F_{cr} and the corresponding value of E_t are substituted into Eq. (4-9) and the value of L/r determined. The curve BD of Fig. 4-8 was obtained in this way.

Later, Engesser's conclusion was challenged on the basis that buckling begins with no increase in load. If this is true, stress on the convex side decreases at the onset of buckling while that on the concave side increases. But stress decreases at the rate E, that is, parallel to the initial loading path (Fig. 4-7a), while stress increases at the rate E_t. Therefore, the stress distribution for a slightly bent column would be that shown in Fig. 4-7*c*, in which the slopes of the stress block are E/ρ on the convex side and E_t/ρ on the concave side, where ρ is the radius of curvature at the section. The location of the center of rotation of the cross section is determined by $\int f\,dA = P$. Following this concept, Engesser presented a second solution to the inelastic-buckling problem in 1895, in which the bending stiffness of the cross section is expressed in terms of a double modulus E_r. For a simplified H cross section consisting of only the two flanges (Ref. 1, p. 178)

$$E_r = \frac{2EE_t}{E + E_t} \tag{4-10a}$$

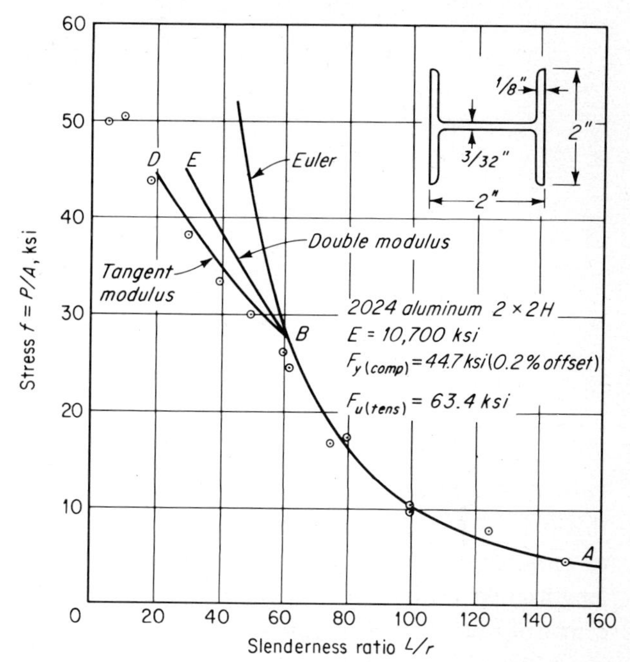

FIGURE 4-8
Column test results. (*From W. R. Osgood and M. Holt, The Column Strength of Two Extruded Aluminum-alloy H-sections, NACA Tech. Rep. 656, 1939.*)

For the rectangular cross section

$$E_r = \frac{4EE_t}{(\sqrt{E} + \sqrt{E_t})^2} \tag{4-10b}$$

The double modulus is also called the reduced modulus. The critical stress according to the double modulus is found by replacing E in Eq. (4-2) by E_r:

$$F_{cr} = \frac{\pi^2 E_r}{(L/r)^2} \tag{4-11}$$

To determine a point on the double-modulus curve for inelastic buckling a value of F_{cr} is assumed, which determines E_t, and E_r is computed. Substituting corresponding values of E_r and F_{cr} into Eq. (4-11) gives a corresponding value of L/r. The curve *BE* of Fig. 4-8 was obtained in this way.

Results of tests on 2 × 2 in. extruded aluminum-alloy H-shaped columns are shown in Fig. 4-8. It will be noted that the tests agree with the tangent-modulus prediction rather than with the double-modulus values. Although this is usually the case, the reason remained unknown for many years. Shanley is credited with having

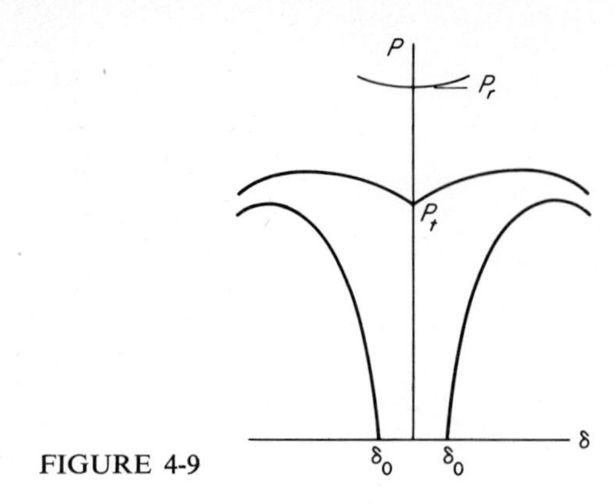

FIGURE 4-9

been the first to offer an explanation. He pointed out in 1947 that the tangent-modulus theory is correct if the load increases at the onset of buckling by an amount sufficient to offset strain reversal due to bending.[5] In this case, the stress distribution at mid-length could be that shown in Fig. 4-7*d*, for which $\int f\,dA = P + \Delta P$. The corresponding center of rotation of the cross section is on the convex edge, so that the strain there is stationary at the instant bending begins.

Rather than comparing the two theories by postulating on the one hand an increase in load at the onset of buckling (tangent modulus) and no increase on the other (double modulus), it is perhaps easier to understand the difference in terms of the order of magnitude of the increase in load. It was pointed out in Art. 4-2 that the increase in load for a small deflection at the onset of elastic buckling is a small quantity of second order. Stresses due to such an increase in load could not offset those due to a first-order moment $P\,\Delta\delta$. Therefore, if inelastic buckling also begins with a second-order increase in load, there must be strain reversal and the double modulus applies. On the other hand, stresses due to a first-order increase in load could be large enough to offset a first-order reversal. Thus, if inelastic buckling begins with a first-order increase in load ΔP, reversal need not take place and the tangent modulus could apply.

A first-order increase in load as bending begins requires that the postbuckling curve have a positive slope at $\delta = 0$, rather than zero slope as in Figs. 4-1 and 4-4. Therefore, the tangent-modulus postbuckling behavior is represented by curves such as those originating at P_t in Fig. 4-9. However, it has been shown that buckling of a perfect column can begin at any load between P_t and the double-modulus load P_r.[6,7] Thus, there is a family of postbuckling curves originating at points lying between P_t and P_r in Fig. 4-9 each of which is a possible behavior curve for the perfect column. In this sense, then, the tangent-modulus load is a lower bound and the double-modulus load an upper bound on the critical load for a perfect column. However, in order to

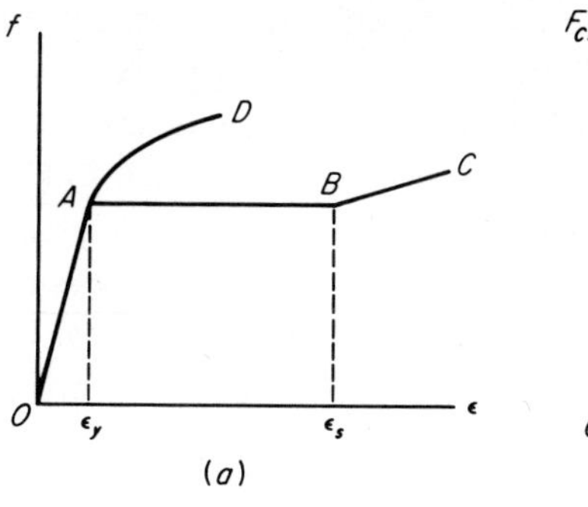

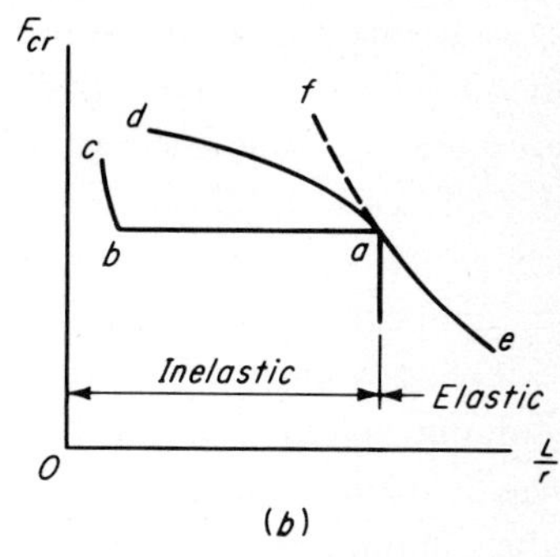

FIGURE 4-10

reach a load higher than P_t, the column must somehow survive the lowest potential instability, namely, P_t.

The load-deflection curve for the crooked column with initial midlength deflection δ_0 is also shown in Fig. 4-9. Conclusive proof that the tangent-modulus load is indeed the critical load for inelastic buckling of a real, perfect column (in the sense that the real column can only approach perfection) was given by Duberg and Wilder.[6] Using a webless cross section, as in Fig. 4-13*b*, to simulate the I, they showed that the tangent-modulus load is the critical load for the perfectly straight column if the straight column is regarded as the limiting case of the crooked column as the initial crookedness vanishes. Furthermore, the peak of the postbuckling curve is only slightly higher than the tangent-modulus load. Therefore, the tangent-modulus load is a good measure of column strength in the inelastic-buckling range and can be considered to be an *upper bound* on the critical load of a *real* column.

Although there is no strain reversal at midlength of the convex side of the column at the beginning of bending at the tangent-modulus load, strain reversal does occur, and begins (on the convex side at midlength) immediately after the column starts to bend.[6,7] Reversal then spreads toward each end of the column with increasing load, until the entire convex side is encompassed when the ultimate load (peak of the postbuckling curve) is reached.

Equation (4-9) shows that the shape of the column F_{cr}-L/r curve in the inelastic range depends on the shape of the f-ε curve since E_t is determined by the latter. Consider the two steels represented by the f-ε curves of Fig. 4-10*a*. *OABC* typifies carbon steels and the high-strength low-alloy steels, which have the yield plateau *AB*, while *OAD* typifies heat-treated high-strength carbon steels, heat-treated constructional alloy steels, and many of the sheet and strip steels used in cold-formed members. *OAD* is also typical of the aluminum alloys, except that for them the slope of *OA* is flatter because of the smaller value of *E*. For the purposes of this discussion, it is assumed that the proportional limit for the curve OAD is the same as the yield point for $OABC$. The buckling behavior of straight, centrally loaded columns made of these two steels is shown in Fig. 4-10*b*. Because the modulus of elasticity is the same for all steels, the elastic-buckling curve *eaf* is the same for all. In the nonproportional range AD of the gradually yielding steel, the slope E_t decreases gradually, so that the column curve pulls away from the Euler curve, beginning at a. This gives the inelastic buckling branch *ad*, corresponding to AD in Fig. 4-10*a*. On the other hand, the column curve for the steel typified by $OABC$ breaks sharply at a, because the column yields before buckling. This gives the horizontal line *ab* in Fig. 4-10*b*. However, if the column is short enough to withstand a strain larger than ε_s without buckling, it can strain harden. This results in the branch *bc* of the inelastic-buckling curve, which corresponds to BC in Fig. 4-10*a*.

4-5 COLUMNS WITH ENDS ROTATIONALLY RESTRAINED

The pin-ended compression member is rare. Rotation of the ends of the columns in building frames is usually limited by the beams connecting to them, while compression members in trusses may have restricted end rotation because of other members connecting at the joints. Of course, such restraint is never complete in the sense that the column ends are prevented from turning. However, some understanding of the effects of complete restraint is helpful in judging the effects of partial restraint.

Figure 4-11*a* shows a column supporting an axial load P with both ends rotationally fixed. If P is the critical load, the column buckles as shown, with the reactive moments M_0 preventing rotation of the ends. The equation of equilibrium is

$$-EI\frac{d^2y}{dx^2} = Py - M_0 \tag{a}$$

which gives the solution

$$y = A \sin kx + B \cos kx + \frac{M_0}{P} \tag{b}$$

where $k^2 = P/EI$. With the boundary conditions $y = 0$ at $x = 0$ and $x = L$ we get

$$y = \frac{M_0}{P}\left(1 - \cos kx - \frac{1 - \cos kL}{\sin kL}\sin kx\right) \tag{c}$$

Since the ends are prevented from turning, we have the additional boundary conditions $dy/dx = 0$ at $x = 0$ and $x = L$. Using the first of these, we get

$$\frac{kM_0}{P}\,\frac{1 - \cos kL}{\sin kL} = 0 \tag{d}$$

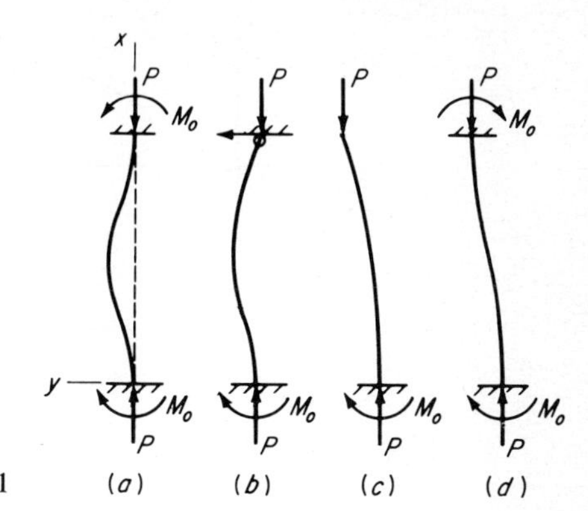

FIGURE 4-11

The smallest nonzero root of Eq. (*d*) is $kL = 2\pi$. Thus, since $k^2 = P/EI$, we have

$$P = \frac{4\pi^2 EI}{L^2} = \frac{\pi^2 EI}{(0.5L)^2} \tag{e}$$

Solutions for other cases are obtained similarly. If only one end is fixed against rotation (Fig. 4-11*b*), we find

$$P = \frac{20.2EI}{L^2} = \frac{2.05\pi^2 EI}{L^2} = \frac{\pi^2 EI}{(0.7L)^2} \tag{f}$$

If one end is fixed and the other completely free (Fig. 4-11*c*),

$$P = \frac{\pi^2 EI}{4L^2} = \frac{\pi^2 EI}{(2L)^2} \tag{g}$$

Finally, if both ends are rotationally fixed but one end is free to translate relative to the other (Fig. 4-11*d*),

$$P = \frac{\pi^2 EI}{L^2} \tag{h}$$

which is the same as the critical load for the pin-ended column. It will be noted that the critical loads given by these equations can be expressed in the general form

$$P = \frac{\pi^2 EI}{(KL)^2} \tag{4-12}$$

KL in this equation is called the *effective length* of the column. The effective length of columns with limited restraint of end rotation is discussed in Art. 4-18.

It should be noted that the reactive moment M_0 in Eq. (*d*) is indeterminate, so that the deflections given by Eq. (*c*) are also indeterminate. Thus, as was the case for

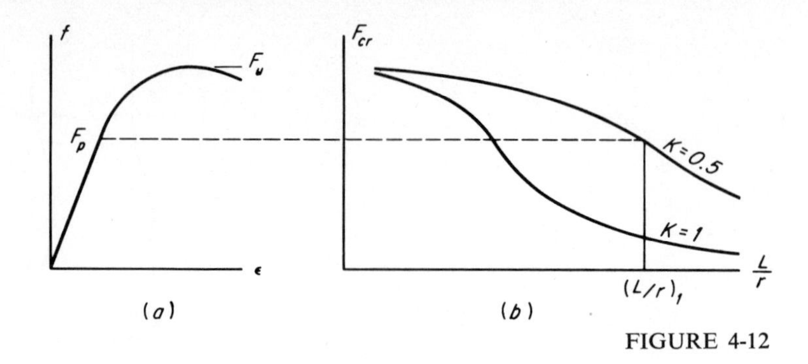

FIGURE 4-12

the pin-ended column, we have determined the load at which a perfect column becomes unstable, but the solution tells us nothing about the postbuckling behavior.

According to the tangent-modulus theory, E_t replaces E in Eq. (4-12) if the stress P/A exceeds the proportional limit. Therefore, with the critical load in terms of the stress F_{cr} as in Eq. (4-9), we get

$$F_{cr} = \frac{\pi^2 E_t}{(KL/r)^2} \tag{4-13}$$

In general, the effect of end restraint on the critical stress is less for inelastic buckling than it is for elastic buckling. This is because the very short column can develop yield stress (and more) whether the ends are pinned or fixed. Thus, column F_{cr}-L/r curves for various values of K converge toward F_u at $L/r = 0$. This is shown in Fig. 4-12*b* for a pin-ended column and a column with fixed ends, each made of a metal with the f-ε variation shown in Fig. 4-12*a*. $(L/r)_1$ in Fig. 4-12*b* is the boundary between elastic and inelastic buckling for the case $K = 0.5$. If $L/r \geqq (L/r)_1$, a column with fixed ends can support four times as much load as a column with pinned ends. However, this benefit is seen to decrease with decreasing L/r, until F_{cr} finally becomes virtually independent of K.

4-6 EFFECT OF RESIDUAL STRESSES

Residual stresses that are usually present in the steel column may reduce its strength significantly. Sources of residual stress and the ways in which it may be distributed were discussed in Art. 3-1.

The effect of residual stress on column strength can be demonstrated by considering the webless H (Fig. 4-13*b*) that was investigated as a tension member in Art. 3-1. The stress-strain curve for the steel coupon is shown in *a* and the residual-

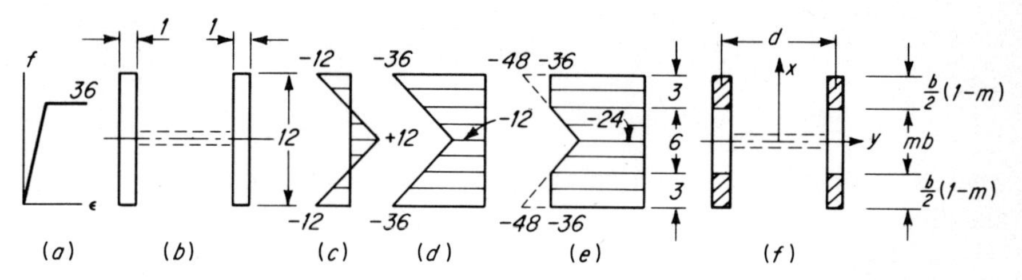

FIGURE 4-13

stress distribution in *c*. Residual stresses also vary across the thickness (Art. 4-7), so that the distribution shown is an average stress. The compressive residual stress, 12 ksi, is the average value found in 21 measurements of residual stresses in wide-flange shapes.[8] The area of the cross section is 24 in.^{2} If the column is loaded with $P = 576$ kips, the stress is $f = P/A = 576/24 = 24$ ksi. This is superimposed on the residual stresses, with the result shown in Fig. 4-13*d*. Should the column be of such a length as to become unstable at this load, initiation of buckling induces bending stresses which must be superimposed on those in *d*. Since it is stressed to yield only at the very edges of the flanges, the entire cross section behaves elastically at the onset of buckling and contributes to the bending resistance. Therefore, the bending stiffness is EI, where I is the moment of inertia of the cross section for the axis about which buckling occurs.

Assume next that the column is short enough to remain stable until the load P is large enough to produce a stress of 24 ksi at the midpoint of each flange. The stress distribution is shown in Fig. 4-13*e*, where the flanges are yielded a distance of 3 in. from each tip. The load is

$$P = 2(36 \times 6 \times 1 + 30 \times 6 \times 1) = 792 \text{ kips}$$

and the *average* stress on the cross section is $f = 792/24 = 33$ ksi. Since it is assumed that the column becomes unstable at this load, bending begins. However, the compressive stress cannot increase in the yielded portions of the flanges on the concave side, and if there is no strain-reversal as bending begins (tangent-modulus theory), there can be no reversal of stress in the yielded portions on the convex side. Thus, only the unyielded portions can develop a resisting moment, and the bending stiffness of the cross section is reduced. Using I_{eff} to denote the *effective* moment of inertia (the moment of inertia of the unyielded portion of the cross section) the bending stiffness is EI_{eff}. Substituting this into Eq. (4-1) gives

$$P = \frac{\pi^2 EI_{\text{eff}}}{L^2} = \frac{\pi^2 EI}{L^2} \frac{I_{\text{eff}}}{I}$$

from which

$$F_{cr} = \frac{\pi^2 E}{(L/r)^2} \frac{I_{\text{eff}}}{I} \tag{4-14}$$

where r is the radius of gyration of the entire cross section for the axis about which buckling occurs.

If the column of Fig. 4-13*f* buckles about its weak (y) axis, the effective moment of inertia is

$$I_{y,\text{ eff}} = \frac{2t(mb)^3}{12} = \frac{tm^3b^3}{6} \tag{a}$$

so that

$$\frac{I_{y,\text{ eff}}}{I_y} = \frac{tm^3b^3/6}{2tb^3/12} = m^3 \tag{b}$$

On the other hand, if the column is supported against weak-axis buckling and buckles about the strong (x) axis,

$$I_{x,\text{ eff}} = 2mbt\left(\frac{d}{2}\right)^2 = \frac{tmbd^2}{2} \tag{c}$$

$$\frac{I_{x,\text{ eff}}}{I_x} = \frac{tmbd^2/2}{2btd^2/4} = m \tag{d}$$

For the stress distribution of Fig. 4-13*e*, $P = 792$ kips and $f = 33$ ksi, as determined previously. Also, $m = 0.5$, which determines I_{eff}/I. Substituting the values of I_{eff}/I and f into Eq. (4-14) gives the corresponding value of L/r. Thus:

$$33 = \frac{30{,}000\pi^2}{(L/r_x)^2} \times 0.5 \qquad \frac{L}{r_x} = 67$$

$$33 = \frac{30{,}000\pi^2}{(L/r_y)^2} \times 0.5^3 \qquad \frac{L}{r_y} = 34$$

The two curves *AB* in Fig. 4-14 were obtained by plotting results obtained in this way. Since the strength of the straight, centrally loaded column with no residual stress would be given by *ACBD*, we see that residual stress may reduce the strength considerably. Furthermore, we note that the strength of the H column also depends on the direction of buckling. Finally, we observe that the juncture of the inelastic buckling curves *AB* with the elastic (Euler) buckling curve *CBD* is determined by the magnitude of the compressive residual stress.

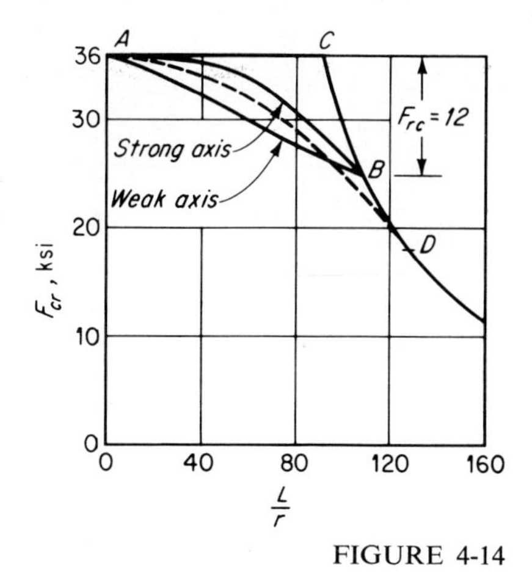

FIGURE 4-14

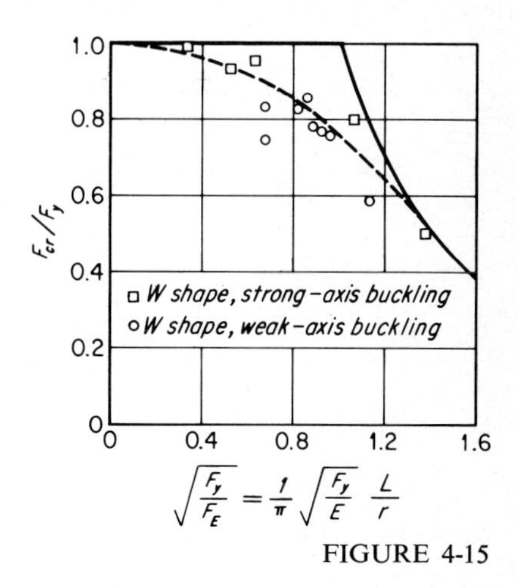

FIGURE 4-15

Results of tests[8] on W shapes are shown in Fig. 4-15. These are tests on relatively thin cross sections (less than 1 in.). The coordinates are nondimensionalized in terms of F_y. It will be noted that abscissas are directly proportional to L/r. These test results confirm the effect of residual stresses in causing weak-axis buckling at loads smaller than those for strong-axis buckling. Although this suggests two column formulas for the steel W, the Column Research Council (CRC) proposed in 1960 that design procedure be simplified by using a parabola beginning with a vertex at $F_{cr} = F_y$ where $L/r = 0$ and terminating at the point $F_{cr} = F_y/2$ where it intersects and is tangent to the Euler hyperbola. The equation of this parabola is

$$F_{cr} = F_y\left[1 - \frac{F_y}{4\pi^2 E}\left(\frac{KL}{r}\right)^2\right] \tag{4-15}$$

where K is the effective-length coefficient discussed in Art. 4-5. The equation can also be put in the form

$$F_{cr} = F_y\left[1 - \frac{1}{2}\left(\frac{KL/r}{C_c}\right)^2\right] \tag{4-16}$$

where $C_c = \pi\sqrt{2E/F_y}$. The dashed curves in Figs. 4-14 and 4-15 are plots of the CRC formula with $K = 1$.

The inelastic critical stress for a column with residual stress can be determined by using the tangent modulus E_t obtained from the stress-strain curve of a *stub-column* test, which is a compression test of a column short enough not to buckle but long enough to contain the residual-stress distribution. This test has been standardized.[9] (The tensile stress-strain curve for a member with residual stress was discussed in

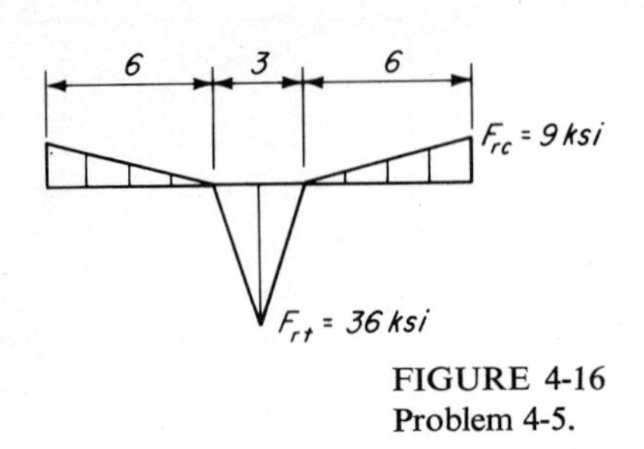

FIGURE 4-16
Problem 4-5.

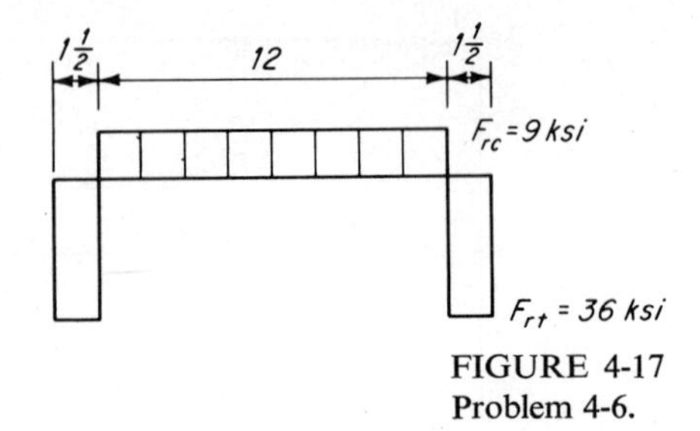

FIGURE 4-17
Problem 4-6.

Art. 3-1.) The manner in which the tangent modulus defines the critical stress must be determined from an analysis such as the one used to obtain Fig. 4-14. Using the notation $\tau = E_t/E$, the critical stresses for the strong (x) axis and the weak (y) axis of an H-shape are given approximately by[10]

$$F_{cr,x} = \frac{\pi^2 E\tau}{(L/r_x)^2} \qquad F_{cr,y} = \frac{\pi^2 E\tau^3}{(L/r_y)^2} \tag{e}$$

Similarly, for solid round bars with polar symmetric residual stress[11]

$$F_{cr} = \frac{\pi^2 E\tau^2}{(L/r)^2} \tag{f}$$

Residual-stress effects on the strength of columns of various cross sections are discussed and compared in Art. 4-7.

PROBLEMS

4-1 Compare the maximum deflection of a pin-ended column whose unloaded shape is parabolic with that given by Eq. (4-5) for the crooked column of Art. 4-3.

4-2 Derive the formulas for the critical loads of the columns shown in Fig. 4-11*b*, *c*, *d*.

Note: To simplify the numerical work in the following problems, assume the area of each cross-sectional element to be concentrated at mid-thickness.

4-3 Plot the column curve for the cross section of Prob. 3-1 for both strong-axis and weak-axis buckling.

4-4 Plot the column curve for the cross section of Prob. 3-2.

4-5 The cross section is the same as that of Prob. 3-2 except that it consists of four 8 × 8 × 1 angles. The residual-stress distribution is shown in Fig. 4-16. Plot the column curve.

4-6 Plot the column curve for the cross section of Prob. 3-2 for the residual-stress distribution shown in Fig. 4-17.

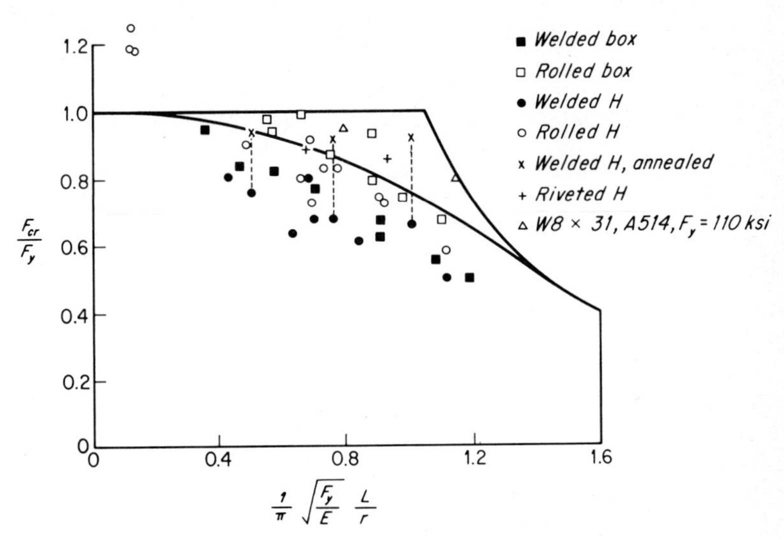

FIGURE 4-18
Column test results. (*From L. Tall, J. Inst. Eng. Aust., vol. 36, no. 12, December 1964.*)

4-7 DETERMINANTS OF COLUMN STRENGTH

Because of the variety of imperfections that influence the behavior of a column, it is not easy to predict its strength. Results of tests on a number of steel columns are shown in Fig. 4-18. It is important to note that test results for perfect columns, i.e., those which are absolutely straight, free of residual stress, supported on frictionless hinges, loaded precisely on the centroidal axis, etc., would fall on the Euler curve to its intersection with the line $F_{cr} = F_y$ and thence on the latter line to the axis of ordinates, provided the steel has a yield plateau. Thus, the deviations in this figure are a result of imperfections. However, it is not always easy to identify the imperfection, and in most cases a combination of imperfections is responsible.

The CRC column curve, Eq. (4-16), is plotted in Fig. 4-18. Except for the welded columns, it appears to be a reasonable, average curve for these tests. Although the test results appear to be more or less randomly scattered, certain patterns emerge when the parameters are studied. The comparison of three welded I's with three identical specimens stress-relieved by annealing (identified by the dashed lines connecting the corresponding points) is convincing evidence of the effect of residual stresses. It is seen that welded columns as a group tend to be less strong than rolled shapes. However, some types of welded column are stronger than others. Thus, the H fabricated

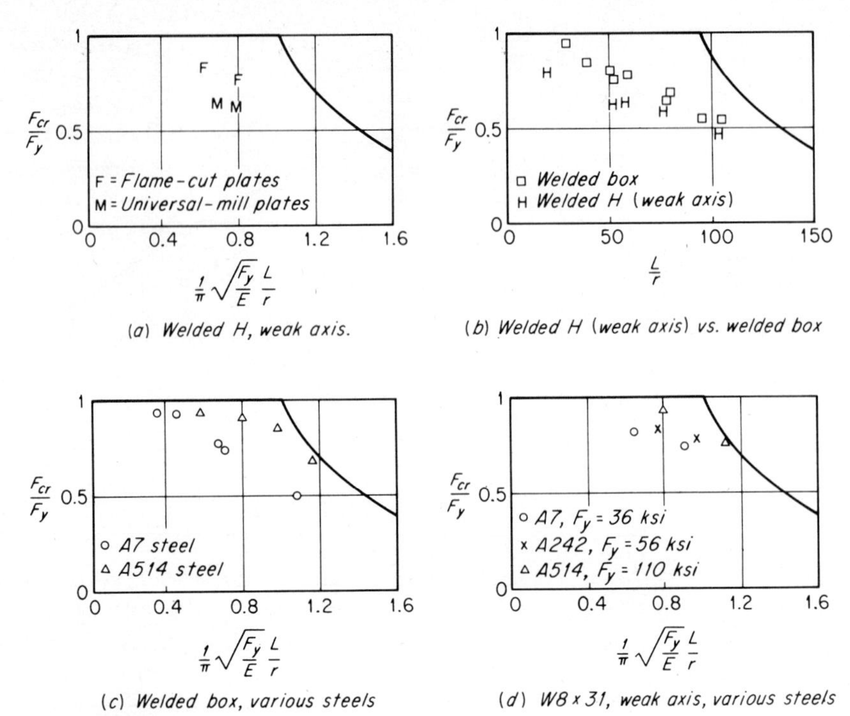

FIGURE 4-19
Effect of residual stress on column strength. (*From L. Tall, Welded Built-up Columns, Lehigh Univ. Fritz Eng. Lab. Rep. 249.29, 1966.*)

from flame-cut plates has a more favorable distribution of residual stress (tension at the flange tips) than the H fabricated from universal-mill plates (Fig. 3-1). As a result, the weak-axis strength of the former may be considerably more than that of the latter. This is shown in Fig. 4-19*a*. However, the strong-axis buckling strengths are more nearly equal. Welded box columns tend to be stronger than the welded H with universal-mill plates (Fig. 4-19*b*). Of course, this is because the former has a more favorable residual-stress pattern (Fig. 3-1).

Residual stresses in rolled shapes tend to be independent of the yield stress. Thus, reduction in strength tends to be smaller for rolled shapes of higher-strength steels. Furthermore, the heat treatment used in the production of A514 steel reduces thermal residual stresses even below the level that would normally exist. These effects are shown in Fig. 4-19*d*. The A514 welded box is also relatively stronger than the A7 box (Fig. 4-19*c*).

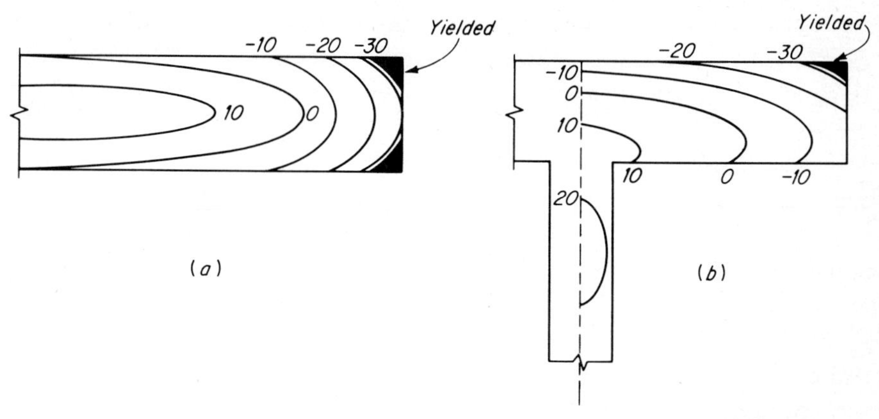

FIGURE 4-20

All the columns for which test results are shown in Figs. 4-18 and 4-19 were composed of relatively thin elements. Residual stress variation across the thickness of such elements is small. This is not the case for thick parts. Figure 4-20 shows the computed residual-stress distribution in one-half of a $24 \times 3\frac{1}{2}$ in. plate and in one-quarter of a W14 × 426.[14] The inner surface of the flange of the W shows smaller residual stresses than the outer surface since it is less favorably situated, because of the web, for cooling. It will be noted that the tips of both the flange of the W and the plate have yielded. The yield stress was assumed to be 31 ksi.

Residual stresses tend to increase in magnitude with increase in size of the element. Thus, the *average* residual compression at the ends of $6 \times \frac{1}{2}$, 20×1, and 12×2 plates and the $20 \times 3\frac{1}{2}$ plate of Fig. 4-20*a* was 10, 13, 21, and 31 ksi, respectively. The average tension at the center was 3, 2, 8, and 9 ksi, respectively. These residual stresses are likely to constitute the major part of the residual stress in the welded, built-up member composed of thick plates. This is because of the proportionately smaller heat input from welding of such members, for which the welding itself may contribute only 5 to 10 ksi of residual stress.[13] Thus, the weakening effect of residual stress can be expected to be larger for welded columns with thick plates. Of course, this is also true for the W with thick flanges, On the other hand, thick-plated columns are likely to have small slenderness ratios, for which the weakening effect of residual stresses is relatively small. However, this favorable condition may be partially offset by the fact that straightness becomes more important with larger residual compression. This is because there is earlier yielding in the region of high residual compression, which causes earlier deterioration of bending stiffness.

The fact that columns in the very short L/r range can be stressed well into the strain-hardening region is shown by the three tests in Fig. 4-18 which average about 20 percent higher than yield stress. This corresponds to the branch *bc* of the column curve *eabc* in Fig. 4-10*b*.

The preceding discussion of column strength as it is affected by differences in cross-sectional shape and in fabrication suggests that a single column-strength curve is inadequate. Of course, questions arise as to practicable groupings and classifications, and compromises are inevitable. Three column curves have been proposed by a task committee of the Column Research Council. A committee of the European Convention of Constructional Steelwork has also recommended three formulas. However, the two committees differ in respect to their classifications of column types.

Strengths of cold-formed compression members also exhibit considerable scatter in the inelastic buckling range. This is partly because many of the sheet and strip steels used in cold-formed steel construction are gradually yielding steels, with stress-strain curves such as *OAD* in Fig. 4-10*a*. On the other hand, the cold-forming operation produces residual stresses which lower the proportional limit, so that even a member cold-formed from steel with a flat-top yield may fall considerably short of the perfect-column curve.

Column strengths of aluminum members usually correspond closely to those predicted by the tangent-modulus formula (Fig. 4-8). Furthermore, the plot of the tangent-modulus formula is very nearly a straight line.[15]

4-8 ALLOWABLE STRESSES FOR STEEL COLUMNS

It is difficult to determine a suitable factor of safety for the design of compression members, although it is clear that (among other things) it must allow for the imperfections that have been discussed in the preceding articles. Columns in the intermediate range of slenderness (say, $30 < L/r < 120$) are sensitive to residual stress (Fig. 4-18), crookedness (Fig. 4-6), and eccentricity of load (Art. 4-12), while columns with small slenderness ratio can be stressed beyond the yield value (Fig. 4-18). This suggests that the factor of safety for columns with small slenderness ratio should be the same as the factor of safety for tension members but that it should increase in some fashion with L/r. However, it should not decrease with the diminishing effect of residual stress, crookedness, and eccentricity in the range of large L/r, because columns in this range of slenderness are sensitive to errors in estimating their effective lengths, as shown in Art. 4-5.

Some authorities recommend a constant factor of safety, pointing out that the strength of columns in the range of large slenderness is determined by Young's modulus, which varies very little, while the strength of those in the short and intermediate ranges is determined by yield stress, which varies considerably. However, the variation in yield stress is already taken into account for steels bought to ASTM specifications, since the yield stress is guaranteed, in the sense that a heat, or lot, of steel may be rejected if the specified yield and tensile strength values are not met.* Therefore, manufacture is so controlled as to give a high probability that the specified properties will be met, so that yield strength, ultimate strength, etc., almost always exceed the specified minimums, sometimes by a considerable margin. Therefore, the arguments would seem to favor the varying factor of safety.

Although the considerable variation in column strength in the inelastic range, as it relates to the shape of the cross section, thickness of the elements, and method of manufacture, suggests that a variety of column formulas is needed in this range, it is the practice in standard specifications to use a single formula. Allowable stress for elastic buckling is usually based on the Euler formula, although the secant formula [Eq. (4-29)] is used in the AASHO specifications.

Allowable-stress formulas from several standard specifications for design will now be discussed. The formulas are not necessarily in the same form as they appear in the specifications but are written so as to facilitate comparison.

The AISC formulas for allowable stress F_a on axially loaded compression members are

$$F_a = \frac{F_y\left[1 - \dfrac{1}{2}\left(\dfrac{KL/r}{C_c}\right)^2\right]}{\dfrac{5}{3} + \dfrac{3}{8}\dfrac{KL/r}{C_c} - \dfrac{1}{8}\left(\dfrac{KL/r}{C_c}\right)^3} \qquad \frac{KL}{r} \lesssim C_c \tag{4-17}$$

$$F_a = \frac{149{,}000}{(KL/r)^2} \qquad \frac{KL}{r} \gtrsim C_c \tag{4-18}$$

where K is the effective length coefficient (Art. 4-5) and

$$C_c = \pi\sqrt{\frac{2E}{F_y}}$$

Comparing Eq. (4-17) with Eq. (4-16), we see that the AISC allowable stress is obtained by dividing the CRC column strength by a factor of safety which depends on KL/r. This factor of safety varies from 1.67 at $KL/r = 0$ to 1.92 at $KL/r = C_c$. Equation (4-18) is the Euler stress [Eq. (4-2)] with $E = 29{,}000$ ksi, divided by a factor of safety of 1.92.

* The specification does allow a retest of one random sample if the results of the original test are within 2,000 psi of the specified tensile strength, 1,000 psi of the required yield point, or within 2 percentage units of the required elongation.

The AASHO formulas are

$$F_a = \frac{0.55F_y}{1.25}\left[1 - \frac{1}{2}\left(\frac{0.75L/r}{C_c}\right)^2\right] \tag{4-19}$$

for members with riveted ends and

$$F_a = \frac{0.55F_y}{1.25}\left[1 - \frac{1}{2}\left(\frac{0.875L/r}{C_c}\right)^2\right] \tag{4-20}$$

for members with pinned ends. These formulas are also derived from the CRC formula, using a factor of safety of $1.25/0.55 = 2.27$ for all values of L/r and specifying effective-length coefficients of 0.75 and 0.875 for members with riveted ends and pinned ends, respectively. Although the upper limit of L/r for which the formulas may be used is approximately C_c, its numerical value is specified instead for each of the various yield stresses covered. Thus, the limit is 130 for $F_y = 36$ ksi, 125 for $F_y = 42$ ksi, etc. The formulas reduce to simple form for specific yield stresses. Thus, with some roundoff, for $F_y = 36{,}000$ psi,

$$F_a = 16{,}000 - 0.30\left(\frac{L}{r}\right)^2 \qquad \frac{L}{r} \not> 130 \tag{4-21}$$

for members with riveted ends and

$$F_a = 16{,}000 - 0.38\left(\frac{L}{r}\right)^2 \qquad \frac{L}{r} \not> 130 \tag{4-22}$$

for members with pinned ends.

The AREA formulas are

$$F_a = 0.55F_y \qquad 0 \le \frac{KL}{r} \le \frac{3{,}388}{\sqrt{F_y}} \tag{4-23}$$

$$F_a = 0.60F_y - \left(\frac{F_y}{1{,}662}\right)^{3/2}\frac{KL}{r} \qquad \frac{3{,}388}{\sqrt{F_y}} \not< \frac{KL}{r} \not< \frac{27{,}111}{\sqrt{F_y}} \tag{4-24}$$

$$F_a = \frac{147{,}000{,}000}{(KL/r)^2} \qquad \frac{KL}{r} > \frac{27{,}111}{\sqrt{F_y}} \tag{4-25}$$

F_a and F_y in these formulas are in pounds per square inch. It will be noted that two straight-line formulas are used for inelastic buckling, while elastic buckling is based on the Euler stress. The factor of safety is $1/0.55 = 1.82$ for $0 \le KL/r \le 3{,}388/\sqrt{F_y}$, after which it begins to increase until it attains the value 1.95 for elastic buckling, based on $E = 29{,}000{,}000$ psi. The specifications prescribe "under usual conditions" the same values of K as the AASHO specifications, namely, 0.75 for

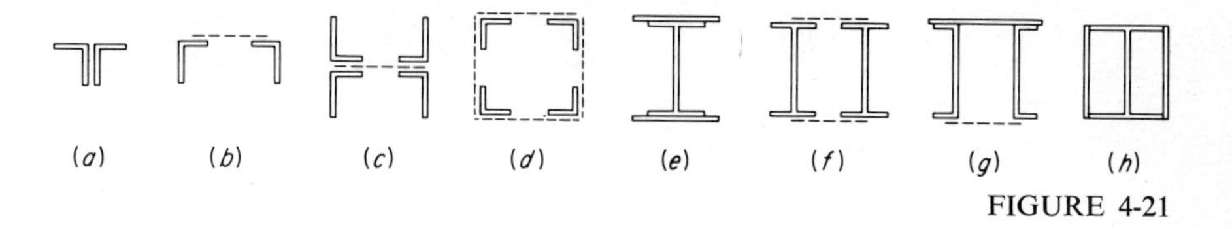

FIGURE 4-21

riveted, bolted, or welded ends and 0.875 for pinned ends. Equations (4-23) and (4-24) reduce to simple form for specific yield stresses. Thus, with some roundoff, for $F_y = 36{,}000$ psi,

$$F_a = 20{,}000 \qquad 0 \leq \frac{KL}{r} \leq 15 \tag{4-26}$$

$$F_a = 21{,}500 - 100\frac{KL}{r} \qquad 15 \leq \frac{KL}{r} \leq 143 \tag{4-27}$$

The American Iron and Steel Institute (AISI) Specification for the Design of Cold-formed Steel Structural Members prescribes compression-member formulas which are similar to Eqs. (4-17) and (4-18) of the AISC specification. However, because such shapes are usually much thinner than hot-rolled shapes, design formulas must take the possibility of local buckling into account. How this is done is discussed in Chap. 9.

Design formulas for aluminum columns are discussed in Art. 4-30.

4-9 TYPICAL SECTIONS FOR COMPRESSION MEMBERS

A single shape, such as the angle, the W, the tee, the tube, or the pipe, may be used as a compression member. Single angles are used extensively in towers, particularly in transmission towers. They are sometimes used in small roof trusses. The W's are used extensively as columns in buildings and as compression members in large trusses. The tee is a suitable compression-chord member for welded roof trusses; web members may be welded directly to the stem of the tee.

Figure 4-21 shows some of the forms of built-up cross sections. The cross section depends to a large extent upon the structure in which the member is to be used, the manner in which the ends are to be connected, and the requirements of other members which connect to it. In this figure, shapes which extend the full length of the member are shown in solid lines. Intermittent connections, such as lacing, tie plates, etc. (Art. 4-24), whose function is to force the several shapes to act as one, are indicated by dashed lines.

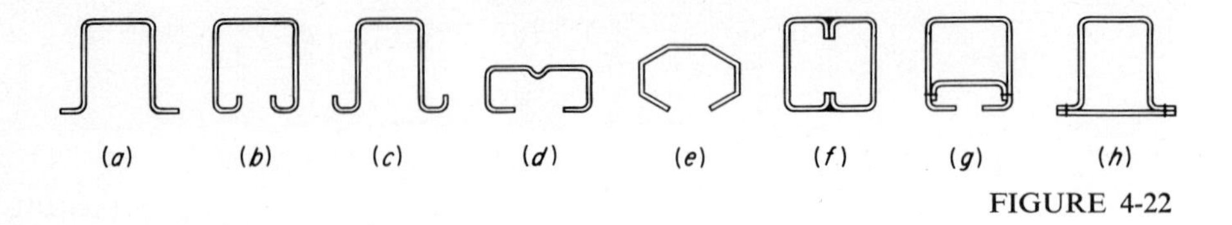

FIGURE 4-22

The double-angle member in Fig. 4-21*a* is widely used in roof trusses. It may be used for both web and chord members of riveted or bolted trusses, with connections made to gusset plates between the vertical legs. It is suitable for web members of welded trusses in conjunction with the tee chord mentioned previously. In some structures the outstanding legs of the double-angle member may be riveted, bolted, or welded to the connecting member. Two angles are sometimes used, as in Fig. 4-21*b*, with the connection made to two gusset plates, one at each outstanding leg. Again, the gusset plates may usually be omitted in welded construction. Four angles arranged as in Fig. 4-21*c* are used occasionally. Connection is usually through the outstanding legs. Four angles arranged as in Fig. 4-21*d* are common for crane booms. Columns for industrial buildings are sometimes made as in Fig. 4-21*f* with one I supporting a crane runway and the other extending past the runway to support the roof.

Occasionally the largest W will have insufficient area, and plates (called *cover* or *flange plates*) may be attached to the flanges as in Fig. 4-21*e*. However, three plates welded to form an I are likely to be a better solution.

Members of the type shown in Fig. 4-21*g* were common in bridge trusses as top-chord members but have been superseded by the W, the welded I, and the welded box. The box section is also used for columns in buildings. It is available as a rolled shape, called structural tubing, in square form up to 12 × 12 in. in size, and in rectangular form to 12 × 6 in. In the form of four plates welded at the corners its size is virtually unlimited. Box sections are sometimes made as in Fig. 4-21*h* by welding plates to the W shape.

Because of the relative ease with which a great variety of cold-formed shapes can be produced, structural sections of this type have not been standardized. However, a number of fabricators have developed individual products. Typical cross sections suitable for compression members are shown in Fig. 4-22. The depth of these shapes ranges from about 3 to 12 in. and the thickness from about 8 to 18 gage (0.164 to 0.048 in.). Sections *a*, *b*, and *c* are suitable for general use. They, and *d* and *e*, are often used as chord members in trusses. Closed sections such as *f*, *g*, and *h* are particularly suitable for columns.

Approximate radii of gyration of a variety of cross sections are given in the Appendix.

4-10 DESIGN OF COLUMNS

The design of a column is necessarily a trial-and-error procedure, since the allowable stress depends upon the distribution of the cross-sectional area while the required area depends in turn upon the allowable stress. One may assume a value of the allowable stress, proportion a section with the corresponding required area, determine the slenderness ratio and the corresponding allowable stress, and, finally, revise the section if the calculated allowable stress differs from the assumed value.

EXAMPLE 4-1 A building column of A36 steel, 15 ft long, supports a load of 250 kips and is to comply with the AISC specification. Assume that the effective length coefficient $K = 1$. With an assumed allowable stress of 15 ksi, the required area is 16.7 in.^{2} A W from either the 8 × 8 or the 10 × 10 series will have sufficient area. Because of its larger radius of gyration, the 10 × 10 will require less area. Noting that the least radius of gyration for the 10 × 10 series is 2.54 in., we find $KL/r = 15 \times 12/2.54 = 71$. From the AISC Manual table of allowable stresses according to Eq. (4-17), $F_a = 16.3$ ksi, so that $A = P/F_a = 250/16.3 = 15.3$ in.^{2} Therefore, the W10 × 54, which has an area of 15.88 in.^{2}, is adequate.

The proportioning of built-up sections is somewhat more difficult, since there are usually many possible combinations of the components. However, with the help of Table A-1 in the Appendix, one can usually contrive a trial section which will need only minor revision.

Tables of allowable loads for commonly used members, such as the W, the double angle, etc., are available in sources such as the AISC Manual.

PROBLEMS

4-7 Choose a W shape for an 18-ft building column which supports an axial load of 510 kips. The column is free to buckle about either axis and has negligible rotational restraints. A36 steel, AISC specification.

4-8 Same as Prob. 4-7 except that the load is 65 kips.

4-9 Choose a W shape for a 16-ft building column which supports an axial load of 1,200 kips. The column is free to buckle about either axis and has negligible rotational restraints. A441 steel, AISC specification.

4-10 Same as Prob. 4-9 except that an attached partition supports the column against weak-axis buckling.

4-11 Choose an A36 W shape for a 12-ft building column which supports an axial load consisting of 310 kips dead load and 270 kips live load. The column is free to buckle about either axis and has negligible rotational restraint. Using the CRC formula, design for load factors of 1.3 on dead load and 1.7 on live load.

4-12 A 14-ft column supports an axial load consisting of 130 kips dead load and 95 kips live load. It is free to buckle about either axis and has negligible rotational restraints. Using the CRC formula, design the column in A36 structural tubing for load factors of 1.25 on dead load and 1.8 on live load.

4-13 Choose a W shape for a 20-ft column for an elevated highway structure. The load is 210 kips axial. The connections are bolted. A36 steel, AASHO specifications.

4-14 Same as Prob. 4-13 except that the column is in a railroad structure. AREA specifications.

4-11 DP4-1: COMPRESSION MEMBERS FOR WELDED TRUSS

Data for the truss of this example are given in DP3-1. In designing the chord, we must decide whether the purlins support the chord against buckling in the plane of the roof. If we decide they do not, the distance between diagonal bracing connections must be taken as the unsupported length for buckling in the plane of the roof and the distance between truss panel points as the unsupported length for buckling in the plane of the truss. In this example we assume that the purlins do support the chord.

The effective-length coefficients K of compression members in trusses do not exceed unity. This is because trusses are usually loaded only at the joints, so that joint displacements are primarily a function of axial deformations. Thus, the lateral displacement of one end of a member relative to the other end is small. However, K should not be taken less than unity, even in trusses with bolted, riveted, or welded joints, if the truss supports a fixed system of loads and is designed so that all its members fail simultaneously. In this case, there is little, if any, rotational restraint at the joints.

If the compression chord of a truss is of the same cross section for two or more panels, there are rotational restraints if the chord forces are unequal. In this case, the effective lengths are less than unity. If such a chord has the same cross section for the entire length of a truss of approximately constant depth, an effective length coefficient of 0.9 may be used if the truss supports fixed loads.[3,9]

Simple rules for the effective length of web members which are rotationally restrained by chord members are difficult to formulate. Bleich gives values of K ranging from 0.57 to 1 (Ref. 3, p. 247). If the truss carries moving or movable loads, rotational restraint may be appreciable. This is because the maximum stress in a web member results from a different position of live load than that which produces maximum stress in the chords. For this case, $K = 0.85$ is suggested as a conservative, approximate rule.[9]

COMPRESSION MEMBERS FOR TRUSS OF DP3-1	DP4-1

U_4U_5
$P = 109.7^k$, $L = 6'$
$A = 109.7/19 = 5.77\ in.^2$
Try WT 6×20, $A = 5.89\ in.^2$
$KL/r = 0.9 \times 72/1.56 = 42$
$F_a = 19.0$ ksi
$A = 109.7/19.0 = 5.77 < 5.89$
Use WT 6×20

U_1U_2, U_2U_3, U_3U_4 same as U_4U_5

U_1L_1
$P = 37.6^k$, $L = 5'$
$A = 37.6/19 = 1.98\ in.^2$
Try 2L 3×2
$L/r = 60/0.89 = 67$
$F_a = 16.7$ ksi
$A = 37.6/16.7 = 2.25\ in.^2$
Use 2L 3×2×1/4][$A = 2.38\ in.^2$

U_2L_2
$P = 21.7^k$, $L = 5.5'$
$A = 21.7/12 = 1.81\ in.^2$
Try 2L 2×2
$L/r = 66/0.60 = 110$
$F_a = 11.7$ ksi
$A = 21.7/11.7 = 1.86\ in.^2$
Use 2L 2×2×1/4][$A = 1.88\ in.^2$

U_3L_3
$P = 13.6^k$, $L = 6'$
$A = 13.6/10 = 1.36\ in.^2$
Try 2L 2×1¼
$L/r = 72/0.59 = 122$
$F_a = 10.0$ ksi
$A = 13.6/10 = 1.36\ in.^2$
Use 2L 2×1¼×1/4][$A = 1.50\ in.^2$

U_4L_4
$P = 7.0^k$, $L = 6.5'$
Try 2L 2×1¼
$L/r = 78/0.59 = 1.32$
$F_a = 8.57$ ksi
$A = 7.0/8.57 = 0.82\ in.^2$
Use 2L 2×1¼×1/4][$A = 1.50\ in.^2$

L_0L_1 (wind force – see below)
*WT 7×17, $A = 5.01$, $L/r = 77/1.52 = 51$
$F_a = 18.3 \times 1.33 = 24.4$ ksi
$P = 24.4 \times 5.00 = 122^k$
$P_{max} = 11.5^k$ (see below)

*See DP3-1

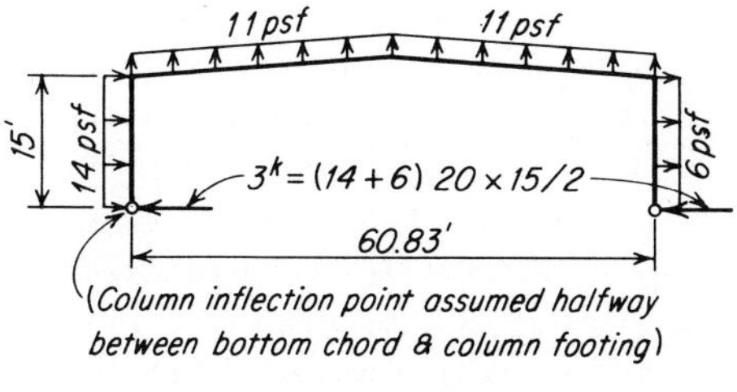

(a) External forces

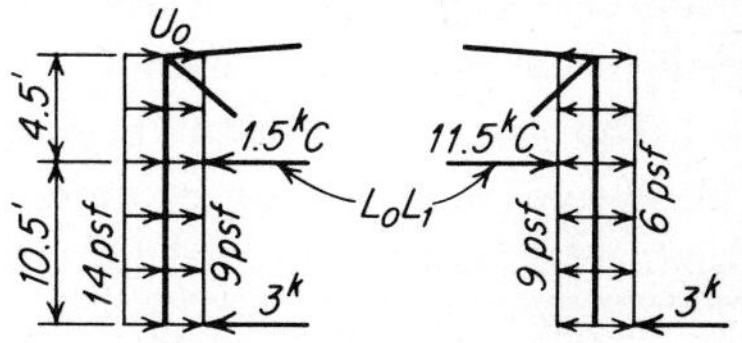

(b) Ext. wind + int. suction

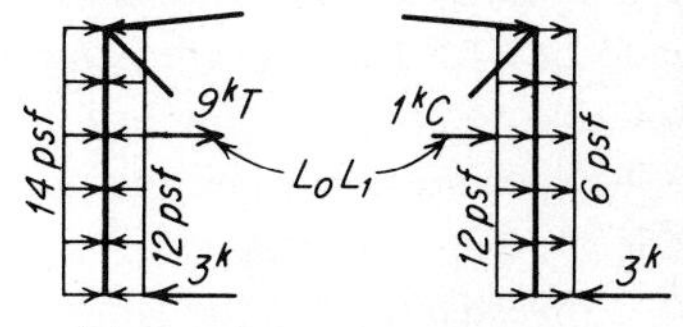

(c) Ext. wind + int. pressure

DP 4-1
Compression members for truss of DP 3-1.

$U_4 U_5$ The section for $U_4 U_5$ should be used for the entire top chord of the truss, or for at least the portion $U_2 U_5$, since any saving from the use of lighter sections for the less heavily loaded members would be offset by the cost of the additional connections. Therefore, $K = 0.9$. The estimated required area is available in tees ranging from 4 to 8 in. in depth. The smaller size can be ruled out as almost certain to be inadequate for suitable connections of web members. The larger size might be ruled out on the basis of its ratio of length to depth ($72/8 = 9$) since it was mentioned in Art. 3-18 that secondary stresses tend to become excessive for trusses with relatively wide members, say, $L/d < 10$. Instead of the calculation in the last line, some designers prefer to compute the "actual" stress and compare it with the allowable stress. A comparison of areas makes it easier to determine whether there is a shape which better meets the requirements of the member.

Web members Unequal-legged angles with the long legs back to back give more nearly equal radii of gyration for the two principal axes than either angles with equal legs or unequal-legged angles with short legs back to back. The radius of gyration for a specific pair of angles, such as $2\frac{1}{2} \times 2$, 3×2, etc., is virtually independent of the thickness. Therefore, the allowable stress can be determined without assuming the thickness.

$L_0 L_1$ Although it is nominally a tension member, $L_0 L_1$ may be subjected to compression due to wind forces. We shall investigate it for the wind forces specified in DP3-1, 20 psf for vertical surfaces and 11 psf suction on the roof. The prescribed 20 psf is shown in Fig. *a* of the design sheet as consisting of 14 psf pressure on the windward side and 6 psf suction on the leeward side. This is in agreement with Art. 1-8, where it was noted that the windward-wall shape factor in Eq. (1-4) is about 0.9 while the shape factor for the building is assumed to be 1.3 in arriving at the wind-design pressures of Table 1-5. Therefore, the windward-wall pressure is about $0.9 \times 20/1.3 = 14$ psf, which leaves 6 psf suction on the leeward wall. The points of inflection in the columns are assumed to lie midway between the bottom chord and the footing, and the wind force above these points is assumed to be divided equally between the two columns.

The possibility of large internal wind pressures was noted in Art. 1-8. A building that is likely to have 30 percent or more of its wall surfaces subject to being opened, or broken open, may experience internal suctions of as much as 9 psf and internal pressures of as much as 12 psf, at a wind velocity corresponding to a resultant external force of 20 psf on vertical surfaces. These are superimposed on the external forces, as shown in Figs. *b* and *c*. The forces in $L_0 L_1$ are then determined by taking moments at U_0. The largest compression turns out to be 11.5 kips. There is no dead-load tension to offset this force. In determining the allowable compression for $L_0 L_1$, the allowable stress is increased by one-third, as provided by the specification when member forces due to wind or combinations of wind and other loads are being investigated.

Investigation of the wind force in $L_0 L_1$ is not customary. It is given here only as a reminder that there may be situations in which it is not enough to make the usual assumption that the only effect of wind is a pressure on the windward face of a structure.

PROBLEMS

4-15 The web tension members for the truss of DP4-1 were designed in DP3-1. Design and detail the joint at U_1.

4-16 Choose a pair of angles for a riveted roof-truss top-chord member 10 ft long whose compressive force is 48 kips. The member is supported at each end in both principal planes. AISC specification, A36 steel.

4-17 Design for a welded roof truss a top-chord member whose unsupported length is 8 ft for buckling in either plane. The load is 112 kips. AISC specification, A36 steel.

4-18 Same as Prob. 4-17, except that the truss is bolted.

4-19 Same as Prob. 4-17, except that the unsupported length is 8 ft in the plane of the truss and 16 ft in the plane of the roof.

4-20 Design I-shaped top-chord and single-angle web compression members for the welded truss of DP4-1. Position the top chord so that its web is in the vertical plane with the web members butting against its bottom flange, and design and detail the connection at U_1 if U_1L_2 is a $3 \times 3 \times \frac{3}{8}$ angle. AISC specification, A36 steel.

4-21 Design two-angle sections for the top-chord and web compression members for bolted construction (A307 bolts) of the truss of DP4-1. Choose the gusset-plate thickness, and design the connections of these members and the tension web members you designed in Prob. 3-19. AISC specification, A36 steel.

4-22 Design the compression chord and web members for the highway-bridge truss of DP3-3.

4-12 COLUMNS WITH ECCENTRIC LOAD

The preceding discussions of compression members are based on the assumption that the load acts at the center of gravity of the cross section. Except for trusses, this is usually not the case in practice. Thus, even in building frames whose beams are joined to the columns with connections having a small moment of resistance, so that the beam is very nearly simply supported, the column itself may be subjected to bending because the shear transferred by the connection is eccentric. A typical connection is shown in Fig. 4-23*a*. Of course, interior columns with beams connecting to opposite flanges experience moment from eccentricity of load only if the beam reactions are unequal. Columns with beams joined by moment-resistant connections, of which Fig. 4-23*b* is typical, are also subjected to bending moments. The essential difference is that the column may be considered continuous on simple supports with connections such as that in Fig. 4-23*a*, while it is rotationally restrained at the joints by the connecting beams if the connections are moment-resistant. Furthermore, the moment is

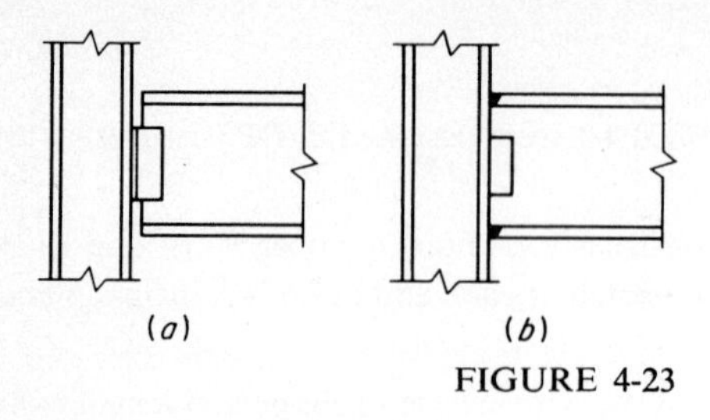

FIGURE 4-23

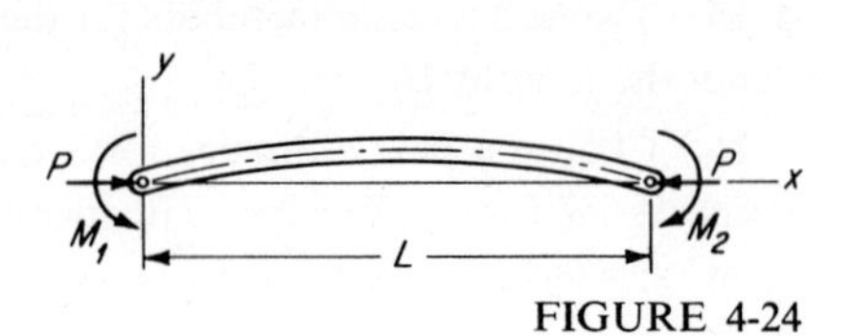

FIGURE 4-24

usually much larger with moment-resistant connections. Members which are subjected to both axial force and moment are called *beam-columns*.

An initially straight pin-ended column subjected to axial forces P and end moments M_1 and M_2, with $M_1 < M_2$, is shown in Fig. 4-24. Equating bending moment to bending resistance, we get

$$-EI\frac{d^2y}{dx^2} = Py + M_1 + (M_2 - M_1)\frac{x}{L} \tag{a}$$

Using the notation $k^2 = P/EI$, Eq. (*a*) becomes

$$\frac{d^2y}{dx^2} + k^2y = \frac{k^2}{P}(M_1 - M_2)\frac{x}{L} - \frac{k^2}{P}M_1 \tag{b}$$

The solution of this equation is

$$y = A\sin kx + B\cos kx + \frac{M_1 - M_2}{P}\frac{x}{L} - \frac{M_1}{P} \tag{c}$$

With the boundary conditions $y = 0$ at $x = 0$ and $x = L$, Eq. (*c*) becomes

$$y = \left(\frac{M_2}{P} - \frac{M_1}{P}\cos kL\right)\frac{\sin kx}{\sin kL} + \frac{M_1}{P}(\cos kx - 1) + (M_1 - M_2)\frac{x}{PL} \tag{d}$$

The bending moment at any point is determined from $M = -EI\,d^2y/dx^2$. The maximum value is found to be

$$M_{\max} = M_2 \csc kL\sqrt{\left(\frac{M_1}{M_2}\right)^2 - 2\frac{M_1}{M_2}\cos kL + 1} \tag{e}$$

Equation (*e*) is valid so long as Hooke's law holds. Assuming the proportional-limit stress to be equal to the yield stress F_y, values of P and M_2 (M_1/M_2 is assumed known) at which yielding begins are found by using $F_y = P/A + M_2 c/I$. The result, with P/EI substituted for k^2, is

$$F_y = \frac{P}{A} + \frac{M_2 c}{I} \sqrt{\left(\frac{M_1}{M_2}\right)^2 - 2\frac{M_1}{M_2}\cos\pi\sqrt{\frac{P}{P_E}} + 1} \quad \csc\pi\sqrt{\frac{P}{P_E}} \tag{4-28}$$

For the case $M_1 = M_2 = M$, Eq. (4-28) reduces to

$$F_y = \frac{P}{A} + \frac{Mc}{I}\sec\frac{\pi}{2}\sqrt{\frac{P}{P_E}} \tag{4-29a}$$

or

$$F_y = \frac{P}{A}\left(1 + \frac{ec}{r^2}\sec\frac{\pi}{2}\sqrt{\frac{P}{P_E}}\right) \tag{4-29b}$$

where $e = M/P$. Equation (4-29) is called the *secant formula*. In the form of Eq. (4-29*b*) it resembles Eq. (4-6) for the crooked column. This and the correspondence (Fig. 4-25) of $\sec[(\pi/2)\sqrt{P/P_E}]$ and $1/(1 - P/P_E)$ show that equal end eccentricities e of the load P for the straight column have much the same effect as crookedness of amplitude $\delta_0 = e$. Thus, Eqs. (4-29) can be put in the simpler, approximate form

$$F_y = \frac{P}{A} + \frac{Mc}{I}\frac{1}{1 - P/P_E} = \frac{P}{A}\left(1 + \frac{ec}{r^2}\frac{1}{1 - P/P_E}\right) \tag{4-30}$$

Equation (4-30) is usually put in different form by dividing through by F_y. The result is

$$\frac{P}{P_y} + \frac{M}{M_y}\frac{1}{1 - P/P_E} = 1 \tag{4-31}$$

where $P_y = F_y A$ and $M_y = F_y I/c$. The equation in this form is called an *interaction formula*. The factor $1/(1 - P/P_E)$ is sometimes called an *amplification factor* since it is the factor by which the end moment M is multiplied to approximate the moment $M + Py$.

It is instructive to determine the extreme values of P and M in Eq. (4-31). To do this, we write the equation in the form

$$\frac{M}{M_y} = \left(1 - \frac{P}{P_y}\right)\left(1 - \frac{P}{P_E}\right) \tag{f}$$

Obviously, $M = M_y$ when $P = 0$. However, there are two extreme values of P. Thus, when $M = 0$, $P = P_y$ or $P = P_E$. These are the values for the perfect column; that is, P is the smaller of P_y and P_E. Therefore, Eq. (4-31) does not allow for the effect of

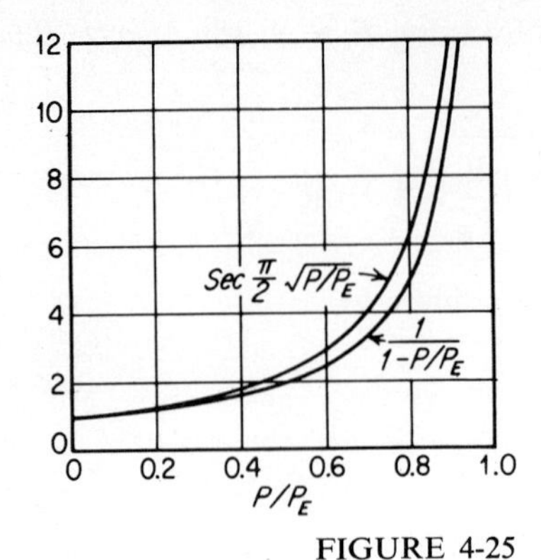

FIGURE 4-25

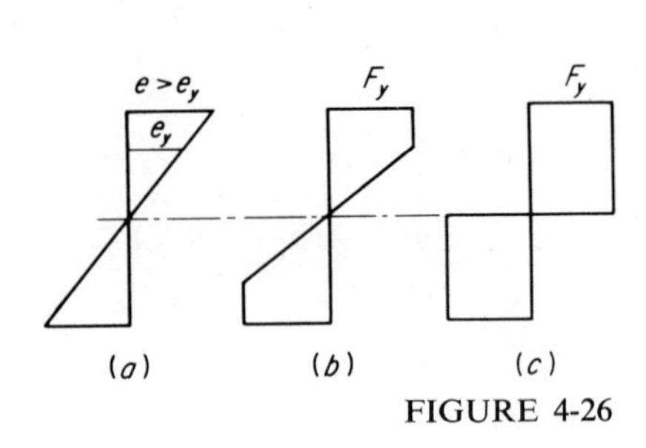

FIGURE 4-26

imperfections such as residual stress, crookedness, etc., on the strength of a column made of flat-yielding steel or for the increase of P beyond P_y for a gradually yielding material. Since the formula gives the correct extreme value of P for elastic buckling, namely, P_E, we change the inelastic extreme P_y by substituting P_{cr} for P_y, where by P_{cr} we mean the strength of the imperfect column of flat-yielding steel or the tangent-modulus load for a column of gradually yielding metal. Thus, Eq. (4-31) becomes

$$\frac{P}{P_{cr}} + \frac{M}{M_y} \frac{1}{1 - P/P_E} = 1 \tag{4-32}$$

This equation gives correct extreme values of P. It is used to predict the concurrent values of P and M which produce yield stress on the extreme fiber of that cross section between the ends of a beam-column at which the moment is a maximum. Therefore, it underestimates the strength of the member. This is because the extreme value $M = M_y$ is not correct for $P = 0$, since bending strength is not fully developed when the extreme fiber begins to yield. Instead, beams can tolerate considerable penetration of yield stress into the cross section. Thus, Fig. 4-26*a* shows the strain over a cross section where the extreme-fiber strain is several times larger than ε_y. The resulting stress distribution is shown in *b*. For large values of extreme-fiber strain the stress distribution can be approximated closely by two rectangles, as shown in *c*. The corresponding moment is called the *plastic moment* and is denoted by M_p. (Plastic bending is discussed further in Art. 5-2.) A formula giving correct extreme values of both M and P is obtained by substituting M_p for M_y in Eq. (4-32). This gives

$$\frac{P}{P_{cr}} + \frac{M}{M_p} \frac{1}{1 - P/P_E} = 1 \tag{4-33}$$

Of course, such an extension of a formula based on proportionality of stress and strain to a form intended to predict strength (which may involve a considerable amount of nonproportional behavior) may not give very good results. Therefore, Eq. (4-33) must be checked by comparing it with results of analysis based on inelastic behavior and of tests. This is discussed in Art. 4-13.

It remains to extend Eqs. (4-32) and (4-33) to the case of unequal end moments. This could be done by using Eq. (4-28). However, a much simpler procedure, based on formulas which approximate data from numerical analyses, consists in replacing M by $C_m M$, where C_m is given by any of the following:

$$C_m = \frac{1}{1.75 - 1.05M_1/M_2 + 0.3(M_1/M_2)^2} \not< 0.4 \tag{4-34a}$$

$$C_m = \sqrt{0.3 + 0.4\frac{M_1}{M_2} + 0.3\left(\frac{M_1}{M_2}\right)^2} \not< 0.4 \tag{4-34b}$$

$$C_m = 0.6 + 0.4\frac{M_1}{M_2} \not< 0.4 \tag{4-34c}$$

M_1 in these equations is the smaller of the end moments M_1 and M_2, and M_1/M_2 is positive when the member bends in single curvature, as in Fig. 4-24.* These formulas are by Salvadori,[16] Massonnet,[17] and Austin,[18] respectively. They give approximately the same results. Although they were derived for the case of the member which fails by bending out of the plane of the end moments (lateral-torsional buckling), which is discussed in Art. 5-7, they are used in practice for the case considered here where bending is confined to the plane of the moments. Finally, then, we have

$$\frac{P}{P_{cr}} + \frac{C_m M}{M_y}\frac{1}{1 - P/P_E} = 1 \tag{4-35}$$

$$\frac{P}{P_{cr}} + \frac{C_m M}{M_p}\frac{1}{1 - P/P_E} = 1 \tag{4-36}$$

where M is understood to be the larger of the two end moments.

Equation (4-35) gives values of P and M at yielding of the extreme fiber, while Eq. (4-36) gives ultimate values of P and M. Equation (4-36) is compared in Art. 4-13 with the results of a detailed analysis of the beam-column of I cross section and with results of tests.

* The opposite sign convention is also used, as in the AISC specification, where M_1/M_2 is positive when the member bends in double curvature. With this sign convention the sign of the M_1/M_2 term in each of Eqs. (4-34) is changed.

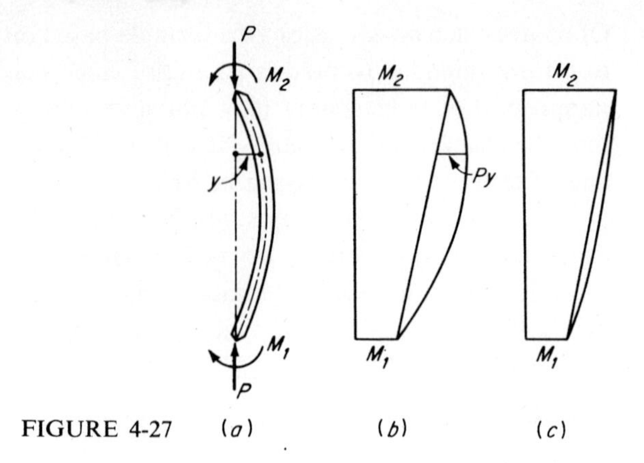

FIGURE 4-27

4-13 COLUMNS WITH MAXIMUM MOMENT AT ONE END

The moment $C_m M/(1 - P/P_E)$ in Eqs. (4-35) and (4-36) is the maximum moment at an interior cross section of the beam-column. This is shown for the member of Fig. 4-27*a* in the moment diagram in *b*. In some cases, however, this maximum moment does not exist. Instead, the largest moment is M_2 itself, as shown in the moment diagram of Fig. 4-27*c*. For this case, beginning of yield on the extreme fiber is given by

$$F_y = \frac{P}{A} + \frac{M_2 c}{I} \tag{a}$$

which gives the interaction formula

$$\frac{P}{P_y} + \frac{M}{M_y} = 1 \tag{4-37}$$

This is the same as Eq. (4-31) with an amplification factor of unity, which is as it should be because the moment Py is zero at the end. The extreme values are correct; that is, $M = M_y$ if $P = 0$ and $P = P_y$ if $M = 0$. The latter is correct because there is no question of stability.

Equation (4-37) does not give the maximum values of P and M which the cross section can support. To determine this we must consider the postyielding behavior. Since there is no question of buckling, extreme fiber strains can increase well beyond the strain ε_y at the beginning of yield, as shown in Fig. 4-28*b*. The corresponding stress distribution for a flat-yielding steel is shown in Fig. 4-28*c*. The interaction formula for this stress distribution acting on the rectangular cross section of Fig. 4-28*a* is easily determined. The stress distribution can be approximated very closely by two

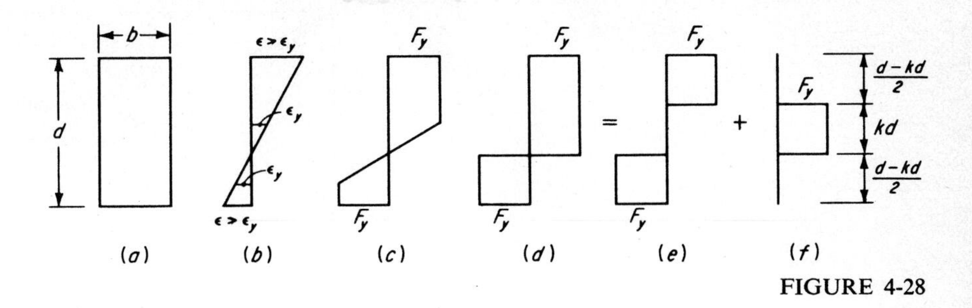

FIGURE 4-28

rectangular stress blocks (Fig. 4-28*d*), which can then be decomposed into the two distributions shown in *e* and *f*. From *e* we get

$$M = F_y \frac{b(d - kd)}{2} \frac{d + kd}{2} = F_y \frac{bd^2}{4}(1 - k^2) \tag{b}$$

while that in *f* gives

$$P = F_y bkd \tag{c}$$

If $k = 0$, $P = 0$ and $M = F_y bd^2/4$, which is the plastic resisting moment M_p of the cross section. Similarly, if $k = 1$, $M = 0$ and $P = F_y bd$, which is the axial load capacity P_y. Therefore, Eqs. (*b*) and (*c*) can be written

$$\frac{M}{M_p} = 1 - k^2 \qquad \frac{P}{P_y} = k \tag{d}$$

which, upon eliminating k, give

$$\left(\frac{P}{P_y}\right)^2 + \frac{M}{M_p} = 1 \tag{e}$$

This equation is plotted in Fig. 4-29.

Analysis of the I leads to similar curves, which, for cross sections of practical proportions, are bounded by the two curves shown in Fig. 4-29. These curves are approximated closely by

$$\frac{P}{P_y} + 0.85 \frac{M}{M_p} = 1 \qquad \frac{P}{P_y} \gtrsim 0.15 \tag{4-38a}$$

$$M = M_p \qquad \frac{P}{P_y} \lesssim 0.15 \tag{4-38b}$$

These formulas are also plotted in Fig. 4-29.

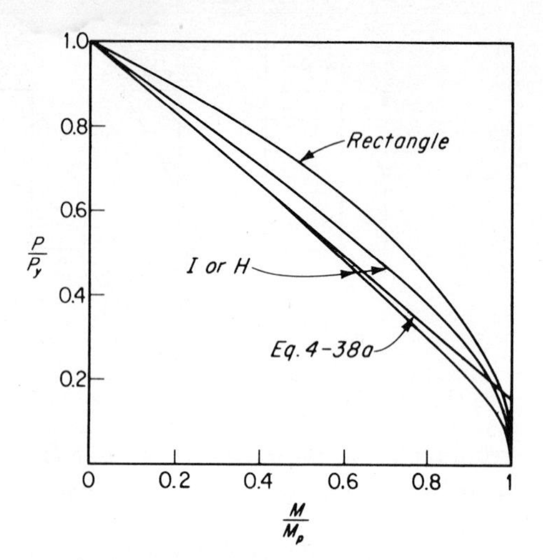

FIGURE 4-29

Both Eq. (4-36) and Eqs. (4-38) must be used to evaluate the strength of the beam-column of I shape. As has already been noted, this is because Eq. (4-36) is based on the existence of a maximum moment between the ends (Fig. 4-27*b*), while with Eqs. (4-38) it is assumed that the largest moment is at an end (Fig. 4-27*c*). These equations are compared in Fig. 4-30*a*, *b*, and *c* with a detailed investigation of a W8 × 31 steel beam-column.[19] Thermal residual stresses were taken into account in this analysis. Deflected shapes were established by an iterative procedure in which elements of the member are fitted together, the curvature of each element being found from predetermined variations of moment with strain for various values of P. (Figure 5-4 shows a typical plot, in this case for $P = 0$.) The procedure involves assuming a deflected shape, computing the corresponding moments in the elements into which the member is divided, determining the curvatures, and fitting the elements together to see whether the resulting deflected shape agrees with the assumed one. The process is repeated until the differences between the assumed deflections and the calculated deflections are negligible.

Results from Ref. 19 are shown in the dashed lines in Fig. 4-30; Eqs. (4-36) and (4-38) are shown in solid lines. The case $M_1 = M_2$ ($C_m = 1$) is shown in Fig. 4-30*a*. Agreement is seen to be very good. Values for $L/r = 0$ agree with Eqs. (4-38), while those for $L/r = 40$, 80, and 120 agree with Eq. (4-36). The case $M_1 = 0$ ($C_m = 0.6$) is shown in Fig. 4-30*b*, where the correspondence is again very good except for columns with $L/r = 120$ and whose M is larger than $0.8M_p$. It will be noted that Eq. (4-36) holds for $L/r = 40$ if M/M_p is smaller than about 0.25, while Eqs. (4-38) apply for larger values. The case $M_1 = -M_2$ ($C_m = 0.4$) is shown in Fig. 4-30*c*. Here the agreement of Eq. (4-36) with the results of Ref. 19 is not very good for $L/r = 80$ and 120. However, results from the equation are on the safe side. Finally, then, since it is shown

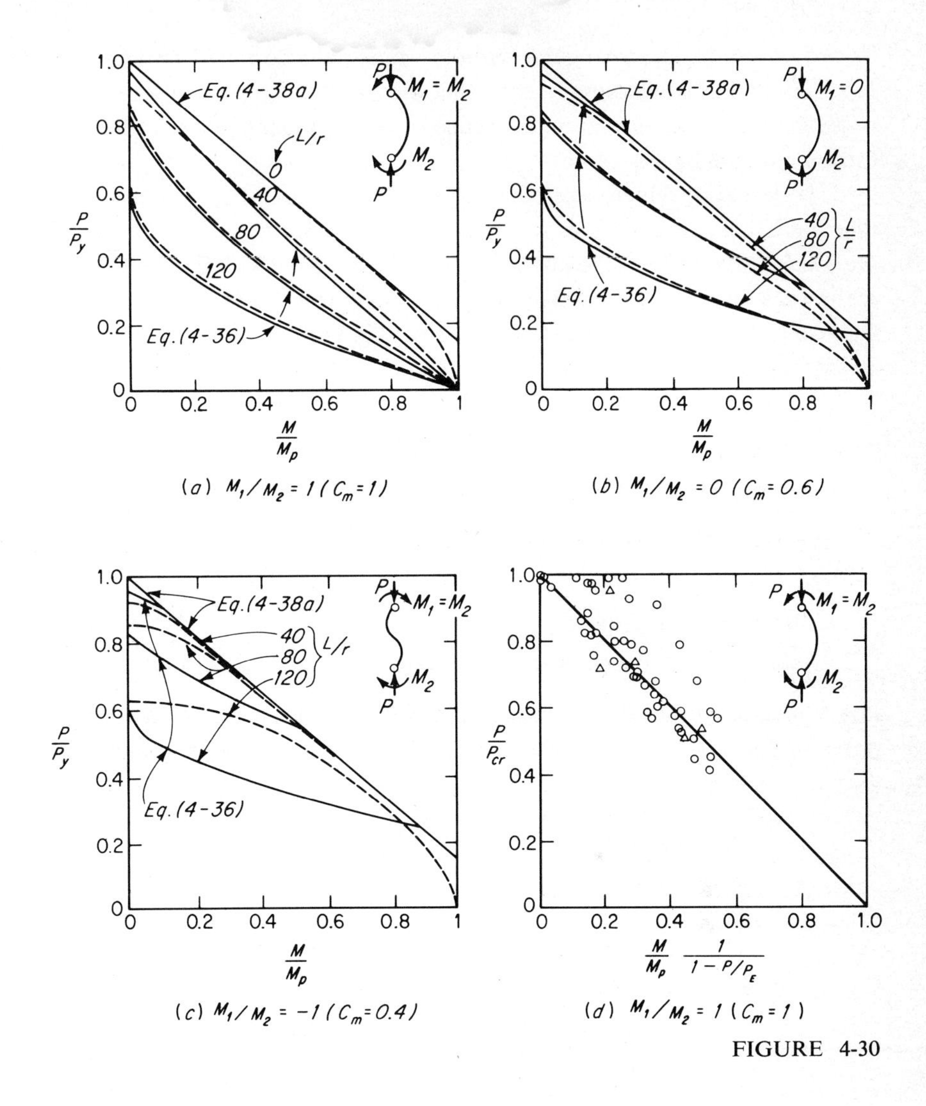

FIGURE 4-30

in Ref. 19 that the predicted strengths are in good agreement with the results of a number of tests on steel beam-columns, we can conclude that Eq. (4-36) and Eqs. (4-38) also give good predictions of beam-column strength except for slender members subjected to reversed-curvature bending, for which they underestimate the strength.

Equation (4-36) is compared in Fig. 4-30*d* with the results of tests of 53 aluminum beam-columns[20] whose slenderness ratios ranged from 20 to 150. Since the eccentricity was the same at both ends, $C_m = 1$. Only five of the specimens were I's, the remainder being rectangular bars, rectangular tubes, and round tubes. Test results for the five

I's are shown by the triangular symbol. It should be noted that the coordinates in Fig. 4-30*d* differ from those in the other plots in the figure, so that Eq. (4-36) plots as a straight line. The test strengths ranged from 11 percent more to 9 percent less than those predicted by Eq. (4-36).

It should be remembered that Eqs. (4-35) and (4-36) are based on the assumption that the beam-column bends in the plane of the end moments. Cross sections which are much stiffer in bending about one axis than about the other, as the I is, may bend in the weak direction, even if they are loaded so as to bend in the strong direction, unless weak-axis bending is prevented by some kind of support. This is called lateral-torsional buckling (Art. 5-7).

4-14 SPECIFICATION FORMULAS FOR BEAM-COLUMNS

The preceding discussions suggest two possible criteria for determining the limiting load on a beam-column:

1 First yielding (on the extreme fiber) or
2 Attainment of the peak load (topmost point on the P-δ curve)

For convenience in the discussion to follow, the corresponding formulas from Arts. 4-12 and 4-13 are repeated here. For first yielding of the extreme fiber

$$\frac{P}{P_{cr}} + \frac{C_m M}{M_y} \frac{1}{1 - P/P_E} = 1 \tag{4-35}$$

$$\frac{P}{P_y} + \frac{M}{M_y} = 1 \tag{4-37}$$

For attainment of peak load

$$\frac{P}{P_{cr}} + \frac{C_m M}{M_p} \frac{1}{1 - P/P_E} = 1 \tag{4-36}$$

$$\frac{P}{P_y} + 0.85 \frac{M}{M_p} = 1 \qquad \frac{P}{P_y} \geqslant 0.15 \tag{4-38a}$$

$$M = M_p \qquad \frac{P}{P_y} \leqslant 0.15 \tag{4-38b}$$

Equations (4-35) and (4-36) check the member for maximum moment at a point between the ends; Eqs. (4-37) and (4-38) check the member for maximum moment at one end.

There are several ways in which these formulas can be used to evaluate the strength of members in frames with moment-resistant beam-to-column connections, as follows:

1 Analyze the frame by a method based on elastic behavior, e.g., slope-deflection equations, and then evaluate the members by Eqs. (4-35) and (4-37). In this case, the evaluation of the frame moments and evaluation of member strengths are consistent, in that both are based on elastic behavior.

2 Analyze the frame as in procedure 1 but evaluate the members by Eqs. (4-36) and (4-38). This corresponds to the "ultimate-strength design" procedure used for reinforced-concrete frames. Of course, it gives larger values of frame capacity than procedure 1. Since we assume elastic behavior to determine frame moments and inelastic behavior to determine member strength, the assumptions in procedure 2 are inconsistent.

3 Analyze the frame by theories of plastic behavior (these take into account certain redistributions of moment which develop with increased loading of a frame after first yielding) and evaluate the members by Eqs. (4-36) and (4-38). Here the assumptions are consistent in that frame moments and assessment of member strength are both based on inelastic behavior. This procedure is called *plastic design* when applied to steel frames and *limit design* when applied to reinforced-concrete frames. In general, this procedure gives larger values of frame capacity than procedure 2. Plastic design is discussed in Chap. 8.

The AISC specification uses Eqs. (4-36) and (4-38) for the beam-column in plastically designed frames. Equations (4-35) and (4-37) are used for structures designed on a first-yield basis, except that they are written in terms of allowable stress. Thus, if P and M are defined as service-load (working-load) values and n is the factor of safety, nP and nM should be substituted for P and M in Eq. (4-35). This gives

$$\frac{nP}{P_{cr}} + \frac{nM}{M_y}\frac{C_m}{1 - nP/P_E} = 1 \qquad (a)$$

Dividing P and P_{cr} in the first term by the cross-sectional area A and, in the second term, dividing M and M_y by the section modulus S and P and P_E by the area A, we can write Eq. (a) in the form

$$\frac{f_a}{F_{cr}/n} + \frac{f_b}{F_y/n}\frac{C_m}{1 - f_a/(F_E/n)} = 1 \qquad (b)$$

where f_a and f_b are service-load stresses. The stresses which are divided by the factor of safety are now allowable stresses, so that

$$\frac{f_a}{F_a} + \frac{f_b}{F_b}\frac{C_m}{1 - f_a/F'_E} = 1 \qquad (4\text{-}39a)$$

where F_a = axial compressive stress that would be allowed if there were only the axial force P

F_b = bending compressive stress that would be allowed if there were only the moment M

F'_E = allowable Euler stress

F_a is given by Eqs. (4-17) and (4-18). Allowable bending stresses F_b are discussed in Chap. 5. If no out-of-plane (lateral) buckling is involved, as is assumed in this chapter, F_b is either $0.6F_y$ or $0.66F_y$, depending on whether certain requirements relative to thickness of the elements of the cross section are met. F'_E is given by Eq. (4-18) for all values of KL/r, so that the factor of safety n for F_E in Eq. (*b*) is the same as it is for columns which buckle elastically.

Equation (4-37) is transformed similarly, the result being

$$\frac{f_a}{0.6F_y} + \frac{f_b}{F_b} = 1 \tag{4-39b}$$

Of course, both equations must be checked, since one corresponds to the case where the maximum moment is at a section between the ends while the other corresponds to the case where the maximum moment is at one end.

Equations (4-39) are written in more general form in the AISC specification so as to include the case of biaxially eccentric load, i.e., a load P together with x-axis end moments M_{1x} and M_{2x} and y-axis end moments M_{1y} and M_{2y}. This is done by adding a bending term, so that

$$\frac{f_a}{F_a} + \frac{f_{bx}}{F_{bx}} \frac{C_{mx}}{1 - f_a/F'_{Ex}} + \frac{f_{by}}{F_{by}} \frac{C_{my}}{1 - f_a/F'_{Ey}} = 1 \tag{4-40a}$$

$$\frac{f_a}{0.6F_y} + \frac{f_{bx}}{F_{bx}} + \frac{f_{by}}{F_{by}} = 1 \tag{4-40b}$$

The accounting in Eq. (4-40*a*) for the effect of the moments M_y is approximate because it neglects the effect of the twisting which accompanies biaxial bending.[21] Nevertheless, the formula is generally conservative. This is because it is based on first yield, rather than attainment of peak load, and this turns out to be more than enough to offset the neglect of twist.[22] Equations (4-40) are also used in the AREA specifications, except that they are based on a larger factor of safety. The AASHO specifications formula is based on Eq. (4-28).

The proportioning of beam-columns is a trial-and-error procedure. One can estimate the cross-sectional area by determining what is required for either the axial load alone or the moment alone or by using the equation

$$A = \frac{P}{F_a(1 - f_b/F_b)} \tag{4-41}$$

This equation derives from Eq. (4-39a) by substituting P/A for f_a and using $C_m/(1 - f_a/F'_E) = 1$. Both F_a and the ratio f_b/F_b must be estimated.

Another equation for preliminary design is obtained by substituting into Eq. (4-39a) P/A for f_a and M/S for f_b, which gives

$$A = \frac{P}{F_a} + \frac{M}{F_b}\frac{A}{S} \tag{4-42}$$

The AISC Manual lists values of A/S, under the name *bending factor B*, for both principal axes of the W, M, and S shapes.

4-15 DP4-2: INTERIOR COLUMN FOR 26-STORY BUILDING

The design of the bottom-tier column B2 of the 26-story building shown in plan in the figure is given in this example. The height from the basement to the lobby is 15 ft; other floor heights are 13 ft. The floor system consists of a 3-in. cellular steel deck with $2\frac{1}{2}$ in. concrete slab supported on steel joists. The joists are 15 ft on centers. Live loads are 100 psf on the first, or lobby, floor, 80 psf on the upper floors, and 20 psf on the roof. X bracing in the longitudinal direction and K bracing (Art. 12-2) in the lateral direction of the service area are used to resist wind forces. This allows the use of moment-free beam-to-column and beam-to-beam connections throughout. In this case, as mentioned in Art. 4-12, the columns may be considered to be continuous on simple supports. Furthermore, because of the service-area bracing, relative joint translation can be considered to be negligible. Therefore, $K = 1$.

Columns to which beams connect with simple framing connections (Fig. 4-23a) are usually designed as concentrically loaded members. Such a connection is not moment-free, but the moment that can be generated is relatively small compared to that of a typical moment-resistant connection (Fig. 4-23b).

The following comments are intended to clarify computations identified by the corresponding letter on the design sheet.

a Felt and gravel roofing weighs about 5 psf and the cellular deck and concrete slab 42 psf. Weight of beams and columns depends upon their spacing and spans and the load to be supported.

b Live-load reduction is discussed in Art. 1-4. The American Standard Building Code reduction is 0.08 percent per ft² of supported area, provided the area is larger than 150 ft², but cannot exceed 60 percent or $R = 100(D + L)/4.33L$ percent. The column weight is averaged for the 26 floors.

c The W shape is ordinarily used for building columns and will usually be cheaper than the welded box used in this example. A W14 × 605 in A36 steel or a W14 × 426 in

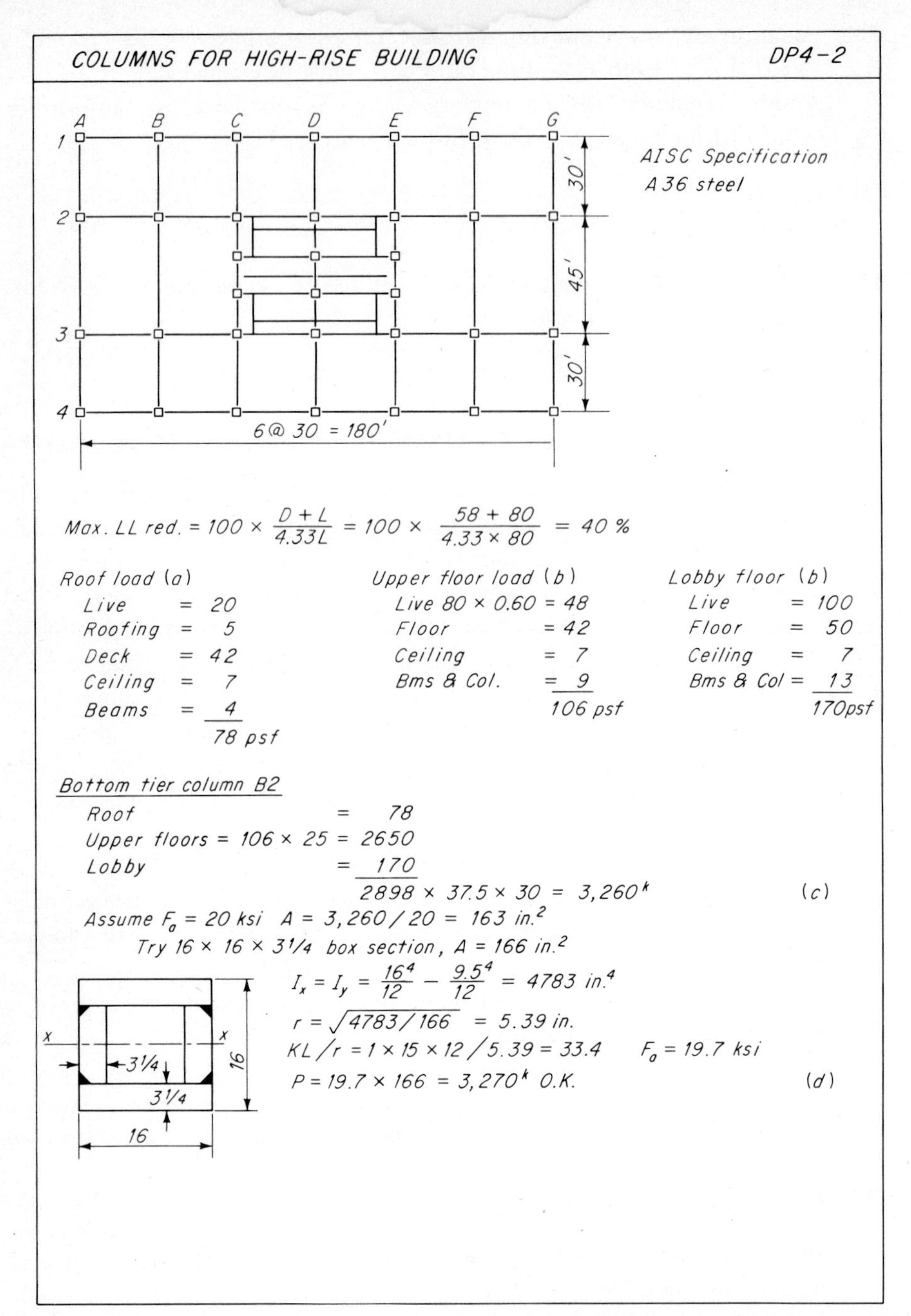

DP 4-2
Columns for high-rise building.

A572 Grade 55 can be used instead. The former is 21 in. deep and has a 17.4-in. flange; the latter is 18.7 in. deep and has a 16.7-in. flange. Thus, in some situations the extra cost of the 16 × 16 in. box might be justified.

Column loads in a building of this magnitude are ordinarily determined by computer, and in many cases sizing of the members is also computerized.

4-16 COLUMN BASES

The base of a column which is supported by a concrete footing must have a bearing plate large enough to distribute the load over an area sufficient to preclude excessive bearing stresses. The base plate may be shop-connected, or it may be shipped loose. Except where end moments are involved, the connection need only hold the parts in line if the ends of the column are finished to a plane surface, since the load is transmitted by bearing at the contact surfaces. It may be cheaper to omit planing the ends of lightly loaded columns and to design the connection between base and shaft for the total load.

The distribution of the pressure of the plate on the footing depends upon the relative stiffnesses of the two. Even if the distribution were known, the resulting stresses in the plate could not be determined easily since bending in two directions is involved. The usual analysis is based on two assumptions: (1) the pressure of the footing on the plate is uniformly distributed, and (2) those portions of the plate which project from the column shaft act as cantilever beams. Since the sections of zero shear, and consequently of maximum moment, in the bearing plates are inside the area of contact between the column shaft and plate, the lengths of the cantilevered portions are usually assumed to be longer than the actual projection of the plate. The AISC recommended analysis for H-shaped columns assumes that the maximum moments occur at sections which are $0.95d$ apart in one direction and $0.8b$ apart in the other direction, where d and b are, respectively, the depth and flange width of the shape. The allowable bending stress for these moments is $0.75F_y$ (Art. 5-5).

Bases of columns in industrial buildings, and those in tier-building construction designed to resist wind forces, may need to be proportioned to resist end moments resulting from lateral forces. This problem is discussed in Art. 7-31.

EXAMPLE 4-2 Base plate for column of DP4-2. The columns of this building are supported on reinforced concrete footings for which $f_c' = 3{,}000$ psi. Assume that the area of the base plate for column B2 exceeds one-third the area of the footing. In this case, the AISC allowable bearing pressure F_p on the concrete is $0.25f_c' = 750$ psi. Therefore, $A = 3{,}270/0.750 = 4{,}360$ in.^{2}

Use a 66 × 66 in. plate, which gives $A = 4{,}356$ in.^{2} The plate projection is $\frac{1}{2}(66 - 0.95 \times 16) = 25.4$ in. Therefore, $M = 0.750 \times 25.4^2/2 = 242$ in.-kips per 1-in. width. The allowable bending stress is $F_b = 0.75F_y = 0.75 \times 36 = 27$ ksi, so that

$$\frac{bt^2}{6} = \frac{242}{27} = 8.95$$

$$t = \sqrt{8.95 \times 6} = 7.35 \text{ in.}$$

Use an 8 × 66 × 66 plate planed to $7\frac{1}{2}$ in. thick.

PROBLEMS

Note: Since lateral-torsional buckling is not discussed in this chapter, columns of box or tubular cross section are suggested in the following problems except where it can be assumed that weak-axis bending is prevented by wall or other construction.

4-23 Columns in tier buildings are usually fabricated in two-story lengths, using the same cross section throughout. Design column B2 for the tenth to twelfth floor of the building of DP4-2, using a box section. Investigate the effect of neglecting eccentricity of load. AISC specification with $F_b = 0.66F_y$.

4-24 Same as Prob. 4-23 except design the topmost segment (twenty-fifth floor to roof) for column B2.

4-25 For the office building of DP4-2 design the bottom-tier exterior column A2. Assume the average weight of the exterior wall is 25 psf of surface. Select a box section made of four A36 steel plates. AISC specification with $F_b = 0.66F_y$.

4-26 Same as Prob. 4-25 except design the topmost segment (twenty-fifth floor to roof) for column A1.

4-27 Same as Prob. 4-25 except use a built-up section consisting of a W14 and two plates welded to form a closed section.

4-28 Choose an A572 box section for the A36 column of DP4-2. If the cost ratios of A572 to A36 are 1.04, 1.08, 1.12, 1.18, and 1.36 for Grades 42, 45, 50, 55, and 60, respectively, compare the material cost of the two designs. See Table 2-2.

4-29 Assume the office building of DP4-2 is 18 stories high. Design the exterior column B1 and its base plate. The weight of the exterior wall is 25 psf of surface. Choose a W shape of A36 steel. The wall supports the column in the weak direction. AISC specification with $F_b = 0.66F_y$ for the column and $0.75F_y$ for the base plate.

4-30 An 18 × 22 in. box column is made of $1\frac{1}{2}$-in. A36 steel plate. The column height is 16 ft. Design an A36 base plate for the allowable concentric load on the column. AISC specification with $F_b = 0.75F_y$. For the footing $f'_c = 3{,}000$ psi.

4-31 Figure 4-31 shows the plan of a one-story industrial building. The roof deck is 2-in. precast concrete plank, made with slag aggregate, which weighs 14 psf. Roofing weighs 6 psf.

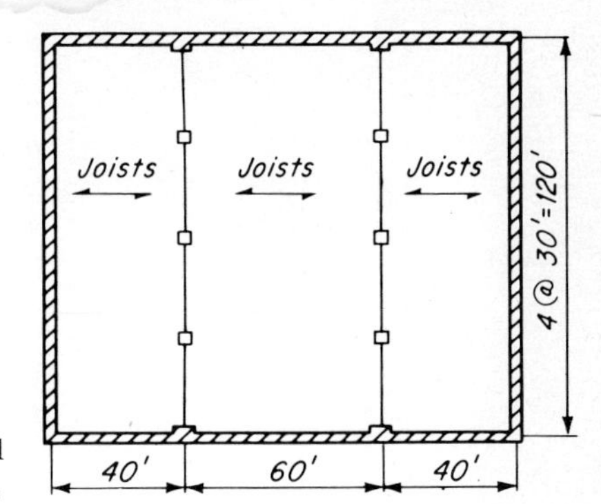

FIGURE 4-31
Problem 4-31.

The deck is supported on standard longspan steel joists spaced 5 ft on centers. The 40-ft joists are 24 in. deep and weigh 19 plf. The 60-ft joists are 32 in. deep and weigh 28 plf. The beams supporting the joists are A36 W21 × 68.

The building is 16 ft high between the floor and the bottom of the beams. Design the columns in A36 steel tubing, and base plates supported on concrete footings with $f'_c = 3{,}000$ psi. AISC specification with $F_b = 0.66F_y$ for the column and $0.75F_y$ for the base plate.

4-32 Design column B2 for the tenth to twelfth floors of the building of DP4-2 for load factors of 1.3 on dead load and 1.7 on live load. Use a box section of A36 steel.

4-33 The reactions on the beams A, B, and C in Fig. 4-32 are 28.2, 21.6, and 36.1 kips, respectively. In addition to the three beams, the column supports a centroidal axial load of 64.2 kips. The beams rest on seats 4 in. wide and clear the adjacent surface of the column 1/2 in. The effective length of the column is 12 ft. What 8 × 8 in. standard A36 structural tubing is required? AISC specification with $F_b = 0.66F_y$.

4-34 Figure 4-33 shows the half elevation of a frame for an industrial building. The frames are 30 ft on centers. The roof dead load is 20 psf and the live load 30 psf. The floor dead load is 60 psf and the live load 100 psf. The crane reaction is 38.6 kips. The building is braced at the ends and at several interior bents, so that B, C, E, and F can be assumed fixed against translation. Beam-to-column connections are not moment-resistant. Design columns ABC and DEF. AISC specification.

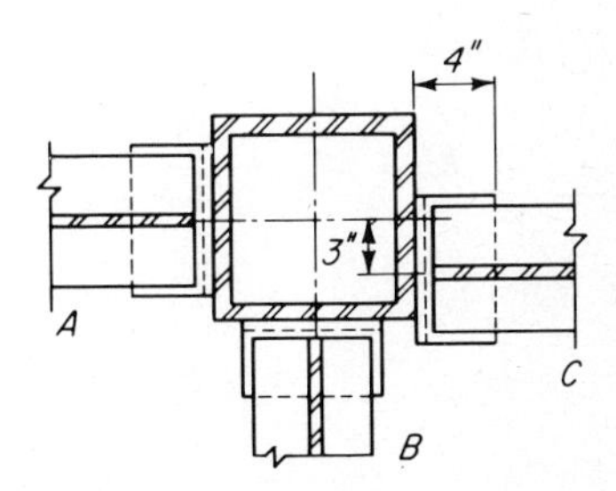

FIGURE 4-32
Problem 4-33.

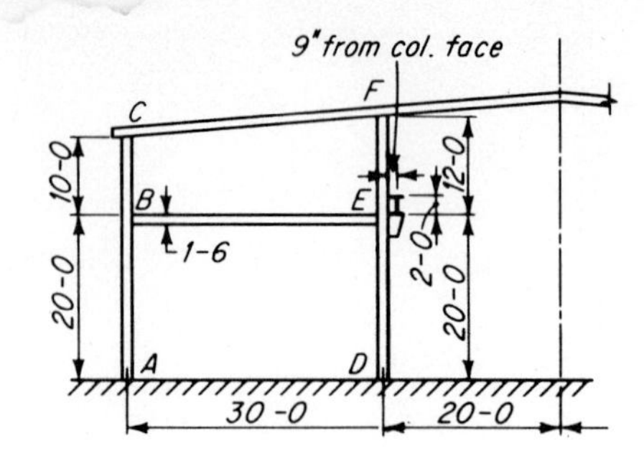

FIGURE 4-33
Problem 4-34.

4-17 BEAM-COLUMNS: GENERAL CASE

In the derivation of Eqs. (4-35) and (4-36) in Art. 4-12 we assumed that the column is acted upon only by end moments, in addition to P, and that there is no lateral displacement of one end relative to the other. In this article, we consider the effects of transverse load and of lateral displacement.

For a simply supported member with transverse loads in addition to P, the moment at any point is given by

$$M = M_0 + Py \tag{a}$$

where M_0 is the moment due to the transverse loads alone. The deflection y can be found by integrating $EI\, d^2y/dx^2 = M$. However, it is given to good approximation by

$$y = \frac{y_0}{1 - P/P_E} \tag{b}$$

where y_0 is the deflection due to the transverse loads alone. This equation is identical in form to Eq. (4-5) for the column with initial crookedness of amplitude δ_0. Substituting from Eq. (b) into Eq. (a) gives

$$M = M_0 + \frac{Py_0}{1 - P/P_E} = M_0 \frac{1 + (P_E y_0/M_0 - 1)(P/P_E)}{1 - P/P_E} \tag{c}$$

which can be written

$$M = M_0 \frac{1 + \Psi P/P_E}{1 - P/P_E} = \frac{C_m M_0}{1 - P/P_E} \tag{4-43a}$$

where

$$C_m = 1 + \Psi \frac{P}{P_E} \qquad \Psi = \frac{P_E y_0}{M_0} - 1 \tag{4-43b}$$

The beam-column with transverse load can now be evaluated by using Eq. (4-35) or Eq. (4-36), provided M is taken to be the maximum value of M_0 and C_m is determined by Eq. (4-43*b*).

If the member and its loads are symmetrical, both M_0 and y_0 are maximum at midlength. Thus, for the pin-ended beam-column of length L with a transverse load W at midlength

$$M_0 = \frac{WL}{4} \qquad y_0 = \frac{WL^3}{48EI} \qquad P_E = \frac{\pi^2 EI}{L^2}$$

from which

$$\Psi = \frac{\pi^2 EI}{L^2}\frac{WL^3}{48EI}\frac{4}{WL} - 1 = 0.822 - 1 = -0.178 \tag{d}$$

$$C_m = 1 - 0.178\frac{P}{P_E} \tag{e}$$

Similarly, for a pin-ended beam-column of length L with a uniformly distributed load W

$$M_0 = \frac{WL}{8} \qquad y_0 = \frac{5WL^3}{384EI} \qquad P_E = \frac{\pi^2 EI}{L^2}$$

$$\Psi = \frac{\pi^2 EI}{L^2}\frac{5WL^3}{384EI}\frac{8}{WL} - 1 = 1.028 - 1 = 0.028 \tag{f}$$

$$C_m = 1 + 0.028\frac{P}{P_E} \tag{g}$$

Values of C_m for the beam-column with lateral displacement of one end are more difficult to determine. An approximate value can be found for the case where there are no transverse loads between supports by considering the frame with an infinitely stiff beam (Fig. 4-34*a*). Each column can be considered to be a simply supported beam of span $2L$ supporting a load $2H$ at midspan if the bases are hinged (Fig. 4-34*b*) or, if the bases are fixed, as a beam of span L supporting a load $2H$ at midspan (Fig. 4-34*c*). In either case, C_m is given by Eq. (*e*). However, a constant value $C_m = 0.85$ gives a good approximation to the multiplier $C_m/(1 - P/P_E)$ of M_0 in Eq. (4-43*a*), as is shown in Table 4-1. For this reason, the AISC specification prescribes $C_m = 0.85$ for all compression members in frames subject to joint translation.

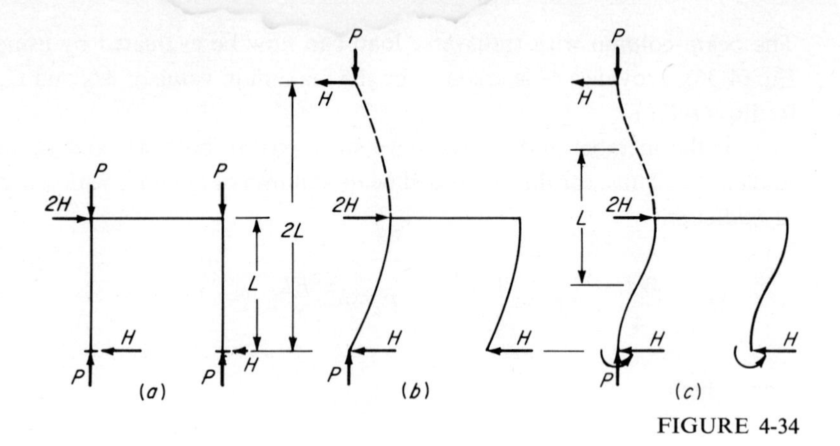

FIGURE 4-34

An approximate value of C_m for the beam-column with transverse loads between the ends can be determined in the same way as for the frame of Fig. 4-34*a*. Thus, if there is a uniformly distributed load on the column, the two cases shown in Fig. 4-35*b* and *c* show that the equivalent uniformly loaded beam is of span $2L$ for hinged bases and L for fixed bases. In either case, C_m is given by Eq. (*g*). A constant value $C_m = 1$ gives a good approximation to $C_m/(1 - P/P_E)$, as is shown in Table 4-2. The AISC specification prescribes $C_m = 0.85$ for this case also.

Table 4-1

$\frac{P}{P_E}$	$\frac{1-0.18P/P_E}{1-P/P_E}$	$\frac{0.85}{1-P/P_E}$
0	1	0.85
0.2	1.20	1.06
0.4	1.55	1.42
0.6	2.23	2.13
0.8	4.28	4.25

Table 4-2

$\frac{P}{P_E}$	$\frac{1+0.028P/P_E}{1-P/P_E}$	$\frac{1}{1-P/P_E}$
0	1	1
0.2	1.26	1.25
0.4	1.69	1.67
0.6	2.54	2.50
0.8	5.11	5.00

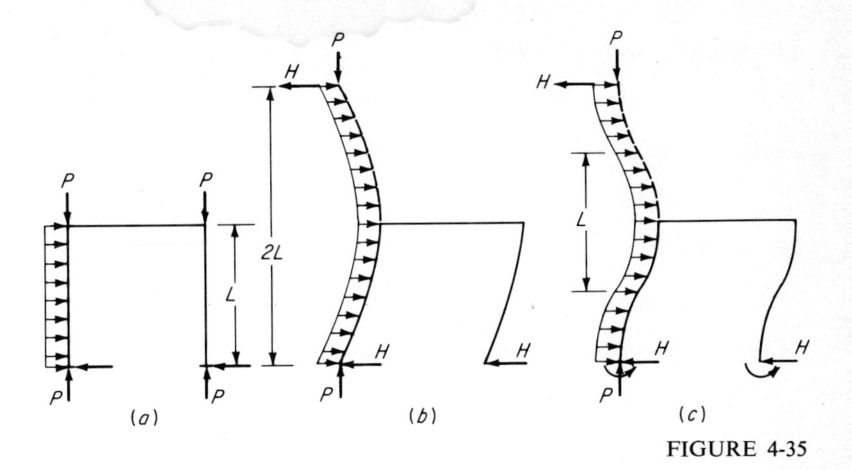

FIGURE 4-35

4-18 EFFECTIVE LENGTH OF COLUMNS IN FRAMES

The column rarely occurs as an isolated member, and its end conditions are influenced by the members to which it connects. For example, the frame shown in Fig. 4-36*a* will take the shape in Fig. 4-36*b* if it buckles under the vertical loads *P*, provided the connections at *B* and *C* are moment-resistant and stiff enough to allow little change in the 90° angles at *B* and *C*. Typical connections that satisfy these requirements are shown in Figs. 7-24 and 7-40. The critical load can be determined as follows. With the coordinate axes shown in *b*, the equation of equilibrium of column *AB* is

$$E_c I_c \frac{d^2 y}{dx^2} = -Py \tag{a}$$

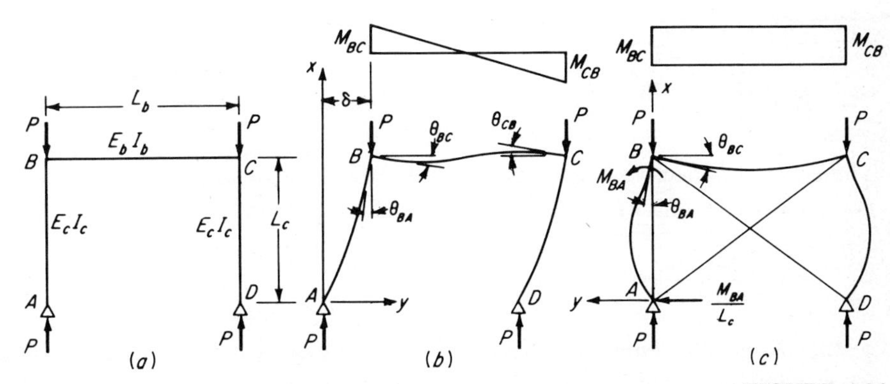

FIGURE 4-36

The solution, with $k^2 = P/E_c I_c$, is

$$y = A \sin kx + B \cos kx \tag{b}$$

Since $y = 0$ at $x = 0$

$$y = A \sin kx \tag{c}$$

The deflection $y = \delta$ at $x = L_c$ is

$$\delta = A \sin kL_c \tag{d}$$

so that the bending moment M_{BA} in the column is

$$M_{BA} = PA \sin kL_c \tag{e}$$

The moment M_{BC} for the beam is

$$M_{BC} = \frac{2E_b I_b}{L_b}(2\theta_{BC} + \theta_{CB}) = \frac{6E_b I_b \theta_{BC}}{L_b} \tag{f}$$

since $\theta_{CB} = \theta_{BC}$. Then, since $M_{BC} = M_{BA}$, Eqs. (*e*) and (*f*) give

$$\theta_{BC} = \frac{M_{BA} L_b}{6E_b I_b} = \frac{L_b PA \sin kL_c}{E_b I_b} \tag{g}$$

The rotation θ_{BA} of the column is

$$\theta_{BA} = \left(\frac{dy}{dx}\right)_{x=L_c} = Ak \cos kL_c \tag{h}$$

Equating θ_{BA} and θ_{BC} and simplifying the result gives

$$kL_c \tan kL_c = 6\frac{(EI/L)_b}{(EI/L)_c} \tag{i}$$

If the beam is infinitely stiff, Eq. (*i*) gives $\tan kL_c = \infty$, from which $kL_c = \pi/2$ and, since $k^2 = P/E_c I_c$,

$$P = \frac{\pi^2 E_c I_c}{4L_c^2} = \frac{\pi^2 E_c I_c}{(2L_c)^2} \tag{j}$$

Therefore, the effective-length coefficient $K = 2$. If beam and column are equally stiff, Eq. (*i*) gives $kL_c \tan kL_c = 6$, from which $kL_c = 1.35$ and

$$P = \frac{1.82E_c I_c}{L_c^2} = \frac{\pi^2 E_c I_c}{(2.33L_c)^2}$$

so that $K = 2.33$. Thus, the effective length increases with decreasing stiffness of the beam and tends to infinity as the beam stiffness approaches zero.

If the frame of Fig. 4-36*a* is restrained against sidesway, as by diagonal bracing, it buckles as shown in Fig. 4-36*c*. The equation of equilibrium is

$$E_c I_c \frac{d^2y}{dx^2} = -Py + M_{BA}\frac{x}{L_c} \tag{k}$$

which gives

$$y = A \sin kx + B \cos kx + \frac{M_{BA}}{P}\frac{x}{L_c} \tag{l}$$

Using $y = 0$ at $x = 0$ and at $x = L_c$, we get

$$y = \frac{M_{BA}}{P}\left(\frac{x}{L_c} - \frac{\sin kx}{\sin kL_c}\right) \tag{m}$$

Then, using the boundary conditions $\theta_{BC} = \theta_{BA} = -(dy/dx)_{x=L_c}$, we get

$$1 - kL_c \cot kL_c = -\frac{(kL_c)^2}{2}\frac{(EI/L)_c}{(EI/L)_b} \tag{n}$$

If the beam is infinitely stiff, Eq. (*n*) gives $kL_c = 4.49$, so that

$$P = \frac{20.2E_c I_c}{L_c^2} = \frac{\pi^2 E_c I_c}{(0.7L_c)^2}$$

If beam and column are equally stiff, Eq. (*n*) gives $kL_c = 3.59$, from which

$$P = \frac{12.9E_c I_c}{L_c^2} = \frac{\pi^2 E_c I_c}{(0.875L_c)^2}$$

and $K = 0.875$. Thus, the effective-length coefficient increases with decreasing stiffness of the beam and becomes unity with zero stiffness.

These examples show that the critical load for a column depends on its stiffness relative to that of the beams framing it and on the presence or absence of restraint against relative lateral displacement of its ends. (Relative joint displacement is usually called *sidesway*.) Critical loads for frames in multibay, multistory frames can be determined as in these examples. They can also be evaluated by using the three-moment equation, the slope-deflection equation, moment distribution, etc., provided certain modifications are made. Thus, distribution factors and carry-over factors in the moment-distribution procedure must be modified to account for the effect of the moment *Py*. To simplify the analysis for frames with many members, the following assumptions can be made:

1 The frame is subjected to vertical loads applied only at the joints.
2 All columns in the frame become unstable simultaneously.

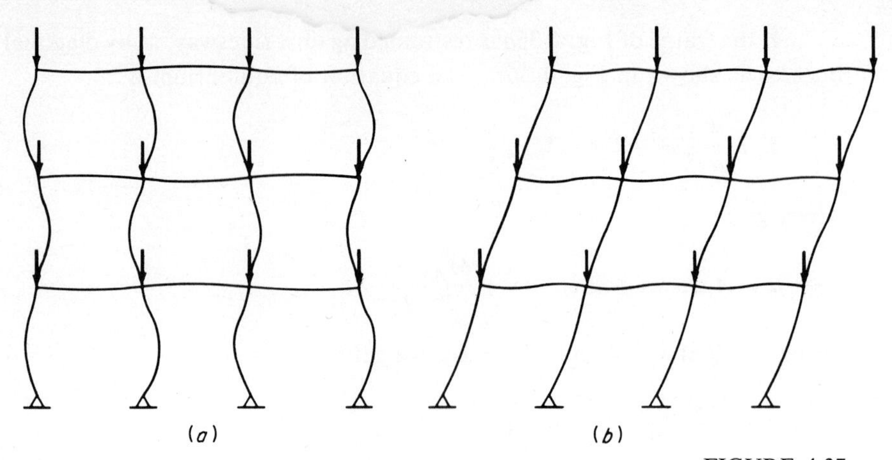

FIGURE 4-37

3 All joint rotations at a floor are equal. They are alternately clockwise and counterclockwise if there is no sidesway (Fig. 4-37*a*) and all in the same direction if there is (Fig. 4-37*b*).
4 The restraining moment exerted by the beams at a joint as the columns begin to buckle is distributed to the columns at the joint in proportion to their stiffnesses EI/L.

The preceding assumptions enable a single column AB of a frame to be analyzed by using Fig. 4-38*a* if there is no sidesway and Fig. 4-38*b* if there is. Using the ordinary slope-deflection equation for beams AC, AE, BD, and BF and the modified (for effect of axial load) slope-deflection equation for column AB, a transcendental equation for the effective-length coefficient K can be established[21] for each of the two cases in

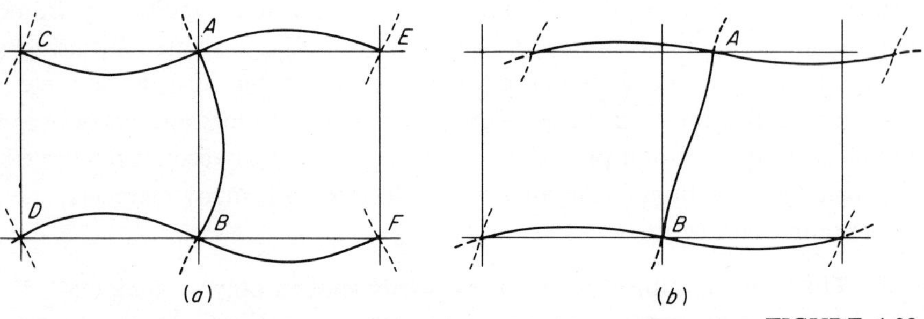

FIGURE 4-38

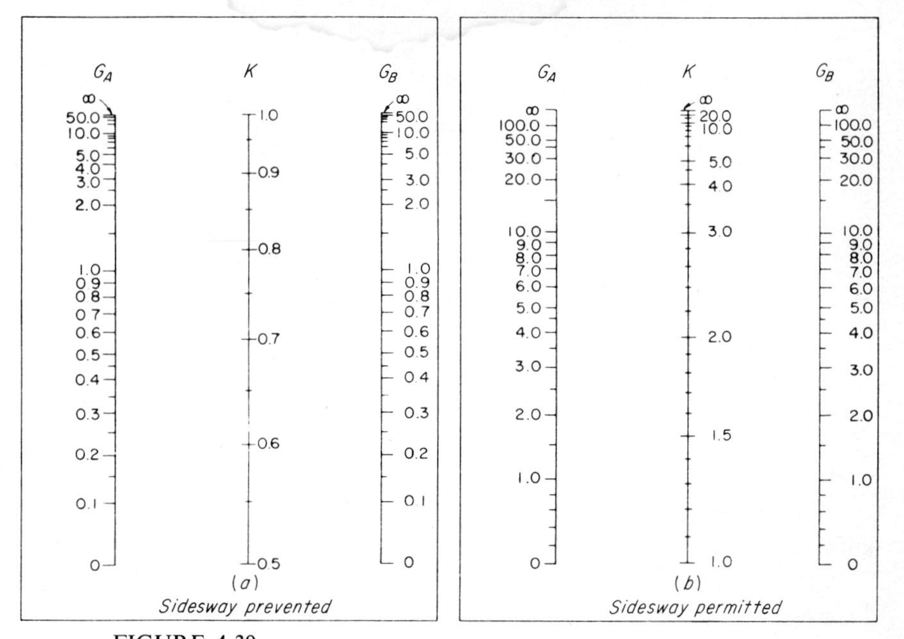

FIGURE 4-39
Nomograph for effective length of columns. (*From O. G. Julian and L. S. Lawrence, unpublished notes, 1959.*)

Fig. 4-38. Solutions of these equations were put in nomographic form in 1959 for use in the Boston Building Code[23] (Fig. 4-39). The symbol G in these diagrams is defined by

$$G = \frac{\sum (EI/L)_c}{\sum (EI/L)_b} \tag{4-44}$$

where $\sum (EI/L)_c$ is the sum of stiffnesses of columns entering the joint and $\sum (EI/L)_b$ is the sum of stiffnesses of beams entering the joint. A straight line connecting G_A and G_B intersects the axis of K at the corresponding value of K.

The nomographs can also be used for certain cases where the end rotations of the beams differ from those described in category 3 above. This is done by taking account of the change in beam stiffness. Thus, for the frame of Fig. 4-40*a*, the moment M_{CB} in the buckled configuration is zero because of the simple support. Using the slope-deflection equations for M_{BC} and M_{CB}, we get $\theta_{BC} = M_{BC} L_b / 3E_b I_b$. But the nomograph for the case where the frame buckles with sidesway is based on the deflected shape of Fig. 4-36*b*, for which the joint rotation is given by Eq. (g). Thus, the rotation at B in the frame of Fig. 4-40*b* is twice that of the frame of Fig. 4-36*b* so that the

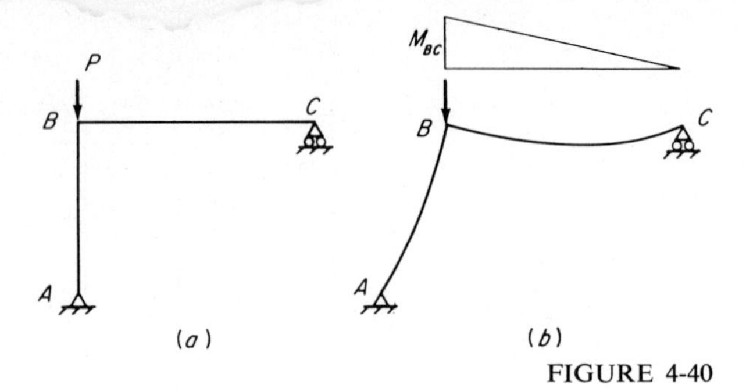

FIGURE 4-40

beam in Fig. 4-40*a* is only half as stiff as the beam in Fig. 4-36*a*. Relative stiffnesses for other cases are determined similarly. The correction factors (by which beam stiffness I/L must be multiplied) for four cases are as shown in Table 4-3.

If a column end is hinged, the value of G at that end is infinite. This is because the hinge is equivalent to zero stiffness of the connecting beams. On the other hand, if a column end is completely restrained rotationally, G at that end is zero. Intermediate values of $G = 10$ and $G = 1$ have been suggested for the practical case of the simple column base and the fixed column base, respectively.[9]

In the case of a column which buckles upon reaching a stress exceeding the proportional limit, the tangent modulus E_t should be used in evaluating column stiffness in Eq. (4-44). However, because the loads are assumed to act at the joints, the beams are straight at the onset of buckling. Therefore, they are free of stress as bending begins and are governed by Young's modulus. For this case, Eq. (4-44) can be written

$$G = \frac{\sum (\tau\, EI/L)_c}{\sum (EI/L)_b} \tag{4-45}$$

where $\tau = E_t/E$. If we assume further that both columns at the joint buckle at the same stress, we have

$$G = \frac{\tau \sum (EI/L)_c}{\sum (EI/L)_b} \tag{4-46}$$

Table 4-3

Condition	Sidesway	No sidesway
Far end of beam hinged	$\frac{1}{2}$	$\frac{3}{2}$
Far end of beam fixed	$\frac{2}{3}$	2

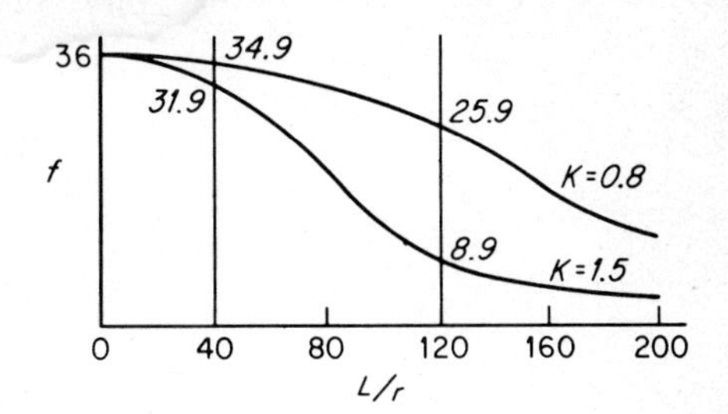

FIGURE 4-41

This requires that G be determined by trial, since the critical stress, which determines τ, and the value of G are interdependent. However, it is always safe to take $\tau = 1$, which, of course, amounts to using Eq. (4-44). This is because the beams are stiffer, relative to the columns, if the columns buckle inelastically, and exert more rotational restraint than they would if the columns buckled elastically.

The effect of end conditions on column strength is influenced to a marked degree by the slenderness of the column. This is demonstrated in Fig. 4-41, where column curves for A36 steel are shown for effective-length coefficients of 0.8 and 1.5, using the CRC formula for the inelastic range. For a column whose $L/r = 40$, the critical stresses are 34.9 and 31.9 ksi for $K = 0.8$ and 1.5, respectively, while for $L/r = 120$ they are 25.9 and 8.9 ksi. Thus, a reasonably good estimate of end conditions is of primary importance in the design of slender columns but of little consequence in the design of stocky members.

It is important to remember that the discussion in this article holds only if (1) the beam-to-column connections are moment-resistant and stiff and (2) the loads are concentrated at the joints. Thus, although the connection consisting of two angles connecting the web of the beam to the column (Fig. 4-23*a*) does develop some moment, it is too flexible to qualify as a stiff or "rigid" connection. The lateral stability of frames in which such connections are used must be assured by walls, partitions, diagonal bracing, shear walls, and the like. Furthermore, the effective-length coefficient of the columns in such a frame is practically unity.

The behavior of frames in which the loads are carried by the beams, rather than by the joints, is considered in Art. 4-20.

EXAMPLE 4-3 Determine the effective-length coefficients for the frame of Fig. 4-42. The stiffnesses are as follows:

Member I/L:		
	AB 110/15 = 7.33	BD 800/30 = 26.7
	CD 110/15 = 7.33	DG 800/20 = 40
	DE 110/12 = 9.17	GJ 800/20 = 40
	FG 110/15 = 7.33	EH 290/20 = 14.5
	GH 110/12 = 9.17	

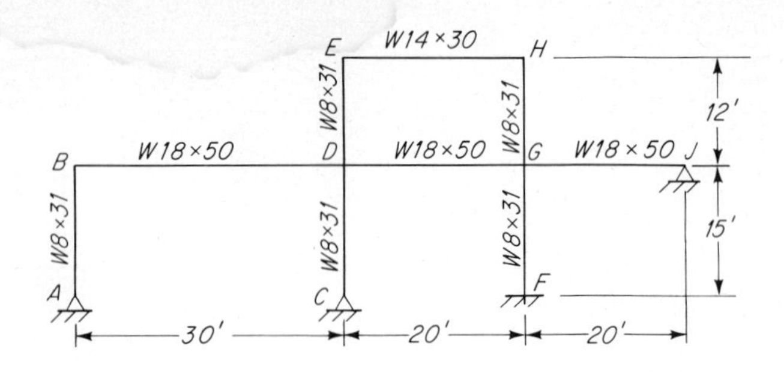

FIGURE 4-42

Column AB: $G_A = 10 \quad G_B = \dfrac{7.33}{26.7} = 0.274 \quad K = 0.77$ (Fig. 4-39a)

Column CD: $G_c = 10 \quad G_D = \dfrac{7.33 + 9.17}{26.7 + 40} = 0.247 \quad K = 0.76$ (Fig. 4-39a)

Column FG: $G_F = 1 \quad G_G = \dfrac{7.33 + 9.17}{40 + \frac{3}{2} \times 40} = 0.165 \quad K = 0.67$ (Fig. 4-39a)

Column DE: $G_D = 0.247 \quad G_E = \dfrac{9.17}{14.5} = 0.633 \quad K = 1.14$ (Fig. 4-39b)

Column GH: $G_G = \dfrac{7.33 + 9.17}{40 + \frac{1}{2} \times 40} = 0.275 \quad G_H = \dfrac{9.17}{14.5} = 0.633 \quad K = 1.15$ (Fig. 4-39b)

EXAMPLE 4-4 Determine the critical load P for the frame of Fig. 4-36a, braced as in Fig. 4-36c, with the following dimensions: $L_b = 40$ ft, $L_c = 20$ ft, $BC =$ W24 × 76, $AB = DC =$ W12 × 106, A36 steel. The columns are placed with their webs in the plane of the frame and are supported against buckling out of the plane. Use the CRC formula, Eq. (4-16), for buckling in the inelastic range. Use $E = 30{,}000$ ksi.

SOLUTION

Column: W12 × 106 $\quad I_x = 931$ in.^{4} $\quad A = 31.2$ in.^{2} $\quad r_x = 5.46$ in.

Beam: W24 × 76 $\quad I_x = 2{,}100$ in.^{4}

$$\text{Column } \frac{I}{L} = \frac{931}{20} = 46.5 \qquad \text{Beam } \frac{I}{L} = \frac{2{,}100}{40} = 52.4$$

$$G_B = \frac{52.4}{46.5} = 1.13 \qquad G_A = 10 \text{ (pinned base)} \qquad K = 0.87 \qquad \text{(Fig. 4-39}a\text{)}$$

$$\frac{L}{r_x} = \frac{20 \times 12}{5.46} = 44 \qquad \frac{KL}{r_x} = 0.87 \times 44 = 38$$

$$C_c = \pi\sqrt{\frac{2E}{F_y}} = \pi\sqrt{\frac{60{,}000}{36}} = 128$$

$$F_{cr} = 36\left[1 - \frac{1}{2}\left(\frac{38}{128}\right)^2\right] = 34.4 \text{ ksi}$$

Since F_{cr} exceeds the proportional-limit stress, the value of G_B determined above should be revised. The tangent modulus corresponding to a given inelastic-buckling stress, $\pi^2 E_t/(L/r)^2$, can be determined by dividing that stress by the Euler stress for the same L/r, $\pi^2 E/(L/r)^2$. Thus, $\tau = E_t/E = F_{cr}/F_E$. Since any stress calculated by the CRC formula is, in effect, a tangent-modulus critical stress, it can be used to determine the corresponding value of τ. Therefore,

$$F_E = \frac{\pi^2 E}{(KL/r)^2} = \frac{296{,}000}{38^2} = 210 \text{ ksi}$$

$$\tau = \frac{34.4}{210} = 0.164 \qquad G_B = 0.164 \times 1.13 = 0.185 \qquad \text{[Eq. (4-46)]}$$

$$K = 0.745 \qquad \frac{KL}{r_x} = 0.745 \times 44 = 33$$

$$F_{cr} = 36\left[1 - \frac{1}{2}\left(\frac{33}{128}\right)^2\right] = 34.8 \text{ ksi}$$

The revised value of F_{cr} is only slightly larger than the value for $\tau = 1$. Therefore, no further refinement is necessary, and

$$P_{cr} = 34.8 \times 31.18 = 1{,}085 \text{ kips}$$

For practical purposes, the first evaluation, $F_{cr} = 34.4$ ksi, is close enough.

PROBLEMS

4-35 Can a buckled form consistent with the buckled forms on which the effective-length nomographs are based be drawn for the frame of Example 4-3? Explain.

4-36 Same as Example 4-3 except with rollers at J to allow horizontal movement.

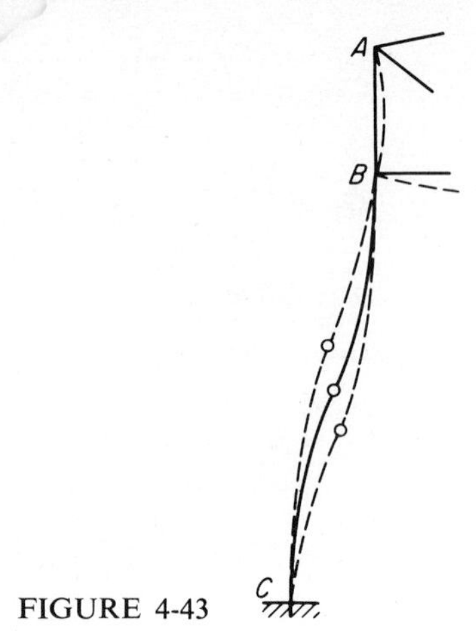

FIGURE 4-43

4-37 Compute the critical load P for the unbraced frame of Fig. 4-36a, using the same dimensions as in Example 4-4.

4-38 Choose a W for the columns of the frame of Fig. 4-36a for the following data: $L_c =$ 15 ft, $L_b = 30$ ft, $BC =$ W16 × 36, $P = 160$ kips. The frame is not braced against sidesway. A36 steel, AISC specification.

4-39 Same as Prob. 4-38 except that the frame is braced against sidesway.

4-19 DP4-3: COLUMN FOR BUILDING OF DP3-1 AND DP4-1

In this example, we make the conventional assumption that the wind force acts entirely on the windward side of the building. The points of inflection of the columns are taken midway between the column base and the bottom of the truss, as in DP4-1. This is equivalent to the assumption that the column bends as shown by the solid line in Fig. 4-43. This condition exists if the column remains straight between A and B and if the base does not rotate. Of course, rotation of the base is a function of the base detail. For example, if the anchor bolts are placed on the centerline of the column which is normal to the web, resistance to rotation around this axis may be relatively small unless the axial load is large. With a moment-resistant base, some designers would take the point of inflection below the midpoint, on the reasonable assumption that the column base is not likely to furnish complete restraint. This condition is shown by the dotted line to the right in Fig. 4-43. On the other hand, the column will rotate at B, accompanied by bending of the bottom chord connecting at B, unless the chord is infinitely stiff and rigidly connected. This rotation tends to pull the point of inflection upward, as indicated by the dotted line at the left in the figure. Since the two rotations offset each other, the point of inflection may actually lie either

COLUMN FOR BUILDING OF DP3-1 and DP4-1 — DP4-3

Building of DP3-1
- AISC specs.
- Bents 20′ c.c.
- Wind 20 psf
- DL 14 psf
- LL 40 psf

(Bents, Wind, DL, LL: See DP3-1)

$20 \times 20 = 0.4$ klf

$\frac{3 \times 15}{4.5} = 10^k$ C

6.4^k 12.3^k 6.4^k

4.5′, 10.5′, 10.5′

$3^k = 15 \times 0.4/2$

3^k

$3 + 0.4 \times 10.5 = 7.2^k$

60.83′

$0.74^k = \frac{0.4 \times 15 \times 7.5}{60.83}$

Shear diagram windward col.: 3, 7.2

Mom. diagram windward col.: $11'^k$; $\frac{7.2+3}{2} \times 10.5 = 53.5'^k$

Mom. diagram leeward col.: $31.5'^k$; $3 \times 10.5 = 31.5'^k$

Column load: $34 \times 20 \times 60.83/2 = 20.7^k$ DL + LL (a)

3.4 siding, DP5-5

12.5 conc. loads

$36.6^k - 0.7^k = 35.9^k$ (windward col.)

$+ 0.7^k = 37.3^k$ (leeward col.)

Design loads with wind:

Windward col.	Leeward col.	
$35.9 \times 3/4 = 26.9^k$	$37.3 \times 3/4 = 28.0^k$	(b)
$53.5 \times 12 \times 3/4 = 480''^k$	$31.5 \times 12 \times 3/4 = 283''^k$	

Windward column

$$A = \frac{26.9}{12(1 - 3/4)} = 8.97 \text{ in.}^2 \quad \text{Min. } r = 21 \times 12/200 = 1.26 \text{ in.} \tag{c}$$

Try W8 × 28; $A = 8.23$, $S_x = 24.3$, $r_x = 3.45$, $F_b = 0.66 \times 36 = 24$ ksi (d)

$f_a = 26.9/8.23 = 3.27$ ksi, $f_b = 480/24.3 = 19.8$ ksi

$$\frac{f_a}{0.6F_y} + \frac{f_b}{F_b} = \frac{3.27}{22} + \frac{19.8}{24} = 0.149 + 0.825 = 0.974 < 1 \tag{e}$$

Leeward column

$KL/r = 1.5 \times 21 \times 12/3.45 = 110$, $F_a = 11.67$ ksi, $F_E' = 12.34$ ksi (f)

$f_a = 28/8.23 = 3.40$ ksi, $f_b = 283/24.3 = 11.6$ ksi

$$\frac{f_a}{F_a} + \frac{f_b}{F_b}\frac{C_m}{1 - f_a/F_E'} = \frac{3.40}{11.67} + \frac{11.6}{24} \times \frac{0.85}{1 - 3.40/12.34} = 0.291 + 0.566 = 0.857 \tag{g}$$

DP 4-3
Column for building of DP 3-1 and DP 4-1.

above or below the midpoint, depending upon the relative rotations at B and C. For this reason, the point of inflection for a column with an adequately designed moment-resistant base might just as well be assumed to lie at midlength.

The wind pressure on the part of the building which lies above the points of inflection is assumed to be divided equally between the columns at these points. The 10-kip compressive force in the bottom chord of the end panel is not needed for this example; it was computed for the sake of completeness, since ordinarily the analysis here would be made, rather than the more detailed study of DP4-1. The force in the corresponding bottom chord on the windward side need not be determined, since it will always be zero if the shears in the columns are assumed to be equal at the points of inflection. These results should be compared with those of DP4-1, Figs. *b* and *c*.

The following comments are identified by letters corresponding to those alongside the computations in DP4-3.

a Although the basic live load is 40 psf, it is assumed here that a snow load of 20 psf is ample in combination with wind load. The unlikelihood of large accumulations of snow together with high-velocity wind was discussed in Art. 1-7.

b For load due to wind alone or the combinations of wind and other forces, the basic allowable stresses may be increased by one-third. This is equivalent to designing for three-quarters of the expected load, using the basic allowable stress.

c Since the axial load is small compared with the moment, bending stresses will be relatively large. This explains the estimate of 3/4 for the ratio of f_b/F_b. The column is a W shape placed with its minor axis in the plane of the frame. The exterior wall prevents bending about the minor axis, so that only in-plane bending need be considered.

d Allowable bending stresses are discussed in Art. 5-5. In this case, the AISC value is $F_b = 0.66F_y$.

e This is the AISC prescribed check [Eq. (4-39*b*)] for maximum stress at the end with the larger moment, in this case, at the column base. Equation (4-39*a*) would be used to check the stress at the interior point of maximum moment. However, this maximum is only 11 ft-kips (see moment diagram), and it is evident that it would not be increased sufficiently by the moment Py to require a check.

f The value of K is an estimate. Even if the bottom chord were connected to the column with a moment connection, Eq. (4-44) could not be used to compute the value of G to be used in Fig. 4-39*b* to determine K. This is because one of the assumptions on which the nomogram is based is that all columns in a frame reach their critical loads simultaneously. In this case, however, the 4.5-ft segment AB (Fig. 4-43) of the continuous column ABC has a much larger critical load than the 21-ft segment BC. Thus, AB exerts rotational restraint on BC. Therefore, this suggests a modification of Eq. (4-44) for this situation; namely, consider AB to be a beam, rather than a column, for the purpose of evaluating G_B. Thus

$$G_B = \frac{(I/L)_{BC}}{0.5(I/L)_{AB}} = \frac{4.5}{0.5 \times 21} = 0.44$$

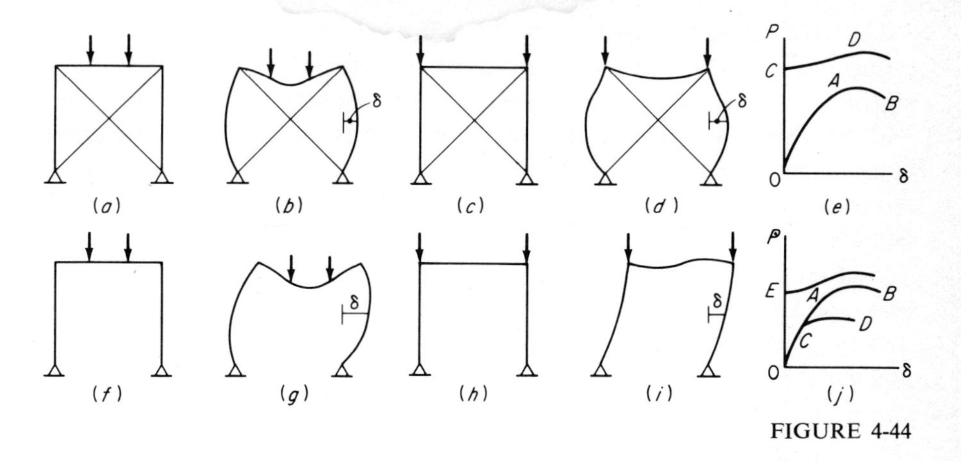

FIGURE 4-44

where the factor 0.5 in the denominator accounts for the assumed hinged end at A. Then, using $K_C = 1$ for the fixed base, rather than the theoretical value zero (Art. 4-18), we get $K = 1.22$ from Fig. 4-39*b*. Then, because we have overestimated the stiffness of AB in neglecting the axial force, we increase K arbitrarily to 1.5. The allowable values F_a and F'_E are from the AISC Manual.

g The leeward column must also be checked by both Eqs. (4-39*a*) and (4-39*b*). Upon examining the result from Eq. (4-39*a*), however, it is clear that Eq. (4-39*b*) does not control. The value $C_m = 0.85$ is discussed in Art. 4-17.

The columns have now been checked for gravity load plus wind load at the permitted increased allowable stress for this loading condition. They must also be checked for gravity load alone at the normal allowable stress. Since we reduced the live load from 40 to 20 psf in combination with wind, the 20.7-kip load computed in *a* must be corrected to $P = 54 \times 20 \times 60.83/2 = 32.9$ kips. Adding the siding and concentrated loads to this gives 48.8 kips, for which $f_a = 48.8/8.23 = 5.9$ ksi. With gravity load alone, K for the column would be less than unity. Therefore, the allowable stress F_a would be larger than the value calculated in *f*, so that the column is adequate for gravity load alone.

4-20 BUCKLING OF FRAMES WITH LOADED BEAMS

The behavior of the braced frame shown in Fig. 4-44*a* and *b*, in which symmetrically placed loads P are carried by the beam, is quite different from that of the same frame in Fig. 4-44*c* and *d*, where the loads act at the joints. Thus, lateral deflection δ of a point on the column (at midheight, say) of the frame in *a* begins at $P = 0$ and increases until a maximum value of P is reached (OAB in Fig. 4-44*e*). On the other hand, there is no lateral deflection of a point on the column of the frame in Fig. 4-44*c* until the critical load is reached, at which buckling begins (C in Fig. 4-44e). Furthermore, the points of inflection are in the beams in the one case and in the columns in the other (Fig. 4-44b and d). Since the columns in the frame with the loaded beam are beam-columns, the peak load can be determined by Eqs. (4-36) and (4-38), provided the frame moments are determined by plastic rather than elastic analysis. The critical load corresponding to point C is determined by the procedures described in Examples 4-3 and 4-4.

If the frame with the loaded beam is not braced (Fig. 4-44f), it will follow, at least for a time, the same load-deflection path as the braced frame (OAB in Fig. 4-44j). However, before point A is reached, it may become unstable in the symmetrical configuration and shift to the asymmetrical one shown in Fig. 4-44g. The subsequent load-deflection curve is CD in Fig. 4-44j. This behavior is the same as that of the frame in Fig. 4-44h and i, which buckles at a load corresponding to point E in Fig. 4-44j; i.e. there is a bifurcation of the equilibrium configuration. The critical load corresponding to point E is determined in the same way as the critical load corresponding to point C in Fig. 4-44e; the only difference in the solution is in the effective-length coefficient K to be used in the column-buckling formula. The problem, then, is to determine the load at which the unbraced frame with loaded beams becomes unstable in respect to lateral movement. As it turns out, this critical load is relatively insensitive to the location of the loads P. In other words, the critical values of P for the frames of Fig. 4-44f and h are practically equal. This was shown first by Chwalla, who investigated the sidesway stability of the one-bay one-story frame which buckles elastically (Ref. 3, p. 227).

The frame shown in Fig. 4-45a has been used to investigate the sidesway stability of multistory frames which buckle elastically.[24] The load $(n - 1)wL_b/2$ at each column top represents the load from the upper stories of an n-story frame. Of course, the simulation is approximate because continuity with the second-story column is neglected. The calculated loads at bifurcation of equilibrium are shown in Fig. 4-45b, together with the corresponding values of the effective-length coefficients K. Results of the analysis were in excellent agreement with tests on two small-scale frames; the ratio of test load to predicted load was 0.98 in one case and 0.96 in the other. OCD shows the variation of P with the horizontal reaction H for a simulated three-story frame. Bifurcation occurs at point C. Thus, OCD corresponds to OCD in Fig. 4-44j. The effective-length coefficient for this case is $K = 2.95$. OAB shows the variation of P with H for the frame loaded only at the column tops. The value $K = 2.92$ for this case is given by the nomograph, Fig. 4-39b. It is clear that the difference between the extreme case where only the beam is loaded ($K = 3.02$) and the other extreme case where only the column tops are loaded is of no consequence. This is an important conclusion because, to the extent that we can expect the behavior shown in Fig. 4-45 to be typical, it means that the nomograph in Fig. 4-39b can be used to determine the

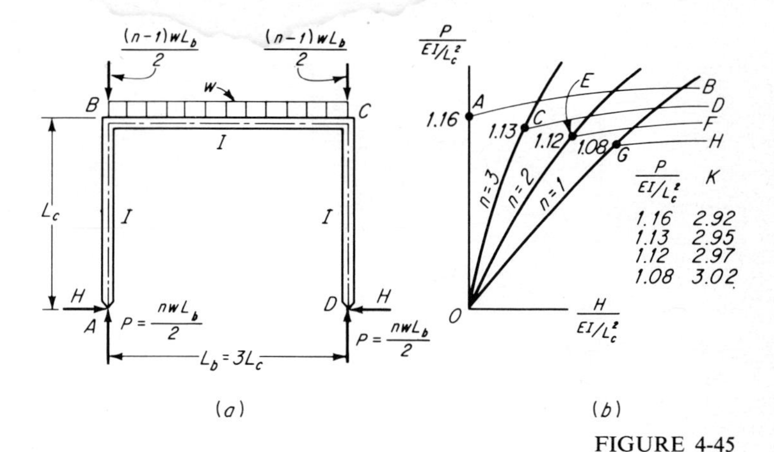

FIGURE 4-45

critical load for the unbraced frame with loaded beams, provided buckling occurs before the beams become inelastic. The case where the beam yields before the frame becomes unstable is discussed in Art. 8-18.

A braced frame loaded to simulate a multistory frame is shown in Fig. 4-46. The columns are W8 × 31, 20 ft long, and the beam a W18 × 50, 30 ft long, all A36 steel. The load nP which produces first yielding is determined by Eqs. (4-35) and (4-37) for the columns, using the CRC formula to compute P_{cr} in the inelastic range, and by $F_y = Mc/I$ for the beam. The results, $nP = 32$, 128, and 184 kips for $n = 1$, 4, and 8, respectively, are shown on the nP-δ plots OB, OC, and OD of Fig. 4-47a. The peaks of these load-deflection curves, $nP = 39$, 143, and 250 kips, are determined by Eqs.(4-36) and (4-38), using frame moments determined by plastic analysis. Also shown is the critical load $nP = 280$ kips for the frame with all the load acting at the column tops.

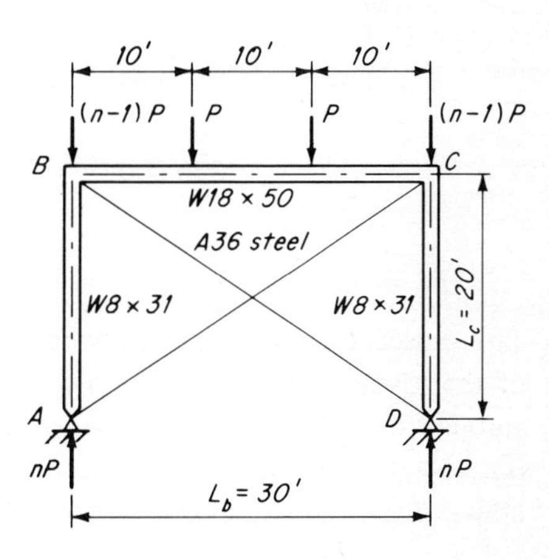

FIGURE 4-46

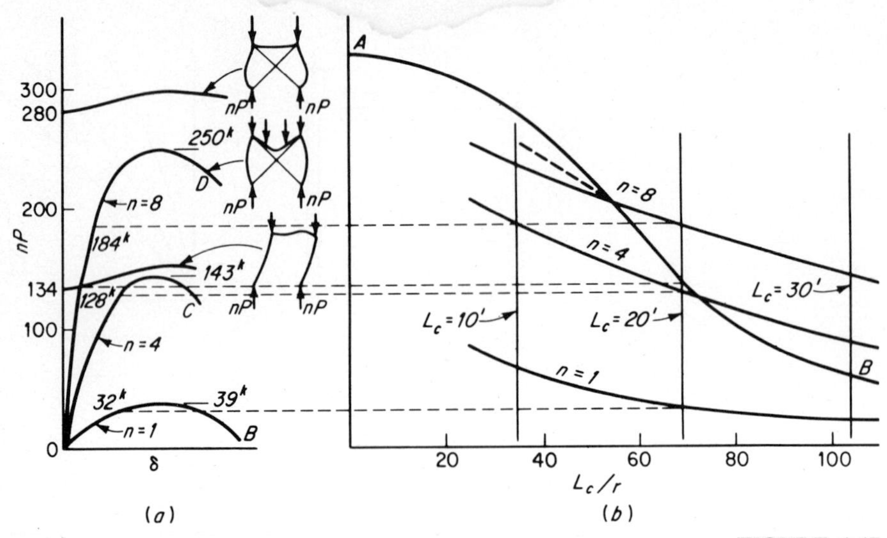

FIGURE 4-47

This load is given by the CRC formula, using the effective-length coefficient K from Fig. 4-39*a*. It is seen that this critical load can be orders of magnitude larger than the load that can be supported when part of it acts on the beam.

The critical load $nP = 134$ kips for the unbraced frame with the loads nP at the column tops is also shown in Fig. 4-47*a*. This load is also easily determined, this time with K from Fig. 4-39*b*. Thus,

$$G_B = \frac{(I/L)_c}{(I/L)_b} = \frac{110/20}{802/30} = 0.205 \qquad G_A = \infty \qquad \text{for hinged base}$$

$$K = 2.05 \qquad \frac{KL}{r} = 2.05 \times 20 \times \frac{12}{3.47} = 142$$

$$F_E = \frac{\pi^2 E}{(KL/r)^2} = \frac{296{,}000}{142^2} = 14.7 \text{ ksi}$$

$$P_{cr} = F_E A = 14.7 \times 9.12 = 134 \text{ kips}$$

This critical load, as has already been shown (Fig. 4-45), is not sensitive to the position of the loads. Therefore, it is also a close estimate of the critical load for sidesway stability of unbraced frames loaded as in Fig. 4-46. From this we can conclude that the one-story frame can attain the peak load 39 kips with or without bracing against lateral displacement. Also, the four-story frame should be able to reach the beginning-of-yield load, 128 kips, without bracing but would be unlikely to reach its peak load, 143 kips, in the unbraced condition. Finally, if $n = 8$, the frame can be expected to buckle laterally at about 134 kips if it is not braced, so that it cannot attain even its yield load unless it is braced.

Figure 4-47*b* shows the variation in first-yield loads with column slenderness. The curves for $n = 1$, 4, and 8 were plotted by computing additional first-yield loads for the frame of Fig. 4-46 with $L_c = 10$ and 30 ft and corresponding beam lengths $L_b = 15$ and 45 ft, so that the ratio of column stiffness to beam stiffness is constant. Also shown is the variation in critical load for the unbraced frame (curve *AB*). According to this figure, the one-story frame would be stable to its first-yield load even if the W8 × 31 columns were longer than 30 ft. Similarly, for $n = 8$, the frame should be stable to its first-yield load for columns shorter than about 15 ft.

Sidesway stability of unbraced frames at loads exceeding first-yield values are not as readily determined because penetration of yield stress into those cross sections of a beam where the bending moment exceeds M_y reduces the beam stiffness. This reduces the rotational restraint of the columns, which increases their effective lengths. Therefore, *AB* in Fig. 4-47*b* is lowered to the left of its intersection with each *n* curve, as is shown by the dashed line at the intersection with the curve for $n = 8$. This problem is discussed further in Art. 8-18.

The AISC specification does not prescribe the above procedure for evaluating stability of unbraced frames. Instead, the interaction formulas are to be used as in the case of the braced frame, with the effective-length coefficient K determined by a "rational method," of which the nomograph of Fig. 4-39*b* is one. It should also be pointed out that bifurcation of the equilibrium configuration does not occur in frames with lateral loads. Thus, this is a first-yield or peak-load problem, which is investigated with the appropriate interaction formulas. For this application, AISC prescribes $C_m = 0.85$ for the unbraced frame (Art. 4-17).

PROBLEMS

4-40 Determine the allowable load P for the frame of Fig. 4-46 with $n = 1$. AISC specification with $F_b = 0.66F_y$ for the beam and $0.6F_y$ for the column. (Reasons for the differences in F_b are discussed in Art. 5-5.) The frame is supported against out-of-plane buckling.

4-41 Same as Prob. 4-40 except that the frame is unbraced.

4-42 Given a braced frame of the type shown in Fig. 4-46 with $L_b = 40$ ft and $L_c = 16$ ft. The frame is to be designed for a uniform load of 1,600 plf, of which 1,000 plf is dead load. Determine member sizes in A36 steel, using first-yield load factors of 1.2 on dead load and 1.6 on live load. The frame is supported against out-of-plane buckling.

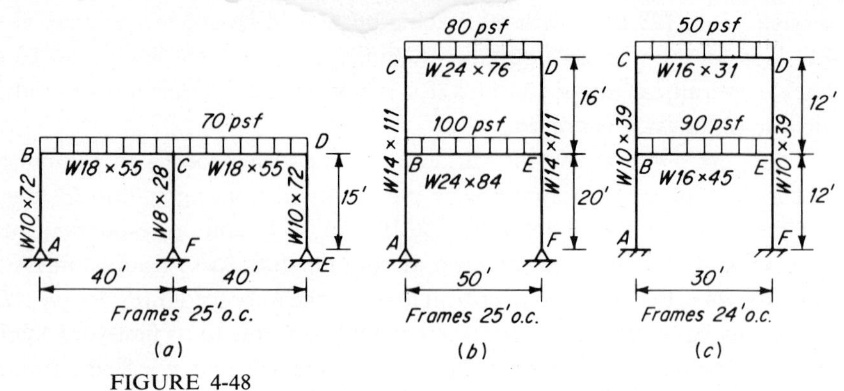

FIGURE 4-48
(*a*) Problem 4-44, (*b*) Problems 4-45 and 4-46, (*c*) Problems 4-47 and 4-48.

4-43 Check the adequacy of the frame of Fig. 4-46, without bracing but supported against out-of-plane buckling, for a uniform load of 1.8 klf together with a wind force of 6 kips at *B*. AISC specification with the same values of F_b as in Prob. 4-40, except that allowable stresses are increased by one-third when wind forces are considered.

4-44 Check the adequacy of the A36 frame shown in Fig. 4-48*a*. The frame is to be considered braced, and supported against out-of-plane buckling. AISC specification with $F_b = 0.66F_y$ for all members.

4-45 Check the adequacy of the A36-steel frame shown in Fig. 4.48*b*. The frame is unbraced but is supported against out-of-plane buckling. AISC specification.

4-46 Same as Prob. 4-45 except that there is a wind load of 20 psf in addition to the specified gravity loads. Assume the wind load to be concentrated at *B* and *C*. Allowable stresses are increased by one-third when wind forces are considered.

4-47 Check the adequacy of the A36-steel frame of Fig. 4-48*c* for a first-yield load factor of 1.6. The frame is to be considered braced and supported against out-of-plane buckling.

4-48 Same as Prob. 4-47. Use the procedure discussed in Art. 4-20 to determine whether the frame can support the loads without in-plane bracing.

4-21 LOCAL BUCKLING

The flanges, webs, and other plate elements of structural members may develop wave formations when they are compressed. This is called *local buckling*. Figure 4-49*a* shows an aluminum H and Fig. 4-49*b* an aluminum cruciform cross section, each tested in uniform axial compression, which have buckled in this fashion.

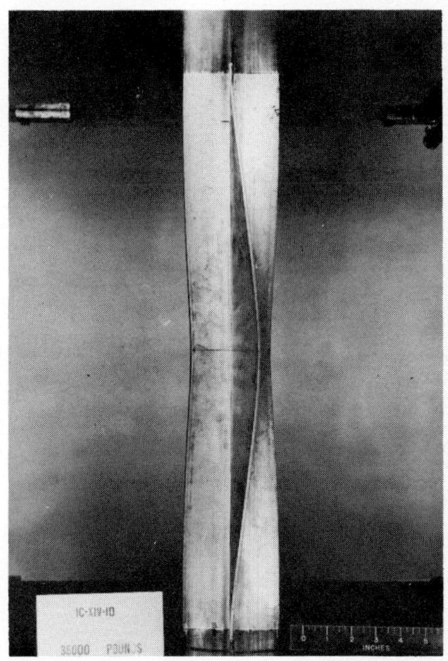

FIGURE 4-49
Local buckling of columns. (*National Aeronautics and Space Administration.*)

The critical stress for rectangular plates with various types of edge support, and with loads in the plane of the plate distributed along the edges in various ways, is given by

$$F_{cr} = \frac{k\pi^2 E}{12(1 - \mu^2)(b/t)^2} \tag{4-47}$$

where k = a constant which depends upon how the edges are supported, upon the ratio of plate length to plate width, and upon the nature of the loading

μ = Poisson's ratio

b = length of loaded edge of plate (except that it is the smaller lateral dimension when the plate is subjected only to shearing forces)

t = plate thickness

The derivation of this equation for the plate shown in Fig. 4-50*a* is given in Chap. 9. In this case, the plate is simply supported on all four edges and is uniformly compressed on two opposite edges of width b. Such a plate buckles in one transverse wave and one or more longitudinal waves. Values of the coefficient k in Eq. (4-47)

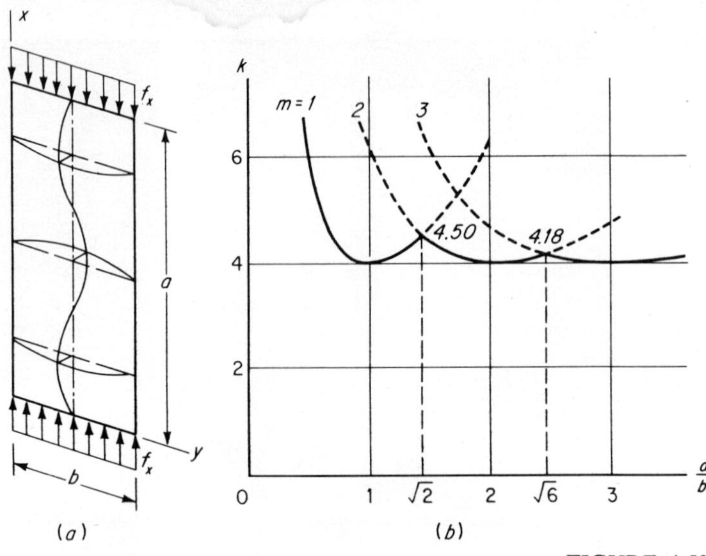

FIGURE 4-50

for this case are given in Fig. 4-50*b*, where *m* denotes the number of longitudinal waves. The ratio a/b of plate length to plate width is called the *aspect ratio*. There is one longitudinal wave if $a/b \lessapprox \sqrt{2}$, two if $\sqrt{2} \lessapprox a/b \lessapprox \sqrt{6}$, etc. The coefficient k has a minimum value of 4 for $a/b = 1, 2, 3$, etc. However, except for the unlikely case of the extremely short plate, the error in using $k = 4$ for all cases is at most about 10 percent. The error decreases with increasing a/b, and in the usual case, for which a/b is likely to be of the order of 10 or more, it is extremely small.

Values of k for five cases are given in Fig. 4-51. Case *a* in this figure is the plate of Fig. 4-50. Behavior of the plates in *b*, *c*, and *d* is similar to that of the plate in *a*; that is, they buckle in one transverse wave and a number of longitudinal waves. In each of these cases the value of k in the figure is the minimum value. On the other hand, the plate with one unloaded edge free and the other simply supported (case *e*) buckles in one longitudinal wave regardless of the aspect ratio. The corresponding value of k approaches the limiting value 0.456 with increasing aspect ratio. However, even for a plate as short as $a = 5b$ the value of k (0.496) is only 9 percent larger than the minimum value. Therefore, except for very short plates, the minimum value is a good approximation. Figure 4-49*b* is an example of this case, except that here the loaded edges are rotationally restrained. Although the member has four plate elements, with one edge each in common, there is no rotational restraint on this edge because all four legs buckle simultaneously since they are identical.

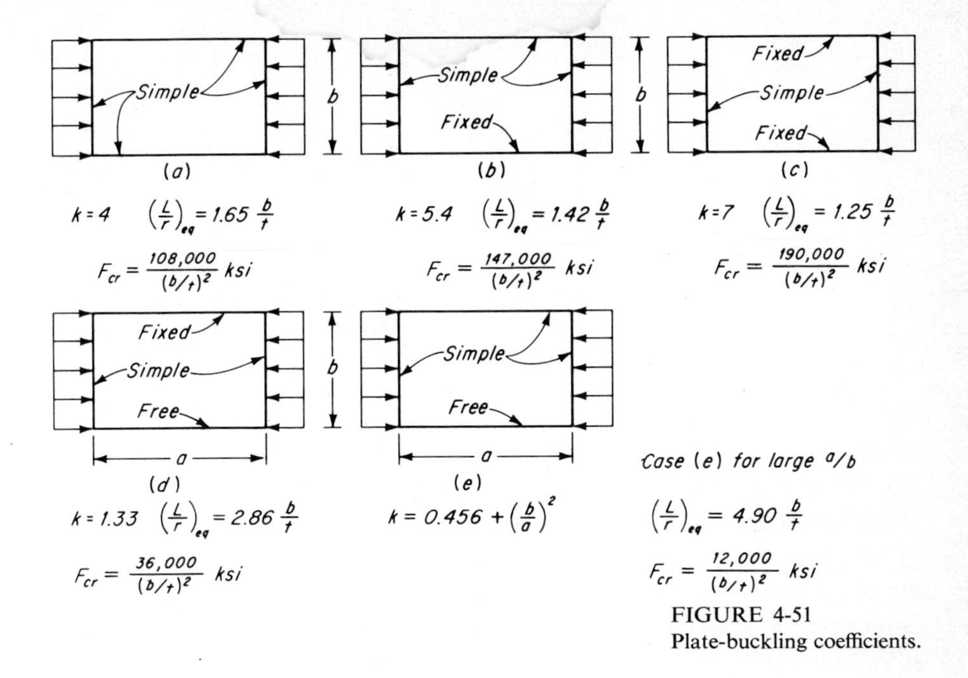

FIGURE 4-51
Plate-buckling coefficients.

The buckling shown in Fig. 4-49*a* is an example of a case between cases *d* and *e* of Fig. 4-51. Each of the four half-flanges is held straight and is rotationally restrained (but not fixed) at the juncture with the web. Of course, the corresponding value of k lies between the values for cases *d* and *e*. Values of k for incomplete rotational restraint have been determined for a number of cases (Ref. 3, p. 346).

Comparison of Eq. (4-47) with Eq. (4-2) shows that the ratio b/t of a plate plays the same role in its buckling behavior as the slenderness ratio of a column does. Also, as for the Euler formula, Eq. (4-47) is correct only if the critical stress does not exceed the proportional limit, but it can be extended to the inelastic range by using a reduced modulus of elasticity. However, the reduced modulus is not the tangent modulus, as in the case of the column. This is because the plate is anisotropic in its resistance to buckling at stresses exceeding the proportional limit. This can be shown by noting that the plate is assumed to be perfectly flat at the onset of buckling. Thus, for the plate loaded as in Fig. 4-50*a*, stresses at the onset of buckling are f_x (compressive) in the x direction and $f_y = 0$ in the y direction. These are shown in Fig. 4-52, where $f_x > F_p$. Now, if the plate begins to bend in the manner shown in Fig. 4-50*a*, bending stresses develop in both the x and the y directions. The bending stresses in the y direction are governed by E, since they begin with $f_y = 0$ (Fig. 4-52). Thus, the plate stiffness for this direction, per unit of width, is $Et^3/12(1 - \mu^2)$. On the other hand,

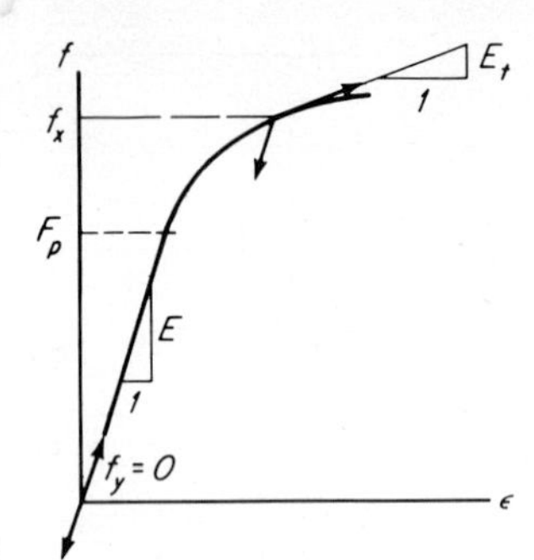

FIGURE 4-52

bending stresses in the x direction are superimposed on the uniform compression f_x. According to the double-modulus theory, compressive bending stresses would initiate at the rate E_t, while tensile bending stresses would initiate at the rate E (Fig. 4-52). In this case, the plate stiffness per unit of width is $E_r t^3/12(1 - \mu^2)$, where E_r is the double modulus (Art. 4-4). According to the tangent-modulus theory, however, there is no reversal of stress at the instant buckling begins (Art. 4-4), in which case the stiffness is $E_t t^3/12(1 - \mu^2)$. Either way, the plate is anisotropic, because the stiffness in the y direction is $Et^3/12(1 - \mu^2)$. It is shown in Ref. 3, p. 354 that, using the tangent modulus, anisotropy for the cases shown in Fig. 4-51 is accounted for with good, conservative approximation by replacing E with $\sqrt{EE_t} = E\sqrt{\tau}$, where $\tau = E_t/E$.

Critical stresses for plate buckling can be evaluated by determining the *equivalent slenderness ratio* for which a column will buckle at the same stress. The equivalent slenderness ratio is found by replacing E in Eq. (4-47) with $E\sqrt{\tau}$ and equating F_{cr} to the value given by Eq. (4-9). The result is

$$\left(\frac{L}{r}\right)_{eq} = \frac{3.3 \sqrt[4]{\tau}}{\sqrt{k}} \frac{b}{t} \qquad (4\text{-}48)$$

The value of τ to be used in this equation depends on the critical stress of the plate, which depends in turn on the value of $(L/r)_{eq}$. Therefore, the inelastic buckling stress should be determined by trial and error. However, it is on the safe side to ignore τ, since this results in a larger value of $(L/r)_{eq}$. The error is not significant, because the error in $(L/r)_{eq}$ is large only for the smaller values of τ, which correspond to stresses close to the yield stress. The critical stress is not very sensitive to L/r in such cases. The resulting equivalent slenderness ratios are given in Fig. 4-51.

Figure 4-53 shows the variation of critical stress with slenderness b/t. For a perfectly flat plate made of steel with a flat-top yield and with no residual stresses and no eccentricities of the edge stresses, the critical stress is given by ABC if strain

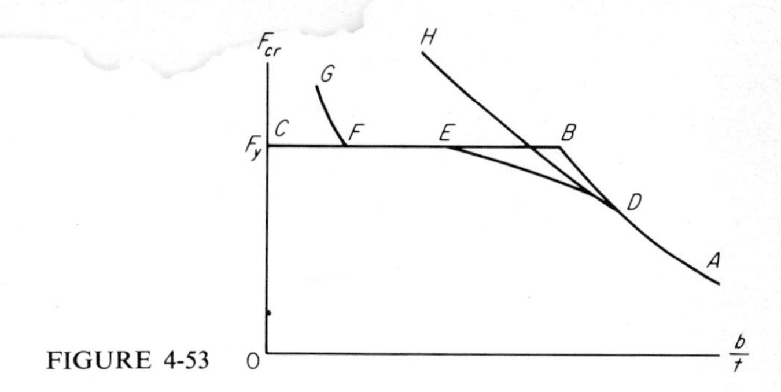

FIGURE 4-53

hardening is neglected and by $ABFG$ if it is not. Of course, as with columns, imperfections reduce the critical stress given by this curve, and the local buckling curve for a real plate is $ADEFG$. On the other hand, DH is typical for inelastic buckling of a plate made of gradually yielding metals, where the ordinate to D is the proportional-limit stress. This curve is determined by Eq. (4-47), using the inelastic modulus $E\sqrt{\tau}$, or by using Eq. (4-2) with the equivalent slenderness ratios from Eq. (4-48).

4-22 LOCAL BUCKLING WITH RESIDUAL STRESSES

The effect of residual stresses on buckling is not the same for local buckling of plates as it is for primary buckling of columns. Thus, in a buckled column, the bending moment at any cross section caused by the residual stresses F_r at an adjacent parallel cross section Δx distant is $\int \Delta\delta \, F_r \, dA$, where $\Delta\delta$ is the increment in deflection in the distance Δx. However, all points of a given cross section deflect equally, so that this moment is $\Delta\delta \int F_r \, dA = 0$. Therefore, Eq. ($g$) of Art. 4-2 remains the same in the presence of symmetrical residual-stress patterns, and their only effect is to reduce the bending stiffness of the cross section through premature yielding (Art. 4-6). On the other hand, deflection of the middle surface of a plate varies transversely as well as longitudinally. In this case, the residual-stress bending-moment increment on a transverse section of a buckled plate generally is not zero. This may reduce the local-buckling strength.

In the case of a plate welded along both longitudinal edges, as in the plates of a welded box or the web of a welded I, the large residual tension (usually of yield-stress intensity) induces a fairly uniform residual compression F_{rc} over most of the width of the plate (Fig. 4-54a). When the plate buckles, the contribution of the residual tension to bending at a cross section is small, since the increments of deflection are smaller at the edge of the plate than they are in the interior. The result is a bending moment which is in addition to the moment from the externally applied compressive

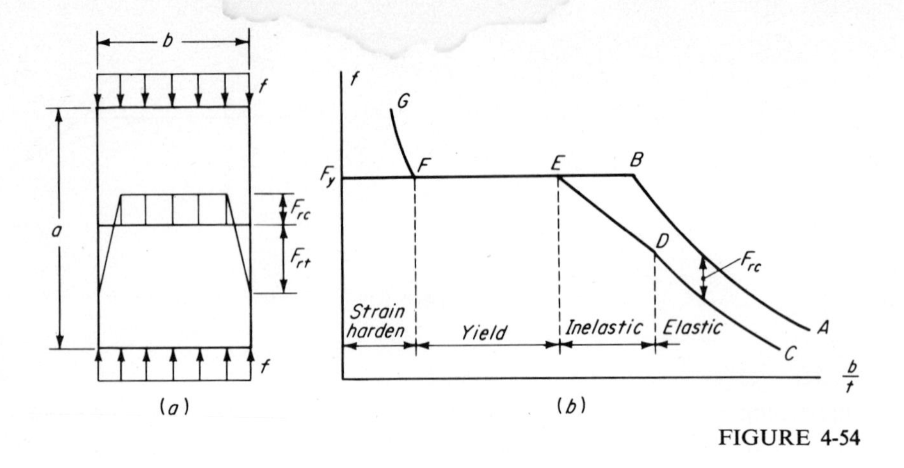

FIGURE 4-54

stress f. This reduces the critical stress. If $F_{rc} \lesssim 0.15F_y$, the reduction is practically equal to F_{rc} if the plate buckles elastically. This is shown in Fig. 4-54*b*, where *CD* and *AB* are the curves for elastic buckling of the plate with and without residual stress, respectively. If $F_{rc} \gtrsim 0.15F_y$, the reduction in critical stress is smaller than F_{rc}. Inelastic buckling is shown by *DE*. In this case, buckling begins *after* some premature yielding of the cross section because of the residual compression F_{rc}. In this case, the resulting reduction in stiffness of the plate must be taken into account. Thus, the problem is now one of buckling of an anisotropic plate. Methods of solution are discussed in Ref. 25.

4-23 DESIGN PROCEDURES FOR LOCAL BUCKLING OF STEEL COLUMNS

Economy of weight of steel columns in bridges, buildings, and similar structures is most likely to be realized if the plate elements do not buckle locally before the critical load for primary buckling of the column is reached. This requires that limiting values of plate slenderness b/t be established.

Equation (4-47) is plotted in Fig. 4-55 for four of the cases given in Fig. 4-51. Two levels of yield stress are shown. According to this figure, a perfect plate of A36 steel loaded in compression on two opposite edges and simply supported on the unloaded edges ($k = 4$) can reach yield stress without buckling if $b/t \lesssim 55$. The corresponding value for a yield stress of 50 ksi is 46. However, these values do not account for the various imperfections which lower the proportional limit. Therefore, smaller values, corresponding to point E in Fig. 4-53, must be used. This is indicated

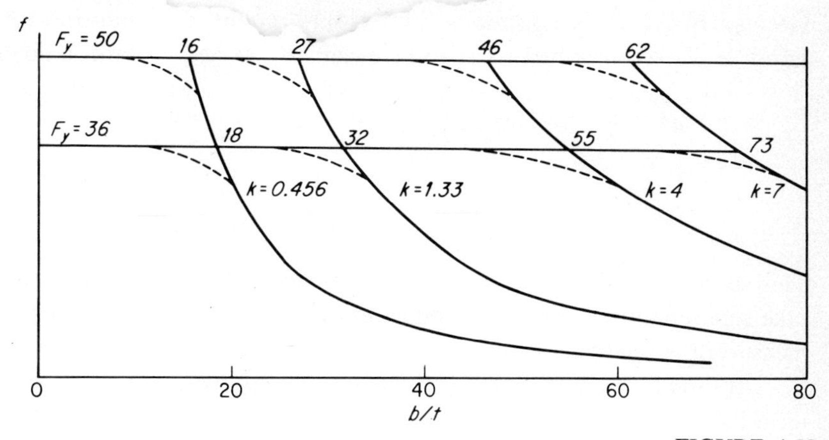

FIGURE 4-55

by the dashed lines in Fig. 4-55. Of course, these values are not easy to determine because of the considerable variation in out-of-flatness, residual stresses, etc. Therefore, they are largely based on judgment and experience. Where there is likely to be rotational restraint, values are interpolated between simple-edge and clamped-edge values.

Table 4-4 compares typical limiting values of b/t for the AISC, AREA, and AASHO specifications. AISC and AREA values are based on yield stress. In the former F_y is in kips per square inch, in the latter, pounds per square inch. The AASHO values are based on the calculated ("actual") stress f_a ksi. Each specification defines the various elements in its own way, so that there is no uniformity. Therefore, the description of the types of elements in this table is a compromise and may differ from that in the specification. Furthermore, the table is for illustrative purposes, and is not complete.

It will be noted that the AISC values suggest precision which is not consistent with the uncertainty in establishing these limits. This is because they were written in round numbers with F_y in pounds per square inch until 1969 and were not rounded when the units were changed. Thus, the limit $253/\sqrt{F_y}$ was $8{,}000/\sqrt{F_y}$ in earlier editions.

The equivalent slenderness ratios of Fig. 4-51 suggest another way in which limiting values of b/t may be determined. Thus, for the plate of Fig. 4-51*a*, we have

$$\frac{b}{t} = \frac{1}{1.65}\,\frac{L}{r} = 0.6\,\frac{L}{r}$$

Primary buckling and local buckling are equally likely at this value of b/t. The

German Buckling Specifications[26] DIN 4114 require the following for column web or cover plates supported so as to be essentially hinged on both unloaded edges:

$$\frac{b}{t} \leqq \begin{cases} 45 & \text{for } \dfrac{L}{r} \leqq 75 \\ 0.6\dfrac{L}{r} & \text{for } \dfrac{L}{r} \geqq 75 \end{cases}$$

This specification prescribes formulas for various other types of edge support. These take rotational restraint into account, so that they are in terms of a parameter which measures the relative stiffnesses of the element and the parts to which it connects.

It should be noted that the critical stress given by Eq. (4-47) is the stress at which local buckling of a perfect plate just begins. Therefore, whether it is a good estimate of the actual strength of a plate can be determined only by investigating the post-buckling behavior, as in the case of the axially loaded column (Fig. 4-3). As it turns out, the strength of the plate under uniform edge compression can be considerably larger than the load at which buckling begins. However, forces in compression

Table 4-4 TYPICAL LIMITING VALUES OF PLATE SLENDERNESS b/t

	Specification			Value for $F_y = 36$ ksi		
Type of element	AISC F_y, ksi	AREA F_y, psi	AASHO f_a, psi	AISC	AREA	AASHO*
Supported on one unloaded edge:						
Single angle	$\dfrac{76}{\sqrt{F_y}}$	...	$\dfrac{1{,}625}{\sqrt{f_a}}$	13	...	12
Projecting element	$\dfrac{95}{\sqrt{F_y}}$	$\dfrac{2{,}300}{\sqrt{F_y}}$	$\dfrac{1{,}625}{\sqrt{f_a}}$	16	12	12
Stem of T	$\dfrac{127}{\sqrt{F_y}}$	$\dfrac{3{,}000}{\sqrt{F_y}}$	...	21	16	
Supported on both unloaded edges:						
Webs	$\dfrac{253}{\sqrt{F_y}}$	$\dfrac{6{,}000}{\sqrt{F_y}}$	$\dfrac{4{,}000}{\sqrt{f_a}}$	42	32	32
Cover plates	$\dfrac{253}{\sqrt{F_y}}$	$\dfrac{7{,}500}{\sqrt{F_y}}$	$\dfrac{5{,}000}{\sqrt{f_a}}$	42	40	40
Perforated cover plates	$\dfrac{317}{\sqrt{F_y}}$	$\dfrac{7{,}500}{\sqrt{F_y}}$	$\dfrac{6{,}000}{\sqrt{f_a}}$	53	40	48

* Based on f_a = maximum allowable stress.

members of steel frames in which hot-rolled shapes are used are usually such that economy in weight requires plate elements thick enough to preclude local buckling at stresses less than the yield stress or, as in the German specifications, less than the primary-buckling stress. Thus, limiting values of slenderness such as those in Table 4-4 cover most situations in the design of hot-rolled compression members. However, there are provisions in the AISC and AREA specifications for relaxing these requirements, discussed in Chap. 9. On the other hand, cold-formed construction usually involves elements which are so thin as to require different procedures. This is also discussed in Chap. 9.

EXAMPLE 4-5 Design a minimum-weight welded H cross section for a pin-ended column which carries a central load of 165 kips. The unsupported length is 18 ft. A36 steel, AISC specification.

SOLUTION Assuming $F_a = 16$,

$$A = \frac{165}{16} = 10.32 \text{ in.}^2$$

From Table A-1, $r_x = 0.435h$ and $r_y = 0.25b$. For $I_x = I_y$, $h \approx 0.6b$. For a flange width of 12 in., the minimum thickness is $6/16 = \frac{3}{8}$ in. The area of the two flanges is 9.00 in.^{2} A $6\frac{1}{2} \times \frac{3}{16}$ in. web, for which $6\frac{1}{2}/\frac{3}{16} = 35 < 42$, gives an additional 1.22 in.^{2} The proposed section is shown in Fig. 4-56.

Part		Area	I_x	I_y
Flanges	$12 \times \frac{3}{8} \times 2$	9.00×3.44^2	$= 106.8$	$\times \frac{12^2}{12} = 108$
Web	$6\frac{1}{2} \times \frac{3}{16}$	$1.22 \times \frac{6.5^2}{12}$	$= 4.3$	
		10.22	111.1	108

$$r_y = \sqrt{\frac{108}{10.22}} = 3.24 \text{ in.}$$

$$\frac{KL}{r_y} = 1 \times 18 \times \frac{12}{3.24} = 66.8 \qquad F_a = 16.76 \text{ ksi}$$

$$A = \frac{165}{16.76} = 9.86 < 10.22$$

From column tables of the AISC Manual, the lightest W for this member is a W8 × 48. The weight of the built-up section is 3.4 × 10.22 = 32.5 lb. However, it

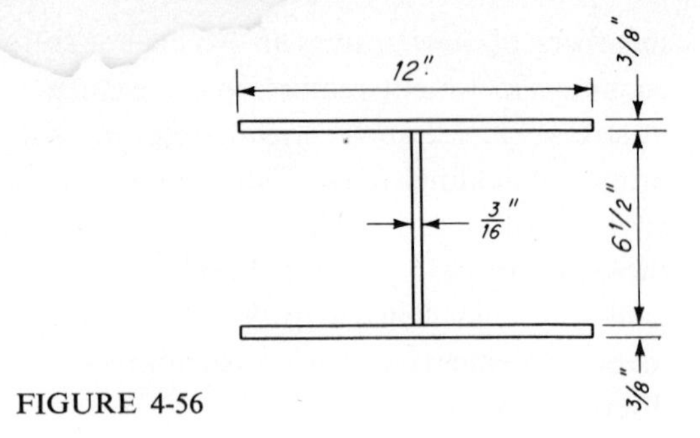

FIGURE 4-56

should be noted that a minimum-weight section is not necessarily a minimum-cost section or a "best" design. Architectural considerations, space limitations, costs of fabrication and erection, etc., all enter into the problem.

PROBLEMS

4-49 Design an A36-steel welded H cross section for a pin-ended column which supports a central load of 500 kips and has an unsupported length of 24 ft. AISC specification. Compare the weight with that of the lightest W column designed for the same load and conditions.

4-50 Design an A36-steel square box pin-ended column to support a central load of 250 kips. The unsupported length is 18 ft. For architectural reasons the outside dimension must not exceed 12 in. AISC specification.

4-51 In the article on members consisting of segments connected by lacing bars or by solid cover plates, the AREA specification limits the ratio b/t of the web plate and cover plate to $32\sqrt{p_c/f}$ and $40\sqrt{p_c/f}$ respectively, where p_c is the allowable stress and f the actual stress in the member. What is the logic in support of the factor $\sqrt{p_c/f}$?

4-24 SHEAR IN COLUMNS

The segments of built-up columns with cross sections such as those shown in Fig. 4-21*b*, *c*, *d*, *f*, and *g* must be interconnected so that they will act as parts of a whole rather than as individual columns. They may be joined by single lacing (Fig. 4-57*a*) or by double lacing (Fig. 4-57*b*). At the ends and at intermediate points where it is necessary to interrupt the lacing (to admit gusset plates, for example), the open sides are connected with *stay plates* (Fig. 4-57*d*). Stay plates are also called *batten plates* or *tie plates*. Lacing is sometimes omitted, with the segments connected by battens (Fig. 4-57*e*). Lacing has been largely supplanted by perforated cover plates (Fig. 4-57*f*) in members of bridge trusses.

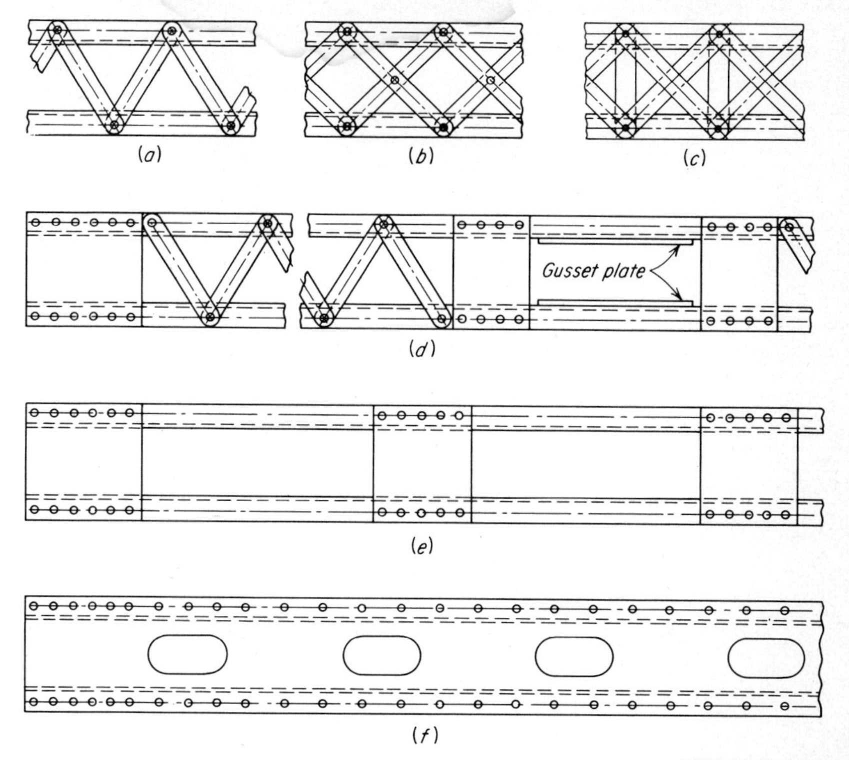

FIGURE 4-57

Lacing and batten plates are important details. A considerable amount of study of their role in column strength followed the collapse in 1907 of the first Quebec Bridge, which was caused, at least in part, by inadequate lacing of compression-chord members.

The lacing or battens in a column must carry the shear forces which develop when the column bends. The shear V is shown in Fig. 4-58, which pictures a portion of a bent column. Since θ is a small angle, we have

$$V = P \sin \theta = P \tan \theta = P \frac{dy}{dx} \qquad (a)$$

The slope dy/dx for a perfect column is indeterminate, even though the deflected shape is known. This is because the deflection δ in $y = \delta \sin \pi x/L$ is indeterminate. Therefore, an evaluation of δ must be attempted or else the shear determined for an imperfect column, such as a crooked column or an eccentrically loaded one. Both schemes have been used.

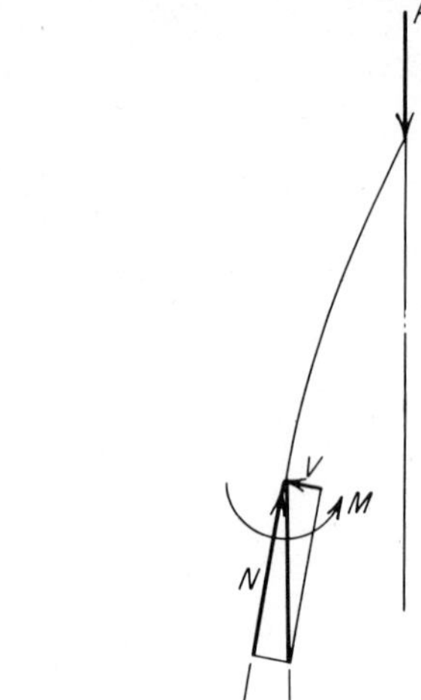

FIGURE 4-58

For a column whose load is eccentric equal amounts e at each end, dy/dx is maximum at the end and can be found from Eq. (d) of Art. 4-12. Equating M_2/P and M_1/P in that equation to e, we get

$$y = e\left(\tan\frac{kL}{2}\sin kx + \cos kx - 1\right) \tag{b}$$

From Eq. (b), $dy/dx = ek\tan(kL/2)$ at $x = 0$. Substituting this into Eq. (a), we get

$$V = Pek\tan\frac{kL}{2} \tag{c}$$

Substituting into this equation $k = \sqrt{P/EI}$, we have

$$\frac{V}{P} = \pi\frac{e}{L}\sqrt{\frac{P}{P_E}}\tan\frac{\pi}{2}\sqrt{\frac{P}{P_E}} \tag{4-49a}$$

Curves A and B of Fig. 4-59 are plots of Eq. (4-49a) for $F_y = 33$ ksi, $E = 30{,}000$ ksi, and $e = 0.25r$ and $0.50r$, respectively, where r is the radius of gyration of the cross section. Values of V/P were found by substituting in the equation corresponding values of P/A and L/r computed from the secant formula [Eq. (4-29b)] assuming $c = r$. It is seen that V/P increases with L/r. However, since P decreases with increase in L/r, the shear itself does not increase indefinitely but reaches a maximum value at about $L/r = 150$.

If the column load is eccentric equal but opposite amounts e at each end, the maximum value of dy/dx is at midlength. Proceeding as above, the solution is found to be

$$\frac{V}{P} = \pi \frac{e}{L} \sqrt{\frac{P}{P_E}} \csc \frac{\pi}{2} \sqrt{\frac{P}{P_E}} \tag{4-49b}$$

Curves C and D of Fig. 4-59 are graphs of this equation for the same values as before, with e/r equal to 0.25 and 0.50, respectively.

Specification requirements are based on empirical formulas which are simpler in form than Eqs. (4-49). AASHO specifies the shear given by

$$\frac{V}{P} = \frac{1}{100} \left(\frac{100}{L/r + 10} + \frac{L/r}{3{,}300{,}000/F_y} \right) \tag{4-50}$$

This equation is plotted for $F_y = 33{,}000$ psi in Fig. 4-59. It will be noted that it gives values which approximate those of Eq. (4-49*b*) for the e/r values for which it was plotted but falls short of the values according to Eq. (4-49*a*). AREA uses the same formula, except that the constant in the second term in the parentheses is 3,600,000. On the other hand, AISC specifies a shear which is independent of L/r, namely

$$V = 0.02P \tag{4-51}$$

This formula is also plotted in Fig. 4-59.

The formulas just discussed are intended to predict the shear component of the axial load P which results from curvature of the column. Both AASHO and AREA require that the shear force due to any other external force and to the weight of the member be taken into account. The AISC does not.

Specifications for aluminum structures suggest the shear given by[27]

$$V = 0.02F_a A + V_t \tag{4-52}$$

where F_a is the specified allowable stress for short columns and V_t is the shear caused by any transverse load on the column.

The considerable variation in specification requirements for shear for which lacing must be designed reflects the uncertainties in predicting the deflected shape of the column and the differing opinions of specification writers regarding the assumptions that must be made.

Although the shearing force is relatively small, the arrangement of lacing to resist it is of considerable importance. Since the laced column is a trussed framework, secondary stresses in the lacing may result from the axial deformation. These stresses can be quite large in double-lacing systems, particularly in that of Fig. 4-57*c* (Ref. 3, p. 183). Lacing as in Fig. 4-57*b* is better in this respect because the accordionlike action permits lateral expansion of the column, which must accompany shortening under

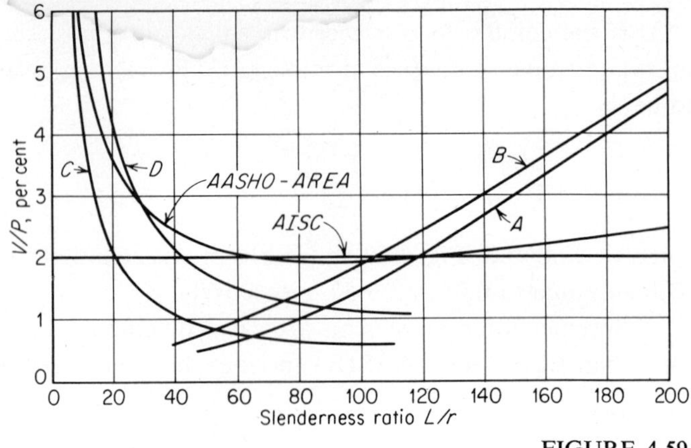

FIGURE 4-59

load if there is to be no shortening of the lacing bars. The lateral ties in Fig. 4-57*c* restrain this adjustment. The Quebec Bridge member which failed was laced in this fashion.

The spacing of the connections of lacing to the column segments is an important consideration. If this distance is too great, individual segments may buckle alternately in and out between lattice connections. This suggests that the slenderness ratio of a column segment, considered as a column whose length is the spacing of lattice connections, should not be more than that of the column as a whole. This is a requirement of the AISC specification. However, AASHO and AREA require that lacing connections be spaced so that the slenderness ratio of the portion of the flange between them be no more than two-thirds the slenderness ratio of the member and not more than 40.

The lacing bar must be designed to resist either tension or compression. Proportions of perforated cover plates are based on results of analyses and tests. Specification requirements are essentially identical and conform to recommendations in Ref. 28.

4-25 EFFECT OF SHEAR ON THE CRITICAL LOAD

The shear component of the axial force P, which was discussed in the preceding article, results in shear deformation which is neglected in the derivation of the Euler formula for the critical load. To determine the effect this has on the predicted critical load, the additional curvature of the column due to shear deformation must be considered. The buckled column is shown in Fig. 4-60*a*, where y_m and y_s are the displacements due to moment and shear, respectively. According to Fig. 4-60*b*, the moment is

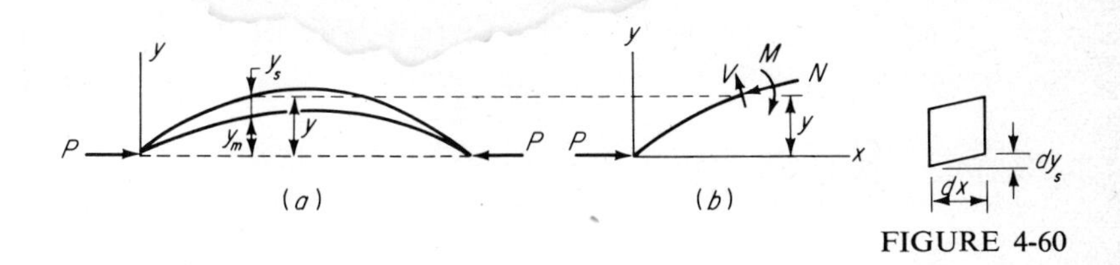

FIGURE 4-60

Py, and the shear is $P\,dy/dx$. Figure 4-60c shows the shear deformation of an element dx long, from which

$$\frac{dy_s}{dx} = \frac{nP}{AG}\frac{dy}{dx} \tag{a}$$

where A = area of cross section
G = modulus of elasticity in shear
n = factor which depends on shape of cross section

The factor n depends on the manner in which the shearing stress is distributed over the cross section. For rectangular cross sections $n = 1.2$, while for an I bent in the plane of the web it is about 2.

The curvature at any point is

$$\frac{d^2y}{dx^2} = \frac{d^2y_m}{dx^2} + \frac{d^2y_s}{dx^2} \tag{b}$$

From Eq. (*a*)

$$\frac{d^2y_s}{dx^2} = \frac{d}{dx}\frac{nP}{AG}\frac{dy}{dx} = \frac{nP}{AG}\frac{d^2y}{dx^2} \tag{c}$$

Furthermore,

$$\frac{d^2y_m}{dx^2} = \frac{M}{EI} = -\frac{Py}{EI} \tag{d}$$

Substituting these two values into Eq. (*b*) gives

$$\frac{d^2y}{dx^2} = -\frac{Py}{EI} + \frac{nP}{AG}\frac{d^2y}{dx^2} \tag{e}$$

from which

$$\frac{d^2y}{dx^2} + \frac{Py}{EI(1 - nP/AG)} = 0 \tag{f}$$

Comparing this equation with Eq. (i) of Art. 4-2, we see that they are the same except for the meaning of k^2. Therefore, the solution is obtained from Eq. (l) of Art. 4-2.

$$P = \frac{\pi^2 EI}{L^2}\left(1 - \frac{nP}{AG}\right) \tag{g}$$

Solving this equation for P gives

$$P = \frac{\pi^2 EI}{L^2} \frac{1}{1 + n\pi^2 EI/L^2 AG} \tag{4-53}$$

This equation can be written in terms of an effective-length coefficient. Using the tangent-modulus concept to extend it to the inelastic range and assuming G/E to be constant so that $G = 0.4E$, we get

$$F_{cr} = \frac{\pi^2 E_t}{(K'L/r)^2} \tag{4-54}$$

in which

$$K' = \sqrt{1 + \frac{25n}{(L/r)^2}} \tag{4-55}$$

Values of K' exceed unity only slightly. Using $n = 2$ for the I, $K' = 1.002$ and 1.060 for $L/r = 120$ and 20, respectively. Thus, it is clear that neglecting shear deformation results in a negligible error in the critical load for I columns.

The effect of shear deformation on built-up columns whose segments are connected by lacing, battens, or perforated cover plates is larger than for solid-webbed columns. The strength of a column with single lacing inclined at 60° to the longitudinal axis of the column or with double lacing at 45° can be determined, as for the solid-webbed column, by using the effective-length coefficient given by[9]

$$K' = \begin{cases} 1.1K & \dfrac{KL}{r} \not> 40 \qquad (4\text{-}56a) \\[2ex] K\sqrt{1 + \dfrac{300}{(KL/r)^2}} & \dfrac{KL}{r} \not< 40 \qquad (4\text{-}56b) \end{cases}$$

where K is the effective-length coefficient for end restraint. Although the increase in effective-length coefficient is 10 percent for stocky columns, the resulting reduction in strength is small because of the relative insensitivity of short columns to variation in L/r. However, the reduction in strength may be as much as 10 percent for slender columns.

The strength of a battened column (Fig. 4-57*e*) can be determined by using the effective-length coefficient given by[3]

$$K' = \sqrt{1 + \frac{\pi^2}{12}\left(\frac{L_0/r_0}{L/r}\right)^2} \tag{4-57}$$

where L/r = slenderness ratio of column
L_0 = center-to-center spacing of battens
r_0 = radius of gyration of one chord for its axis normal to plane of batten

The strength of a steel column with $L/r = 110$ is reduced about 10 percent for $L_0/r_0 = 40$.

4-26 DP4-4: DESIGN OF BATTENED COLUMN

In this example the channels are spaced 9 in. in order to make the radii of gyration about equal for both axes. After the cross section has been established, the size and spacing of battens is determined. This is largely a question of judgment and sense of proportion, except that the batten plate must accommodate the required number of fasteners or weld to connect it to the column flange and the spacing must not be so large as to make L_0/r_0, the slenderness ratio of one channel, more than the slenderness ratio of the member.

The battened column resists buckling by frame action. Points of inflection are assumed to be located midway between battens in each channel and midway between connections of the batten to the channels. The specified shear, 2 percent of the axial load, is divided equally between channels, and the bending moment in one channel at the edge of the batten determined. The resulting stress $P/A + M/S$ should not exceed the allowable stress for $L/r = 0$.

The shear Q midway in the batten plate is determined next. This enables the bending moment in the batten at the fastener line (in this case, at the end where it is welded to the channel flange) to be computed. The required plate thickness is $\frac{1}{4}$ in.

Procedures for determining the weldment to connect the batten plate to the channel are discussed in Chap. 7.

PROBLEMS

4-52 Design the column of DP4-4 for single lacing instead of batten plates.

4-53 Design the column of DP4-4 using four angles with battens. AISC specification, A36 steel.

4-54 Design a pin-ended column, battened on two opposite sides, 24 ft long for $P = 300$ kips. AISC specification, A36 steel.

4-55 Design the column of Prob. 4-54 with perforated cover plates instead of battens.

DESIGN OF BATTENED COLUMN — DP4-4

Design data:

$P = 130^k$

$L = 20'$ pin-ended, lateral dimensions not restricted

A36 steel, AISC Specification

Try [ ] with 12" channels, $r_x = 0.36 \times 12 = 4.3$ (Table A-1)

$L/r_x = 240/4.3 = 56, \quad F_a = 17.8$ ksi

$$A = \frac{130}{17.8} = 7.3 \text{ in.}^2$$

Lightest C12 is 6.03 in.², so try two C10 × 15.3, $A = 2 \times 4.47 = 8.94$ in.²

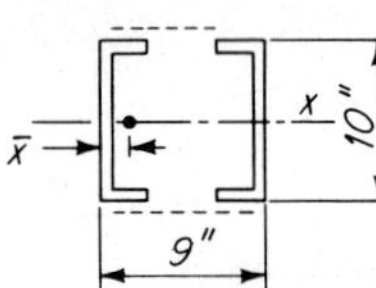

C10 × 15.3	$A = 4.47$	$\bar{x} = 0.64$
	$I_x = 66.9$	$I_y = 2.3$
	$r_x = 3.87$	$r_y = 0.72$

$$I_y = 2 \times 2.3 + 2 \times 4.47 \times 3.86^2 = 4.6 + 133 = 138 \text{ in.}^4$$

$$r_y = \sqrt{\frac{138}{8.94}} = 3.93$$

$L/r_x = 240/3.87 = 62, \quad F_a = 17.2, \quad P = 17.2 \times 8.94 = 154^k > 130^k$ O.K.

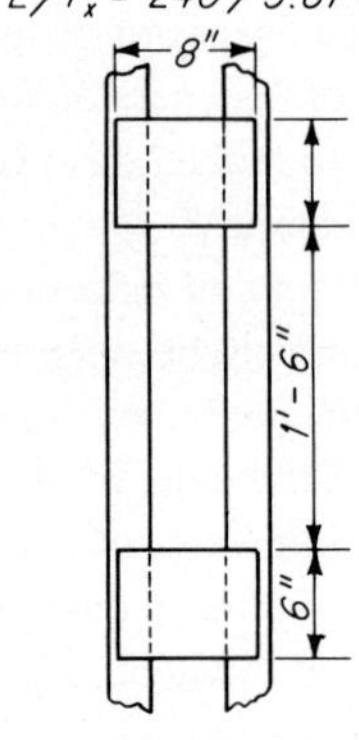

Try 6" × 8" battens 2' o.c.

$$\frac{L_o}{r_o} = \frac{24}{0.72} = 33, \qquad \frac{L}{r_y} = \frac{240}{3.93} = 61$$

$$K = \sqrt{1 + \frac{\pi^2}{12}\left(\frac{33}{61}\right)^2} = \sqrt{1 + 0.247} = 1.12 \qquad \text{[Eq. (4-57)]}$$

$KL/r_y = 1.12 \times 61 = 68, \quad F_a = 16.6$

$P = 16.6 \times 8.94 = 149^k > 130^k$ O.K.

65ᵏ
1.3ᵏ
20"
Q = 6.8ᵏ
4"
1.3ᵏ
0.64
3.86
65ᵏ

Check shear

$V = 0.02 \times 130 = 2.6^k = 1.3^k$ per channel

Channel at edge of batten

$M = 1.3 \times 9 = 11.7''^k$

$$f = \frac{P}{A} + \frac{M}{S} = \frac{65}{4.47} + \frac{11.7}{1.2} = 14.5 + 9.8 = 24.3 > 22 \text{ ksi}$$

Reduce spacing of battens to 20" O.C.

$M = 1.3 \times 7 = 9.1''^k$

$$f = \frac{65}{4.47} + \frac{9.1}{1.2} = 14.5 + 7.6 = 22.1 \text{ ksi OK}$$

Check batten plate in bending at weld

$$Q = \frac{1.3 \times 20}{3.86} = 6.8^k$$

$$S = \frac{t \times 6^2}{6} = 6t, \quad M = 4'' \times 6.8^k = 27.2''^k$$

$$\frac{M}{S} = \frac{27.2}{6t} = 22; \quad t = 0.21''$$

Use 1/4" plate

DP 4-4
Design of battened column.

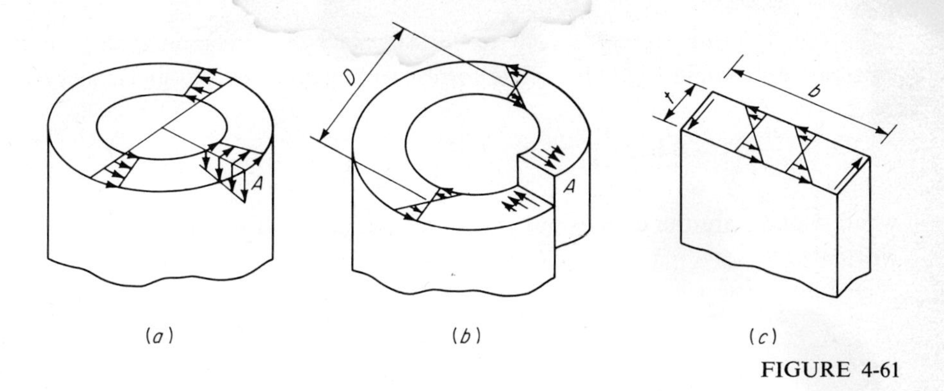

FIGURE 4-61

4-27 STRUCTURAL MEMBERS IN TORSION

The shearing stresses which result when a circular tube is twisted are shown in Fig. 4-61*a*. The variation in stress is linear, as shown, if the proportional limit is not exceeded. The angle of twist θ per unit length is

$$\theta = \frac{T}{GI_p} \tag{a}$$

where T = torsional moment
G = shearing modulus of elasticity
I_p = polar moment of inertia

Of course, Eq. (*a*) also applies to the solid circular cross section.

If the tube of Fig. 4-61*a* is cut on a longitudinal section at A, the shearing stresses which result when it is twisted are those shown in Fig. 4-61*b*. Because the shearing stresses on the longitudinal section of the tube of Fig. 4-61*a* cannot exist in the split tube, there is a relative vertical displacement at the split and the distribution of shearing stresses is altogether different. However, the variation in stress is still linear if the proportional limit is not exceeded. The relative stiffnesses of the two tubes can be inferred by noting that the torsional resistance of the closed tube consists of couples with an average moment arm about equal to the diameter of the tube, while in Fig. 4-61*b* the average moment arm is less than the thickness.

The angle of twist per unit length of a noncircular cross section, solid or tubular, is given by

$$\theta = \frac{T}{GJ} \tag{4-58}$$

where J is the torsion constant of the cross section. Of course, $J = I_p$ if the cross section is circular.

The shearing stresses which result when a solid rectangular cross section is twisted are shown in Fig. 4-61*c*. The corresponding constant J is given very closely by

$$J = \frac{bt^3}{3}\left(1 - 0.630\,\frac{t}{b}\right) \qquad b > t \tag{b}$$

where b and t are the dimensions shown in the figure. If b/t is large, Eq. (*b*) may be written

$$J = \frac{bt^3}{3} \tag{4-59}$$

For example, if $b/t = 6$, the error in using Eq. (4-59) is only 10 percent.

The torsion constant of the split tube of Fig. 4-61*b* is given by Eq. (4-59) with b equal to the circumference. Thus

$$J = \tfrac{1}{3}\pi D t^3$$

For the tube of Fig. 4-61*a*

$$I_p = \pi D t\left(\frac{D}{2}\right)^2 = \frac{\pi D^3 t}{4}$$

Then

$$\frac{I_p}{J} = \frac{\pi t D^3/4}{\pi D t^3/3} = \frac{3}{4}\left(\frac{D}{t}\right)^2$$

Thus, a 10-in. tube with a 1-in. wall is 75 times as stiff as a split tube with the same dimensions.

The torsion constant of any shape composed of rectangular and/or curved elements for which b/t is sufficiently large can be determined by adding the quantities $bt^3/3$ for all the elements, provided no part of the cross section is closed. Such a section is called an *open section*. Pipes, tubes, box sections, etc., are called *closed sections*. In Fig. 4-21, *a*, *b*, *c*, and *e* are open sections. The section in *h* is closed, and *d*, *f*, and *g* are also if the open sides are adequately laced.

The torsion constant of the single-cell closed section is given by

$$J = \frac{4A^2}{\int ds/t} \tag{4-60}$$

where A = area enclosed by midline of wall
ds = element of circumference of wall
t = thickness of wall

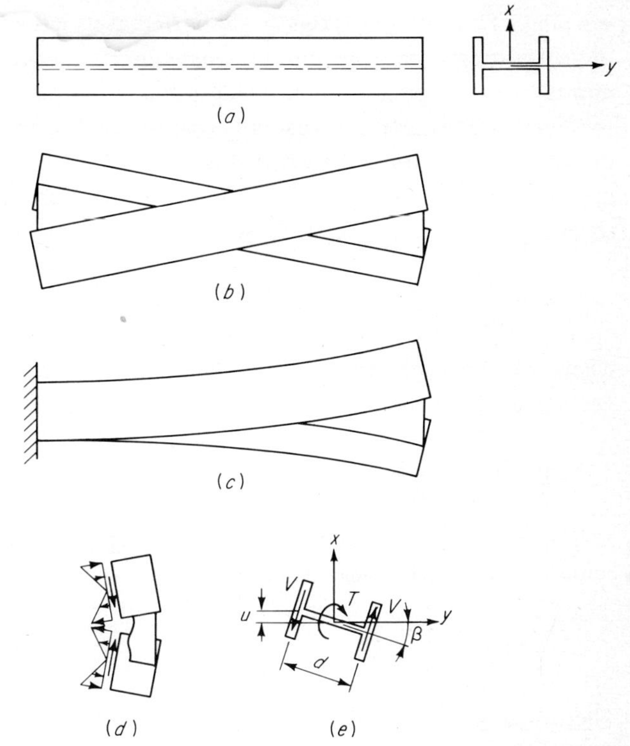

FIGURE 4-62

Integration is around the entire periphery. Thus, for a square box section 20 × 20 in. outside dimension by $\frac{1}{2}$ in. thick,

$$J = \frac{4(19.5 \times 19.5)^2}{4 \times 19.5/0.5} = 3{,}707 \text{ in.}^4$$

Values of J for the multicell cross section can be determined by methods given in Ref. 30.

Displacements in the longitudinal direction of the split tube of Fig. 4-61*b* are called *warping displacements*, and the cross section, which does not remain plane, is said to have *warped*. If there is nothing to restrain such warping, it is uniform throughout its length and the stress distributions, such as those in Fig. 4-61*b* and *c*, are also uniform throughout the length. Torsion with uniform warping is usually called *St. Venant torsion*, he having been the first to develop the theory for the general case.

Uniform warping of the I of Fig. 4-62*a* is shown in Fig. 4-62*b*. However, structural members are usually supported in such a manner as to prevent uniform warping. Thus, if the I is rigidly supported at its left end, so that warping is prevented there, twisting is accompanied by *nonuniform warping*, as shown in Fig. 4-62*c*. Such nonuniform warping results in additional shearing stresses and an increase in the torsional stiffness. In this case, the oppositely directed bending of the flanges of the I produces the shears V shown in Fig. 4-62*d*, which constitute a couple opposing the applied torque T (Fig. 4-62*e*). Of course, there are also bending stresses; these are shown in the figure. The torsional resistance is evaluated as follows. Since

$$V = \frac{dM}{dz} \qquad M = -EI_f \frac{d^2u}{dz^2} \tag{a}$$

where u is the displacement of one flange (Fig. 4-62*e*) and I_f the moment of inertia of *one* flange about the y axis of the I, we have

$$V = -EI_f \frac{d^3u}{dz^3} \tag{b}$$

In Fig. 4-62*e*, $u = \beta d/2$, where β is the angle of twist and d the distance between centerlines of the flanges. Therefore,

$$V = -EI_f \frac{d}{2} \frac{d^3\beta}{dz^3} \tag{c}$$

The resisting couple is Vd. Therefore

$$T_w = -EI_y \frac{d^2}{4} \frac{d^3\beta}{dz^3} \tag{4-61}$$

where T_w is the torsional resistance due to nonuniform warping and I_y $(= 2I_f)$ is the moment of inertia of the I.

The St. Venant torsional resistance itself is assumed to be unaffected by nonuniform warping. In other words, the St. Venant resistance is supplemented by the nonuniform-warping resistance T_w. Therefore, using Eq. (4-58) and noting that θ in that equation is the twist per unit of length while β is the angle of twist at any cross section, so that $d\beta = \theta\, dz$, we have

$$T = T_v + T_w = GJ \frac{d\beta}{dz} - EI_y \frac{d^2}{4} \frac{d^3\beta}{dz^3} \tag{4-62}$$

Of course, the nonuniform-warping torsional resistance, which for the I is measured by $d^2 I_y/4$ as determined above, depends upon the shape of the cross section. Therefore, to put Eq. (4-62) in a more general form, we write

$$T = GJ \frac{d\beta}{dz} - EC_w \frac{d^3\beta}{dz^3} \tag{4-63}$$

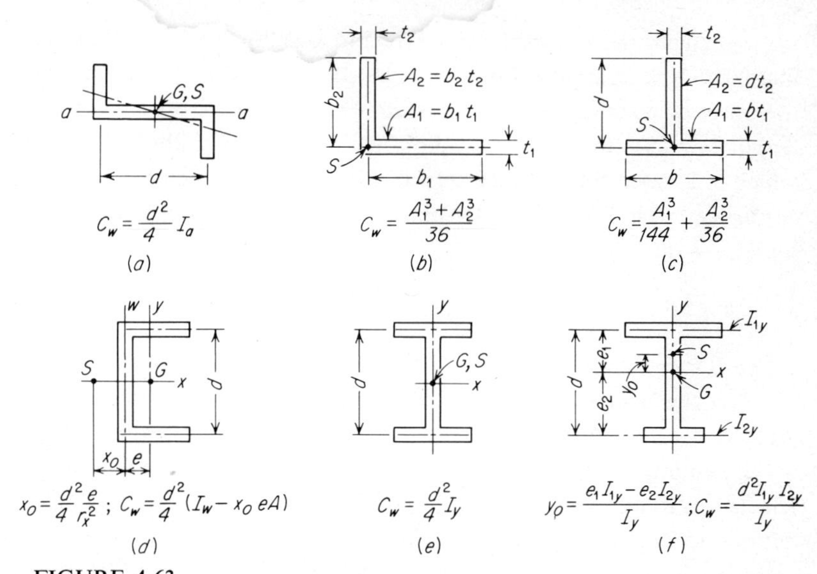

FIGURE 4-63
Values of warping constant C_w. (C_w for the angle and the tee is small enough to be neglected in most applications.)

where C_w is the warping constant of the cross section. Values of C_w for various open cross sections are given in Fig. 4-63. A procedure for determining C_w is given in Ref. 3, p. 218.

PROBLEMS

4-56 Compute the angle of twist of one end of an A36 W8 × 31, 10 ft long, loaded and supported as in Fig. 4-62*b*. $T = 2$ ft-kips.

4-57 Same as Prob. 4-56 except that the member is loaded and supported as in Fig. 4-62*c*.

4-28 TORSIONAL BUCKLING OF COMPRESSION MEMBERS

It was shown in Art. 4-27 that the torsional stiffness of open cross sections is quite small. Because of this, open-section columns may buckle in a torsional mode, rather than in a bending mode. The column of cruciform cross section shown in Fig. 4-49*b* is an example, even though this was called a local-buckling failure in Art. 4-21. For this particular cross section, the two modes of collapse are identical. In this article, we consider the failure as a torsional one.

Figure 4-64*a* shows the cross section and Fig. 4-64*b* a differential length dz of the member. The element dA in area is acted upon by the force $f\,dA$, where $f = P/A$. At the onset of buckling, this stress is uniform over the cross section and throughout the length because the column is supporting a centrally applied load P. The element, which is located the distance z from the end of the member, is shown in its buckled configuration in Fig. 4-64*c*. Displacements measured from the unbuckled position are u and $u + du$. From Fig. 4-64*a*

$$u = r\beta \tag{a}$$

where β is the angle of twist at z and r is the distance from the shear center to dA.

In the buckled state, the element is acted upon by the shears Q and $Q + dQ$ and the bending moments M and $M + dM$ (Fig. 4-64*d*), where Q is the shear and M is the bending moment, both per unit of length in the direction of r. These are the effects of nonuniform warping. These stresses are accompanied by the stresses associated with uniform warping (St. Venant torsion). Therefore, the shearing stresses acting on the cross section are those shown in Fig. 4-64*e*. The resultant of the shear stresses due to uniform warping is denoted by T_v (Fig. 4-64*b*) and is the same as T_v in Eq. (4-62).

Summing moments about the z axis in Fig. 4-64*b* gives

$$dT_v + \int_A r\,dQ\,dr = 0 \tag{b}$$

Summing moments in Fig. 4-64*d* gives

$$dM\,dr + Q\,dr\,dz + f\,dA\,du = 0 \tag{c}$$

Solving Eq. (*c*) for $Q\,dr$ and differentiating the result with respect to z gives

$$\frac{dQ}{dz}\,dr = -\frac{d^2M}{dz^2}\,dr - f\,dA\,\frac{d^2u}{dz^2} \tag{d}$$

Dividing Eq. (*b*) by dz and substituting for dQ/dz from Eq. (*d*) gives

$$-\frac{dT_v}{dz} + \int_A \frac{d^2M}{dz^2}\,r\,dr + f\int_A \frac{d^2u}{dz^2}\,r\,dA = 0 \tag{e}$$

Since M is the moment per unit of r, the moment on the element $dA = t\,dr$ is $M\,dr$. Therefore,

$$M\,dr = EI\,\frac{d^2u}{dz^2} = E\,\frac{t^3\,dr}{12}\,\frac{d^2u}{dz^2} \tag{f}$$

where $I = t^3\,dr/12$ is the moment of inertia of the element. Differentiating Eq. (f) twice with respect to z and substituting d^2M/dz^2 into Eq. (e) gives

$$-\frac{dT_v}{dz} + E\frac{t^3}{12}\int_A \frac{d^4u}{dz^4}\,r\,dr + f\int_A \frac{d^2u}{dz^2}\,r\,dA = 0 \tag{g}$$

The St. Venant torsion resultant T_v is found from Eq. (4-58). Since $\theta = d\beta/dz$, we have

$$T_v = GJ\frac{d\beta}{dz} \qquad \frac{dT_v}{dz} = GJ\frac{d^2\beta}{dz^2} \tag{h}$$

Substituting the values of dT_v/dz from Eq. (h) and u from Eq. (a) into Eq. (g) gives

$$-GJ\beta'' + \frac{Et^3}{12}\beta^{\text{iv}}\int_A r^2\,dr + f\beta''\int_A r^2\,dA = 0 \tag{i}$$

But

$$\int_A r^2\,dr = 4\frac{r^3}{3}\bigg|_0^b = \frac{4b^3}{3}$$

where b = width of leg (Fig. 4-64a). Furthermore, $\int_A r^2\,dA = I_p$. Therefore,

$$E\frac{b^3t^3}{9}\beta^{\text{iv}} + (fI_p - GJ)\beta'' = 0 \tag{4-64}$$

The factor $b^3t^3/9$ in Eq. (4-64) is the warping constant C_w, which was discussed in Art. 4-27. The corresponding shear stresses are the nonuniform-warping shear stresses pictured in Fig. 4-64e, which are analogous to the nonuniform-warping shears V in Fig. 4-62d. Thus, the two types of resistance to torsion are represented by $(Eb^3t^3/9)\beta^{\text{iv}}$ and $GJ\beta''$, respectively, in Eq. (4-64).

Using the notation

$$k^2 = \frac{fI_p - GJ}{EC_w} \tag{j}$$

Eq. (4-64) becomes

$$\beta^{\text{iv}} + k^2\beta'' = 0 \tag{4-65}$$

The solution of this equation is

$$\beta = A\sin kz + B\cos ky + Cz + D \tag{k}$$

The constants of integration in Eq. (k) will be evaluated for the boundary conditions of the column in Fig. 4-49b. It will be noted in that figure that plane cross sections remain plane at the ends; i.e., there is no warping at the ends. Consequently,

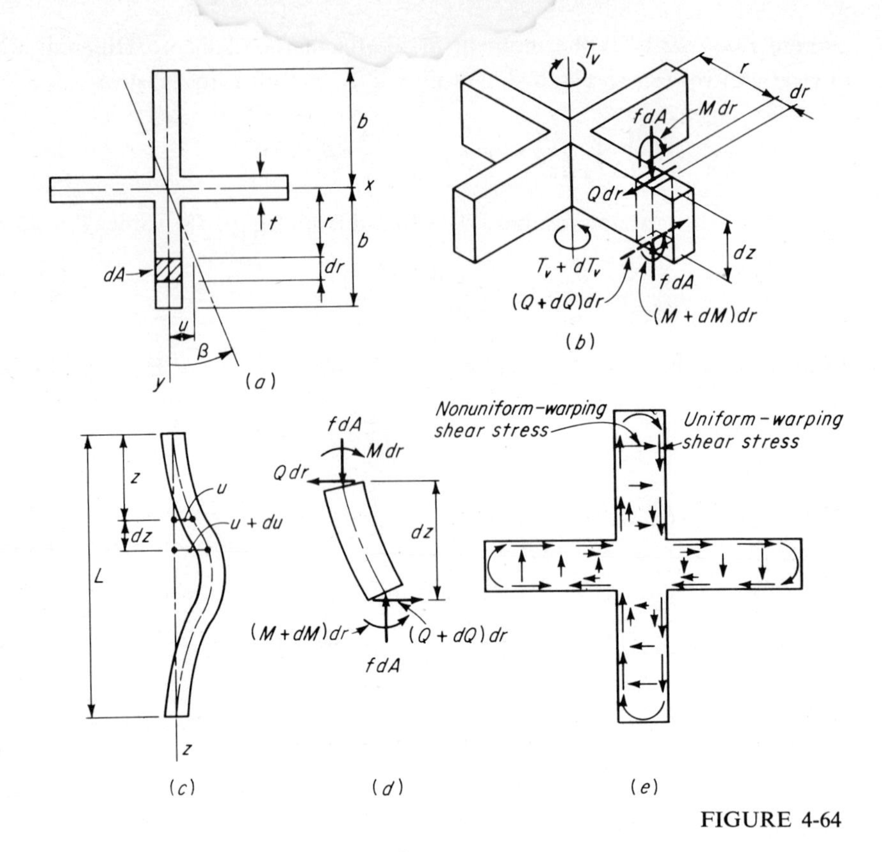

FIGURE 4-64

$du/dz = 0$ at each end (Fig. 4-64c), which, because of Eq. (a), gives $d\beta/dz = 0$. In addition, $\beta = 0$ at each end. Using these four boundary conditions with Eq. (k) gives

$$\beta_{z=0} = 0 \qquad 0 = B + D$$

$$\beta_{z=L} = 0 \qquad 0 = A \sin kL + B \cos kL + CL + D$$

$$\left(\frac{du}{dz}\right)_{z=0} = 0 \qquad 0 = Ak + C$$

$$\left(\frac{du}{dz}\right)_{z=L} = 0 \qquad 0 = Ak \cos kL - Bk \sin kL + C$$

Eliminating C and D from these equations gives

$$A(\sin kL - kL) + B(\cos kL - 1) = 0$$

$$A(\cos kL - 1) - B \sin kL = 0$$

These are homogeneous equations in A and B. One solution is $A = B = 0$, in which case $C = D = 0$, and Eq. (k) reduces to $\beta = 0$. Nonzero (indeterminate) values of A and B exist only if the determinant of the coefficients vanishes. This gives

$$\sin \frac{kL}{2} \left(2 \sin \frac{kL}{2} - kL \cos \frac{kL}{2} \right) = 0$$

which is satisfied by $\sin kL/2 = 0$ or $\tan kL/2 = kL/2$. The corresponding smallest roots are $kL/2 = \pi$ and $kL/2 = 4.49$. Substituting the smaller of these roots into Eq. (j) gives the smallest critical stress:

$$F_{cr} = \frac{GJ}{I_p} + \frac{4\pi^2}{L^2} \frac{EC_w}{I_p} \tag{4-66}$$

This equation gives the stress F_{cr} at which torsional buckling begins, provided the column is perfectly straight, free of residual stress, etc. Although it was derived for the cruciform cross section of Fig. 4-49b, it holds for any cross section for which the shear center and the centroid coincide.

If the ends of the column are free to warp, each of the four legs of the cruciform in Fig. 4-49b will bend in single curvature. Since the ends are free to warp only if there is no rotational restraint of the legs, this is equivalent to saying that the moment M in Fig. 4-64d is zero at each end. Thus, $d^2u/dz^2 = 0$ at $z = 0$ and $z = L$. Using these boundary conditions together with $\beta = 0$ at each end gives

$$F_{cr} = \frac{GJ}{I_p} + \frac{\pi^2}{L^2} \frac{EC_w}{I_p} \tag{4-67}$$

Comparing Eqs. (4-66) and (4-67), we see that the boundary conditions relating to warping can be expressed in terms of an effective-length coefficient, as in the case of bend buckling. Thus,

$$F_{cr} = \frac{GJ}{I_p} + \frac{\pi^2}{(KL)^2} \frac{EC_w}{I_p} \tag{4-68}$$

where $K = 1$ if the end cross sections are free to warp and $K = \frac{1}{2}$ if warping of the end cross sections is completely restrained.

Equation (4-68) holds only for buckling which begins when the stress F_{cr} is less than the proportional-limit stress. The following arguments enable us to modify the formula to apply to inelastic buckling as well. The second term on the right is associated with the bending stresses caused by nonuniform warping (Fig. 4-62d). When the column starts to twist, these bending stresses are superimposed on the uniform stress F_{cr}, which is above the proportional limit. Therefore, it is reasonable to expect the tangent modulus or double modulus to govern the bending behavior.

Furthermore, the same arguments as in Art. 4-4 suggest that the tangent modulus is the correct one. On the other hand, there are no shearing stresses on the cross section at the instant buckling begins, so that the elastic modulus G would be expected to hold. Tests on two circular tubes which were twisted after having been compressed to the strain ε_s are reported in Ref. 31. The shearing modulus at the beginning of twist was practically equal to the elastic value G. However, it is on the safe side to assume a reduced value of G in the inelastic range, and it is convenient to take the ratio G/E the same for both elastic and inelastic behavior.

The concept of an equivalent radius of gyration, which was used in local buckling of plates, is also useful in torsional buckling. If we replace E and G in Eq. (4-68) with tangent-modulus values and equate the stress F_{cr} to the tangent-modulus bend-buckling stress $\pi^2 E_t/(KL/r_t)^2$, where r_t is the equivalent radius of gyration for torsional buckling, we get

$$r_t^2 = \frac{C_w + 0.04J(KL)^2}{I_{pS}} \tag{4-69}$$

The change in notation (I_{pS} in place of I_p, where I_{pS} means the polar moment of inertia with respect to the shear center S) is for the purpose of generalizing the formula for an application to be discussed in Art. 4-29. For the member of this article, the shear center S and centroid G coincide, so that $I_{pS} = I_{pG}$. Equation (4-69) enables the critical load for any column for which the shear center and centroid coincide to be determined. We need only compute r_t and compare it with r_x and r_y. The smallest of the three determines the buckling mode. The critical load is computed by substituting the corresponding KL/r into the appropriate bend-buckling formula, elastic or inelastic.

EXAMPLE 4-6 Compute the critical load for the column with the cruciform cross section shown in Fig. 4-65. The column is of A36 steel, 15 ft long, and supported so that warping at the ends is prevented, as in Fig. 4-49*b*. Use the CRC formula [Eq. (4-16)], for the inelastic range.

SOLUTION

$$I_x = I_y = 2\frac{tb^3}{3} = \frac{2 \times 0.5 \times 6^3}{3} = 72 \text{ in.}^4$$

$$I_{pS} = I_{pG} = I_x + I_y = 144 \text{ in.}^4 \qquad J = 4\frac{bt^3}{3} = \frac{4 \times 6 \times 0.5^3}{3} = 1 \text{ in.}^4$$

$$C_w = \frac{b^3t^3}{9} = \frac{(6 \times 0.5)^3}{9} = 3 \text{ in.}^6 \qquad A = 4 \times 6 \times 0.5 = 12 \text{ in.}^2$$

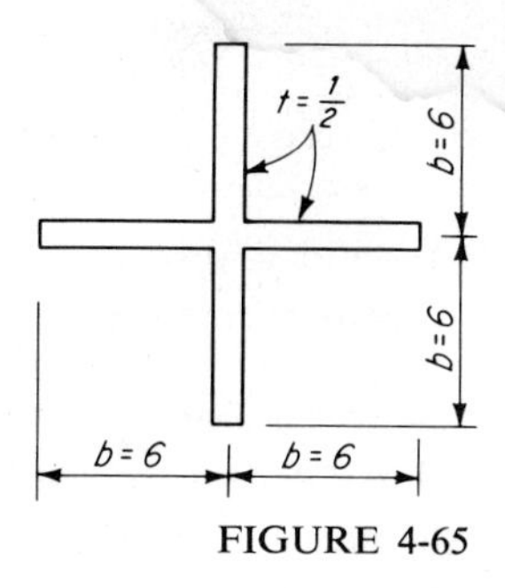

FIGURE 4-65

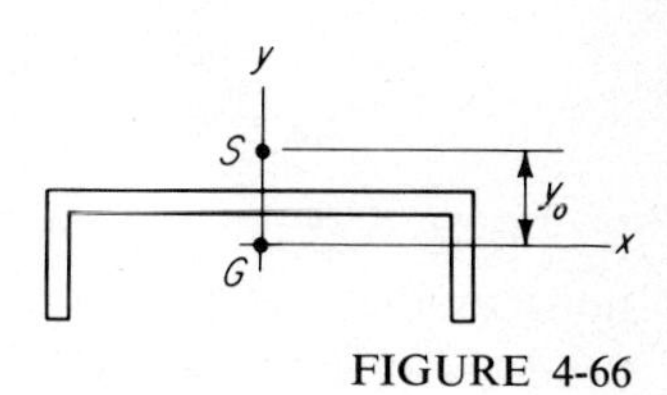

FIGURE 4-66

$$r_x = r_y = \sqrt{\frac{72}{12}} = 2.45 \text{ in.}$$

$$r_t^2 = \frac{3 + 0.04 \times 1 \times (0.5 \times 180)^2}{144} = \frac{3 + 324}{144} = 2.27 \qquad r_t = 1.51$$

Since $r_t < r_x = r_y$, the column fails by twist buckling

$$\frac{KL}{r_t} = \frac{0.5 \times 180}{1.51} = 60 \qquad C_c = \pi\sqrt{\frac{2E}{F_y}} = \pi\sqrt{\frac{30{,}000}{36}} = 128$$

$$F_{cr} = F_y\left[1 - \frac{1}{2}\left(\frac{KL/r}{C_c}\right)^2\right] = 36\left[1 - \frac{1}{2}\left(\frac{60}{128}\right)^2\right] = 32.0 \text{ ksi}$$

$$P_{cr} = 32.0 \times 12 = 384 \text{ kips}$$

It will be noted that the warping stiffness C_w for the cruciform is so small that its contribution to torsional resistance can be neglected. The warping stiffness of certain other cross sections, such as the angle and the tee, is also small enough to be neglected.[3]

4-29 COLUMNS WITH ONE AXIS OF SYMMETRY

It was shown in Art. 4-28 that columns for which the shear center and the centroid coincide fail in one of three independent buckling modes. This is not the case if the cross section has only one axis of symmetry, as for the channel, the angle, and the tee. In this case, the equations of equilibrium are[3]

$$EI_x v^{iv} + Pv'' = 0 \qquad (4\text{-}70a)$$

$$EI_y u^{iv} + Pu'' + Py_0 \beta'' = 0 \qquad (4\text{-}70b)$$

$$Py_0 u'' + EC_w \beta^{iv} + (fI_{ps} - GJ)\beta'' = 0 \qquad (4\text{-}70c)$$

where y = axis of symmetry
y_0 = distance from centroid G of the cross section to its shear center S (Fig. 4-66)
u, v = displacements in direction of x and y, respectively
I_{pS} = polar moment of inertia with respect to shear center S

If G and S coincide, Eq. (4-70c) becomes identical to Eq. (4-64). Also, the three equations are then independent, and the solution of Eqs. (4-70a) and (4-70b) gives the two bend-buckling modes, while Eq. (4-70c) gives the twist-buckling mode.

For the case of one axis of symmetry, which Eqs. (4-70) represent, it is seen that Eq. (4-70a) is independent and can be solved without reference to the other two. On the other hand, Eqs. (4-70b) and (4-70c) are simultaneous equations. This means that the bend-buckling mode corresponding to Eq. (4-70a) is independent, while the other two are coupled. Therefore, the column fails in one of *two* buckling modes, i.e. bend buckling about the x axis or a combination of twisting and bending about the y axis.

The critical load for the coupled buckling mode can be found by determining the equivalent radius of gyration r_{tb} given by

$$\left(1 - \frac{y_0^2}{r_{pS}^2}\right) r_{tb}^4 - (r_y^2 + r_t^2) r_{tb}^2 + r_y r_t^2 = 0 \qquad \text{(4-71)}$$

where r_{pS}, the polar radius of gyration, equals $\sqrt{I_{pS}/A}$ and r_t is the equivalent radius of gyration for torsional buckling, given by Eq. (4-69).

To determine the critical load for a column whose cross section has one axis of symmetry (y axis), we compute r_{tb} from Eq. (4-71) and compare it with r_x. The smaller of the two identifies the buckling mode and determines the corresponding critical stress. The effective-length coefficient K can be used in evaluating r_t and r_y, provided the boundary conditions for β and u are the same.

EXAMPLE 4-7 Compute the critical load for the column with the tee cross section shown in Fig. 4-67. The column is of A36 steel, 10 ft long, and supported so that warping, y-axis bending, and x-axis bending are all prevented at each end. Use the CRC formula [Eq. (4-16)] for the inelastic range.

SOLUTION

$$A = 0.5(8 + 12) = 10 \text{ in.}^2$$

$$I_x = 12 \times 0.5 \times 1.6^2 + \frac{0.5 \times 1.6^3}{3} + \frac{0.5 \times 6.4^3}{3} = 59.7 \text{ in.}^4$$

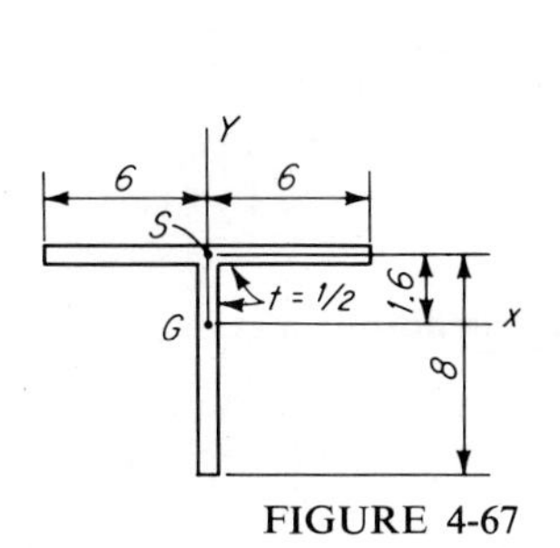

FIGURE 4-67

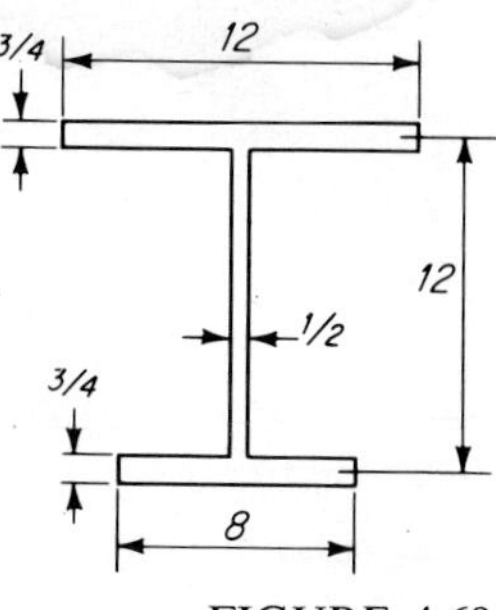

FIGURE 4-68
Problem 4-61.

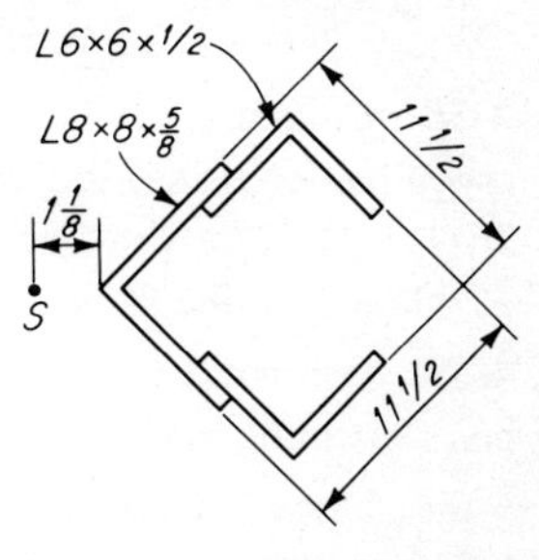

FIGURE 4-69
Problem 4-62.

$$I_y = \frac{0.5 \times 12^3}{12} = 72 \text{ in.}^4 \qquad I_{pS} = 59.7 + 72 + 10 \times 1.6^2 = 157.3 \text{ in.}^4$$

$$J = \frac{12 \times 0.5^3}{3} + \frac{8 \times 0.5^3}{3} = 0.833 \text{ in.}^4 \qquad C_w = 0$$

$$r_x^2 = \frac{59.7}{10} = 5.97 \qquad r_y^2 = \frac{72}{10} = 7.2 \qquad r_{pS}^2 = \frac{157.3}{10} = 15.7$$

$$r_t^2 = \frac{C + 0.04J(KL)^2}{I_{pS}} = \frac{0 + 0.04 \times 0.833 \times (0.5 \times 120)^2}{157.3} = 0.763 \text{ in.}^2$$

$$\left(1 - \frac{1.6^2}{15.7}\right) r_{tb}^4 - (7.2 + 0.763)r_{tb}^2 + 7.2 \times 0.763 = 0$$

$$0.837r_{tb}^4 - 7.96r_{tb}^2 + 5.49 = 0$$

$$r_{tb}^2 = \frac{7.96 \pm \sqrt{63.3 - 18.4}}{1.674} = 0.751 \text{ in.}^2$$

$$r_{tb} = \sqrt{0.751} = 0.87 \text{ in.} \qquad r_x = \sqrt{5.97} = 2.44 \text{ in.}$$

Since $r_{tb} < r_x$, the column fails by twist-bend buckling.

$$\frac{KL}{r_{tb}} = \frac{0.5 \times 120}{0.87} = 69 \qquad C_c = \pi\sqrt{\frac{2E}{F_y}} = \pi\sqrt{\frac{60{,}000}{36}} = 128$$

$$F_{cr} = F_y\left[1 - \frac{1}{2}\left(\frac{KL/r}{C_c}\right)^2\right] = 36\left[1 - \frac{1}{2}\left(\frac{69}{128}\right)^2\right] = 30.8 \text{ ksi}$$

$$P_{cr} = 30.8 \times 10 = 308 \text{ kips}$$

PROBLEMS

4-58 Compare the local-buckling stress [Eq. (4-47)] of a long column with equal-legged cruciform cross section with the twist-buckling stress of Eq. (4-67). Use the relation $E = 2G(1 + \mu)$. The value of k in case e of Fig. 4-51 is based on $\mu = 0.25$. The two values of F_{cr} should be the same.

4-59 For what length of pin-ended column of equal-legged cruciform cross section are bend buckling and twist buckling equally likely?

4-60 For what length of pin-ended column whose cross section is a W10 × 49 are bend buckling and twist buckling equally likely?

4-61 Compute the critical load for an A36 steel pin-ended column 12 ft long whose cross section is shown in Fig. 4-68. Use the CRC formula for buckling in the inelastic range.

4-62 The cross section shown in Fig. 4-69 was used for the bottom legs of a 437-ft tower supporting a transmission line crossing the Sacramento River (*Civ. Eng.*, January 1954, p. 41). The unsupported length of the leg is 20 ft. The Bureau of Reclamation specification allowable stress for A7 steel columns was $P/A = 22{,}000 - 98L/r$, in pounds per square inch, assuming failure by bend buckling. The warping constant C_w of the cross section is 6,300 in.^{6} What is the factor of safety?

4-30 ALUMINUM COLUMNS

The tangent-modulus formula (*DB* of Fig. 4-8) may be approximated by a straight line for the aluminum alloys, rather than by a parabola as suggested by CRC for steel. For alloy 6061-T6 the column strength is given by[29]

$$F_{cr} = 39.4 - 0.246\frac{KL}{r} \qquad \frac{KL}{r} \leq 66 \tag{4-72}$$

In the elastic range the Euler equation (4-2) becomes

$$F_{cr} = \frac{\pi^2 E}{(KL/r)^2} = \frac{\pi^2 \times 10{,}100}{(KL/r)^2} = \frac{100{,}000}{(KL/r)^2} \qquad \frac{KL}{r} > 66 \tag{4-73}$$

A factor of safety must be applied to the column strength determined from Eqs. (4-72) and (4-73). The ASCE Task Committee on Lightweight Alloys suggests factors of safety of 2.2 for bridges and 1.95 for buildings.[27]

Local-buckling strength of elements of compression members of alloy 6061-T6 can be calculated from the following formulas.

For one unloaded edge free and the other simply supported

$$F_{cr} = \begin{cases} 45.0 - 1.54\dfrac{b}{t} & \dfrac{b}{t} \leq 12 \qquad (4\text{-}74a) \\ \left(\dfrac{61.5}{b/t}\right)^2 & \dfrac{b}{t} > 12 \qquad (4\text{-}74b) \end{cases}$$

For both unloaded edges simply supported

$$F_{cr} = \begin{cases} 45.0 - 0.49\dfrac{b}{t} & \dfrac{b}{t} \leq 37 \qquad (4\text{-}75a) \\ \left(\dfrac{169}{b/t}\right)^2 & \dfrac{b}{t} > 37 \qquad (4\text{-}75b) \end{cases}$$

Plates that buckle in the inelastic range [Eqs. (4-74*a*) and (4-75*a*)] do not develop appreciable postbuckling strength, so that their allowable stresses should be determined by using the same factor of safety as for the allowable primary buckling stresses. On the other hand, plates that buckle in the elastic range [Eqs. (4-74*b*) and (4-75*b*)] may have appreciable postbuckling strength. Methods for evaluating this extra load-carrying capacity are discussed in Chap. 9.

Suggested specifications for the design of structures of aluminum alloys 6061-T6, 6062-T6, 6063-T5, and 6063-T6 have been published by the ASCE Task Committee mentioned above. Buckling formulas for columns and plates of typical structural aluminum alloys are given in Ref. 29, p. 10-6.

4-31 DP4-5: ALUMINUM-TRUSS COMPRESSION MEMBERS

In this article we design the compression members of the roof truss of DP3–4 for riveted construction in aluminum 6061-T6, using a factor of safety of 2. The comments which follow are identified by corresponding letters on the design sheet.

a The value of the column strength as determined from Eq. (4-72) is multiplied by one-half to provide the factor of safety of 2.

b Since the two angles are joined by rivets at rather long intervals, with washers as fillers, we assume that the long legs are free to buckle locally.

c Since these angles are of the same series chosen for member U_2U_5, we use as an initial guess a value approximately equal to the allowable for U_2U_5 but adjusted for $K = 1$.

d The web members will be small, and so we assume the size of the angles (except for thickness) instead of an allowable stress.

e For the remaining members $L/r > 66$, and so we use Eq. (4-73).

f Since the value of $b/t = 11$ is the same as for member U_2L_2, $F = 14.0$.

g Equation (4-74*b*) applies since $b/t > 12$.

ALUMINUM TRUSS COMPRESSION MEMBERS — DP4-5

Same truss as DP3-1
Material: aluminum 6061-T6
3/4 rivets: aluminum 6061-T6
5/16-in. gusset plates

$\underline{U_2U_5}$: $P = 109.7^k$, $L = 6'$

$A = 109.7/15 = 7.31$

Try 2L $5 \times 3\frac{1}{2}$, ⌉⌈

$KL/r = 0.9 \times 72/1.47 = 44$

$F_a = \frac{1}{2}(39.4 - 0.246 \times 44) = 14.3$ (a)

$A = 109.7/14.3 = 7.68$

Try 2L $5 \times 3\frac{1}{2} \times \frac{1}{2}$ $A = 8.00$

$\frac{b}{t} = \frac{5 - 1/2}{1/2} = 9$ (b)

$F = \frac{1}{2}(45.0 - 1.54 \times 9) = 15.57 > 13.67$

Use 2L $5 \times 3\frac{1}{2} \times \frac{1}{2}$ ⌉⌈

$\underline{U_0U_2}$: $P = 87.3^k$, $L = 6'$

$A = 87.3/13.5 = 6.48$ (c)

Try 2L $5 \times 3\frac{1}{2} \times \frac{7}{16}$ $A = 7.06$

$L/r = 72/1.44 = 50$

$F_a = \frac{1}{2}(39.4 - 0.246 \times 50) = 13.55$

$A = 87.3/13.55 = 6.44$

$\frac{b}{t} = \frac{5 - 7/16}{7/16} = 10.4$

$F = \frac{1}{2}(45.0 - 1.54 \times 10.4) = 14.5 > 13.55$

Use 2L $5 \times 3\frac{1}{2} \times \frac{7}{16}$ ⌉⌈

$\underline{U_1L_1}$: $P = 37.6^k$, $L = 5'$

Try 2L $3 \times 2\frac{1}{2}$ (d)

$L/r = 60/0.94 = 64$

$F_a = \frac{1}{2}(39.4 - 0.246 \times 64) = 11.8$

$A = 37.6/11.8 = 3.18$

Try 2L $3 \times 2\frac{1}{2} \times \frac{5}{16}$, $A = 3.24$

$\frac{b}{t} = \frac{3 - 5/16}{5/16} = 8.6$

$F = \frac{1}{2}(45.0 - 1.54 \times 8.6) = 15.9 > 11.8$

Use 2L $3 \times 2\frac{1}{2} \times \frac{5}{16}$ ⌉⌈

$\underline{U_2L_2}$: $P = 21.7^k$, $L = 5.5'$

Try 2L 3×2

$L/r = 66/0.86 = 72$

$F_a = \frac{1}{2} \times 100{,}000/72^2 = 9.64$ (e)

$A = 21.7/9.64 = 2.25$

Try 2L $3 \times 2 \times \frac{1}{4}$ $A = 2.38$

$\frac{b}{t} = \frac{3 - 1/4}{1/4} = 11$

$F = \frac{1}{2}(45.0 - 1.54 \times 11) = 14.0 > 9.64$

Use 2L $3 \times 2 \times \frac{1}{4}$ ⌉⌈

$\underline{U_3L_3}$: $P = 13.6^k$, $L = 6.04'$

Try 2L $3 \times 2 \times \frac{1}{4}$ $A = 2.38$

$L/r = 72.5/0.86 = 84.3$

$F_a = \frac{1}{2} \times 100{,}000/84.3^2 = 7.03 < 14.0$ (f)

$A = 13.6/7.03 = 1.94$

Use 2L $3 \times 2 \times \frac{1}{4}$ ⌉⌈

$\underline{U_4L_4}$: $P = 7.0^k$, $L = 6.54'$

Try 2L $2\frac{1}{2} \times 2 \times \frac{3}{16}$ $A = 1.62$

$L/r = 78.5/0.76 = 103$

$F_a = \frac{1}{2} \times 100{,}000/103^2 = 4.71$

$\frac{b}{t} = \frac{2\frac{1}{2} - 3/16}{3/16} = 12.3$

$F = \frac{1}{2}\left(\frac{61.5}{12.3}\right)^2 = 12.5 > 4.71$ (g)

$A = 7.0/4.71 = 1.49$

Use 2L $2\frac{1}{2} \times 2 \times \frac{3}{16}$ ⌉⌈

DP 4-5
Aluminum truss compression members.

PROBLEMS

4-63 Design for aluminum 6061-T6 a two-angle top-chord member for a riveted roof truss. The design load is 40 kips. The unsupported length is 5 ft for buckling in the plane of the truss and 10 ft for buckling in the plane of the roof. Gusset plates are $\frac{1}{4}$ in. thick. Use a factor of safety of 2.

4-64 Same as Prob. 4-63, except that the unsupported length is 5 ft for buckling in either direction.

4-65 Design an aluminum 6061-T6 column 20 ft long to support an axial load of 120 kips. The section is to consist of four angles and a plate in the form of an H. Space requirements are such that neither the depth nor the width of the member can exceed 12 in. Assume $K = 1$ and a factor of safety of 2.

4-66 Same as Prob. 4-65, except that the column is 10 ft long.

4-67 Same as Prob. 4-65, except that the column is 30 ft long.

REFERENCES

1 Timoshenko, S., and J. M. Gere: "Theory of Elastic Stability," 3d ed., McGraw-Hill, New York, 1969.

2 Horne, M. R., and W. Merchant: "The Stability of Frames," Pergamon, New York, 1965.

3 Bleich, F.: "Buckling Strength of Metal Structures," McGraw-Hill, New York, 1952.

4 Osgood, W. R., and M. Holt: The Column Strength of Two Extruded Aluminum-alloy H-sections, *NACA Tech. Rep.* 656, 1939.

5 Shanley, F. R.: Applied Column Theory, *Trans. ASCE*, vol. 115, 1950.

6 Duberg, J. E., and T. W. Wilder: Inelastic Column Behavior, *NACA Tech. Note* 2267, January 1951.

7 Johnston, B. G.: Buckling Behavior above the Tangent Modulus Load, *J. Eng. Mech. Div. ASCE*, December 1961.

8 Beedle, L. S., and L. Tall: Basic Column Strength, *Trans. ASCE*, vol. 127, p. 138, 1962.

9 Column Research Council, Engineering Foundation: "Guide to Design Criteria for Metal Compression Members," 2d ed., B. G. Johnston (ed.), Wiley, New York, 1966.

10 Huber, A. W., and L. S. Beedle: Residual Stresses and the Compression Strength of Steel, *Weld. J.*, vol. 33, p. 589, December 1954.

11 Nitta, A., and B. Thürlimann: Ultimate Strength of High Yield Strength Constructional-alloy Circular Columns; Effect of Thermal Residual Stresses; Effect of Cold Straightening, *Pub. IABSE*, 1962.

12 Tall, L.: Recent Developments in the Study of Column Behavior, *J. Inst. Eng. Aust.*, vol. 36, no. 12, December 1964.

13 Tall, L.: Welded Built-up Columns, *Lehigh Univ. Fritz Eng. Lab. Rep.* 249.29, 1966.

14 Alpsten, G. A.: Thermal Residual Stresses in Hot-rolled Steel Members, *Lehigh Univ. Fritz. Eng. Lab. Rep.* 337.3, 1968.

15 Clark, J. W., and R. L. Rolf: Buckling of Aluminum Columns, Plates, and Beams, *J. Struct. Div. ASCE*, June 1966.

16 Salvadori, M.: Lateral Buckling of Eccentrically Loaded I Columns, *Trans. ASCE*, vol. 121, p. 1163, 1956.

17 Massonnet, C.: Stability Considerations in the Design of Steel Columns, *J. Struct. Div. ASCE*, September 1959.

18 Austin, W. J.: Strength and Design of Metal Beam-Columns, *J. Struct. Div. ASCE*, April 1961.

19 Ketter, R. L.: Further Studies of the Strength of Beam-Columns, *J. Eng. Mech. Div. ASCE*, August 1961.

20 Hill, H. N., E. C. Hartmann, and J. Clark: Design of Aluminum Alloy Beam-Columns, *Trans. ASCE*, vol. 121, p. 1, 1956.

21 Galambos, T. V.: "Structural Members and Frames," p. 185, Prentice-Hall, Englewood Cliffs, N.J., 1968.

22 Sharma, S. S., and E. H. Gaylord: Strength of Steel Columns with Biaxially Eccentric Load, *J. Struct. Div. ASCE*, December 1969.

23 Julian, O. G., and L. S. Lawrence: Notes on J and L Nomograms for Determination of Effective Lengths, unpublished, 1959.

24 Lu, Le-Wu: Stability of Frames under Primary Bending Moments, *J. Struct. Div. ASCE*, June 1969.

25 Ueda, Y., and L. Tall: Inelastic Buckling of Plates with Residual Stresses, *Lehigh Univ. Fritz Eng. Lab. Rep.* 290.2, 1964.

26 German Buckling Specifications, T. V. Galambos and J. Jones (trans.), Column Research Council, June 1957.

27 Suggested Specifications for Structures of Aluminum Alloys 6061-T6 and 6062-T6, Report of Task Committee on Lightweight Alloys, *J. Struct. Div. ASCE*, December 1963.

28 White, M. W., and B. Thürlimann: Study of Columns with Perforated Cover Plates, *AREA Bull.* 531, September-October 1956.

29 Clark, J. W.: Aluminum Structures, sec. 10 in E. H. Gaylord and C. N. Gaylord (eds.), "Structural Engineering Handbook," McGraw-Hill, New York, 1968.

30 McGuire, W.: "Steel Structures," Prentice-Hall, Englewood Cliffs, N.J., 1968.

31 Thürlimann, B.: New Aspects Concerning Elastic Stability of Steel Structures, *Trans. ASCE*, vol. 127, 1962.

5

BEAMS

5-1 INTRODUCTION

Although there are no hard-and-fast rules, the following names are in common use to describe beams with reference to their function:

Floor beam In buildings, a major beam usually supporting joists; a transverse beam in bridge floors

Girder In buildings, the same meaning as (and more commonly used than) floor beam; also any major beam in a structure

Girt A horizontal member fastened to and spanning the wall columns of industrial buildings, used to support wall covering, such as corrugated metal

Header A beam framed to two beams at right angles to it, and usually supporting joists on one side of it; used at openings such as stairwells

Joist A beam supporting floor construction but not major beams

Lintel A beam spanning door, window, or other wall opening and supporting wall immediately above

Purlin A roof beam, usually supported by trusses

Rafter A roof beam, usually supported by purlins

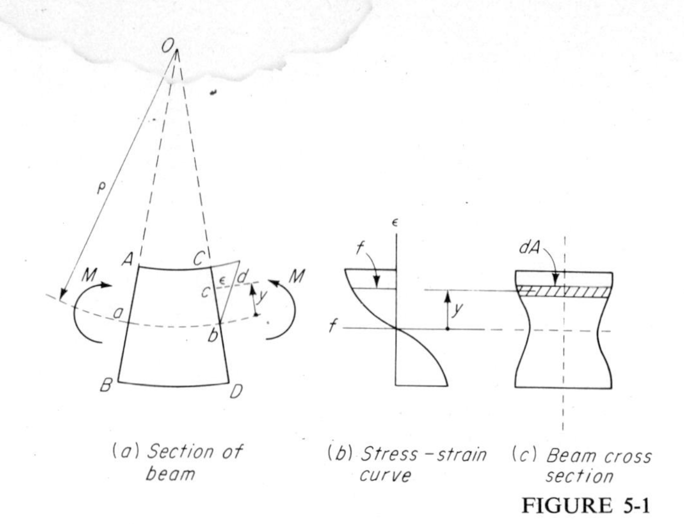

FIGURE 5-1

Spandrel beam A beam at the outside wall of a building, supporting its share of the floor and also the wall up to the floor above

Stringer In bridge floors, a longitudinal beam supported by floor beams (sometimes called a joist); in buildings, a beam supporting stair steps

Trimmer One of the beams or joists supporting a header

5-2 BENDING BEHAVIOR OF BEAMS

Figure 5-1*a* represents a length of originally straight beam which has been bent to the radius ρ by couples M; that is, the segment is subjected to pure bending. It is assumed that plane cross sections normal to the length of the unbent beam are still plane after the beam is bent. Therefore, considering two cross sections *AB* and *CD* a unit distance apart, similar sectors *Oab* and *bcd* give

$$\varepsilon = \frac{y}{\rho} \tag{a}$$

where y is measured from the axis of rotation (neutral axis). Thus, strains are proportional to distance from the neutral axis. The corresponding variation in stress over the cross section is given by the stress-strain diagram of the material, rotated 90° from the conventional orientation, provided the strain axis ε is scaled through Eq. (*a*) with the distance y (Fig. 5-1*b*). The bending moment M is given by

$$M = \int_A yf \, dA \tag{b}$$

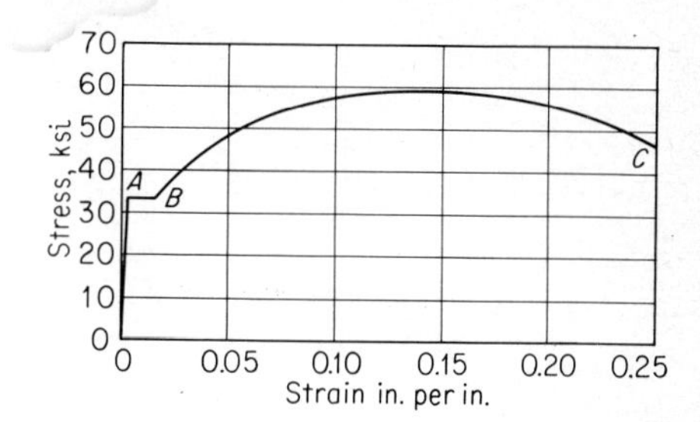

FIGURE 5-2

where dA is an element of area at the distance y (Fig. 5-1c). Thus, the moment M can be determined if the relation between stress and strain shown in Fig. 5-1b is known. If stress is proportional to strain, so that $f = E\varepsilon$, Eqs. (a) and (b) give

$$M = \frac{E}{\rho} \int_A y^2 \, dA = \frac{EI}{\rho} \tag{c}$$

or, eliminating ρ through Eq. (a),

$$M = \frac{EI\varepsilon}{y} = \frac{fI}{y} \tag{d}$$

The bending behavior of a beam of rectangular cross section made of a sharply yielding steel (Fig. 5-2) will now be investigated. Equation (d) holds so long as the stress is given by OA of Fig. 5-2; i.e., if $f \lesssim F_y$. When the extreme-fiber strain attains the value ε_y, the strain distribution and stress distribution are given by Fig. 5-3b and c. The corresponding moment is

$$M_y = \frac{F_y I}{d/2} = F_y \frac{bd^2}{6} \tag{e}$$

where b is the width and d the depth of the cross section (Fig. 5-3a). For $M < M_y$, moment is proportional to extreme-fiber strain. This is shown by OA in Fig. 5-4, where the ratio M/M_y is plotted against ε.

As the load on the beam increases beyond the load corresponding to M_y, strain continues to increase in proportion to distance from the neutral axis, but the stress distribution consists of OA and a portion of AB of Fig. 5-2, provided the extreme-fiber strain is less than ε_s. Thus, if the extreme-fiber strain is $2\varepsilon_y$, as in Fig. 5-3d, the stress distribution is that shown in Fig. 5-3e. The corresponding resisting moment is

$$M = F_y \frac{bd}{4} \frac{3d}{4} + F_y \frac{bd}{8} \frac{d}{3} = \frac{11}{48} F_y bd^2 \tag{f}$$

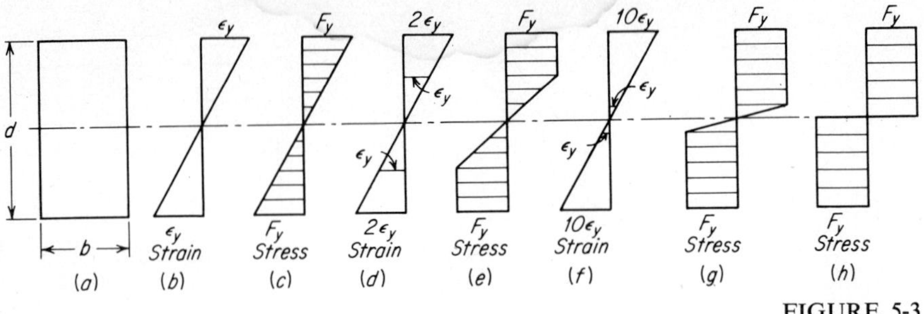

FIGURE 5-3

which gives point B in Fig. 5-4. This moment is only 37.5 percent more than the yield moment M_y, even though the extreme-fiber strain is doubled. Still further deformation is shown in Fig. 5-3f, where 90 percent of the cross section has yielded. The corresponding moment from Fig. 5-3g is $0.249F_y bd^2$, which gives point C in Fig. 5-4.

It will be noted in Fig. 5-4 that the rate of increase in moment falls off rapidly soon after M_y is exceeded, so that M appears to be approaching a limiting value. This limit is determined by the distribution of stress shown in Fig. 5-3h, for which

$$M = F_y \frac{bd}{2}\frac{d}{2} = F_y \frac{bd^2}{4} \tag{g}$$

This is only 0.4 percent larger than the moment for an extreme-fiber strain of $10\varepsilon_y$. Thus, although the assumed stress distribution cannot exist, it determines a practical limiting value with negligible error. However, even this limit is exceeded after strain hardening begins, since stresses at and near the extreme fiber now exceed F_y. This is shown by DE in Fig. 5-4.

The moment given by Eq. (g) is called the *plastic moment* of resistance, denoted by M_p. It is usually taken as the limiting value; i.e., the benefits of strain hardening are neglected. The ratio of the plastic moment to the yield moment for the rectangular cross section is given by

$$\frac{M_p}{M_y} = \frac{F_y bd^2/4}{F_y bd^2/6} = 1.5$$

This ratio is called the *shape factor.*

A beam element that has been strained plastically is left with some permanent deformation after it is unloaded. Thus, if the moment reaches the value at F in Fig. 5-4 and the beam is then gradually unloaded, the resisting moment decreases along FG. Upon reloading it will increase along GF and will continue to behave elastically in subsequent loading and unloading, provided the moment corresponding to F is not exceeded. This is the same phenomenon that is observed in the tension test.

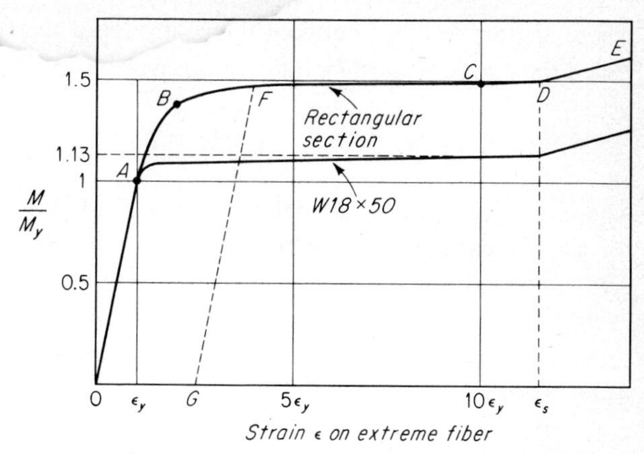

FIGURE 5-4

For the unequal-flanged I cross section of Fig. 5-5a, the stress distribution at the beginning of yield on the extreme fiber is shown in b. The moment M_y is given by

$$M_y = F_y \frac{I}{c} = F_y S$$

where S is the elastic section modulus. To satisfy the condition $\int_A f\,dA = 0$, the neutral axis must pass through the center of gravity of the cross section. The fully plastic stress distribution is shown in c. To satisfy the condition $\int_A F_y\,dA = 0$ in this case requires that the neutral axis divide the area A of the cross section into two equal parts. The moment M_p is given by

$$M_p = F_y \frac{A}{2} a = F_y Z$$

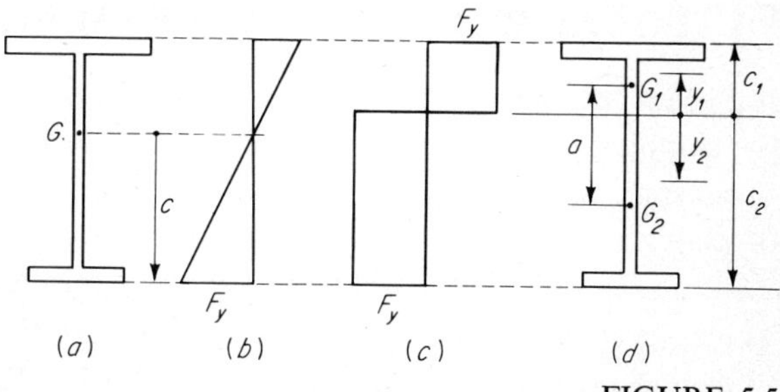

FIGURE 5-5

where a is the distance between the centroids of the two equal areas (Fig. 5-5d). Z is called the plastic section modulus. In general, the value of M_p can be determined from

$$M_p = F_y\left(\int_0^{c_1} y_1\, dA + \int_0^{c_2} y_2\, dA\right) = F_y Z$$

where y_1, y_2, c_1, and c_2 are defined in Fig. 5-5d. Values of the section moduli S and Z for the standard shapes commonly used for beams and columns are tabulated in the AISC Manual.

The variation of moment with extreme-fiber strain for the I or H shape is similar to that of the rectangular cross section. However, the ratio of the plastic moment to the yield moment is much smaller. Thus, for the W18 × 50

$$\frac{M_p}{M_y} = \frac{Z}{S} = \frac{101}{89.1} = 1.13$$

This is typical of the W shapes, for which the shape factor ranges from 1.10 to 1.18. The moment-strain plot for the W18 × 50 is shown in Fig. 5-4. Shape factors for certain other cross sections are the following:

Circular	1.70
Thin-walled circular tube	1.27
Thin-walled rectangular tube	1.12

PROBLEMS

5-1 Prove that the shape factor for the circular cross section is 1.70.

5-2 Compute the shape factor for the cross section consisting of two equal isosceles triangles whose common base constitutes the neutral axis.

5-3 Compute the shape factor for the tee of Fig. 4-67.

5-3 LOCAL BUCKLING OF BEAM FLANGES

If a beam cross section such as the I, the box, etc., is to develop the yield moment M_y, the compression flange must be able to reach yield stress without buckling. Of course, this means that it must be capable of accepting a compressive strain ε_y. Local buckling of plates is discussed in Arts. 4-21 and 4-22. Values of plate slenderness b/t which must not be exceeded if yield stress is to be reached are determined in Art. 4-23, and Table 4-4 gives typical specification values. Thus, according to AISC, the thickness of the flange of an I must be such that $b/t \lesssim 95/\sqrt{F_y}$, where b is the distance from web

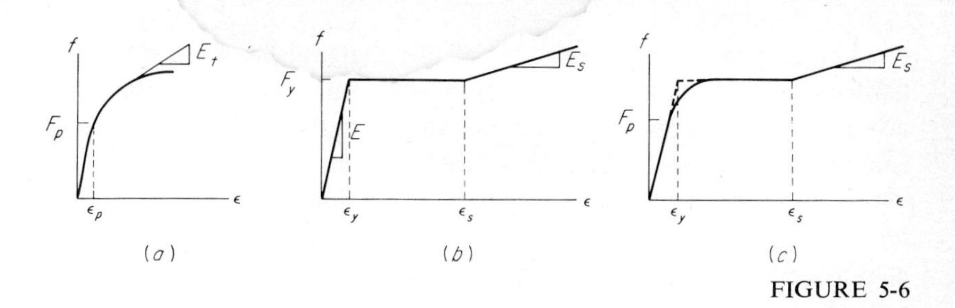

FIGURE 5-6

centerline to flange tip and F_y is in kips per square inch. Similarly, the flange of a box section must satisfy $b/t \lessapprox 253/\sqrt{F_y}$, where b is the distance between the lines of fasteners or welds.

More restrictive limits of plate slenderness must be observed if a beam cross section is to attain the plastic moment M_p. This is because the compressive strain in the flange must be several times the yield strain to develop a moment essentially equal to the fully plastic moment (Figs. 5-3 and 5-4). For this case, point F of Fig. 4-53 gives a conservative limiting value since it allows the strain-hardening strain ε_s, which may be as much as 12 times ε_y, to develop.

Proportions of plates which can be strained beyond the yield value ε_y are difficult to determine. This is partly because of the anisotropic behavior of plates that buckle after yielding has begun and partly because of the discontinuous nature of the yielding process itself. Local buckling of plates in the nonproportional range of stress was discussed in Art. 4-21, where it was shown that, for a gradually yielding material (Fig. 5-6*a*), the elastic modulus E in Eq. (4-47) should be replaced by $\sqrt{EE_t}$ to extend the formula to the inelastic range. However, in the case of steel with a flat-top yield (Fig. 5-6*b*), it appears that the modulus changes instantaneously from E to zero at yield. Even in the curved-knee diagram of average stress vs. strain, which results if there are initial residual stresses (Fig. 5-6*c*), the tangent modulus appears to reach zero at strains much less than the strain-hardening value. Thus, in this situation, it would seem that resistance to local buckling would vanish at strains too small to allow the plastic moment to develop fully. However, plates accept much larger strains than this without buckling. The reason appears to be that yielding actually develops in small yield planes, or slip bands, in which the strain jumps suddenly from the elastic-limit value to the value ε_s at the beginning of strain hardening. This phenomenon was discussed in Art. 2-4. These slip bands form one after another after having initiated at a weak point, such as an inclusion or a point of stress concentration. Thus, there is no material, in the length over which strain is measured, for which $\varepsilon_y < \varepsilon < \varepsilon_s$. Instead, some portions are strained to ε_y while the remainder is strained to ε_s. The *measured* strain averages these values. During this stage, the material is not homogeneous. However, after all the material has been strained to the strain-hardening value, it becomes homogeneous again and the stress begins to increase according to the strain-hardening modulus E_s in Fig. 5-6*b*. Nevertheless, the material is anisotropic because of changes caused by slip.

Discontinuous yielding has been taken into account to determine values of b/t corresponding to strains of the order of strain hardening.[1,2] For the uniformly compressed plate simply supported on one unloaded edge and free on the other, the critical stress according to Ref. 1 is

$$F_{cr} = \frac{G_t}{(b/t)^2} + \frac{\pi^2 E}{12(1 - \mu_x \mu_y)} \frac{t^2}{L^2} \tag{a}$$

where G_t = tangent modulus in shear
μ_x, μ_y = Poisson's ratio in direction of x, y
L = length of plate
b = width of plate
t = thickness of plate

This equation is identical in form to Eq. (4-67) for twist buckling of the cruciform cross section. The identity of twist buckling of the cruciform and local buckling of the plate has already been mentioned (Art. 4-28). Substituting $I_p = 4tb^3/3$, $J = 4bt^3/3$, and $C_w = b^3t^3/9$ (Art. 4-28) into Eq. (4-67) gives

$$F_{cr} = \frac{G}{(b/t)^2} + \frac{\pi^2 E}{12} \frac{t^2}{L^2} \tag{b}$$

Thus, the only difference between inelastic buckling and elastic buckling is that $E/(1 - \mu_x \mu_y)$ replaces E and G_t replaces G in the buckling formula. Furthermore, plates used in structural members are long enough to warrant neglecting the second term of Eq. (*a*), so that

$$F_{cr} = \frac{G_t}{(b/t)^2} \tag{5-1}$$

Since the value of b/t which permits a plate to reach the beginning of strain hardening without buckling is needed, G_t in Eq. (5-1) must be evaluated accordingly. Tests on two circular tubes which were compressed to the strain ε_s and then twisted are reported in Ref. 3. The shearing modulus at the beginning of twist was practically equal to the elastic value G. However, it dropped rapidly at small values of shear strain. Then, at a value of 2,000 or 3,000 ksi, it began to decrease more slowly, so that a value of this order of magnitude would seem to be reasonable for the buckling problem. Based on torsional buckling tests on single angles, the value $G_t = 2{,}400$ ksi was suggested.[1] Therefore, from Eq. (5-1)

$$\frac{b}{t} = \frac{\sqrt{2{,}400}}{\sqrt{F_y}} = \frac{49}{\sqrt{F_y}} \tag{5-2}$$

where F_y is in kips per square inch. An analysis of the I flange under compression, taking into account rotational restraint from the web, showed that this value is increased by only 2 or 3 percent by such restraint.[4]

Analysis of the uniformly compressed plate supported on all four edges leads to an equation similar to Eq. (*a*). If the supports are simple, the plate can reach strain hardening if

$$\frac{b}{t} \lessapprox \frac{192}{\sqrt{F_y}} \tag{5-3}$$

For A36 steel, Eqs. (5-2) and (5-3) give $b/t = 8.2$ and 32, respectively.

Summarizing the results of this article, we have the following limits of plate slenderness which preclude premature local buckling of compression flanges of beams:

Projecting element: $\quad \frac{b}{t} \lessapprox \frac{95}{\sqrt{F_y}}$ for M_y $\quad \frac{b}{t} \lessapprox \frac{49}{\sqrt{F_y}}$ for M_p

Flange of box: $\quad \frac{b}{t} \lessapprox \frac{253}{\sqrt{F_y}}$ for M_y $\quad \frac{b}{t} \lessapprox \frac{192}{\sqrt{F_y}}$ for M_p

Since these limits are not well defined, they differ somewhat from one specification to another.

5-4 BEAM CROSS SECTIONS

Typical beam cross sections are shown in Fig. 5-7. Two types of I cross section are rolled in a wide variety of sizes. The American Standard Beam (S), which was the first steel beam section rolled in the United States, ranges in depth from 3 to 24 in. (Fig. 5-7*a*). Increase in section modulus for a given depth is achieved by spreading the rolls to increase the flange width and web thickness while maintaining the same depth. The wide-flange shapes (W) give more section modulus per pound (Fig. 5-7*b*). They range in depth from 4 to 36 in., with increase in section modulus achieved by increasing flange and web thicknesses and flange width while maintaining a constant depth *inside* the flanges. Light beams and miscellaneous shapes (M) are of the same

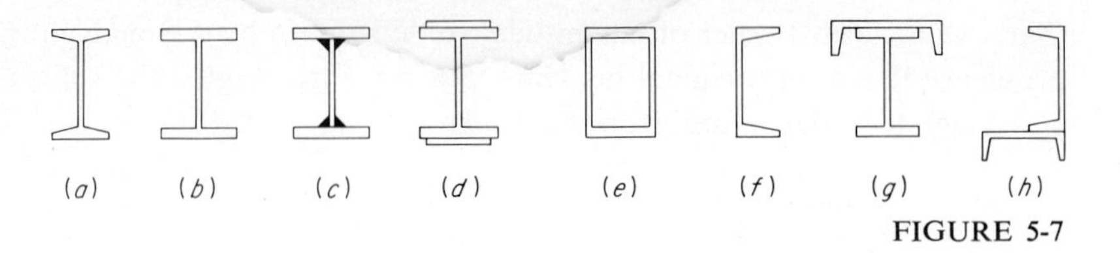

FIGURE 5-7

shape as the W but of lighter weight (and smaller section modulus) for the same depth. Some producers weld three plates to form standard shapes of the same dimensions as the deeper W's (Fig. 5-7*c*). The section modulus of the W may be increased by welding plates to the flanges (Fig. 5-7*d*).

Since the web of an I contributes only a small part of the bending resistance, it is sometimes economical in welded beams of high-strength steel to use a lower-strength steel for the web. Such beams, called *hybrid* beams, are discussed in Chap. 6.

Box sections (Fig. 5-7*e*) are also efficient beam sections. They are available as rolled shapes, called *structural tubing*, in rectangular form ranging from 3×2 to 12×6 in. The four-plate welded box is also used extensively.

Channels (Fig. 5-7*f*) are used occasionally, usually as purlins, girts, eave struts, lintels, and as trimmers and headers for stairwells and other openings. They are sometimes used with the S or W for crane-runway girders as in Fig. 5-7*g*. Two channels arranged as in *h* are common as eave struts in industrial buildings.

Typical cold-formed cross sections suitable for beams are shown in Fig. 5-8. The channel in *a* may be used for short spans. Local-buckling resistance of the thin flanges is increased if they are stiffened with lips, as in *b*. Two channels spot welded back to back are used as floor joists (*c* and *d*). Another form of I cross section is shown in *e*. The crimped web of the I in *f* is designed to receive nails. The cross sections in *g* and *h* can be used as girts and eave struts respectively.

5-5 DESIGN FOR SIMPLE BENDING

The selection of a standard shape to resist at a specified allowable stress the bending due to loads in a plane of symmetry is one of the most common problems in the design of metal beams. The W, which is used almost exclusively in this situation, is proportioned so that the moment of inertia about the major principal axis is considerably larger than that about the minor principal axis. This is done to produce shapes that make economical beams. As a result, they are relatively weak in resistance to torsion and to bending about the minor axis, and if not held in line by attached construction, they may become unstable under load. The instability manifests itself as a sidewise

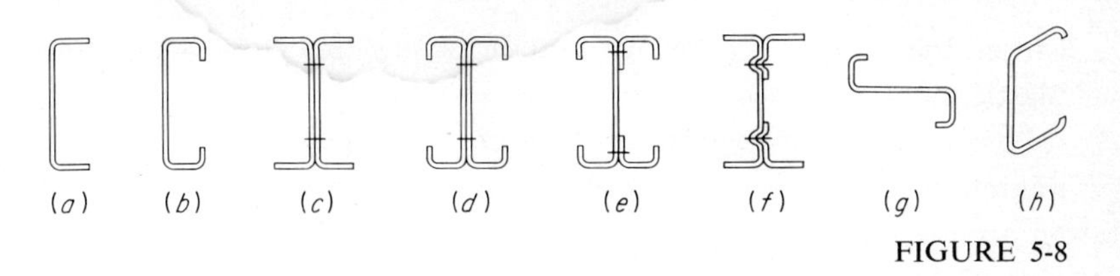

FIGURE 5-8

bending accompanied by twist, and is called *lateral buckling* or *lateral-torsional buckling*. Lateral buckling is discussed in Art. 5-7.

If the beam cannot buckle because of support furnished by floor or other construction, it is necessary only to compute the maximum bending moment and then choose a shape which has the required section modulus. Since structural metals are sold by weight, it is desirable to require further that the shape be the lightest one that will furnish the necessary section modulus. Occasionally, limited headroom or the necessity of matching the depth of adjacent beams will make it impossible to use the lightest shape. Also, as will be shown later, deflection limitations, and occasionally shearing stresses, will sometimes dictate the choice. The AISC Manual contains a table of section moduli which makes it a simple matter to choose the lightest available steel shape.

For beams in bridges designed on the basis of elastic behavior, the AASHO and AREA allowable tensile bending stress is $0.55F_y$. The allowable compressive bending stress is also $0.55F_y$, provided there is continuous lateral support. This gives a factor of safety of 1.82 with respect to the yield moment M_y. The corresponding AISC allowable stress for beams in buildings is $0.60F_y$, which gives a factor of safety of 1.65 on the yield moment. However, this allowable is increased by 10 percent, to $0.66F_y$, if the beam flange is thick enough to permit the plastic moment to develop. This corresponds to the smallest value of the shape factor for W's. Shapes which can develop the plastic moment are called *compact sections*. Of course, the increased allowable stress is to be used with the elastic section modulus S rather than the plastic section modulus Z. Thus, both allowable stresses give the same factor of safety with respect to the bending strength of the beam.

Limiting values of flange slenderness b/t to permit the allowable stresses discussed above are given in Table 5-1. Values for $F_b = 0.66F_y$ are based on Eqs. (5-2) and (5-3).

The AISC specification also permits an increased allowable bending stress, $0.75F_y$, for I shapes bent about the minor axis, provided they are compact sections, i.e., provided the flange projection satisfies the slenderness limit $52.2/\sqrt{F_y}$ of Table 5-1. The same allowable stress applies to round and square bars and to rectangular cross sections bent about the minor axis. As for the allowable stress $0.66F_y$ discussed earlier, the increase is in recognition of the fact that compact sections can develop the plastic moment.

The allowable bending stress of the AISI specification is $0.60F_y$, where F_y is the minimum specified yield stress before forming to shape. The corresponding limits of b/t are given in Table 5-1.

The limiting values given in Table 5-1 for design in hot-rolled shapes cover most situations, since it usually is not economical to use thinner elements that require reduction in the allowable stress because of local buckling. Nevertheless, there are provisions in the AISC and AREA specifications for relaxing these requirements, discussed in Chap. 9. On the other hand, b/t values of elements of cold-formed members often exceed the values in Table 5-1. Design procedures for this situation are also discussed in Chap. 9.

Because of the nonlinear stress-strain curve for aluminum, the bending stress computed by $f = M/S$ can exceed the yield strength (0.2 percent offset) appreciably before the beam yields significantly. This can be taken into account in computing the yield moment M_y by using a shape factor. Thus, the shape factor for the I bent about the major axis is 1.07, while that for bending about the minor axis is 1.30. Shape factors for other cross sections are given in Ref. 5. It should be noted that these are yield-moment shape factors rather than plastic-moment shape factors. However, aluminum beams also develop the plastic moment, provided they are proportioned to preclude local buckling and lateral buckling. Shape factors for plastic moment are the same as those discussed in Art. 5-2. Allowable bending stresses to be used in $M = FS$, which are suggested in Ref. 6, take the plastic-moment shape factor into account.

Table 5-1 LIMITING VALUES OF b/t FOR FLANGES OF BEAMS

Type of element	AASHO	AREA	AISC		AISI
	$F_b = 0.55F_y$	$F_b = 0.55F_y$	$F_b = 0.60F_y$	$F_b = 0.66F_y$	$F_b = 0.60F_y$
Flange of I shape	$\dfrac{3{,}250^{a,b}}{\sqrt{f_{,\text{psi}}}}$	$\dfrac{2{,}300^{c}}{\sqrt{F_{y,\text{ psi}}}}$	$\dfrac{95^{d}}{\sqrt{F_{y,\text{ ksi}}}}$	$\dfrac{52.2^{d}}{\sqrt{F_{y,\text{ ksi}}}}$	$\dfrac{63.3^{e}}{\sqrt{F_{y,\text{ ksi}}}}$
Flange of box	$\dfrac{5{,}000^{a}}{\sqrt{f_{,\text{ psi}}}}$	$\dfrac{7{,}500}{\sqrt{F_{y,\text{ psi}}}}$	$\dfrac{253^{f}}{\sqrt{F_{y,\text{ ksi}}}}$	$\dfrac{190}{\sqrt{F_{y,\text{ ksi}}}}$	$\dfrac{184^{a}}{\sqrt{f_{,\text{ ksi}}}}$

[a] f = service-load stress.
[b] b = width of flange.
[c] b = distance from free edge of flange to fillet.
[d] b = half width of flange.
[e] b = flat projection of flange from web, exclusive of fillets and stiffening lip.
[f] 238 for boxes of uniform thickness.

5-6 DP5-1: FLOOR FRAMING

Beams for a typical floor of the building of DP4-2 are designed in this example. A complete floor plan is given in DP4-2 and a general description of the framing in Art. 4-15. A portion of the floor-framing plan is shown on the design sheet. A sketch of the floor construction shows the 3-in. cellular steel decking with $2\frac{1}{2}$-in. concrete slab, which spans the 15 ft between joists. The frame is designed for simple-beam framing connections; wind bracing is provided in the walls surrounding the service area. The following comments are intended to clarify computations identified by the corresponding letters on the design sheet.

a Live-load reduction is discussed in Art. 1-4. The American Standard Building Code reduction is 0.08 percent per square foot of supported area, provided the area is larger than 150 ft², but cannot exceed 60 percent or $R = 100(D + L)/4.33L$ percent. In this case, the maximum reduction of 40 percent controls except for the 30-ft joists.

b The AISC specification requires that beam spans be taken as the distance center to center of the supporting members.

c Beams must be stiff enough to limit deflection. AISC suggests $L/d = 800/F_y$ as a minimum for fully stressed beams (Art.5–22) which gives $800/36 = 22$ for A36 steel.

d The 30-ft girder carries a concentrated load at midspan, which is the sum of the two 30-ft joist reactions. Therefore, $M = PL/4$.

e The spandrel beam supports one story of wall, which weights 25 psf, in addition to its share of floor.

PROBLEMS

5-4 A simply supported beam spanning 15 ft carries a uniformly distributed load of 3 klf, not including the weight of the beam. Floor construction restrains it against lateral buckling. What size beam of A36 steel is required? AISC specification.

5-5 Purlins 20 ft long and spaced 6 ft on centers support a roof which is on a slope of 1 in 12. The roof is decked with 2-in.-thick precast slabs made of lightweight concrete weighing 65 pcf and securely fastened to the purlins. The built-up roofing weights 6 psf. For the snow load recommended in Table 1-3, what size A36 joist is required if the building is located in (*a*) Alabama, (*b*) Wisconsin? AISC specification.

5-6 A simply supported beam spanning 30 ft carries uniformly distributed loads of 1.2 klf dead and 1.6 klf live and a concentrated live load of 8 kips 12 ft from one end. Headroom limits the depth to not more than 22 in. Determine the laterally supported A36-steel beam required for load factors of 1.2 on dead load and 1.6 on live load, based on the plastic moment.

5-7 The beam shown in Fig. 5-9 has continuous lateral support. What size A36-steel beam is required? AISC specification.

5-8 Same as Prob. 5-7 except that, instead of the fixed concentrated loads shown, the beam supports a moving load consisting of two wheel loads of 10 kips each spaced 6 ft on centers.

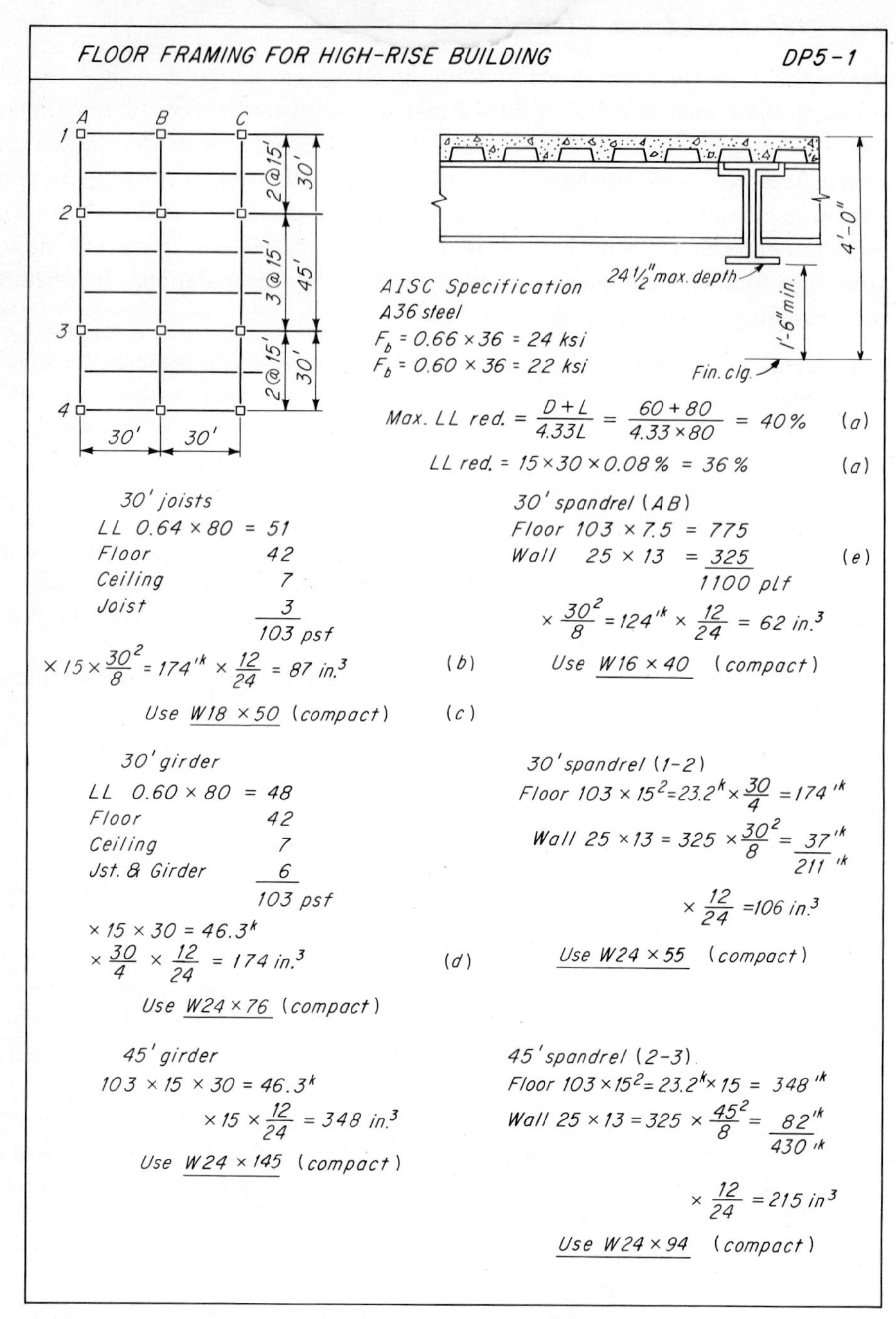

DP 5-1
Floor framing for high-rise building.

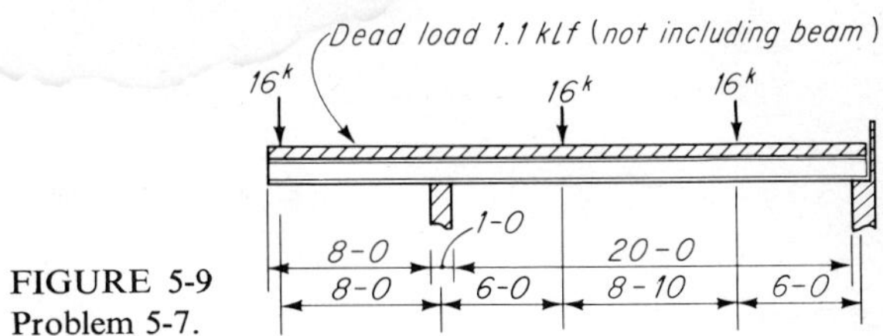

FIGURE 5-9
Problem 5-7.

5-9 Design the joists and girders for the floor of the building shown in plan in Fig. 5-10. The live load is 80 psf uniform. The floor is a 4-in. concrete slab. A36 steel, AISC specification.

5-10 A cross section of the roadway of a highway beam-bridge is shown in Fig. 5-11. The span is 40 ft. The bridge is to be designed for the AASHO HS20-44 loading (Fig. 1-2). Since a beam cannot deflect unless there is a transverse bending of the roadway slab, the result is that part of the wheel directly over a beam may be supported by adjacent beams. Therefore the load on any one beam may be more or less than the load on a wheel directly over it, depending upon the stiffness of the floor and the spacing of the beams. According to the AASHO specifications, for concrete floor slabs the fraction of a wheel load supported by an interior beam directly underneath it is $S/5.5$, where S is the spacing of the beams in feet, while the load on an outside beam is computed on the assumption that the flooring acts as a simple beam.

Design the beams in A36 steel. The clearance diagram for the truck loading is shown in Fig. 1-2, and the impact requirement is given by Eq. (1-3).

5-11 The framing plan for an office building is shown in Fig. 5-12. Design the beams in A36 steel, AISC specification. Given:

Floor $2\frac{1}{2}$-in. stone concrete on cellular steel, 40 psf
Ceiling Acoustical plaster and hung ceiling, 10 psf
Exterior walls Aluminum window wall 10 ft high at 20 psf
Live load Office area 70 psf (includes 20 psf for movable partitions)

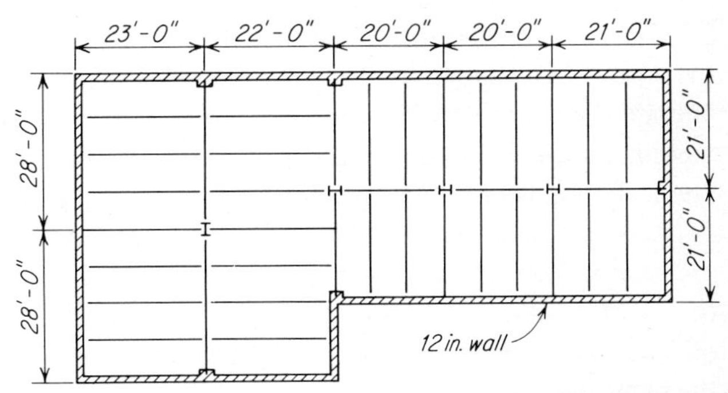

FIGURE 5-10
Problem 5-9.

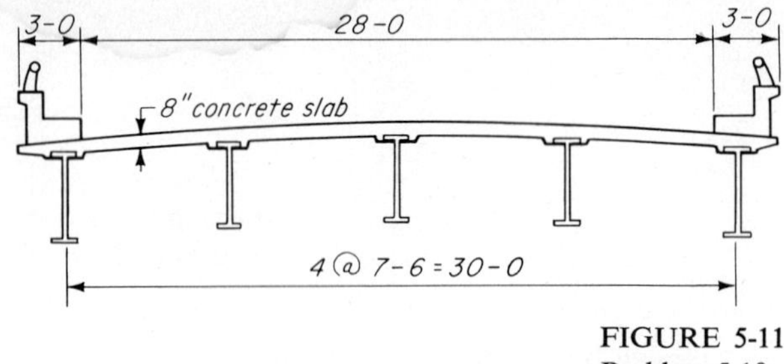

FIGURE 5-11
Problem 5-10.

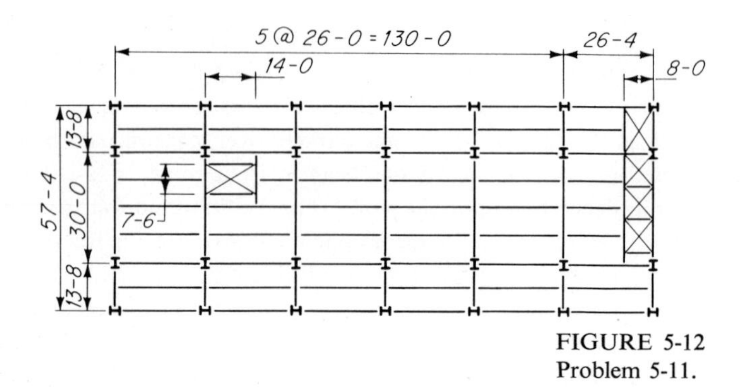

FIGURE 5-12
Problem 5-11.

5-7 LATERAL-TORSIONAL BUCKLING

The possibility of lateral-torsional buckling of beams which are not supported against lateral displacement was mentioned in Art. 5-5. Any beam not so supported may buckle in this manner if its torsional resistance and moment of inertia for the weak axis are small compared with the moment of inertia for the strong axis.

Figure 5-13*a* shows a doubly symmetrical prismatic I beam, both ends of which are simply supported with respect to the x and y axes but held against rotation about the z axis. The beam is subjected to pure bending by moments M_x at each end. In the laterally bent position the center of gravity of the cross section is displaced by the amounts u and v in the direction of x and y, respectively, and there is a rotation β about the z axis. In the projection on the xz plane in Fig. 5-13*b*, the external moment M_x and the resisting moment at z are shown as vectors. The components of the resisting moment are $M_x \cos \theta$ in the plane of the deflected cross section and $M_x \sin \theta$ in the direction of the normal to it. However, if we restrict ourselves to the determination of the value of the end moments M_x at which a perfect beam just begins to bend out of the plane of these moments, u, v, β, and θ can be considered to be infinitesimal and we may take $\cos \theta = 1$ and $\sin \theta = \tan \theta = du/dz$. This gives M_x and $M_x\, du/dz$ for the

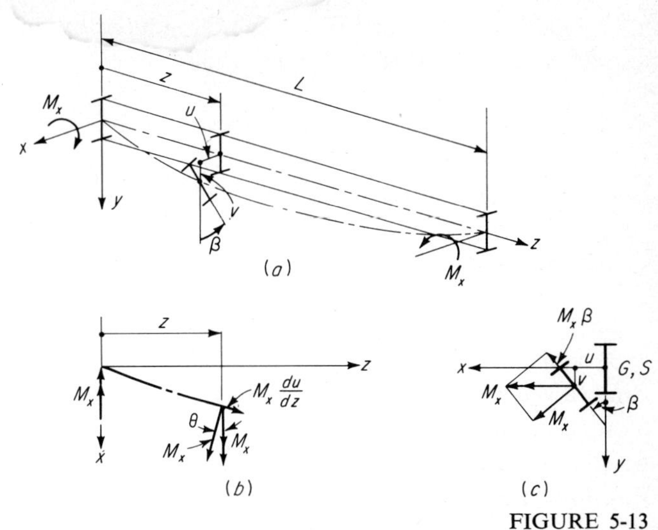

FIGURE 5-13

components of moment in the deflected position, as shown in Fig. 5-13*b*. Similarly, in the projection on the *xy* plane shown in Fig. 5-13*c*, the components in the directions of the principal axes in the deflected position are $M_x \cos \beta = M_x$ on the strong axis and $M_x \sin \beta = M_x\beta$ on the weak axis.

Equations of equilibrium are formulated by equating the components of M_x to the corresponding resistances. Since the displacements are infinitesimal, curvatures in the principal planes of the deflected cross section may be taken as d^2v/dz^2 and d^2u/dz^2. Therefore, the bending resistances are $EI_x\, d^2v/dz^2$ and $EI_y\, d^2u/dz^2$. The torsional resistance is given by Eq. (4-63). In this way we obtain the following equations

$$-EI_x \frac{d^2v}{dz^2} = M_x \tag{5-4a}$$

$$-EI_y \frac{d^2u}{dz^2} = M_x \beta \tag{5-4b}$$

$$GJ \frac{d\beta}{dz} - EC_w \frac{d^3\beta}{dz^3} = M_x \frac{du}{dz} \tag{5-4c}$$

The negative signs are needed in Eqs. (5-4*a*) and (5-4*b*) because the second derivatives are negative. Equation (5-4*a*) is independent and can be solved directly for the displacement *v*. However, *b* and *c* are coupled, which means that the displacements *u* and β cannot exist independently of one another.

Equations (5-4*b*) and (5-4*c*) can be reduced by differentiating (5-4*c*) once with respect to z and eliminating d^2u/dz^2 through (5-4*b*). This gives

$$EC_w \frac{d^4\beta}{dz^4} - GJ \frac{d^2\beta}{dz^2} - \frac{M_x{}^2}{EI_y} \beta = 0 \tag{5-5}$$

A solution to this equation can be obtained by taking

$$\beta = \beta_{L/2} \sin \frac{\pi z}{L} \tag{a}$$

where $\beta_{L/2}$ is the angle of twist at midspan. This gives $\beta = 0$ at each end of the beam and, because of Eq. (5-4*b*), $EI_y\, d^2u/dz^2 = 0$ at each end. Thus, the beam is restrained against twisting at the supports but is free to rotate about the y axis. Furthermore, because $d^2\beta/dz^2$ is also zero at each end, the end cross sections are free to warp (Art. 4-28). Thus, the boundary conditions which were assumed in formulating Eqs. (5-4) are satisfied. Substituting from Eq. (*a*) into Eq. (5-5) gives

$$\left(\frac{\pi^4}{L^4} EC_w + \frac{\pi^2}{L^2} GJ - \frac{M_x{}^2}{EI_y}\right)\beta_{L/2} \sin \frac{\pi z}{L} = 0$$

This equation is satisfied by $\beta_{L/2} = 0$, which means that both β and u are everywhere zero, or by

$$M_{x,\,cr}^2 = \frac{\pi^2}{L^2} EI_y GJ + \frac{\pi^4}{L^4} EI_y EC_w \tag{5-6}$$

This gives the value of M_x at which lateral-torsional buckling begins. It will be noted that, as in the case of the Euler column, the deflection is indeterminate. Thus, Eq. (5-6) identifies the beginning of lateral-torsional buckling, but gives no information about the postbuckling behavior.

Since it was derived for pure bending, Eq. (5-6) is limited in scope. It is restricted further by the assumed freedom to warp and to rotate about the y axis at the supports. The first restriction is removed by rewriting Eqs. (5-4) for whatever variation in bending moment is to be considered. Of course, the bending moment is now dependent on z. Solutions of these equations can be obtained by using infinite series.[11] The second restriction is removed by using appropriate boundary conditions. For this case, however, an end-restraint moment M_y (which is an infinitesimal of the same order as β) must be added to the right member of Eq. (5-4*b*). The resulting critical moments for beams with end moments and beams with transverse loads acting through the shear center can be put in the form of Eq. (5-6) as follows:

$$M_{x,\,cr}^2 = C_b{}^2\left(\frac{\pi^2}{(KL)^2} EI_y GJ + \frac{\pi^4}{(KL)^4} EI_y EC_w\right) \tag{5-7}$$

where C_b is a coefficient which depends on the variation in moment along the span and

Table 5-2 COEFFICIENTS IN EQ. (5-7)*

Case	Loading	y-axis support at ends	K	C_b†	C_1	C_2
1	M M	Simple Fixed	1 0.5	1 1		1 1
2	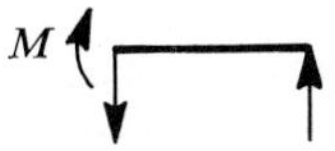 M	Simple Fixed	1 0.5	1.77–1.86 1.78–1.85		6.5
3	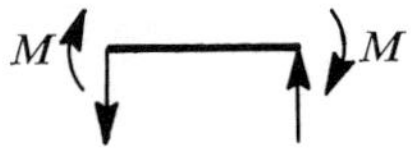 M M	Simple Fixed	1 0.5	2.56–2.74 2.23–2.58		
4	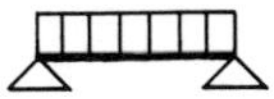	Simple Fixed	1 0.5	1.13 0.97	0.45 0.29	
5	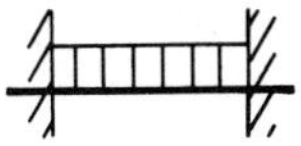	Simple Fixed	1 0.5	1.30 0.86	1.55 0.82	
6	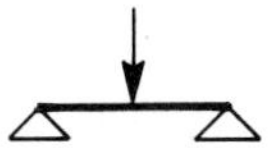	Simple Fixed	1 0.5	1.35 1.07	0.55 0.42	2.5
7	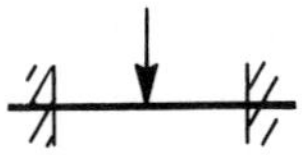	Simple Fixed	1 0.5	1.70 1.04	1.42 0.84	
8	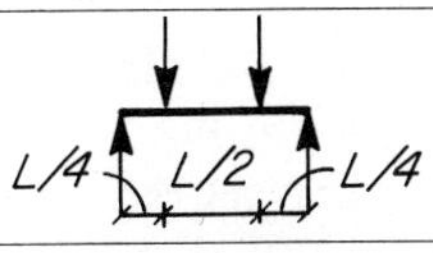 $L/4$ $L/2$ $L/4$	Simple	1	1.04	0.42	
9	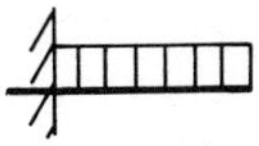	Fixed	1	2.05–3.42		
10	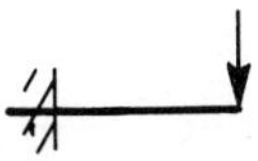	Fixed	1	1.28–1.71	0.64	

* From Ref. 7.

† Where a range of values is given, the smaller value corresponds to beams with negligible warping resistance.

K is an effective-length coefficient which depends on the conditions of restraint at the supports. Values of C_b and K for a number of cases of practical interest are given in Table 5-2. Values of K in this table are based on identical boundary conditions for both warping and y-axis rotation. In other words, the simple support means one for which both warping and y-axis rotation are permitted at each end, while the fixed support means one for which both are prevented. If boundary conditions are mixed, values of K for y-axis rotation and for warping restraint are unequal. Approximate values for such cases are tabulated in Ref. 27. However, mixed boundary conditions are not likely to occur in practice.

Values of C_b for the first three cases in Table 5-2 are approximated closely by the reciprocal of C_m given by Eqs. (4-34). Therefore, C_b for the beam acted upon only by end moments can be evaluated from any of the following equations:

$$C_b = \frac{1}{C_m} = 1.75 - 1.05\frac{M_1}{M_2} + 0.3\left(\frac{M_1}{M_2}\right)^2 \quad \not> 2.5^* \tag{5-8a}$$

$$C_b = \frac{1}{C_m} = \frac{1}{\sqrt{0.3 + 0.4M_1/M_2 + 0.3(M_1/M_2)^2}} \not> 2.5^* \tag{5-8b}$$

$$C_b = \frac{1}{C_m} = \frac{1}{0.6 + 0.4M_1/M_2} \not> 2.5^* \tag{5-8c}$$

where M_1 is the smaller of the two end moments M_1 and M_2, and where M_1/M_2 is positive when the member bends in single curvature. The opposite sign convention is also used, as in the AISC Specification, where M_1/M_2 is negative for single-curvature bending, in which case the sign of the M_1/M_2 term is changed. Since C_b for the beam acted upon only by end moments is very nearly independent of the conditions at the support, as is shown by the values in Table 5-2, Eqs. (5-8) can be used for simple, fixed, or mixed conditions.

Equation (5-6) applies to the symmetrical channel and the point-symmetrical zee as well as to the symmetrical I for which it was derived. Equation (5-7) can be used for an approximate solution for channels and zees supporting transverse load acting through the shear center.[7] Beams whose loads do not act through the shear center are discussed in the next article.

5-8 INELASTIC LATERAL-TORSIONAL BUCKLING

Equations (5-4) are based on proportionality of stress and strain. Therefore, the value of M_x from Eq. (5-7) must not exceed the moment which gives proportional-limit stress on the extreme fiber of the beam. Thus, in the case of steel with a flat-top yield,

* This limiting value is 2.3 in the AISC specification.

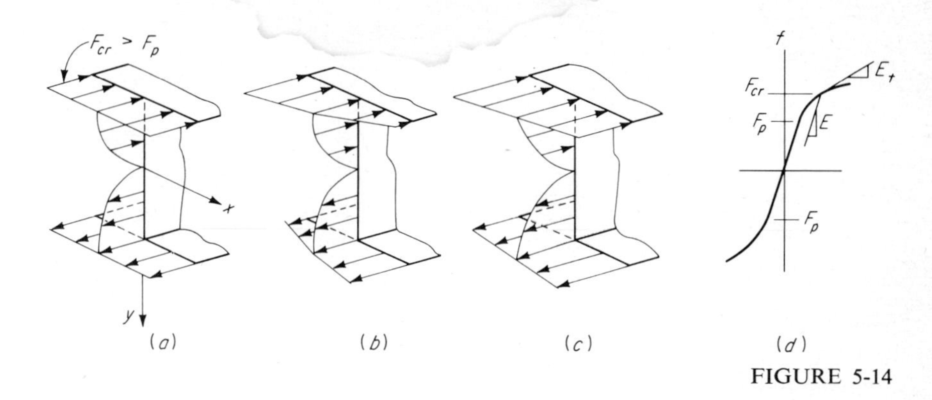

FIGURE 5-14

a solution $M_x > M_y$ is invalid. In this case, behavior of the beam is given by Eq. (5-4*a*) if there are no residual stresses, which means that the beam develops the yield moment with no lateral-torsional buckling.

Equation (5-7) can be extended to the inelastic range of buckling of beams made of gradually yielding metals. Figure 5-14*a* shows the bending stresses at a cross section of a beam bent in the yz plane, with an extreme-fiber stress $f = F_{cr}$ greater than the proportional-limit stress F_p. If the displacement u at the onset of buckling is in the positive direction of x, the resulting bending about the y axis puts tension on the $+x$ side of the flanges and compression on the $-x$ side. If the moment increment is of higher order than the corresponding displacement, the stress distribution changes to that shown in Fig. 5-14*b*, where increase in stress is governed by the tangent modulus E_t and decrease by Young's modulus E (Fig. 5-14*d*). On the other hand, there need be no stress reversal if the moment increment is a first-order effect (Fig. 5-14*c*). This is analogous to the inelastic-buckling behavior of columns, which was discussed in Art. 4-4. The stress distribution in Fig. 5-14*b* changes the bending stiffness EI_y in Eq. (5-7) to $E_r I_y$, where E_r is a double modulus of elasticity, while the distribution in *c* changes it to $E_t I_y$, where E_t is the tangent modulus.

The torsional displacement β produces shear stresses (St. Venant torsion) and, because of the nonuniform warping, cross bending of the flanges. The stresses due to cross bending are shown in Fig. 4-62*d*. Since these are also superimposed on the bending stresses of Fig. 5-14*a*, EC_w in Eq. (5-7) should be replaced by $E_r C_w$ or $E_t C_w$, in accordance with the assumption made for the bending stiffness EI_y. The St. Venant shear stresses are superimposed on any shear stresses which may exist at the onset of buckling. However, shear stresses due to bending are usually relatively small over most of the cross section, and are not likely to exceed the proportional limit. Furthermore, the shearing modulus governing twist at the beginning of buckling is practically

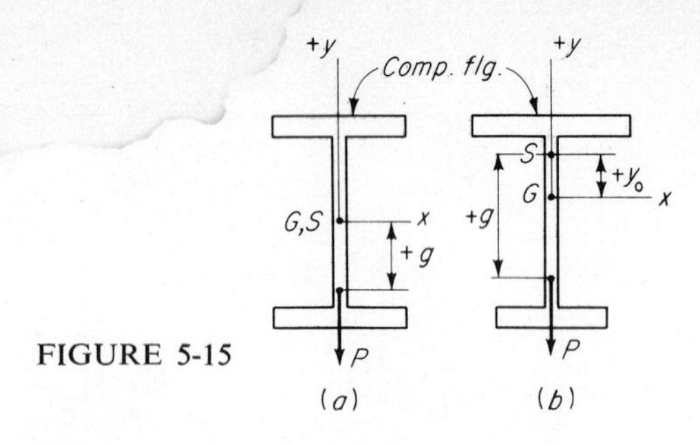

FIGURE 5-15

equal to the elastic value even in the presence of axial stresses well beyond the proportional limit (Art. 5-3). Thus, it would appear that the elastic value G should be retained in extending Eq. (5-7) to the inelastic range. However, it is on the safe side, and results in a simpler equation, if G is replaced by G_t. Thus, Eq. (5-7) becomes

$$M_{x,\,cr}^2 = C_b{}^2\tau^2\left[\frac{\pi^2}{(KL)^2}EI_yGJ + \frac{\pi^4}{(KL)^4}EI_yEC_w\right] \tag{5-9}$$

where $\tau = E_t/E = G_t/G$. Since τ and M_x are interrelated, this equation must be solved by trial.

The solution of Eq. (5-9) can be simplified by using an equivalent radius of gyration which is obtained by equating the critical bending stress to the tangent-modulus critical stress for the column. Thus,

$$F_{cr}{}^2 = \frac{M_{x,\,cr}^2}{S_x{}^2} = \left[\frac{\pi^2\tau E}{(L/r_{eq})^2}\right]^2 \tag{a}$$

Substituting $M_{x\,,\,cr}^2$ from Eq. (5-9) into Eq. (a) gives

$$r_{eq}{}^2 = C_b{}^2\frac{\sqrt{I_y}}{S_x}\sqrt{C_w + 0.04J(KL)^2} \tag{5-10a}$$

Since Eq. (5-7) is based on the assumption that transverse loads act through the shear center of the beam cross section, Eqs. (5-9) and (5-10a), which are derived from it, are equally limited in application. Figure 5-13c shows that a load P applied below the shear center S (at the bottom flange, say) exerts a restoring moment as the beam begins to buckle. Of course, this increases resistance to buckling. On the other hand, a load P applied above S tends to increase the twist once the beam begins to buckle. Procedures for evaluating these effects are discussed in various books.[11] The effect on the equivalent radius of gyration is accounted for in the following equation:[7]

$$r_{eq}{}^2 = C_b{}^2\frac{\sqrt{I_y}}{S_x}\left[C_1g\sqrt{I_y} + \sqrt{(C_1g\sqrt{I_y})^2 + C_w + 0.04J(KL)^2}\right] \tag{5-10b}$$

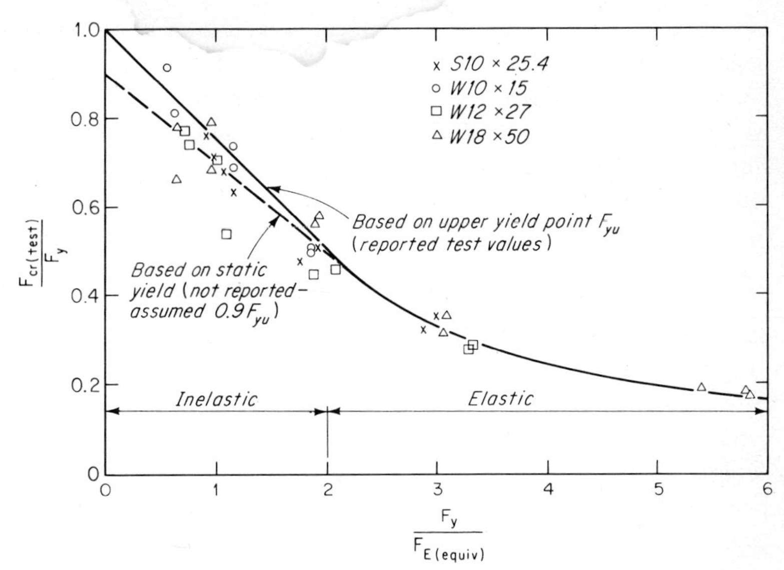

FIGURE 5-16
Comparison of predicted lateral-buckling strengths with test results from Ref. 8.

where g is the distance from the shear center to the point of application of the load, to be taken positive if the load is below the shear center and negative if it is above (Fig. 5-15*a*). Values of C_1 are given in Table 5-2. This equation reduces to Eq. (5-10*a*) if $g = 0$.

In I-shaped beams made of steel with a flat-top yield, residual stresses cause premature yielding of portions of the cross section, just as in the case of columns (Art. 4-6). This causes the beam to behave as though it were made of a gradually yielding steel. The corresponding lateral-buckling resistance can be evaluated by using an effective cross section (the unyielded portion) as is done for columns. The effective cross section is not symmetrical, however, because premature yielding begins in the tips of the compression flange, just as it does in the column, while the tension flange yields first at its juncture with the web, where the residual stress is tensile. Thus, the beam is reduced to the equivalent of one with unequal flanges (Art. 5-9). For practical applications, however, a solution based on the properties of the unreduced cross section is to be preferred. Since the proportional limit in compression is the same for lateral buckling of beams as it is for columns, namely, $F_p = F_y - F_{rc}$, the CRC column formula, with the equivalent radius of gyration from Eqs. (5-10), suggests itself. A comparison of the results of tests on rolled A7-steel beams, reported in Ref. 8, with values computed in this way is shown in Fig. 5-16. The beams were simply supported and ranged in span from 5 to 10 ft for the W10 × 15 and 10 to 50 ft for the W18 × 50. They were tested with concentrated loads acting on the top flange at the quarter points. A lateral-buckling formula for this loading is given in Ref. 8. Since the yield points of the beams ranged from 33.5 to 42.5 ksi, the results are nondimensionalized in Fig. 5-16 by plotting F_{test}/F_y against $F_y/F_{E(\text{equiv})}$, where $F_{E(\text{equiv})}$ denotes the Euler stress for the beam slenderness ratio based on r_{eq} from Eqs. (5-10). The CRC parabola plots as a straight line in these coordinates, since it reduces to

$$F_{cr} = F_y\left(1 - \frac{1}{4}\frac{F_y}{F_{E(\text{equiv})}}\right) \tag{b}$$

as is seen from Eq. (4-15). The upper yield point for each beam coupon was reported in Ref. 8. The static yield level, which was not reported, may be as much as 10 to 15 percent smaller (Art. 2-4). Two plots of Eq. (*b*), one based on the reported upper yield point F_{yu} and one on $0.9F_{yu}$, are shown in Fig. 5-16. It will be noted that there is more scatter of test results in the inelastic range, as is to be expected, and that the procedure described above gives a good average prediction of the test results if it is based on the static yield.

EXAMPLE 5-1 Using a factor of safety of 1.75, determine the allowable load for an A36-steel W10 × 21 spanning 10 ft and supporting a concentrated load at midspan. The load is applied at the shear center. The ends are simply supported in the x and y directions, and there is no lateral support at midspan.

SOLUTION From the AISC Manual, $I_y = 10.8$ in.^{4}, $S_x = 21.5$ in.^{3}, $J = 0.210$ in.^{4}, $C_w = 246$ in.^{6}

From Table 5-2, $C_b = 1.35$, $K = 1$. Then, from Eq. (5-10*a*)

$$r_{eq}^2 = 1.35^2 \frac{\sqrt{10.8}}{21.5}\sqrt{246 + 0.04 \times 0.210 \times 120^2} = 5.32$$

$$r_{eq} = 2.32 \qquad \frac{KL}{r_{eq}} = \frac{120}{2.32} = 52$$

Using the CRC formula [Eq. (4-16)] for the inelastic range,

$$C_c = \pi\sqrt{\frac{2E}{F_y}} = \pi\sqrt{\frac{60{,}000}{36}} = 128$$

$$F_{cr} = 36\left[1 - \frac{1}{2}\left(\frac{52}{128}\right)^2\right] = 33.1 \text{ ksi}$$

$$M_x = F_{cr}S_x = 33.1 \times \frac{21.5}{12} = 59.4 \text{ ft-kips}$$

$$M_x = \frac{PL}{4} = 2.5P \qquad P = \frac{59.4}{2.5} = 23.8 \text{ kips}$$

$$P_{\text{all}} = \frac{23.8}{1.75} = 13.6 \text{ kips}$$

PROBLEMS

5-12 Determine the allowable load for the beam of the example in Art. 5-8 if the concentrated load acts (*a*) on the top flange, (*b*) on the bottom flange.

5-13 Determine the allowable load for the beam of the example in Art. 5-8 if there is lateral support at midspan.

5-14 A cantilever beam 10 ft long with no lateral support carries a concentrated load of 10 kips acting on the top flange at the free end. Determine the A36-steel beam required for a factor of safety of 2.

5-15 A simply supported beam spanning 24 ft supports a concentrated load of 20 kips acting on the top flange at midspan. Determine the A441-steel beam required for a factor of safety of 1.8. The beam has no lateral support except at the ends.

5-16 Same as Prob. 5-15 except that the beam is also supported laterally at midspan by a beam framing at right angles.

5-17 A simply supported W24 $\times$ 68 beam of A572 Grade 45 steel spans 36 ft. It supports a concentrated load P 10 ft from the left end and another concentrated load of $1.5P$ 20 ft from the left end. The beam is supported laterally at each end as well as at the load points, where beams frame in at right angles. Determine the value of P at which lateral-torsional buckling can be predicted.

5-18 Same as Prob. 5-17 except that there is no lateral support at the points of load. The loads act on the top flange.

5-9 LATERAL BUCKLING OF UNEQUAL-FLANGED BEAMS

Built-up beams are sometimes made with unequal flanges. The compression flange is usually larger. The critical bending stress for lateral-torsional buckling of unequal-flanged beams can be determined by analysis similar to that of Art. 5-7. The results can be put in terms of the equivalent radius of gyration given by Eq. (5-11), provided the cross section is symmetrical about the axis of the web (y axis)[7]

$$r_{eq}^2 = C_b^2 \frac{\sqrt{I_y}}{S_{xc}} \left[C_2 k \sqrt{I_y} + \sqrt{(C_2 k \sqrt{I_y})^2 + C_w + 0.04J(KL)^2} \right] \tag{5-11}$$

in which

$$k = y_0 - \frac{1}{2I_x}\int_A (x^2 + y^2)y\,dA \tag{5-12}$$

where y_0 = distance from centroid to shear center, positive when shear center lies between centroid and compression flange (Fig. 5-15*b*)
y = ordinate with origin at center of gravity, positive when directed toward compression flange
S_{xc} = section modulus for compression flange

For pure bending, $C_2 = 1$. Values for two other cases are given in Table 5-2. The critical stress is determined by using KL/r_{eq} in the column formula, as with Eq. (5-10*a*).

Except for pure bending, Eq. (5-11) applies only if the transverse loads act through the shear center. When this is not the case, the equivalent radius of gyration can be determined by replacing $C_2 k\sqrt{I_y}$ where it occurs by $C_1 g + C_2 k\sqrt{I_y}$, where $C_1 g$ is defined in Eq. (5-10*b*).

An approximate solution for the case of pure bending ($C_b = 1$) with $K = 1$ was developed in Ref. 9. It is equivalent to using for k in Eq. (5-11) the distance from the shear center to the mid-depth as an approximation to the value from Eq. (5-12). Expressed as an equivalent radius of gyration, the result is

$$r_{eq}^{\ 2} = \frac{\sqrt{I_y}\sqrt{C'_w + 0.04JL^2} + (I_c - I_t)\,d/2}{S_{xc}} \tag{5-13}$$

where I_c = y-axis moment of inertia of compression flange
I_t = y-axis moment of inertia of tension flange
$C'_w = (d^2/4)(I_c + I_t)$
d = center-to-center distance of flanges

The warping constant is denoted by C'_w, rather than C_w, because it differs from the theoretically correct value for a beam with unequal flanges (Fig. 4-63). This formula tends to overestimate the buckling stress if $I_c > I_t$. Of course, it reduces to Eq. (5-10*a*) if $I_c = I_t$.

Another approximate solution[10] consists in equating the integral in Eq. (5-12) to zero, so that $k = y_0$. This solution tends to underestimate the buckling stress if $I_c > I_t$ and to overestimate it if $I_c < I_t$. Still another approximate solution, which gives good results, is obtained by assuming the tension flange to be identical with the compression flange, so that the critical stress is based on r_{eq} from Eq. (5-10*a*).

EXAMPLE 5-2 Compute the critical pure-bending moment for a simply supported A36 WT9 × 35 spanning 15 ft. The moment is in the plane of the web, with the flange in compression. Compare the results according to Eq. (5-11) and the approximate

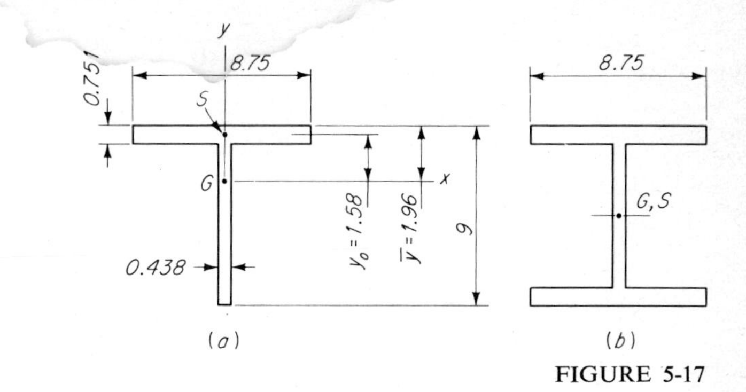

FIGURE 5-17

solutions by Eqs. (5-13) and (5-10*a*). Assume the proportional limit to be $F_y/2$ and use the CRC column formula [Eq. (4-16)] for inelastic buckling.

The cross section is shown in Fig. 5-17*a*. The following properties are from the AISC Manual:

$$I_x = 68.2 \text{ in.}^4 \qquad S_{xt} = 9.68 \text{ in.}^3$$

$$I_y = 42.0 \text{ in.}^4 \qquad S_{xc} = \frac{68.2}{1.96} = 34.7 \text{ in.}^3$$

$$\bar{y} = 1.96 \text{ in.}$$

The shear center S is at the intersection of the flange and web centerlines.

SOLUTION ACCORDING TO EQ. (5-11) $K = 1$ for simple y-axis support and $C_b = C_2 = 1$ for pure bending (Table 5-2). The warping stiffness C_w for the tee is small enough to be neglected (Fig. 4-63). The value of J is one-half the value for the W18 × 70 from which the tee is cut. Therefore, $J = 3.13/2 = 1.57$ in.^{4}. To evaluate k in Eq. (5-12), we need $\int x^2 y\, dA + \int y^3\, dA$:

Flange: $\int x^2 y\, dA = y_0 \int x^2\, dA = y_0 I_y = 1.58 \times 42.0 \quad = \quad 66$

Flange: $\int y^3\, dA = y_0{}^3 \int dA = 1.58^3 \times 8.75 \times 0.751 \quad = \quad 26$

Web: $\int x^2 y\, dA = \text{neglible}$

Web: $\int y^3\, dA = t \int y^3\, dy = t \dfrac{y^4}{4}\Big]_{-7.04}^{1.20} = \dfrac{0.438}{4}(2 - 2{,}456) = -269$

Total $= -177$

$$k = 1.58 - \frac{-177}{2 \times 68.2} = 1.58 + 1.30 = 2.88$$

$$k\sqrt{I_y} = 2.88\sqrt{42.0} = 18.7$$

$$r_{eq}{}^2 = \frac{\sqrt{42.0}}{34.7}\left[18.7 + \sqrt{18.7^2 + 0 + 0.04 \times 1.57 \times 180^2}\right] = 12.67$$

$$r_{eq} = 3.57$$

$$\frac{KL}{r_{eq}} = \frac{180}{3.57} = 50$$

$$C_c = \pi\sqrt{\frac{2E}{F_y}} = \pi\sqrt{\frac{60{,}000}{36}} = 128$$

$$F_{cr} = 36\left[1 - \frac{1}{2}\left(\frac{50}{128}\right)^2\right] = 33.2 \text{ ksi}$$

$$M_{cr} = 33.2S_{xc} = 33.2 \times 34.7 = 1{,}150 \text{ in.-kips}$$

$$M = 36S_{xt} = 36 \times 9.67 = 348 \text{ in.-kips}$$

The stress on the tension flange governs.

APPROXIMATE SOLUTION BY EQ. (5-13)

$$C'_w = \frac{8.62^2}{4} \times 42.0 = 780 \text{ in.}^6$$

$$r_{eq}{}^2 = \frac{\sqrt{42.0}\sqrt{780 + 0.04 \times 1.57 \times 180^2} + 42.0 \times 8.62/2}{34.7} = 15.2$$

$$r_{eq} = 3.91$$

$$\frac{KL}{r_{eq}} = \frac{180}{3.91} = 46$$

$$F_{cr} = 36\left[1 - \frac{1}{2}\left(\frac{46}{128}\right)^2\right] = 33.7 \text{ ksi}$$

$$M_{cr} = 33.7S_{xc} = 33.7 \times 34.7 = 1{,}168 \text{ in.-kips}$$

APPROXIMATE SOLUTION BY EQ. (5-10*a*) Here we compute r_{eq} for the symmetrical shape of Fig. 5-17*b*. The following properties of the cross section are needed:

$$I_x = 2 \times 8.75 \times 0.751 \times 4.12^2 + 0.438 \times \frac{7.5^3}{12} = 223 + 15.4 = 238 \text{ in.}^4$$

$$I_y = 2 \times 42.0 = 84.0 \text{ in.}^4$$

$$J = 2 \times \tfrac{1}{3} \times 8.75 \times 0.751^3 + \tfrac{1}{3} \times 7.5 \times 0.438^3 = 2.470 + 0.210 = 2.68 \text{ in.}^4$$

$$C_w = \frac{8.25^2}{4} \times 84.0 = 1{,}430 \text{ in.}^6$$

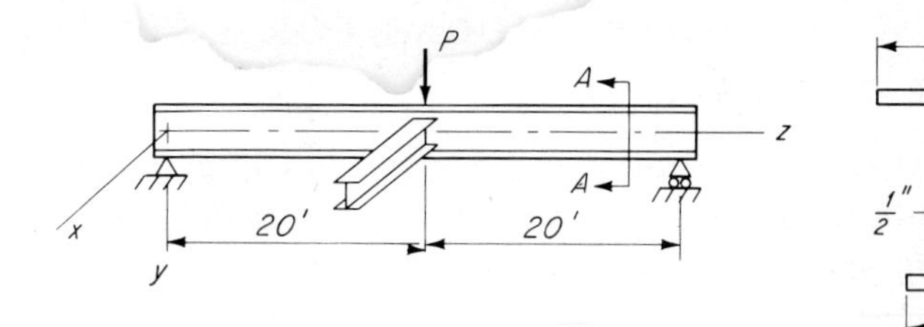

FIGURE 5-18
Problem 5-20.

$$S_x = \frac{238}{4.5} = 52.8 \text{ in.}^3$$

$$r_{eq}^{\ 2} = \frac{\sqrt{84.0}}{52.8}\sqrt{1{,}430 + 0.04 \times 2.68 \times 180^2} = 12.2$$

$$r_{eq} = 3.50$$

$$\frac{KL}{r_{eq}} = \frac{180}{3.50} = 51$$

$$F_{cr} = 36\left[1 - \frac{1}{2}\left(\frac{51}{128}\right)^2\right] = 33.1 \text{ ksi}$$

$$M_{cr} = 33.1 S_{xc} = 33.1 \times 34.7 = 1{,}150 \text{ in.-kips}$$

It should be noted that once the critical stress is found by using the value of r_{eq} for the cross section with equal flanges, the critical moment is found by multiplying this critical stress by the section modulus for the given shape. A disadvantage of this approximate procedure is that the compression-flange section modulus must be computed for both the given shape and its equal-flanged counterpart. However, this is simpler than calculating k.

PROBLEMS

5-19 Compute the critical pure-bending moment for the tee of the example in Art. 5-9 with the flange in tension.

5-20 The A36-steel beam shown in Fig. 5-18 is supported laterally at the ends and also at midspan where the beam which delivers the load P is attached. Determine the critical value of P.

5-10 SIMPLIFIED FORMULAS FOR LATERAL BUCKLING OF BEAMS

In addition to the concept of equivalent radius of gyration, another scheme, based on neglecting the smaller of the terms in brackets in Eq. (5-9), has been used to simplify formulas for lateral buckling. Usually, one of these terms predominates. In the worst case, when they are equal, the critical moment for elastic buckling is underestimated by $(\sqrt{2}-1)/\sqrt{2} = 29$ percent if either term is omitted. However, this error tends to be smaller in the inelastic-buckling range because of the decreasing sensitivity of critical stress to L/r as L/r decreases.

If we neglect the second term in brackets in Eq. (5-9), the critical bending stress is

$$F_{cr} = \frac{M_{x,cr}}{S_x} = \tau C_b \frac{\pi}{KL} \frac{\sqrt{EI_y GJ}}{S_x} \tag{a}$$

Since the webs of I shapes are thin compared to the thickness of their flanges, we can obtain simple and fairly accurate approximations to I_y, J, and S_x of the equal-flanged beam by considering only the flanges. Thus,

$$I_y = \frac{2tb^3}{12} = \frac{tb^3}{6}$$

$$J = \frac{2bt^3}{3}$$

$$S_x = \frac{I_x}{c} = \frac{2bt(d/2)^2}{d/2} = btd$$

where b = flange width
t = flange thickness
d = beam depth

Substituting these values into Eq. (a) and using $E = 2(1 + \mu)G$ with $\mu = 0.25$ gives

$$F_{cr} = \frac{0.21\pi\tau E C_b}{KLd/A_f} \tag{5-14a}$$

where $A_f = bt$ = area of one flange. For steel, using $E = 30{,}000$ ksi, Eq. (5-14a) gives

$$F_{cr} = \frac{20{,}000\tau C_b}{KLd/A_f} \tag{5-14b}$$

This formula is plotted in Fig. 5-19a. For buckling at stresses below the proportional limit, $\tau = 1$. The behavior of a perfect beam of steel with a flat-top yield is given by ACE if there are no residual stresses and by $ABDE$ if there are.*

* If plastic-bending strength is considered, DB continues upward to the left of B, instead of horizontally as shown by BA.

If we neglect the first term in brackets in Eq. (5-9), the critical bending stress is

$$F_{cr} = \frac{M_{x,cr}}{S_x} = \tau C_b \frac{\pi^2}{(KL)^2} \frac{\sqrt{EI_y EC_w}}{S_x} \tag{b}$$

This equation can be simplified for the equal-flanged I by substituting for C_w its value $d^2 I_y/4$:

$$F_{cr} = \tau C_b \frac{\pi^2 E}{(KL)^2} \frac{I_y d}{2S_x} \tag{5-15a}$$

Substituting for S_x its value

$$S_x = \frac{I_x}{d/2} = \frac{2A_f(d/2)^2 + t_w d^3/12}{d/2} = d\left(A_f + \frac{A_w}{6}\right) \tag{c}$$

we get

$$F_{cr} = \tau C_b \frac{\pi^2 E}{(KL)^2} \frac{I_y/2}{A_f + A_w/6} \tag{d}$$

The last fraction in this equation can be written simply as r_y^2, where r_y is the y-axis radius of gyration of the T consisting of one flange and one-sixth the web. Therefore,

$$F_{cr} = \frac{\pi^2 \tau E C_b}{(KL/r_y)^2} \tag{5-15b}$$

For steel, using $E = 30{,}000$ ksi, Eq. (5-15*b*) gives

$$F_{cr} = \frac{296{,}000 \tau C_b}{(KL/r_y)^2} \tag{5-15c}$$

This formula is plotted in Fig. 5-19*b*. As in Fig. 5-19*a*, $\tau = 1$ for buckling at stresses below the proportional limit, and the behavior of a perfect beam of steel with a flat-top yield is given by *ACE* or *ABDE*, depending on the situation in respect to residual stresses.

It should be noted that the larger critical stress given by Eqs. (5-14) and (5-15) should be used, since each underestimates the critical stress because it is obtained by omitting one of the terms in the radical of Eq. (5-9). Although the equations are derived for equal-flanged beams, they can be used for unequal-flanged beams provided A_f in Eq. (5-14) and I_y and r_y in Eqs. (5-15) are based on the compression flange. This is equivalent to substituting for the unequal-flanged beam one with equal flanges, each of which is identical to the compression flange of the unequal-flanged beam. It was shown in Art. 5-9 that this procedure gives a good approximation to the critical compressive bending stress.

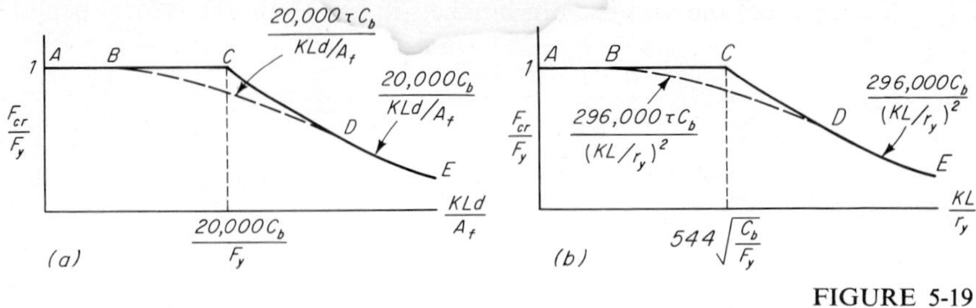

FIGURE 5-19

It was pointed out in Art. 5-7 that Eq. (5-9) applies to any cross section symmetrical about the axis perpendicular to the plane of the loading. Thus, it applies to the channel. Equations (5-14) and (5-15) can also be used for the channel, even though they were derived from Eq. (5-9) for the equal-flanged I. The results are conservative.

5-11 ALLOWABLE BENDING STRESSES

Specification allowable bending stresses are based on Eqs. (5-14) and (5-15), together with various empirical formulas for the inelastic-buckling curves *BD* in Fig. 5-19. To facilitate comparison in the discussion to follow, formulas are written in forms which are not necessarily the same as in the specification.

AISC Allowable bending stresses for noncompact shapes not continuously supported against lateral-torsional buckling are based on the bending strengths given by *ACE* of Fig. 5-19*a* and *ABDE* of Fig. 5-20, assuming the effective-length coefficient $K = 1$. *ABDE* of Fig. 5-20 corresponds to *ABDE* of Fig. 5-19*b* except that r_T replaces r_y, where r_T is the radius of gyration of the section consisting of the compression flange and one-third the compression web area. This definition of radius of gyration permits the formula to be used for beams of unequal flanges, as explained in Art. 5-9. *ABDE* of Fig. 5-20 consists of the portion *DE* of the elastic-buckling hyperbola, which is given by Eq. (5-15*b*) with $\tau = K = 1$, the portion *BD* of an inelastic-buckling parabola *GBD* with vertex at *G* and tangent to *DE* at *D*, and the horizontal line *AB* at ordinate F_y. The abscissa of point *D* is found by equating the Euler stress to $\frac{5}{9}F_y$, which gives $\sqrt{0.9}C_c\sqrt{C_b} = 0.949C_c\sqrt{C_b}$, where $C_c = \pi\sqrt{2E/F_y}$, the abscissa of the corresponding point of the curve for column critical stress vs. slenderness ratio. Similarly, the abscissa of point *B* is found to be $\sqrt{0.18}C_c\sqrt{C_b} = 0.424C_c\sqrt{C_b}$. The allowable

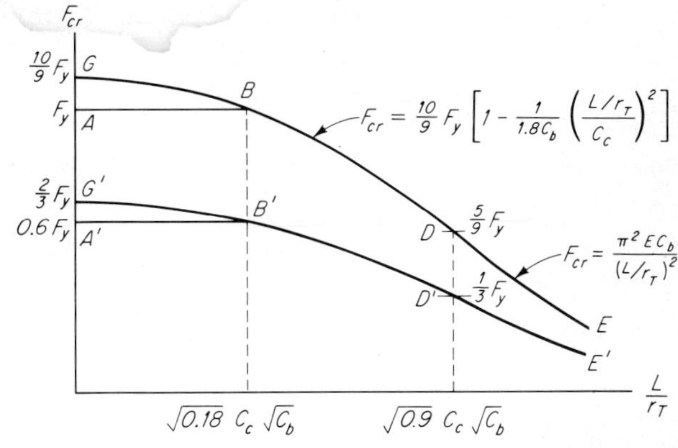

FIGURE 5-20

stress $A'B'D'E'$ is obtained by multiplying the ordinates to $ABDE$ by 0.6 to give a factor of safety of about 1.65. The result is*

$$F_b = \begin{cases} 0.6F_y & 0 \lessdot \dfrac{L}{r_T} \lessdot 0.424C_c\sqrt{C_b} \quad \text{(5-16}a\text{)} \\[2ex] \dfrac{2}{3}F_y\left[1 - \dfrac{1}{1.8C_b}\left(\dfrac{L/r_T}{C_c}\right)^2\right] & 0.424C_c\sqrt{C_b} \lessdot \dfrac{L}{r_T} \lessdot 0.949C_c\sqrt{C_b} \quad \text{(5-16}b\text{)} \\[2ex] \dfrac{170{,}000C_b}{(L/r_T)^2} & 0.949C_c\sqrt{C_b} \lessdot \dfrac{L}{r_T} \quad \text{(5-16}c\text{)} \end{cases}$$

Using the same factor of safety with Eq. (5-14b) and taking $K = 1$ and $\tau = 1$, we get the allowable-stress formula for elastic buckling corresponding to CDE of Fig. 5-19a:

$$F_b = \frac{12{,}000C_b}{Ld/A_f} \lessdot 0.6F_y \tag{5-17}$$

Since the only restriction in Eq. (5-17) is $F_b \lessdot 0.6F_y$, we see that the inelastic-buckling curve BD of Fig. 5-19a is not taken into account.

The larger of the values given by Eq. (5-17) and the applicable Eq. (5-16) should be used. This was explained in Art. 5-10, where it was also mentioned that the formulas can also be used for unequal-flanged I's and for channels. However, the AISC specification restricts Eq. (5-17) to cross sections for which the compression flange is larger, and does not allow Eq. (5-16) to be used for channels.

* Numerical values in the AISC specification are based on $E = 29{,}000$ ksi.

According to the specification, C_b in Eqs. (5-16) and (5-17) is to be determined by

$$C_b = 1.75 + 1.05 \frac{M_1}{M_2} + 0.3 \left(\frac{M_1}{M_2}\right)^2 \not> 2.3 \tag{5-18}$$

where M_1 is the smaller and M_2 the larger bending moment at the ends of the unbraced length and M_1/M_2 is positive for reversed-curvature bending and negative for single-curvature bending. This is Eq. (5-8*a*) except that the sign convention for M_1/M_2 is opposite. If the bending moment at any point within the length L is larger than that at both ends of L, C_b is to be taken as unity. It will be noted that, because of Eq. (5-18), the specification allows the designer the benefit of C_b for beams subjected only to end moments, as in the first three cases of Table 5-2, but prescribes $C_b = 1$ for the remainder except for cases 9 and 10.

Compact sections With $C_b = 1$, the upper limit of L for Eq. (5-16*a*) is $L = 0.424 C_c r_y$. With $E = 29{,}000$ ksi, this gives

$$L = \frac{320 r_y}{\sqrt{F_y}} \tag{a}$$

which gives the abscissa of points B' and B of Fig. 5-20. This limiting length allows the beam to just develop the moment $M_y = F_y S$ without buckling laterally. Thus, it is clear that compact sections not continuously braced must be braced at closer intervals, to allow the plastic moment $M_p = F_y Z$ to develop, if the increased allowable stress $F_b = 0.66 F_y$ of the AISC specification is used. Methods for determining this length are discussed in Chap. 8. According to the specification, the length L of such intervals must satisfy the following formulas:

$$L \not> \frac{20{,}000 A_f}{F_y d} \tag{5-19a}$$

$$L \not> \frac{76b}{\sqrt{F_y}} \tag{5-19b}$$

where b is the width of the flange. Equation (5-19*a*) is Eq. (5-14*b*) with $\tau = C_b = K = 1$ and $F_{cr} = F_y$. Thus, it corresponds to the abscissa of point C in Fig. 5-19*a*, so that the inelastic-buckling curve BD is not taken into account in this formula. Equation (5-19*b*) is equivalent to a limiting value of L/r_y, since, for the I, $r_y = b\sqrt{12}$. Thus,

$$L \not> \frac{76b}{\sqrt{F_y}} \not> \frac{263 r_y}{\sqrt{F_y}} \tag{b}$$

Comparing this limiting value with that of Eq. (*a*) above, we see that it gives us a point on $G'B'$ in Fig. 5-20, as it should.

AASHO The AASHO allowable bending stress for beams not continuously supported laterally is

$$F_b = 0.55F_y\left[1 - \frac{1}{2}\left(\frac{L/r_y}{C_c}\right)^2\right] \tag{5-20}$$

where r_y is the radius of gyration of the compression flange. This is the CRC formula for columns [Eq. (4-16)] with a factor of safety $1/0.55 = 1.82$. Thus, it corresponds to ABD of Fig. 5-19*b*, with a proportional limit of $F_y/2$. These specifications set a limiting length for beams, shown below, such that the elastic-buckling curve DE of Fig. 5-19*b* is not needed. Furthermore, there is no formula corresponding to Fig. 5-19*a*, so that only Eq. (5-20) is used. Using $r_y{}^2 = b^2/12$, where b is the width of the flange of an I, Eq. (5-20) is put in terms of L/b. For example, for A36 steel

$$F_b = 20{,}000 - 7.5\left(\frac{L}{b}\right)^2 \text{ psi} \qquad 0 \lessapprox \left(\frac{L}{b}\right) \lessapprox 36 \tag{5-21}$$

The upper limit $L/b = 36$ is found by substituting $b = r\sqrt{12}$ into $L/r = C_c$. Thus, $L/r = \pi\sqrt{2E/F_y} = 126$ for $E = 29{,}000$ ksi and $F_y = 36$ ksi, and $L/b = 126/\sqrt{12} = 36$.

AREA Allowable bending stresses are based on the elastic-buckling form of Eq. (5-14*a*) and a parabola representing the inelastic buckling range of Eq. (5-15*a*). With $\tau = K = C_b = 1$, Eq. (5-15*a*) can be written

$$F_{cr} = \frac{\pi^2 E}{L^2}\frac{I_y d}{2S_x} = \frac{\pi^2 E}{L^2}\frac{d^2 I_y}{4I_x} = \frac{\pi^2 E}{4}\frac{r_y{}^2}{L^2}\frac{d^2}{r_x{}^2} \tag{c}$$

The ratio r_x/d is fairly constant for the I and can be taken at 0.4. Thus, Eq. (*c*) becomes

$$F_{cr} = \frac{\pi^2 E}{0.64(L/r_y)^2} \tag{d}$$

The inelastic-buckling parabola is taken tangent to Eq. (*d*) at $F_y/2$ and with its vertex at F_y at $L/r_y = 0$. The abscissa of the point of tangency is found by equating F_{cr} in Eq. (*d*) to $F_y/2$. This gives

$$\frac{L}{r_y} = \pi\sqrt{\frac{2E}{0.64F_y}} = 1.25C_c \tag{e}$$

where $C_c = \pi\sqrt{2E/F_y}$, the abscissa of the corresponding point of the column critical-stress curve. Using a factor of safety of 1.8, the allowable stress F_b becomes

$$F_b = 0.55F_y\left[1 - 0.32\left(\frac{L/r_y}{C_c}\right)^2\right] \tag{5-22a}$$

The specification defines r_y as the radius of gyration of a section comprising the compression flange and the part of the web in compression.

With $E = 29{,}000$ ksi and $\tau = C_b = K = 1$ in Eq. (5-14*a*), a factor of safety of 1.8 gives the allowable stress

$$F_b = \frac{10{,}500}{Ld/A_f} \not> 0.55F_y \tag{5-22b}$$

A_f in this formula is defined as the area of the smaller flange. Both Eqs. (5-22*a*) and (5-22*b*) can be used for unequal-flanged beams no matter which flange is in compression. As in the AISC specification, the larger value is used because each is based on an underestimation of the critical stress. Since AREA limits L/r_y to not more than the value by Eq. (*e*), an allowable-stress formula corresponding to Eq. (*d*) is not needed. Thus, the specification is based on *ACE* of Fig. 5-19*a* [Eq. (5-22*b*)] and *ABD* of Fig. 5-19*b* [Eq. (5-22*a*)].

AISI In this specification for design of cold-formed steel structural members, the allowable compressive bending stress for beams is given by formulas which are equivalent to the AISC formulas [Eqs. (5-16)]. A formula corresponding to Eq. (5-17) is not prescribed because the torsional stiffness J is so small for the thin-walled shapes used in cold-formed members that the $EI_y EC_w$ term in the lateral-buckling formula [Eq. (5-9)] usually exceeds the $EI_y GJ$ term. The AISI formulas differ from the AISC formulas only in that they are written in terms of the parameter $I_y d/2L^2 S_x$ of Eq. (5-15*a*), instead of the alternative form, L/r_y of Eq. (5-15*b*). However, I_y is replaced by its equivalent, $2I_{yc}$, the moment of inertia of the compressive part of the cross section. Thus, the AISC formulas [Eqs. (5-16)] and the AISI formulas become identical if L/r_y in the former is replaced by $L^2 S_{xc}/I_{yc}\, d$.

Aluminum The lateral-buckling strength of aluminum beams can be determined from column-buckling formulas by using the equivalent radius of gyration r_{eq}, Eq. (5-10). Thus, the lateral-buckling strength of 6061-T6 beams is found by substituting the equivalent slenderness ratio into Eqs. (4-72) and (4-73) of Art. 4-30.

A good approximation to r_{eq} for the I and the channel is given by[5]

$$\frac{KL}{r_{eq}} = \frac{L}{1.2r_y} \tag{5-23}$$

where r_y is radius of gyration of the cross section about the y axis. This approximation becomes quite conservative for KL/r_{eq} greater than about 50, so that Eq. (5-10) should be used for such cases.

The ASCE Task Committee on Lightweight Alloys suggests factors of safety of 2.2 for bridges and 1.95 for buildings.[6]

It is not always easy to decide whether a beam has adequate support against lateral buckling. Embedment of the top flange in a concrete slab provides support except when the beam is a cantilever so that the compression flange is at the bottom. A completely encased beam is supported no matter which flange is in compression. Wood flooring spiked to nailing strips fastened to the top flange should furnish lateral support. Corrugated sheet-metal roofs are sometimes attached to purlins by metal straps or clips. It is questionable whether such connections provide dependable lateral support.

Lateral bracing must be adequate to hold the braced beam in position. Thus, stiffness as well as strength is required. As a general rule, bracing will be adequate if each lateral brace is designed for 2 percent of the compressive force in the flanges of the beam it braces. This rule of thumb is based on observations from laboratory tests.

Box sections Because of the superior torsional stiffness of the box, lateral-torsional buckling of box-section beams is not usually a problem. Since the effect of nonuniform warping is small, the warping constant C_w may be neglected in computing the equivalent radius of gyration [Eq. (5-10)]. The value of J is given by Eq. (4-60)

$$J = \frac{4A^2}{\int ds/t}$$

The AREA specification gives for the box an equivalent slenderness ratio L/r which is based on Eq. (5-10*a*). The AISC specification gives two limiting values of unsupported length, $L = 2{,}500b/F_y$ if $F_b = 0.60F_y$ and $L = 76b/\sqrt{F_y}$ if $F_b = 0.66F_y$. The latter is probably too conservative if the former is reasonable.

EXAMPLE 5-3 Using a factor of safety of 1.6, determine the allowable moment for an A36 $12 \times 6 \times \frac{1}{4}$ structural tube spanning 24 ft and supporting a concentrated load P at midspan. The ends are simply supported in the x and y directions, and there is lateral support at midspan.

SOLUTION From the AISC Manual, $I_y = 54.2$ in.^{4} and $S_x = 26.2$ in.^{3} The torsional stiffness is

$$J = \frac{4 \times 11.75^2 \times 5.75^2}{2(11.75/0.25 + 5.75/0.25)} = 130 \text{ in.}^4$$

Since there is lateral support at midspan, each half of the beam corresponds to case 2 of Table 5-2, with $K = 1$. Thus, $1.77 \lessapprox C_b \lessapprox 1.86$. [Using $M_1/M_2 = 0$ in Eq. (5-8), we get $C_b = 1.75$.] Then, from Eq. (5-10*a*), with $C_w = 0$,

$$r_{eq}^2 = 1.75^2 \frac{\sqrt{54.2 \times 0.04 \times 130 \times 144^2}}{26.2}$$

$$r_{eq} = 16.6 \text{ in.} \qquad \frac{KL}{r_{eq}} = \frac{1 \times 144}{16.6} = 8.7$$

This is so small that no reduction in stress for lateral buckling need be made. Therefore,

$$M_{\text{all}} = \frac{F_y S_x}{1.6} = \frac{36 \times 26.2}{1.6} = 590 \text{ in.-kips}$$

Beam-columns The formulas for allowable bending stress which have been discussed in this article apply only to beams. If there is axial load (either tension or compression) in addition to bending moments, the additional stress must, of course, be considered. In the case of the beam-column, this is done by extending the interaction formulas discussed in Art. 4-14 to cover lateral-torsional buckling. Thus, in the AISC formula [Eq. (4-40*a*)] F_a is computed for the largest effective slenderness ratio of any unbraced segment of the member, taking into account the possibility of buckling about either principal axis of the cross section. Similarly, F_b is defined as the larger value from Eq. (5-16) or (5-17), rather than $0.60F_y$ or $0.66F_y$ as for the beam-column with continuous lateral support. However, either C_m in the interaction formula or C_b in the formulas for F_a must be taken equal to unity. This is because each is a factor that accounts for the distribution of load on the member, and they are in fact reciprocal, as was pointed out in Art. 5-7. The same procedure is used in both the AISI and the AREA specifications. However, both AISC and AISI require that C_b, rather than C_m, be taken as unity, while AREA uses unity for C_b and C_m for all cases.

The procedure described above is not correct theoretically because the interaction formulas in Chap. 4 are derived for the case where bending is in the plane of the moments. However, the results appear to be conservative, as shown in Art. 5-27, where lateral-torsional buckling of beam-columns is discussed further.

5-12 DP5-2: CRANE FOR SHOP BUILDING

The hoist shown in the figure is supported by a trolley riding a bridge supported on trucks that can be moved along the crane girder. The following comments clarify the correspondingly lettered computations on the design sheet.

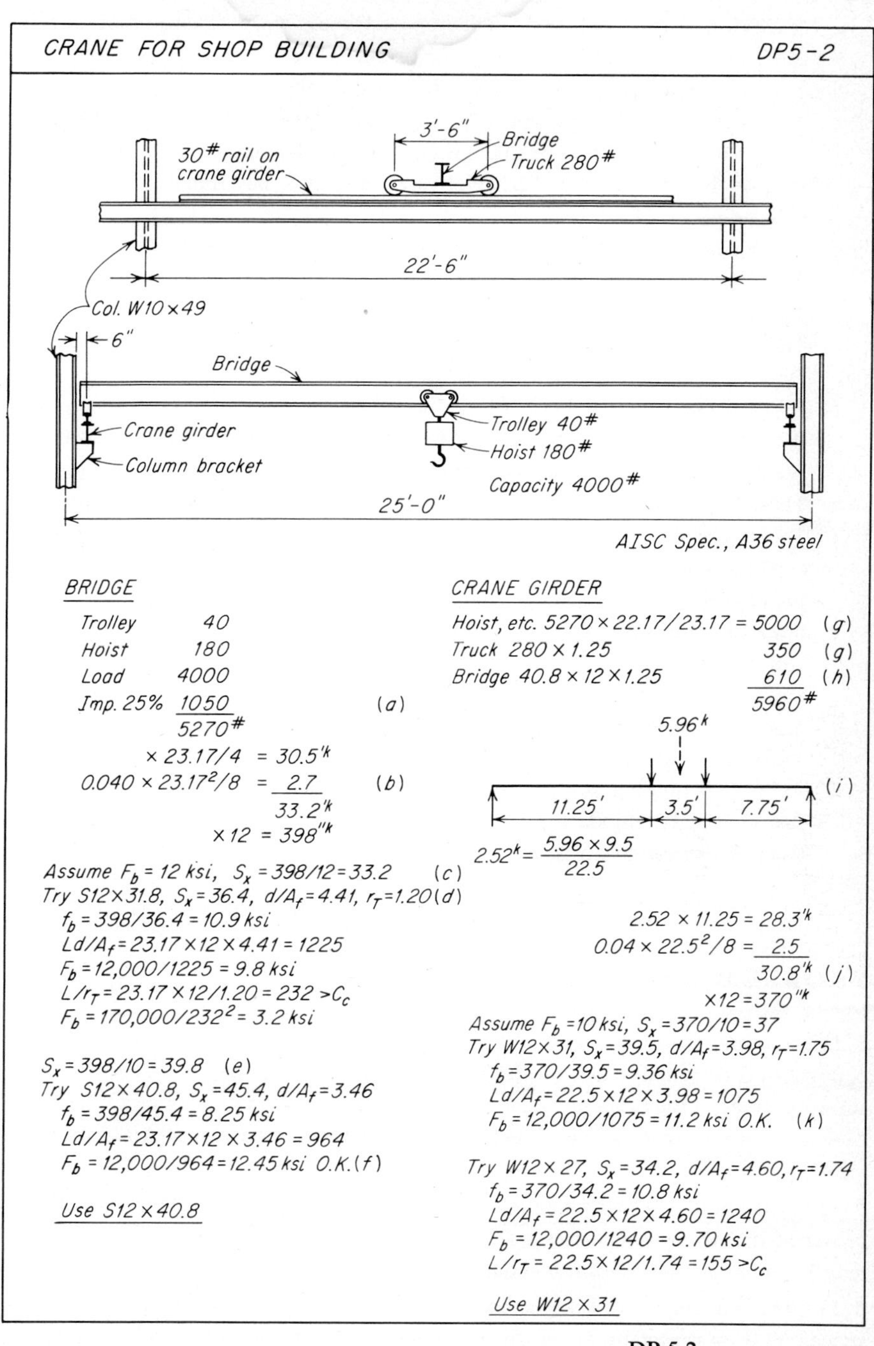

DP 5-2
Crane for shop building.

a AISC 1.3.3 prescribes 25 percent impact for the support girders of traveling cranes. The same impact is used here for the bridge.

b The estimated weight of the bridge beam is 40 plf.

c Since there is no lateral support for the bridge, the reduction in strength, if any, because of lateral buckling must be taken into account. The load is light for a span as long as this, so that the beam will be relatively small and may therefore have a considerably reduced allowable stress.

d The S shape is chosen, rather than a W, because standard trolley wheels are made to fit the 1:6 slope of the inside face of the S. W's have slopes ranging from 0 to 1:20, depending on the manufacturer.

e The S12 × 31.8 was found to be inadequate. We use its allowable stress (the larger of the two values of F_b) as a guide in choosing the next section to try.

f It is unnecessary to check AISC formula 1.5-6*b* [Eq. (5-16*c*)], since formula 1.5-7 [Eq. (5-17)] was found to give the larger value of F_b for the first trial section.

g We assume that the trolley can be no closer to the runway than 1 ft.

h These loads are increased 25 percent for the prescribed impact allowance.

i The loads are positioned for maximum moment at midspan, rather than for the absolute maximum, which occurs at a point 10.5 in. from midspan. The difference between the two maximums is negligible in this case.

j Since the trolley is not motor-driven, no lateral forces on the runway due to acceleration and deceleration of the load on the bridge are assumed. An example where such forces are considered is given in Art. 5-24.

k Since this value of F_b is larger than f_b, it is not necessary to compute the value of F_b by the other formula. However, we try the next lighter shape, the W12 × 27. It turns out to be inadequate according to the Ld/A_f formula, but we check L/r_T before adopting the W12 × 31. Since L/r_T is considerably in excess of C_c, we know that the L/r_T formula cannot give a larger value of F_b.

PROBLEMS

5-21 A monorail hoist weighing 850 lb has a lifting capacity of 5 tons. The trolley weighs 120 lb and is designed to run on the bottom flange of a standard S shape. The trolley beams run lengthwise of the building and are attached through their top flanges to the bottom chords of a series of roof trusses spaced 20 ft on centers. Design the beam. AISC specification, A36 steel.

5-22 A beam in a power plant spans 40 ft. It supports only a column, which stands at the center of the beam and carries a load of 120 kips. There is no lateral support in the 40-ft length. What size A36 beam is required? AISC specification.

5-23 The mezzanine walkway shown in Fig. 5-21 is to be installed along a row of columns spaced 22 ft on centers in an existing building. The floor is to be steel grating spanning 4 ft and weighing 10 psf. The design live loads are 80 psf for the floor and 50 plf laterally at the

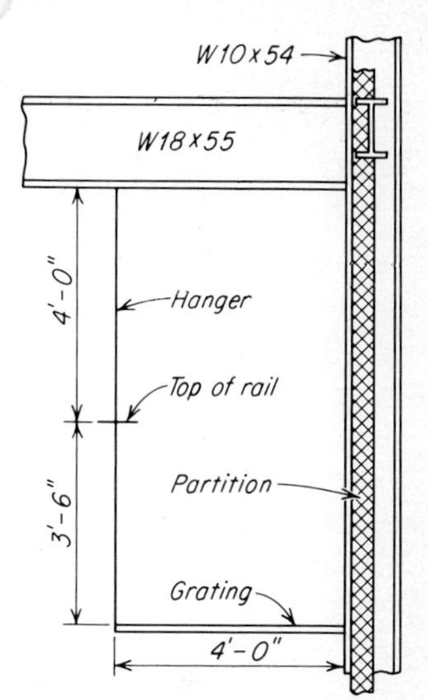

FIGURE 5-21
Problem 5-23.

top of the railing. Design the railing, the supporting structure for the floor, and the hanger to the beam overhead. Design welded connections for the hanger. Design a welded connection of the beam to the column. AISC specification, A36 steel.

5-24 Figure 5-22 shows a part transverse section of a one-story shop building which is 96 ft long inside its 8-in. end walls. The purlins are wall-bearing at the ends of the building. The local building code specifies a roof live load of 30 psf. Each monorail is equipped with a hoist of 2,000 lb capacity weighing 125 lb, suspended from a 50-lb trolley designed to run on the bottom flange of a standard S shape. Choose a spacing for the roof beams, and design the monorail, purlins, and beams. AISC specification, A36 steel.

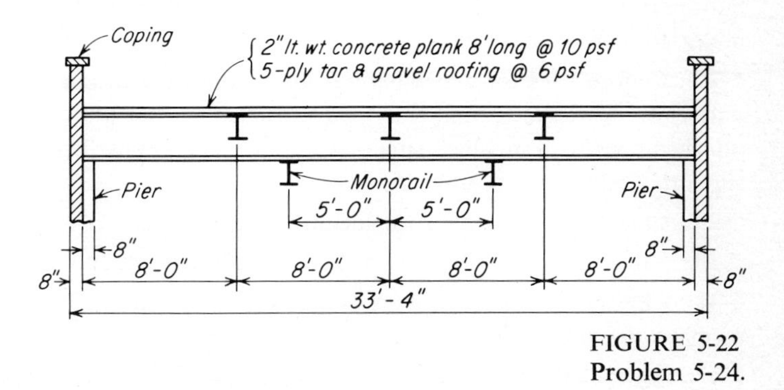

FIGURE 5-22
Problem 5-24.

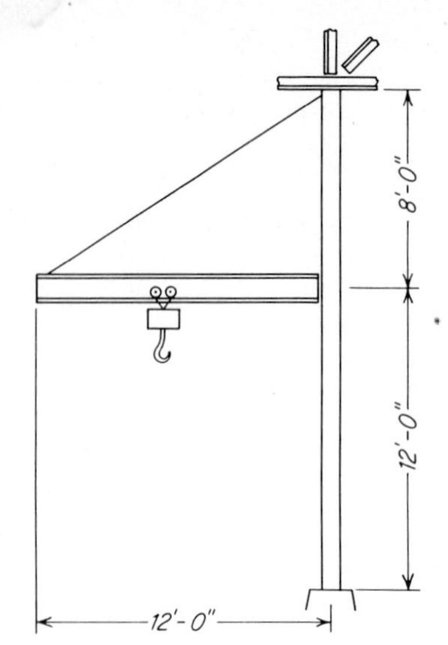

FIGURE 5-23
Problem 5-34.

5-25 Compute the factor of safety for the bridge beam and the crane girder of DP5-2 taking into account the position of the load relative to the shear center.

5-26 Choose a pair of equal-legged angles, separated $\frac{3}{8}$ in. for gusset plates, for a member 8 ft long supporting an axial tension of 40 kips and two transverse loads of 2 kips each located 2.5 ft from either end. There are no holes at the points of attachment of the transverse loads, but there is a single line of $\frac{3}{4}$-in. bolts connecting the member at its ends. Assume simply supported ends. AISC specification, A36 steel.

5-27 A fixture to support a shaft bearing is to be welded to the bottom chord of the truss of DP3-1, midway between L_2 and L_3. The shaft load is 600 lb. What change is required in the member? Assume simply supported ends.

Can you suggest a more rigid support for this bearing (other than a larger shape for the bottom chord) that might be desirable if alignment of the shaft is critical and if vibration is an important consideration?

5-28 Choose a tee for a member 9 ft long supporting an axial compression of 50 kips and a uniformly distributed transverse load of 500 plf. The uniform load is delivered through attached construction which supports the member against lateral buckling. Assume simple supports. AISC specification, A36 steel.

5-29 Choose a W shape for the member of Prob. 5-28.

5-30 Same as Prob. 5-28 except that the member is not supported against lateral buckling.

5-31 Choose a pair of aluminum 6061-T6 angles separated by $\frac{3}{8}$ in. for gusset plates for the member of Prob. 5-28.

5-32 Same as Prob. 5-31 except that the member is not supported against lateral buckling.

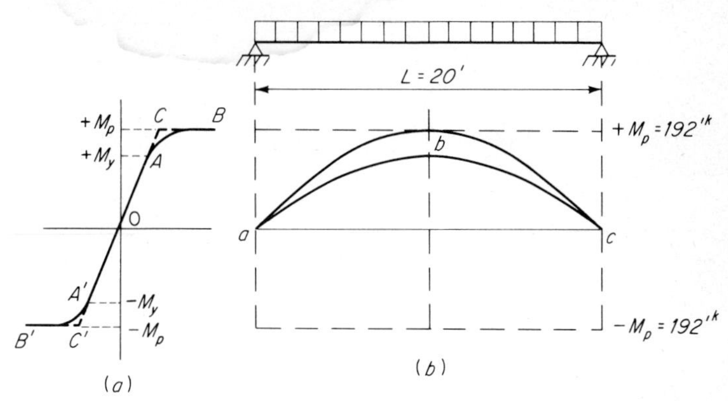

FIGURE 5-24

5-33 The horizontal top chord of a welded roof truss is to be a W shape with the web members butted against its bottom flange and welded to it. The chord is continuous over panel points. Panel lengths are 16 ft and the purlin spacing 8 ft. The purlin reaction at each truss is 8.2 kips, and the axial compression in the chord is 78 kips. Dependable lateral support exists only at the panel points. Design the chord. AISC specification, A36 steel.

5-34 The jib crane shown in Fig. 5-23 is to be installed in an existing building. The mast is to be supported on a new concrete footing and is to be welded at its top to the existing roof truss, whose bottom chord consists of two angles $5 \times 3\frac{1}{2} \times \frac{1}{2}$, with their long legs vertical. The lifting capacity of the hoist is 3,000 lb. The trolley weighs 500 lb, and its wheels are designed to run on the flange of a standard S-shape beam. The boom must be free to rotate about the mast, approximately 90° in either direction from its mid-position. AISC specification, A36 steel.

(*a*) Design the boom, the mast, and the tie.
(*b*) Design the welded connection of the mast to the truss.
(*c*) Design the connections of the tie.
(*d*) Design the connection of the boom to the mast.

5-35 Same as Prob. 5-33 except use A501 structural tubing.

5-13 CONTINUOUS BEAMS

An important difference between the behavior of statically determinate beams and statically indeterminate beams is discussed in this article. Bending behavior of beam cross sections was discussed in Art. 5-2, where it was shown that, except for strain hardening, the plastic moment is the bending strength of the section. The variation of moment with strain on the extreme fiber is shown in Fig. 5-4. Figure 5.24*a* shows this variation for the I cross section for both positive and negative bending. For the

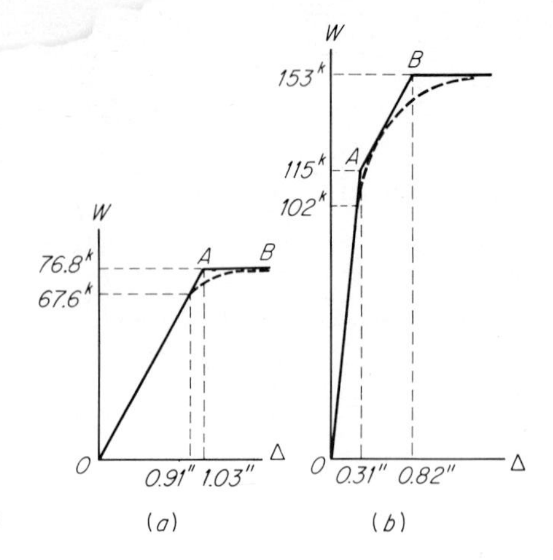

FIGURE 5-25

S or W of the proportions used for beams, the plastic moment M_p averages about 12 percent larger than the moment M_y at the beginning of yield on the extreme fiber. Although residual stresses and other imperfections of the beam lower the proportional limit, so that nonlinear behavior begins at a moment less than M_y, $B'C'OCB$ is a good approximation to the bending behavior.

Figure 5-24*b* shows a simply supported, uniformly loaded A36-steel beam spanning 20 ft. The cross section is a W16 × 36. The beam is supported against lateral buckling. It is clear that, under gradually increasing load, the parabolic moment diagram *abc* increases in amplitude until the moment at midspan reaches the value M_p. The corresponding load W is given by

$$M_p = ZF_y = 64.0 \times \frac{36}{12} = 192 \text{ ft-kips}$$

$$W = \frac{8 \times 192}{20} = 76.8 \text{ kips}$$

Any attempt to increase the load beyond this value fails because, except for strain hardening, the moment cannot exceed M_p. Thus, if deformation of the beam increases, rotation at midspan continues with no increase in moment, along CB in Fig. 5-24*a*, and the beam is said to have developed a *plastic hinge*. The plastic hinge differs from a real hinge in that the latter allows free rotation at no load, whereas the former allows free rotation only after the plastic moment has been attained. When the beam develops the plastic hinge at midspan, it is unstable. In this form, it behaves like a linkage, and because of this analogy it is called a *mechanism*.

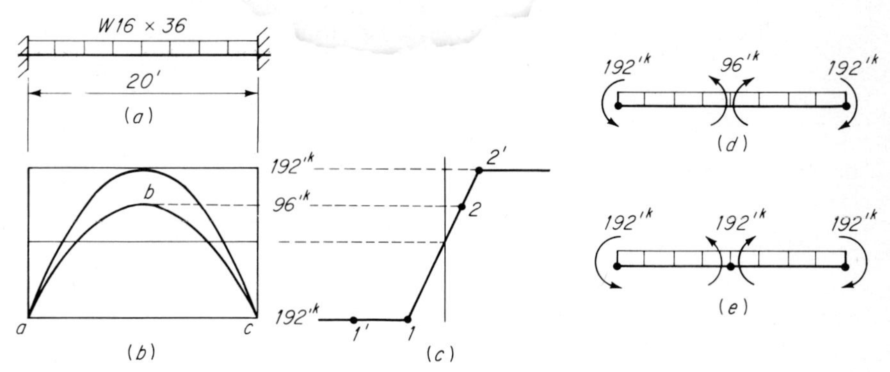

FIGURE 5-26

The midspan deflection of the beam at the instant the mechanism develops is given by

$$\Delta = \frac{5}{384}\frac{WL^3}{EI} = \frac{5 \times 76.8 \times 240^3}{384 \times 30{,}000 \times 447} = 1.03 \text{ in.}$$

OAB in Fig. 5-25*a* shows the assumed variation of load with deflection. In the actual beam, nonlinear behavior begins at the yield moment, which, for the W16 × 36, is $M_y = SF_y = 56.5 \times 36/12 = 169$ ft-kips. The corresponding load W is 76.8 × 169/192 = 67.6 kips, for which $\Delta = 0.91$ in.

Figure 5-26*a* shows the beam of Fig. 5-24*b* with both ends fixed. As long as stress is proportional to strain, the bending moments at the ends are double the moment at midspan. Thus, under uniformly increasing load the moment diagram *abc* of Fig. 5-26*b* will eventually be reached, where the end moments are represented by point 1 in Fig. 5-26*c* and the midspan moment by point 2. At this stage the beam has a plastic hinge at each support (Fig. 5-26*d*) and corresponds to a simply supported beam carrying the uniformly distributed load and a moment M_p at each end. Further increase in load causes the moment at midspan to increase while the end moments hold at M_p. Thus, the vertex of the parabolic moment diagram in Fig. 5-26*b* continues to rise until the midspan moment attains the value M_p. At this stage, the midspan moment is represented by point 2′ and the end moments by point 1′ in Fig. 5-26*c*. It is obvious that these moments can increase no further. Therefore, the beam now has three plastic hinges (Fig. 5-26*e*) and is a mechanism, so that it has reached its load capacity. This peak load is called the *limit load* or *collapse load.* The collapse-load moment diagram in Fig. 5-26*b* is said to result from a *redistribution* of the moments *abc* corresponding to elastic behavior.

The load W_1 at which plastic hinges form at the supports is given by

$$W_1 = \frac{12 \times 192}{20} = 115 \text{ kips}$$

The corresponding deflection is

$$\Delta_1 = \frac{W_1 L^3}{384EI} = \frac{115 \times 240^3}{384 \times 30{,}000 \times 447} = 0.31 \text{ in.}$$

The additional positive moment, 96 ft-kips, which produces the third plastic hinge, corresponds to the additional load W_2 on a simply supported beam

$$W_2 = \frac{8 \times 96}{20} = 38 \text{ kips}$$

The corresponding increment of deflection is

$$\Delta_2 = \frac{5}{384} \frac{W_2 L^3}{EI} = \frac{5 \times 38 \times 240^3}{384 \times 30{,}000 \times 447} = 0.51 \text{ in.}$$

The load-deflection diagram is shown in Fig. 5-25*b*. As is the case for the simply supported beam, nonlinear behavior begins at the yield moment M_y, rather than at the development of the first plastic hinge. Therefore, the load-deflection plot begins to curve at the load 102 kips, as shown in the figure.

It will be noted that the load which initiates yielding of the fixed-end beam is 50 percent larger than that for the simply supported beam. On the other hand, the collapse load of the fixed-end beam is twice that of the simple beam.

Since the W16 × 36 is a compact section in A36 steel, the AISC allowable bending stress is $0.66F_y$. As pointed out in Art. 5-5, this is 10 percent larger than the allowable stress $0.60F_y$ for noncompact sections and is an allowance for the difference between the yield moment and the plastic moment. In the case of the simple beam of Fig. 5-24, it accounts for the fact that the collapse load is larger than the yield load in the ratio M_p/M_y. In the case of the fixed-end beam, Fig. 5-26*a*, the ratio of the collapse load to the yield load is 153/102 = 1.50, part of which is due to the difference between the plastic moment and the yield moment and part to the redistribution of moment. To provide for redistribution of moment, the AISC allows compact continuous beams and compact beams rigidly framed to columns to be designed for 90 percent of the negative moments at points of support if these moments are calculated by elastic analysis, provided the maximum positive moment in the span is increased by 10 percent of the average of the negative moments at the ends of the span. While this reduction in moment is not sufficient to account fully for the 33 percent redistribution of the fixed-end beam of Fig. 5-26, the reduction must be small enough to

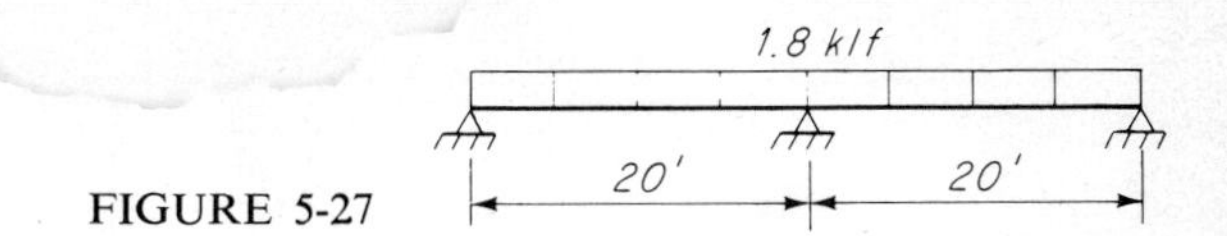

FIGURE 5-27

allow for the fact that the effect of redistribution is smaller in many cases. For example, there is no redistribution of moment in the fixed-end beam with a concentrated load at midspan, since the midspan moment and the two end moments are equal ($WL/8$) from the beginning of loading. Plastic design, which takes rational account of redistribution of moment, is discussed in Chap. 8.

EXAMPLE 5-4 Determine the A36 W required for the continuous beam of Fig. 5-27. The beam is supported against lateral buckling. AISC specification.

SOLUTION

$$-M = \frac{WL^2}{8} = \frac{1.8 \times 20^2}{8} = 90 \text{ ft-kips}$$

$$+M = \frac{9}{128} WL^2 = \frac{9 \times 1.8 \times 20^2}{128} = 50.6 \text{ ft-kips}$$

The beam may be designed for $0.9 \times 90 = 81$ ft-kips negative moment and $50.6 + 9 = 59.6$ ft-kips positive moment.

$$F_b = 0.66 F_y = 0.66 \times 36 = 24 \text{ ksi}$$

$$S = \frac{81 \times 12}{24} = 40.5 \text{ in.}^3$$

Use a W14 × 30, which is compact in A36 steel.

PROBLEMS

5-36 A laterally supported beam built-in at one end and simply supported at the other spans 20 ft and carries a concentrated load at midspan. The cross section is an A36 W12 × 31. Determine its limit load, and plot the load-deflection curve.

5-37 A laterally supported two-span continuous beam carries a concentrated load at the middle of each 20-ft span. The cross section is an A572 Grade 50 W16 × 45. Determine the limit load and plot a load-deflection curve.

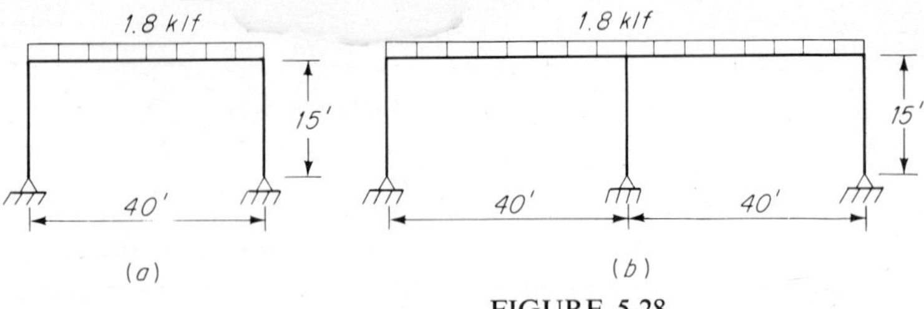

FIGURE 5-28
(*a*) Problem 5-40, (*b*) Problem 5-41.

5-38 The end spans of a three-span continuous beam are each 30 ft long, and the central span is 40 ft long. There is continuous lateral support. The design load is 2.5 klf not including the weight of the beam itself. What W shape of A36 steel is required? AISC specification.

5-39 One of the end spans of a three-span continuous beam is 16 ft long; the other is 20 ft. The central span is 24 ft long. There is a concentrated load of 20 kips at the center of each span. The beam is supported laterally at the supports and at the points of load. What W shape of A441 steel is required? AISC specification.

5-40 The frame shown in Fig. 5-28*a* is unbraced. The design load is 1.8 klf. Design the frame in A36 steel. AISC specification.

5-41 The frame shown in Fig. 5-28*b* is unbraced. The design load is 2.0 klf. Design the frame in A36 steel. AISC specification.

5-14 HOLES IN BEAM FLANGES

Sometimes there must be open holes in beams for piping, conduit, reinforcing steel, etc., or to receive brick and stone anchors and bolts for wood nailing strips. Holes for rivets or bolts to attach reinforcing plates to the flanges, or stiffener angles to the webs, are required occasionally. The effect of such holes on the strength of a beam is not easy to evaluate.

The question of location of the neutral axis for beams with holes in only one flange has been widely discussed. Some writers argue that its position is the same as it would be if there were no holes, while others contend that the beam-flexure formula requires that it lie at the center of gravity of the net section. The latter argument is questionable, since it is the linear distribution of stress assumed by the flexure formula that places the neutral axis at the center of gravity. Stress concentrations at the edges of holes, which were discussed in Art. 3-5, alter the linear distribution of stress and therefore also invalidate the statement $\int y\, da = 0$.

If the neutral axis of a beam does not shift at a cross section containing one or more holes, it is a simple matter to compute either the moment of inertia or the section modulus. From the gross moment of inertia we need only subtract $\sum Ay^2$, where A is the area to be deducted for each hole and y the distance from the neutral axis to the hole. If the holes are only in the flanges, the deduction in section modulus is $\sum Ay^2/c$, but since y is in this case practically the equal of c, we can take the deduction as $\sum Ac$. Some designers consider both flanges to have holes even though there may be holes in only one.

The effect of a hole in the tension flange of a beam cannot be deduced from the behavior of tension members. The strength of a steel beam is usually determined by the strength of the compression flange. Thus, it is possible that an open hole in the compression flange affects the strength of a steel beam more than one in the tension flange does. On the other hand, if a hole in the compression flange contains a rivet or bolt, the weakening effect is probably reduced, since the fastener can transmit compression through the hole. Therefore it would seem reasonable up to a point to neglect holes in steel beams, provided there is no question of fatigue. The AISC specifications allow the designer to neglect reduction in area of beam and girder flanges up to 15 percent of the gross area, provided the holes are for rivets or bolts. If the reduction in area exceeds 15 percent, only the excess need be considered. On the other hand, AASHO and AREA require full deduction for holes.

Holes in beam webs are discussed in the next article.

5-15 SHEAR IN BEAMS

Shearing stress is seldom a factor in the design of a steel beam, and it is usually calculated, if at all, only as a check after the beam has been designed for bending. Shear may determine the design of beams which support heavy concentrated loads near the reaction points and of very short (small values of L/d) beams uniformly loaded.

Figure 5-29*a* shows a differential length dz of a beam which is symmetrical about the y axis. Figure 5-29*b* shows a portion of this element obtained by separation on the plane *abcd*. The resultant forces on the vertical faces of this portion are T and $T + dT$. An equilibrating shear force dT acts in the plane of separation. Division of dT by dz gives the shear q per lineal inch: $q = dT/dz$. But

$$T = \int_{ab}^{e} f\,dA = \int_{ab}^{e} \frac{M_x y}{I_x}\,dA = \frac{M_x Q_x}{I_x}$$

where $Q_x = \int_{ab}^{e} y\,dA$, so that

$$q = \frac{dT}{dz} = \frac{dM_x}{dz}\frac{Q_x}{I_x} = \frac{V_y Q_x}{I_x} \qquad (5\text{-}24)$$

Note that the section *abcd* may be passed through the beam in any direction so long as it is parallel to the z axis and that it may be a zigzag or curved section. The

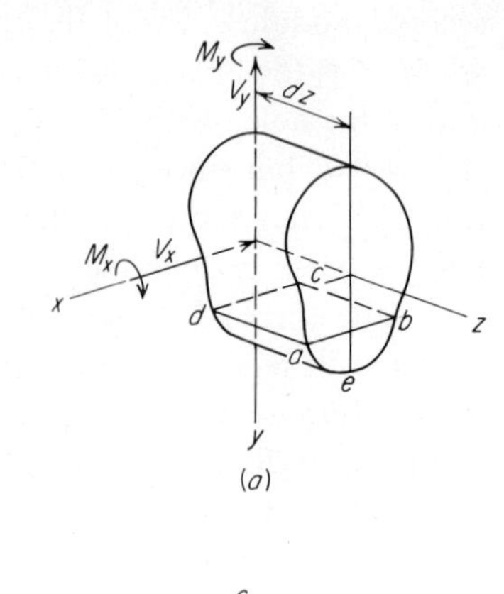

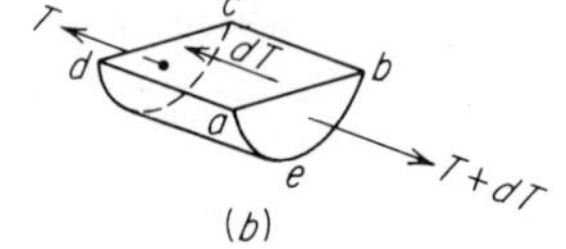

FIGURE 5-29

shearing stress at any point on *abcd* cannot be determined from this analysis. However, if we assume the shear force dT to be uniformly distributed along the width *ab*, the shearing stress f_v is found by dividing q by *ab*. By virtue of the equality of horizontal and vertical shearing stresses at a point, Eq. (5-24) also determines the shearing stresses in the surface *abe* along the line *ab*.

The maximum shearing stress in I beams loaded in the plane of the web can be approximated satisfactorily by assuming the total shear to be uniformly distributed over the area of the web. The explanation lies in the fact that the web furnishes most of the shear resistance, and with very little variation in shearing stress throughout its depth. For example, if the shearing force on a W18 × 50 loaded in the plane of the web is 50 kips, the maximum shearing stress is

$$f_v = \frac{V_y Q_x}{I_x t} = \frac{50 \times 7.36 \times 6.87}{802 \times 0.358} = 8.8 \text{ ksi}$$

The value of Q_x in the above calculation is from the AISC Manual's table of properties of structural tees. The shearing stress at the juncture of the flange and web, neglecting the fillet, is

$$f_v = \frac{50 \times 7.50 \times 0.57 \times 8.72}{802 \times 0.358} = 6.5 \text{ ksi}$$

The shearing force on the web is found by integration to be 48.5 kips, which is 97 percent of the shearing force.

If we assume the shearing force to be distributed uniformly over an area equal to the product of the depth of the beam and the thickness of the web, we get $f_v = 50/(18 \times 0.358) = 7.8$ ksi. This value is 11.4 percent less than the true maximum value. Design-specification allowable stresses are based on the assumption that the shearing stress will be computed in this way.

Holes are often needed in webs of beams in buildings to accommodate ducts, conduits, and other services. In some cases the beam must be reinforced in the vicinity of the hole by welding doubler plates or stiffening angles to the web or by welding flats or bars to the periphery of the hole, while in other cases no reinforcement is required. Procedures have been developed for both elastic and plastic design of beams with unreinforced holes with depths not exceeding six-tenths the depth of the beam.[24] Procedures for the design of reinforcements have also been published.[25] Fatigue tests of beams with rectangular holes in the web have also been made.[26]

Since the web of an I beam is a flat plate, it may buckle at shearing stresses less than the shearing yield strength of the metal. This must be considered in establishing allowable shear stresses. Shear buckling of webs is discussed in the next article.

5-16 SHEAR BUCKLING OF BEAM WEBS

Figure 5-30 shows a flat plate acted upon by shear stresses distributed uniformly along the four boundaries. Because this is a state of pure shear, the shear stresses are equivalent to principal stresses of the same magnitude, one tension and one compression, acting at 45° to the shear stresses. These are shown on an interior element of the web in the figure. Thus, it can be seen that buckling in the form of waves or wrinkles inclined at about 45° may develop, as shown. Such shear buckling of the web of an aluminum beam is shown in Fig. 5-31.

The shear stress $F_{v,cr}$ at which buckling of a perfect plate begins is given by

$$F_{v,cr} = \frac{k\pi^2 E}{12(1-\mu^2)(b/t)^2} \tag{5-25}$$

This is the same as Eq. (4-47), which is derived in Art. 9-1. Values of k are defined by a system of intersecting curves similar to those of Fig. 4-50*b*. However, the common tangent to these curves is itself a curve, rather than a straight line as in Fig. 4-50*b*. This tangent curve for a plate with all four edges simply supported is given to good approximation by[11]

$$k = \begin{cases} 4 + \dfrac{5.34}{(a/b)^2} & \dfrac{a}{b} \lessapprox 1 \qquad (5\text{-}26a) \\[2ex] 5.34 + \dfrac{4}{(a/b)^2} & \dfrac{a}{b} \gtrapprox 1 \qquad (5\text{-}26b) \end{cases}$$

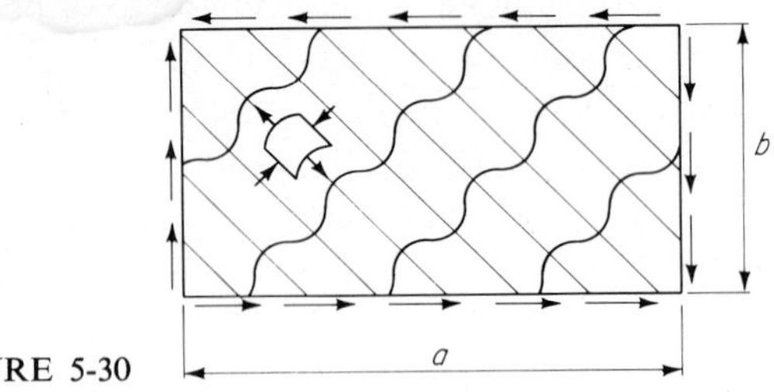

FIGURE 5-30

where a/b is the aspect ratio of the plate (Fig. 5-30). More recently, a single formula, valid for all a/b, has been suggested[12]

$$k = 5\left[1 + \left(\frac{b}{a}\right)^2\right] \tag{5-27}$$

If the four edges are clamped,

$$k = \begin{cases} 5.60 + \dfrac{8.98}{(a/b)^2} & \dfrac{a}{b} \lesssim 1 \quad \text{(5-28}a\text{)} \\[2ex] 8.98 + \dfrac{5.60}{(a/b)^2} & \dfrac{a}{b} \gtrsim 0 \quad \text{(5-28}b\text{)} \end{cases}$$

We note from Eq. (5-26b) that k is only slightly larger than 5.34 for a hinged-edge plate with large aspect ratio. Thus, if $a/b = 5$, $k = 5.50$. Therefore, the shear-buckling behavior of the webs of rolled beams can be conservatively evaluated by using $k = 5.34$ in Eq. (5-25). In any case, some approximation is involved in using Eq. (5-25), since bending stresses will always be present. However, at the ends of simply supported beams these stresses are small enough to be neglected in investigating shear stresses. Substituting $k = 5.34$, $E = 30{,}000$ ksi, and $\mu = 0.30$ into Eq. (5-25) gives

$$F_{v,cr} = \frac{144{,}000}{(b/t)^2} \text{ ksi} \tag{5-29}$$

ABC of Fig. 5-32 is a graph of Eq. (5-29). When $b/t = 83$, $F_{v,cr} = 21$ ksi, the shearing yield stress of A36 steel. *ABD* shows the behavior of a perfect plate. Of course, imperfections lower the proportional limit to a point such as *E*. Moreover, for small b/t, beam webs can strain harden, as is shown by the test results plotted in the figure.[13] These were tests on welded beams with web slenderness b/t ranging from 50 to 70. One group consisted of five beams made of steel for which

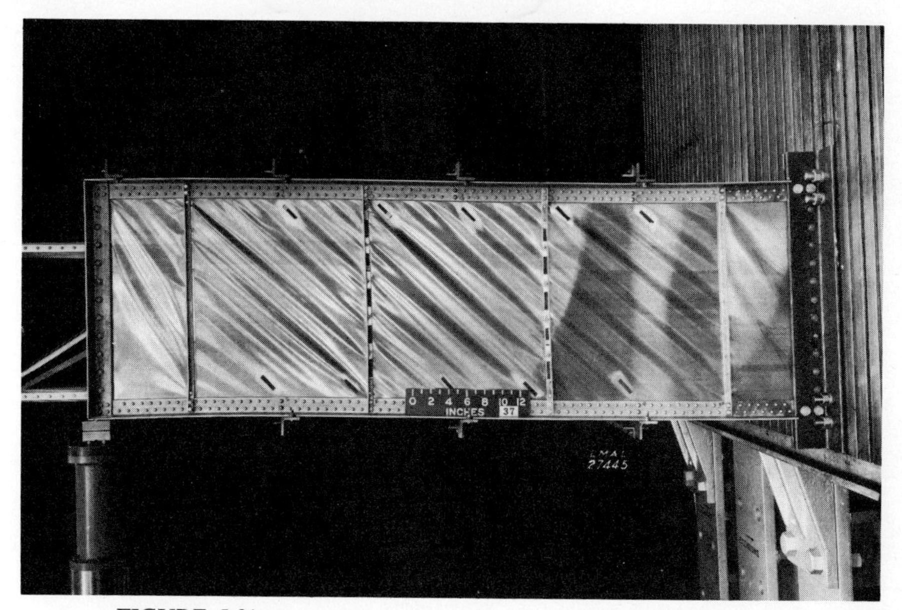

FIGURE 5-31
Shear postbuckling of aluminum-beam web. (*National Aeronautics and Space Administration, Langley Research Center.*)

$29.7 < F_y < 33.7$ ksi. For the second group (three beams) $43.3 < F_y < 49.6$ ksi. Assuming the proportional limit stress F_{vp} to be $0.8F_{vy}$, where F_{vy} is the yield stress in shear, the critical shear stress $F_{v(cr)i}$ in the inelastic and strain-hardening ranges can be expressed by[14]

$$F_{v(cr)i} = \sqrt{0.8F_{vy}F_{v,cr}} \tag{5-30}$$

Line *EF* in Fig. 5-32 is the plot of Eq. (5-30) for $F_y = 36$ ksi. *GH* is a plot for $F_y = 32$ ksi, which is the average yield stress for the group of five test specimens mentioned above, while *IJ* is a plot for $F_y = 47$ ksi, the average yield stress for the group of three.

It will be noted that A36-steel beams with webs for which b/t is less than about 74 can be counted on to develop shear yield. The webs of all rolled-steel I sections are of such thickness that b/t is less than 70; in fact, it is less than 50 for most of them. Therefore, web buckling due to shear is eliminated as a design consideration for rolled beams of A36 steel. On the other hand, the inelastic-buckling curve for steel with $F_y = 100$ ksi intersects the corresponding yield stress, $F_{vy} = 58$ ksi, at $b/t = 50$. Therefore, the shear strength of the thinner webs of beams of high-strength steels may be somewhat less than yield strength.

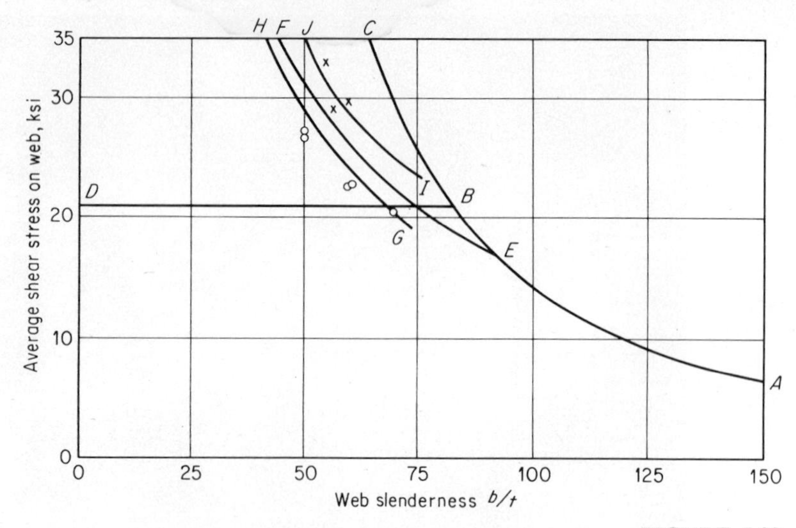

FIGURE 5-32

The allowable shear stresses for beam webs and the corresponding factors of safety of the AISC, AASHO, and AREA specifications are:

Specification	Allowable shear	Factor of safety
AISC	$0.4F_y$	1.45
AASHO	$0.33F_y$	1.75
AREA	$0.35F_y$	1.65

The factors of safety are based on the shear yield stress $F_{vy} = F_y/\sqrt{3}$. For plate girders, however, these allowable stresses usually must be reduced for shear buckling, since the slenderness of plate-girder webs is generally much larger than that of rolled beams. This is discussed in Chap. 6.

The shear-buckling strength of aluminum beams is given by Eq. (5-25) in the elastic range and by the tangent-modulus equivalent of that equation for the inelastic range. For alloy 6061-T6, the values are[5]

$$F_{v(cr)i} = 25.8 - 0.164\frac{b}{t} \qquad \frac{b}{t} \not> 65 \tag{5-31a}$$

$$F_{v,cr} = \frac{63{,}100}{(b/t)^2} \qquad \frac{b}{t} > 65 \tag{5-31b}$$

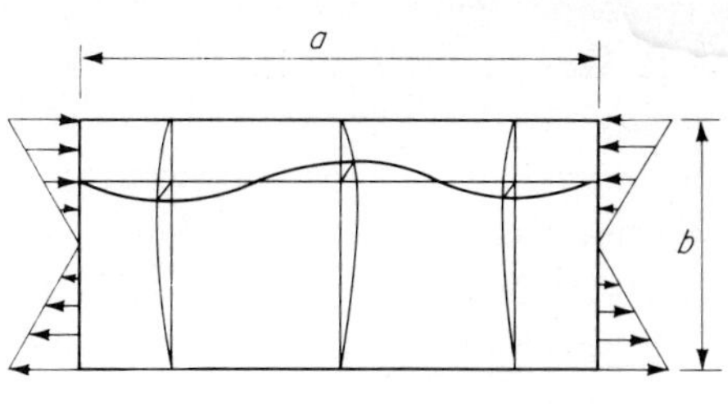

FIGURE 5-33

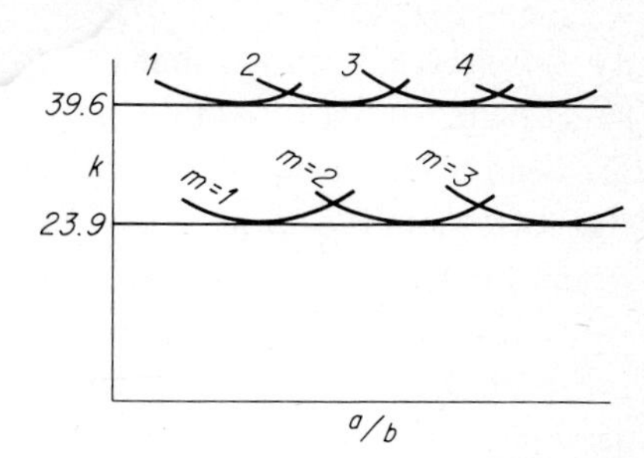

FIGURE 5-34

These stresses are based on edge conditions about halfway between hinged and fixed. The ASCE Task Committee[6] recommends factors of safety of 2.2 and 1.95 for bridge structures and building structures, respectively.

5-17 BEND BUCKLING OF BEAM WEBS

Since bending stresses are compressive over part of the depth of a beam, they may cause local buckling of the web. Figure 5-33 shows a flat plate, simply supported on all four edges, which has buckled due to the bending stresses shown. As in the case of plates under uniform edge compression, beam webs bend-buckle in a single transverse wave and multiple lengthwise waves. Figure 5-34 shows values of k in Eq. (5-25) for a plate with the four edges simply supported and for one with the loaded edges simply supported and the unloaded edges clamped.[11] The number of longitudinal waves is shown by $m = 1$, etc. It will be noted that the minimum values of k, 23.9 for simply supported edges and 39.6 for the unloaded edges clamped, are the same for all values of m. The corresponding critical stresses for $E = 30{,}000$ ksi and $\mu = 0.3$ are

$$F_{b,cr} = \frac{650{,}000}{(b/t)^2} \quad \text{ksi} \tag{5-32a}$$

$$F_{b,cr} = \frac{1{,}080{,}000}{(b/t)^2} \quad \text{ksi} \tag{5-32b}$$

Tests show that beam webs are partially restrained against rotation by the flanges, to the extent that the critical stress is likely to be at least 30 percent higher than that given by Eq. (5-32a). Thus, we may use

$$F_{b,cr} = \frac{850{,}000}{(b/t)^2} \quad \text{ksi} \tag{5-32c}$$

Figure 5-34 shows that the critical stress is independent of a/b except for very short plates (a/b less than about 0.5). From Eq. (5-32c) we find that the critical stress for bend buckling equals the yield stress for A36 steel at $b/t = 154$. For $F_y = 100$ ksi, the corresponding value is 92. Thus, even with some allowance for a lower proportional limit because of imperfections, there is no likelihood of bend buckling of the webs of rolled-steel I shapes since, as mentioned in Art. 5-16, their webs are such that $b/t < 70$. However, plate girders usually have webs much thinner than those of rolled sections, so that bend buckling must be taken into account. This is discussed in Chap. 6.

Since the web slenderness of cold-formed beams is usually much larger than that of standard rolled shapes, allowable bending stresses must provide for bend buckling. The AISI specifications allowable-bending stress for webs is

$$F_b = \frac{520{,}000}{(b/t)^2} \not> 0.60F_y \tag{5-33}$$

This formula gives a factor of safety of 1.25 with respect to bend buckling for hinged-edge panels, Eq. (5-32a). This small factor of safety is justified because thin beam webs have considerable postbuckling strength. This is discussed in Chap. 9.

Bend-buckling critical stresses can be calculated by using an equivalent slenderness ratio in column-buckling formulas, as in the case of the flat plate under uniform edge compression (Art. 4-21). Thus, from Eq. (4-48), with $\tau = 1$, and using $k = 23.9$ for simply supported edges, we get

$$\left(\frac{L}{r}\right)_{eq} = \frac{3.3}{\sqrt{23.9}}\frac{b}{t} = 0.67\,\frac{b}{t} \tag{5-34a}$$

Similarly, with $k = 39.6$ for unloaded edges clamped we have

$$\left(\frac{L}{r}\right)_{eq} = 0.52\,\frac{b}{t} \tag{5-34b}$$

The equivalent slenderness for the partially restrained plate corresponding to Eq. (5-32c) is

$$\left(\frac{L}{r}\right)_{eq} = 0.59\,\frac{b}{t} \tag{5-34c}$$

The value $(L/r)_{eq} = 0.60b/t$ has been recommended for calculating the bend-buckling stress for webs of aluminum beams. Because of the considerable post-buckling strength, factors of safety of 1.2 for buildings and 1.35 for bridges are suggested.[6]

Bend buckling of the webs of beams proportioned to develop the plastic moment

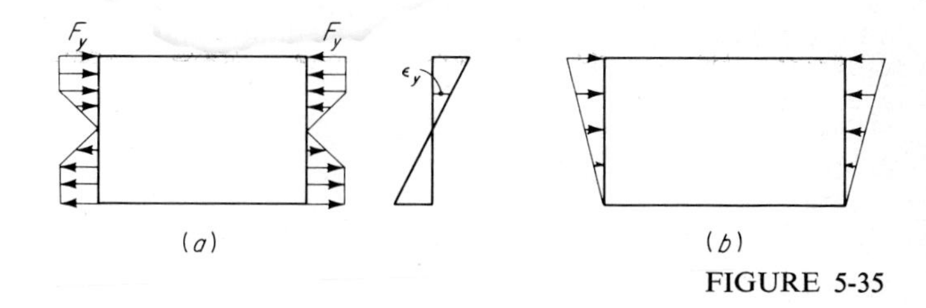

FIGURE 5-35

must also be considered. It is clear that the bending stress distribution of the web of Fig. 5-35*a* involves an extreme-fiber strain which is several times the yield strain. In this case, the web is anisotropic in its resistance to buckling; i.e., its properties are direction-dependent. This is because the longitudinal bending stresses resulting from buckling begin in a direction in which the web is already strained inelastically while the transverse bending stresses begin in a direction in which it is unstressed. This phenomenon is discussed in Art. 4-21. Such bend buckling is analyzed in Ref. 2, assuming that the extreme fiber of the web is strained inelastically ($\varepsilon > \varepsilon_y$) in the longitudinal direction at the onset of buckling, and that there is no strain reversal as buckling begins. It was found that b/t for a web of A36 steel must not exceed about 60 if the web is to develop an extreme-fiber strain $\varepsilon = 4\varepsilon_y$ in pure bending. The AISC has adopted the value $b/t \lessapprox 412/\sqrt{F_y}$ (which gives $b/t \lessapprox 69$ for A36 steel) as the limit in plastic design and in elastic design when the increased allowable bending stress $0.66F_y$ for compact sections is used.

The preceding discussion of bend buckling of beam webs is limited to pure bending. The plate-buckling coefficient k for a beam-column web hinged on its four edges will lie between the value $k = 23.9$ for pure bending and $k = 4$ for uniform compression. Values for various combinations of moment and axial force have been determined.[11] For example, $k = 7.8$ for a plate with the distribution of stress shown in Fig. 5-35*b*. For such a web in A36 steel, a consistent slenderness limit would be $b/t = 42\sqrt{7.8/4} = 59$, where 42 is the AISC prescribed value for uniformly compressed plates (Table 4-4). Such limits are not ordinarily used in elastic design. However, they are essential in plastic design and as compact-section limitations in elastic design. The following AISC formulas, which are based on the research noted above (Ref. 2), are requirements for compact sections:

$$\frac{b}{t} \lessapprox \begin{cases} \dfrac{412}{\sqrt{F_y}}\left(1 - 2.33\dfrac{f_a}{F_y}\right) & \dfrac{f_a}{F_y} \lessapprox 0.16 \qquad (5\text{-}35a) \\[2ex] \dfrac{257}{\sqrt{F_y}} & \dfrac{f_a}{F_y} > 0.16 \qquad (5\text{-}35b) \end{cases}$$

In these formulas, $f_a = P/A$ is the service-load axial compression. The value $412/\sqrt{F_y}$ is the beam-web limit mentioned earlier in this article. Thus, the limiting values for A36-steel webs range from $b/t = 412/\sqrt{36} = 69$ for bending alone to $b/t = 257/\sqrt{36} = 43$ for axial compression alone. The same formulas are prescribed, in slightly different form, for plastic design [Eqs. (8-6)].

5-18 COMBINED SHEAR AND BENDING OF WEBS

Shear buckling and bend buckling of webs were discussed in the preceding articles as independent buckling modes. While bend buckling can occur alone (where pure bending is involved), shear cannot exist without some bending. Furthermore, webs at the interior supports of continuous beams are subjected concurrently to large bending moments and shears. Therefore, the effect of these combinations of stress must be considered.

Figure 5-36 shows a flat plate with hinged edges subjected to bending stresses and to shear stresses distributed uniformly along the edges.[11] The curve in the figure gives, for a square plate, the relation between the ratios $F'_{b,cr}/F_{b,cr}$ and $F'_{v,cr}/F_{v,cr}$, where $F_{b,cr}$ is the critical stress on the extreme fiber for pure bending, $F_{v,cr}$ is the critical stress for pure shear, and $F'_{b,cr}$ and $F'_{v,cr}$ are the bending and shearing stresses which, acting simultaneously, cause buckling. It is seen that the effect of shearing stress on the critical bending stress is small if $F'_{v,cr}/F_{v,cr}$ is small. Thus, if $F'_{v,cr}/F_{v,cr} = 0.4$, a bending stress of almost $0.9F_{b,cr}$ can be resisted. Similarly, the effect of bending stress on the critical shear stress is small if $F'_{b,cr}/F_{b,cr}$ is small.

The interaction curve of Fig. 5-36 is very nearly circular. Furthermore, curves for other aspect ratios a/b differ from it only slightly. Therefore, the interaction of shear stress and bending stress can be written in the form

$$\left(\frac{F'_{v,cr}}{F_{v,cr}}\right)^2 + \left(\frac{F'_{b,cr}}{F_{b,cr}}\right)^2 = 1 \tag{5-36}$$

This equation is based on critical stresses. It can be expressed in terms of actual stresses and allowable stresses by dividing the numerator and denominator of each term by the factor of safety. Thus

$$\left(\frac{f_v}{F_v}\right)^2 + \left(\frac{f_b}{F_b}\right)^2 = 1 \tag{5-37}$$

where f_v, f_b = concurrent actual shearing and bending stresses
F_v = allowable shear stress in absence of bending
F_b = allowable bending stress in absence of shear

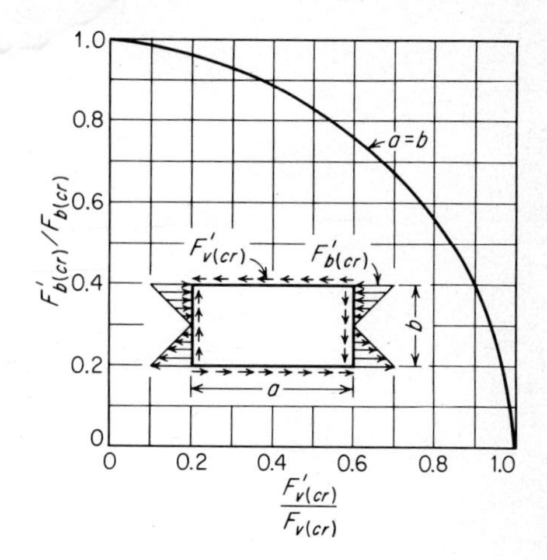

FIGURE 5-36

According to the AISI specification, combined shear and bending must be evaluated according to Eq. (5-37). The same formula is also recommended for webs of aluminum beams.[5] The AISC specification makes no provision for combined shear and bending except where the postbuckling shear strength of thin webs is taken into account. Since webs of standard rolled shapes are thick enough not to buckle, except possibly for the thinner webs in the highest-strength steels, this means that interaction need be considered in the AISC specification only for plate girders. This is discussed in Chap. 6.

Ths AASHO and AREA specifications make no provision for interaction of shear and bend buckling of beam webs. However, at sections where maximum shear and bending occur simultaneously, AREA specifies $0.55F_y$ as the allowable diagonal tension in the web. This principal stress f_t is determined from

$$f_t = \frac{f_b}{2} + \sqrt{\left(\frac{f_b}{2}\right)^2 + f_v^{\,2}}$$

where $f_b = M/S$ and $f_v = V/A_w$.

5-19 CRIPPLING AND VERTICAL BUCKLING OF WEBS

In addition to shearing and bending stresses in the web of a beam, there are compressive stresses in the vertical direction because of bearing of the loads on the flanges. In the beam shown in Fig. 5-37, we see that there must be vertical compression immediately above the bearing plate. Furthermore, since there is no load on the top flange above the support, the vertical compression must diminish at successively

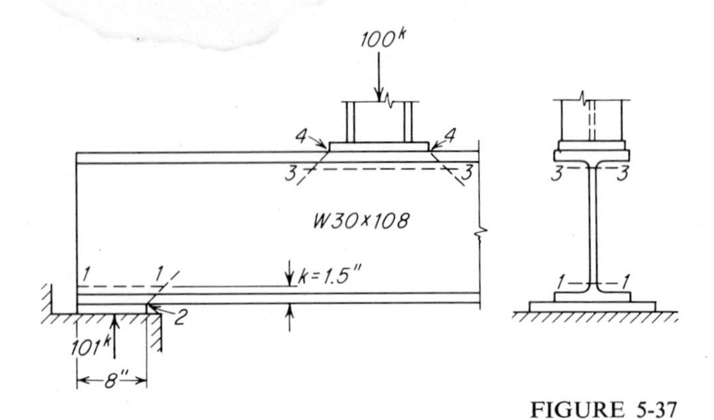

FIGURE 5-37

higher horizontal sections, until it becomes zero at the top flange. To estimate this stress, we isolate a portion of the beam below the section 1-1 at the edge of the fillet and to the left of section 1-2 drawn at an inclination of 45° from the right end of the bearing plate. The inclination of section 1-2 is arbitrary and is an attempt to account for the dispersion of the compressive stresses. The section is sometimes taken vertically. Neglecting the vertical component of any force which may exist on section 1-2 and assuming the compressive stress to be uniformly distributed on section 1-1, we have

$$f = \frac{101}{9.5 \times 0.548} = 19.4 \text{ ksi}$$

The magnitude of this stress obviously can be controlled by varying the length of the bearing plate. Similar compression exists below the top flange at the supported column, and, for consistency with the assumption about the distribution at the support, we assume it to be distributed over the area at section 3-3 which terminates at sections 3-4 drawn at 45° from each end of the bearing plate.

The failure which may result at concentrated loads if vertical compression in webs is excessive is called by various names, such as *direct compression*, *web crimpling*, and *web crippling*. The AISC allowable stress for web crippling is $0.75F_y$, where the web-crippling stress is intended to mean the stress on sections 1-1 and 3-3 in Fig. 5-37. This is a protection against localized yielding. If the allowable stress is exceeded, web stiffeners must be provided. Stiffeners supporting a concentrated load on the top flange of a beam are shown in Fig. 5-38. The stiffeners are plates welded to the web of the beam. They relieve the web of the local crippling stresses. Neither AASHO or AREA specifies an allowable stress for this situation. However, AREA requires web

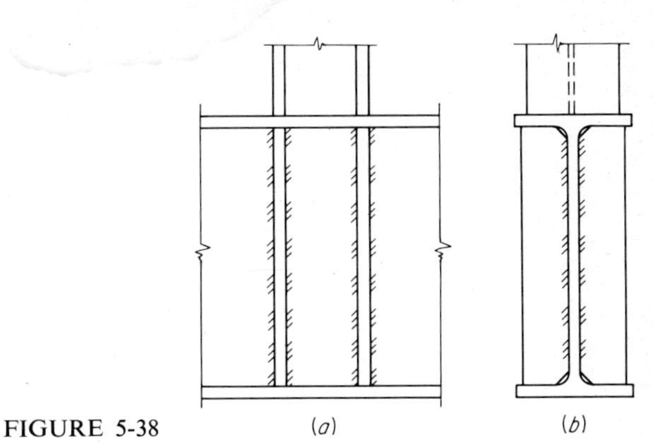

FIGURE 5-38

stiffeners at end bearings and at points of bearing of interior concentrated loads for all beams and girders. AASHO also requires web stiffeners at points of bearing in all plate girders but in rolled beams only if the shearing stress in the beam web exceeds 75 percent of the allowable. The design of stiffeners is discussed in Chap. 6.

Vertical compression in beam webs may also result in a buckling failure (Fig. 5-39). An exact solution of this problem requires a stability analysis of the entire web with unlike load systems on the two horizontal edges. However, a good approximation can be developed for the beam supporting a uniformly distributed load w on its top flange. In this case, the vertical compression in the web varies from w at the top edge to zero at the bottom. Since the shearing stress on a vertical section of the web is distributed almost uniformly, as shown in Art. 5-15, the variation in vertical compression will be very nearly linear. Thus, the web can be treated as a plate with axial compression distributed uniformly along its length. It is shown in Ref. 11, p. 107, that a pin-ended column can support about twice as much load distributed uniformly along its length as it can if the load is concentrated at the ends. The same conclusion can be shown to hold for a plate loaded in the same manner. Thus, the critical vertical compressive stress for the web of a beam supporting a uniformly distributed load is twice the critical stress for a plate uniformly compressed on two opposite edges. This buckling problem is discussed in Art. 9-1, where it is shown that the critical stress for a plate with all four edges simply supported is given by Eq. (9-8), provided the plate buckles in a single half wave in the direction of the compression. This requires that the aspect ratio a/b be less than $\sqrt{2}$ (Fig. 4-50). For the case of the beam web, a in Eq. (9-8) is the depth of the beam and b its length, so that a/b is very small (b will rarely be less than $10a$). Therefore, Eq. (9-8) applies. Furthermore, a/b can be dropped because it is small, which reduces the formula to the simpler form of Eq. (9-4). Thus, the vertical-

FIGURE 5-39
Buckling failure of web of rolled I. (*University of Illinois Experiment Station.*)

buckling stress for the web of a beam loaded uniformly on its top flange is twice the value from Eq. (9-4). Therefore,

$$F_{cr} = \frac{2\pi^2 E}{12(1-\mu^2)(b/t)^2} \tag{5-38a}$$

where b is the depth of web and t the thickness of web. However, the critical stress from Eq. (9-4) must *not* be doubled if the beam is also uniformly compressed on the opposite flange, as in the grillage discussed in Art. 5-20, because in this case the web compression is uniform throughout the depth. Thus

$$F_{cr} = \frac{\pi^2 E}{12(1-\mu^2)(b/t)^2} \tag{5-38b}$$

These equations should not be used if the web has transverse stiffeners. This is discussed in Chap. 6. Furthermore, it must be remembered that Eqs. (5-38) give the critical stress, to which an appropriate factor of safety must be applied.

The critical stresses of Eqs. (5-38) are increased if the web is rotationally restrained by the flanges. For example, if the web is prevented from turning at either flange, the effective-length coefficient is 0.7 for the uniformly compressed web of Eq. (5-38*b*) and F_{cr} is doubled. On the other hand, for the web in which the vertical compression varies from a maximum at the compression flange to zero at the tension

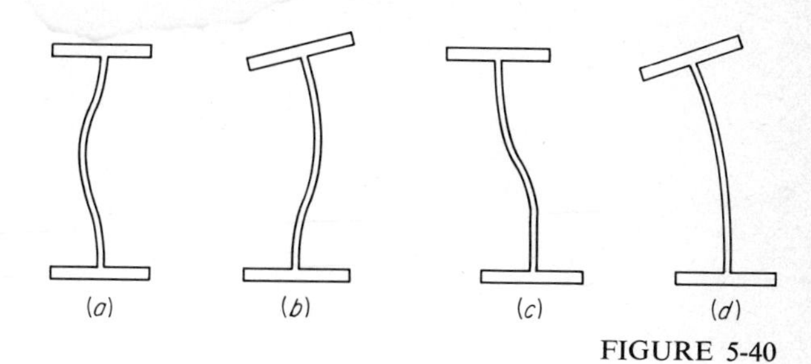

FIGURE 5-40

flange, the critical stress is 2.75 times the value from Eq. (5-38*a*). The larger increase results from the fact that the largest reduction in lateral displacement of the web is in the region where the vertical compression is largest. Of course, there are other possible buckling modes. For example, there may be a lateral displacement of one flange relative to the other, as in Fig. 5-39. Various other possible forms of web buckling are shown in Fig. 5-40.

Web buckling due to concentrated loads is harder to evaluate. Figure 5-41 shows the vertical stresses on three horizontal sections of a beam of rectangular cross section of unit width and depth d supporting a concentrated load P. It will be noted that, at all three levels, the stress is compressive over a length about equal to d. The stress at mid-depth varies from zero at each end of the length d to $0.91P/d$ at the center.[15] The average stress on the area is about $0.5P/d$. Thus, in terms of average stress, the decrease in compression with depth is the same as that for a uniformly distributed load. There is no simple way to evaluate web stability under these conditions. A procedure which was used for many years and which was reported to have been based on tests, consisted in assuming vertical compression in the web to be distributed uniformly over the length of bearing of the concentrated load plus $d/2$ for an interior load and $d/4$ for an end reaction. The allowable value of this compression was computed from a formula based on Eq. (5-38*b*).

Standard design specifications do not cover vertical buckling of beam webs except to the extent that it may be precluded by provisions for web crippling discussed previously. However, the AISC specification does require investigation of vertical buckling of plate-girder webs (Art. 6-12).

Vertical-buckling resistance is reduced in any portion of a beam web which is subjected to bending. Thus, according to Ref. 11, p. 379, the critical uniform vertical compression for a square steel plate simply supported on all four sides is only one-third the value according to Eq. (5-38*b*) if the plate is also subjected to a bending stress 75 percent of the critical stress according to Eq. (5-32*a*). Of course, this value is increased

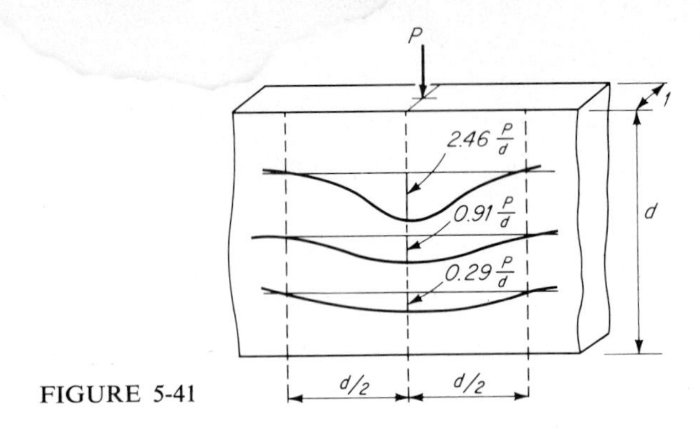

FIGURE 5-41

if the vertical compression varies through the depth as in the beam web. Standard specifications do not require a check of this combination of stresses.

Cold-formed beams Web crippling of cold-formed beams is aggravated by bending due to eccentric application of load which results from the curved transition at the juncture of the flange and web and by the fact that such webs are often so thin that buckling is also involved. Formulas which are based largely on test results are given in the AISI specification.

5-20 DP5-3: GRILLAGE BASE FOR COLUMN

Loads from heavily loaded columns are sometimes distributed to a footing or pile foundation through one or more tiers of closely spaced beams. Such an arrangement is called a *grillage*. Grillages are usually encased in concrete, except those used for temporary support during construction. Web stresses are large in grillage beams and can result in collapse if they are excessive. Two spans of the Second Narrows Bridge in Vancouver, B.C., collapsed during construction in 1958 because of web failure of the W36 × 160 beams in the upper tier of a grillage which supported an erection bent. An error in calculating web stresses appears to have been a major factor: the erector's designers inadvertently used the flange thickness, 1.02 in., instead of the web thickness, 0.653 in. The beam webs buckled in the manner shown in Fig. 5-39. The end of the span supported by the bent fell, pulling the pier at its other end out of plumb so that the adjacent span also fell.[16,17]

In DP5-3 we design a grillage for a column consisting of two 20 × 4½ in. flange plates and a 13 × 2 in. web. The load is 3,800 kips. The footing is 9 × 9 ft. The comments which follow clarify the correspondingly lettered computation on the design sheet.

GRILLAGE FOR COLUMN — DP5-3

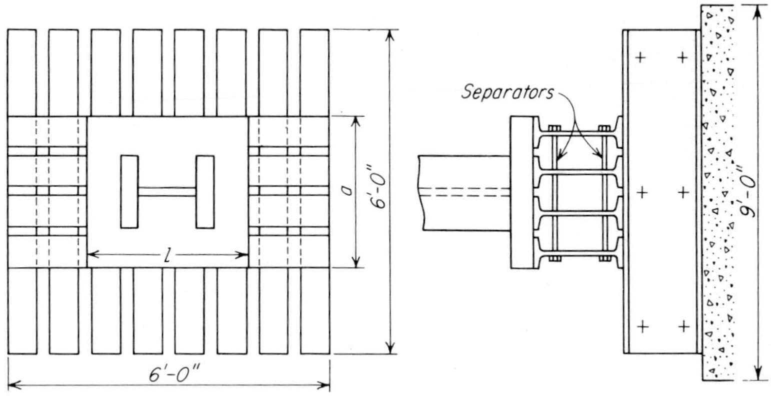

AISC spec. — $P = 3800^k$

Col. { Web 2×13 — $L = 20'$
Flg. 4½ × 20 — A36 steel

Footing 750 psi: 3800/0.75 = 5070 in.² (a)
72" × 72" = 5184 in.²

Upper tier beams

Try $l = 40''$ (b)

$M = \frac{3800\,(72-40)}{8} = 15{,}200''^k$ (c)

÷ 24 = 633 I/c

Try 4 beams ÷ 4 = 158

S20×95, $I/c = 160$, web $t = 0.80''$ (d)

Web crippling: 3800/4 = 950^k per bm.

$f = 950/(43.8 \times 0.80) = 27.1$ ksi (e)

Try $l = 42''$

$f = 27.1 \times 43.8/45.8 = 26.0$ ksi < 27 (f)

$a = 4 \times 7.2 + 3 \times 2 = 34.8$ say 36" (g)

Check shear

$V = \frac{950^k}{72''} \times \frac{72''-42''}{2} = 198^k$ (h)

$f_v = \frac{198}{20 \times 0.8} = 12.4$ ksi < 14.5

Use 4-S20×95

Base plate

$0.8b = 0.8 \times 20 = 16''$ (i)

Cantilever = (36 − 16)/2 = 10"

$0.95d = 0.95 \times 22 = 20.9''$ (i)

Cantilever = (42 − 21)/2 = 10.5"

Base pressure: $p = \frac{3800}{42 \times 36} = 2.52$ ksi

$M = 2.52 \times 10.5^2/2 = 139''^k$

$bt^2/6 = 139/(0.75 \times 36) = 5.15$

$t = \sqrt{6 \times 5.15} = 5.6''$

Use 6 × 36 × 3'-6" base pl. (j)

Lower tier beams

$M = 3800\,(72-36)/8 = 17{,}100''^k$

÷ 24 = 713 I/c

Try 8 beams ÷ 8 = 89.1

S18×54.7, $I/c = 89.4$

(72 − 8×6)/7 = 3" + bet. flgs.

Web crippling 3800/8 = 475^k

$f = 475/(38 \times 0.46) = 26.5$ ksi < 27

$V = \frac{475^k}{72''} \times \frac{72''-36''}{2} = 119^k$

$f_v = \frac{119}{18 \times 0.46} = 14.4$ ksi < 14.5

Use 8-S18 × 54.7

DP 5-3
Grillage for column.

a. The AISC allowable bearing stress on concrete is the same as that of the ACI code, $0.25f'_c$ if the load bears on the entire area and $0.375f'_c$ if it bears on one-third the area. Assuming $f'_c = 3{,}000$ psi, the allowable bearing stress is $0.25 \times 3{,}000 = 750$ psi.

b. This dimension may vary from about one-third to somewhat more than half the length of the upper tier of beams. The preliminary guess may need revision later.

c. The load is assumed to be distributed over the 40-in. length of contact on the upper tier and the 72-in. contact of the upper tier on the lower. For such a distribution the maximum moment is $WL/8 - WL'/8$, where W is the total load and L and L' are the loaded lengths.

d. Web stresses tend to be large in grillage beams. Therefore, the S is likely to be more efficient than the W because S's have thicker webs.

e. Vertical compression is assumed to be distributed over the 40-in. length plus the distance at each end shown in Fig. 5-37. Buckling of the web is not considered, since grillages for columns in buildings are usually encased in concrete, as mentioned above. Some designers argue that a concrete fill also makes it unnecessary to consider web crippling.

f. Since the vertical compression computed in *e* exceeds the allowable, l is increased.

g. The distance a is chosen so as to give clearance between beam flanges for placing concrete. A clearance of 2 in. is usually considered to be minimum.

h. The maximum shear in the beam is at the edge of the base plate.

i. These are the AISC recommendations for base-plate analysis discussed in Art. 4-16.

j. Bearing plates over 4 in. thick must be planed to furnish a true bearing surface in contact with the column. About $\frac{3}{8}$ in. extra thickness is required to allow for planing.

The beams in each tier must be interconnected with *separators*, which usually consist of bolts or tie rods passing through lengths of pipe. Details of this and other forms of separator are given in the AISC Manual. It is evident that several trials may be required to achieve a satisfactory arrangement for a grillage. If the base plate is too large relative to the required area in contact with the concrete, the grillage loses its advantage compared with a base plate large enough to function alone. On the other hand, too small a base plate results in excessively large beams because of requirements for shear and web crippling.

Since vertical buckling of the web is an important consideration in grillages not encased in concrete, we shall illustrate the analysis for an upper-tier beam of this example. The compressive stress is uniform through the depth, so that the critical stress is given by Eq. (5-38*b*), as explained in Art. 5-19. Lateral support of the top flange is unlikely in this situation, so that buckling as in Fig. 5-39 should be assumed. Therefore, the effective length is the depth of the beam. With $E = 30{,}000$ ksi and $\mu = 0.3$, Eq. (5-38*b*) becomes

$$F_{cr} = \frac{27{,}000}{(b/t)^2} \quad \text{ksi} \tag{a}$$

This equation holds only if the critical stress is less than the proportional limit. Assuming $F_p = F_y/2$ to allow for web crookedness, possible residual stresses, etc., the lower limit of b/t for which Eq. (*a*) applies is

$$\frac{b}{t} = \sqrt{\frac{27{,}000}{18}} = 39$$

For the S20 × 95 of the grillage, $b/t = 20/0.80 = 25$. Therefore, vertical buckling is inelastic. We shall assume inelastic-buckling stresses to be given by a parabola, as in the CRC column formula. Therefore,

$$F_{cr} = 36\left[1 - \frac{1}{2}\left(\frac{25}{39}\right)^2\right] = 28.4 \text{ ksi}$$

Assuming the 950-kip load to be distributed over a length $d/2$ longer than the length of the base plate (Art. 5-19), we get

$$f = \frac{950}{0.8(42 + 10)} = 22.8 \text{ ksi}$$

The factor of safety is $28.4/22.8 = 1.25$. This is much too low. Therefore, if the beams of this grillage were not encased in concrete, web stiffeners should be used or some other means taken to relieve the web of the large vertical compressive stress.

PROBLEMS

5-42 Investigate the sufficiency of the lower-tier beams of the grillage of DP5-3 if they are not encased in concrete.

5-43 Design a base plate to replace the grillage of DP5-3. Allow extra thickness for planing. Compare its weight with that of the grillage steel.

5-44 Design an A36-steel grillage to support an A572 ($F_y = 50$ ksi) square-box column consisting of two plates 5 × 24 and two plates 4 × 14. The load is 10,000 kips. The concrete footing ($f'_c = 3$ ksi) is 15 ft square.

5-45 An existing column in a building is to be underpinned temporarily during alteration of the building. It is to be supported on one or more parallel beams, which will be supported in turn on temporary concrete pedestals. Each pedestal furnishes a 12-in. bearing length whose center is 3 ft from the column. The base plate of the column is 20 × 24 in., the 20-in. dimension being parallel to the underpinning beam. The load on the column is 310 kips. Design the supporting beam (or beams). AISC specification, A36 steel.

5-46 The four lines of columns in an existing building are 24 ft on centers (Fig. 5-42). The columns are 18 ft on centers in each line. Floors are concrete slabs 4 in. thick, and the design live load is 80 psf. An alteration requires that the width of a part of the central aisle at the

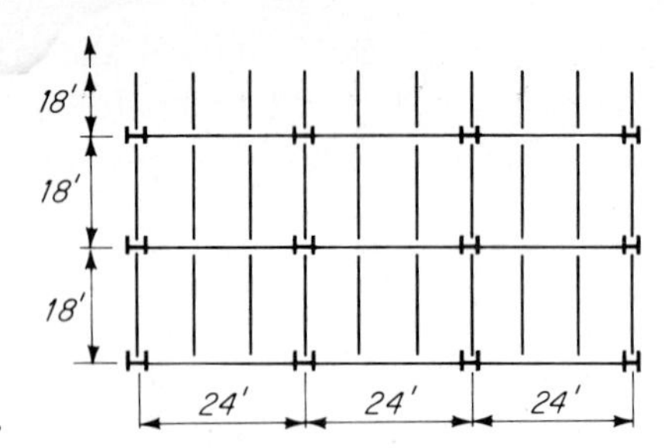

FIGURE 5-42
Problem 5-46.

lowest level be increased to 28 ft. New columns will support a girder 28 ft long, which because of headroom limitations cannot be more than 25 in. deep. In addition to the existing upper-floor columns, the new girder will support the existing W12 × 27 joists, which are 8 ft on centers. Each column load, not including the joists framing at that point, is 210 kips. The girder will be supported against lateral buckling only at the points where the four lines of joists connect to it. Design the girder. AISC specification, A36 steel.

5-47 A beam 20 ft between supports overhangs 2 ft on one end. Between the supports is a uniformly distributed load of 2 klf. A W8 column stands on the overhang, its centerline 6 in. from the tip. The column load is 240 kips. Therefore, excluding the negligible weight of the beam itself, the maximum moment is 360 ft-kips, and the maximum shear 240 kips. Since the beam is supported laterally, the designer plans to use a W24 × 94, which is adequate for moment but not for shear. His idea is to reinforce the overhang by welding to the beam two plates 20 × $\frac{1}{4}$, one on each side of the web. He says the web area will be $24 \times 0.516 + 2 \times 20 \times 1/4 = 22.4$ in.^{2} and the shearing stress $240/22.4 = 10.7$ ksi as compared with $240/12.4 = 19.4$ ksi without the plates.

Why is his analysis wrong? What is the magnitude of the largest shearing stress on the vertical section at the support? Suggest a way out of the difficulty other than using a larger beam or thicker reinforcing plates.

5-21 BEAM BEARING PLATES

A beam supported on masonry or concrete must usually be provided with bearing plates which, in addition to having enough length along the beam to control vertical compression in the web, have bearing area sufficient to give an adequate factor of safety with respect to crushing of the supporting material. Since they can be placed in advance of the beams and grouted level at the required elevation, bearing plates also facilitate erection. For this reason they are sometimes advisable even if the beam itself has contact area sufficient to distribute its reaction.

Although the bearing plate is a simple enough structural element, it is virtually impossible to determine the distribution of the forces acting on it. As a result of

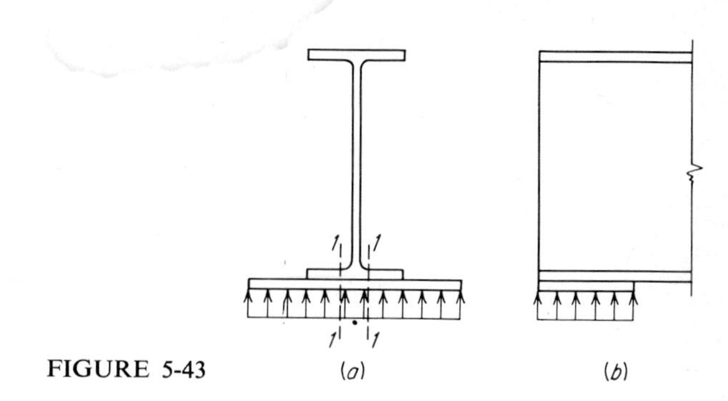

FIGURE 5-43

deflection of the beam, bearing pressures are larger at the edge of the plate nearest the center of the beam than they are at the end. Pressures at the other two edges are relieved by bending of the plate in the direction normal to the beam, and as a consequence the distribution of pressure in this direction also is not uniform.

Various assumptions as to the distribution of the bearing stresses might be made, but it is customary to consider them to be uniform, which amounts to saying that the plate is designed for an average bearing stress. With this assumption, the design procedure resolves itself into one of choosing a plate which has (1) sufficient dimension lengthwise of the beam to keep the compressive stresses in the web of the beam within allowable limits, (2) sufficient area to distribute the reaction over the masonry at the specified allowable stress, and (3) sufficient thickness to satisfy requirements for bending normal to the beam. With respect to the last of these requirements, there is the question of the location of the section of maximum bending moment. There is little to say except that it lies somewhere between the centerline of the plate and the edge of the beam flange. If the beam flange has no stiffness, the maximum moment in the plate is at its centerline, but if the flange is infinitely stiff, it is at the edge of the flange. Since in order for it to perform its function of distributing load the plate requires stiffness as well as strength, it is advisable to take the section of maximum moment somewhere near the center of the plate and to neglect the pressures exerted on it by the flange. The AISC specification recommends that the design be based on the moment at the section 1-1 at the edge of the flange fillet (Fig. 5-43*a*).

EXAMPLE 5-5 Design a bearing plate for an A36 W16 × 40 beam spanning 24 ft supporting a uniformly distributed load of 1.7 klf. The beam is supported on a 12-in. brick-faced masonry wall. AISC specification.

SOLUTION $R = 1.7 \times 12 = 20.4$ kips. The allowable bearing pressure $F_p = 250$ psi gives

$$A = \frac{20.4}{0.25} = 81.6 \text{ in.}^2$$

To allow 4 in. for the brick facing, use a bearing plate $7\frac{1}{2} \times 11$, for which $A = 82.5$ in.^{2} The allowable compression (web crippling) is $F_a = 0.75 \times 36 = 27$ ksi. Then,

$$f_a = \frac{R}{(7.5 + k)t} = \frac{20.4}{(7.5 + 1)0.307} = 7.8 < 27 \text{ ksi}$$

The bearing pressure is $p = 20.4/82.5 = 0.248$ ksi. The bearing plate projection is $\frac{11}{2} - \frac{5}{8} = 4.9$ in. Therefore,

$$M = 0.248 \times \frac{4.9^2}{2} = 2.98 \text{ in.-kips}$$

With the allowable bearing-plate bending stress $0.75F_y = 27$ ksi

$$\frac{bt^2}{6} = \frac{2.98}{27} = 0.11 \qquad t = 0.81$$

Use $1 \times 7\frac{1}{2} \times 11$ plate.

5-22 DESIGN FOR LIMITED DEFLECTION

Although a beam is unsuitable if it cannot support its loads without excessive deflection, it is not easy to set a dividing line between reasonable and unreasonable deflection. Excessive deflection in floor construction is objectionable not only because of the feeling of softness but also because of undesirable vibration characteristics and the possibility of damage to attached construction such as plaster. Excessive deflection in floor construction supporting machinery may result in misalignments as well as dangerous vibration. Excessive deflection in purlins may cause damage to roofing materials and, on flat roofs, accumulation of water during rainstorms which, under certain conditions, can cause collapse. Retention of water due to the deflection of flat-roof framing is called *ponding*.

The maximum deflection Δ of a simply supported beam uniformly loaded in a principal plane is given by

$$\Delta = \frac{5}{384} \frac{WL^3}{EI} \qquad (a)$$

where W denotes the total load on the span. But since the maximum bending moment $M = WL/8$, we may eliminate W from Eq. (a) to get

$$\Delta = \frac{5}{48}\frac{ML^2}{EI} \tag{b}$$

Substituting $M/I = f/c$ into Eq. (b) gives

$$\Delta = \frac{5}{48}\frac{fL^2}{Ec} \tag{c}$$

The length of a beam in structural work is usually given in feet, while its depth is measured in inches. Let L_{ft} stand for length in feet. Substituting $L = 12L_{ft}$ into Eq. (c), we have

$$\Delta = 15\frac{f}{E}\frac{{L_{ft}}^2}{c_{in}} \tag{5-39a}$$

This equation reduces to a very simple form for steel beams. Since $E = 30{,}000$ ksi and $c = d/2$, we have, if f is expressed in kips per square inch,

$$\Delta = \frac{f}{1{,}000}\frac{{L_{ft}}^2}{d_{in}} \tag{5-39b}$$

Similar equations may be derived for other load distributions. However, Eq. (5-39a) will predict deflection with sufficient accuracy for practically any vertical load. Table 5-3 gives ratios of the true deflection to the value given by Eq. (5-39a) for 13 different load distributions. The first six entries show that the equation is satisfactory for almost

Table 5-3 RATIO OF TRUE DEFLECTION TO VALUE GIVEN BY EQ. (5-39a)

Case	Load	Ratio
1	Uniformly distributed on each end quarter of span	1.15
2	Uniformly distributed on each end third of span	1.11
3	Increases uniformly from zero at one end to maximum at other	0.98
4	Increases uniformly from zero at each end to maximum at center	0.96
5	Uniformly distributed over middle half of span	0.96
6	Uniformly distributed over middle quarter of span	0.89
7	Concentrated at center of span	0.80
8	Concentrated at one of the quarter points	0.75
9	Concentrated at one of the eighth points	0.69
10	Concentrated in two equal parts at $\frac{1}{8}$ span from each support	1.17
11	Concentrated in two equal parts at $\frac{1}{4}$ span from each support	1.10
12	Concentrated in two equal parts at $\frac{1}{3}$ span from each support	1.02
13	Concentrated in three equal parts at equidistant points	0.95

any case of distributed load except possibly the first. The next three suggest that the maximum deflection for a single concentrated load is, closely enough, 80 percent of the value given by Eq. (5-39*a*) unless the load is very near a support. The remaining entries show that the formula can be used without significant error for multiple concentrated loads, with the possible exception of the rather extreme location of loads in case 10.

Permissible deflection of a beam is usually relative to the span, since a deflection of, say, 1 in. in a span of 30 ft will not ordinarily be more objectionable than a deflection of $\frac{1}{2}$ in. in a span of 15 ft. Occasionally, however, permissible deflection is independent of span. For example, if a lintel is placed to clear a glass-block panel by $\frac{1}{2}$ in., its permissible deflection is somewhat less than $\frac{1}{2}$ in. regardless of its span. The most frequently quoted deflection limit prohibits live-load deflections in excess of 1/360 of the span for beams supporting plastered ceilings. The source of this rule seems to be unknown. Presumably it was given originally as a safe limit with respect to cracking of plastered ceilings. This deflection limit is a requirement of the AISC specifications. AASHO limits deflection due to live load plus impact to not more than 1/800 of the span, while AREA limits it to 1/640.

The ratio L/d of beam span to beam depth which corresponds to a specific ratio Δ/L of deflection to span can be determined from Eq. (*c*)

$$\frac{L}{d} = \frac{24}{5}\frac{E}{f}\frac{\Delta}{L} \tag{d}$$

If we wish to limit deflection to, say, 1/300 of the span for steel beams designed for $f = 0.6F_y$, we find from Eq. (*d*)

$$\frac{L}{d} = \frac{24}{5}\frac{30{,}000}{0{\cdot}6F_y}\frac{1}{300} = \frac{800}{F_y}$$

The commentary to the AISC specification suggests this as a guide to deflection control of beams in floors. For purlins (except those in flat roofs) the value $1{,}000/F_y$ is suggested. More detailed recommendations, to guard against ponding, are covered in the specification.

PROBLEMS

5-48 Design a bearing plate for a laterally supported A36 W24 × 76 which spans 15 ft and carries its allowable uniformly distributed load. The bearing plates rest on concrete piers for which $f'_c = 3{,}000$ psi. AISC specification.

5-49 Design bearing plates for a laterally supported A36 W18 × 35 which spans 15 ft and carries its allowable concentrated load at 3 ft from the center of one support. The bearing plates rest on concrete piers for which $f'_c = 3{,}000$ psi. AISC specification.

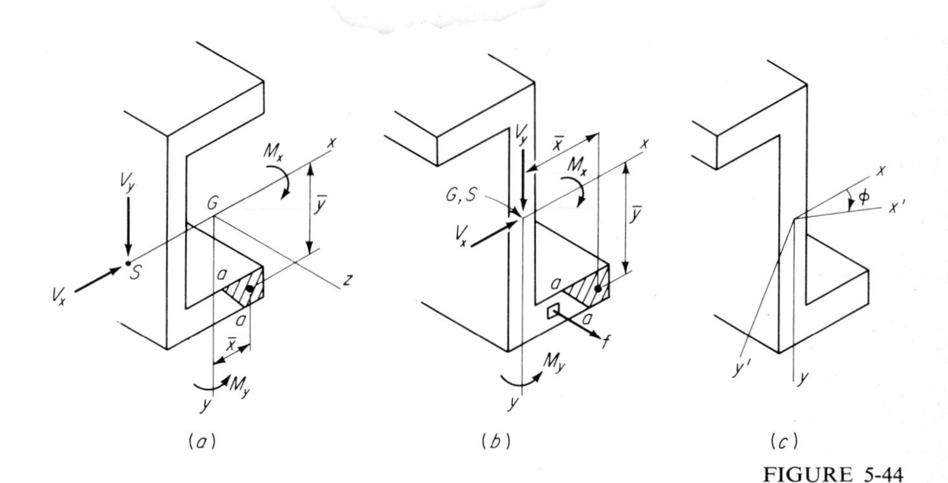

FIGURE 5-44

5-50 A simply supported, laterally restrained beam for a machine-shop floor spans 24 ft. It supports a distributed load of 600 plf and a concentrated load of 3 kips located 6 ft from one end. Impact for the concentrated load is 30 percent. Clearance requirements limit the depth of the beam to 12 in. What A36 W shape is required? AISC specification.

5-23 BIAXIAL BENDING

The design of beams which have cross sections with at least one axis of symmetry and which are loaded through the shear center parallel to one of the two principal axes has been considered in preceding articles. If the load on the beam is not parallel to a principal axis, it can be resolved into components in the directions of the principal axes. The resulting shears and moments are shown in Fig. 5-44*a*. The bending stress f_b is given by

$$f_b = \frac{M_x y}{I_x} + \frac{M_y x}{I_y} \tag{5-40}$$

The shear q per lineal inch of beam on a plane through *a-a* parallel to the axis is

$$q = \frac{V_y A\bar{y}}{I_x} + \frac{V_x A\bar{x}}{I_y} \tag{5-41}$$

where A is the shaded area in Fig. 5-44*a* and $\bar{x}$, $\bar{y}$ are the coordinates of its center of gravity.

If x, y are not principal axes, as in the case of the zee shown in Fig. 5-44b, f_b and q are given by

$$f_b = \frac{M_x I_y - M_y I_{xy}}{I_x I_y - I_{xy}^2} y + \frac{M_y I_x - M_x I_{xy}}{I_x I_y - I_{xy}^2} x \tag{5-42}$$

$$q = \frac{V_y I_y - V_x I_{xy}}{I_x I_y - I_{xy}^2} A\bar{y} + \frac{V_x I_x - V_y I_{xy}}{I_x I_y - I_{xy}^2} A\bar{x} \tag{5-43}$$

where I_{xy} is the product of inertia $\int xy\, dA$. Taking tensile stress positive, the moments M_x and M_y in Eq. (5-42) are positive when they produce tension in the first quadrant (Fig. 5-44b). If x, y are principal axes, $I_{xy} = 0$ and Eqs. (5-42) and (5-43) reduce to Eqs. (5-40) and (5-41). If the beam cross section has an axis of symmetry, that axis is a principal axis. If there is no axis of symmetry, the direction of the principal axes x', y' is given by

$$\tan 2\phi = \frac{2I_{xy}}{I_y - I_x} \tag{5-44}$$

where x, y are orthogonal axes originating at the centroid of the cross section (Fig. 5-44c). The moments of inertia about the principal axes are

$$I_{x'} = I_x \cos^2 \phi + I_y \sin^2 \phi - I_{xy} \sin 2\phi \tag{5-45}$$

$$I_{y'} = I_x \sin^2 \phi + I_y \cos^2 \phi + I_{xy} \sin 2\phi \tag{5-46}$$

The calculations required to compute the principal moments of inertia offset most if not all of the advantage of Eq. (5-40) over Eq. (5-42).

A beam that must support loads that are not in one of its principal planes can be designed only by trial and error. If an approximate range of values of the ratio

Table 5-4 APPROXIMATE VALUES OF S_x/S_y

Shape	Depth d, in.	$\frac{S_x}{S_y}$
W	8–16	3–8
W	16–24	5–10
W	24–36	7–12
S	6–8	d
S	10–18	$0.75d$
S	20 and 24	$0.6d$
C	7 and under	$1.5d$
C	8–10	$1.25d$
C	12 and 15	d

S_x/S_y for a particular shape or combination of shapes is known, the following rearrangement of Eq. (5-40) will often give a fairly close design at the first trial:

$$S_x = \frac{M_x}{f_b} + \frac{S_x}{S_y}\frac{M_y}{f_b} \tag{5-47}$$

Approximate values of S_x/S_y are given in Table 5-4 for W, S, and C shapes. The smaller numbers in each group of W's are for those shapes with relatively wide flanges, i.e., those which are more nearly square. Values for the S and C are related roughly to the depth. For an S12, for example, S_x/S_y is about $0.75 \times 12 = 9$.

A variation of Eq. (5-47) which is convenient for the W may be obtained by using approximate values of moment of inertia. If we neglect the web, the section moduli are given by

$$S_x = \frac{2bt(d/2)^2}{d/2} = btd$$

$$S_y = \frac{2tb^3/12}{b/2} = \frac{tb^2}{3}$$

from which $S_x/S_y = 3d/b$. The error in this approximation due to the omission of the moment of inertia of the web is offset closely by increasing the coefficient of d/b from 3 to 3.5. Substituting this value into Eq. (5-47), we get

$$S_x = \frac{M_x}{f_b} + 3.5\frac{d}{b}\frac{M_y}{f_b} \tag{5-48}$$

Values of d/b for the W's range about as follows:

Shapes 8 to 16 in. deep, 1 to 2
Shapes 16 to 24 in. deep, 1.5 to 2.5
Shapes 24 to 36 in. deep, 2 to 3

The strength of beams subjected to bending moments in both principal planes depends upon their lateral-buckling characteristics. If, in Fig. 5-13*a*, in addition to the couples M_x at each end there are couples M_y producing single curvature in the xz plane, each of Eqs. (5-4) takes on an additional term involving M_y.[11] The solution to these equations is

$$\frac{M_x{}^2}{EI_y} + \frac{M_y{}^2}{EI_x} = \frac{\pi^2}{L^2}GJ + \frac{\pi^4}{L^4}EC_w \tag{a}$$

This equation reduces to Eq. (5-6) if $M_y = 0$. It holds only if the bending stress does not exceed the proportional limit. Therefore, M_x and M_y must also satisfy

$$\frac{M_x}{S_x} + \frac{M_y}{S_y} \leqq F_p \tag{b}$$

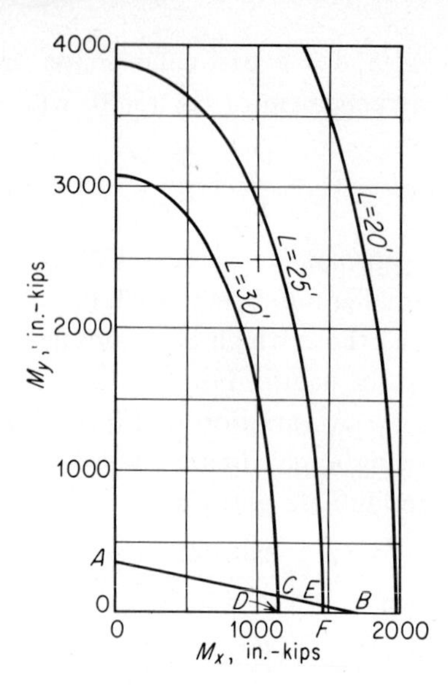

FIGURE 5-45

It is instructive to interpret Eqs. (*a*) and (*b*) graphically. Graphs of Eq. (*a*) for a steel W12 × 40 on spans of 20, 25, and 30 ft, with $E = 30{,}000$ ksi and $G = 12{,}000$ ksi, are shown in Fig. 5-45. Assuming $F_p = F_y = 36$ ksi, Eq. (*b*) gives the line *AB* in the figure. The intersection *C* of *AB* and the ellipse for the 30-ft span gives simultaneous values of M_x and M_y for which the combined bending stress is 36 ksi. No point higher than *C* on the graph for the 30-ft span is significant because it represents values of M_x and M_y that do not satisfy Eq. (*b*). No point on *AB* to the right of *C* is valid because it represents values of M_x and M_y that do not satisfy Eq. (*a*). Similarly, combinations for a 25-ft span lie on the line *AEF*, while only points on the line *AB* are valid for the 20-ft span.

Since the abscissas of points *C* and *E* are nearly equal to those for points *D* and *F*, respectively, *CD* and *EF* are for all practical purposes vertical lines. Furthermore, points *D* and *F* correspond to the case $M_y = 0$, so that they represent critical values of M_x for pure bending in the plane of the web; i.e., they are given by Eq. (5-6). Therefore, the following design procedure is justified. If the span of the beam is such that the allowable stress for bending in the plane of the web must be reduced for lateral buckling, the beam is adequate provided (1) the stress $f = M_x/S_x$ is equal to or less than the allowable stress for bending in the plane of the web and (2) the combined bending stress $f = M_x/S_x + M_y/S_y$ is equal to or less than the allowable bending stress without regard to lateral buckling. These two conditions are fulfilled by any point on the line *ACD* or the line *AEF*. On the other hand, if the span of the beam is such that the allowable stress for bending in the plane of the web need not be reduced for lateral buckling, only condition 2 applies.

The AISC specification does not cover biaxial bending specifically, except as it may be considered to be a particular case of the beam-column with $P = 0$, for which Eqs. (4-40) apply. In this case, F_{bx} and F_{by} may differ. Thus, $F_b = 0.6F_y$ (or, if the section is compact, $0.66F_y$) for major-axis bending of the I and $0.75F_y$ for minor-axis bending (Art. 5-5). If the procedure suggested above is used, F_b should be taken to be $0.6F_y$ or $0.66F_y$, whichever is applicable.

It will be recalled that lateral buckling is aggravated if vertical load acts on the top flange of a beam but relieved if it acts on the bottom flange (Art. 5-8). If lateral load acts on either flange, there is a twisting moment which is not accounted for in Eqs. (*a*) and (*b*). This twisting moment is likely to be adequately compensated for by the greater resistance to lateral buckling resulting from vertical load acting on the bottom flange, but if lateral load acts on either flange in combination with vertical load on the top flange, the effect is cumulative. Therefore, some provision must be made for the latter combination. It is quite common to make an allowance for twisting moment by assuming that the lateral load is resisted only by the flange on which it acts. For symmetrical cross sections this is equivalent to taking the section modulus for the minor axis of the I at half its actual value or to doubling the bending moment due to the lateral loads.

5-24 DP5-4: CRANE RUNWAY

An illustration of design for biaxial bending is given in this example. Except for the lateral force on the runway, the problem is similar to DP5-2. The comments which follow are intended to clarify the correspondingly lettered computations in the example.

a The estimated weight of the girder includes 20 plf for the weight of the rail.

b The lateral forces due to accelerations of the trolley and load are resisted by contact of the truck wheels with the rails, so that the position of the wheels for maximum lateral moment is the same as that for maximum moment in the vertical plane. The specified lateral force is 20 percent of the sum of the weights of the lifted load and of the crane trolley. According to the AISC specification, this force may be assumed to be divided equally between the two girders of the runway, which explains the factor 0.10 in this calculation. The lateral force gives rise to twisting moments on the crane girder, as well as bending moments in the horizontal plane, and since the vertical loads act on the top flange, some provision must be made for the twisting moment. In this example we use the approximation mentioned in Art. 5-23, namely, that only the top flange of the girder resists the lateral force.

CRANE RUNWAY — DP5-4

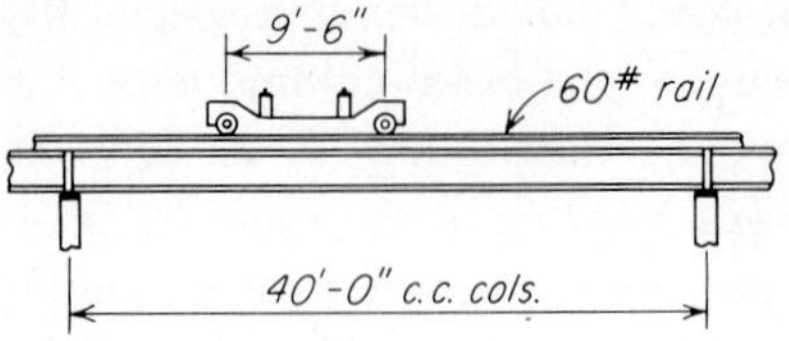

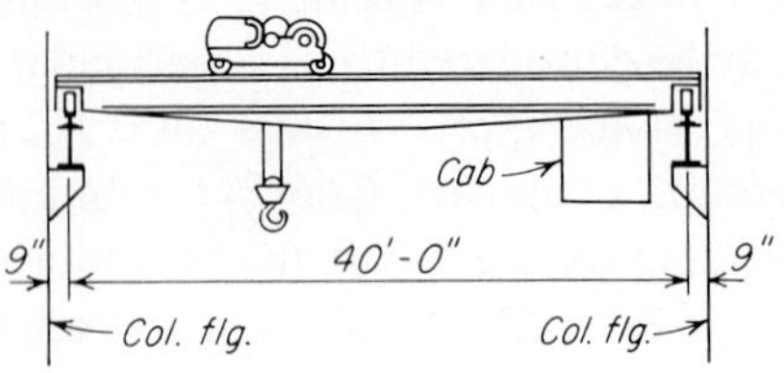

9'-6" — 60# rail — 40'-0" c.c. cols. — Cab — 9" — 40'-0" — 9" — Col. flg. — Col. flg.

Crane capacity 15 tons
Max. end-truck wheel load 26.8^k
Weight of trolley 8.0^k
AISC spec. A36 steel

Crane girder

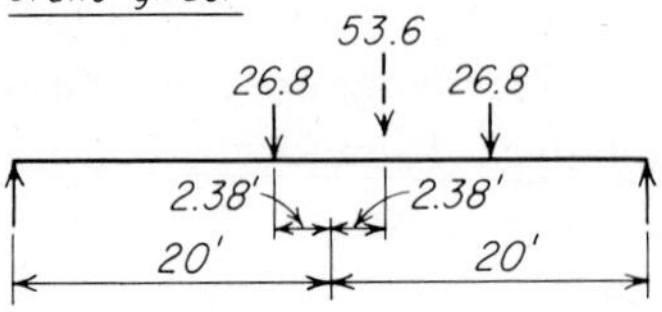

53.6 — 26.8 — 26.8 — 2.38' — 2.38' — 20' — 20'

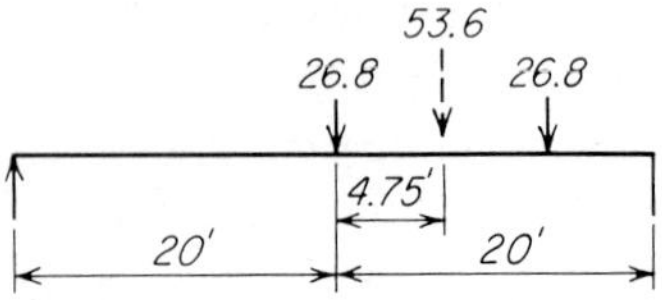

53.6 — 26.8 — 26.8 — 4.75' — 20' — 20'

$$M = \frac{53.6 \times 17.62 \times 17.62}{40} = 416'^k \qquad M = \frac{53.6 \times 15.25 \times 20}{40} = 409'^k$$

Impact 25% $104'^k$
Girder $0.275 \times 40^2/8$ $55'^k$
$575'^k = 6900''^k$ (a)

$$\text{Moment from lateral force} = 0.10 \times \frac{30+8}{53.6} \times 416'^k = 29.5'^k = 354''^k \quad (b)$$

$$S_x = \frac{6900}{24} + 3.5 \times 2 \times \frac{2 \times 354}{24} \quad (c)$$

$$= 287 + 206 = 493$$

Try W30 × 172, $S_x = 530$, $S_y = 80$, $d/A_f = 1.87$, $r_T = 3.98$ (d)

$$\frac{Ld}{A_f} = 480 \times 1.87 = 900, \quad F_{bx} = \frac{12{,}000}{900} = 13.3 \text{ ksi}$$

$$\frac{L}{r_T} = \frac{480}{3.98} = 120; \quad \sqrt{\frac{510{,}000}{36}} = 119; \quad F_{bx} = \frac{170{,}000}{120^2} = 11.8 \text{ ksi}$$

$f_{bx} = 6900/530 = 13.0$ ksi < 13.3 O.K.
$f_{by} = 2 \times 354/80 = 8.9$
21.9 ksi < 24 O.K. (e)

$$\Delta = \frac{13.0}{1000} \times \frac{40^2}{30} = 0.71'' \quad \frac{\Delta}{L} = \frac{0.71}{480} = \frac{1}{680} \quad (f)$$

DP 5-4
Crane runway.

c A crane girder 40 ft long is certain to be more than 2 ft deep. Therefore, in Eq. (5-48) we take the most favorable value of d/b for W shapes in the 24- to 36-in. range, which is 2. Furthermore, we double the moment due to the lateral force, which is equivalent to the assumption that it is resisted only by the top flange. It is assumed that the beam section will be compact, so that $F_b = 0.66 \times 36$, which is rounded to 24 ksi in the specification.

d The W30 × 172 is the lightest shape for which d/b is about 2 and whose section modulus S_x is close to the estimated required value. Values of S_x, S_y, d/A_f, and r_T are from the AISC Manual.

e The W30 × 172 satisfies the criteria for a compact section, so that $F_b = 0.66F_y = 24$ ksi. Therefore, it is adequate according to this analysis, for which the justification was given in Art. 5-23. If the AISC formulas 1.6-1 for the beam-column [Eqs. (4-40)] are used instead, it is not adequate. The two formulas become identical, since $C_m = 1$ is prescribed for this situation, and $f_a = 0$. Therefore, with $F_{bx} = 13.3$ ksi, as already determined, and $F_{by} = 0.75F_y = 27$ ksi, we get

$$\frac{f_{bx}}{F_{bx}} + \frac{f_{by}}{F_{by}} = \frac{13.0}{13.3} + \frac{8.9}{27} = 0.98 + 0.33 = 1.31$$

Thus, with this interpretation of the AISC clause for the beam-column, the W30 × 172 is inadequate; the W30 × 210 would be required instead.

f Since stiffness is an important consideration in crane girders, deflection should be checked. Equation (5-39*b*) is used here. According to Table 5-3, this value would be multiplied by 1.02 if the two wheels were at the third points of the 40-ft girder. However, the wheels are only 9 ft 6 in. apart, so that the correction factor is somewhat smaller. Dunham[18] recommends that deflection of crane girders be limited to not more than $L/1{,}000$. Therefore, a larger beam is required if this criterion is accepted.

Cranes of the type considered in this example also exert longitudinal forces on their runways. The AISC specifies a longitudinal force equal to 10 percent of the maximum wheel loads, acting at the top of the rail. In this case, the force is $0.10 \times 53.6 = 5.36$ kips. The cross-sectional area of the W30 × 172 is 50.7 in.^{2}, so that the axial-force stress is only $5{,}360/50.7 = 106$ psi.

5-25 DP5-5: EAVE STRUT

In this example we consider the analysis of the eave strut for the building of DP3-1. The strut in this case consists of one 12-in. channel with its web in the vertical plane and one 7-in. channel with its web in the horizontal plane. The 12-in. channel serves principally as a purlin, the 7-in. channel as a girt. The eave strut may be subjected to axial load resulting from the wind forces on the ends of the building (hence the name *eave strut*), to lateral load resulting from wind forces on the side of the building, and

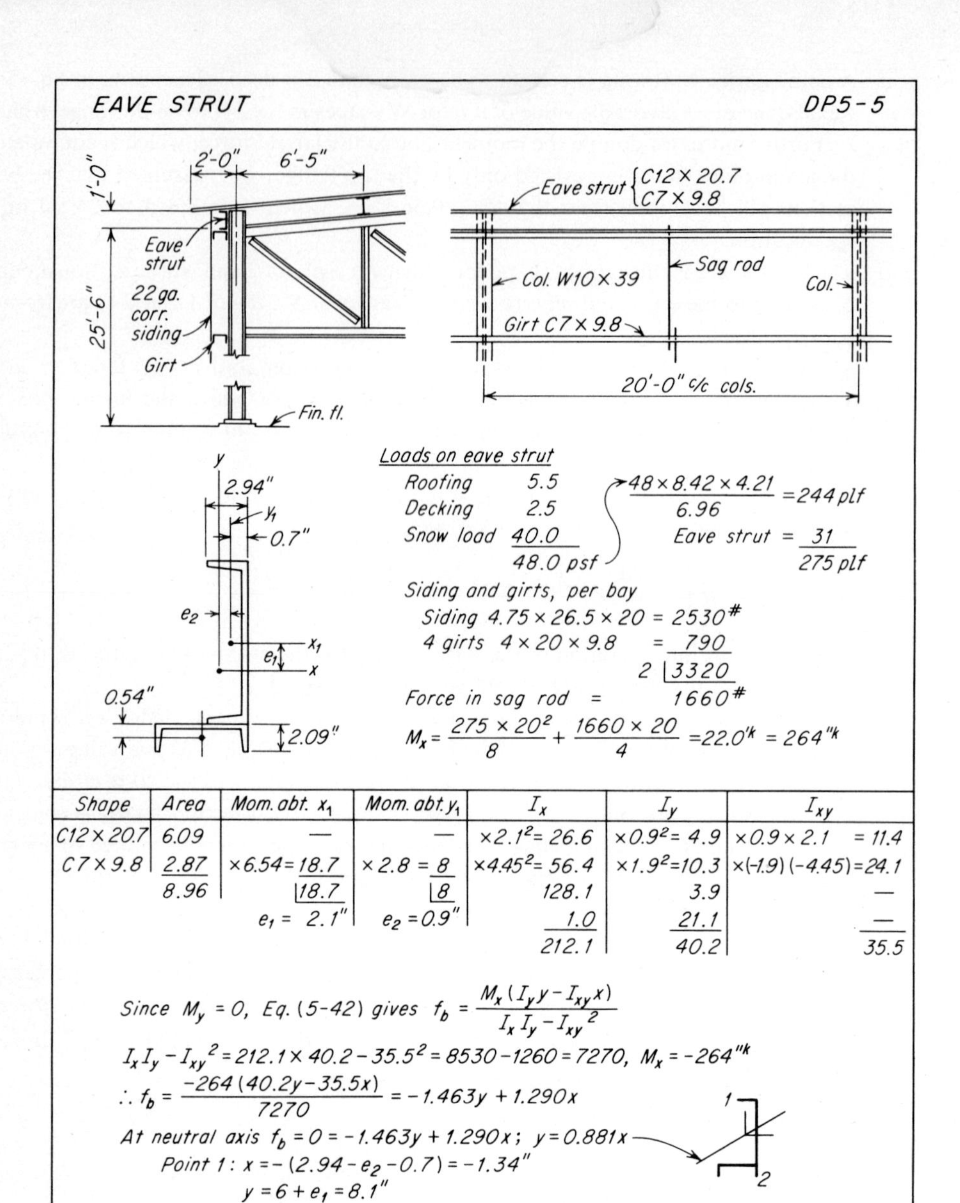

Loads on eave strut

Roofing	5.5	
Decking	2.5	
Snow load	40.0	
	48.0 psf	

$$\frac{48 \times 8.42 \times 4.21}{6.96} = 244\ plf$$

Eave strut = 31

275 plf

Siding and girts, per bay

Siding $4.75 \times 26.5 \times 20 = 2530^{\#}$

4 girts $4 \times 20 \times 9.8 = 790$

$2 \,|\, 3320$

Force in sag rod $= 1660^{\#}$

$$M_x = \frac{275 \times 20^2}{8} + \frac{1660 \times 20}{4} = 22.0'^k = 264''^k$$

Shape	Area	Mom. abt. x_1	Mom. abt. y_1	I_x	I_y	I_{xy}
C12 × 20.7	6.09	—	—	$\times 2.1^2 = 26.6$	$\times 0.9^2 = 4.9$	$\times 0.9 \times 2.1 = 11.4$
C7 × 9.8	2.87	$\times 6.54 = 18.7$	$\times 2.8 = 8$	$\times 4.45^2 = 56.4$	$\times 1.9^2 = 10.3$	$\times(-1.9)(-4.45) = 24.1$
	8.96	18.7	8	128.1	3.9	—
		$e_1 = 2.1''$	$e_2 = 0.9''$	1.0	21.1	—
				212.1	40.2	35.5

Since $M_y = 0$, Eq. (5-42) gives $f_b = \dfrac{M_x(I_y y - I_{xy} x)}{I_x I_y - I_{xy}^2}$

$I_x I_y - I_{xy}^2 = 212.1 \times 40.2 - 35.5^2 = 8530 - 1260 = 7270$, $M_x = -264''^k$

$\therefore f_b = \dfrac{-264(40.2y - 35.5x)}{7270} = -1.463y + 1.290x$

At neutral axis $f_b = 0 = -1.463y + 1.290x$; $y = 0.881x$

Point 1: $x = -(2.94 - e_2 - 0.7) = -1.34''$

$y = 6 + e_1 = 8.1''$

$f_b = -1.463 \times 8.1 + 1.290(-1.34) = -11.8 - 1.7 = -13.5$ ksi

Point 2: $x = e_2 + 0.7 = 1.6''$

$y = -(6 + 2.09 - e_2) = -6.0''$

$f_b = -1.463(-6.0) + 1.290 \times 1.6 = 8.8 + 2.1 = +10.9$ ksi

DP 5-5
Eave strut.

to vertical load resulting from snow and from the weight of roofing and siding. Only vertical load is considered in this example.

The snow load and the weights of roofing and decking are taken from DP3-1. To determine the load on the eave strut, the decking is assumed to act as a simple beam whose span is 6 ft 5 in. plus 5 in. for half the width of the column plus 1.5 in. to the center of the flange of the channel, or 6.96 ft. Windows, weighing 8 psf, occupy 50 percent of the wall area, and since the 22-gage corrugated-steel siding weighs 1.51 psf, the average weight of the wall is 4.75 psf. We assume that the sag rod fastened to the eave strut delivers a load equal to half the weight of the wall and girts in one bay.

The center of gravity of the eave strut is computed by taking moments of area about the axes x_1y_1 of the 12-in. channel. The moments of inertia I_x and I_y are determined next. The third and fourth entries in each column are the moments of inertia for the 12-in. channel and the 7-in. channel, respectively. In computing the product of inertia, we note that the product of inertia for each channel about its own principal axes is zero, so that only the term Axy in the transfer theorem contributes to I_{xy}.

According to the sign convention established in Art. 5-23 for bending (positive moment produces tension on an element of area in the first quadrant), M_x is negative. Equation (5-42) reduces to a simple form. The neutral axis is located in order to determine the points of maximum stress, following which the bending stresses are easily computed.

There are two implied assumptions in this analysis: (1) the loads are applied in such a way that no twisting moments are created, or that we can neglect any stresses resulting from twist; and (2) the eave strut is free to bend in any direction. The conditions which must be fulfilled if no twisting moments exist are discussed in Art. 5-28. With respect to the second assumption, if the decking is rigid enough and secured to the eave strut so as to prevent sidewise bending, the stresses reduce to those resulting from bending in the vertical plane, i.e.,

$$f_{b1} = \frac{-264 \times 8.1}{212} = -10.1 \text{ ksi} \qquad f_{b2} = \frac{-264(-6.0)}{212} = 7.5 \text{ ksi}$$

These values are, respectively, 25 and 31 percent less than those previously computed. Whether lateral bending actually exists in any situation depends upon the nature of the attached construction and the manner in which it is fastened to the beam. In any case, it is always conservative to assume no restraint.

Simplifying assumptions are usually made in order to arrive at the proportions of sections such as that discussed in this example. This question is considered in Art. 5-26. The analysis given in DP5-5, although it is correct if the beam is free to bend in any direction, would not be used in practice on an ordinary eave strut, even though the strut were not restrained against sidewise bending. It is important to remember, however, that an analysis which neglects sidewise bending always underestimates the bending stresses, sometimes by a wide margin.

It is of interest to show that the restraining force which must be exerted on unsymmetrical sections to prevent lateral bending under vertical load can be determined from Eq. (5-42). Bending occurs in the vertical plane only if the second term in that equation vanishes. Therefore, $M_y/M_x = I_{xy}/I_x$, which means that the ratio of the restraining force to the load is I_{xy}/I_x.

PROBLEMS

5-51 Choose a shape for the runway girder of a 25-ton crane which operates on a 40-ft bridge. The trolley weighs 10 kips. The end-truck wheels are 11 ft on centers and run on 60-lb rails. The maximum wheel load is 38.2 kips. Columns are 25 ft on centers. AISC specification, A36 steel.

5-52 Choose a shape for the runway girder of a 40-ft 20-ton crane with a 5-ton auxiliary hoist. The trolley weighs 14 kips. The end-truck wheels are 10 ft 6 in. on centers and run on 60-lb rails. The maximum wheel load is 35.2 kips. Columns are 40 ft on centers. AISC specification, A36 steel.

5-53 A strut lying in a plane normal to the longitudinal axis of a 20-ft simply supported beam intersects the centroid of the cross section at an angle of 60° with the horizontal. The force in the strut is 10 kips, and there are no other loads. Design the beam, assuming no lateral support except at the ends. AISC specification, A36 steel.

5-54 A W21 × 96 simply supported on a span of 16 ft supports at its midpoint a vertical concentrated load of 35 kips. The web of the beam is on a slope of 6 vertical in 1 horizontal. Compute the maximum bending stresses if the beam is free to bend in any direction.

5-55 What forces would the roofing exert on the eave strut of DP5-5 in order to have bending only in the vertical plane? In what direction would these forces act? If the roofing exerted a uniformly distributed force, would the eave strut bend in the vertical plane at all sections?

5-56 Eave struts are sometimes made as in Fig. 5-46*e* with an angle in place of the smaller channel in DP5-5. Assume a $6 \times 3\frac{1}{2} \times \frac{1}{4}$ angle with its long leg attached to the bottom flange of the 12-in. channel and the outside face of its short leg 6 in. from the back of the channel. What will be the bending stresses due to the loads considered in DP5-5 if the strut is (*a*) free to bend in any direction, (*b*) free to bend only in the vertical plane?

5-57 An $8 \times 6 \times \frac{1}{2}$ angle positioned with the long leg pointing vertically downward is used as a beam on a span of 12 ft. If it supports a uniform load of 400 plf and is free to bend in any direction, what is the maximum bending stress? Compare this value with the maximum stress if the angle can bend only in the vertical plane.

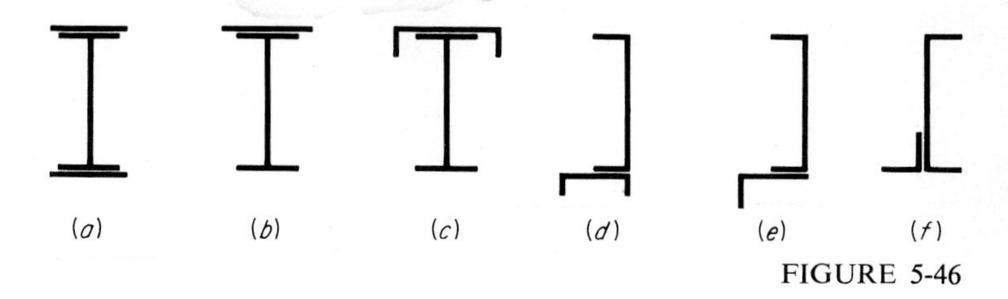

FIGURE 5-46

5-26 BUILT-UP BEAMS

Built-up sections are often required for special-purpose beams such as crane bridge beams and runway girders, spandrel beams, eave struts, lintels, and the like. Sometimes it is a question of reinforcing a shape, as in Fig. 5-46*a*, to increase its capacity to support vertical loads. Crane runway girders are sometimes made as in Fig. 5-46*b* and *c*. The reinforcement of the top flange is intended to increase resistance to lateral forces. A top flange plate may be required as in Fig. 5-46*b* to serve as a shelf, rather than reinforcement, when the section is used as a lintel. In the case of lintels and eave struts it is usually a question of devising a section to suit the requirements of supported construction, as in Fig. 5.46*d* to *f*.

Assumptions are sometimes made to simplify the design of these sections. For example, an eave strut such as that of Fig. 5-46*d* may be designed as two independent members, the vertical channel to support gravity loads and the horizontal channel to support lateral loads. Similarly, when Fig. 5-46*e* and *f* are used as lintels, the shelf angle may be ignored and the channel designed to support the entire load. A crane runway built as in Fig. 5-46*c* may be designed on the assumption that the I supports the vertical load while the channel resists the lateral force.

The required size of reinforcing plates of equal area on both flanges as in Fig. 5-46*a* can be determined very closely by noting that the moment of inertia of the compound beam is

$$I = I_i + 2A_f \left(\frac{d}{2}\right)^2 \qquad (a)$$

where I_i = moment of inertia of primary shape (the I)
A_f = area of one flange plate
d = distance between centers of gravity of flange plates

Since the distances from the neutral axis to the outside flange surface and to the center of gravity and outside surface of the flange plate are very nearly the same, we may set

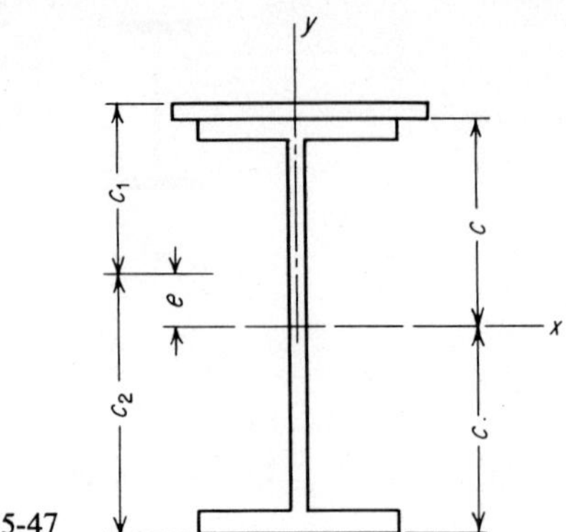

FIGURE 5-47

them all equal to $d/2$. With this assumption, division of each term of Eq. (*a*) by $d/2$ gives

$$S = S_i + A_f d \tag{5-49}$$

where S and S_i are the section moduli of the compound section and the primary shape, respectively.

Although the design of unsymmetrical built-up beams is necessarily a trial-and-error procedure, it is possible to derive useful and fairly simple formulas to obtain close approximations for sections of the type shown in Fig. 5-46*b* and *c*. In Fig. 5-47, let A_i be the area of the primary shape and A_f the area of the flange plate. The eccentricity e is found by taking moments around the x axis of the primary shape

$$A_f c = (A_f + A_i)e$$

from which

$$e = \frac{pc}{1 + p} \tag{b}$$

where $p = A_f/A_i$. Then

$$c_1 = c - e = \frac{c}{1 + p} \tag{c}$$

$$c_2 = c + e = \frac{c(1 + 2p)}{1 + p} \tag{d}$$

The moment of inertia of the compound section is

$$I = I_i + A_i e^2 + pA_i c_1{}^2 \tag{e}$$

Substituting into this equation the values of e and c_1 from Eqs. (b) and (c) and simplifying, we find

$$I = I_i + A_i \frac{pc^2}{1+p} \tag{f}$$

The section moduli $S_1 = I/c_1$ and $S_2 = I/c_2$ are found from Eqs. (c), (d), and (f) to be

$$S_1 = S_i(1+p) + A_i pc \tag{g}$$

$$S_2 = \frac{1}{1+2p}[S_i(1+p) + A_i pc] = \frac{S_1}{1+2p} \tag{h}$$

If the primary shape is an S or W, Eq. (g) can be put into a more convenient form. Substituting for S_i in Eq. (g) the value

$$\frac{I_i}{c} = \frac{A_i r_i^2}{c}$$

we get

$$S_1 = S_i + A_i p\left(\frac{r_i^2}{c} + c\right)$$

But, for I sections, $r^2 = 2c^2/3$ very nearly, so that finally

$$S_1 = S_i + \tfrac{5}{3} A_f c \tag{5-50}$$

$$S_2 = \frac{S_1}{1+2p} \tag{5-51}$$

These equations will give close approximations for sections of the type shown in Fig. 5-46b and c. For design purposes, Eqs. (5-50) and (5-51) may be written

$$A_f = 0.6\,\frac{S_1 - S_i}{c} \tag{5-52a}$$

$$A_f = 0.6\,\frac{S_2 - S_i}{c - 1.2S_2/A_i} \tag{5-52b}$$

It is important to remember that c in this equation refers to the primary shape and not to the compound section (Fig. 5-47).

EXAMPLE 5-6 A number of beams in an existing building must be strengthened because of remodeling. The beams are A36 W24 × 55, supported laterally. The increased load produces a moment of 270 ft-kips. The construction is such that reinforcement is limited to a plate welded to the bottom flange. What A36 plate can be used?

The W24 × 55 is compact in A36 steel. Therefore, $F_b = 0.66 \times 36 = 24$ ksi, and

$$S_x = \frac{270 \times 12}{24} = 135 \text{ in.}^3$$

The section modulus of the W24 is 114 in.³ The required size of the plate can be approximated by Eq. (5-52*b*):

$$A_f = 0.6 \frac{135 - 114}{23.55/2 - 1.2 \times 135/16.2} = \frac{0.6 \times 21}{11.78 - 10.00} = 7.1 \text{ in.}^2$$

The flange of the W24 is 7 in. wide. A 9-in. plate can be welded to it downhand. Therefore, try a $\frac{13}{16} \times 9$ plate, for which $A = 7.32$ in.²

The eccentricity e (Fig. 5-47) is

$$e = \frac{7.32 \times 12.19}{7.32 + 16.2} = 3.80 \text{ in.}$$

The moment of inertia of the reinforced section is

$$I_x = 1{,}340 + 16.2 \times 3.80^2 + 7.32 \times 8.39^2 = 2{,}090 \text{ in.}^3$$

$$\frac{I_x}{c} = \frac{2{,}090}{15.58} = 134 \text{ in.}^3 \qquad M = \frac{134 \times 24}{12} = 268 \text{ ft-kips}$$

This moment is about 0.7 percent less than the required value, 270 ft-kips.

PROBLEMS

5-58 Design a symmetrical cross section with welded cover plates for a laterally supported beam the maximum bending moment of which is 600 ft-kips. Because of requirements for clearance, the depth of the beam cannot exceed 26 in., and its width is limited to 10 in. AISC specification, A36 steel.

5-59 Design for the crane runway of DP5-4 a girder of the type shown in Fig. 5-46*c*.

5-60 Design a 25-ft crane-runway girder of the type shown in Fig. 5-46*c* for a 30-ton crane operating on a 40-ft bridge. The truck wheels are 12 ft on center and run on 80-lb (per yard) rails. The maximum wheel load is 46 kips. Compare the section with the required W shape without a reinforced flange. AISC specification, A36 steel.

5-61 An old building in a town which has no building code is being remodeled. A 20-ft clear opening is required in an exterior 13-in. brick wall. It is 2 ft from the top of the opening to the bottom of 3 × 14 wood joists which support a 1-in. pine subfloor with a $\frac{7}{8}$-in. hardwood finish. There are no openings above the floor. Design a lintel with a welded bottom reinforcing plate which will also serve as a shelf to support a 4-in. brick veneer.

Because of the arching action of masonry over openings, lintels need support only the

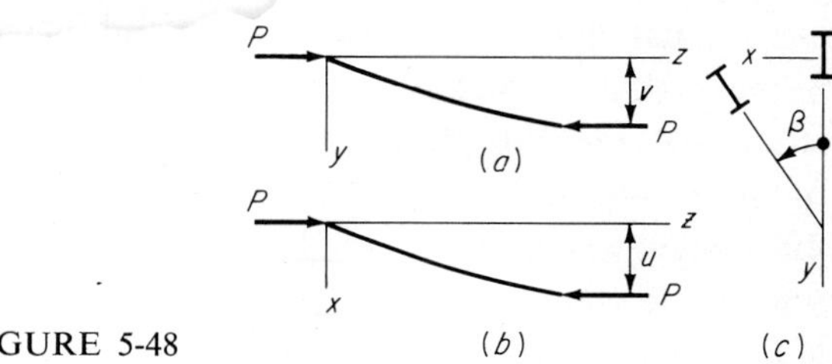

FIGURE 5-48

masonry beneath the "arch." It is usual to assume that the lintel supports a section of wall in the shape of a triangle whose altitude is half the span of the beam. However, if the bottom of the next higher opening falls within the triangle, it is better to assume that the lintel supports all the masonry between the two openings. Furthermore, if the wall is load-bearing and a floor falls within the triangle, it is safer to design the lintel to support the floor load and the masonry between it and the lintel, in addition to a triangular section of masonry above the floor.

5-27 LATERAL-TORSIONAL BUCKLING OF BEAM-COLUMNS

In Art. 5-7 we discussed the three differential equations of equilibrium for lateral-torsional buckling of beams [Eqs. (5-4)]. These equations can be extended to include the effect of a centrally applied compression force P by adding to the components of the end moments M_x the moments due to P. Figure 5-48 shows that the load P produces a moment Pv about the beam's x axis and a moment Pu about its y axis. These must be added to the right side of Eqs. (5-4*a*) and (5-4*b*), respectively. Thus,

$$-EI_x \frac{d^2v}{dz^2} = M_x + Pv \tag{5-53a}$$

$$-EI_y \frac{d^2u}{dz^2} = M_x \beta + Pu \tag{5-53b}$$

Because the resultant internal force P is not perpendicular to the cross section in Fig. 5-48, it also produces a twisting moment. It was shown in Art. 4-30 that the rate of change of this moment with respect to z is given by $fI_p\, d^2\beta/dz^2$, where $f = P/A$ [see Eq. (4-64) and Eq. (*h*) of Art. 4-28]. Therefore, the moment itself at any cross section is $fI_p\, d\beta/dz$. This must be added to the right side of Eq. (5-4*c*), which gives

$$-EC_w \frac{d^3\beta}{dz^3} + GJ \frac{d\beta}{dz} = M_x \frac{du}{dz} + \frac{P}{A} I_p \frac{d\beta}{dz} \tag{5-53c}$$

Equation (5-53*a*) is independent and, when integrated, gives the deflected shape in the yz plane, which is the plane in which the moment $M_x + Pv$ acts. On the other hand, the lateral displacement u and the twist β are not independent and must be found by solving Eqs. (5-53*b*) and (5-53*c*) simultaneously. The boundary conditions for a beam which is supported at each end ($u = 0$ and $\beta = 0$ at each end) and which also has no y-axis rotational restraint and is free to warp at each end ($d^2u/dz^2 = 0$ and $d^2\beta/dz^2 = 0$ at each end) are satisfied by

$$u = A \sin \frac{\pi z}{L} \qquad \beta = B \sin \frac{\pi z}{L}$$

If these displacement functions are solutions to the problem, they must satisfy Eqs. (5-53*b*) and (5-53*c*). Substituting into the equations gives

$$M_x^{\,2} = \frac{I_p}{A}(P_{Ey} - P)(P_T - P) \tag{a}$$

where I_p = polar moment of inertia
P_{Ey} = Euler load for y-axis buckling
P_T = twist-buckling load determined by Eq. (4-67)

Equation (*a*) can also be expressed in terms of the critical moment for pure bending, given by Eq. (5-6). Thus,

$$M_{x,cr}^2 = \frac{\pi^2}{L^2} EI_y GJ + \frac{\pi^4}{L^4} EI_y EC_w = \frac{\pi^2}{L^2} EI_y\left(GJ + \frac{\pi^2}{L^2} EC_w\right) = \frac{I_p}{A} P_{Ey} P_T \tag{b}$$

Dividing Eq. (*a*) by Eq. (*b*) gives

$$\left(\frac{M_x}{M_{x,cr}}\right)^2 = \left(1 - \frac{P}{P_{Ey}}\right)\left(1 - \frac{P}{P_T}\right) \tag{5-54}$$

Equation (5-54) is not used in design specifications. Instead, as pointed out in Art. (5-11), the interaction formulas derived for the beam which does not buckle laterally are used. These interaction formulas are based on Eq. (4-32), which is derived from the secant formula. The secant formula is the solution to Eq. (5-53*a*). Equation (4-32) is arbitrarily extended to cover lateral-torsional buckling by redefining P_{cr} as the smaller of $P_{cr,\,x}$ and $P_{cr,\,y}$ and M_y as the critical moment for lateral-torsional buckling. Thus, Eq. (4-32) becomes

$$\frac{P}{P_{cr(x \text{ or } y)}} + \frac{M}{M_{cr}} \frac{C_m}{1 - P/P_{Ex}} = 1 \tag{5-55}$$

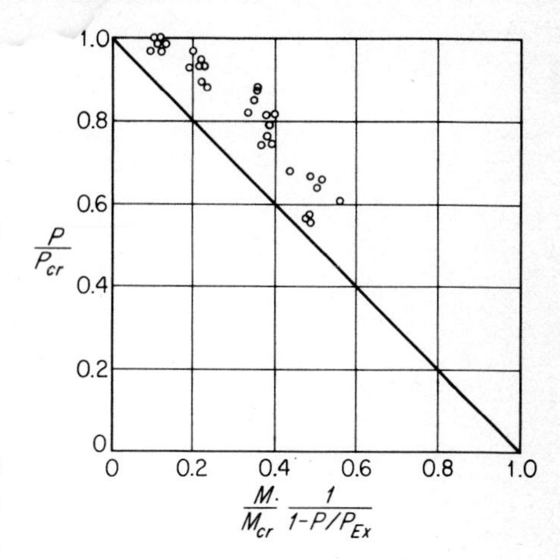

FIGURE 5-49
Comparison of Eq. (5-55) with tests on laterally unsupported aluminum beam-columns reported in Ref. 19.

The interaction formula Eq. (4-33) can be extended similarly. In this case, however, M_{cr} must approach M_p, rather than M_y, as a limiting value. The following formula is based on tests:

$$\frac{M_{cr}}{M_p} = 1.07 - \frac{\sqrt{F_y}}{3{,}160}\frac{L}{r_y} \not> 1 \tag{5-56}$$

This formula is used for plastic design in the AISC specification, except that M_{cr} is denoted by M_m.

A comparison of Eq. (5-55) with tests of 36 aluminum beam-columns that failed by lateral-torsional buckling is shown in Fig. 5-49. These tests were of the same type as those described in Art. 4-12, Fig. 4-30*d*, except that all the cross sections were of I shape.[19] The actual strengths ranged from 4 to 22 percent more than the predicted values.

5-28 THE SHEAR CENTER

If the load on a beam is in a plane parallel to a principal plane (the *yz* plane, say) and passes through the shear center, the beam bends, without twisting, about the *x* axis. The stresses are given by $f = M_x y/I_x$ and $q = V_y A\bar{y}/I_x$. For any other condition of loading there are additional stresses because of twisting. Such twisting is not a primary consideration so long as the load passes through the shear center even if it is not in a principal plane. This is because the twisting moments in this case result from eccentricities of the end reactions due to the component of deflection normal to the

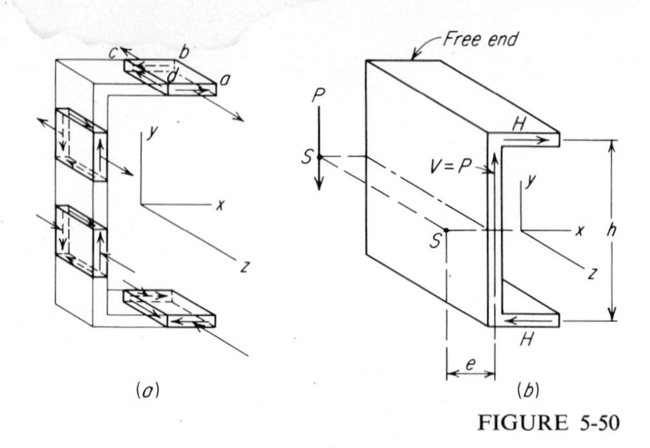

FIGURE 5-50

plane of loading. If the load does not pass through the shear center, the resulting twisting moments are more significant. However, in many cases attached construction forces the beam to bend in the plane of the loads even when that plane does not intersect the shear center. The eave strut in Art. 5-25 is an example. Shearing stresses resulting from twist and the bending stresses caused by nonuniform warping can be determined by methods discussed in Art. 4-27.

The shear center can be located by considering the z-axis moment of the shear stresses caused by bending. In Fig. 5-50*a* are shown four elements cut from the channel in Fig. 5-50*b* which supports a load *P*, parallel to the web, acting through the shear center *S*. The bending stresses increase in the direction of the positive z axis. The difference between the tensile forces acting on the element *abcd* must be balanced by a shearing force on the vertical face *cd*, since there is no force on the external vertical face *ab*. This shearing force and the resultant of the tensile forces form a couple which is balanced by a couple consisting of shearing forces on the vertical surfaces *ad* and *bc*. The same reasoning proves the existence of the shearing forces shown on the other three elements. The resultant internal shearing forces are shown in Fig. 5-50*b*. Since the horizontal shearing forces *H* form a couple, equilibrium exists only if the load *P* and the vertical shearing force $V = P$ constitute an equal and opposite couple.

The shearing forces on several other types of cross section are shown in Fig. 5-51. In all but Fig. 5-51*e* bending is assumed to be in the vertical plane. The horizontal shears for the symmetrical I in Fig. 5-51*a* form self-equilibrating couples, so that the shear center *S* coincides with the center of gravity *G*. If the bottom flange is smaller than the top flange, *S* lies above *G*, at the distance given in Fig. 4-63*f*. In the

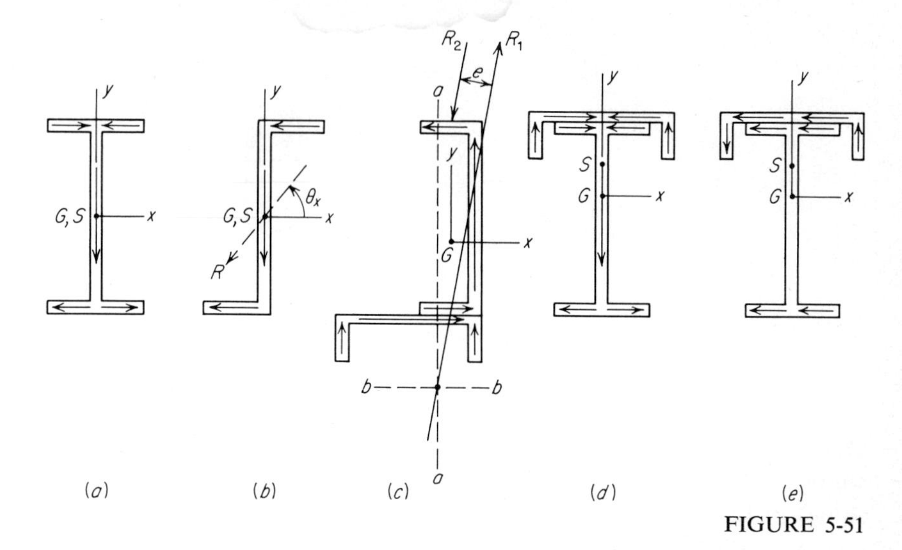

FIGURE 5-51

limiting case where the section becomes a tee, the shear center falls at the intersection of the flange and web centerlines (Fig. 4-63*c*).

The shear center of the equal-legged zee in Fig. 5-51*b* is at the centroid, since the resultant of the equal horizontal shears intersects the vertical shear at that point. The inclination θ_x of the resultant R of these shear forces can be found from Eq. (5-42) as follows. Since the cross section is bent around the x axis, the second term on the right in that equation must vanish. Thus, $M_x/M_y = I_x/I_{xy}$. But the resultant R must lie in the plane of the loading, so that

$$\tan \theta_x = \frac{M_x}{M_y} = \frac{I_x}{I_{xy}} \tag{a}$$

The shearing forces for a section consisting of an I and a channel are shown in Fig. 5-51*d* for bending in the vertical plane and in Fig. 5-51*e* for bending in the horizontal plane. The resultant of the forces in Fig. 5-51*d* is evidently a vertical force at the centerline of the web. The resultant in Fig. 5-51*e* is horizontal, but it lies nearer the top flange than the bottom. The intersection of this resultant with that of Fig. 5-51*d* is the shear center. For a section with this kind of symmetry, the position of the horizontal resultant can be determined by observing that if the girder bends sidewise without twisting, the two flanges have equal radii of curvature at the neutral (y) axis. Then since $M = EI/\rho$, the lateral moment must be resisted by the flanges in proportion to their moments of inertia. Since the shearing forces on the two flanges are in the

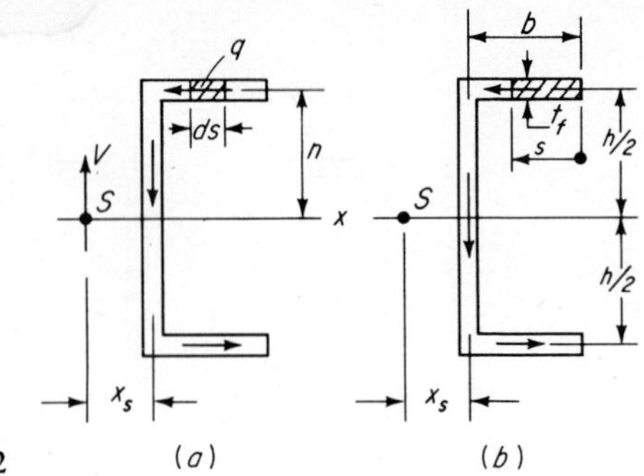

FIGURE 5-52

same ratio as the moments, and since their sum is the resultant horizontal shear, a simple equation of moments locates the resultant. The position of the shear center shown in Fig. 4-63*f* can be verified in this way. The higher shear center of the cross section in Fig. 5-51*d* as compared with that in Fig. 5-51*a* demonstrates the effectiveness of the reinforced compression flange in reducing the twisting moments from lateral forces on crane girders.

The eave strut of DP5-5 is shown in Fig. 5-51*c*. The shearing forces shown result if the strut is constrained to bend in the vertical plane. The resultant of the vertical shearing forces lies on a line *aa*, while that of the horizontal shearing forces is on *bb*. The resultant shearing force R_1 acts at the intersection of *aa* and *bb* and makes with the horizontal the angle θ_x given by Eq. (*a*). If the strut supports a gravity load on its top flange, the resultant R_2 of this load and of a lateral restraining force supplied by roof construction attached to the top flange acts on the flange. The corresponding eccentricity e is shown in the figure. The shear center for this cross section is not at the intersection of *aa* and *bb*. Its position can be determined by locating the resultant shearing force for bending in the horizontal plane and finding the intersection of this resultant with R_1.

If the cross section has an axis of symmetry, the shear center S lies on that axis, as in the channel of Fig. 5-52. If x is the axis of symmetry, the location of S is given by

$$Vx_s = \int qr\, ds \tag{5-57}$$

where x_s = distance of S from center of moments
q = shear-flow intensity
r = moment arm of q
ds = element of length

The center of moments is arbitrary and is shown here at the web centerline to eliminate the web shear in the equation of moments.

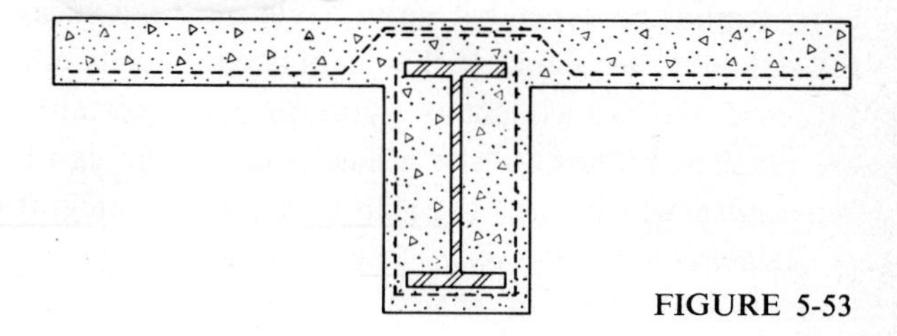

FIGURE 5-53

To evaluate x_s for the channel, we compute q at the distance s from the flange tip (Fig. 5-52*b*),

$$q = \frac{VA\bar{y}}{I_x} = \frac{V}{I_x} t_f s \frac{h}{2}$$

The moment arm r of q is $h/2$. Substituting these values into Eq. (5-57) and multiplying by 2 to account for the moment of the shear force in the opposite flange, we get

$$Vx_s = 2\int_0^b \frac{V}{I_x} t_f s \frac{h}{2}\frac{h}{2}\, ds = 2\frac{V}{I_x}\frac{t_f h^2}{4}\frac{b^2}{2}$$

$$x_s = \frac{t_f b^2 h^2}{4I_x}$$

If the cross section has no axis of symmetry, the shear center is located by evaluating Eq. (5-57) twice, once for a shear V_y and once for a shear V_x. If x, y are principal axes, the shear intensity q for V_y is given by the first term on the right in Eq. (5-41), while that for V_x is given by the second term. If x, y are not principal axes, Eq. (5-43) must be used.

PROBLEMS

5-62 Locate the shear centers for the cross section of Figs. 4-68 and 4-69.

5-63 A C15 × 33.9 spans 20 ft and supports a uniformly distributed load of 1.4 klf. The line of action of the load is centered on the top flange. If the vertical reactions are centered on the web, compute the twisting moments exerted by the channel on its supports.

5-29 COMPOSITE BEAMS

Floor construction in buildings and bridges often consists of a reinforced-concrete slab supported on steel beams. In the preceding discussions in this chapter, it has been assumed that the beams act independently of the floor slab. This is because a natural bond cannot be depended upon to develop the shear VQ/I on the interface between the slab and the beam. If the beam is completely encased in concrete, however, as in Fig. 5-53, the two will act as a unit, provided certain conditions (to be discussed later) are satisfied. Alternatively, a mechanical bond can be established by means of *shear connectors*, which force the slab to act as an integral part of the beam. Such beams are called *composite beams*. They offer advantages in the form of possible reductions in depth, reductions of 20 to 30 percent in the weight of the steel beam, and increased stiffness of the floor system. (In the elastic range, composite beams are two to three times as stiff as noncomposite beams.)

Two types of construction are used. Commonly, formwork for the concrete slab is supported on the steel beams, so that the weight of the concrete is carried by the steel beam alone. However, if the steel beam is supported by shores that are not removed until the concrete has attained sufficient strength, both dead load and live load are resisted by composite action.

5-30 DESIGN OF COMPOSITE BEAMS

A composite floor is assumed to consist of a series of T beams. The beams are analyzed for elastic behavior by transforming the effective cross-sectional area of the concrete slab into an equivalent area of steel. For concrete weighing 145 pcf, $E_c = 57{,}400\sqrt{f_c'}$, where f_c' is in pounds per square inch, so that $n = 500/\sqrt{f_c'}$. Alternatively, n can be determined by using $E_c = 1{,}000f_c'$. The width of slab assumed to be effective as flange varies with the specification. The AISC effective flange widths are the same as those of ACI Committee 315 for buildings, as follows.

The effective flange width for an interior beam is the smallest of:

1 One-fourth the span of the beam
2 The distance center to center of beams
3 Sixteen times the least thickness of the slab, plus the width of the top flange of the steel section

In the case of an edge beam, the overhanging width of flange is the smallest of:

1 One-twelfth the span of the beam
2 One-half the clear distance to the adjacent beam
3 Six times the thickness of the slab

Effective widths of flanges of composite beams in bridges are given in Art. 10-13.

Encased beams Since there can be no local or lateral buckling of an encased steel beam, it can be treated as a compact section, for which $F_b = 0.66F_y$ in the AISC

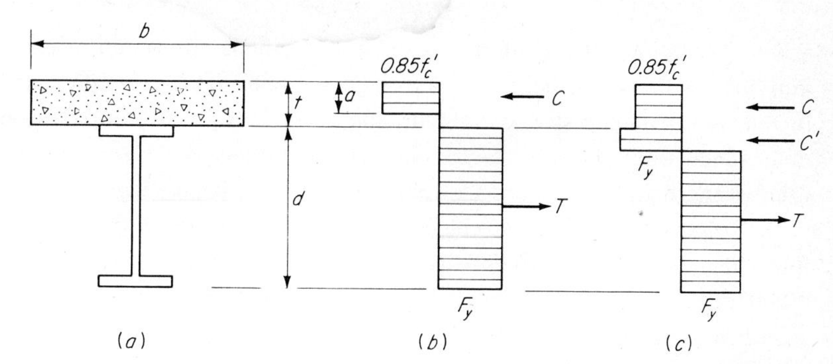

FIGURE 5-54

specification. In unshored beams the steel beam must be proportioned to support all dead load acting on the structure prior to hardening of the concrete, while live load and any dead load applied after hardening of the concrete is assumed to be supported by composite action. The AISC specification allows a simplified analysis, in which composite action is taken into account approximately by proportioning the steel beam alone to support all the load at an allowable stress $F_b = 0.76F_y$. This is conservative enough to cover the case of unshored construction.

In order for the encasement to be fully effective, the concrete cover of the steel section must be at least 2 in. thick and reinforced with wire mesh (Fig. 5-53). The AISC specification requires the top of the steel beam to be at least $1\frac{1}{2}$ in. below the top of the slab and 2 in. above the bottom.

Beams with shear connectors Most composite beams are built with shear connectors. The stress distribution at the ultimate bending strength of such a beam is shown in Fig. 5-54. The neutral axis may be in the slab, as in *b*, or in the steel beam, as in *c*. In either case, the moment is easily determined. Thus, for the case shown in *b*,

$$C = 0.85f_c'ba \qquad T = F_y A_s \tag{5-58}$$

where *a* and *b* are defined in the figure and A_s is the area of the steel section. Since $C = T$ and f_c', b, F_y, and A_s are known, *a* can be computed to determine the moment arm of the couple. For the case shown in *c*,

$$C_1 = 0.85f_c'bt \qquad T = F_y A_s - C_2 \qquad T = C_1 + C_2 \tag{5-59a}$$

Eliminating T from the second and third of these equations and substituting into the result the value of C_1 from the first gives

$$2C_2 = A_s F_y - 0.85f_c'bt \tag{5-59b}$$

Since everything on the right is known, C_2 can be computed, after which T and C_1 and the resulting moment are easily determined. Thus, whether the neutral axis is in the slab or in the steel beam, the adequacy of a given composite section is readily determined once the load factor has been established. A load factor of 2, which is used in the British Standard Code of Practice, has been suggested.[20]

Tests show that the strength of a composite beam with shear connectors is not affected by the manner of construction. Thus, Eqs. (5-58) and (5-59) can be used for either type. In unshored construction, however, this could result in overstress of the steel beam during construction or at service loads. Therefore, limiting values of these stresses must be established.

Composite design by the AISC and AASHO specifications is based on elastic analysis, using the transformed section. In the AISC specification, the allowable stress in the steel section is $0.66F_y$ if the section is compact and $0.6F_y$ if it is not. Lateral-torsional buckling need not be considered for the completed structure, but it must be guarded against during construction. The allowable stress in the concrete is the same as for reinforced-concrete beams, $0.45f'_c$. However, the neutral axis is close to the top of the section, so that the stress in the steel is usually the determining factor. Because of this, a lighter (but not necessarily cheaper) design can be achieved by using a cover plate on the bottom flange.

To design a composite beam, we compute the required transformed-section modulus, based on the allowable stress in the steel, to determine the cross section needed. This procedure is facilitated by tables in the AISC Manual and other sources.[20, 21] To protect the steel beam in unshored construction against the possible overstress mentioned above, the following two precautionary checks are recommended[22] (and required by the AISC specification): (1) to protect the steel beam against overstress during (unshored) construction, the stress in the steel beam acting alone should not exceed the AISC allowable value before the concrete has attained 75 percent of its 28-day strength, and (2) the section modulus used in calculations should not exceed the value

$$S_{tc} = \left(1.35 + 0.35\frac{M_L}{M_D}\right) S_{ts} \tag{5-60}$$

where S_{tc}, S_{ts} = section moduli of tension flange for transformed section and steel beam, respectively

M_D = moment due to dead load applied before concrete has attained 75 percent of its 28-day strength

M_L = moment due to loads applied afterward

Equation (5-60) is a device for limiting the service-load stress on the steel beam in unshored construction to not more than 1.35 times the standard allowable stress for the steel beam alone. This can be shown as follows. The stress on the tension flange of the steel beam, assuming composite action for the total load, is

$$f_{t1} = \frac{M_D + M_L}{S_{tc}} \tag{a}$$

However, because of the absence of shores during construction, the true stress is

$$f_{t2} = \frac{M_D}{S_{ts}} + \frac{M_L}{S_{tc}} \tag{b}$$

Setting f_{t2} in Eq. (*b*) equal to $1.35 f_{t1}$ and then eliminating f_{t1} between Eqs. (*a*) and (*b*) gives Eq. (5-60). Therefore, if Eq. (5-60) is satisfied, the tension-flange steel stress in a composite beam built without shores will not exceed $1.35 \times 0.60F_y = 0.81F_y$ for a noncompact steel section or $1.35 \times 0.66F_y = 0.89F_y$ for a compact steel section. [It would seem simpler to use Eq. (*b*) itself, i.e., to compute f_{t2} and compare it with an allowable value $1.35F_b$, where F_b is the nominal allowable bending stress for the steel beam.]

Shear in composite beams is assumed to be resisted by the steel beam alone. Deflection is computed by the usual formulas. Because of creep of the concrete, however, the long-term deflection will be larger than the predicted value based on the properties of the transformed section used to evaluate stresses. A good estimate of long-term deflection can be made by using a transformed section based on double the value of n used to compute stresses.[20]

EXAMPLE 5-7 Figure 5-55*a* shows the layout of beams and girders for a typical warehouse floor. The live load is 200 psf. The slab thickness is 4 in., concrete strength f_c' is 3,000 psi, and steel is A36. Determine the steel beam required according to the AISC specification. Assume the beam to be simply supported. Construction is to be without shores.

The allowable stresses and the modular ratio are $F_b = 24$ ksi (assuming the section will be compact), $f_c = 0.45f_c' = 1.35$ ksi, $n = 500/\sqrt{3{,}000} = 9$. The dead-load and live-load moments are:

Load		Weight, klf	Moment
Dead:	Slab = 7.5 × 150 × 4/12 =	0.375	
	Beam (estimated)	0.050	
		0.425 × 30² × 12/8	575
Live:	7.5 × 200	= 1.500 × 30² × 12/8 =	2,025
		1.925 klf	2,600 in.-kips

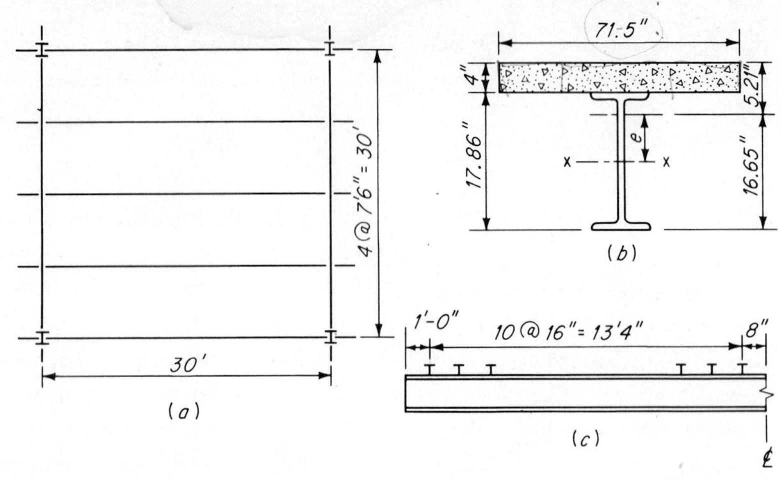

FIGURE 5-55

A good estimate of the required area A_s of the steel beam, based on the ultimate moment M_u according to Fig. 5-54, is given by

$$A_s = \frac{M_u}{(0.5d + 0.8t)F_y}$$

where d is the depth of the steel beam and t is the slab thickness.[20] The AISC allowable-stress design gives a factor of safety of about 2.2 on ultimate strength. Therefore, assuming a 16-in. beam,

$$A_s = \frac{2.2 \times 2{,}600}{(0.5 \times 16 + 0.8 \times 4)(36)} = 14.2 \text{ in.}^2$$

A W16 × 50 gives 14.7 in.^{2} Assuming an 18-in. beam,

$$A_s = \frac{2.2 \times 2{,}600}{(0.5 \times 18 + 0.8 \times 4)(36)} = 13.1 \text{ in.}^2$$

A W18 × 45 gives 13.2 in.^{2} The lightest 21-in. W weighs 44 plf. This would give a slightly lighter beam but at a cost of 3 in. of headroom. Therefore, the W18 × 45 appears to be a good choice. It is a compact section, as was assumed.

The effective flange width is the smallest of (*a*) $L/4 = 30/4 = 7.5$ ft, (*b*) spacing of beams (=7.5 ft), and (*c*) $2 \times 8t + b_f = 2 \times 8 \times 4 + 7.5 = 71.5$ in. The transformed effective flange width is $b/n = 71.5/9 = 7.94$ in. The properties of the transformed section (Fig. 5-55*b*) are:

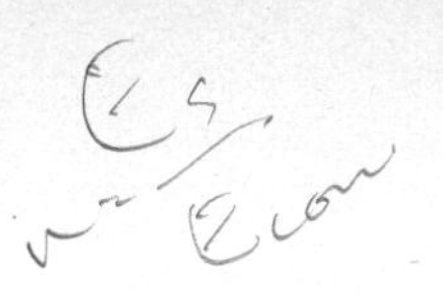

Section	Area	Moment about x	I_x	
W18 × 45 =	13.2		× 7.72² =	788
			I_0 =	706
4 × 7.94 =	31.8	× 10.93 = +348	31.8 × 3.21² =	328
	45.0	+348	7.94 × 4³/12 =	42
		e = +7.72 in.		1,864 in.⁴

The tension-flange moduli are 79.0 in.³ for the W18 × 45 and 1,864/(7.72 + 8.93) = 112 in.³ for the transformed composite section. Since construction is to be without shores, however, the latter value must be compared with the value from Eq. (5-60) and the smaller of the two used to compute the bottom-fiber stress. The equation gives

$$S_{tc} = \left(1.35 + 0.35\,\frac{2{,}025}{575}\right)79.0 = 204 \text{ in.}^3$$

Therefore, $S_{tc} = 112$ in.³ The bending stresses are

$$\text{Bottom fiber:} \qquad f_t = \frac{M}{S_{tc}} = \frac{2{,}600}{112} = 23.2 < 24 \text{ ksi}$$

$$\text{Top of slab:} \qquad f_c = \frac{Mc}{nI} = \frac{2{,}600 \times 5.21}{9 \times 1{,}864} = 0.81 < 1.35 \text{ ksi}$$

The stress in the unshored steel beam due to construction dead load is

$$f = \frac{M_D}{S_s} = \frac{575}{79} = 7.3 < 24 \text{ ksi}$$

The deflections are given by $\Delta = 5ML^2/48EI$. With L in feet this gives

$$\Delta = 5M(12L)^2/48EI = 15ML^2/EI.$$

Therefore,

$$\Delta = \frac{15 \times 575 \times 30^2}{30{,}000 \times 706} = 0.367 \text{ in.} \qquad \text{dead load}$$

$$\Delta = \frac{15 \times 2{,}025 \times 30^2}{30{,}000 \times 1{,}864} = 0.490 \text{ in.} \qquad \text{live load}$$

The live-load deflection is well within the limiting value $L/360 = 30 \times 12/360 = 1$ in. There is no allowance for creep in this calculation, however, and if the live load is likely to be sustained for long periods of time, as in a warehouse, $n = 2 \times 9 = 18$ should be used to compute the moment of inertia.

The ultimate resisting moment of the beam designed above will now be determined. This is for illustrative purposes, since it is not required by the AISC specification.

Assuming the neutral axis to be in the slab, Eq. (5-58) gives

$$C = T = F_y A_s = 36 \times 13.2 = 475 \text{ kips}$$

$$a = \frac{C}{0.85 f'_c b} = \frac{475}{0.85 \times 3 \times 71.5} = 2.60 \text{ in.}$$

Since $a < t$, the neutral axis is in the slab, as was assumed. The moment arm of the resisting couple is

$$\frac{d}{2} + t - \frac{a}{2} = \frac{17.86}{2} + 4 - \frac{2.60}{2} = 11.63 \text{ in.}$$

Therefore, $M = 475 \times 11.63 = 5{,}520$ in.-kips. Since the maximum service-load moment is 2,600 in.-kips, the factor of safety is $5{,}520/2{,}600 = 2.12$.

An example of the design of a composite-beam highway bridge is given in Chap. 10.

5-31 SHEAR CONNECTORS

The stud shear connector is a short length of round steel bar with a round head to provide anchorage. The diameter of the head is $\frac{1}{2}$ in. larger than that of the stud. The other end of the stud is welded to the beam flange. The usual diameters are $\frac{1}{2}$, $\frac{5}{8}$, $\frac{3}{4}$, and $\frac{7}{8}$ in., and the most commonly used lengths are 3 and 4 in. Longer studs are needed if the slab is haunched over the beam, however, since connectors should penetrate not less than 2 in. into the slab. Furthermore, the concrete cover on connectors should be at least 1 in. in any direction.

The channel shear connector is a short length of rolled channel with one flange welded to the beam and the other providing an anchorage for the slab. Channels 3 in. at 4.1 lb and 4 in. at 5.4 lb are most commonly used. These are usually welded with continuous fillet welds in front and back of the channel.

The function of the shear connector is to transfer the horizontal shear at the slab-beam interface, so that the required spacing of connectors at any cross section is determined by dividing the shear VQ/I at that section by the resistance of one connector (two if they are in pairs, etc.). This suggests that the spacing should vary along the span continuously with V. For practical purposes, however, it is changed at larger intervals and held constant within the interval. Where fatigue must be considered, as in bridges, connector spacing must be based on the appropriate fatigue strength of the connector and the expected range of the shear V. This is discussed in Chap. 10.

The static strength q_u of a stud shear connector is given by

$$q_u = 930 d_s^{\,2} \sqrt{f'_{c,\text{psi}}} \tag{5-61a}$$

where d_s is the diameter of the stud.[23] For a channel connector

$$q_u = 550(h + 0.5t)w\sqrt{f'_{c,\text{psi}}} \tag{5-61b}$$

where h = average thickness of flange
t = web thickness
w = length of channel[23]

These formulas are based on results of tests. Equation (5-61a) should not be used for studs shorter than $4d_s$.

Tests have shown that the ultimate strength of a composite beam under static load is largely independent of the spacing of the shear connectors. The strength of beams in which the connectors were spaced uniformly between the points of maximum moment and zero moment was the same as for beams with the same number of connectors spaced according to the intensity of the shear.[23] This is because deformation of the concrete and the more heavily stressed shear connectors redistributes the horizontal shear among the less heavily stressed connectors. This action is analogous to that which takes place in connections having a number of fasteners in line (Art. 3-11).

The number of connectors required between the point of maximum moment and the point of zero moment is determined by dividing the compression-flange force at the point of maximum moment by the connector strength. If the neutral axis is in the slab (Fig. 5-54b), the flange compression C is easily found by determining the equal force $T = A_s F_y$. If the neutral axis is in the steel beam (Fig. 5-54c), the flange compression is $C_1 = 0.85 f'_c bt$. This procedure is prescribed by the AISC specification, except that the flange compression is reduced to a service-load value by dividing by a factor of safety of 2, so that the horizontal shear is given by

$$V_h = \begin{cases} \dfrac{A_s F_y}{2} & \text{neutral axis in slab} \quad (5\text{-}62a) \\[2ex] \dfrac{0.85 f'_c bt}{2} & \text{neutral axis in beam} \quad (5\text{-}62b) \end{cases}$$

The corresponding allowable shear q per connector is given in Table 5-5. These values are based on Eqs. (5-62), using a factor of safety of 2.5.

If a composite section is proportioned on the basis of its bending strength (Fig. 5-54), the location of the neutral axis is known, so that the applicable Eq. (5-62) is also known. But proportioning is on an allowable-stress basis in the AISC specification, so that the location of the neutral axis at ultimate load is not known. However, instead of determining its position, it is simpler to compute the connector requirement for both of Eqs. (5-62), since it turns out that the smaller value is the correct one. This can be proved by noting in Fig. 5-54 that T is less than C_1 if the neutral axis is in the slab, but C_1 is less than T if the neutral axis is in the beam. The second statement is true because C_1 is less than T_1, which itself is less than T. Therefore, the correct number of connectors is always given by the smaller value of V_h by Eqs. (5-62).

In cases where standard sizes result in a larger cross section than is needed, some economy can be achieved by using fewer connectors than the number required for full composite action. Tests have shown that the bending strength in the case of partially developed composite action can be determined by linear interpolation between the case of no composite action (the steel beam alone) and the case of complete composite action.[23] Therefore, denoting the required transformed section modulus by S_{eff}, the corresponding horizontal shear V_h' is given by

$$\frac{V_h'}{V_h} = \frac{S_{\text{eff}} - S_s}{S_{tc} - S_s} \tag{5-63}$$

EXAMPLE 5-8 Determine the shear-connector requirement for the beam of Art. 5-30. From Eqs. (5-62)

$$V_h = \frac{13.2 \times 36}{2} = 238 \text{ kips} \qquad V_h = \frac{0.85 \times 3 \times 71.5 \times 4}{2} = 365 \text{ kips}$$

Table 5-5 AISC ALLOWABLE LOADS FOR SHEAR CONNECTORS

Type and size of connectors*	Allowable load q, kips		
	$f'_c = 3$ ksi	$f'_c = 3.5$ ksi	$f'_c = 4$ ksi
$\frac{1}{2}$-in. stud × 2 in.	5.1	5.5	5.9
$\frac{5}{8}$-in. stud × 2.5 in.	7.9	8.6	9.2
$\frac{3}{4}$-in. stud × 3 in.	11.5	12.5	13.3
$\frac{7}{8}$-in. stud × 3.5 in.	15.6	16.8	18.0
3-in. channel 4.1 lb	$4.3w$	$4.7w$	$5.0w$
4-in. channel 5.4 lb	$4.6w$	$5.0w$	$5.3w$
5-in. channel 6.7 lb	$4.9w$	$5.3w$	$5.6w$

* Studs longer than those shown have the same allowable loads.

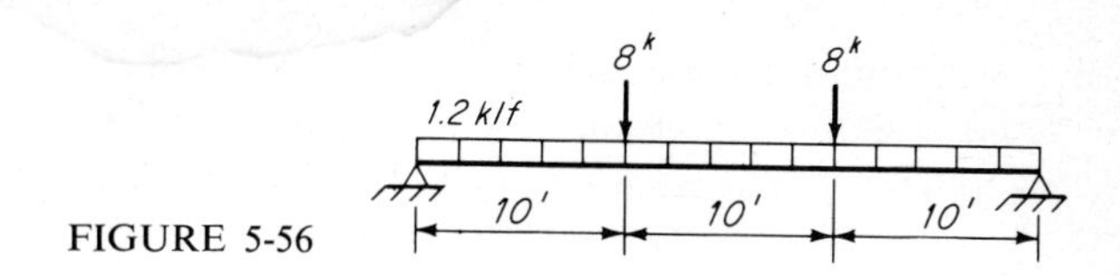

FIGURE 5-56

Therefore, connectors must be provided to develop 238 kips. Using $\frac{3}{4}$-in. studs, the allowable shear per connector is 11.5 kips (Table 5-5). Therefore, $n = 238/11.5 = 21$ for each half of the beam. Using two studs at 16-in. intervals uniformly along the beam gives 22 for each half (Fig. 5-55*c*).

Since the beam has some excess bending capacity, Eq. (5-63) will be used to see what saving in connectors is possible. The required section modulus S_{eff} is $M/F_b = 2{,}600/24 = 108$ in.^{3} The modulus S_{tc} furnished is 112 in.^{3}. Therefore,

$$\frac{V'_h}{238} = \frac{108 - 79}{112 - 79} \qquad V'_h = 209 \text{ kips}$$

The corresponding connector requirement is $n = 209/11.5 = 18$. These can be furnished by connectors in pairs at equal 20-in. spaces along the beam.

Concentrated loads The abrupt changes in shear in beams with concentrated loads suggest a modification of the uniform connector spacing discussed above. This is because the moment at concentrated loads near midspan may be almost as large as the midspan moment, so that the difference in flange compression at these cross sections may be small. Thus, relatively few of the required connectors are needed in this portion of the span. Since the distribution of the connectors is proportional to the horizontal shears, it can be determined by Eq. (5-63) by defining S_{tc} and S_{eff} as the section moduli required by the maximum moment and the moment at the concentrated load, respectively.

EXAMPLE 5-9 The cross section of the beam of Fig. 5-56 is the same as that of the beam in Art. 5-30 (Fig. 5-55*b*). Determine the connector requirements.

The bending moment and the required section modulus at midspan are

$$M = 8 \times 10 + \frac{1.2 \times 30^2}{8} = 215 \text{ ft-kips} \qquad S = \frac{215 \times 12}{24} = 107.5 \text{ in.}^3$$

The moment and required section modulus at the 10-kip load are

$$M = 8 \times 10 + \frac{1.2 \times 10 \times 20}{2} = 200 \text{ ft-kips} \qquad S = \frac{200 \times 12}{24} = 100 \text{ in.}^3$$

From Example 5-7, $S_{tc} = 112$ in.^{3} and $S_s = 79$ in.^{3} The number of $\frac{3}{4}$-in. connectors for full composite action is $n = 21$. This number can be reduced because the full bending capacity of the beam is not needed. From Eq. (5-63)

$$\frac{n}{21} = \frac{V_h'}{V_h} = \frac{107.5 - 79}{112 - 79} = \frac{28.5}{33} \qquad n = 18$$

Thus, 18 connectors each side of midspan are required. To determine the number of these which should be put in the outer 10-ft segment of the beam, Eq. (5-63) is used again, this time with the required section moduli at midspan and at the concentrated load:

$$\frac{n}{18} = \frac{100 - 79}{107.5 - 79} = \frac{21}{28.5} \qquad n = 13.3$$

Therefore, 14 of the 18 connectors should be located between the end of the beam and the 10-kip load, with the remaining four between the 10-kip load and midspan.

The extension of this procedure to the case of additional concentrated loads involves similar successive applications of Eq. (5-63).

5-32 CONTINUOUS COMPOSITE BEAMS

The composite beams of the floor shown in Fig. 5-55*a* were designed in Art. 5-30 as simply supported beams. If the slab is cast continuously over the girders to which the beams connect, deflection of the beams will cause cracking of the concrete over the girder unless reinforcement or an expansion joint is used. This suggests that it may pay to make the composite beam continuous at the supports. Tests show that negative reinforcement in the slab acts compositely with the steel beam if shear connectors are used. The composite behavior results from the bond of the reinforcing bars to the slab and thence through the connectors to the steel beam. All the reinforcement within the effective width can be considered effective, provided it is anchored beyond the points of inflection as is required for reinforced-concrete beams.

Since the shear connectors are the link between the reinforcing bars and the steel beam, they must be capable of developing a shear equal to the tensile force in the bars within the effective width. Therefore, the number of connectors is obtained by dividing $F_{yr} A_r$, where A_r is the total area of reinforcing, by the strength q_u of one connector. In the AISC specification, this is put on a service-load basis by dividing $F_y A_r/2$ by the allowable shear from Table 5-5.

PROBLEMS

5-64 Design the composite beam of Art. 5-30 using a steel beam with $F_y = 50$ ksi.

5-65 Design the composite beam of Art. 5-30 for continuous construction, with $F_y = 36$ ksi, $f_c' = 3{,}000$ psi, and $F_{yr} = 40$ ksi.

5-66 A floor system with steel beams 8 ft on centers spans 40 ft, simply supported. The 4-in. reinforced-concrete slab is made of 4,000-psi concrete. The live load is 125 psf. Design a composite beam in A36 steel for a load factor of 2 on ultimate strength.

REFERENCES

1 Haaijer, G.: Plate Buckling in the Strain-hardening Range, *Trans. ASCE*, vol. 124, 1959.

2 Haaijer, G., and B. Thürlimann: Inelastic Buckling in Steel, *Trans. ASCE*, vol. 125, 1960.

3 Thürlimann, B.: New Aspects Concerning Elastic Instability of Steel Structures, *Trans. ASCE*, vol. 127, 1962.

4 Lay, M.: Flange Local Buckling in Wide-flange Shapes, *J. Struct. Div. ASCE*, December 1965.

5 Clark, J. W.: Aluminum Structures, sec. 10 in E. H. Gaylord and C. N. Gaylord (eds.), "Structural Engineering Handbook," McGraw-Hill, New York, 1968.

6 Suggested Specifications for Structures of Aluminum Alloys 6061-T6 and 6062-T6, Report of Task Committee on Lightweight Alloys, *J. Struct. Div. ASCE*, December 1962.

7 Clark, J. W., and H. N. Hill: Lateral Buckling of Beams, *J. Struct. Div. ASCE*, July 1960.

8 Hechtman, R. A., J. S. Hattrup, E. F. Styer, and J. L. Tiedemann: Lateral Buckling of Rolled Steel Beams, *Proc. ASCE*, vol. 81, pap. 797, 1955.

9 Winter, G.: Lateral Stability of Unsymmetrical I-beams and Trusses in Bending, *Trans. ASCE*, vol. 108, p. 247, 1943.

10 Hill, H. N.: Lateral Stability of Unsymmetrical I-beams, *J. Aeronaut. Sci.*, vol. 9, no. 5, March 1942.

11 Timoshenko, S., and J. M. Gere: "Theory of Elastic Stability," 3d ed., McGraw-Hill, New York, 1969.

12 Vincent, G. S.: Tentative Criteria for Load Factor Design of Steel Highway Bridges, *AISI Bull.* 15, March 1969.

13 Lyse, I., and H. J. Godfrey: Investigation of Web Buckling in Steel Beams, *Trans. ASCE*, vol. 100, 1935.

14 Basler, K.: Strength of Plate Girders in Shear, *Trans. ASCE*, vol. 128, pt. II, 1963.

15 Timoshenko, S. P., and J. N. Goodier: "Theory of Elasticity," 3d ed., p. 119, McGraw-Hill, New York, 1970.

16 Masters, F. M., and J. R. Geise: "Findings in Second Narrows Bridge Collapse at Vancouver," *Civ. Eng.*, February 1969.

17 Hrennikoff, A.: "Lessons of Collapse of Vancouver Second Narrows Bridge," *Trans. ASCE*, vol. 126, 1961.

18 Dunham, C. W.: "Planning Industrial Structures," McGraw-Hill, New York, 1948.

19 Hill, H. N., E. C. Hartmann, and John Clark: Design of Aluminum Alloy Beam-Columns, *Trans. ASCE*, vol. 121, p. 1, 1956.

20 Viest, I. M.: Composite Construction, sec. 14 in E. H. Gaylord and C. N. Gaylord (eds.), "Structural Engineering Handbook," McGraw-Hill, New York, 1968.

21 "Properties of Composite Sections for Buildings," Bethlehem Steel Corporation, 1962.

22 Tentative Recommendations for the Design and Construction of Composite Beams and Girders for Buildings, Progress Report of Joint ASCE-ACI Committee on Composite Construction, *J. Struct. Div. ASCE*, December 1960.

23 Slutter, R. G., and G. C. Driscoll: Flexural Strength of Steel-Concrete Composite Beams, *J. Struct. Div. ASCE*, April 1965.

24 Bower, J. E.: Design of Beams with Web Openings, *J. Struct. Div. ASCE*, March 1968.

25 Segner, E. P., Jr.: Reinforcement Requirement for Girder Web Openings, *J. Struct. Div. ASCE*, June 1964.

26 Frost, R. W., and R. E. Leffler: Fatigue Tests of Beams with Rectangular Web Holes, *J. Struct. Div. ASCE*, February 1971.

27 Galambos, T. V.: "Structural Members and Frames," Prentice-Hall, Inc., Englewood Cliffs, N.J., 1968.

6

PLATE GIRDERS

6-1 INTRODUCTION

In its simplest form the plate girder is a built-up beam consisting of two flange plates welded to a web plate to form an I (Fig. 6-1*a*). Box girders (Fig. 6-1*b*) are also used. Prior to the development of modern welding techniques, plate girders usually consisted of four angles riveted to the web as in Fig. 6-1*c*. Plates, called cover plates, were riveted to the angles to increase the flange area (Fig. 6-1*d*).

Plate girders (or trusses) are used in buildings where long spans are needed over large assembly areas such as auditoriums. They are also used extensively in bridges for spans ranging from 60 to 1,000 ft. The principal differences between the design of a rolled beam and the design of a plate girder are that the designer has greater freedom in proportioning the cross section of a plate girder and that the larger depth of the plate girder often results in relatively thin webs which make web-buckling problems more important.

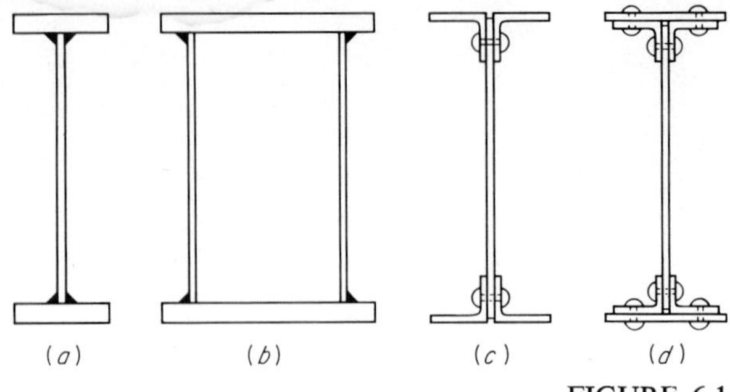

FIGURE 6-1

6-2 BEND BUCKLING OF PLATE-GIRDER WEBS

Bend buckling of webs is discussed in Art. 5-17. The extreme-fiber bending stress at which a perfectly flat web buckles is given by Eq. (5-32*a*) for webs not rotationally restrained by the flanges and by Eq. (5-32*b*) for webs that are completely restrained. It was shown that webs of standard rolled beams reach yield stress on the extreme fiber without buckling. This is not usually the case for plate girders, and if bend buckling is to be avoided, upper limits of web slenderness must be determined. Such limits can be established from Eqs. (5-32). Thus, assuming the edges of a web of depth h and thickness t to be hinged at the junctures with the flanges, and using a factor of safety of 1.25 with respect to the service-load bending stress f_b, Eq. (5-32*a*) gives

$$1.25 f_b = \frac{650 \times 10^6}{(h/t)^2} \quad \text{psi}$$

from which

$$\frac{h}{t} = \frac{23{,}000}{\sqrt{f_{b,\,\text{psi}}}} \tag{6-1}$$

This is the AASHO specifications' slenderness limit for plate-girder webs without longitudinal stiffeners. According to the AREA specifications, $h/t = 32{,}500/\sqrt{F_y}$. Using the AASHO allowable stress $f_b = 0.55F_y$, Eq. (6-1) gives $h/t = 165$ for A36 steel. The corresponding AREA limit is 170.

Except for the relatively small factor of safety, slenderness limits derived as above make no allowance for the fact that thin-webbed beams and plate girders can develop a considerable additional bending resistance beyond that at which bend buckling of the web occurs. This postbuckling behavior is discussed in Art. 6-3. The bend-buckling resistance of beam webs can be increased considerably by longitudinal stiffeners. Since this means that webs much thinner than those given by Eq. (6-1) can be used, longitudinal stiffeners may be economical for deep girders.

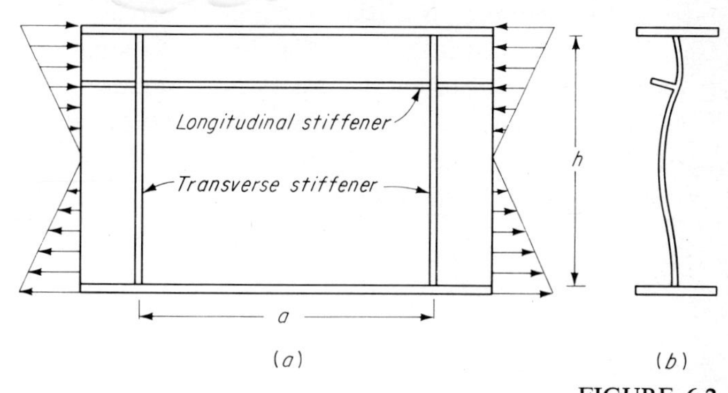

FIGURE 6-2

Thin webs also require transverse stiffeners to increase their resistance to shear (Fig. 6-2*a*). Stiffeners usually consist of rectangular bars welded to the web. Transverse stiffeners may be in pairs, one each side of the web, or they may be placed on one side of the web. Longitudinal stiffeners are usually placed on one side. The effect of a longitudinal stiffener on bend buckling of a web plate is shown in Fig. 6-2*b*. It increases resistance to bend buckling by enforcing a longitudinal node in the buckled configuration. The figure suggests that the efficiency of the stiffener is a function of its location in the compression zone. A stiffener midway between the neutral axis and the compressive edge of the web increases the value of k in the plate-buckling formula Eq. (4-47) to 101 for hinged edges, as compared with 23.9 for unstiffened webs. The optimum location for a longitudinal stiffener has been determined to be at the distance $h/5$ from the compressive edge of the web, in which case $k = 129$.[1] The corresponding allowable web slenderness (for the same factor of safety) is found by multiplying the slenderness limit for webs with no longitudinal stiffeners by $\sqrt{129/23.9} = 2.32$. Thus, the slenderness limit 165, given by Eq. (6-1) for the unstiffened web in A36 steel, becomes 380. However, according to the AASHO specifications this limit is 330, and, in general,

$$\frac{h}{t} = \frac{46{,}000}{\sqrt{f_{b,\,\mathrm{psi}}}} \not> 340 \tag{6-2}$$

The longitudinal stiffener must be stiff enough to produce the higher buckling mode shown in Fig. 6-2*b*. In doing so, it acts as a beam supported at the ends where the vertical stiffeners hold the web in line. It must also resist axial compression because of its location in the compressive zone of the web. Thus, it is a beam-column and must be proportioned in terms of both cross-sectional area and moment of inertia. Furthermore, as a beam-column, its strength is influenced by the distance a between the vertical stiffeners (Fig. 6-2a). Consequently, the buckling coefficient $k = 129$ is attained only with certain combinations of these factors. The AASHO requirement for longitudinal stiffeners, which is independent of the ratio of the area A_s of the stiffener to the web area ht, gives an acceptable upper bound to the required moment of inertia of the stiffener given in Ref. 1. The AASHO formula is

$$I_s = ht^3\left[2.4\left(\frac{a}{h}\right)^2 - 0.13\right] \tag{6-3}$$

According to Massonnet,[2] results of tests suggest that the theoretical values of I_s on which Eq. (6-3) is based should be multiplied by about 7 to obtain a stiffener which will remain practically straight up to collapse of the web. He also notes that a longitudinal strip of the web acts as a part of the stiffener and suggests that the width of the strip be taken equal to $20t$. However, according to the AASHO specifications, I_s is the moment of inertia of the stiffener alone, taken at the edge in contact with the web.

The numerical data from the analysis on which Eq. (6-3) is based covered values of the ratio A_s/ht of stiffener area to web area ranging from 0 to 0.1 and panel aspect ratios a/h from 0.5 to 1.5. The equation should not be used for values outside this range. This requirement is met in the AASHO specifications since they limit a/h to not more than unity.

The stiffener must also be proportioned to resist local buckling, according to procedures discussed in Art. 4-23. For plates supported on only one longitudinal edge, AASHO requires $b/t \lessapprox 1{,}625/\sqrt{f_{a,\text{psi}}}$ (Table 4-4), where f_a is the service-load stress. For a stiffener at $h/5$, $f_a = 0.6f_b$, where f_b is the bending stress on the compression flange. Substituting this into the equation above gives $b/t \lessapprox 2{,}100/\sqrt{f_{b,\text{psi}}}$; AASHO uses $2{,}250/\sqrt{f_{b,\text{psi}}}$.

A single longitudinal stiffener is not likely to be sufficient to produce economical webs for the large depths required for long-span girder bridges, where haunch depths as much as 40 ft have been used. Thus, for a 30-ft web with $F_y = 60$ ksi, the AASHO requirement is $h/t = 46{,}000/\sqrt{0.55 \times 60{,}000} = 253$ [Eq. (6-2)], so that $t = 360/253 = 1.42$ in. Multiple longitudinal stiffeners are used in such cases.

According to tests reported in Ref. 9, the linear distribution of stress corresponding to beam theory is preserved in longitudinally stiffened webs with slendernesses even greater than those discussed above. However, for this to be assured, a longitudinal stiffener at $h/5$ must satisfy an additional requirement, namely, that its critical stress as a column of length equal to the transverse-stiffener spacing be not less than 0.6 times the critical stress for the flange. However, it is pointed out that one or more additional longitudinal stiffeners may be needed for webs with $h/t > 450$.

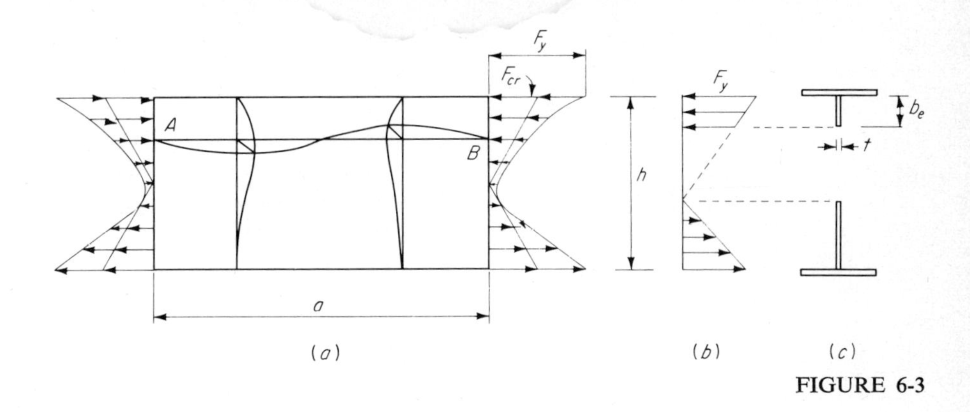

FIGURE 6-3

Longitudinal stiffeners are not covered by the AISC specification because it attains large web slenderness by accepting bend buckling of the web and taking its postbuckling strength into account. This is discussed in Art. 6-4.

6-3 POSTBUCKLING BENDING STRENGTH OF GIRDER WEBS

Figure 6-3*a* shows a beam-web panel for which the critical bending stress F_{cr} is less than the yield stress F_y. If the bending strain increases after F_{cr} is reached, the upper edge of the panel shortens and the bottom edge lengthens. Provided no lateral-torsional buckling of the girder occurs, the edges remain straight and the extreme-fiber stresses continue to increase. If the web were to remain flat, proportionate increases in stress would develop in the remainder of the web. Because the web has buckled, however, the increase in stress will be nonlinear in the compression zone, as shown in the figure. This is because some of the shortening in a length AB will be taken up by an increase in the amplitude of the buckle. During the increase in moment beyond that corresponding to F_{cr}, the neutral axis moves down. The maximum moment for a symmetrical cross section is reached at an extreme-fiber stress F_y in the compression flange if strain hardening is neglected.

The bending behavior discussed above is an idealization of the actual behavior. Since beam webs are not perfect, there will be some waviness even in the unloaded beam. Therefore, lateral deflection of the web begins at the beginning of loading. However, the rate of increase of these deflections increases rapidly when F_{cr} is reached. Thus, the behavior is analogous to that of the axially loaded column.

Since the variation in stress in the postbuckled state is unknown, simplifying assumptions must be made to determine the maximum moment. The one shown in

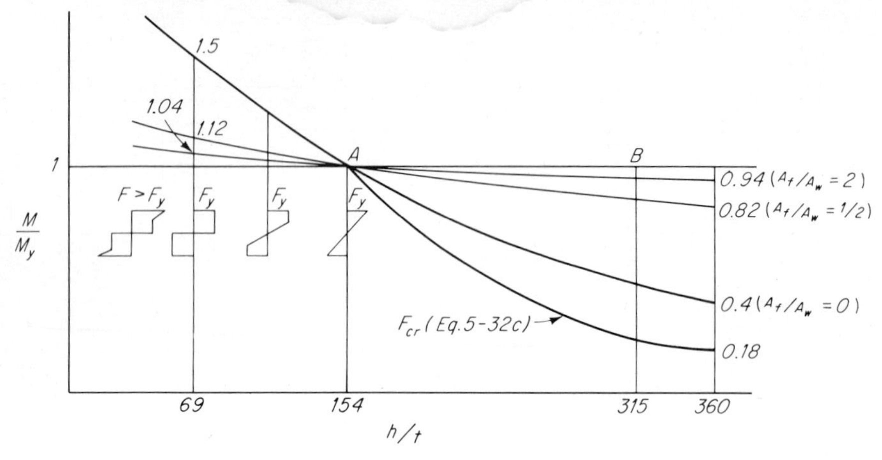

FIGURE 6-4

Fig. 6-3*b* was proposed in Ref. 3. Here the nonlinear compression of Fig. 6-3*a* is replaced with a linear distribution according to elastic-behavior beam theory, acting on an effective depth of web b_e (Fig. 6-3*c*). This enables the moment to be calculated, provided b_e is known. In Ref. 3 it was assumed that $b_e = 30t$ for girders with $h/t = 360$. Using this value and $F_y = 36$ ksi, values of M are computed for $A_f/ht = 0$, 0.5, and 2, and plotted, in the nondimensional form M/M_y, in Fig. 6-4. Also shown is the point $M_{cr}/M_y = 0.18$, where M_{cr} is the moment based on the critical bending stress for the web given by Eq. (5-32*c*). We can also use Eq. (5-32*c*) to determine the web slenderness h/t that enables a girder to reach its full yield moment $M_y = F_y I/c$. The result is

$$\frac{h}{t} = \sqrt{\frac{850{,}000}{F_y}} = \frac{925}{\sqrt{36}} = 154$$

which gives point A in Fig. 6-4. Also shown are the ratios 1.5, 1.12, and 1.04, of the plastic moment M_p to M_y, for $A_f/A_w = 0$, 0.5, and 2, respectively. In order for these moments to be developed, web slenderness must not exceed $412/\sqrt{F_y}$ (Art. 5-17). For A36 steel, this gives $h/t = 69$.

If the girder has a longitudinal stiffener at $h/5$ from the compression flange, $k = 129$ in the plate-buckling formula, so that

$$F_{cr} = \frac{3{,}510{,}000}{(h/t)^2} \qquad \frac{h}{t} = \sqrt{\frac{3{,}510{,}000}{36}} = 315$$

which gives point B in Fig. 6-4. This shows the effectiveness of longitudinal stiffeners in increasing the bending strength of thin-webbed girders.

The variation of M/M_y in the regions between $h/t = 0$, 69, 154, and 360 in Fig. 6-4 is not determined by the analysis above, so that the curves shown in the figure are approximate. The variation where web efficiency is reduced by buckling was assumed in Ref. 3 to be linear. Furthermore, the slenderness at which reduction in web effectiveness was assumed to begin was taken to be $980/\sqrt{F_y}$, rather than the value corresponding to point A of Fig. 6-4. This was because the AISC specification used 170 as the upper limit of slenderness of girders of A7 steel ($F_y = 33$ ksi) prior to the revision in which postbuckling strength was taken into account. The resulting equation connecting the revised point A with the points corresponding to $h/t = 360$ in Fig. 6-4 is

$$\frac{M}{M_y} = 1 - 0.0005 \frac{A_w}{A_f}\left(\frac{h}{t} - \frac{980}{\sqrt{F_y}}\right) \tag{6-4}$$

Equation (6-4) is compared in Ref. 3 with results of nine tests. The ratio of test moment to computed moment ranged from 0.96 to 1.11.

To simplify the computations for design purposes, Eq. (6-4) is expressed in terms of an equivalent stress $f_{eq} = Mc/I$, where I is the moment of inertia of the gross cross section. Thus, dividing M and M_y in Eq. (6-4) by I/c, we get

$$\frac{f_{eq}}{F_y} = 1 - 0.0005 \frac{A_w}{A_f}\left(\frac{h}{t} - \frac{980}{\sqrt{F_y}}\right) \tag{6-5}$$

Using a factor of safety of 1.65, the allowable bending stress $F_b' = f_{eq}/1.65$ is

$$F_b' = 0.6F_y\left[1 - 0.0005 \frac{A_w}{A_f}\left(\frac{h}{t} - \frac{980}{\sqrt{F_y}}\right)\right] \tag{6-6}$$

But $0.6F_y$ is the allowable bending stress F_b for a plate girder with a fully effective web. Therefore, substituting $F_y = 1.65F_b$ we get

$$F_b' = F_b\left[1 - 0.0005 \frac{A_w}{A_f}\left(\frac{h}{t} - \frac{760}{\sqrt{F_b}}\right)\right] \tag{6-7}$$

Finally, to extend Eq. (6-7) to the case where the girder fails by lateral-torsional buckling at a reduced stress, F_b is defined as the allowable bending stress according to Eq. (5-16) if there is no lateral support.

Figure 6-4 shows that a girder with large h/t and no longitudinal stiffener can develop almost as much moment as the same girder with a longitudinal stiffener. Thus, the stiffener appears to offer little advantage. However, tests show that the fatigue life of the unstiffened girder with large h/t is less than that of the stiffened girder.[4–6] Several girders with $h/t > 200$ failed at fewer than 2 million cycles of load because of excessive lateral deflection of the web. This is not a significant consideration for girders in buildings, so that no provision is made for it in the AISC specification.

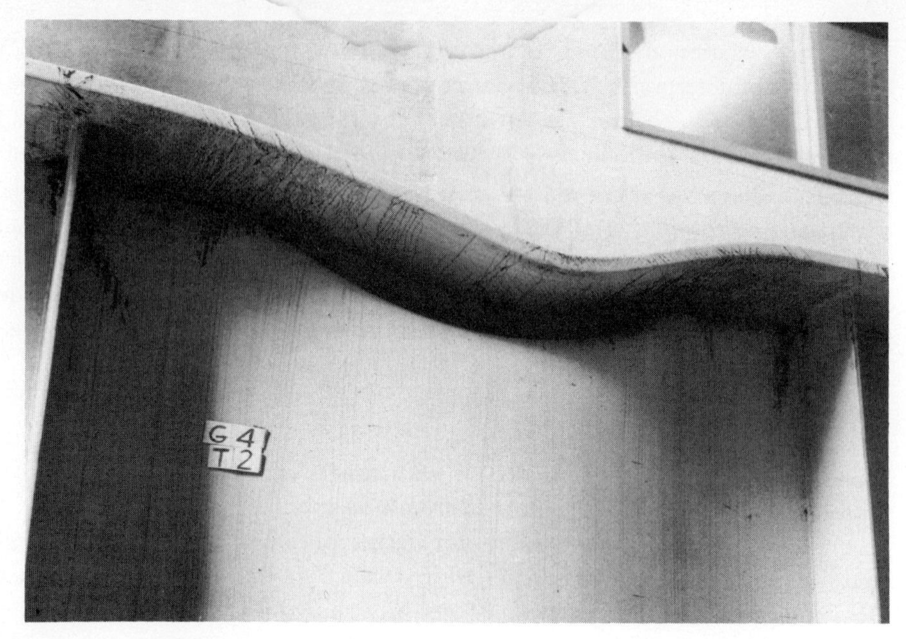

FIGURE 6-5
Vertical buckling of plate-girder flange. (*Fritz Engineering Laboratory, Lehigh University.*)

However, in proposed specifications for load-factor design of steel highway bridges,[7] web slenderness for girders without longitudinal stiffeners is limited by

$$\frac{h}{t} \lessapprox \frac{36{,}500}{\sqrt{F_{y,\mathrm{psi}}}} \tag{6-8}$$

For A36 steel, this gives $h/t \lessapprox 190$. This is somewhat larger than the limit $23{,}000/\sqrt{f_b}$ discussed in Art. 6-2, which gives $h/t \lessapprox 165$ for A36 steel.

6-4 VERTICAL BUCKLING OF THE COMPRESSION FLANGE

If a plate-girder web is too slender, the compression flange may buckle in the vertical plane at a stress less than yield stress. This failure mode is shown in Fig. 6-5. A study of this photograph shows that the compression flange is a beam-column, continuous over the vertical stiffeners as supports, whose stability depends on the stiffener spacing and the relative stiffnesses of the flange and the web. The difficulty in analyzing this phenomenon is compounded by the bending stresses in the web. The following analysis of the problem was presented in Ref. 3.

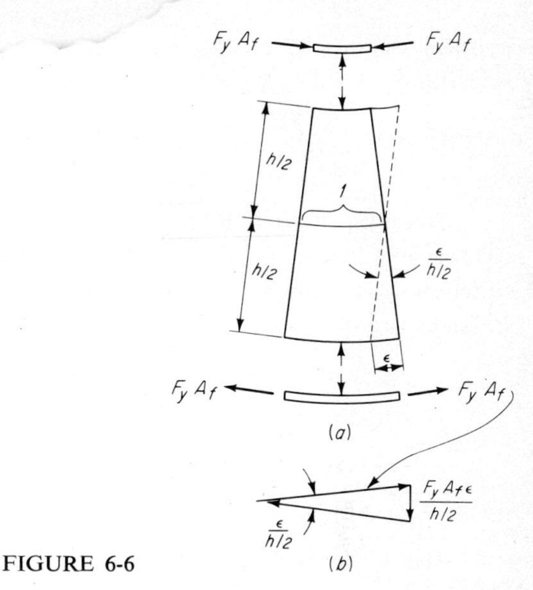

FIGURE 6-6

Figure 6-6 shows a vertical element, of unit width, of a girder which is stressed to its yield moment. The extreme-fiber strains ε produce a relative rotation $\varepsilon/(h/2)$ of the vertical edges of the element. Equilibrium of the flange forces $F_y A_f$ requires a vertical compression $F_y A_f \varepsilon/(h/2)$ in the element, which the web must be able to support without buckling. The stability of webs in vertical compression was discussed in Art. 5-19, where it was shown that the critical stress for a web without transverse stiffeners is given by Eq. (5-38*b*):

$$F_{cr} = \frac{\pi^2 E}{12(1-\mu^2)(h/t)^2} \tag{a}$$

For equilibrium of the element in Fig. 6-6*a*, we have

$$F_{cr} \times 1 \times t = \frac{F_y A_f \varepsilon}{h/2} \tag{b}$$

Equations (*a*) and (*b*) give

$$\left(\frac{h}{t}\right)^2 = \frac{\pi^2 E}{12(1-\mu^2)} \frac{A_w}{A_f} \frac{1}{\varepsilon F_y} \tag{c}$$

Because of residual stresses in the flanges, the extreme-fiber strain ε must exceed ε_y by an amount sufficient to offset the residual strain of opposite sign, in order to obtain uniform yielding of the flanges. Therefore, $\varepsilon = \varepsilon_y + \varepsilon_r = (F_y + F_r)/E$, which, upon substitution in Eq. (*c*), yields

$$\left(\frac{h}{t}\right)^2 = \frac{\pi^2 E^2}{12(1-\mu)^2} \frac{A_w}{A_f} \frac{1}{F_y(F_y + F_r)} \tag{6-9}$$

Since Eq. (6-9) is based on the vertical-buckling strength of a web with no transverse stiffeners, it will give conservative results for a web with closely spaced stiffeners. According to tests reported in Ref. 8, the slenderness of webs with vertical stiffeners can be taken conservatively at

$$\frac{h}{t} \lessapprox \begin{cases} \dfrac{2,500}{\sqrt{F_y}} & \text{for } \dfrac{a}{h} \lessapprox 1 \qquad (a) \\[2ex] \dfrac{2,000}{\sqrt{F_y}} & \text{for } \dfrac{a}{h} \lessapprox 1.5 \qquad (b) \end{cases}$$

The AISC specification is based on Eqs. (6-9) and (*b*). The ratio A_w/A_f in Eq. (6-9) is taken at 0.5, which is a lower limit for girders of practical proportions, and the residual stress F_r at 16.5 ksi. The result is

$$\frac{h}{t} \lessapprox \frac{14,000}{\sqrt{F_y(F_y + 16.5)}} \tag{6-10}$$

except that h/t need not be less than $2,000/\sqrt{F_y}$ for $a/h \lessapprox 1.5$.

Although Eq. (6-9) involves the ratio of the areas of the flange and the web, it contains no parameter involving their stiffnesses. Therefore, it gives conservative results for flanges with large bending stiffnesses.

PROBLEMS

6-1 Given an A36-steel girder with a $75 \times \frac{1}{4}$ in. web and with each flange consisting of a $12 \times 1\frac{1}{2}$ in. plate. Compute the ultimate resisting moment.

6-2 Same as Prob. 6-1 but with a longitudinal stiffener 15 in. from the compression flange. Compute the ultimate resisting moment and determine the stiffener cross section.

6-5 PROPORTIONING OF GIRDERS

The depth of a plate girder is influenced by many factors, and it is impossible to lay down workable rules to which no exception will occur. In many cases the determining factor is headroom or, in the case of deck bridges, clearance for high water or for traffic passing beneath. Even when no such limitations exist, the range of depths for economical results is considerable. Although the depth of the average girder is one-tenth to one-twelfth of its span, depths of as much as one-eighth to one-sixth of the span may occasionally be necessary, particularly if there are heavy concentrated loads such as occur for girders supporting columns in buildings. On the other hand, lighter loads may be accommodated economically with depths as small as one-fourteenth or one-fifteenth of the span.

In proportioning the cross section, one determines a web which has the necessary capacity in shear, and flanges which, together with the web, have the required section modulus. Since it is more convenient to proportion by area than by moment of inertia, a procedure that enables a tentative determination of the flange is desirable.

Let A_f be the area of one flange, h_g the distance between the centers of gravity of the two flanges, h_w and t the height and thickness, respectively, of the web plate, and h_o the overall depth. Except for the small moment of inertia of each flange about its own gravity axis, the moment of inertia of the cross section is

$$I = 2A_f\left(\frac{h_g}{2}\right)^2 + \frac{th_w^{\,3}}{12}$$

But h_w and h_g are approximately equal, and $th_w = A_w$, the area of the web. Therefore,

$$I = \frac{h_g^{\,2}}{2}\left(A_f + \frac{A_w}{6}\right) \tag{6-11}$$

Substituting this value of I into the equation $I/c = M/f$, we get

$$A_f = \frac{M}{fh_g^{\,2}h_o} - \frac{A_w}{6}$$

But fh_g/h_o is the bending stress at the center of gravity of the flange, which we denote by f_g. Therefore,

$$A_f = \frac{M}{h_g f_g} - \frac{A_w}{6} \tag{6-12}$$

Although plate-girder components are usually welded, bolted field splices are sometimes used. Opinion differs as to the effect of bolt holes on the bending strength. This question was discussed in Art. 5-14. Both AASHO and AREA specifications require that the tensile bending stress be based on the moment of inertia of the net section, i.e., holes on both sides of the neutral axis are deducted, while the compressive bending stress is based on the moment of inertia of the gross cross section. The AISC specification requires a reduction in area of the flange of only the area lost in excess of 15 percent of the gross area of the flange.

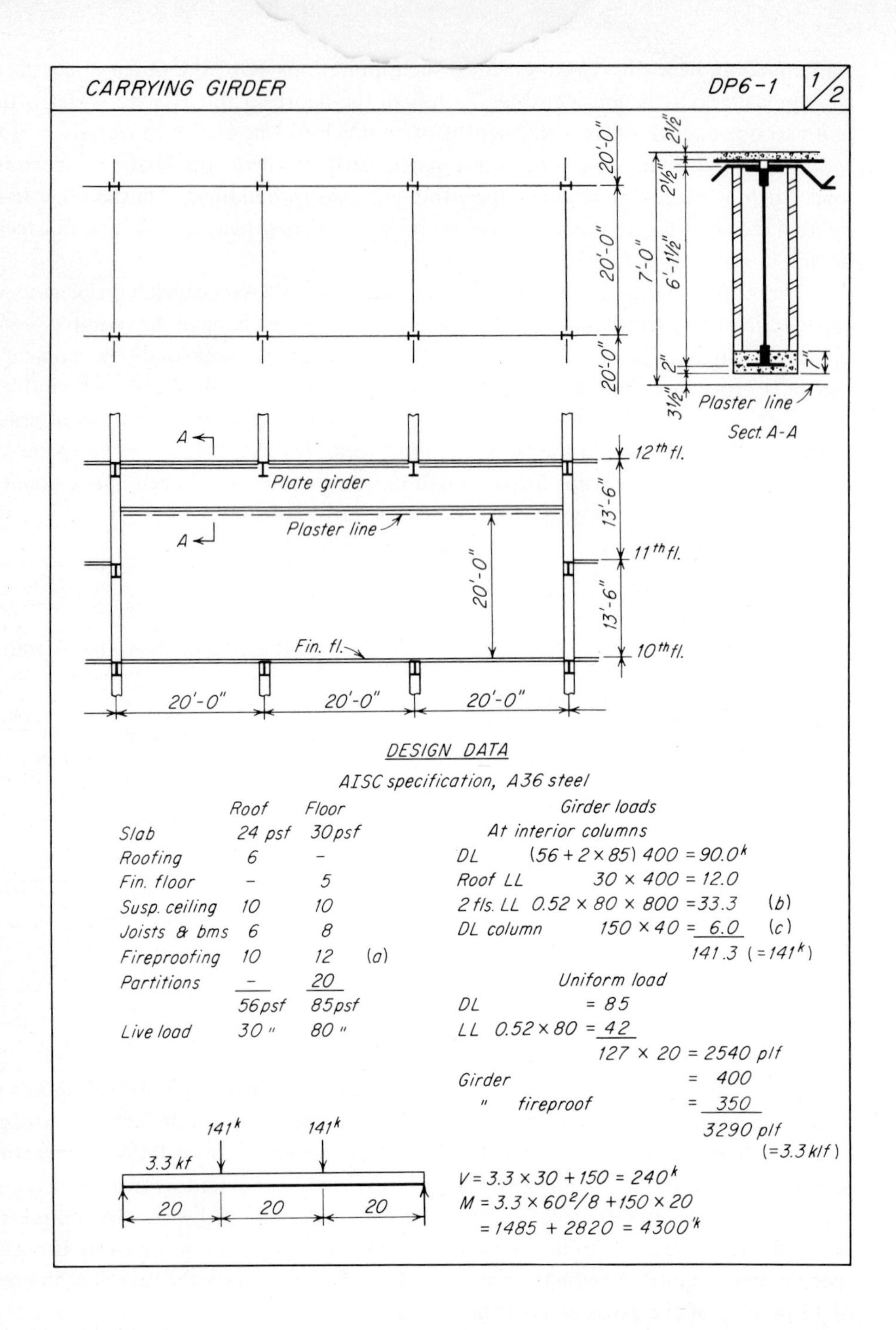

CARRYING GIRDER
DP6-1
1/2
20'-0"
20'-0"
20'-0"
2½"
2½"
7'-0"
6'-11½"
2"
7"
3½"
Plaster line
Sect. A-A
A
A
12th fl.
Plate girder
13'-6"
Plaster line
20'-0"
11th fl.
13'-6"
Fin. fl.
10th fl.
20'-0"
20'-0"
20'-0"
DESIGN DATA
AISC specification, A36 steel
Roof Floor
Slab 24 psf 30 psf
Roofing 6 -
Fin. floor - 5
Susp. ceiling 10 10
Joists & bms 6 8
Fireproofing 10 12 (a)
Partitions - 20
56 psf 85 psf
Live load 30 " 80 "
Girder loads
At interior columns
DL (56 + 2 × 85) 400 = 90.0k
Roof LL 30 × 400 = 12.0
2 fls. LL 0.52 × 80 × 800 = 33.3 (b)
DL column 150 × 40 = 6.0 (c)
141.3 (= 141k)
Uniform load
DL = 85
LL 0.52 × 80 = 42
127 × 20 = 2540 plf
Girder = 400
" fireproof = 350
3290 plf
(= 3.3 klf)
141k
141k
3.3 kf
20
20
20
V = 3.3 × 30 + 150 = 240k
M = 3.3 × 60²/8 + 150 × 20
= 1485 + 2820 = 4300'k

CARRYING GIRDER — DP6-1 — 2/2

WEB PLATE

Max. depth girder 6-1½ (Sect. AA)

Try 68 × 3/8 web (d)

$A = 68 \times 3/8 = 25.5\ \text{in.}^2$

$$\frac{h}{t} = \frac{68}{3/8} = 181$$

$$\frac{h}{t} = \frac{14{,}000}{\sqrt{36(36+16.5)}} = 322 \qquad (e)$$

$$\leqq \frac{2{,}000}{\sqrt{36}} = 333 \text{ if } a/h \leqq 1.5 \qquad (e)$$

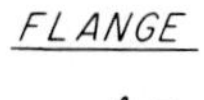

FLANGE

$$A = \frac{4300 \times 12}{70 \times 21.4} = 34.5 \qquad (f)$$

$$A_w/6 = 25.5/6 = \underline{4.2}$$

$$30.3\ \text{in.}^2$$

1 ℞ $14 \times 2\tfrac{1}{4} = 31.5\ \text{in.}^2$

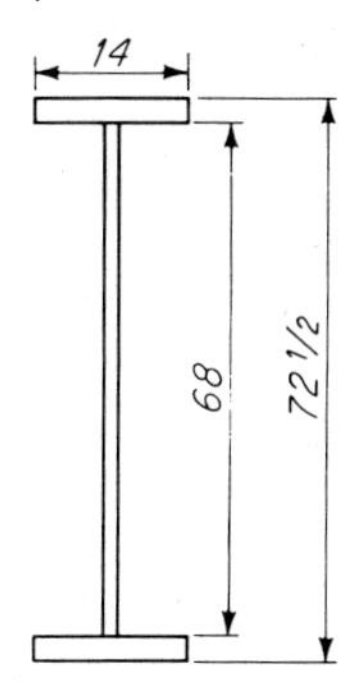

Moment of Inertia		Weight
Web $3/8 \times 68^3/12$	9830	87
Flg. $2 \times 31.5 \times 35.12^2$ =	77750	214
	87,580 (g)	301

$$f = \frac{4300 \times 12 \times 36.25}{87{,}580} = 21.4 \text{ ksi}$$

$$760/\sqrt{F_b} = 760/\sqrt{22} = 162$$

$$F_b' = 22\left[1 - 0.0005\,\frac{25.5}{31.5}\,(181-162)\right] \quad \text{Eq. (6-7)}$$

$$= 21.7 > 21.4 \text{ ksi O.K.}$$

6-6 DP6-1: CARRYING GIRDER

The girder to be designed is one of a series over an assembly hall in a tier building. In addition to its share of the twelfth floor, each girder supports two columns which carry the thirteenth and fourteenth floors and the roof. The design live loads are 30 psf for the roof and, except for the assembly room, 80 psf for the floors. The concrete slab for the roof is 2 in. thick, while that for the floors is $2\frac{1}{2}$ in. thick. Both are supported on 12-in. truss joists spaced 2 ft on centers. The girder is fireproofed, as shown in section *AA* of the figure, with $1\frac{3}{4}$ in. of gypsum plaster on metal lath. The girder cannot exceed 6 ft $1\frac{1}{2}$ in. in depth. The comments which follow are intended to clarify the correspondingly lettered computations.

a This item of dead load is the weight of fireproofing for the beams. The suspended ceiling fireproofs the joists.

b Since the girder supports a floor area in excess of 150 ft², the basic live load may be reduced in accordance with the ASBC (see AISC Manual or Art. 1-4). The allowable reduction is 8 percent per 100 ft² of supported area but not to exceed 60 percent or the value given by Eq. (1-1), namely, $R = (D + L)/4.33L$. In our case, the latter limitation controls, since $D = 85$ psf, $L = 80$ psf, and $R = 165/(4.33 \times 80) = 0.48$, while the supported area is 20×60 ft at the twelfth floor and 20×20 ft at each column for each of the two floors above, or 2,800 ft² altogether.

c Each column together with its fireproofing weighs 150 plf.

d The 68-in. web allows flanges $2\frac{1}{2}$ in. thick within the 73 in. allowable overall depth. The $\frac{3}{8}$-in. thickness is a preliminary judgment which will be checked in the investigation for shear. (In practice, it should be checked, at least roughly, before proceeding further. However, discussion of shear is postponed to Art. 6-8.)

e These are the maximum allowable values of web slenderness, which were discussed in Art. 6-4.

f The joists, which are attached to the top flange of the girder, provide adequate lateral support. Therefore, $F_b = 0.6 \times 36 = 22$ ksi. Assuming 2-in. flange plates, $h_g = 70$ in. and $h_o = 72$ in. Therefore, $f_g = 22 \times 70/72 = 21.4$ ksi in Eq. (6-12).

g Tables of moment of inertia in the AISC Manual can be used to facilitate this computation.

6-7 LENGTH OF FLANGE PLATES

Some economy may be achieved by varying the area of the flange plate to suit the variation in moment. This can be done for welded girders by flanking the flange plate required for maximum moment by one or more successively thinner plates butt welded end to end. The theoretical cutoff point of a plate is found by calculating the moment capacity of the reduced cross section and locating the corresponding point on the moment diagram. In the case of simple load systems, cutoff points can be determined by formula. Thus, since the moment diagram for a uniformly loaded, simply supported beam is a parabola,

$$\frac{L_1}{L} = \sqrt{\frac{M_1 - M_2}{M_1}} = \sqrt{\frac{A_1 - A_2}{A_1 + A_w/6}} \tag{6-13}$$

where A_1 = area of center plate
M_1 = corresponding moment
A_2 = area of adjacent plate
M_2 = corresponding moment
L_1 = length of center plate
L = span of girder

Substitution of the areas A for the corresponding moments M in Eq. (6-13) is approximate. It is based on the assumption that $h_g f_g$ in Eq. (6-12) does not change with the change in cross section of the girder. In any case, a strict interpretation of standard specifications requires that the cross section at the cutoff be checked by $f = Mc/I$. Equation (6-13) can be extended to a third pair of plates by substituting L_2, M_3, and A_3 for L_1, M_2, and A_2.

If the plate girder supports moving loads, as in the case of highway and railway bridges, the envelope of maximum moments is needed. Since the maximum moment at each point results from a different position of the moving live load, the envelope is not the same thing as a moment diagram. However, a satisfactory approximate method which avoids the construction of the envelope has been developed.[10] It is based on the assumption that the envelope is a straight line through the point of maximum moment, extending $0.05L$ on either side of the center point of the span, flanked by parabolas tangent at its ends and passing through the ends of the span. Since each parabolic segment has a base $0.45L$ in length, cover-plate lengths can be determined by adding $0.1L$ to the value of L_1, L_2, from Eq. (6-13), provided L in that equation is replaced with $0.9L$.

In some cases (and in particular in the case of continuous spans) plotting the various resisting moments on a plot of the maximum-moment envelope helps to determine cutoff points.[11]

Cover plates used to vary flange area, as in the case of the cover plate added to the flange of a rolled shape to increase its moment capacity, must extend beyond the theoretical cutoff a distance sufficient to develop the capacity of the plate; i.e., the allowable fastener strength beyond the cutoff must equal the allowable axial force in the plate.

It is highly unlikely that the saving in the cost of the steel would offset the additional cost of fabrication if a reduction in flange-plate area is made in the girder of DP6-1. However, to illustrate the procedure, the point at which a $1\frac{1}{2}$-in. flange plate is adequate will be determined. Since Eq. (6-13) cannot be used in this case, because of the concentrated loads, we proceed as follows:

$$I = \frac{3}{8}\frac{68^3}{12} + 2 \times 14 \times 1.5 \times 34.75^2 = 9{,}830 + 50{,}700 = 60{,}530 \text{ in.}^4$$

$$M = \frac{22 \times 60{,}530}{35.5 \times 12} = 3{,}120 \text{ ft-kips}$$

From the data in DP6-1

$$M = 240x - \frac{3.3x^2}{2}$$

where x is the distance from either end to the point where the $1\frac{1}{2}$-in. plate is adequate. Equating the two moments and solving the equation, we get $x = 14.4$ ft.

PROBLEMS

6-3 Redesign the girder of DP6-1 using a $\frac{1}{4}$-in. web. Compare the weights per foot of cross section.

6-4 A simply supported plate girder spanning 70 ft supports a uniformly distributed load, not including its own weight, of 4 klf. There is no fireproofing, and the girder is supported against lateral buckling. Design a cross section. AISC specification, A36 steel.

6-5 A simply supported plate girder spans 90 ft and supports a uniformly distributed load, not including its own weight, of 8 klf. There are lateral supports at intervals of 18 ft. The depth of the girder cannot exceed 7 ft. Design a cross section. AISC specification, A36 steel.

6-6 Columns in a 14-story building similar to that of DP6-1 are 22 ft on centers in each direction. The floor-to-floor height is 13 ft. In one area of the building a row of columns between the tenth and twelfth floors is to be omitted in an assembly room, with the columns that support the twelfth, thirteenth, and fourteenth floors to be carried on a 44-ft plate girder. The design live load and the weights of slab, roofing, etc., are the same as for DP6-1. Design a plate girder to allow a 20-ft ceiling in the assembly room. AISC specification, A36 steel.

6-7 Given a simply supported welded plate girder whose flange consists of one central flange plate flanked by two thinner plates butt welded to it. If the girder supports a load concentrated at midspan and the weight of the girder itself is negligible, prove that the weight of the flange plates is a minimum if the length of the central plate is half the span of the girder. Assume that the girder is supported against lateral buckling.

6-8 Determine the cutoff for the flange plate of the girder of DP6-1 if the abutting flange plates are to be $1\frac{1}{4}$ in. thick.

6-8 SHEAR BUCKLING OF PLATE-GIRDER WEBS

Shear buckling of beam webs was discussed in Art. 5-16, where it was shown that the critical shear stress is given by the plate-buckling formula

$$F_{v,cr} = \frac{k\pi^2 E}{12(1-\mu^2)(h/t)^2} \tag{6-14}$$

where

$$k = 4 + \frac{5.34}{(a/h)^2} \qquad \text{if } \frac{a}{h} \lessgtr 1 \tag{6-15a}$$

$$k = 5.34 + \frac{4}{(a/h)^2} \qquad \text{if } \frac{a}{h} \gtrless 1 \tag{6-15b}$$

in which a and b are defined in Fig. 6.2. A simpler formula for k, valid for all values of a/h, has also been proposed [Eq. (5-27)]. If the design of plate-girder webs is based on shear buckling, these equations can be used to determine the required spacing of transverse stiffeners. With the proportions of the web known, the service-load shear stress can be determined, multiplied by the desired factor of safety, and substituted for $F_{v,cr}$ in Eq. (6-14) to determine k. The stiffener spacing is then found by the appropriate Eq. (6-15). A formula for this procedure can be developed. Thus, for a factor of safety of 1.2, we get, for $a/h \gtrless 1$,

$$1.2f_v = F_{v,cr} = \left[5.34 + \frac{4}{(a/h)^2}\right]\frac{\pi^2 E}{12(1-\mu^2)(h/t)^2}$$

With units of pounds per square inch, this gives, for steel

$$a = \frac{11{,}000t}{\sqrt{f_v - \left(\dfrac{9{,}500}{h/t}\right)^2}} \qquad \frac{a}{h} \gtrless 1 \tag{6-16a}$$

Similarly, with Eq. (6-15a), we get

$$a = \frac{9{,}500t}{\sqrt{f_v - \left(\dfrac{11{,}000}{h/t}\right)^2}} \qquad \frac{a}{h} \lessgtr 1 \tag{6-16b}$$

Simpler, approximate forms of Eq. (6-16) are used in the AASHO and AREA specifications. Thus

$$a_{\text{AASHO}} = \frac{11{,}000t}{\sqrt{f_v}} \qquad a_{\text{AREA}} = \frac{10{,}500t}{\sqrt{f_v}} \tag{6-17}$$

It will be noted that these simpler formulas increase the factor of safety, since they result in closer spacing of the stiffeners. Furthermore, AASHO limits the spacing of stiffeners by specifying $a/h \lesssim 1$, while according to AREA a must not exceed 6 ft.

The AISC specification uses the same formula as the AASHO, except that it is in units of kips per square inch, which gives

$$a = \frac{348t}{\sqrt{f_v}} \tag{6-18}$$

However, this formula is used only for the end panels of a girder, for reasons which are explained in Art. 6-9.

Since transverse stiffeners support the boundaries of a panel by keeping them straight, they must be stiff enough to prevent their buckling with the web. Several investigators have developed theoretical solutions of this problem. Stein and Fralich devised a solution for the infinitely long, simply supported plate reinforced with equally spaced stiffeners, which shows fair agreement with laboratory tests on 20 specimens with various sizes of stiffener.[13] They give numerical results for three different values of a/h, namely, 0.2, 0.5, and 1. Bleich developed from these data a formula for the required moment of inertia I_s of the transverse stiffener,[14] which can be put in the following form

$$I_s = 2.5ht^3\left(\frac{h}{a} - 0.7\frac{a}{h}\right) \qquad a \lesssim h \tag{6-19}$$

If the web is stiffened on only one side, I_s is to be computed at the face of the stiffener which is in contact with the web.

The AASHO formula is

$$I_s = \frac{Jat^3}{10.92} \tag{6-20a}$$

in which

$$J = 25\left(\frac{h}{a_1}\right)^2 - 20 \gtrsim 5 \tag{6-20b}$$

where a_1 is the stiffener spacing according to Eq. (6-17), while a is the actual stiffener spacing. For $a = a_1$, Eqs. (6-20) reduce to

$$I_s = 2.29ht^3\left(\frac{h}{a} - 0.8\frac{a}{h}\right)$$

which is essentially the same as Eq. (6-19). In addition, the width of the stiffener b must be not less than $\frac{1}{30}$ the depth of the girder, and the thickness t must satisfy $b/t \lesssim 16$.

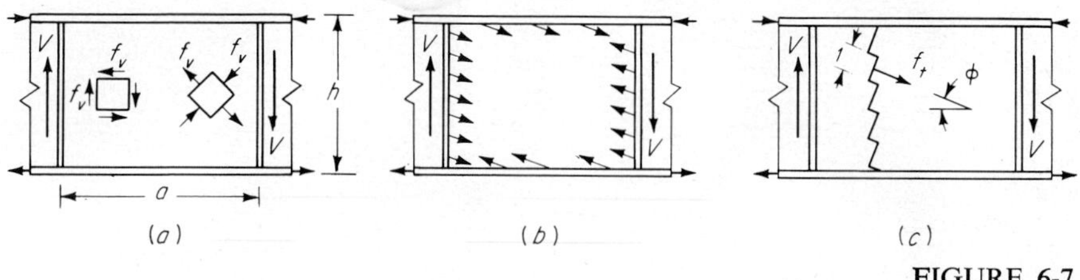

FIGURE 6-7

The AISC provision is $I_s = (h/50)^4$, which gives values of I_s that are the same for all webs of a given height, regardless of their thickness. Since Eq. (6-19) shows that the required moment of inertia is proportional to the cube of the web thickness for given values of a and h, it is clear that this formula is deficient.

The AREA specifies the same dimensions b and t of the stiffener as AASHO but makes no provision for moment of inertia except to say that stiffeners on one side of a web must have the same moment of inertia as that of the minimum allowable pair.

6-9 POSTBUCKLING SHEAR STRENGTH OF GIRDER WEBS

Figure 6-7*a* shows a panel of a plate girder acted upon by shear forces V. Bending stresses are assumed to be small enough to be neglected. The resulting shearing stresses f_v on an element of the web are equivalent to principal stresses f_v, one tensile and one compressive, at 45° to the shear stresses. Values of f_v at which buckling occurs were discussed in Art. 6-8. The direction of the principal compression indicates that these buckles will be in the form of waves or wrinkles running in the direction of the principal tension. Shear buckling in an aluminum girder is shown in Fig. 5-31; shear buckling of a steel girder with longitudinal stiffeners is shown in Fig. 6-8.

After the web buckles in the manner just described, the compressive principal stress can increase very little. On the other hand, the tensile principal stress continues to increase with increase in strain in the diagonal direction. This suggests that such a panel may have a considerable postbuckling strength, since the increase in tension is limited only by the yield stress, provided the flanges and stiffeners framing the panel are stiff enough.

Figure 6-7*b* shows the action of the incremental tension on the boundary members and suggests that the tension will be uniform if they are infinitely stiff. The shearing resistance that can be developed by a uniform tension field, inclined at ϕ with the horizontal, can be determined by considering equilibrium of the stresses on

FIGURE 6-8
Shear buckling of plate-girder web. (*Fritz Engineering Laboratory, Lehigh University.*)

the serrated cross section of Fig. 6-7*c*. Summation of vertical forces gives

$$V_t = f_t ht \cos\phi \sin\phi = \tfrac{1}{2} f_t ht \sin 2\phi \tag{6-21}$$

If the tension field develops at the optimum angle $\phi = 45°$,

$$V_t = \tfrac{1}{2} f_t ht$$

When f_t reaches F_y, $V_{ty} = \frac{1}{2}F_y ht$. Comparing this with the shear capacity of a web which is thick enough to develop the shearing yield stress F_{vy}, for which $V_y = F_{vy} ht$, we get

$$\frac{V_{ty}}{V_y} = \frac{\frac{1}{2}F_y ht}{F_{vy} ht} = \frac{F_y}{2F_{vy}}$$

According to the commonly used distortion-energy theory (also called the Mises yield criterion) $F_{vy} = F_y/\sqrt{3}$. Therefore,

$$V_{ty} = \frac{\sqrt{3}}{2} V_y = 0.87 V_y$$

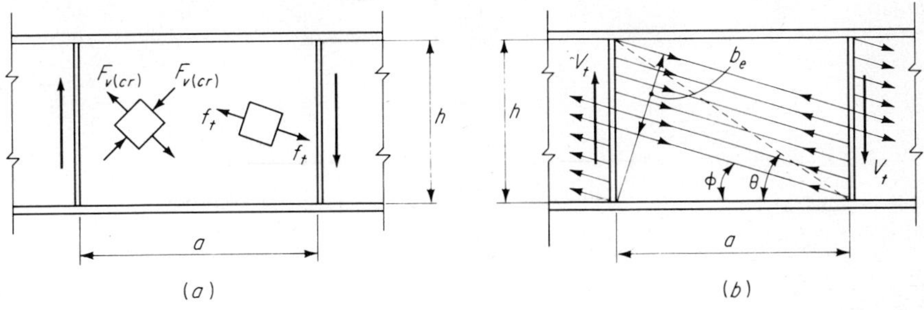

FIGURE 6-9

Thus, a panel which resists shear entirely by tension-field action may be nearly as strong as one which resists it entirely by shear-field action.

The shear strength V_u of a panel whose web buckles before reaching shear yield is the sum of the shear resistance V_{cr} at buckling and the shear component V_t of the tension field (Fig. 6-9*a*),

$$V_u = V_{cr} + V_t = F_{v,cr}ht + V_t$$

Assuming that the panel develops a uniform tension field, V_t is given by Eq. (6-21)

$$V_u = F_{v,cr}ht + \tfrac{1}{2}f_t ht \sin 2\phi$$

from which

$$F_{vu} = \frac{V_u}{ht} = F_{v,cr} + \tfrac{1}{2}f_t \sin 2\phi \tag{6-22}$$

According to Basler,[15] the tension field in a welded girder with plate flanges will be incomplete because the bending strength of the flanges is insufficient to develop the necessary anchorage (Fig. 6-7*b*). He assumes instead the partial field of width b_e shown in Fig. 6-9*b*. Here the flanges are free and the tension field in the panel is reacted by the fields in the adjacent panels. From the geometry of the panel

$$b_e = h \cos \phi - a \sin \phi$$

The shear component V_t of the tension field is

$$V_t = f_t b_e t \sin \phi = f_t t(h \cos \phi - a \sin \phi) \sin \phi$$

$$= f_t t\left[\frac{h}{2} \sin 2\phi - \frac{a}{2}(1 - \cos 2\phi)\right]$$

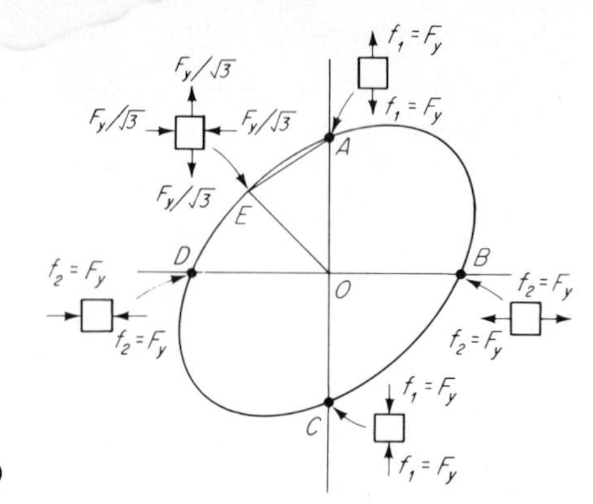

FIGURE 6-10

Assuming that the tension develops at the angle ϕ which maximizes V_t,

$$\frac{dV_t}{d\phi} = f_t t(h \cos 2\phi - a \sin 2\phi) = 0$$

so that

$$\tan 2\phi = \frac{h}{a}$$

But $h/a = \tan \theta$, where θ is the angle of the panel diagonal with the horizontal (Fig. 6-9*b*). Therefore $2\phi = \theta$. Substituting this into Eq. (6-22) gives

$$F_{vu} = F_{v,cr} + \tfrac{1}{2} f_t \sin \theta \tag{6-23}$$

There is an obvious inconsistency in determining ϕ for the partial tension field of Fig. 6-9*b* and assuming that the complete tension field of Eq. (6-22) forms at the same angle. However, with f_t determined as explained in the next paragraph, Eq. (6-23) (which was developed differently in Ref. 15) is in satisfactory agreement with results of tests.

The ultimate shear is assumed to be reached when the two stress fields in Fig. 6-9*a* produce general yielding of the panel. The Mises yield criterion, which was used in Ref. 15, is given by

$$f_1^2 - f_1 f_2 + f_2^2 = F_y^2 \tag{a}$$

where f_1 and f_2 are principal stresses. The ellipse in Fig. 6-10 is the plot of this equation. The directions of the principal stresses represented by points A, B, C, and D are as shown. The coordinates of E on a line OE at 45° are $f_1 = F_y/\sqrt{3}$ and $f_2 = -F_y/\sqrt{3}$. This is a condition of pure shear. Therefore, the combination of the stress fields in Fig. 6-9*a* is represented by a point on *EA*. To simplify the result, Basler took instead the chord *EA*, whose equation is

$$f_1 = F_y + (\sqrt{3} - 1)f_2 \tag{b}$$

Furthermore, instead of using Mohr's circle to get the true principal stresses, he assumed that the difference in direction of the principal tensions $F_{v,cr}$ and f_t could be neglected. Therefore,

$$f_1 = F_{v,cr} + f_t \qquad f_2 = -F_{v,cr}$$

Substituting these values into Eq. (*b*), we get

$$F_{v,cr} + f_t = F_y + (\sqrt{3} - 1)(-F_{v,cr})$$

from which

$$f_t = F_y - \sqrt{3}\,F_{v,cr} \tag{c}$$

Substituting f_t from Eq. (*c*) into Eq. (6-23) gives

$$F_{vu} = F_{v,cr} + \tfrac{1}{2}(F_y - \sqrt{3}\,F_{v,cr}) \sin\theta \tag{6-24}$$

where

$$\sin\theta = \frac{h}{\sqrt{a^2 + h^2}} = \frac{1}{\sqrt{1 + (a/h)^2}}$$

The effect of the simplifying assumptions used in deriving Eq. (6-24) was investigated in Ref. 15 by using Mohr circle stresses in Eq. (*a*). The results differed by less than 10 percent.

The critical stress to be used in Eq. (6-24) is given by Eq. (6-14) provided the value of $F_{v,cr}$ so determined does not exceed F_{vy} or, if the proportional limit is taken into account, provided it does not exceed F_{vp}. The inelastic and strain-hardening ranges of shear behavior were discussed in Art. 5-16, where it was shown that the nonproportional range of stress can be determined by Eq. (5-30). Thus,

$$F_{v(cr)i} = \sqrt{0.8F_{vy}F_{v,cr}} = \sqrt{F_{vp}F_{v,cr}} \tag{6-25}$$

where $0.8F_{vy}$ is the proportional limit F_{vp}, and $F_{v,cr}$ is given by Eq. (6-14). For this case, Eq. (6-24) becomes

$$F_{vui} = F_{v(cr)i} + \tfrac{1}{2}(F_y - \sqrt{3}\,F_{v(cr)i}) \sin\theta \tag{6-26a}$$

It will be noted that the factor in parentheses in this equation vanishes if $F_{v(cr)i} = F_{vy} = F_y/\sqrt{3}$. This means that the panel yields as a shear field, so that no tension field can develop and

$$F_{vui} = F_{v(cr)i} \tag{6-26b}$$

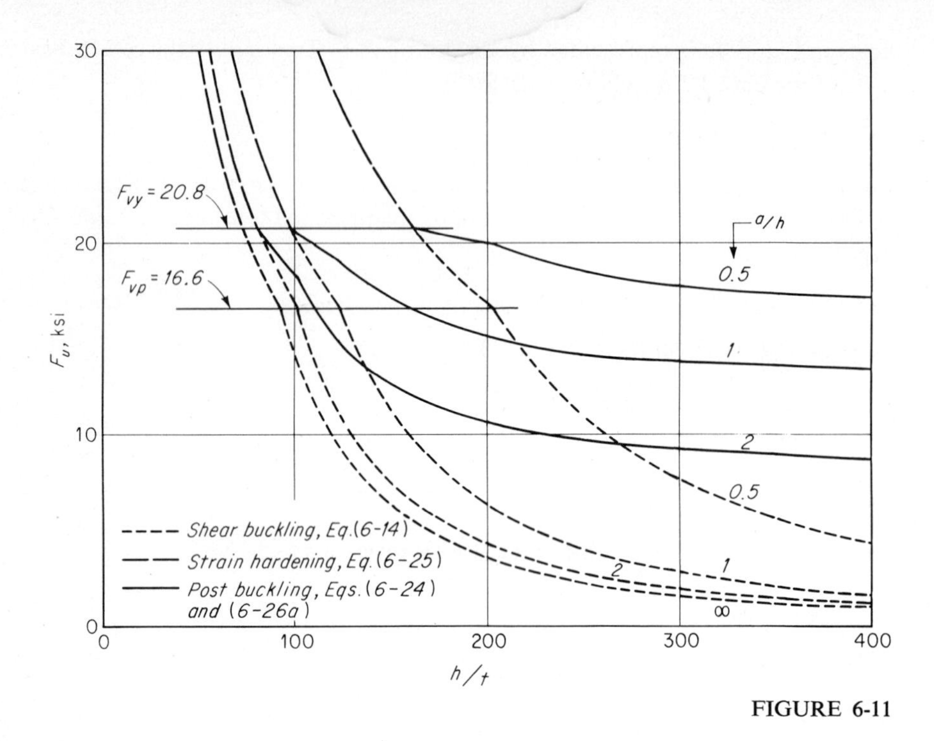

FIGURE 6-11

Figure 6-11 shows plots of the shear-buckling stress, the strain-hardening stress, and the postbuckling stress of webs of A36 steel for four values of a/h. With no transverse stiffeners a/h is infinite, in which case Eqs. (6-24) and (6-26) reduce to $F_{vu} = F_{v,\,cr}$ and $F_{vui} = F_{v(cr)i}$, and there is no postbuckling strength. It will be noted that the postbuckling strength of slender webs is considerable. Thus, with $h/t = 300$ and $a/h = 1$, the shear-buckling stress is 2.8 ksi while the postbuckling strength is 13.8 ksi.

Equations (6-24) and (6-26*a*) are compared in Ref. 15 with results of tests on girders with 50-in. webs whose slenderness h/t ranged from 131 to 382. The ratio of the test shears to the predicted values ranged from 0.88 to 1.12. A comparison of the strain-hardening shear strength with results of tests on beams without transverse stiffeners, whose web slenderness ranged from 50 to 70, is shown in Fig. 5-32.

EXAMPLE 6-1 Compute V_u for an A36-steel plate-girder panel, for which $a =$ 84 in. and $h = 60$ in., for the following thicknesses: (*a*) $t = \frac{1}{4}$ in., (*b*) $t = \frac{5}{8}$ in., (*c*) $t =$ 1 in.

SOLUTION

$$\frac{a}{h} = \frac{84}{60} = 1.4 \qquad k = 5.34 + \frac{4}{1.4^2} = 7.38$$

$$\sin\theta = \frac{1}{\sqrt{1+1.4^2}} = 0.577 \qquad F_{vy} = \frac{36}{\sqrt{3}} = 21 \text{ ksi}$$

$$F_{vp} = 0.8 \times 21 = 16.8 \text{ ksi}$$

$$F_{v,\,cr} = \frac{k\pi^2 E}{12(1-\mu^2)(h/t)^2} = \frac{27{,}000k}{(h/t)^2} = \frac{196{,}000}{(h/t)^2}$$

Panel a: $\quad \dfrac{h}{t} = \dfrac{60}{0.25} = 240 \qquad F_{v,\,cr} = \dfrac{196{,}000}{240^2} = 3.42 \text{ ksi}$

From Eq. (6-24)

$$F_{vu} = 3.42 + \tfrac{1}{2}(36 - 3.42\sqrt{3}\,)\,0.577 = 3.42 + 8.66 = 12.08 \text{ ksi} < F_{vp}$$

$$V_u = F_{vu}ht = 12.08 \times 60 \times 0.25 = 181 \text{ kips}$$

Panel b: $\quad \dfrac{h}{t} = \dfrac{60}{0.625} = 96 \qquad F_{v,\,cr} = \dfrac{196{,}000}{96^2} = 18.1 \text{ ksi} > F_{vp}$

From Eq. (6-25)

$$F_{v(cr)i} = \sqrt{0.8F_{vy}\,F_{v,cr}} = \sqrt{16.8 \times 18.1} = 17.5 \text{ ksi} < F_{vy}$$

From Eq. (6-26a)

$$F_{vui} = 17.5 + \tfrac{1}{2}(36 - 17.5\sqrt{3}\,)\,0.577 = 17.5 + 1.7 = 19.2 \text{ ksi}$$

$$V_u = 19.2 \times 60 \times 0.625 = 720 \text{ kips}$$

Panel c: $\quad \dfrac{h}{t} = 60 \qquad F_{v,\,cr} = \dfrac{196{,}000}{60^2} = 54.5 \text{ ksi} > F_{vp}$

From Eq. (6-25)

$$F_{v(cr)i} = \sqrt{0.8F_{vy}F_{vcr}} = \sqrt{16.8 \times 54.5} = 30.4 \text{ ksi} > F_y$$

From Eq. (6-26b)

$$F_{vui} = F_{v(cr)i} = 30.4 \text{ ksi}$$

$$V_u = 30.4 \times 60 \times 1 = 1{,}820 \text{ kips}$$

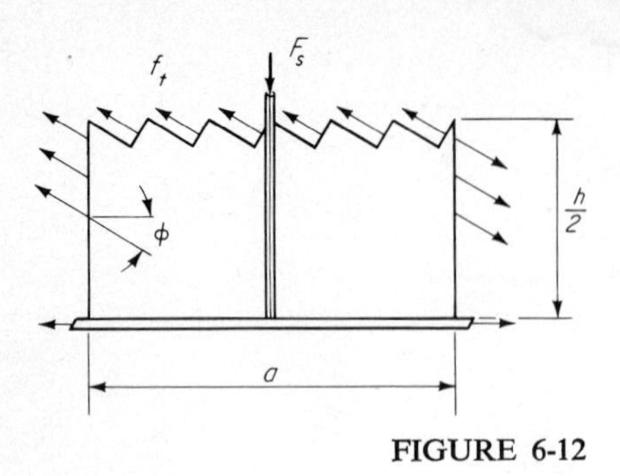

FIGURE 6-12

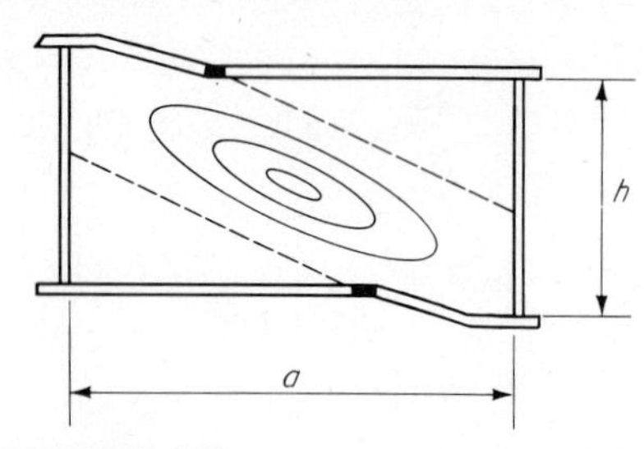

FIGURE 6-13
Shear buckling of plate-girder panel. (*From K. C. Rockey, Br. Constr. Steelwork Assoc. Conf. on Steel Bridges, Inst. Civ. Eng., London, June 1968, Sess. 1, pap. 3.*)

Figure 6-7*b* shows that the transverse stiffeners of a tension-field plate girder serve as supports for the flanges, which are subjected to transverse loading. In the case of the partial tension fields shown in Fig. 6-9*b*, the stiffeners can be thought of as the vertical compression members of a truss in which the web panels act as diagonal tension members and the flanges as chords. In either case, the stiffener forces F_s can be determined by summing vertical components in Fig. 6-12. Thus,

$$F_s = f_t \, at \sin \phi \cos \phi = f_t \frac{at}{2}(1 - \cos 2\phi) = \tfrac{1}{2} f_t \, at \,(1 - \cos \theta)$$

which gives, using Eq. (*c*),

$$F_s = \tfrac{1}{2} at(F_y - \sqrt{3} F_{v,\,cr})(1 - \cos \theta) \tag{6-27}$$

For panel a of the example above:

$$\cos \theta = \frac{a}{\sqrt{a^2 + h^2}} = \frac{a/h}{\sqrt{1 + (a/h)^2}} = \frac{1.4}{\sqrt{1 + 1.4^2}} = 0.813$$

$$F_s = \tfrac{1}{2} \times 84 \times 0.25(36 - 3.42\sqrt{2})(1 - 0.813) = 59 \text{ kips}$$

As has been mentioned, the tension-field formulas in this article are based on the assumption that the flanges are flexible. According to Rockey,[16] the collapse mechanism of a shear panel is that shown in Fig. 6-13, where plastic hinges develop in the flanges. With increasing flexibility of the flanges each hinge moves toward the near stiffener, so that the limiting case for flanges with no stiffness is as shown in Fig. 6-9*b*. In tests on two longitudinally stiffened girders with webs 50 in. deep and 0.0666 in. thick ($h/t = 750$), one with flanges whose moment of inertia was $7.35 \times 10^{-6} a^3 t$ and the other with flanges whose moment of inertia was $2.54 \times 10^{-4} a^3 t$, he found that the shear capacity of the panel with the stiffer flanges was 75 percent greater than that of the one with the more flexible flanges. Equation (6-24) would predict the same shear capacity for both.

6-10 SPECIFICATION PROVISIONS FOR TENSION-FIELD SHEAR

The AISC specification allowable shear on plate-girder webs is based on Eqs. (6-24), (6-25), and (6-26*a*). Equation (6-24) can be written in the form

$$F_{vu} = F_{vy}\left[\frac{F_{v,cr}}{F_{vy}} + \frac{1}{2}\left(\frac{F_y}{F_{vy}} - \sqrt{3}\,\frac{F_{v,cr}}{F_{vy}}\right)\frac{1}{\sqrt{1+(a/h)^2}}\right]$$

Substituting $F_{vy} = F_y/\sqrt{3}$ and using the notation

$$C_v = \frac{F_{v,cr}}{F_{vy}} \tag{6-28}$$

we get

$$F_{vu} = \frac{F_y}{\sqrt{3}}\left[C_v + \frac{\sqrt{3}}{2}\,\frac{1-C_v}{\sqrt{1+(a/h)^2}}\right]$$

Using a factor of safety of 1.65, the allowable shear stress F_v is

$$F_v = \frac{F_{vu}}{1.65} = \frac{F_y}{2.89}\left[C_v + \frac{1-C_v}{1.15\sqrt{1+(a/h)^2}}\right] \tag{6-29}$$

If the value of C_v from Eq. (6-28) exceeds 0.8, the equation is invalid. This is because the proportional limit was taken at $0.8F_{vy}$, as mentioned in Art. 6-9. For this case, we must use Eq. (6-25). With the C_v notation, this gives

$$C_v = \frac{F_{v(cr)i}}{F_{vy}} = \frac{\sqrt{0.8F_{vy}F_{v,cr}}}{F_{vy}} = \sqrt{\frac{0.8F_{v,cr}}{F_{vy}}} \tag{6-30}$$

Finally, if C_v from Eq. (6-30) is unity, $F_{v(cr)i} = F_{vy}$, which means that the web yields as a shear field and no tension field develops. However, if strain hardening in shear is taken into account, C_v may exceed unity. The corresponding shear strength is given by Eq. (6-26*b*), which, with the same factor of safety as in Eq. (6-29), gives

$$F_v = \frac{F_{v(cr)i}}{1.65} = \frac{F_{v(cr)i}}{F_{vy}}\,\frac{F_{vy}}{1.65} = C_v\,\frac{F_y}{1.65\sqrt{3}} = C_v\,\frac{F_y}{2.89} \tag{6-31}$$

The AISC specification makes only limited use of the strain-hardening shear field by limiting F_v by Eq. (6-31) to not more than $0.4F_y$, which corresponds to $C_v = 1.16$. It will be noted that Eq. (6-29) reduces to Eq. (6-31) when $C_v = 1$.

The AISC specification also allows the design of plate-girder webs to be based on shear buckling. In this case, only Eq. (6-31) is needed. This results in stiffener spacing which gives a factor of safety of 1.65 with respect to shear buckling. Furthermore, Eq. (6-29) should not be used for hybrid girders (Art. 6-16) because test data indicate that, under some conditions, such girders may have very little postbuckling shear strength. Therefore, the AISC specification prescribes Eq. (6-31) for webs of hybrid girders.

The AISC formula for the transverse-stiffener area A_{st} is given by Eq. (6-27), put in the form

$$A_{st} = \frac{1 - C_v}{2}\left[\frac{a}{h} - \frac{(a/h)^2}{\sqrt{1 + (a/h)^2}}\right] YDht \tag{6-32}$$

where Y is the ratio of F_y for the web steel to F_y for the stiffener steel and D is a factor to account for eccentricity of the diagonal tension in the web relative to the centroid of stiffeners. Thus $D = 1$ for a pair of stiffeners, 1.8 for a single-angle stiffener, and 2.4 for a single-plate stiffener. In addition, the stiffener is required to have a moment of inertia not less than $(h/50)^4$, the value for transverse stiffeners in shear-field webs, which was mentioned in Art. 6-8.

If a web panel which has developed a tension field adjoins panels with similar fields, as in Fig. 6-9*b*, it is assumed that the stiffeners need furnish only the axial force according to Eq. (6-32). This is only approximately true, since it is clear from the figure that there will be some bending of the stiffener if it alone equilibrates the adjacent tension fields. However, the stiffener at an end of a girder resists diagonal tension on only one side, and if it is not designed accordingly, the tension field may not develop fully because of premature bending failure of the stiffener. Instead of prescribing a design procedure for this situation, the AISC specification requires that an end panel be designed as a shear panel, which allows it to anchor the tension field in the adjacent panel. The required dimensions of the end panel are based on Eq. (6-18), except that the smaller of the panel dimensions a and h must not exceed $348t/\sqrt{f_v}$, rather than the distance a between stiffeners as stated by Eq. (6-18). It would be more consistent to use Eq. (6-31), which is also based on shear-field behavior, and which gives a uniform factor of safety of 1.65, rather than the variable factor of safety of Eq. (6-18).

The AREA specifications do not permit plate-girder design based on tension-field webs. However, it is taken into account in the AASHO provisions for load-factor design of steel bridges.[7]

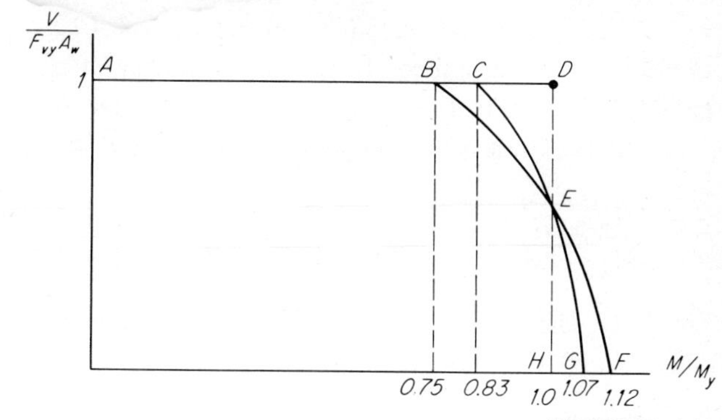

FIGURE 6-14

6-11 COMBINED SHEAR AND BENDING IN WEBS

An interaction formula based on elastic behavior of beam webs under combined shear and bending was discussed in Art. 5-18. In this article, an interaction formula for beams with tension-field webs is discussed. It is based on the following reasoning. If the web of a beam or plate girder is completely yielded in shear, any accompanying moment must be resisted entirely by the flanges. The largest moment that can be developed in this case is $M = F_y A_f h$, where h is the distance between the flange centroids. This situation is represented by

$$V_y = F_{vy} A_w$$

$$0 \lesseqgtr M \lesseqgtr F_y A_f h$$

These equations can be written

$$\frac{V_y}{F_{vy} A_w} = 1$$

$$0 \lesseqgtr \frac{M}{M_y} \lesseqgtr \frac{F_y A_f h}{F_y(A_f + A_w/6)h} = \frac{A_f}{A_f + A_w/6} = \frac{1}{1 + A_w/6A_f}$$

A plot of these equations is shown in Fig. 6-14. AB represents the case $A_w/A_f = 2$, for which the abscissa of point B is

$$\frac{M}{M_y} = \frac{1}{1 + 2/6} = 0.75$$

Similarly, AC represents $A_w/A_f = 1$, for which the abscissa of point C is $M/M_y = 0.83$.

If M/M_y exceeds 0.75 for a girder with $A_w/A_f = 2$, the web must contribute part of the bending resistance. In this case, the shear V must be less than V_y. A curve such as BEF represents this situation. If the girder proportions allow the fully plastic moment M_p to develop, the abscissa of point F is given by

$$\frac{M_p}{M_y} = \frac{F_y(A_f + A_w/4)h}{F_y(A_f + A_w/6)h} = \frac{1 + A_w/4A_f}{1 + A_w/6A_f} = \frac{1 + \frac{1}{2}}{1 + \frac{1}{3}} = 1.12$$

Similarly, the abscissa of point G on the curve CEG for a girder with $A_w/A_f = 1$ is found to be 1.07. In Ref. 17 these curves are assumed to be parabolas with vertices on the M/M_y axis. They have a common intersection E at $V/F_{vy}A_w = 1/\sqrt{3}$. However, since beams with thin webs can develop little or no moment in excess of M_y, $ABEH$ is taken as the interaction curve, instead of $ABEG$, etc. Results of tests on 11 girders show good agreement with the interaction envelope.[17]

The AISC specification provision is based on a straight line connecting B and E. The equation of this line is

$$\frac{M}{M_y} = 1.375 - 0.625\frac{V}{F_{vy}A_w}$$

In terms of allowable stresses $F_b = 0.6F_y$ and F_v $(= 0.4F_y)$

$$\frac{f_b}{0.6F_y} = 1.375 - 0.625\frac{f_v}{F_v}$$

from which

$$f_b = \left(0.825 - 0.375\frac{f_v}{F_v}\right)F_y \not> 0.6F_y \tag{6-33}$$

This is the AISC interaction formula.

6-12 CRIPPLING AND VERTICAL BUCKLING OF PLATE-GIRDER WEBS

The distinction between crippling and web buckling, due to loads on the flange, was discussed in Art. 5-19. Crippling in webs of plate girders is no different from crippling of webs in rolled beams, so that the procedures explained in Art. 5-19 apply also to plate girders. Vertical buckling is also the same phenomenon for both, but because the plate girder may have transverse stiffeners, Eq. (5-38a) must be modified. This equation was derived from Eq. (9-8) for a plate simply supported on all four edges by assuming a^2/b^2 to be small enough, compared to unity, to be dropped. Therefore, it applies to the plate girder with wide stiffener spacing. With the notation of this chapter Eq. (5-38*a*) becomes

$$F_{cr} = \frac{2\pi^2 E}{12(1-\mu^2)(h/t)^2} \tag{a}$$

Basler[18] suggests that the critical stress for closer stiffener spacing should approach the value given by the plate-buckling formula [Eq. (9-11)] with $k = 4$. With the notation of this chapter, this gives

$$F_{cr} = \frac{4\pi^2 E}{12(1-\mu^2)(a/t)^2} \tag{b}$$

This is equivalent to assuming uniformly distributed compression on both flanges, rather than on one as in Eq. (*a*). Basler's formula is

$$F_{cr} = \frac{2\pi^2 E}{12(1-\mu^2)(h/t)^2}\left(1 + 2\frac{h^2}{a^2}\right) \tag{c}$$

The value of F_{cr} from this equation approaches that of Eq. (*a*) if h/a is small and that of Eq. (*b*) if h/a is large. With a factor of safety of 2.7 Eq. (*c*) gives

$$F = \left[2 + \frac{4}{(a/h)^2}\right]\frac{10{,}000}{(h/t)^2} \quad \text{ksi} \tag{6-34a}$$

which is the AISC formula for girders with rotationally free compression flanges. In Art. 5-19, it was mentioned that the critical stress for a web in a beam with a rotationally restrained compression flange is 2.75 times the value for a rotationally free compression flange. AISC takes this into account by specifying

$$F = \left[5.5 + \frac{4}{(a/h)^2}\right]\frac{10{,}000}{(h/t)^2} \quad \text{ksi} \tag{6-34b}$$

for girders with rotationally restrained compression flanges.

Since Eqs. (6-34) are derived for a uniformly distributed load, provision must be made for extending them to the case of a concentrated load or a load distributed over a distance less than the panel length a. According to the AISC specification, such a load is divided by the smaller of the areas ht or at to obtain an equivalent uniformly distributed load.

If the web-crippling and vertical-buckling requirements discussed above are not satisfied, either the web must be made thicker or bearing stiffeners must be provided.

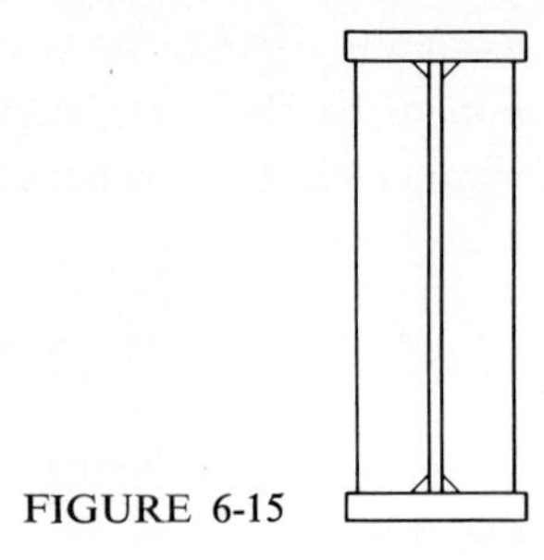

FIGURE 6-15

6-13 BEARING STIFFENERS

Figure 6-15 shows bearing stiffeners consisting of plates welded to the web of a girder. They must fit tightly against the loaded flange. There must be enough area of contact between the stiffeners and the flange to deliver the load in bearing, the stiffeners must be adequate against buckling, and the connection to the web must be sufficient to transmit the load.

The bearing stress on the contact area between stiffener and flange is analogous to the compressive stress at the junction of web and flange of rolled beams subjected to concentrated loads. Because it is a bearing stress, the allowable value can be relatively large. The AISC specification allows $0.90F_y$. Both the AASHO and AREA require milled or ground contact surfaces or a full-penetration groove-weld connection, for which the allowable bearing stresses are $0.80F_y$ and $0.83F_y$, respectively.

Since buckling of bearing stiffeners is analogous to buckling of webs at points of concentrated load, the required moment of inertia of the stiffener is not easy to evaluate. The buckled stiffener may take any of the forms pictured for beam webs in Fig. 5-40, depending on the manner in which the flanges are restrained. In most cases, the compression flange of the girder will be supported laterally at points of concentrated load by bracing or by beams framing into it, so that buckling will approximate the form of an end-fixed column. Even if the flanges are free to rotate, the stiffeners need not be considered as end-hinged columns because the load concentrated on one end of the stiffener is resisted by forces distributed along its connection to the web instead of by a force concentrated at the opposite end, as in columns.

The AISC specifies that the effective column length of a pair of stiffeners be taken at not less than three-fourths the depth of the girder. A strip of web of width not more than $25t$, where t is the thickness of the web, is considered to be a part of the cross section if the stiffeners are at an interior point of the girder. For stiffeners at the end of the girder, the strip is taken to be $12t$ in width. AASHO also requires that bearing stiffeners be designed as columns but does not specify the effective length. The effective strip of web is $18t$. AREA specifies an allowable compression of $0.55F_y$.

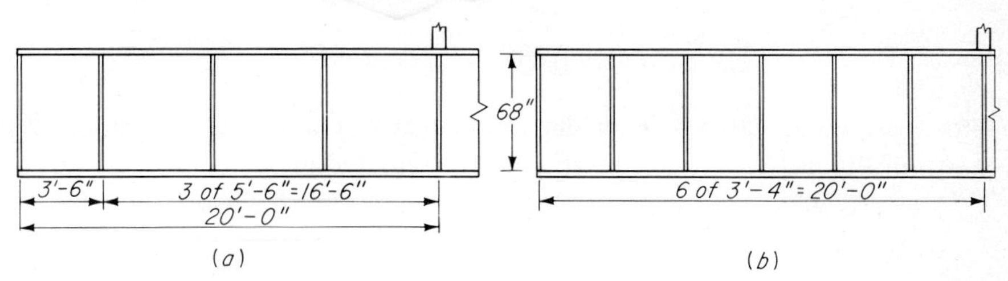

FIGURE 6-16

Slenderness limits for bearing stiffeners are $b/t \lessapprox 95/\sqrt{F_{y,\,\text{ksi}}}$ for AISC, $2{,}300/\sqrt{F_{y,\,\text{psi}}}$ for AREA, and $12\sqrt{33{,}000/F_{y,\text{psi}}}$ for AASHO.

The connection to the web is merely a matter of providing sufficient welding to transmit the calculated load on the stiffener.

In order to distinguish between transverse stiffeners required at points of concentrated load (bearing stiffeners) and those required to develop the necessary shear resistance of a panel, the latter are sometimes called intermediate transverse stiffeners.

6-14 STIFFENERS FOR GIRDER OF DP6-1

To determine the intermediate stiffeners, we have, from the data of DP6-1,

$$h = 68 \text{ in.} \qquad t = 0.375 \text{ in.} \qquad A_w = 25.5 \text{ in.}^2 \qquad V = 240 \text{ kips}$$

$$f_v = \frac{240}{25.5} = 9.06 \text{ ksi}$$

From Eq. (6-18)

$$a = \frac{348t}{\sqrt{f_v}} = \frac{348 \times 0.375}{\sqrt{9.06}} = 43 \text{ in.}$$

This is the minimum distance from the end of the girder to the first transverse stiffener (Fig. 6-16*a*). If tension-field action is considered in determining the spacing of the remaining stiffeners, Eqs. (6-29) and (6-30) must be used. Tables of the allowable shear F_v by these equations are given in the AISC Manual. The shear at the end of the first panel is

$$V = 240 - 3.3 \times \frac{42}{12} = 228 \text{ kips}$$

$$f_v = \frac{228}{25.5} = 8.94 \text{ ksi} \qquad \frac{h}{t} = \frac{68}{0.375} = 181$$

With these data, Table 3-36 in the Manual gives $a/h = 1$, or $a = 68$ in. The layout shown in Fig. 6-16*a* satisfies the stiffener-spacing requirement.

The required area of the stiffener is given by Eq. (6-32), in which C_v is given by Eq. (6-28). Thus,

$$C_v = \frac{F_{v,cr}}{F_{vy}} = \frac{1}{F_y/\sqrt{3}} \frac{\pi^2 E}{12(1-\mu^2)(h/t)^2} = \frac{45{,}000k}{F_y(h/t)^2}$$

as in the AISC specification. Therefore, for $a/h = 66/68 = 0.97$

$$k = 4 + \frac{5.34}{0.97^2} = 9.67$$

$$C_v = \frac{45{,}000 \times 9.67}{36 \times 181^2} = 0.368$$

$$A_{st} = \frac{1 - 0.368}{2}\left(1 - \frac{2}{\sqrt{1 + 0.97^2}}\right) \times 1 \times 2.4 \times 68 \times 0.375 = 5.47 \text{ in.}^2$$

for a single-plate stiffener. The required moment of inertia I_s and slenderness b/t are

$$I_s = \left(\frac{h}{50}\right)^4 = \left(\frac{68}{50}\right)^4 = 3.18 \text{ in.}^4$$

$$\frac{b}{t} = \frac{95}{\sqrt{F_y}} = 16$$

A stiffener 1 × 6 in. gives $A = 6$ in.^{2} and $I_s = 1 \times 6^3/3 = 72$ in.^{4}

Since the stiffener spacing requirements have been based on shear alone, the combined effect of shear and bending must be checked in the panel adjacent to the interior column.

$$V = 240 - 20 \times 3.3 = 174 \text{ kips} \qquad f_v = \frac{174}{25.5} = 6.83 \text{ ksi}$$

$$F_v = 0.4 \times 36 = 14.4 \text{ ksi} \qquad \frac{f_v}{F_v} = \frac{6.83}{14.4} = 0.474$$

From Eq. (6-33)

$$f_b = (0.825 - 0.375 \times 0.474)F_y = 0.647F_y \qquad (\text{use } 0.6F_y)$$

Since the allowable value of the bending stress concurrent with the shear stress is $0.6F_y$, the 66-in. panel length is satisfactory.

If the AISC option of neglecting tension-field action is taken, the stiffeners must be spaced according to Eq. (6-31). Thus,

$$F_v = \frac{F_y}{2.89} C_v = \frac{F_y}{2.89} \frac{45{,}000k}{F_y(h/t)^2} = \frac{45{,}000k}{2.89 \times 181^2} = 0.474k$$

$$f_v = 9.06 \text{ ksi} = 0.474k \qquad k = 19.1$$

$$4 + \frac{5.34}{(a/h)^2} = 19.1 \qquad \frac{a}{h} = 0.596$$

$$a = 0.596 \times 68 = 40 \text{ in.}$$

The layout shown in Fig. 6-16*b* satisfies this requirement. (The decrease in shear in the 20-ft distance to the concentrated load is not large enough to save a stiffener by increasing the spacing as the shear decreases.) The only requirements for the stiffener are the values of I_s and b/t calculated above. A stiffener $\frac{1}{4}$ in. × 4 in. gives $I_s = 0.25 \times 4^3/3 = 5.3$ in.^{4} and $b/t = 16$.

Bearing stiffeners at interior column The allowable bearing stress on the stiffeners is $F_p = 0.90 \times 36 = 32.4$ ksi. Therefore,

$$A = \frac{141}{32.4} = 4.35 \text{ in.}^2$$

Use two 5 × $\frac{1}{2}$ in. stiffeners: $b/t = 10$; $A = 2 \times 4.5 \times 0.5 = 4.5$ in.^{2} ($\frac{1}{2}$ in. is deducted from the 5-in. width to allow for the fillet weld connecting flange and web).

The two stiffeners together with a strip $24t$ of the web are checked as a column of effective length $L = 0.75 \times 68 = 51$ in.

$$24t = 24 \times 0.375 = 9 \text{ in.}$$

$$A = 5 + 9 \times 0.375 = 8.4 \text{ in.}^2$$

$$I = \frac{2 \times 0.5 \times 5^3}{3} = 42 \text{ in.}^4$$

$$r = \sqrt{\frac{42}{8.4}} = 2.24 \qquad \frac{L}{r} = \frac{51}{2.24} = 23$$

$$F_a = 20.4 \text{ ksi} \qquad P = 20.4 \times 8.4 = 171 \text{ kips} > 141$$

PROBLEMS

6-9 Check the stiffener requirements and the combined bending and shear for the girder of Prob. 6-3.

6-10 Same as Prob. 6-9 for the girder of Prob. 6-4.

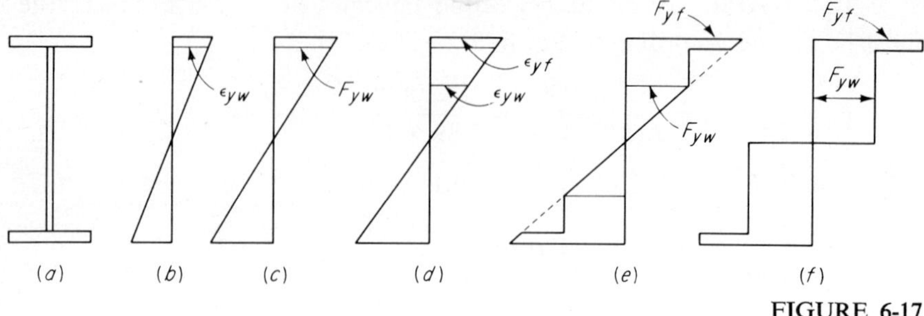

FIGURE 6-17

6-11 Same as Prob. 6-9 for the girder of Prob. 6-6.

6-12 An A572 Grade 42 plate girder consists of a $\frac{7}{16} \times 84$ in. web with 2×16 in. flange plates. The girder is simply supported on a 72-ft. span and has continuous support against lateral buckling. It carries a concentrated load of 220 kips at midspan and a load of 4 klf distributed over the full span. Check the stiffener requirements and the combined bending and shear. AISC specification.

6-13 A simply supported plate girder spans 120 ft. There are concentrated loads at midspan and at each quarter point; each is 120 kips dead and 80 kips live. The uniformly distributed load is 1.8 klf dead and 1.2 klf live. There is continuous lateral support. Design the girder and its stiffeners for a load factor of 1.3 on dead load and 1.7 on live load.

6-14 A railroad plate-girder deck bridge spans 100 ft center to center of bearings. The dead load of track, etc., including an allowance for the weight of the girder is 2.1 klf. There are no conditions at the site that restrict the depth. Live load is Cooper E80 (Fig. 1-3). Design the girder and its stiffeners. AREA specifications.

6-15 HYBRID PLATE GIRDERS

Since the web of a beam or plate girder contributes only a small part of the bending resistance, and since its strength in shear depends on its slenderness h/t, it may be economical to make the web of a lower-strength steel than the flange. Studies have shown that this is often the case.[19] Thus, with the price of A514 steel ($F_y = 100$ ksi) about twice that of A36 steel, a $\frac{1}{2}$-in. A36 web costs no more than a $\frac{1}{4}$-in. A514 web. Furthermore, for a given depth, the web slenderness of the $\frac{1}{2}$-in. A36 web would be only one-half that of the A514 web. Therefore the A36 shear-field web may be able to carry a larger shear than the A514 shear-field web. On the other hand, the tension-field strength of a $\frac{1}{4}$-in. A514 web may be more than that of the $\frac{1}{2}$-in. A36 web.

Beams with stronger steel in the flanges than in the web are called *hybrid beams.* The bending-stress distributions of a hybrid beam at various stages of loading are

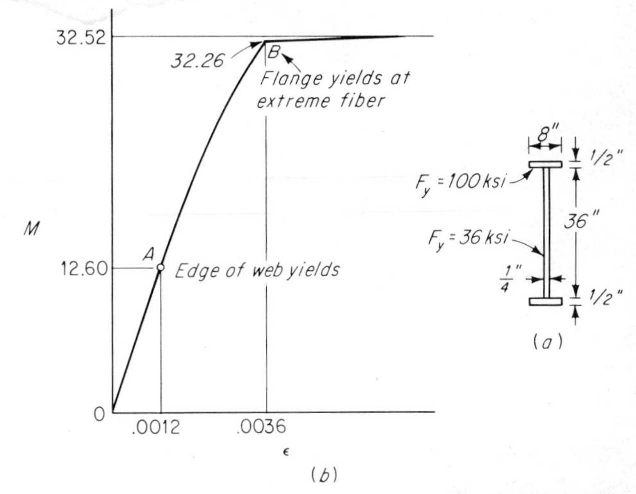

FIGURE 6-18

shown in Fig. 6-17. Beginning of yield in the web is shown in *b* and *c*. If the load is increased beyond that at this stage to the beginning of yielding of the flange, the strain distribution shown in Fig. 6-17*d* is reached. The corresponding distribution of stress is shown in *e*. Continued increase in strain leads eventually to the plastic-moment distribution shown in Fig. 6-17*f*.

Variation in moment with strain on the extreme fiber for the cross section of Fig. 6-18*a* is shown in Fig. 6-18*b*. Although the variation is linear only to the stage where the edge of the web yields (point *A*), the curvature of the segment *AB* is so small that *OAB* is practically straight. After yielding of the flange the rate of increase of moment falls off rapidly. Thus, the bending behavior of the hybrid beam is virtually the same as that of the homogeneous beam.

The ratio of test moment to the predicted moment at beginning of yield of the flange in a series of tests on eight hybrid girders with A514 flanges and A36 webs ranged from 0.88 to 1.01. Web slenderness h/t ranged from 95 to 298. The slenderness b/t of the flange projection was about 7.6. Three girders (with $h/t = 298$) failed by vertical buckling of the compression flange, four by local buckling of the flange, and one by lateral-torsional buckling.[20] Failure in the latter case was caused by failure of the lateral supports.

EXAMPLE 6-2 Design a hybrid cross section for the girder of DP6-1, using an A514 flange and an A36 web.

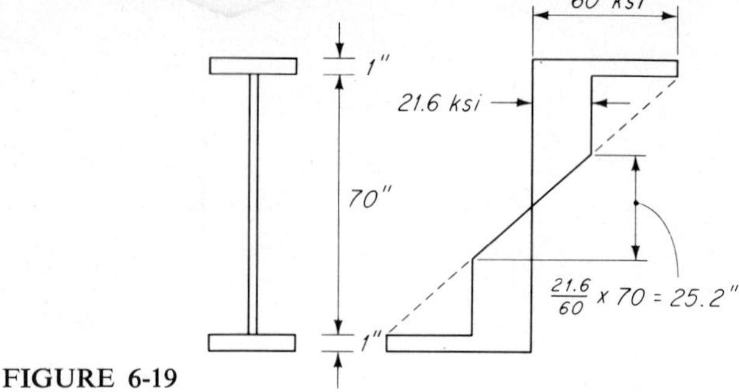

FIGURE 6-19

SOLUTION Since the flange will need roughly only 36/100 times the flange area for the A36 girder, a flange about 1 in. thick should be sufficient. Therefore, the web can be made 70 in. deep without exceeding the allotted depth of girder of 73.5 in.

Assuming the web to be about half as effective as for a homogeneous girder, the required flange area is given by

$$A_f = \frac{M}{F_b h} - \frac{A_w}{12}$$

The allowable stress is 0.6 × 100 = 60 ksi. Using a web $\frac{3}{8} \times 70 = 26.25$ in.^{2}, we get

$$A_f = \frac{4{,}300 \times 12}{60 \times 71} - \frac{26.25}{12} = 12.1 - 2.2 = 9.9 \text{ in.}^2$$

Using a 1 × 10 flange plate, the allowable moment is given by (Fig. 6-19)

$$\text{Flange:} \qquad 60 \times 10 \times 71 \qquad = 42{,}600$$

$$\text{Web:} \qquad 21.6 \times \frac{3}{8} \times \frac{70^2}{4} \qquad = 9{,}920$$

$$\text{Web:} \qquad -\frac{1}{2} \times 21.6 \times \frac{3}{8} \times \frac{25.2^2}{6} = -430$$

$$52{,}090 \text{ in.-kips}$$

Therefore, the allowable moment is 4,340 ft-kips; the required value is 4,300 ft-kips.

An approximate, equivalent extreme-fiber stress has been developed,[8] which, when applied to a homogeneous girder of the same steel as the flange, gives the allowable moment for the hybrid girder. This reduced stress is used in the AISC specification, which specifies the following allowable stress on hybrid girders:

$$F_b' = F_b \frac{12 + (3\alpha - \alpha^3)A_w/A_f}{12 + 2A_w/A_f}$$

where α is the ratio of the yield stresses, F_{yw}/F_{yf}. Using this formula for the girder in Example 6-2 gives

$$\frac{A_w}{A_f} = \frac{26.25}{10} = 2.625 \qquad \alpha = \frac{36}{100} = 0.36$$

$$F_b' = 0.6 \times 100 \frac{12 + 2.625(1.08 - 0.047)}{12 + 5.25} = 51.2 \text{ ksi}$$

The moment of inertia is

$$I = 2 \times 10 \times 35.5^2 + \frac{3}{8}\frac{70^3}{12} = 25{,}210 + 10{,}720 = 35{,}930 \text{ in.}^4$$

This gives

$$M = \frac{51.2 \times 35{,}930}{36 \times 12} = 4{,}260 \text{ ft-kips}$$

AASHO uses a similar formula for a reduced allowable bending stress on hybrid girders.

The AISC increased allowable bending stress $0.66F_y$ for compact sections should not be used for hybrid girders even though the flange and web b/t ratios satisfy the requirements for compact sections. This is because the 10 percent increase is based on the fact that the ratio of the plastic moment to yield moment of a homogeneous girder is about 1.10. This ratio will be smaller for a hybrid beam.

6-16 WEBS OF HYBRID GIRDERS

Shear tests on two hybrid girders with $h/t = 191$ and one combined shear and bending test on a girder with $h/t = 147$ are described in Ref. 20. The experimental values for the two shear tests were 86 and 92 percent of the value predicted by Eq. (6-24), upon which the AISC formula [Eq. (6-29)] is based. On the other hand, in six tests in combined shear and bending reported in Ref. 8, the shear strength in some cases was significantly less than that predicted by Eq. (6-24), and in two cases it was less than the critical stress computed by Eqs. (6-14) and (6-25). For this reason, it has been recommended that webs of hybrid beams be designed as shear fields and that stiffener spacing be determined accordingly.[8] Hence, with the AISC specification, Eq. (6-31) would be used, while for the AASHO specifications Eq. (6-17) is applicable. Accordingly, the stiffener requirements for the hybrid girder designed in the preceding article would be the same as for the A36 girder of DP6-1. These requirements were determined in Art. 6-14, where it was found that a stiffener spacing of 40 in. was needed (Fig. 6-16*b*).

The zone of yielding in the compression portion of the web of hybrid girders raises a question about web crippling under transverse loads not carried by bearing stiffeners. Tests performed to answer this question showed that crippling loads equal to or greater than the value obtained by multiplying the effective bearing area shown in Fig. 5-37 by the yield stress of the web can be supported.[21] These tests also showed that the bending strength of the girder is slightly reduced by the transverse load but that the reduction is negligible.

The yielded compression zone may also reduce vertical-buckling resistance of webs compressed by transverse load on the flanges. Based on results of one test, it has been suggested that only Eq. (6-34*a*) be used to evaluate the allowable vertical-buckling stress and that $(F_y - 50{,}000)/75$ be subtracted from the value given by the formula.[8]

6-17 FLANGE BUCKLING IN HYBRID GIRDERS

Tests have shown that the yielding in the compression zone of hybrid girders does not significantly affect the ability of the web to support the compression flange against vertical buckling[8] (Fig. 6-5). The tests also show that the prescribed slenderness limits b/t according to AISC and AASHO are adequate to control local buckling of the compression flange. Thus, according to AISC, $b/t \lessapprox 95/\sqrt{F_{y,\,\text{ksi}}}$ while for AASHO $b/t \lessapprox 1{,}625/\sqrt{f_{a,\,\text{psi}}}$ (Table 4-4).

Regarding lateral-torsional buckling in hybrid girders, it will be recalled that the torsional resistance of the girder as well as its lateral bending resistance are involved in this phenomenon (Art. 5-7) and that the torsional resistance consists of two parts, the St. Venant torsion and the additional contribution of nonuniform warping. The latter, which is given by $EC_w\, d^3\beta/dz^3$, is contributed almost entirely by the flanges. Furthermore, since $J = bt^3/3$, most of the St. Venant resistance is also contributed by the flanges. Thus, it would appear that lateral-torsional buckling should be affected only slightly by yielding of the compression zone of the web. However, it is suggested in Ref. 8 that only the nonuniform-warping torsion be considered. This would mean, for example, that J would be taken zero in Eq. (5-10*a*) in using the equivalent radius of gyration procedure for evaluating lateral-torsional buckling or, where specification provisions are followed, only Eqs. (5-16) would be used in the AISC specification.

6-18 WELDING OF GIRDER COMPONENTS

Fillet welds connecting the flange to the web may be as small as the specifications permit for the thickness of the flange, provided the allowable stress is not exceeded. The effect of flange thickness on the size of the weld should be kept in mind in proportioning the girder, since the minimum permissible size of weld increases with thickness of the flange (Table 2-14). A $\frac{5}{16}$-in. fillet is usually the largest that can be placed in one pass without special equipment. Although modern shop practice makes continuous welds very desirable, intermittent welds may be economical. Replacement of large intermittent fillet welds, say, $\frac{3}{8}$-in., with $\frac{1}{4}$-in. continuous fillet welds made by automatic welding equipment would probably result in a considerable saving in cost. However, it is questionable whether any savings in cost would ensue from substituting a $\frac{1}{4}$-in. continuous weld made with automatic equipment for $\frac{1}{4}$-in. intermittent fillet welds.

Transverse stiffeners may be stopped short of the tension flange but should have a tight fit at the compression flange. If the stiffeners are on one side of the web, they should be welded to the compression flange. Bearing stiffeners must have a tight fit against the flange through which they receive their load and, according to AASHO, must have a milled or ground surface or be attached to the flange by full-penetration groove welds. Stiffeners may be connected to the web by intermittent or continuous fillet welds. Because of the difficulty of starting and stopping automatic welding machines, fabricators often prefer the continuous weld.

EXAMPLE 6-3 The minimum size fillet weld for the $2\frac{1}{4}$-in. flange of the girder of DP6-1 is $\frac{3}{8}$ in. (Table 2-14). The allowable shear per inch for two $\frac{3}{8}$-in. E60 fillet welds is

$$q = 2 \times 18 \times 0.707 \times 0.375 = 9.5 \text{ kli}$$

The actual shear is

$$q = \frac{VQ}{I} = \frac{240 \times 31.5 \times 35.12}{87{,}580} = 3.0 \text{ kli}$$

The yield strength of the weld metal need not equal the yield strength of the flange. The yield strength of the weld metal was less than that of the flange in most of the hybrid beams tested in Ref. 8, and no problems were encountered even in the beams which were loaded to their shear strength.

If intermittent welds are used, their longitudinal clear spacing connecting the tension flange must not exceed 24 times the thickness of the thinner plate or 12 in.

For the compression flange the clear spacing must not exceed $127/\sqrt{F_y}$ times the thickness of the thinner plate or 12 in. The length of intermittent welds should be at least four times their nominal size and not less than 1.5 in.

For 3-in. $\frac{3}{8}$-in. E60 fillet welds the spacing is

$$s = 3 \times \frac{9.5}{3.0} = 9.5 \text{ in.}$$

The clear spacing is 6.5 in., which satisfies the above requirements.

6-19 SHOP AND FIELD SPLICES

The locations of web shop splices usually depend upon available lengths of plates, and the actual location is often left to the fabricator. Since they are usually connected by full-penetration groove welds, no computations are needed to verify the adequacy of the splice.

The splice of abutting flange plates of different thicknesses must be sloped to avoid an abrupt change in cross section. According to the AWS specifications, this slope must not exceed 1 vertical in $2\frac{1}{2}$ horizontal. The transition may be provided by chamfering the thicker plate, by sloping the weld itself, or by a combination of chamfering and weld sloping (Fig. 6-20).

Field splices are required in continuous spans and in single spans too long to ship in one piece. Field splices in bridge girders may be bolted because of the difficulty of welding large girders in the field. In some cases, the flanges may be groove welded and the web bolted. Flanges which must be welded from both sides present a problem on the inner face because it is difficult to get a sound weld at the juncture of the flange and the web. Coping the web facilitates complete welding of the flange and gives access for cleaning, chipping, or gouging (Fig. 6-20). However, the resulting stress concentration decreases the fatigue life of the girder.[22] The coped area may be filled with weld metal after the flange has been welded.

A bolted web splice is made with two splice plates (Fig. 6-21*b*). To determine the bolt pattern, we consider a strip of web plate *p* in. high, where *p* is the distance between adjacent horizontal rows of bolts (Fig. 6-22). This portion of the web is acted upon by the bending stresses and the shearing forces shown. Figure 6-22*b* shows half the corresponding portion of the splice plates, the right edges of which are assumed to be subjected to the forces that would exist in the web plate were it continuous at the splice. The upper edges and the left edges are free of forces, but there may be a shearing force on the lower edges. If we neglect the latter force, the two bolts shown must furnish horizontal forces sufficient to balance the resultant of the bending

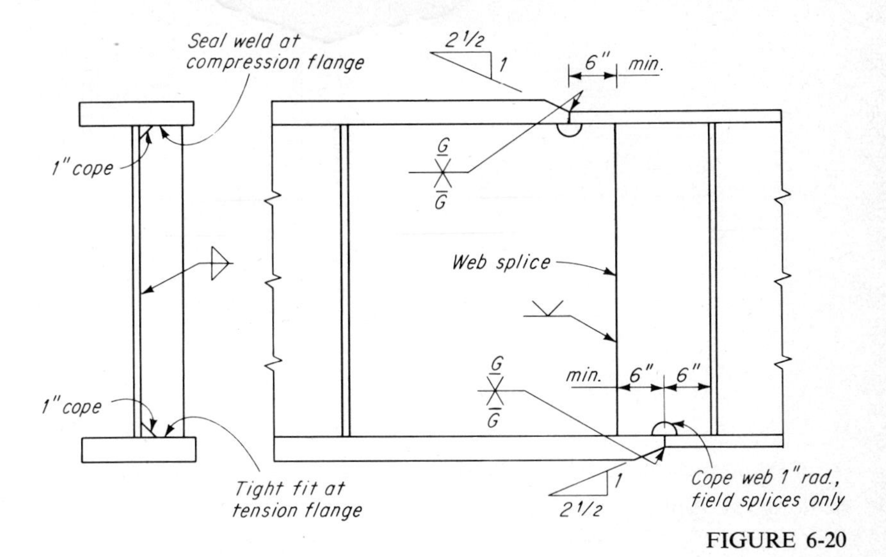

FIGURE 6-20

stresses and, in addition, vertical forces sufficient to balance the vertical shearing force. Let f_b be the average bending stress, f_v the average shearing stress, and t the thickness of the web plate. Then, if R is the allowable force on one bolt, and if the load is shared equally by n bolts, we have

$$\left(\frac{f_b pt}{n}\right)^2 + \left(\frac{f_v pt}{n}\right)^2 = R^2$$

whence

$$p = \frac{Rn}{t\sqrt{f_b^{\,2} + f_v^{\,2}}} \tag{6-35}$$

The shearing stress at any section is practically constant throughout the depth of the web. Therefore, since the bending stress depends upon the distance from the neutral axis, Eq. (6-35) would seem to indicate that the pitch p may increase as we approach the neutral axis or that the number of bolts in each row may decrease if p is held constant. This would be true if R were the same for all rows. But the resisting force that can be developed by any bolt depends upon the amount of its deformation. This deformation must depend on the distance from the neutral axis and must in fact vanish at the neutral axis since deformations in the horizontal direction are reversed on the tension side of the girder. Therefore, if we assume behavior of the bolts consistent with the behavior of the girder itself, the bending stress f_b and the horizontal component of the bolt force obey the same law. Therefore, the horizontal rows in the splice plates should be equidistant, or nearly so.

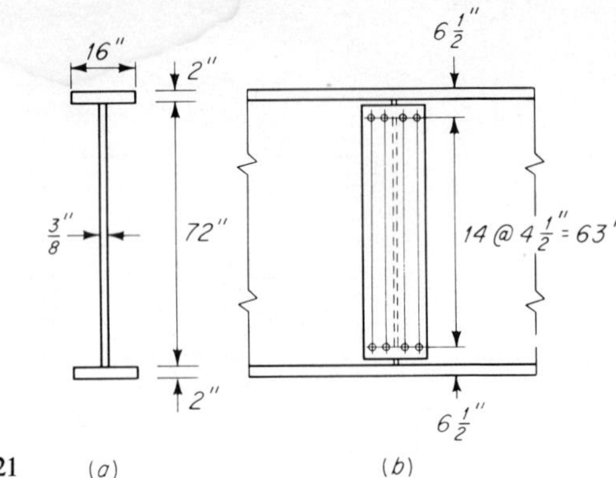

FIGURE 6-21

Opinions differ as to the shear and moment for which a bolted web splice should be designed. Splices are often located at points where the web has excess strength, in which case it would seem reasonable to design the splice for the most unfavorable combination of shear and moment at the point of splice. This is required by the AISC specification. However, the AASHO specifications require that the splice be designed to resist a shear and moment each equal to the average of the calculated service-load value and the allowable value but not less than 75 percent of the latter. The AREA specifications require design for the allowable shear on the gross cross section of the web and for the combined effect of the allowable moment on the net section and the maximum service-load shear at the point of splice.

EXAMPLE 6-4 Design a friction-type bolted splice for the web of the A36 girder shown in Fig. 6-21*a*. The concurrent values of the shear and moment on the girder at the section to be spliced are 210 kips and 320 ft-kips. The moment of inertia of the cross section is 99,260 in.^{4} AISC specification, A325 bolts.

SOLUTION Assuming the first row of bolts to be 4 in. from the inside face of the flange, the bending stress at the outer row of bolts in the splice is

$$f_b = \frac{320 \times 12 \times 32}{99{,}260} = 12.4 \text{ ksi}$$

The shearing stress on the web is

$$f_v = \frac{210}{72 \times 0.375} = 7.8 \text{ ksi}$$

With $\frac{3}{4}$-in. A325 bolts, the allowable double shear per bolt is $2 \times 15 \times 0.44 = 13.2$ kips.

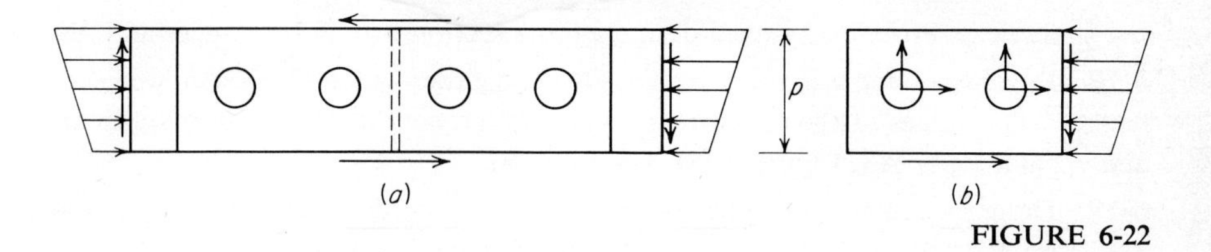

FIGURE 6-22

Since the splice is a friction-type connection, bearing on the bolts is not a factor. For one vertical row of bolts on each side of the splice, Eq. (6-35) gives

$$p = \frac{13.2 \times 1}{0.375\sqrt{12.4^2 + 7.8^2}} = 2.4 \text{ in.}$$

Use two rows with $p = 4\frac{1}{2}$ in. This gives 15 bolts per vertical row (Fig. 6-21*b*). The splice can be checked by computing the moment of inertia of the bolt group by $\sum Ay^2$. However, it is simpler to determine the section modulus S by Eq. (7-4):

$$S = \frac{np(n+1)}{6} = \frac{15 \times 4.5 \times 16}{6} = 180 \text{ per row}$$

$$R_x = \frac{320 \times 12}{2 \times 180} = 10.7 \text{ kips} \qquad R_y = \frac{210}{2 \times 15} = 7.0 \text{ kips}$$

$$R = \sqrt{10.7^2 + 7.0^2} = 12.8 \text{ kips}$$

Since the allowable shear per bolt is 13.2 kips, the bolt layout is satisfactory. Although two $\frac{3}{16}$-in. plates would be equivalent to the $\frac{3}{8}$-in. web, most designers would probably use $\frac{1}{4}$-in. plates. Since the splice is often located at points where the cross section has excess strength, as mentioned above, the thickness of each splice plate may sometimes be less than half the thickness of the web.

PROBLEMS

6-15 A simply supported plate girder spans 98 ft and supports a uniform load, not including its own weight, of 8 klf. It has continuous lateral support. The depth of the girder cannot exceed 7 ft. Design a cross section with A514 flanges and A36 web, and determine the transverse stiffener requirements. AISC specification.

6-16 A simply supported plate girder spans 72 ft and supports a uniformly distributed load, not including its own weight, of 4.2 klf. There are lateral supports at intervals of 18 ft. Design a cross section with A572 flanges and A36 web, and determine the transverse-stiffener requirements. AISC specification.

6-17 Same as Prob. 6-13 except design a cross section with A514 flanges and A36 web.

6-18 Design a bolted web splice for an A36 plate girder with a $\frac{1}{2} \times 100$ in. web and 3×18 in. flanges. The values of the concurrent shear and moment at the point of splice are 410 kips and 7,200 ft-kips. A325 bolts, AISC specification.

6-19 Design a bolted web splice for a hybrid plate girder with $\frac{1}{2} \times 100$ in. A36 web and 3×18 in. A514 flanges. The values of the concurrent shear and moment at the point of splice are 520 kips and 21,200 ft-kips. A325 bolts, AISC specification.

6-20 Design a bolted flange splice for the girder of Prob. 6-18 to be located at the same cross section as the web splice.

6-21 Same as Prob. 6-20 for the girder of Prob. 6-19.

REFERENCES

1 Dubas, C.: A Contribution to the Buckling of Stiffened Plates, *Prelim. Rep. 3d Congr. IABSE Liege*, 1948.

2 Massonnet, C. E. L.: Stability Considerations in the Design of Steel Plate Girders, *Trans. ASCE*, vol. 127, 1962.

3 Basler, K., and B. Thürlimann: Strength of Plate Girders in Bending, *Trans. ASCE*, vol. 128, pt. II, 1963, p. 655.

4 Yen, B. T., and J. A. Mueller: Fatigue Tests of Large-size Welded Plate Girders, *WRC Bull.* 118, November 1966.

5 Lew, H. S., and A. A. Toprac: Fatigue Strength of Hybrid Plate Girders under Constant Moment, *HRB Proc. 40th Annu. Meet.*, 1967.

6 Mueller J. A., and B. T. Yen: Girder Web Boundary Stresses and Fatigue, *WRC Bull.* 127, January 1968.

7 Vincent, G. S.: Tentative Criteria for Load Factor Design of Steel Highway Bridges, *AISI Bull.* 15, March 1969.

8 Report of Subcommittee 1, Joint ASCE-AASHO Committee on Flexural Members, Design of Hybrid Steel Beams, *J. Struct. Div. ASCE*, June 1968.

9 Cooper, P. B.: Strength of Longitudinally Stiffened Plate Girders, *J. Struct. Div. ASCE*, April 1967.

10 Kuntz, F. C.: "Design of Steel Bridges," p. 156, McGraw-Hill, New York, 1915.

11 Elliott, A. L.: Bridges, sec. 18 in E. H. Gaylord and C. N. Gaylord (eds.), "Structural Engineering Handbook," McGraw-Hill, New York, 1968.

12 Report of Task Committee on Lightweight Alloys, Suggested Specifications for Structures of Aluminum Alloys 6061-T6 and 6062-T6, *J. Struct. Div. ASCE*, December 1962.

13 Stein, M., and R. W. Fralich: Critical Shear Stress of Infinitely Long, Simply Supported Plate with Transverse Stiffeners, *NACA Tech. Note* 1851, 1949.

14 Bleich, F.: "Buckling Strength of Metal Structures," McGraw-Hill, New York, 1952.

15 Basler, K.: Strength of Plate Girders in Shear, *Trans. ASCE*, vol. 128, pt. II, 1963.

16 Rockey, K. C.: Factors Influencing Ultimate Behaviour of Plate Girders, *Br. Constr. Steelwork Assoc. Conf. on Steel Bridges, Inst. Civ. Eng., London*, June 1968, sess. 1, pap. 3.

17 Basler, K.: Strength of Plate Girders under Combined Bending and Shear, *Trans. ASCE*, vol. 128, pt. II, 1963.

18 Basler, K.: New Provisions for Plate Girder Design, *Proc. 1961 Natl. Eng. Conf. AISC*, New York.

19 Haaijer, G.: Economy of High Strength Steel Structural Members, *J. Struct. Div. ASCE*, December 1961.

20 Lew, H. S., M. Natarajan, and A. A. Toprac: Static Tests of Hybrid Girders, *Weld. J.*, vol. 48, no. 2, February 1969.

21 Schilling, C. G.: Web Crippling Tests on Hybrid Beams, *J. Struct. Div. ASCE*, February 1967.

22 Stallmeyer, J. E., W. H. Munse, and B. J. Goodal: Behavior of Welded Built-up Beams under Repeated Loads, *Weld. J.*, January 1957.

7

CONNECTIONS

7-1 INTRODUCTION

The design of the structural members discussed in the preceding chapters is based on theories which usually can be depended upon to produce satisfactory results. On the other hand, the behavior of connections is so complex that, in many cases, it is impossible to describe in terms of simple formulas, or for that matter any formula. For this reason, formulas which are derived analytically often require modification to bring them into agreement with test results. Design of connections is sometimes left to the fabricator, and designers do not always give them the attention they deserve. Investigations of structural failures often show connections or other details, rather than the members, to be the origin of the failure.

7-2 BOLTED AND RIVETED CONNECTIONS FOR BEAMS

A riveted or bolted connection between two beams framing at right angles is usually made with two angles placed as shown in Fig. 7-1*a* to form a *framed* beam connection. If the flanges of the two beams are at the same elevation, the connecting beam is cut to clear (Fig. 7-1*b*). The same connection is suitable for beam-to-column connections

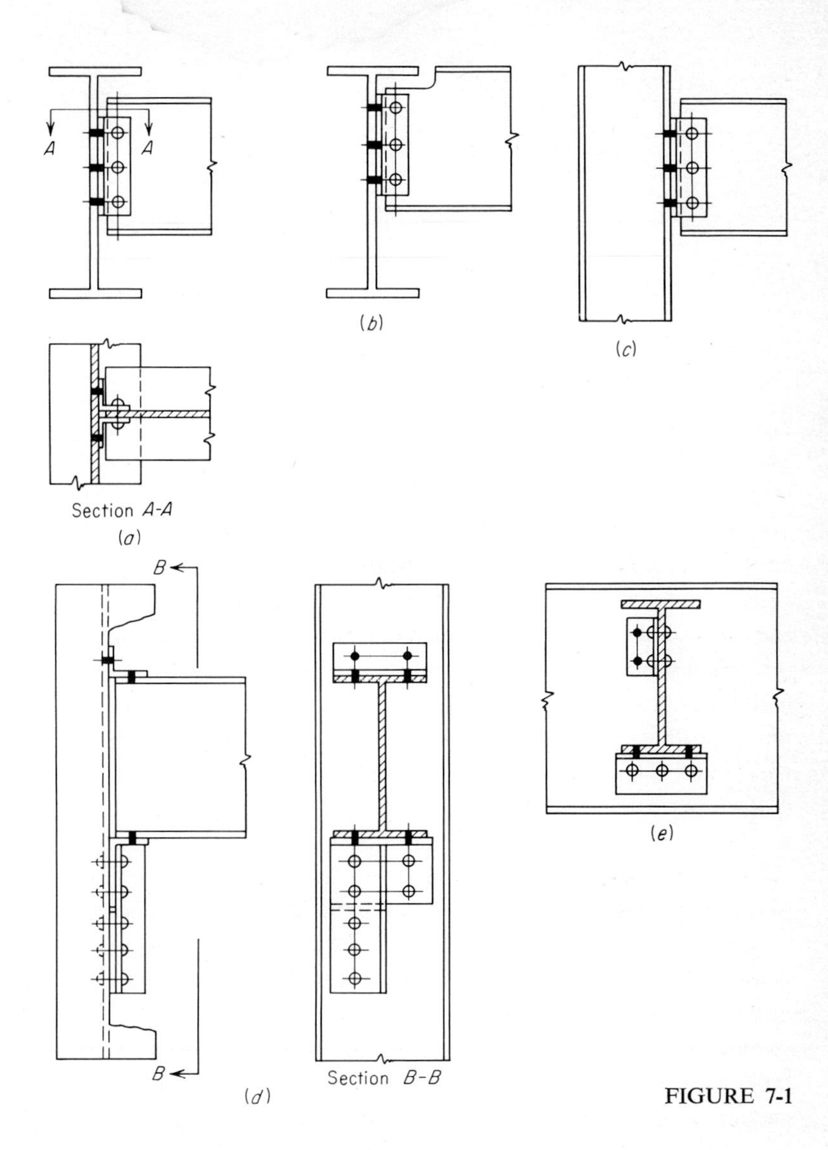

FIGURE 7-1

(Fig. 7-1*c*). If the single row of fasteners connecting the angles to the web in Fig. 7-1*c* is insufficient, angles with legs wide enough for two rows can be used. The outstanding legs of the connecting angles can also be made wide enough to accommodate two rows each. The connections in Fig. 7-1*a* and *b* can of course be similarly modified.

The *seated* beam connection shown in Fig. 7-1*d* is often used when the beam connects to the web of a column. It may also be used to connect the beam to a column flange. In its simplest form the seat consists of a single angle (Fig. 7-1*e*). When the

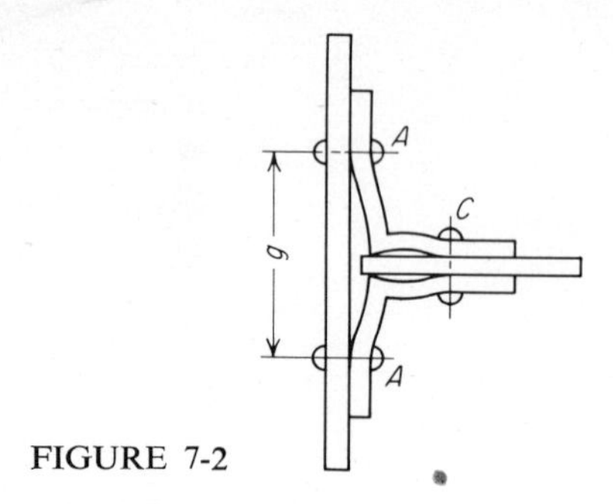

FIGURE 7-2

number of fasteners that can be placed in the leg connected to the column is not enough to support the load, the angle stiffener shown in Fig. 7-1*d* can be added. A second stiffener may be placed alongside the single one shown. The top angle is an essential part of the connection, since it holds the top flange in position and therefore contributes significantly to the resistance of the web to vertical buckling. If the beam rests on a seat connected to the web of another beam, as in Fig. 7-1*e*, the side angle shown should be used if clearance does not permit a top angle.

Conventional design procedure for the framed and seated connections of Fig. 7-1 assumes that they offer no resistance to rotation of the end of the beam in the vertical plane, so that the beam reaction is the only force to be considered. It is evident that rotation of the end of the beam exerts some moment on the connection. The rotation is accommodated in part by an elongation of the upper fasteners of framed connections, and in part by a distortion of the upper portion of the framing angles somewhat as pictured in Fig. 7-2. This suggests that framing angles should be relatively thin and that the gage g should be reasonably large in order to relieve the top fasteners of the framing angles. For the same reason, the top angle of the seat connection should be relatively flexible. Ordinarily, framing angles need be only thick enough to develop in bearing the shear value of the fastener.

Framed connections and seated connections are used so extensively that standard types have been developed. The AISC Manual gives data for a variety of such connections, which are adequate for all but exceptional cases of span and loading.

In Fig. 7-1, the parts of the connections which are shown to be shop-connected (denoted by the open circles) are more likely to be welded, as in Figs. 7-33 and 7-37.

7-3 UNSTIFFENED SEATED CONNECTIONS

The action which takes place when an unstiffened seat angle supports a beam may be inferred from a study of the photographs of Fig. 7-3. It will be noted that the angle acts approximately as a cantilever beam and that the thicker seat in Fig. 7-3*a* tends

FIGURE 7-3
Failure of riveted beam-to-column connections. (*Final Rep. Steel Struct. Res. Comm., Dept. Sci. Ind. Res., H.M. Stationery Office, London, 1936.*)

to concentrate the reaction at the toe of the outstanding leg, while the thinner seat in Fig. 7-3*b* tends to distribute it somewhat more. On the other hand, the web thickness of the beam and the stiffness of its flange both influence the distribution of the reaction. It is virtually impossible to take all these variables into account in an analysis. It is common practice to assume the reaction to be uniformly distributed over a length, measured from the end of the beam, just sufficient to satisfy the web-crippling requirement of the beam. This is obviously a compromise which makes some allowance for the effect of web thickness but none for the relative stiffnesses of the beam and seat. The calculated length of bearing b is given by

$$b = \frac{R}{F_c t} - k$$

where R = reaction
F_c = allowable compression in web at toe of fillet
t = web thickness
k = distance from flange outside face to toe of fillet (Fig. 7-4)

For a large beam with a small reaction, this equation may actually yield a negative value of b. This inconsistency is disposed of by specifying half of $R/F_c t$ as a minimum length.

If the beam were not connected to the seat, the critical section for bending of the seat would be at the top row of fasteners in the vertical leg. However, as the bottom flange of the beam elongates under load, the angle is forced against the column so that the moment in the vertical leg is relieved. Furthermore, the outstanding leg cannot bend as a simple cantilever because of the restraining action of the beam flange to which it is attached. Consequently, it is not easy to determine where the maximum bending moment occurs. It is usually assumed to be at the toe of the fillet in the horizontal leg. The shape factor, which is 1.5 for a rectangular cross section, should be taken into consideration in establishing an allowable stress. The AISC specification makes a partial allowance by specifying $F_b = 0.75F_y$ for bending of solid rectangular cross sections.

EXAMPLE 7-1 A bolted beam seat to connect a W16 × 40 beam to the web of a W8 × 35 column (A36 steel and A325 bolts) is designed in this example, using AISC allowable stresses. The beam reaction is 19.3 kips. The allowable compression F_c is $0.75F_y = 27$ ksi. For the W16 × 40, $k = 1\frac{1}{8}$ in., so that

$$b = \frac{19.3}{27 \times 0.307} - 1.12 = 2.33 - 1.12 = 1.21 > \frac{2.33}{2}$$

The value of k for angles $\frac{1}{2}$ to $\frac{5}{8}$ in. thick ranges from about $\frac{7}{8}$ to $1\frac{1}{4}$ in., depending on the size of the angle. Using $k = 1$ in., the critical section for bending is 1 in. from the face of the column (Fig. 7-4). Beams are usually detailed to clear the column by $\frac{1}{2}$ in., but to allow for underrun in the length of the beam the clearance will be taken $\frac{3}{4}$ in. Therefore, the moment at the critical section is

$$M = 19.3\left(\frac{1.21}{2} + 0.75 - 1\right) = 6.85 \text{ in.-kips} \tag{a}$$

A 6-in. length of seat is the largest that can fit the web of the column. The allowable bending stress is $F_b = 0.75F_y = 27$ ksi. Then, with $F_b = M/S = 6M/bt^2$, we get

$$t = \sqrt{\frac{6M}{bF_b}} = \sqrt{\frac{6 \times 6.85}{6 \times 27}} = 0.503 \text{ in.} \tag{b}$$

Assuming a bearing connection with threads excluded from the shear plane, the allowable shear stress is $F_v = 22$ ksi. For $\frac{3}{4}$-in. bolts,

$$R = 22 \times 0.44 = 9.68 \text{ kips} \qquad n = \frac{19.3}{9.68} = 2$$

A $4 \times 4 \times \frac{1}{2}$ angle would accommodate these bolts and has the thickness calculated in Eq. (*b*). However, $k = \frac{7}{8}$ in. for this angle, while the moment in Eq. (*a*) was computed for $k = 1$ in. Since a smaller k increases M (and t), the next thicker angle, $4 \times 4 \times \frac{5}{8}$, is chosen. For this angle, $k = 1$. A check of the required bolt length shows that threads will be excluded from the shear plane for the thicknesses to be joined. Therefore, the $4 \times 4 \times \frac{5}{8}$ angle, 6 in. long, satisfies the requirements.

The procedure just described is extremely sensitive to small changes in the clearance at the end of the beam, the beam-web thickness, and the values of k for the beam and the seat angle. For example, if the beam clearance or the value of k for the angle is changed by only $\frac{1}{16}$ in., the moment computed in Eq. (*a*) is changed by 18 percent. This sensitivity can lead to inconsistent results which raise some doubts about the validity of the analysis.

EXAMPLE 7-2 Framed connections for the beams of Fig. 7-5 are designed in this example, using AISC allowable stresses. The beams are of A36 steel, and the connections are made with $\frac{3}{4}$-in. A307 bolts. The beam reactions are 22 kips on the W12 × 27 and 35 kips on the W16 × 40. The allowable bearing stress is $1.35F_y = 48.6$ ksi, so that the bearing values on the beam webs are

$$48.6 \times 0.75 \times 0.237 = 8.65 \text{ kips} \qquad \text{W12}$$

$$48.6 \times 0.75 \times 0.307 = 11.2 \text{ kips} \qquad \text{W16}$$

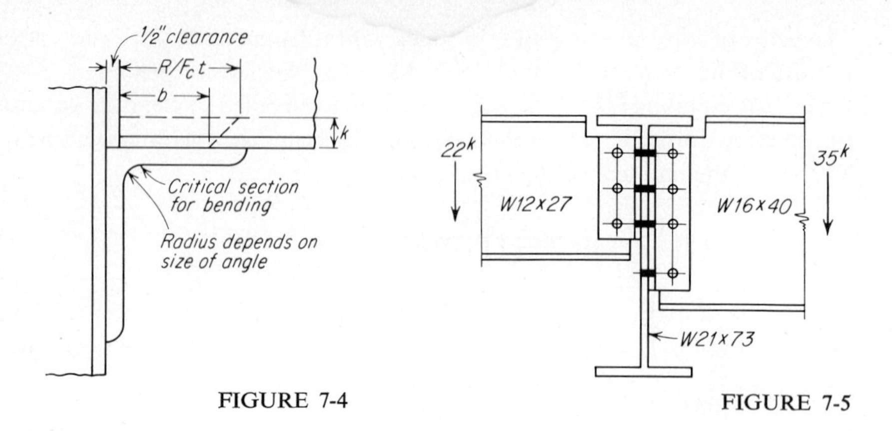

FIGURE 7-4

FIGURE 7-5

The allowable shear is 10 ksi, and the double-shear value per bolt is $2 \times 10 \times 0.44 = 8.8$ kips. The number of bolts required to connect the framing angles to the webs of the W12 and W16 is

$$n = \frac{22}{8.65} = 2.5 \qquad \text{use 3 (W12} \times 27)$$

$$n = \frac{35}{8.8} = 4.0 \qquad \text{use 4 (W16} \times 40)$$

The bolts in the W21 × 73 beam must be sufficient to transmit the reactions of the two beams, or $22 + 35 = 57$ kips. The allowable bearing on the web is $48.6 \times 0.75 \times 0.455 = 16.6$ kips, so that shear controls. The required number of bolt-shear areas is $57/4.4 = 13$. With the arrangement shown in the figure there are six bolts in double shear and two in single shear, which gives 14 shear areas. Framing angles $3\frac{1}{2} \times 3\frac{1}{2} \times \frac{3}{8}$ are large enough to accommodate the bolts with the necessary clearances for tightening.

PROBLEMS

7-1 Design a seat angle to support a W10 × 21 beam on the web of a W12 × 65 column. The beam reaction is 30 kips. There is no beam on the opposite side of the web of the column. Use $\frac{3}{4}$-in. A325 bolts in a bearing-type connection with threads excluded from the shear plane. A36 steel. AISC specification.

7-2 Design seat angles to support two W12 × 50 beams, one on each side of the web of a W12 × 72 column. Each beam reaction is 16 kips. Use $\frac{3}{4}$-in. A502 Grade 2 rivets. A36 steel. AISC specification.

7-3 A W21 × 55 beam is supported on the flange of a W8 × 31 column. The beam reaction is 55 kips. Detail a framed connection using A325 bolts in a bearing-type connection with threads excluded from the shear planes. A36 steel. AISC specification.

7-4 A W12 × 40 beam and a W18 × 55 beam frame on opposite sides of the web of a W21 × 62 girder. The top flanges of the beams and the girder are at the same elevation. The W12 beam spans 12 ft and supports dead and live loads of 1.5 and 4.0 klf, respectively. The W18 spans 13 ft and supports dead and live loads of 2.3 and 6.2 klf, respectively. Design a connection using A325 bolts. A36 steel. AISC specification.

7-5 Design the connections for the stringers to the floor beams of Prob. 5-10. Use $\frac{3}{4}$-in. A325 bolts. AASHO specification.

7-4 ECCENTRICALLY LOADED BOLTED OR RIVETED CONNECTIONS—ELASTIC ANALYSIS

The lines of action of the forces in connected members should in general pass through the centroid of the fastener group. However, it is often impracticable to arrange members in this way, and as a consequence the fasteners are subjected to eccentric forces. The connections of the double-angle members of the truss of DP3-4 are eccentric because the centroids of the angles are not on the gage lines of the rivets. For example, the centroid of the $2\frac{1}{2} \times 1\frac{1}{2} \times \frac{3}{16}$ angles of U_2L_3 is 0.85 in. from the back of the angles, while the gage line of the fasteners is at a distance of $1\frac{3}{8}$ in. Eccentricities of this magnitude are usually neglected in the design of connections. Beams and girders which connect to columns but which cannot be located on or near the centerlines of the columns are sometimes supported on brackets, as shown in Fig. 7-6. The moment resulting from this eccentricity must be considered in the design of the connection.

It is helpful to discuss the connection subjected to eccentric forces by considering first the connection which supports only a couple. Let Fig. 7-7*a* be any arrangement of fasteners in a plate supporting the couple M. If the plate is rigid and the fasteners elastic, rotation of the plate produces shearing deformations in the fasteners which are proportional to and normal to radii from the center of rotation O. Then if stress is proportional to strain, the shearing stress f_v on any fastener is proportional to and normal to the radius r; that is, $f_v = kr$, where k is a constant. Then, if A is the cross-sectional area of the fastener, the force R is

$$R = f_v A = krA \tag{a}$$

In Fig. 7-7*b* the force R is resolved into components R_x and R_y, where x, y are axes originating at O. By similar triangles these components are

$$R_x = \frac{Ry}{r} \qquad R_y = \frac{Rx}{r} \tag{b}$$

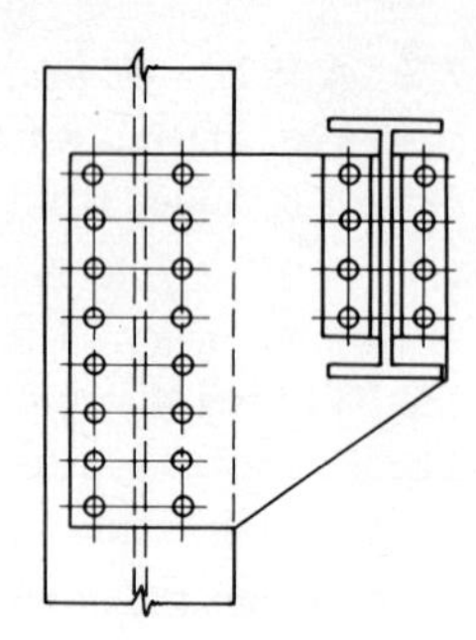

FIGURE 7-6

Substitution of the value of R from Eq. (a) into Eqs. (b) gives

$$R_x = kAy \qquad R_y = kAx \tag{c}$$

Applying the equations of equilibrium to the forces in Fig. 7-7a, and using Eqs. (a) and (c), we have

$$\begin{aligned} \sum R_x &= k \sum Ay = 0 \\ \sum R_y &= k \sum Ax = 0 \\ \sum Rr &= k \sum Ar^2 = M \end{aligned} \tag{d}$$

From the first two of these equations we see that the center of rotation O is at the centroid of the group of fasteners. From the third equation we find $k = M/\sum Ar^2$, which when substituted into Eq. (a) gives

$$R = \frac{MrA}{\sum Ar^2} \tag{e}$$

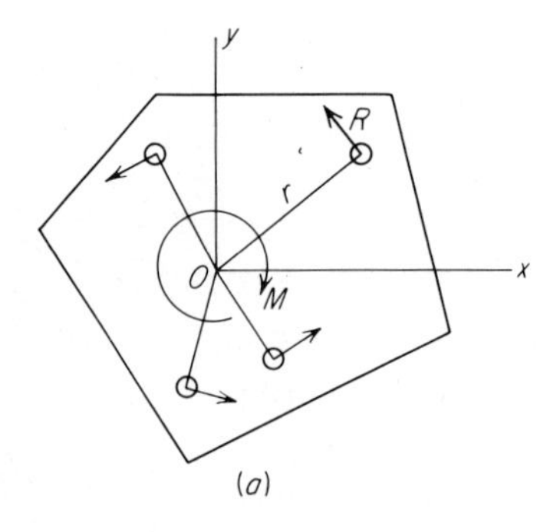

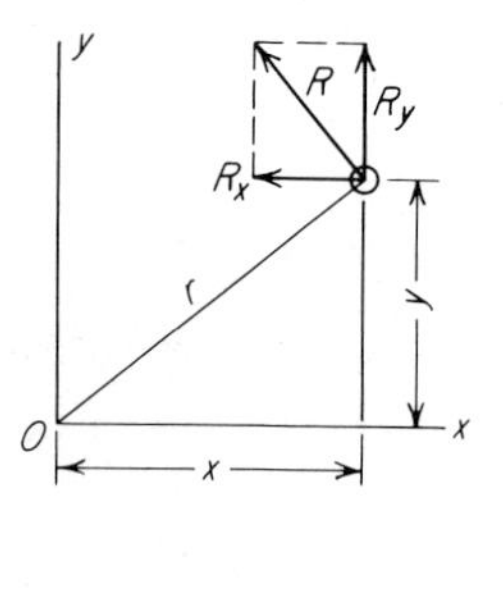

FIGURE 7-7

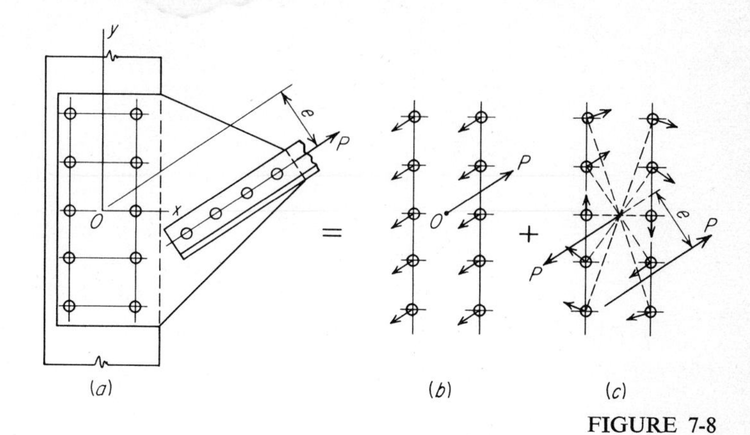

FIGURE 7-8

If the fasteners are of uniform size, as is usually the case, $\sum Ar^2 = A \sum r^2$ and Eq. (*e*) becomes

$$R = \frac{Mr}{\sum r^2} \tag{7-1}$$

Substituting the value of R from Eq. (7-1) into Eqs. (*b*), we get

$$R_x = \frac{My}{\sum r^2} \qquad R_y = \frac{Mx}{\sum r^2} \tag{7-2}$$

A connection supporting a force P which is eccentric with respect to the centroid of a group of fasteners (Fig. 7-8*a*) can be analyzed by replacing P by an equal force at the centroid (Fig. 7-8*b*) and a couple with the moment $M = Pe$ (Fig. 7-8*c*). Assuming again that the plate is rigid and the fasteners elastic, it follows that the fastener forces in Fig. 7-8*b* are equal, while those in Fig. 7-8*c* are given by Eq. (7-1) or Eqs. (7-2). The force on any fastener in Fig. 7-8*a* is the resultant of the force due to P in Fig. 7-8*b* and the force due to $M = Pe$ in Fig. 7-8*c*.

The following example illustrates the analysis of an eccentrically loaded connection (Fig. 7-9). The moment of the 20-kip load with respect to the centroid of the fasteners is $20 \times \frac{3}{5} \times 8 = 96$ in.-kips. The force in fasteners A and B due to the moment is

$$R_x = \frac{96 \times 8}{2(4^2 + 8^2)} = 4.8 \text{ kips}$$

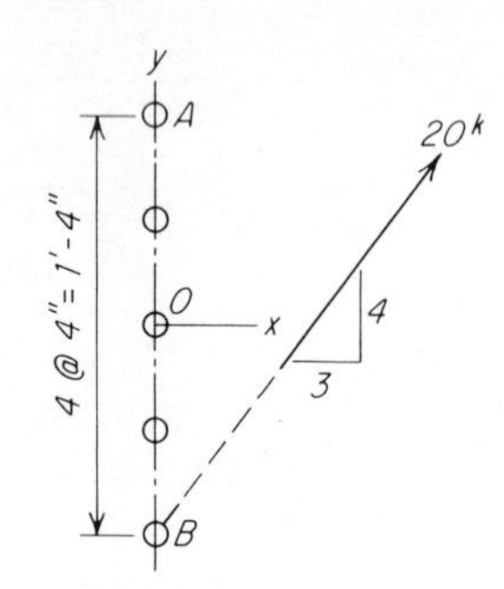

FIGURE 7-9

The components of the force in each fastener due to a 20-kip force applied at the centroid are

$$R_x = \frac{3/5 \times 20}{5} = 2.4 \text{ kips} \qquad R_y = \frac{4/5 \times 20}{5} = 3.2 \text{ kips}$$

The resultant force on the fastener at B is

$$R = \sqrt{(2.4 + 4.8)^2 + 3.2^2} = 7.88 \text{ kips}$$

7-5 ECCENTRICALLY LOADED BOLTED OR RIVETED CONNECTIONS—PLASTIC ANALYSIS

The analysis in Art. 7-4 is based on the assumption that the fasteners are elastic and the plate rigid and that the shearing stress on the most highly stressed fastener does not exceed the proportional limit. If the connection is loaded beyond this point, the rate of increase of shear on the fastener farthest from the center of rotation will decrease, while the forces on fasteners closer to the center of rotation continue to increase at the linearly elastic rate. Therefore, the assumed linear variation in fastener force no longer obtains.

If all fasteners reach their ultimate shearing strength F_{vu}, the ultimate moment M_u is given by

$$M_u = F_{vu} \sum Ar$$

The location of the center of rotation must be assumed and checked by equations of equilibrium. The following example illustrates the procedure (Fig. 7-10). From symmetry, the center of rotation is on the x axis. Its location with respect to the y axis is determined by trial and error. Assuming it to be 3 in. to the left of the y axis, the distances r, given by $r^2 = 3^2 + y^2$, are

$$r_1 = 8.08 \text{ in.} \qquad r_2 = 5.41 \text{ in.} \qquad r_3 = 3.35 \text{ in.}$$

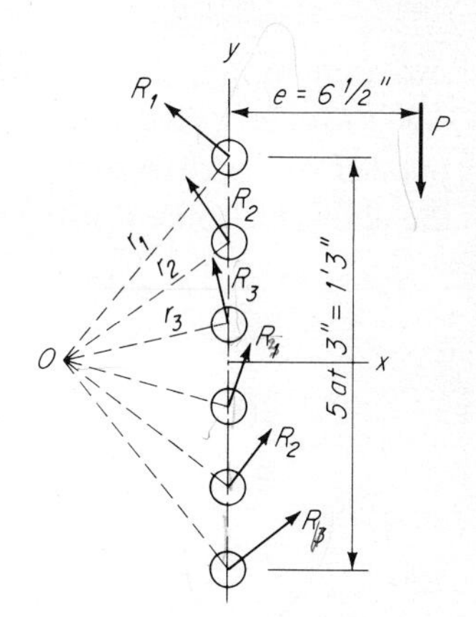

FIGURE 7-10

From summation of vertical components

$$2R\left(\frac{3}{8.08} + \frac{3}{5.41} + \frac{3}{3.35}\right) = P$$

$$P = 3.64R$$

Also, $\sum M_0 = P(e + 3)$, which gives

$$2R(8.08 + 5.41 + 3.35) = P(6.5 + 3)$$

$$P = 3.55R$$

The two values of P do not agree, so that the calculation must be repeated. Since the first value exceeds the second, the assumed distance to the center of rotation is too large. A value of 2.80 in. yields $P = 3.54R$ and $P = 3.52R$, which is close enough. If we assume elastic behavior, as in the previous example, $P = 2.85R$. The ratio of the plastic value of P to the elastic value is 1.24.

Based on the results of tests, the AISC recommends that inelastic behavior be accounted for, in part, in an elastic analysis by substituting the following effective eccentricity e_e for the actual eccentricity e:

$$e_e = e - \frac{1 + 2n}{4}$$

for fasteners equally spaced in a single gage line, and

$$e_e = e - \frac{1 + n}{2}$$

for fasteners equally spaced in two or more gage lines, where n is the number of the fasteners per row.[1] These formulas may give negative values of the effective eccentricity. In this case, it is assumed to be zero. One of the test specimens consisted of two framing angles fastened to a $\frac{3}{8}$-in. A36 steel plate by a single row of six $\frac{3}{4}$-in. A502 Grade 1 rivets at 3-in. spacing. The connection was tested under a load P parallel to the rivet row with an eccentricity of 6.5 in., as in Fig. 7-10. To compute the allowable force P using the effective eccentricity, we have

$$e_e = 6.5 - \frac{1 + 12}{4} = 3.25$$

$$y^2 = 2(1.5^2 + 4.5^2 + 7.5^2) = 158$$

$$R_x = \frac{3.25P \times 7.5}{158} = 0.154P \qquad R_y = \frac{P}{6} = 0.167P$$

$$R = \sqrt{(0.154P)^2 + (0.167P)^2} \qquad P = 4.40R$$

It will be noted that this value of P exceeds even the value based on plastic analysis ($P = 3.53R$). However, the allowable value of R according to the AISC specification is $R = 2 \times 15 \times 0.44 = 13.2$ kips. Therefore, the AISC allowable load is $P = 4.40 \times 13.2 = 58.2$ kips. In the test, the first rivet failed in shear at $P = 181$ kips. Thus, the factor of safety based on P determined by the effective eccentricity is $181/58.2 = 3.1$. If the actual eccentricity, 6.5 in., is used, the allowable P is 37.7 kips, which gives a factor of safety of 4.80.

PROBLEMS

7-6 Refer to Fig. 7-8. How would deformation of the plate affect the distribution of the fastener forces in Fig. 7-8*b*? How would this effect depend upon the length of the fastener rows? How would deformation of the plate affect the distribution of the fastener forces in Fig. 7-8*c*?

7-7 A line of W16 × 50 beams is offset 14 in. from a line of columns. The beam-to-column connection at the end of the last beam in the line is shown in Fig. 7-11. The beam is fastened to the plate through one $4 \times 3\frac{1}{2} \times \frac{3}{8}$ angle. Steel is A36 and the $\frac{7}{8}$-in. rivets are A502 Grade 1. Determine the allowable beam reaction P according to the AISC specification.

7-8 Determine the allowable value of P for the connection shown in Fig. 7-12. A36 steel, $\frac{3}{4}$-in. A307 bolts, AISC specification.

7-6 DESIGN OF ECCENTRICALLY LOADED BOLTED OR RIVETED CONNECTIONS

Since the design of a connection requires a determination of the number of fasteners, it is usually necessary to use trial-and-error methods. Fortunately, most eccentric connections contain one or more parallel lines of fasteners at a uniform pitch, and the

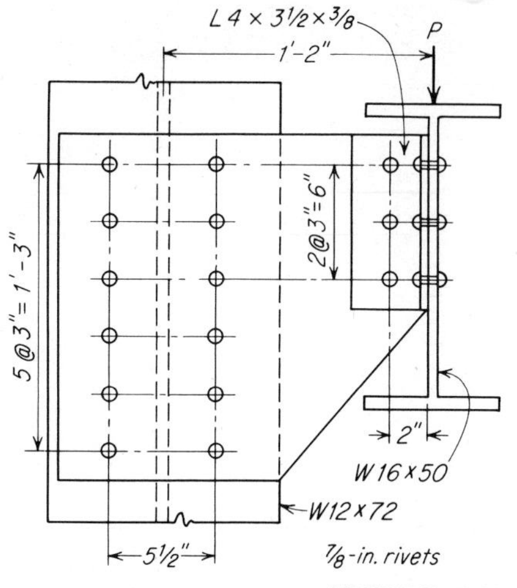

FIGURE 7-11
Problem 7-7.

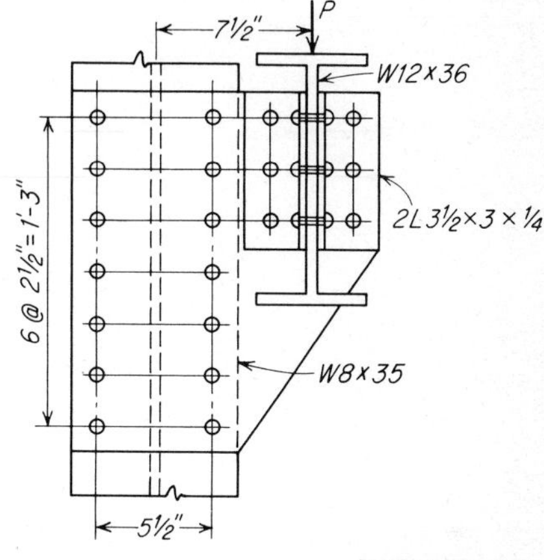

FIGURE 7-12
Problem 7-8.

equation to be developed in this article enables one to design such a connection in one or two trials.

In Fig. 7-13, n is the number of fasteners of uniform pitch p. If the connection is acted upon only by a couple M, then from Eq. (7-1)

$$R = \frac{My}{\sum y^2} \tag{a}$$

For an odd number of fasteners (Fig. 7-13a)

$$\begin{aligned} y^2 &= 2[p^2 + (2p)^2 + (3p)^2 + \cdots + (kp)^2] \\ &= 2p^2(1^2 + 2^2 + 3^2 + \cdots + k^2) \end{aligned}$$

The sum of the k terms in parentheses is $(k/6)(k + 1)(2k + 1)$. From the figure, $k = (n - 1)/2$, and therefore

$$y^2 = 2p^2 \frac{n-1}{2 \times 6}\left(\frac{n-1}{2} + 1\right)(n - 1 + 1) = \frac{np^2(n^2 - 1)}{12} \tag{b}$$

Similarly, for an even number of fasteners (Fig. 7-13b)

$$\begin{aligned} y^2 &= 2\left[\left(\frac{p}{2}\right)^2 + \left(\frac{3p}{2}\right)^2 + \left(\frac{5p}{2}\right)^2 + \cdots + \left(\frac{2k-1}{2}\,p\right)^2\right] \\ &= \frac{p^2}{2}\left[1^2 + 3^2 + 5^2 + \cdots + (2k-1)^2\right] \end{aligned} \tag{c}$$

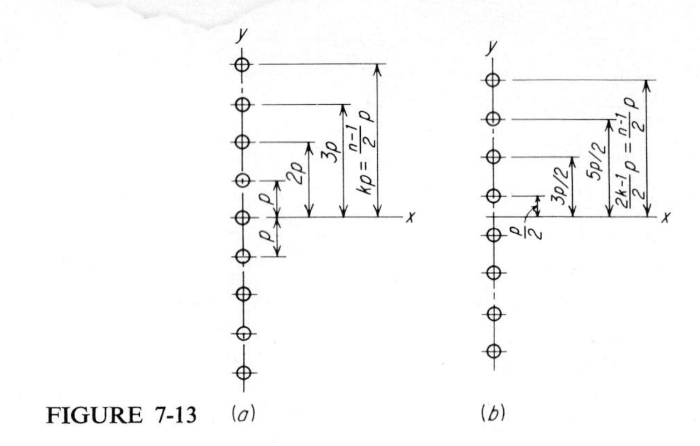

FIGURE 7-13 (a) (b)

The sum of the terms in the bracketed series is $(k/3)(2k+1)(2k-1)$. In this case $k = n/2$, and therefore

$$\sum y^2 = \frac{p^2}{2}\frac{n}{6}(n+1)(n-1) = \frac{np^2(n^2-1)}{12} \tag{7-3}$$

Since it is identical with Eq. (*b*), Eq. (7-3) may be used for either an odd number of fasteners or an even number. In each case the distance from the centroid to the extreme fastener is $p(n-1)/2$. Substituting this value and the value of $\sum y^2$ into Eq. (*a*), we get

$$R = \frac{Mp(n-1)/2}{np^2(n^2-1)/12} = \frac{6M}{np(n+1)} \tag{d}$$

It is evident from the form of this equation that we may define the section modulus S of the group of fasteners by the equation

$$S = \frac{np(n+1)}{6} \tag{7-4}$$

Solving Eq. (*d*) for n, we get

$$n = \sqrt{\frac{6M}{pR} + \frac{1}{4}} - \frac{1}{2} \tag{e}$$

This equation gives the number of fasteners, provided they are in a single line. The error is slight if we omit the fractions $\frac{1}{4}$ and $\frac{1}{2}$,

$$n = \sqrt{\frac{6M}{pR}} \tag{7-5}$$

If a connection contains two or more parallel lines of fasteners, Eq. (7-5) can be used to determine the number of fasteners in each line simply by dividing the number of lines into the moment to be resisted. It is obvious that this procedure involves an error which is not present when the fasteners are in a single line, namely, the omission of $\sum x^2$ in the evaluation of $\sum r^2$ and the substitution of y for r as the distance from the center of rotation to the farthest fastener. The number of fasteners calculated by Eq. (7-5) for multiple-line connections will always be somewhat larger than the number required.

Equation (7-5) gives the number of fasteners required for a connection acted upon only by a couple. For a connection supporting an eccentric force P, the number of fasteners predicted by the equation, using Pe for M, will usually be sufficient provided P is small relative to Pe. For larger ratios of P to Pe, the equation underestimates the number. This requires a trial solution. The number of fasteners must be guessed, using Eq. (7-5) as a guide, and their sufficiency checked by determining the resultant of the fastener forces due to P and to Pe.

It is possible to derive a formula for the number of fasteners required to support an eccentric force P which is parallel to a single row of uniformly spaced fasteners. The equation is quadratic in n^2.

7-7 DP7-1: COLUMN BRACKET FOR CRANE RUNWAY

A girder to support a hand crane is to be installed in an existing shop building. Since the girders must be set out from the columns, they must be supported on brackets. To provide the necessary clearance, the center of the rail must be 6 in. from the edge of the column flange. The weights of the bridge, truck, girder, rail, and lifted load produce a reaction of 16 kips on each bracket. A bearing-type connection with threads excluded from the shear planes is assumed.

Since Eq. (7-5) overestimates the number of bolts required for moment, and since the reaction is relatively small, a connection with three bolts in each row is assumed. The analysis is based on the effective eccentricity discussed in Art. 7-5.

Some saving might be made by using a connection with only one row of bolts. Since the eccentricity is reduced, fewer bolts are required. However, the alternate design is less resistant to longitudinal forces. This would be a significant consideration if the crane were power-operated.

COLUMN BRACKET FOR CRANE RUNWAY — DP7-1

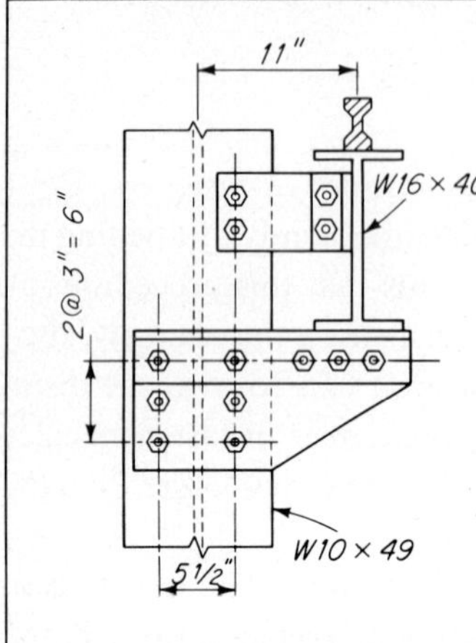

Connection to column

AISC Spec.

$P = 16^k$, $e = 11''$

3/4-in. A325 bolt, s.s. $= 9.72^k$

$$n = \sqrt{\frac{6 \times 11 \times 16}{2 \times 3 \times 9.72}} = 4.25 \quad \text{Try } 3$$

$$e_e = 11 - \frac{1+3}{2} = 9'' \qquad M = 16 \times 9 = 144''^k$$

$$\Sigma r^2 = \Sigma x^2 + \Sigma y^2 = 4 \times 3^2 + 6 \times 2.75^2 = 81.4$$

$$R_x = \frac{144 \times 3}{81.4} = 5.31^k$$

$$R_y = \frac{144 \times 2.75}{81.4} = 4.87^k$$

$$R_y = \frac{16}{6} = 2.67^k$$

$$R = \sqrt{(4.87 + 2.67)^2 + 5.31^2} = 9.22^k < 9.72$$

Alternate design with one row of bolts

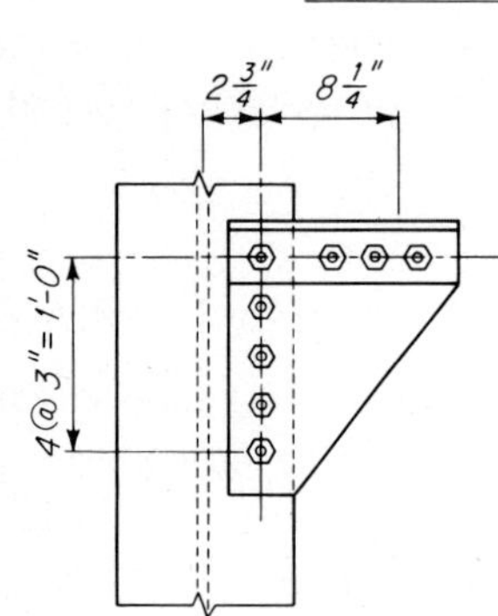

$$n = \sqrt{\frac{6 \times 16 \times 8.25}{3 \times 9.72}} = 5.2 \quad \text{Try } 5$$

$$e_e = 8.25 - \frac{1+10}{4} = 5.50'', \; M = 16 \times 5.5 = 88''^k$$

$$\Sigma y^2 = \frac{5 \times 9}{12}(25 - 1) = 90$$

$$R_x = \frac{88 \times 6}{90} = 5.87^k$$

$$R_y = \frac{16}{5} = 3.2^k$$

$$R = \sqrt{5.87^2 + 3.2^2} = 6.7^k < 9.72$$

PROBLEMS

7-9 For a single row of uniformly spaced fasteners supporting an eccentric force P parallel to the row, the fastener force due to the moment $M = Pe$ can be found from Eq. (7-5), while that due to P is P/n. Let the resultant of these forces equal the allowable shear R on a fastener, and develop a formula for n.

7-10 Determine the number and spacing of A325 $\frac{3}{4}$-in. bolts required to connect the angle to the plate and column in the first bracket of DP7-1.

7-11 Repeat Prob. 7-10 for the alternate design of the bracket.

7-12 In DP7-1 two of the bolts connect both the angle and plate to the column. Does this make any difference?

7-13 Discuss the effect on the column of the bracket with the single row vs. the bracket with the double row of bolts in DP7-1.

7-14 A C12 × 25 is connected to the flange of a W8 × 40 column as shown in Fig. 7-14. Design the connection for a moment of 330 in.-kips due to wind force. Use $\frac{3}{4}$-in. A502 Grade 1 rivets. A36 steel. AISC specification. (Allowable stresses may be increased for wind loading.)

7-15 Design data for a 10-ton electric crane operating in a shop building are as follows: span of the bridge 20 ft, wheelbase 9.5 ft, minimum clearance to center of rail 8 in., maximum load on each wheel (not including impact) 17.4 kips, weight of girder and rail 0.17 klf. Columns are W10 × 60 spaced 30 ft on centers. The webs of the columns are parallel to the crane girder. Design a bracket connection using A325 bolts. Use a friction-type connection since it is subjected to repeated stress. A36 steel. AISC specification.

7-16 A W12 × 50 spandrel beam is connected to the flange of a W10 × 49 column as shown in Fig. 7-15. Design the connection for a beam reaction of 36 kips, using A502 Grade 1 rivets on the standard gage line for the column. A36 steel. AISC specification.

7-8 STIFFENED BEAM SEATS

The capacity of an unstiffened seat angle is limited by the number of fasteners (usually four) that can be accommodated by the leg which connects to the column. Larger reactions require one or two stiffeners which fit tightly against the underside of the seat angle (Fig. 7-1*d*). The width, parallel to the beam, of the horizontal leg of the seat angle will be determined by web-crippling requirements of the beam unless the web of the beam is provided with stiffeners. Since the seat stiffeners relieve the seat of the bending which was discussed in Art. 7-3, it is good practice to have them extend as near to the edge of the seat as practicable. Their outstanding legs must have sufficient area in contact with the seat to support the load at a safe bearing stress. Because it is impracticable to cut the stiffener to fit tightly against the fillet of the seat angle, the effective width of the outstanding leg in bearing against the seat is less than its actual width by at least the radius of the fillet of the seat. This radius varies from $\frac{3}{8}$ to $\frac{5}{8}$ in. for angles ordinarily used as seats, but it is customary for design purposes to use $\frac{1}{2}$ in.

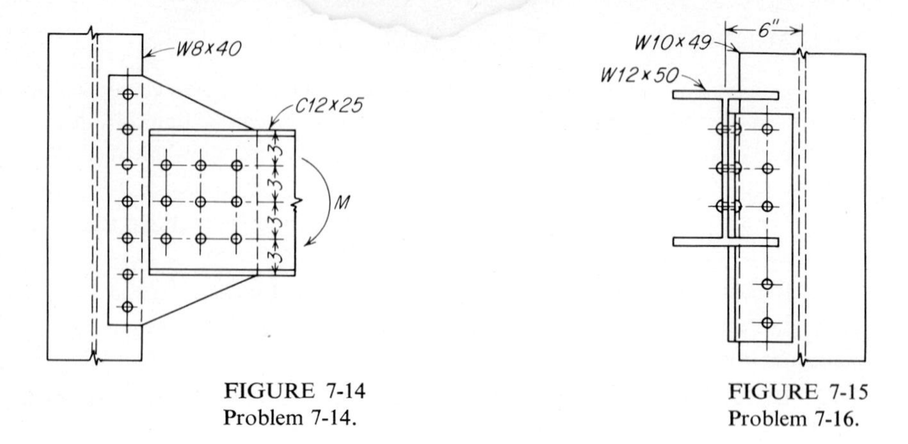

FIGURE 7-14
Problem 7-14.

FIGURE 7-15
Problem 7-16.

Since the outstanding leg of the stiffener is in compression, it may fail by local buckling. Except for the fact that the beam reaction which is applied at the upper end is resisted along the supported edge, instead of at the edge opposite the load, the stiffener leg is a compressed element free on one unloaded edge. Therefore it should be on the safe side to use the same limiting thickness as for outstanding elements of compression members, as discussed in Art. 4-23. For example, this would require the b/t ratio for steel stiffeners designed according to the AISC specification to be not more than $95/\sqrt{F_y}$.

The design of a seat to connect a W18 × 64 beam to the web of a W8 × 40 column will illustrate this discussion. The beam reaction is 68 kips. The A325 bolts are $\frac{3}{4}$ in., steel is A36, AISC specification. To prevent crippling of the beam web, the necessary length of bearing is

$$b = \frac{68}{27 \times 0.403} - 1.38 = 4.9 \text{ in.}$$

Allowing $\frac{3}{4}$ in. from the face of the column to the end of the beam, a 6-in. seat is required. The allowable bearing stress is $0.90 \times 36 \approx 33$ ksi, so the required bearing area is $68/33 = 2.06$ in.^{2} For two stiffeners with 5-in. outstanding legs, $t = 2.06/(2 \times 4.5) = 0.23$ in. But since the ratio b/t for a $\frac{1}{4}$-in. thickness exceeds $95/\sqrt{36} = 15.8$, two angles $5 \times 3 \times \frac{5}{16}$ are chosen. The width of the connected leg (3 in.) is determined by the width between fillets of the column ($6\frac{3}{8}$ in.). Since the beam flange is too wide to clear the column flanges, it must be cut to fit.

The allowable load for each bolt of the group which connects the stiffeners to the web of the column is $R = 22 \times 0.44 = 9.68$ kips in single shear for a bearing-type connection, and $R = 1.35 \times 36 \times 0.75 \times 0.312 = 11.3$ kips in bearing on the leg of the stiffener. The required number of bolts is therefore $n = 68/9.68 = 7$, or 4 in each

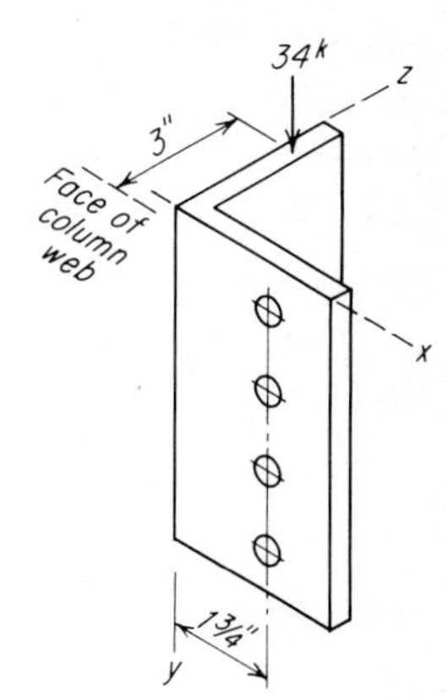

FIGURE 7-16

stiffener. However, since these bolts pass through a loose filler (Fig. 7-1*d*) and as a consequence tend to bend, one should consider the advisability of increasing their number. The AISC specification requires that "when rivets or bolts carrying computed stress pass through fillers thicker than $\frac{1}{4}$ inch, except in friction-type connections assembled with high strength bolts, the filler shall be extended beyond the splice material and the filler extension secured by enough rivets or bolts to distribute the total stress in the member uniformly over the combined sections of the member and fillers." In this example, one might extend the filler beyond the lower end of the stiffener a distance sufficient to receive one additional bolt in each line. It should be mentioned, however, that this is not always done in the case of stiffened beam seats, and in fact the specification is ignored in determining the allowable loads on certain standard beam seats listed in the AISC Manual.

Since the beam reaction acts on the outstanding leg of the stiffener, it is eccentric with respect to the bolts connecting the stiffener to the column. This condition is shown in Fig. 7-16. If the gage on the 3-in. connected leg is the standard $1\frac{3}{4}$ in., the bolt group is subjected to a moment

$$M_{xy} = 34 \times 1\tfrac{3}{4} = 59.5 \text{ in.-kips}$$

and an additional moment

$$M_{yz} = 34 \times 3 = 102 \text{ in.-kips}$$

With the required four bolts per stiffener spaced 3 in.,

$$\sum y^2 = 2(1.5^2 + 4.5^2) = 45$$

so that the shear on the extreme bolt resulting from M_{xy} is $R = 59.5 \times 4.5/45 = 5.95$ kips. Combining this with the shear in the direction of the y axis, $R = 34/4 = 8.5$ kips, the resultant is 10.4 kips. This exceeds the allowable value 9.68 kips. However, the effect of this eccentricity is usually ignored. Of course, if the outstanding legs of the stiffeners are fastened together, the bolts are relieved of the additional shear just computed. Usually, this is not done.

The moment M_{yz} tends to rotate the seat in the yz plane. Resistance to this moment must therefore come from tensile forces in the upper bolts of the connection, together with bearing pressures between the web of the column and the lower end of the stiffener. This question is discussed in Art. 7-9.

7-9 FASTENERS IN TENSION

The behavior of a connection that is loaded in such a way as to produce tension on the fasteners must be studied in terms of the tension which exists in the fasteners in the unloaded state. Since hot-driven rivets are not free to contract during cooling, they attain large tensile forces upon reaching normal temperatures. Tests indicate that an initial tensile stress of the order of 24 ksi may be expected. The initial tension in high-strength bolts is much greater (Art. 2-8). The initial tension in the rivets or bolts of a connection produces a compressive force between the faying surfaces. If an external moment tends to rotate the connection, as in the stiffener angle shown in Fig. 7-16, the initial tension in the fasteners will increase on one side of the neutral axis while the initial compression will decrease. Similarly, on the opposite side of the neutral axis, the initial tension in the fastener decreases while the initial compression increases. Two parts connected by a single rivet or bolt are shown in the unloaded state in Fig. 7-17*a*. A free-body sketch of one of the parts is shown in Fig. 7-17*b*. T_0 is the initial tension in the fastener, and C_0 is the resultant of the initial compressive stresses. This distribution of the latter is unknown but is assumed to be uniform. Then for equilibrium,

$$T_0 = C_0 \tag{a}$$

If a pull P is applied to the parts, as shown in Fig. 7-17*c*, the tension in the fastener and the resultant of the compressive stresses become T and C, respectively, as shown in Fig. 7-17*d*. Then

$$P + C = T \tag{b}$$

If the increase in the length of the fastener due to the increment of tension equals the increase in the thickness of the plates due to the decrease in compression,

$$\frac{T - T_0}{A_f E_f} = \frac{C_0 - C}{A_p E_p} \tag{c}$$

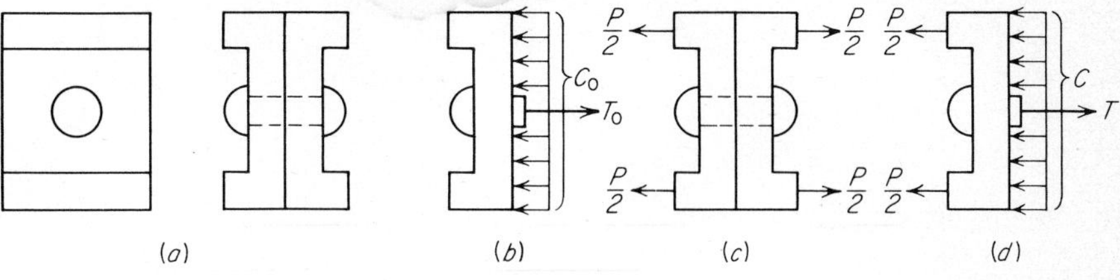

FIGURE 7-17

where the subscripts f and p refer to the fastener and the plate, respectively. Substituting the values of C_0 from Eq. (a) and C from Eq. (b), we get

$$(T - T_0)\frac{A_p E_p}{A_f E_f} = T_0 - (T - P) \qquad (d)$$

From this equation we find, for T,

$$T = T_0 + \frac{P}{1 + A_p E_p/A_f E_f} \qquad (e)$$

Equation (e) is valid only if the faying surfaces remain in contact. This condition obtains until $C = 0$, after which the connection opens up and $P = T$. To find the value of P at which contact is lost, we put $C = 0$ in Eq. (c). Since $C_0 = T_0$, this gives

$$T = T_0\left(1 + \frac{A_f E_f}{A_p E_p}\right) \qquad (f)$$

If the lateral dimensions of the connected part are as small as specifications permit for the size of the fastener (usually three diameters), then $A_p = (3d)^2 - \pi d^2/4 = 8.2d^2$ and $A_f = \pi d^2/4$. Then, if the fastener and connected part are of the same material, so that $E_p = E_f$, Eq. (e) gives

$$T = T_0 + \frac{P}{1 + 8.2/0.79} \approx T_0 + 0.09P \qquad (g)$$

while Eq. (f) gives

$$T \approx 1.1P \qquad (h)$$

These results are shown in Fig. 7-18a. This behavior is confirmed in Fig. 7-18b, which shows results of a test on a connection consisting of a tee cut from a W36 × 300 connected by four $\frac{7}{8}$-in. A325 bolts to a heavy plate assembly.[2]

According to Eq. (g) the increase in tensile force in the fastener is 9 percent of the load P. This increment is reduced if we increase the contact area of the connected

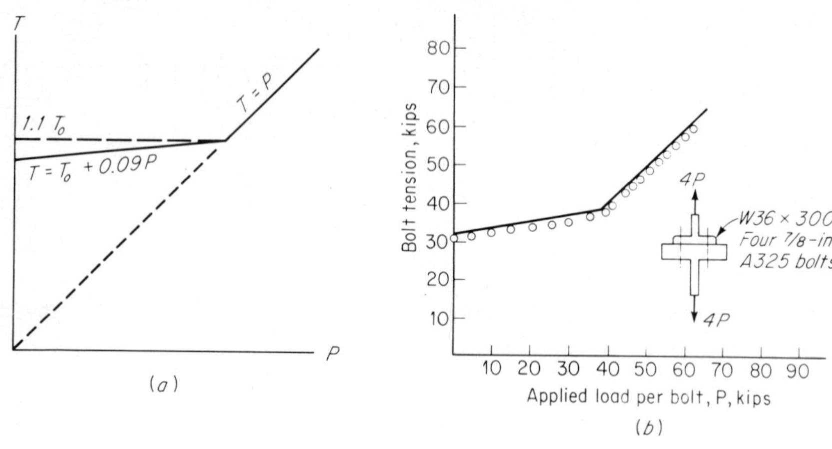

FIGURE 7-18

parts. However, it is obvious that the distribution of the compressive stresses tends to become less and less uniform as the contact area increases. However, Eqs. (*c*) and (*f*) are still applicable, provided a reduced, or effective, area is used in place of the actual contact area.

7-10 CONNECTIONS WITH FASTENERS IN TENSION

Let Fig. 7-19*a* represent an unloaded tee, with two rows of fasteners (one on each side of the stem) connecting it to a stiff member. In Fig. 7-19*b* a couple M acts on the tee. If the tee is also stiff, the distribution of the altered tensile forces and compressive stresses will be linear as shown, so long as the connection does not open up at the top. If M in Fig. 7-19*b* is large enough to release the compression on the upper portion of the faying surfaces, the forces acting on the connection may be assumed to be distributed as shown in Fig. 7-19*c* if the extreme fastener has not yielded. With progressive yielding of the fasteners the distribution of the forces approaches that shown in Fig. 7-19*d*.

The distributions of forces in Fig. 7-19*b* and *c* are both consistent with the fundamental assumptions of elastic design, while that in Fig. 7-19*d* is consistent with plastic or limit design. If the analysis is based on the distribution of Fig. 7-19*b*, the couple M can be shown in the equivalent form of two pairs of forces, P and Q, each acting on a length of tee equal to the pitch of the fasteners (Fig. 7-19*e*). The topmost section is analogous to the connection of Fig. 7-17, as is the section immediately below. We have already shown that T is not significantly greater than T_0. Therefore

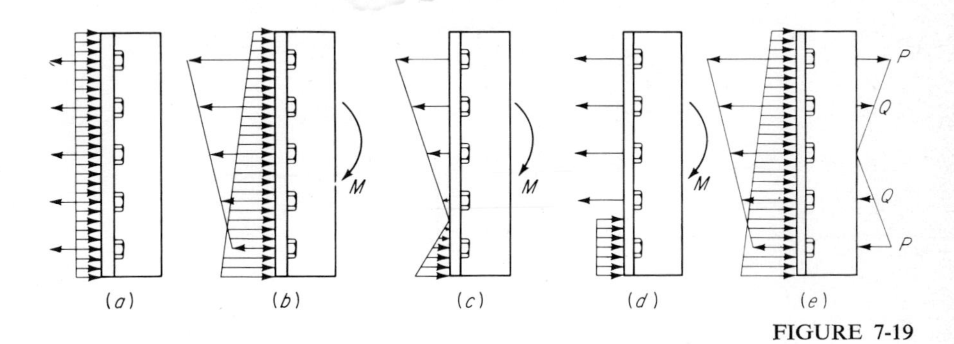

FIGURE 7-19

we may design the connection on the basis of a prescribed allowable tensile stress on the fastener as though the maximum tension in the fastener were equal to P. Since the distribution of the forces P and Q in Fig. 7-19*e* is identical with the distribution assumed for the shearing forces in deriving Eq. (7-1), we may use that equation for the analysis of a connection whose fasteners are subjected to tension. Similarly, we may use Eq. (7-5) to determine the number of fasteners required to resist a known moment.

The neutral axis for the force distribution shown in Fig. 7-19*c* lies at the center of gravity of a cross section consisting of the fastener areas on one side of the neutral axis and the bearing area of the tee on the other. The distance y to the neutral axis will be one-sixth to one-seventh the length of the connection. Its value may either be assumed and checked or solved for directly by equating moments of the areas. The tensile force can then be determined from the flexure formula. A formula has been derived from which one can determine the number of fasteners required to resist a given moment.[3] This number is about 80 percent of the number given by Eq. (7-5).

A conservative estimate of the ultimate moment capacity of a connection using high-strength bolts can be determined by assuming (1) the bolt tensions in Fig. 7-19*d* to be equal to the minimum required tension at installation, which is 70 percent of the specified minimum tensile strength of the bolt, and (2) the area in bearing to be stressed to the yield point of the connected material.[4] Since the compressive force equals the sum of the tensions in the bolts on the tension side of the neutral axis, the bearing area and the location of the neutral axis are easily determined. Of course, the moment is the moment of the couple consisting of the compressive force and the sum of the tensile forces.

Cold-driven rivets are sometimes used in structural work. The absence of initial tension invalidates the distribution of forces of Fig. 7-19*b*, so that the distribution of Fig. 7-19*c* should be used for elastic design and that of Fig. 7-19*d* for plastic design.

The bracket of Fig. 7-20*a* will be used to illustrate the methods of analysis discussed in this article. The AISC allowable tension in a $\frac{3}{4}$-in. A325 bolt is $R = 0.442$

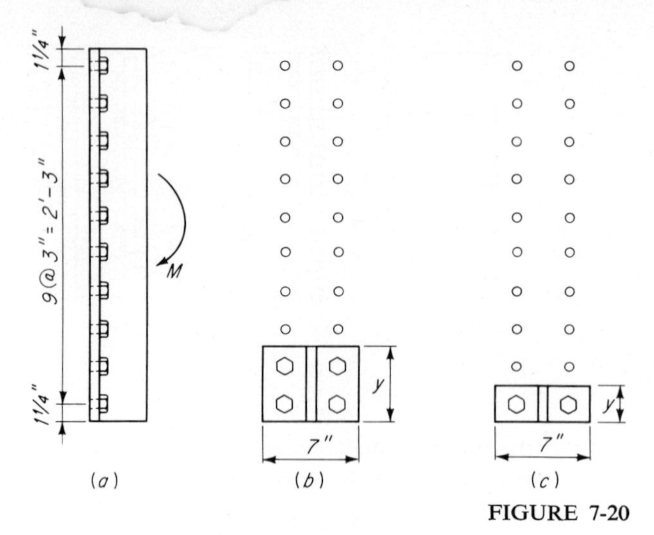

FIGURE 7-20

$\times$ 40 = 17.68 kips. For the distribution of forces of Fig. 7-19*e*, we use Eq. (7-5) to find, for two rows,

$$M = \frac{2n^2pR}{6} = \frac{2 \times 100 \times 3 \times 17.68}{6} = 1{,}768 \text{ in.-kips}$$

For the distribution of forces in Fig. 7-19*c*, knowing that y is one-sixth to one-seventh the length of the connection, the cross section will be as shown in Fig. 7-20*b*. Equating the moment of the contact area below the neutral axis to the moment of the bolt areas above the neutral axis, we have

$$\frac{7y^2}{2} = 16 \times 0.442(17.75 - y)$$

$$y = 5.06 \text{ in.}$$

The moment of inertia of the 16 bolt areas above the neutral axis can be found most easily by using Eq. (7-3) to find their moment of inertia about their own centroidal axis and then using the transfer theorem. Adding to this the moment of inertia of the contact area below the neutral axis, we have

$$I = 2 \times 0.442 \times 8 \times 3^2 \times \frac{64 - 1}{12} + 2 \times 0.442 \times 8 \times 12.69^2 + 7 \times \frac{5.06^3}{3}$$

$$= 1{,}774 \text{ in.}^4$$

The allowable moment is

$$M = \frac{40 \times 1{,}774}{23.19} = 3{,}060 \text{ in.-kips}$$

For the distribution of forces shown in Fig. 7-19*d*, assuming the neutral axis to be located between the bottom two rows, we have the cross section shown in Fig. 7-20*c*. The specified minimum installation tension of the $\frac{3}{4}$-in. bolt is 28 kips (Table 2-7). Equating the compressive force on the contact area below the neutral axis to the tension in the bolts above the neutral axis, we have, for A36 steel,

$$36 \times 7y = 18 \times 28$$
$$y = 2.0 \text{ in.}$$

The ultimate moment is

$$M_u = 18 \times 28 \times 15.25 = 7{,}180 \text{ in.-kips}$$

The ratios of the ultimate moment to the two allowable moments are

$$\frac{7{,}180}{1{,}768} = 4.06 \qquad \text{and} \qquad \frac{7{,}180}{3{,}060} = 2.35$$

This difference suggests that design specifications should state the assumption to be used in the analysis of connections of this type, to be consistent with the specified allowable fastener tension. The distribution of forces shown in Fig. 7-19*c*, which gives the smaller factor of safety, would appear reasonable with the allowable tensions presently specified.

The analyses in this article neglect the effect of certain forces that may develop because of the flexibility of the connected parts. These forces are discussed in Art. 7-13.

7-11 COMBINED TENSION AND SHEAR

Many connections support loads which produce both tension and shear on the fasteners. Tests indicate that the shearing strength of hot-driven rivets is 0.75 times their tensile strength.[5] This is also the ratio of the AISC allowable shear on rivets to the allowable tension. The tests also show that the ultimate strength of rivets under combined tension and shear is given closely by (Fig. 7-21)

$$\left(\frac{f_v}{0.75F_u}\right)^2 + \left(\frac{f_t}{F_u}\right)^2 = 1 \tag{7-6a}$$

Tests on high-strength bolts subjected to shear and tension show that their ultimate strength may also be approximated by ellipses.[6] For A325 bolts with a single shear plane through the shank, the equation is

$$\left(\frac{f_v}{0.83F_u}\right)^2 + \left(\frac{f_t}{F_u}\right)^2 = 1 \tag{7-6b}$$

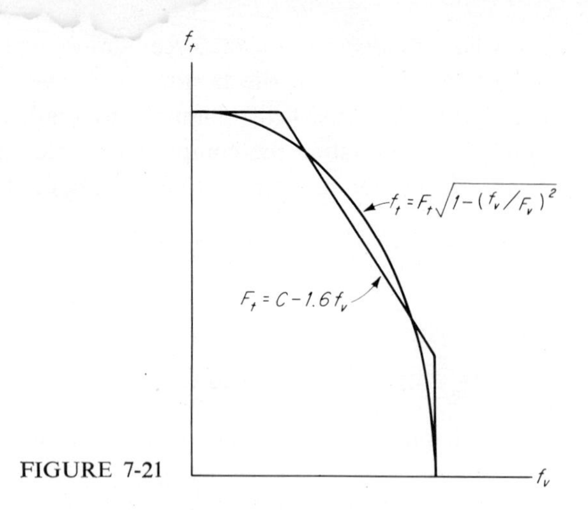

FIGURE 7-21

For A325 bolts with shear planes through the threads the equation is

$$\left(\frac{f_v}{0.64F_u}\right)^2 + \left(\frac{f_t}{F_u}\right)^2 = 1 \tag{7-6c}$$

where f_v, f_t are, respectively, simultaneous shearing stress and tensile stress at failure, based on nominal area, and F_u is the tensile strength of fastener subjected only to tension, based on nominal area.

According to the AISC specification, the allowable tensile stress F_t in rivets and bolts subjected to combined shear and tension in bearing-type connections is (Fig. 7-21)

$$F_t = C - 1.6f_v \tag{7-7}$$

where f_v is the calculated shear stress (not to exceed F_v). Maximum values of F_t and F_v, and values of C are given in Table 7-1.

Table 7-1 VALUES FOR EQ. (7-7)

Fastener type	F_t, ksi	F_v, ksi	C, ksi
A502 Grade 1 rivet	20	15	28
A502 Grade 2 rivet	27	20	38
A307 bolt	20*	10	28
A325 bolt, threads not excluded	40	15	50
Threads excluded	40	22	50
A490 bolt, threads not excluded	54	22.5	70
Threads excluded	54	32	70

* Allowable tension on the area $\frac{\pi}{4}\left(D - \frac{0.9743}{n}\right)^2$, where D = major thread diameter and n = number of threads per inch. All other values of F_t are allowable stresses on the nominal (unthreaded) area.

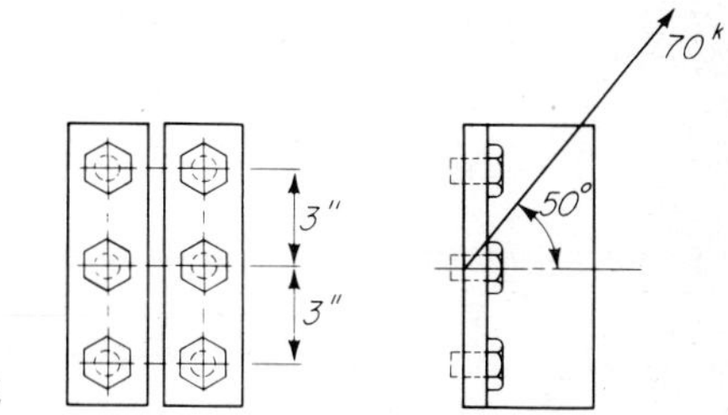

FIGURE 7-22

For bolts in friction-type connections AISC specifies the allowable shear stress

$$F_v \gtrless \begin{cases} 15\left(1 - \dfrac{f_t A_b}{T_b}\right) & \text{for A325 bolts} \quad (7\text{-}8a) \\ 20\left(1 - \dfrac{f_t A_b}{T_b}\right) & \text{for A490 bolts} \quad (7\text{-}8b) \end{cases}$$

where f_t is the tensile stress due to applied load and T_b the specified pretension load of bolt.

For friction-type connections with bolts in shear and tension, AASHO specifies a maximum shear stress f_v of

$$f_v = F_v - 0.22 f_t \tag{7-9}$$

while for rivets and high-strength bolts in bearing-type connections the AASHO formula is

$$\left(\frac{f_v}{F_v}\right)^2 + \left(\frac{k f_t}{F_v}\right)^2 \leq 1 \tag{7-10}$$

where k is a constant equal to 0.75 for rivets and for A325 bolts with threads in shear planes, and 0.555 for A325 bolts with threads excluded from the shear plane. Equation (7-10) is similar to Eqs. (7-6).

The formulas just discussed will be used to determine the number of $\frac{3}{4}$-in. A325 bolts needed for the bracket of Fig. 7-22. If it is a bearing-type connection, with threads excluded from the shear plane, the AISC allowable shear per bolt is $22 \times 0.442 = 9.72$ kips. The number of bolts required to resist the shear component of the 70-kip load is

$$n = \frac{70 \times 0.766}{9.72} = 5.5$$

Assume three bolts per row. The shear stress on each bolt is

$$f_v = \frac{70 \times 0.766}{6 \times 0.442} = 20.2 \text{ ksi}$$

The tensile stress on each bolt is

$$f_t = \frac{70 \times 0.643}{6 \times 0.442} = 17.0 \text{ ksi}$$

From Eq. (7-7), the allowable tension is

$$F_t = 50 - 1.6 \times 20.2 = 17.7 > 17.0 \text{ ksi}$$

The number of bolts is sufficient.

If the bracket is designed as a friction-type connection, the number of bolts required is

$$n = \frac{70 \times 0.766}{15 \times 0.442} = 8.1$$

Since this number is an underestimate, we shall assume five bolts per row. The tensile stress on each bolt is

$$f_t = \frac{70 \times 0.643}{10 \times 0.442} = 10.2 \text{ ksi}$$

From Eq. (7-8*a*), the simultaneous allowable shear stress is

$$F_v = 15\left(1 - \frac{10.2 \times 0.442}{28}\right) = 12.6 \text{ ksi}$$

The actual shear stress is

$$f_v = \frac{70 \times 0.766}{10 \times 0.442} = 12.1 < 12.6 \text{ ksi}$$

Therefore a connection with 10 bolts is adequate.

If the bracket is designed according to the AASHO specifications as a bearing-type connection with threads excluded from the shear plane, the allowable shear per bolt is $20 \times 0.442 = 8.84$ kips. The number of bolts required for the shear component of the 70-kip load is

$$n = \frac{70 \times 0.766}{8.84} = 6.1$$

For a connection with eight bolts, the shearing stress is

$$f_v = \frac{70 \times 0.766}{8 \times 0.442} = 15.2 \text{ ksi}$$

The tensile stress is

$$f_t = \frac{70 \times 0.643}{8 \times 0.442} = 12.7 \text{ ksi}$$

From Eq. (7-10), with $k = 0.555$,

$$\left(\frac{15.2}{20}\right)^2 + \left(\frac{0.555 \times 12.7}{20}\right)^2 = 0.70 < 1$$

The effect of prying forces, which are discussed in Art. 7-13, has been neglected in this example.

PROBLEMS

7-17 Compute the allowable load, in terms of the allowable load R per rivet, for a seat stiffener whose 3-in. connected leg has three rivets at 3-in. pitch at $1\frac{3}{4}$-in. gage, (*a*) neglecting the torsional effect discussed in Art. 7-8 and (*b*) considering the torsional effect. Also compare the allowable loads for four rivets and for five rivets at the same gage and for three, four, and five rivets if the gage is $2\frac{1}{2}$ in. on a 4-in. connected leg. What would you conclude about the advisability of ignoring the torsional effect?

7-18 Compute the allowable load for the following stiffened beam seats, (1) neglecting the moment M_{yz} discussed in Art. 7-8 and (2) considering the moment M_{yz}.

(*a*) Seat angle $6 \times 4 \times \frac{5}{8}$ with the 4-in. leg outstanding, two $3\frac{1}{2} \times 3 \times \frac{1}{2}$ stiffener angles with the $3\frac{1}{2}$-in. legs outstanding, and two rows of $\frac{3}{4}$-in. A325 bolts in a bearing-type joint with six bolts in each row at 3-in. pitch.

(*b*) Seat angle $6 \times 6 \times \frac{5}{8}$, two $5 \times 3\frac{1}{2} \times \frac{1}{2}$ stiffener angles with the 5-in. legs outstanding, and two rows of $\frac{7}{8}$-in. A325 bolts in a bearing-type joint with six bolts in each row at 3-in. pitch.

(*c*) Seat angle $8 \times 6 \times \frac{7}{16}$ with the 8-in. leg outstanding, two $7 \times 4 \times \frac{7}{16}$ stiffener angles with the 7-in. legs outstanding, and two rows of $\frac{7}{8}$-in. A490 bolts in a bearing-type joint with six bolts in each row at 3-in. pitch.

Assume that the reaction is concentrated at the center of the contact area between the beam and the seat angle. What would you conclude about the advisability of ignoring the moment M_{yz}?

7-19 The load on a laterally supported beam spanning 10 ft is 12 klf. Design the beam and a seat to connect it to the web of a W10 × 29 column. Use $\frac{3}{4}$-in. A502 Grade 1 rivets. AISC specification, A36 steel.

7-20 A W21 × 127 beam connects to the web of a W16 × 58 column. The beam reaction is 142 kips. Design a beam seat using $\frac{7}{8}$-in. A325 bolts. AISC specification, A36 steel.

7-21 Design a bracket to connect the beam to the flange of the column of Fig. 7-23, using $\frac{7}{8}$-in. A502 Grade 2 rivets. $R = 80$ kips, $d = 12$ in. AISC specification, A36 steel.

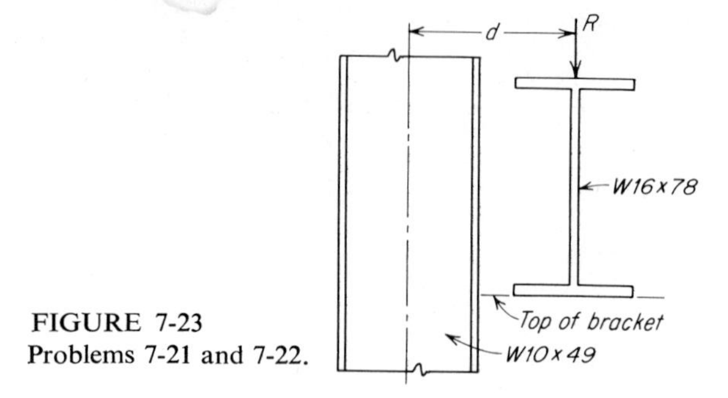

FIGURE 7-23
Problems 7-21 and 7-22.

7-22 Design a bracket to connect the beam to the flange of the column of Fig. 7-23, using $\frac{3}{4}$-in. A325 bolts. $R = 70$ kips, $d = 10$ in. AISC specification, A36 steel.

7-23 Design a bracket to connect the crane girder of DP5-4 to the flange of a W24 × 100 column, using $\frac{7}{8}$-in. A502 Grade 1 rivets.

7-24 A W16 × 36 spans 15 ft and supports a uniform load of 3.2 klf. The top flange of the beam is 6 in. above the top flange of the W16 × 64 supporting girder. There is no beam on the other side of the girder. Design the connection for 1-in. A502 Grade 1 rivets. AISC specification, A36 steel.

7-25 Design a bracket for the connection of the crane girder to the column in DP5-2. Use $\frac{7}{8}$-in. A325 bolts.

7-12 MOMENT-RESISTANT CONNECTIONS WITH FASTENERS IN TENSION

Many connections must be designed to develop moment because the beams they support are parts of a rigid frame. For example, beams which are parts of the wind-bracing system of a tier building must resist end moments resulting from both wind forces and gravity loads. In the usual case, moment resistance of riveted or bolted connections in such frameworks depends upon tension in the fasteners.

The simplest connection of this type, which is satisfactory for small moments, is illustrated in Fig. 7-24*a*. On the assumption that they are unlikely to resist their proportional share of the end shear, the fasteners in this connection which connect the angles at the flanges to the column are usually designed for their allowable value in tension, no consideration being given to a reduction because of accompanying shear. Except for the relatively small contribution of the web angles, the resisting moment of the connection is limited by the number of fasteners it is possible to put in the vertical legs of the angles at the flanges. For reasons which follow, this number should probably not exceed two. If the column flange is wide enough to accommodate four fasteners in a single gage line on the angle, the innermost fasteners may take a disproportionate share of the load (Fig. 7-25*a*). The same argument applies if two

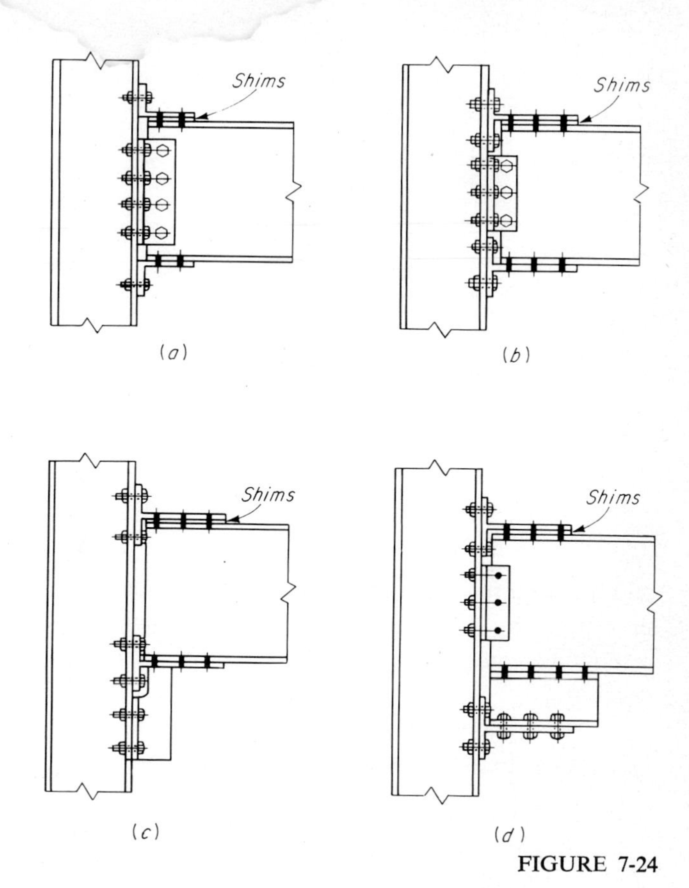

FIGURE 7-24

fasteners are used in each of two gage lines, as in Fig. 7-25*b*. The actual distribution of the fastener tensions is obviously a function of the stiffnesses of the angle and of the flanges of the beam and the column.

Figure 7-24*b* illustrates a connection in which tees connect the beam flanges to the column. This is sometimes called a *T-stub connection*. Again, the moment resisted by the web angles is relatively small and is usually neglected, in which case the fasteners connecting these angles are designed to carry only the end shear. The end moment is then assumed to be resisted by a couple furnished by the tees. The framing angles may be omitted if the fasteners in the T-stubs are adequate for both moment and shear. A seat may be used instead of framing angles (Fig. 7-24*c*).

Figure 7-24*d* shows a modification of the T-stub connection which may be used if the end moment produces a pull greater than the value of four fasteners in tension. The piece which is inserted between the tee and the flange of the beam is called a *beam stub*. The forces acting on the stub are shown in Fig. 7-26. The couple formed by C

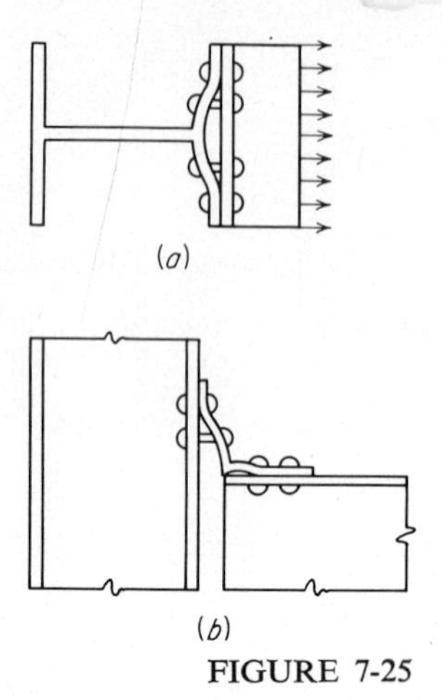

FIGURE 7-25

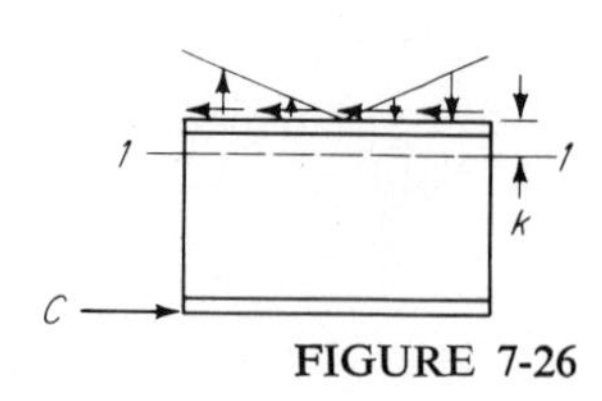

FIGURE 7-26

and the shearing forces on the fasteners connecting the stub to the beam is assumed to be resisted by the couples consisting of fastener tensions at the one end and corresponding compressive forces at the other. The stress on section 1-1 should also be checked.

The parts of the connections shown in Fig. 7-24 which are attached in the fabricating shop are usually welded, rather than riveted or bolted.

7-13 PRYING FORCES IN MOMENT-RESISTANT CONNECTIONS

In Art. 7-9 we discussed initial tension in the fasteners of a connection and the resulting contact pressure between the connected parts. In this article we shall discuss the effect of flexibility of the connected parts. Figure 7-27 shows T-stubs which were connected with four A325 bolts to a thick plate and tested in tension.[2] In each case, failure resulted from fracture of one or more of the bolts. It will be noted that there was considerable bending of the relatively thin flange of the specimen at the left and virtually none in the thick flange of the specimen at the right. In the latter case, the analysis of Art. 7-9 could be used to determine the relationship between bolt tension and applied load. Of course, an assumption would have to be made as to the effective area over which the contact pressure is distributed.

Behavior of the more flexible flange is shown in Fig. 7-28. The unloaded tee is shown in *a*, with the initial tensions T_0 and the equal compressive forces C_0. The latter are distributed nonuniformly over a relatively small area of contact. Figure 7-28*b* shows the tee acted upon by a force $2P$. The bolt tension is now T. The reduced compressive force, denoted by Q, has migrated toward the flange tip. This behavior has

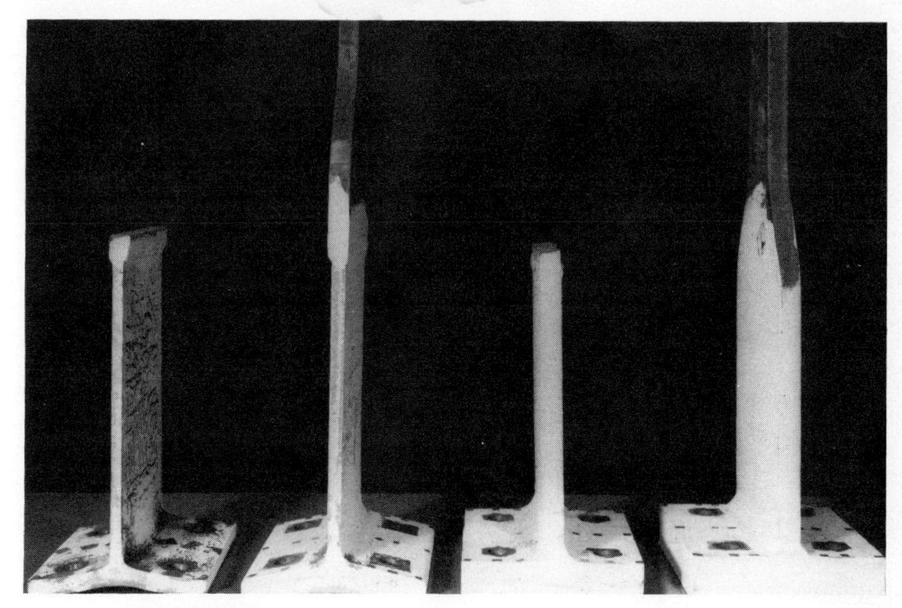

FIGURE 7-27
T-stubs tested in tension. (*From R. T. Douty and W. McGuire, Proc. ASCE, J. Struct. Div., vol. 91, no. ST2, April 1965.*)

been studied in detail by a finite-element analysis of a T-stub connection, where it was found that Q may eventually be so distributed as to be virtually concentrated at the flange tip if the flange is fairly flexible.[7]

The compressive force Q is usually called a *prying force*. An approximate analysis can be made by assuming it to be concentrated at the flange tip (Fig.7-28*c*). The procedure followed in Art. 7-9 can then be used; i.e., the deflection δ of the flange can be equated to bolt elongation to get the relationship between the bolt tension and the applied load. Assuming the flange to remain elastic, δ is easily computed by moment-area principles. The following equation was obtained by making some adjustments in the resulting formula for the prying force Q, to simplify it and to bring it into better agreement with test results:[2]

$$\frac{Q}{P} = \frac{15ab^2 A_b - wt^4}{10a^2(a + 3b)A_b + 5wt^4} \tag{7-11}$$

where A_b = area of bolt
w = length of flange tributary to bolt
t = thickness of flange
a,b = dimensions shown in Fig. 7-28*c*

[Equation (7-11) is written in a form different from that in Ref. 2.]

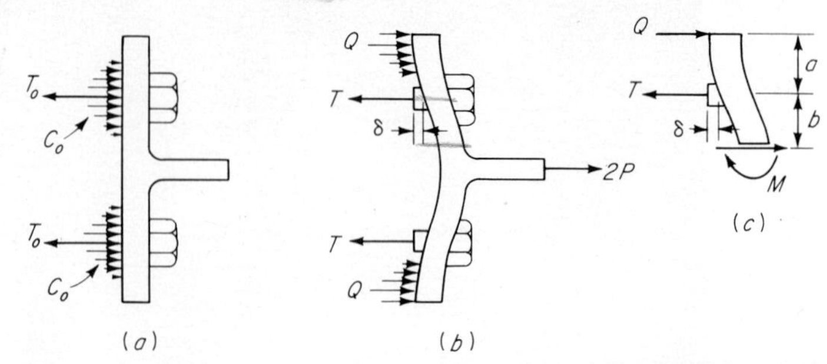

FIGURE 7-28

The solution to this problem is somewhat simpler if we assume that the ultimate load is reached with the development of a plastic hinge at the junction of the flange and web, that is, $M = M_p$ in Fig. 7-28c. In this case, we have the two equations of equilibrium

$$P + Q = T \tag{a}$$

$$Pb - Qa = M_p \tag{b}$$

Dividing Eq. (a) by Eq. (b) and solving the result for Q/P gives

$$\frac{Q}{P} = \frac{b - M_p/T}{a + M_p/T} \tag{c}$$

But $M_p = F_y wt^2/4$. Then, with the bolt tension at ultimate load $T_u = F_u A_b$, M_p/T in Eq. (c) becomes

$$\frac{M_p}{T} = \frac{F_y wt^2}{4} \frac{1}{F_u A_b} = \frac{F_y}{F_u} \frac{wt^2}{\pi d^2} = \beta \frac{wt^2}{d^2}$$

where d is the diameter of the bolt. Substituting this result into Eq. (c) gives[7]

$$\frac{Q}{P} = \frac{bd^2 - \beta wt^2}{ad^2 + \beta wt^2} \tag{d}$$

If the bolt fractured at or after the formation of the plastic hinge in the flange, β would be known and the prying force Q could be determined from Eq. (d). However, the finite-element analysis in Ref. 7 showed that while a substantial amount of yielding had developed in most cases, the plastic hinge was completely formed in only a few of the connections when the bolt failed. Therefore, the coefficients of the terms

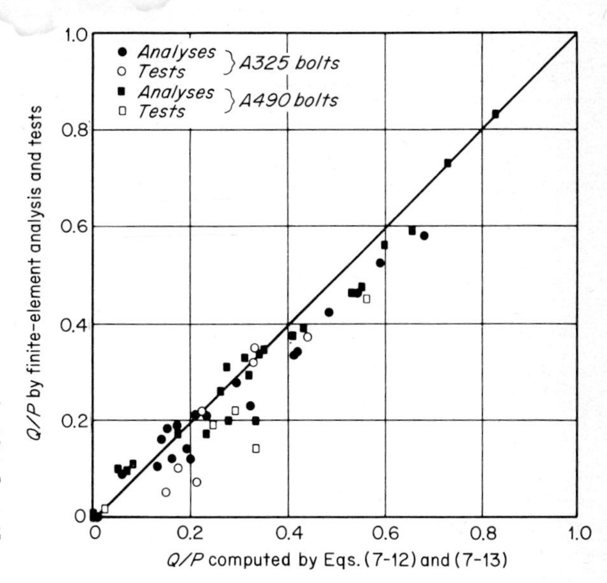

FIGURE 7-29
Prying force Q in T-stub connections. (*From R. S. Nair, P. C. Birkemoe, and W. H. Munse, Behavior of Bolts in Tee-connections Subjected to Prying Action, Univ. Ill. Civ. Eng. Stud., Struct. Res. Ser. 1969.*)

in Eq. (d) were adjusted to give better agreement with the analysis and with tests, as follows:

$$\text{Connections with A325 bolts:} \qquad \frac{Q}{P} = \frac{100bd^2 - 18wt^2}{70ad^2 + 21wt^2} \tag{7-12}$$

$$\text{Connections with A490 bolts:} \qquad \frac{Q}{P} = \frac{100bd^2 - 14wt^2}{62ad^2 + 21wt^2} \tag{7-13}$$

Figure 7-29 compares the predictions of these equations with results of the finite-element analyses and of tests. The agreement is good. In general, values of Q by the approximate formulas are larger than those of the tests and values by the analysis. This is conservative, since the bolt tension T is increased by increase in Q.

According to Ref. 7, if the dimension a in Fig. 7-28c exceeds $2t$, the value $2t$ should be used for a in Eqs. (7-12) and (7-13). Furthermore, the flange should be analyzed for bending at the bolt line and at a section $\frac{1}{16}$ in. from the face of the web.

The proportioning of a T-stub connection must be by trial because the ratio of the prying force Q to the applied load P depends on the dimensions of the connection. The procedure is illustrated in the next article.

7-14 T-STUB AND END-PLATE MOMENT CONNECTIONS

DP7-2 illustrates the design of a T-stub connection for a beam which supports both gravity and wind loads. For loads due to wind alone or to combinations of wind and other loads, allowable stresses may be increased by one-third. Since this is equivalent

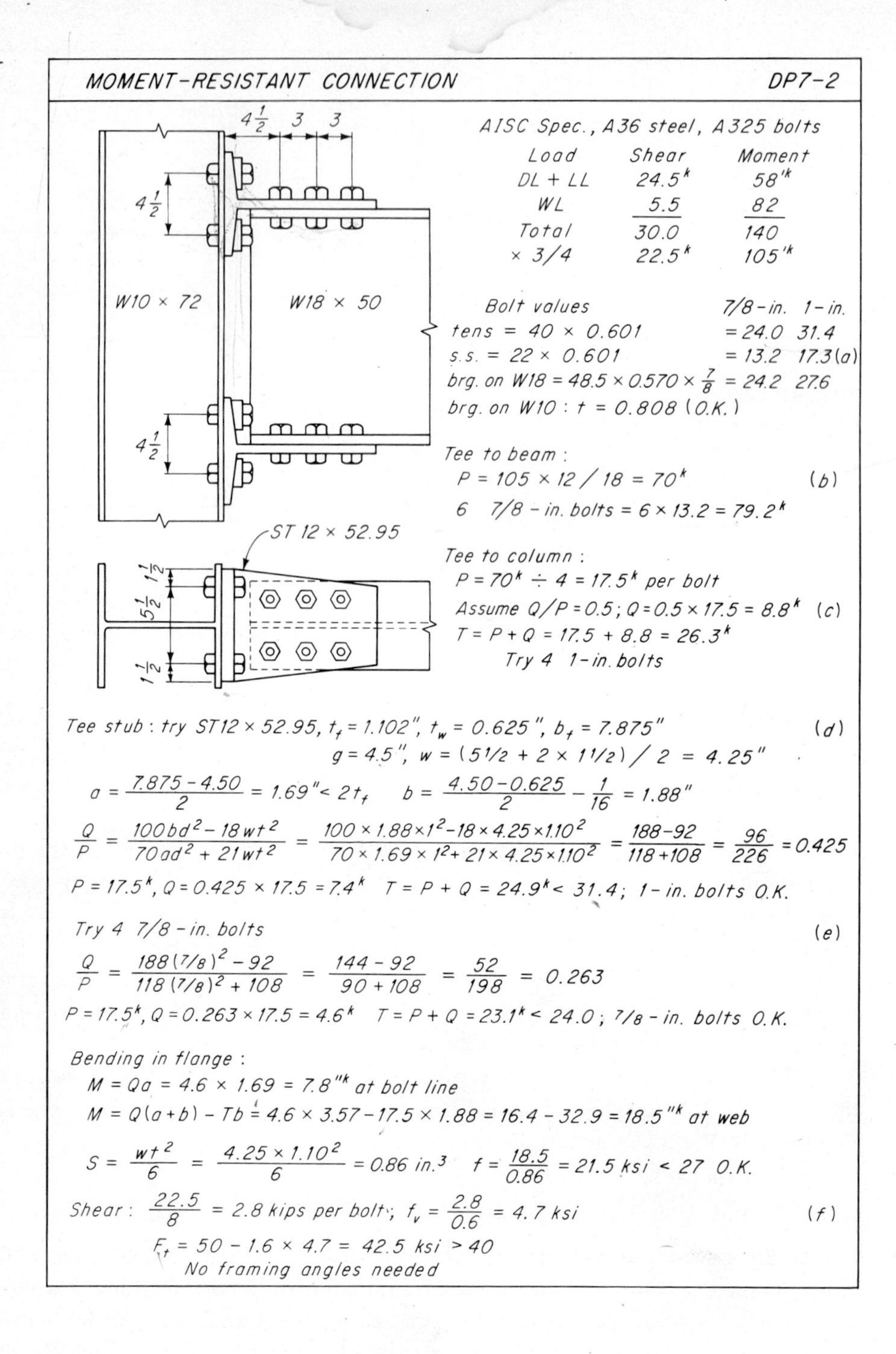

MOMENT-RESISTANT CONNECTION DP7-2

AISC Spec., A36 steel, A325 bolts

Load	Shear	Moment
DL + LL	24.5^k	$58'^k$
WL	5.5	82
Total	30.0	140
× 3/4	22.5^k	$105'^k$

Bolt values	7/8-in.	1-in.
tens = 40 × 0.601	= 24.0	31.4
s.s. = 22 × 0.601	= 13.2	17.3 (a)
brg. on W18 = 48.5 × 0.570 × $\frac{7}{8}$	= 24.2	27.6

brg. on W10 : t = 0.808 (O.K.)

Tee to beam : (b)

$P = 105 \times 12 / 18 = 70^k$

6 7/8-in. bolts $= 6 \times 13.2 = 79.2^k$

Tee to column : (c)

$P = 70^k \div 4 = 17.5^k$ per bolt

Assume $Q/P = 0.5$; $Q = 0.5 \times 17.5 = 8.8^k$

$T = P + Q = 17.5 + 8.8 = 26.3^k$

Try 4 1-in. bolts

Tee stub : try ST12 × 52.95, $t_f = 1.102''$, $t_w = 0.625''$, $b_f = 7.875''$ (d)

$g = 4.5''$, $w = (5\frac{1}{2} + 2 \times 1\frac{1}{2}) / 2 = 4.25''$

$$a = \frac{7.875 - 4.50}{2} = 1.69'' < 2t_f \qquad b = \frac{4.50 - 0.625}{2} - \frac{1}{16} = 1.88''$$

$$\frac{Q}{P} = \frac{100bd^2 - 18wt^2}{70ad^2 + 21wt^2} = \frac{100 \times 1.88 \times 1^2 - 18 \times 4.25 \times 1.10^2}{70 \times 1.69 \times 1^2 + 21 \times 4.25 \times 1.10^2} = \frac{188 - 92}{118 + 108} = \frac{96}{226} = 0.425$$

$P = 17.5^k$, $Q = 0.425 \times 17.5 = 7.4^k$ $T = P + Q = 24.9^k < 31.4$; 1-in. bolts O.K.

Try 4 7/8-in. bolts (e)

$$\frac{Q}{P} = \frac{188(7/8)^2 - 92}{118(7/8)^2 + 108} = \frac{144 - 92}{90 + 108} = \frac{52}{198} = 0.263$$

$P = 17.5^k$, $Q = 0.263 \times 17.5 = 4.6^k$ $T = P + Q = 23.1^k < 24.0$; 7/8-in. bolts O.K.

Bending in flange :

$M = Qa = 4.6 \times 1.69 = 7.8''^k$ at bolt line

$M = Q(a + b) - Tb = 4.6 \times 3.57 - 17.5 \times 1.88 = 16.4 - 32.9 = 18.5''^k$ at web

$$S = \frac{wt^2}{6} = \frac{4.25 \times 1.10^2}{6} = 0.86 \text{ in.}^3 \quad f = \frac{18.5}{0.86} = 21.5 \text{ ksi} < 27 \text{ O.K.}$$

Shear : $\frac{22.5}{8} = 2.8$ kips per bolt; $f_v = \frac{2.8}{0.6} = 4.7$ ksi (f)

$F_t = 50 - 1.6 \times 4.7 = 42.5$ ksi > 40

No framing angles needed

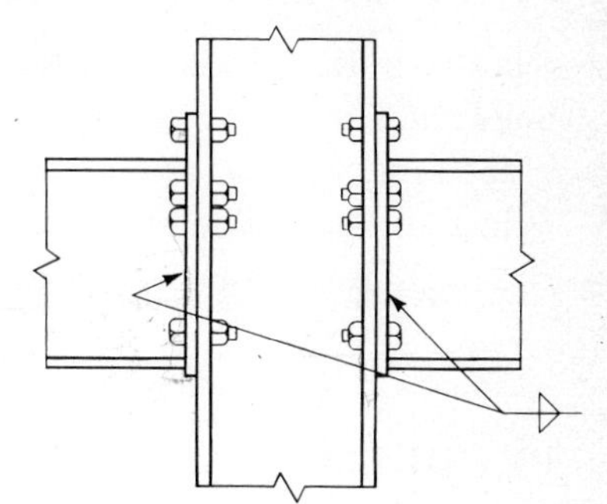

FIGURE 7-30

to designing for three-fourths of the sum of dead load, live load, and wind load at the normal allowable stresses, we may tabulate $DL + LL$ and $\frac{3}{4}(DL + LL + WL)$ and design for the larger of these two combinations, using the normal allowable stress. The following clarifications are identified by letters corresponding to those alongside certain computations in DP7-2.

a The connection is bearing-type, with threads excluded from the shear planes.

b The force *P* on the T-stub is determined by dividing the end moment on the beam, 105 ft-kips, by the beam depth.

c An estimate of the prying force must be made in order to determine the first-trial bolt size.

d The tee used here is cut from the S24 × 105.9. A 12-in. tee is needed to accommodate the six bolts connecting it to the flange of the beam. A table* in the AISC Manual facilitates the selection of the tee by giving an estimate of the required thickness in terms of the force per inch and the moment arm *b* of the bolt line. In this example, the length of the tee is $8\frac{1}{2}$ in., so that the force is $70/8.5 = 8.25$ kli. With the $4\frac{1}{2}$-in. gage shown, *b* will be about 2 in., in which case the AISC table shows that a 1-in. thickness is needed. The tee chosen has an average thickness $t_f = 1.10$ in.

e Although the 1-in. bolts are adequate, the bolt tension is only slightly larger than the allowable tension for the $\frac{7}{8}$-in. bolt. The smaller bolt turns out to be adequate.

f The shear force per bolt turns out to be small enough to permit the full allowable bolt tension, 40 ksi, concurrently. Therefore, framing angles are not needed to provide a shear connection.

The $4\frac{1}{2}$-in. gage in the flange of the tee is not large enough to give the clearance, relative to the bolts in the stem of the tee, to allow the bolts in the flange to be tightened with an impact wrench. They can be tightened with a spud wrench, however. Connections of this type can be made with the T-stub web shop-welded to the beam. This would allow the impact wrench to be used with the bolts at $4\frac{1}{2}$ in.

Field-bolted moment-resistant connections can also be made by using an end plate which is shop-welded to the beam (Fig. 7-30). Tests on such connections are

* See 7th ed., p. 4-80.

reported in Ref. 2, where for the bolt arrangement of Fig. 7-30 it was suggested that the bolts and a portion of the end plate symmetrical about the tension flange be analyzed in the same way as for a T-stub connection. The flanges of the beam can be groove-welded or fillet-welded to the plate. The web will usually be fillet-welded. The design and analysis of an end-plate connection are illustrated in the AISC Manual.

PROBLEMS

7-26 A W21 × 62 beam connects to the flange of a W14 × 111 column, A36 steel. The end shears and moments are as follows:

Load	Shear, kips	Moment, ft-kips
DL	8	26.7
LL	16	53.3
WL	6	80.0

Design a T-stub connection (Fig. 7-24*b*), bearing type with no threads in the shear planes, using A325 or A490 bolts. AISC specification. Note the increased allowable stresses for wind forces.

7-27 A W24 × 84 beam connects to the flange of a W14 × 127 column, A36 steel. The end shears and moments are as follows:

Load	Shear, kips	Moment, ft-kips
DL	14	70
LL	29	175
WL	10	200

Design a T-stub connection of the type shown in Fig. 7-24*d*, bearing type with no threads in the shear planes, using A325 or A490 bolts. Note the increased allowable stresses for wind forces.

7-28 Same as Prob. 7-27, except use a connection of the type shown in Fig. 7-24*b*.

7-15 RIVETS AND A307 BOLTS IN TENSION

Hot-driven rivets develop tension upon cooling, as mentioned in Art. 7-9, which may be of a high order of magnitude. This suggests that the behavior of hot-driven riveted connections in tension can be expected to be about the same as that of high-strength bolted connections in tension. A series of 28 tests on T-stub connections with $\frac{3}{4}$-in. A325 bolts and identical specimens with $\frac{3}{4}$-in. A141 rivets (now designated A502 Grade 1) showed that this is indeed the case.[19] Each specimen consisted of two

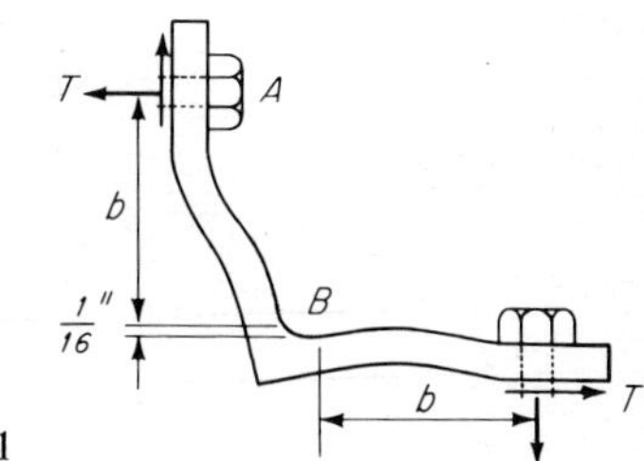

FIGURE 7-31

T-stubs connected through the flanges, so that it was similar to the bolted specimen discussed in Art. 7-13. The efficiencies of the bolted connections and their riveted counterparts were almost identical, in spite of the fact that the separation of the riveted stubs was considerably greater than that of the bolted stubs because of the greater ductility of the rivets. (Efficiency was defined as the ratio of the test load to the sum of the strengths of the individual fasteners in the connection.) Nevertheless, it does not follow that Eqs. (7-11) and (7-12) can be used to determine prying forces in hot-riveted connections, because there was some adjustment of the coefficients of the terms of both equations to bring the predicted values into better agreement with the results of tests on high-strength bolted connections. Therefore, in connections in which rivets will be subjected to calculated tension, some attention should be given to proportioning to control prying, e.g., by using relatively thick connection angles, T-stubs, and the like, or else by computing prying forces conservatively.

A307 bolts are rarely used in connections which subject them to calculated tension. If they are so used, there may be situations in which prying forces develop even though there may be little or no initial tension in the bolts. However, any such prying forces are likely to be small. Furthermore, no design information concerning them is available.

7-16 ANGLES IN MOMENT-RESISTANT CONNECTIONS

Proportioning of the angles in connections such as that of Fig. 7-24*a* can be based on an analysis similar to that of the T-stub. Deformation of such an angle is shown in Fig. 7-31. If the fasteners are high-strength bolts, Eq. (7-12) can be used to approximate the prying force, and the bolt tension and bending moments in the angle evaluated as in Art. 7-14. If rivets or A307 bolts are used and prying forces are neglected, analysis based on linearly elastic behavior, assuming complete rotational restraint at the fastener lines, gives $M_A = 0.6Tb$ and $M_B = 0.4Tb$. However, if plastic hinges develop at A and B, the moments are equal and $M_A = M_B = 0.5Tb$. It would seem reasonable to take the section at B for this case at the same position as for the T-stub, as shown in the figure.

MOMENT-RESISTANT CONNECTION — DP7-3

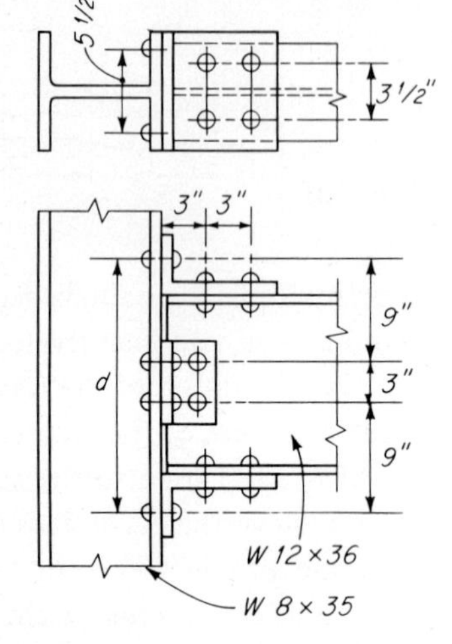

AISC Specification
A 36 steel
3/4-in. A 502 Grade 1 rivets

End moment $= 26'^k$
End shear $= 15^k$

Rivet value

s.s. $= 15 \times 0.44 = 6.6^k$
d.s. $= 13.2^k$
brg. web $= 48.5 \times 3/4 \times 0.305 = 11.1^k$
brg. flg $= 48.5 \times 3/4 \times 0.540 = 19.6^k$
tens. $= 0.44 \times 20 = 8.8^k$

$$d = \frac{M}{2T} = \frac{26 \times 12}{2 \times 8.8} = 17.7''$$
Use 18"

gage on top angle $= \frac{1}{2}(18 - 12.24) = 2.88$
Use 4" leg

Beam-flange connection

Rivets to beam flg. $\frac{2 \times 8.8}{6.6} = 2.7$, use 4

Try $8 \times 4 \times \frac{3}{4}$ L

$a = 2.88 - \frac{3}{4} - \frac{1}{16} = 2.07$

$M = 0.5 \times 2 \times 8.8 \times 2.07 = 18.2''^k$

$b = \frac{6 \times 18.2}{27 \times (3/4)^2} = 7.2$

Use $8 \times 4 \times \frac{3}{4} \times 7\,1/2$ L

Web connection

2L $4 \times 3\,1/2 \times 3/8 \times 5\,1/2$

$R = 2 \times 11.1 = 22.2^k$ to web $> 15^k$

$R = 4 \times 6.6 = 26.4^k$ to column $> 15^k$

DP7-3 illustrates the design of a moment connection using angles instead of tees. Such connections can be used where moments are small. However, they are classed as semirigid connections (Art. 7-17).

PROBLEMS

7-29 A W14 × 34 beam, whose end shear and end moment are 16 kips and 40 ft-kips, connects to the flange of a W10 × 54 column. Design a semirigid connection such as that of DP7-3, using A325 or A490 bolts. A36 steel, AISC specification.

7-30 A W14 × 53 beam, whose end shear and end moment are 12 kips and 70 ft-kips, connects to the flange of a W10 × 60 column. Design a semirigid connection such as that of DP7-3, using A502 Grade 1 rivets. A36 steel, AISC specification.

7-31 A W18 × 50 beam, whose end shear and end moment are 26 kips and 90 ft-kips, connects to the flange of a W10 × 72 column. Design a semirigid connection such as that of DP7-3, using A325 or A490 bolts. A36 steel, AISC specification.

7-17 TESTS OF CONNECTIONS

The beam connections that have been discussed so far in this chapter fall naturally into two groups: (1) those which are designed to resist only the end shear of a beam and (2) those which are designed to resist end moment in addition to shear. But beams are free of end moments only if they have complete freedom of rotation at the supports, and, conversely, they have end moments only if end rotation is prevented either wholly or in part. Since neither the completely free support nor the perfectly fixed one exists in fact, it is important to know how closely a given connection approximates the performance we assign to it for design purposes.

Results of three of a series of tests on the moment-resistant properties of various types of riveted connections are shown in Fig. 7-32. The three graphs OA, OB, and OC show the relation between the rotation ϕ and the corresponding moment M for the three types of connection shown at the right of the figure, the beam being an S12 × 31.8 (A7 steel) in each case. The graphs are only a partial record, since ultimate moments for connections A and C were 403 and 1,845 in.-kips, respectively. Rotations at the ultimate moment were not reported, nor was the ultimate moment for connection B. It is helpful to interpret these graphs by observing that the moment-rotation graph for a simply supported beam coincides with the axis of ϕ, since M remains zero for any rotation. Similarly, the graph for a perfectly fixed beam coincides with the axis of M, since there is no rotation at any moment. Thus we see that the framing connection A can be classed as fairly flexible, while the T-stub connection C is quite stiff.

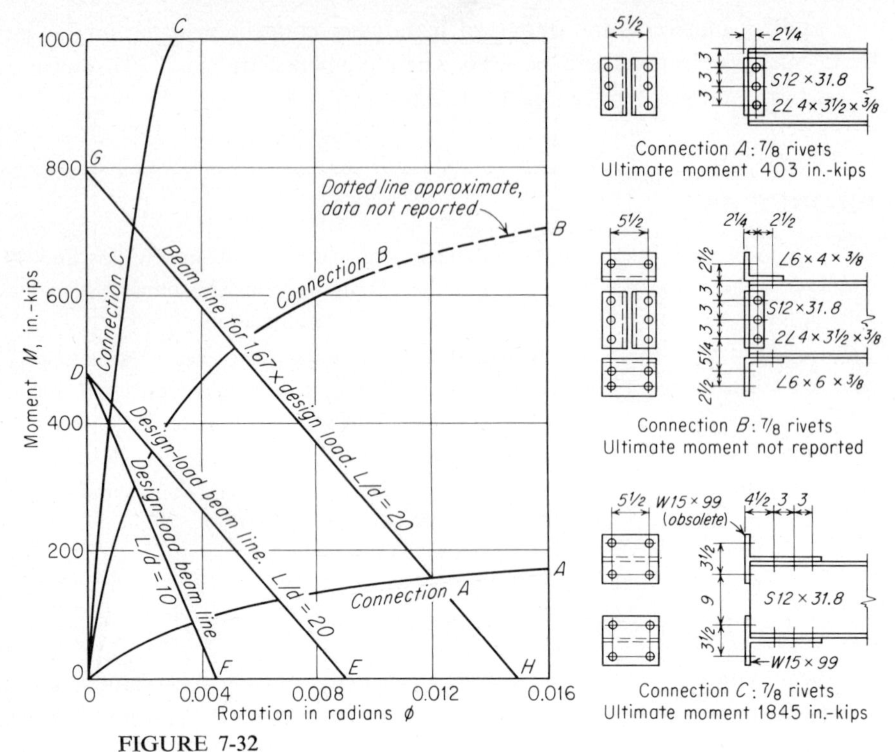

FIGURE 7-32
Comparative rigidity of connections. (*From J. C. Rathbun, Elastic Properties of Riveted Connections, Trans. ASCE, vol. 101, 1936.*)

It is to be noted that the variation of M with ϕ is curvilinear from the beginning. Although this complicates the problem of determining the moment that will be developed by a particular connection when it is used on a beam whose span and load are given, a simple graphical procedure has been suggested.[8] If the stresses do not exceed the yield point of the material, the rotation at the end of a beam of span L supporting the uniform load W and equal end moments M is given by

$$\phi = \frac{WL^2}{24EI} - \frac{ML}{2EI} \tag{a}$$

But since the values of ϕ and M for the beam must equal the corresponding values for the connection, the intersection of the graph of this equation with that of the connection determines the solution. Furthermore, since Eq. (a) is linear in M and ϕ, two points are sufficient to obtain its graph. If $\phi = 0$, then $M = WL/12$, which is the end moment for a beam with built-in ends. Also, if $M = 0$, $\phi = WL^2/24EI$, the end rotation of a simply supported beam. This formula for ϕ can be put into a convenient form by substituting, in turn, M for $WL/8$ and fI/c for M to get, for a simply supported beam,

$$\phi = \frac{2}{3}\frac{f}{E}\frac{L}{d} \tag{b}$$

Thus we see that, for a given stress f, ϕ is directly proportional to the ratio L/d.

As a numerical example, consider the shape used in the tests, namely, the S12 × 31.8, as a beam spanning 20 ft, so that $L/d = 20$. If this beam is supported against lateral buckling and is designed on the assumption that its end supports are simple, then, using an allowable bending stress of 20 ksi, $W = 8M/L = 8 \times 20 \times 36.4/240 = 24$ kips. Therefore, $M_F = WL/12 = 24 \times 240/12 = 480$ in.-kips, which, as we have seen, corresponds to $\phi = 0$. Plotting this value on Fig. 7-32, we obtain the point D. Also, from Eq. (b), $\phi = (\frac{2}{3})(20/30{,}000)(20) = 0.0089$ radian, which gives point E on the figure. The straight line connecting D and E intersects the graph of any one of the connections at the point whose coordinates are the end moment and the corresponding end rotation *at the design load* if that connection is used on the beam we have chosen. The line DE is called the *beam line*.

The variation of end moment with the ratio L/d is easily shown. If we consider the S12 × 31.8 on a span of 10 ft, so that $L/d = 10$, the allowable load $W = 8M/L$ is double that for the 20-ft span. The fixed-end moment remains the same, however, since $M_F = WL/12$. We have already seen that the end rotation for the simply supported condition is directly proportional to L/d. Thus a straight line from D to the point F midway between O and B gives us the beam line for the S12 × 31.8 on a span of 10 ft.

A significant aspect of the typical moment-rotation graph should be mentioned. Connections unload linearly, parallel to the initial slope of the loading curve. Thus, the unloaded connection retains a permanent deformation. Subsequent loadings and unloadings are elastic in the sense that they follow the original unloading line so long as the initial load is not exceeded. This behavior is analogous to that observed in the simple tension test.

7-18 BEHAVIOR OF STANDARD CONNECTIONS

The information in Fig. 7-32 enables us to determine how closely certain types of connections approximate the assumptions we make for design purposes. For example, we see that the framed-beam connection develops an end moment of about 120 in.-kips when it is used on an S12 × 31.8 supporting the uniform load for an allowable stress of 20 ksi and considered as a simple beam spanning 20 ft. This is 25 percent of the end moment required to completely fix the ends of the beam. Of course, this is on the safe side as far as the beam is concerned, since the bending moment at the center is reduced by the amount of the end moment. However, it is clear that it results in uncomputed stresses in the connection itself, since it is standard procedure to design framed-beam connections for only the end shear, in this case 12 kips. The analysis for a bolt or rivet group resisting torsion indicates that the end fasteners of the line of three connecting the angles to the web must develop a force of 20 kips to resist a moment of 120 in.-kips. This combines with the vertical force of 4 kips due to the 12-kip reaction to give a resultant of 20.4 kips. Thus, the uncomputed stress is considerably larger than the computed one. Furthermore, there is a tension on the fasteners which is not considered in the usual procedure for designing such a connection.

Assuming that the beam line *DE* in Fig. 7-32 is based on a factor of safety of 1.67 on yielding of the extreme fiber, the beam line at yielding is *GH* if the beam is simply supported. This line intersects the curve for connection *A* at 160 in.-kips. This is only 33 percent greater than the end moment at the design load. Thus the end moment increases at a slower rate than does the moment at the center of the span.

The beam line *DE* intersects the graph for the T-stub connection *C* at the point $\phi = 0.0008$ radian, $M = 430$ in.-kips. The latter value is 90 percent of the end moment for a fixed support, and the corresponding rotation is less than 3 minutes. Thus the uniformly loaded S12 × 31.8 spanning 20 ft is practically end-fixed if connection *C* is used.

7-19 THREE TYPES OF CONSTRUCTION

Compared with connections *A* and *C*, connection *B* of the tests reported in Fig. 7-32 is of intermediate stiffness. It is clear that it should not be assumed to effect a rigid connection between a beam and its supporting member. On the other hand, it is uneconomical to assume it to be a moment-free connection, since it develops a resisting moment large enough to reduce the bending moment at the center well below the simple-beam value, $WL/8$.

The two common assumptions as to the behavior of a building frame are (1) that its beams are free to rotate at their connections or (2) that its members are so connected that the angles they make with each other do not change under load. It is obvious that the behavior of a structure which employs connections of intermediate stiffness will be intermediate between these two extremes. Frameworks with connections of intermediate stiffness are usually called *semirigid frames*, as contrasted to *simple* framing and *rigid-frame* or *continuous construction*. Semirigid frames can be analyzed by any of the methods for statically indeterminate structures, provided that the rotations of the connections can be predicted and accounted for. The AISC specification allows semirigid framing only if the performance of the proposed connections can be demonstrated. Stiff connections and flexible connections have been sufficiently standardized and verified in performance to make it unnecessary to impose such restrictions on the other two types of construction.

7-20 WELDED FRAMED BEAM CONNECTIONS

Figure 7-33 shows various ways in which a welded framed beam connection can be made. In *a* the beam web is welded directly to another member, e.g., the flange of a column. The connection can be by groove weld or by two fillet welds. The seat, which may be a plate, as shown, or an angle, is for purposes of erection. Although this is the most direct connection that can be made, it has a number of disadvantages. It requires that the gap between the end of the beam and the adjacent surface be small, otherwise an adequate connection is assured only if the size of the weld is increased. The AISC specification requires that the size of fillet welds be increased by the amount of the separation if it is $\frac{1}{16}$ in. or more and that in no event shall the separation exceed $\frac{3}{16}$ in. But the rolling mill allows itself a tolerance of $\pm\frac{3}{8}$ to $\frac{1}{2}$ in. or more, depending on the depth and length, on the ordered length of a beam and also considers acceptable an end which may be out of square by not more than $\frac{1}{64}$ in. per inch of depth of the shape. Thus a 12-in. beam might show a gap which is $\frac{12}{64} = \frac{3}{16}$ in. greater at one flange than at the other. These deficiencies can be corrected by flame cutting, but this is an expensive operation. Furthermore, the member to which the beam connects will be subject to tolerances. For example the depth of a W-shape column may vary $\pm\frac{1}{8}$ in. from its specified value. Also, vertical welding in the field is costly and usually should be avoided. Finally, the stiffness of such a connection may be greater than can be tolerated. Line *OC* in Fig. 7-34 shows the results of one test on a W18 × 55 beam connected by $\frac{1}{2}$-in. fillet welds 11 in. long, one on each side of the web. Initial yielding of the web at the lower end of the welds was observed at a moment of 360 in.-kips. Yielding of the web along the full length of the welds occurred at 660 in.-kips, and at 870 in.-kips the welds cracked slightly at the top of the connection. It will be noted that for $L/d = 20$ (span 30 ft) the design-load beam line intersects curve *OC* at 750 in.-kips. Since the web yielded full length at 660 in.-kips, yielding of the connection could be expected at service loads.

The connection shown in Fig. 7-33*b* eliminates the problem of fit-up of the connection of Fig. 7-33*a*. Here, a connection plate shop-welded to the column allows field adjustment for mill tolerances in the length of the beam and depth of the column.

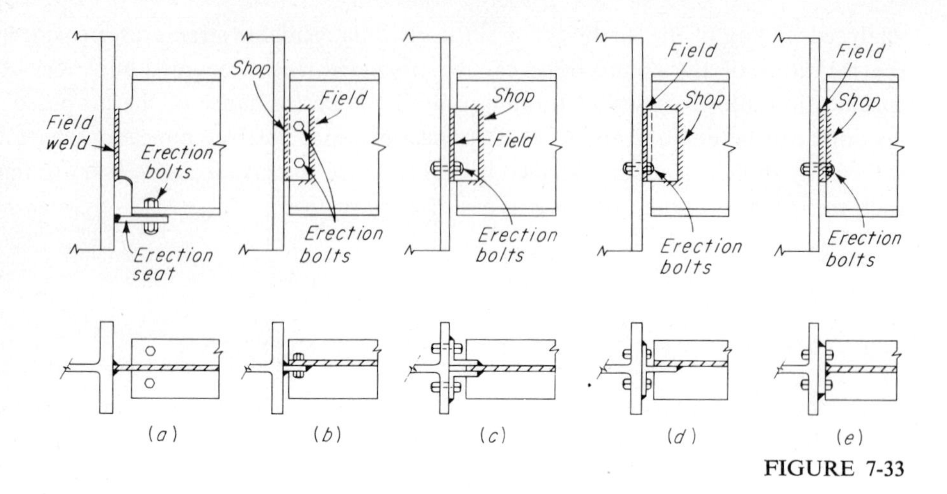

FIGURE 7-33

Two side plates, one on each side of the web, may also be used instead of the single plate shown. Erection bolts take the place of a seat. This connection does not eliminate the costly vertical welding in the field, however, and is of the same order of stiffness as the connection of Fig. 7-33*a*. Furthermore, the plates are harder to hold in alignment during welding than angles are.

The framing angles shown in Fig. 7-33*c* are shop-welded to the beam web and field-welded to the column. This connection also allows for mill tolerance in the length of the beam. To provide flexibility in the connection, the angles are field-welded on the vertical edges, with short returns at the top. The returns are needed to assure a good weld with no crater at the end. The partial moment-rotation curve *OA* of Fig. 7-34 shows the average of data from tests on two W18 × 55 beams connected by $3\frac{1}{2} \times 2\frac{1}{2} \times \frac{1}{4}$ angles 10 in. long, with the $3\frac{1}{2}$-in. legs outstanding.[10] The maximum moment attained was 68 in.-kips, at a rotation of about 0.03 radian. A comparison of this curve with the curve *OB* for an S18 × 54.7 connected by $4 \times 3\frac{1}{2} \times \frac{3}{8}$ angles, with $\frac{7}{8}$-in. rivets,[9] shows the relative flexibility of the two connections. The design-load beam line for the W18 × 55 on a span of 30 ft indicates that, for this case, the welded connection develops an end moment of about 40 in.-kips compared with 400 in.-kips for the riveted connection.

The connection shown in Fig. 7-33*d* is similar to the framing-angle connection and comparable in flexibility. The end-plate connection shown in Fig. 7-33*e* also gives comparable flexibility and uses less connection material. However, it is more difficult to provide for mill tolerances.

The connections shown in Fig. 7-33*c* and *e* can also be bolted to the columns, rather than welded. This has the advantage of eliminating vertical welding in the field.

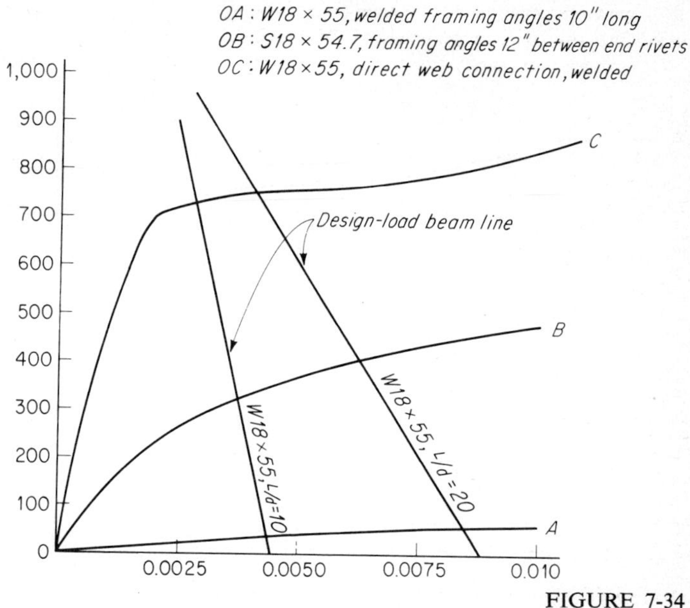

FIGURE 7-34

7-21 DESIGN OF WELDED FRAMED CONNECTIONS

Figure 7-35*a* shows two framing angles separated from the beam which they support. The beam reaction R is assumed to be shared equally by the two field welds. The corresponding forces on the shop welds as they act on the angles and on the web are as shown, their lines of action intersecting the centroids of the shop welds.

If we assume the beam to be simply supported at the face of the outstanding legs of the connection angles, then the field welds can exert no moment in a plane parallel to xz. This leaves on each angle an unbalanced moment $Ra/2$, which we must then assume to be resisted by the shop weld. It is customary to compute the resulting stresses by using the torsion formula, $f_v = Tr/J$, even though the formula applies only to circular cross sections. The shearing stress due to torsion must be combined with the shearing stress due to the reaction $R/2$.

There is also on each angle an unbalanced moment $Rb/2$, which tends to produce rotation about an axis parallel to x. Although this moment is undoubtedly shared by the shop weld and the field weld, it is the usual practice to assume that it is resisted by the latter. The resisting couple is taken to be in the form of bearing pressures between the angles and the web at the upper ends of the angles and between the field weld and the angles over the remaining length, as is indicated in Fig. 7-35*b*. The analogy between this situation and that of the angle connected by rivets or bolts suggests that

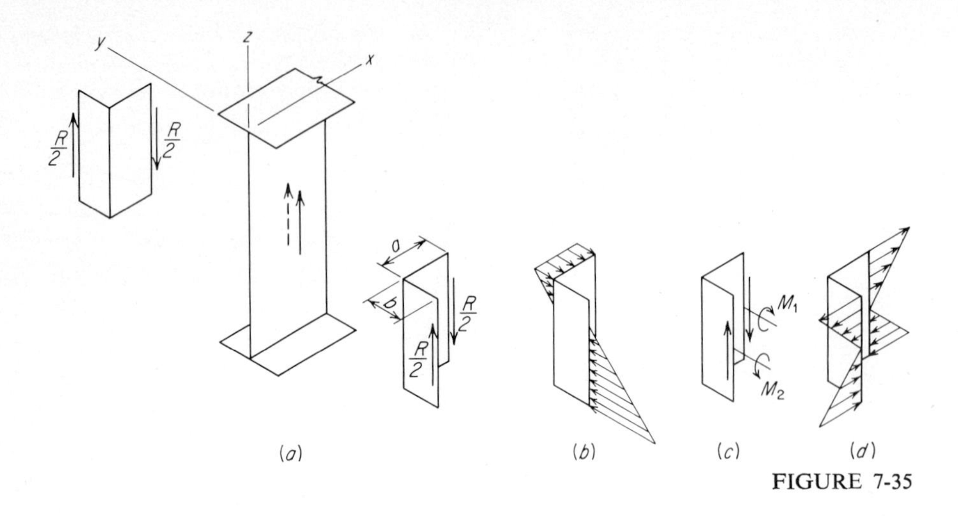

FIGURE 7-35

contact between the angles and the web be assumed to extend for about one-sixth the length of the angle. The assumption of linear distribution of stress, as shown in Fig. 7-35*b*, leads to a simple solution for the maximum bending stress in the weld. This stress is readily combined with the vertical shearing stress due to $R/2$, since the two are at right angles and fillet-weld stresses are assumed to be shearing stresses irrespective of their direction with respect to the cross section through the throat.

For the assumption that the shop welds furnish the couples which balance the couples $Ra/2$ in Fig. 7-35*a* to be true, the beam must rotate counterclockwise with respect to the angles. But this could happen only if the angles turned clockwise relative to the beam, which would mean that they help, rather than retard, the end rotation of the beam. In other words, any torsional shearing stresses on these welds result from couples which are *clockwise* on the angles and which therefore tend to increase the unbalance of moments around the y axis. Consequently, the unbalanced moment $Ra/2$ *plus the neglected clockwise moment* can be resisted only by the field weld. The situation is as shown in Fig. 7-35*c*, where M_1 is the clockwise moment exerted on the angle by the beam and M_2 is the balancing couple supplied by the field weld. Obviously, $M_2 = M_1 + Ra/2$. The couple M_2 is in reality the end moment which we ignored when we assumed in the beginning that the beam is simply supported at the outstanding legs of the connection angles.

Since the framing angles will not ordinarily fit snugly against the face of the member to which they connect, because of variation in sizes of members, necessary erection clearances, etc., it would seem reasonable to assume that the couple M_2 is provided by the stress distribution shown in Fig. 7-35*d*. Therefore, we may expect the field weld to be subjected to a combination of this stress and that of Fig. 7-35*b*, as well as a vertical shearing stress due to $R/2$.

One may ask how we achieve acceptable design with a procedure that omits consideration of the factors we have just discussed. The answer lies in the high degree of flexibility of the connection. For example, we found in Art. 7-20 an end moment of only 40 in.-kips at the design load for a W18 × 55 on a span of 30 ft, so that for each angle $M_2 = 20$ in.-kips. Since the uniformly distributed design load for this beam is 44 kips for an allowable stress of 20 ksi, $R/2 = 11$ kips. Therefore, since a is of the order of 2 in., we see that the couple $Ra/2$ is of the same order of magnitude as is M_2. If this were not true, as would be the case if the connection were much stiffer, then the moment $Ra/2$ for which the shop weld is designed would not be adequate, nor would it be safe to neglect the effect of M_2 on the field weld.

EXAMPLE 7-3 In this example we illustrate the usual design procedure for welded framed connections, using the AISC allowable stresses for A36 steel and E70 electrodes. The beam is a W21 × 62 spanning 21 ft and supporting a uniform load of 92 kips, and connected at each end to the flange of a W12 × 65 column. There are several ways to arrive at a trial design. The length of this type of connection usually ranges from one-half to two-thirds the depth of the beam. Therefore, we might assume the length of the connection and then compute the required size of the weld. But we may also assume the size of the weld, which is likely to range from $\frac{3}{16}$ in. for smaller beams to $\frac{3}{8}$ in. for the large sizes, and then determine its length.

We shall try $3\frac{1}{2} \times 2\frac{1}{2}$ connection angles, with the $3\frac{1}{2}$-in. legs outstanding. The end shear is $92/2 = 46$ kips. Investigating first the connection to the web, we try a $\frac{3}{16}$-in. shop weld. The AISC allowable shear for a $\frac{3}{16}$-in. fillet weld is $q = 21 \times 0.707 \times \frac{3}{16} = 2.78$ kli. The length of weld required to resist the end shear is $l = 23/2.78 = 8.27$ in., where 23 kips is the shearing force per weld ($R/2$ in Fig. 7-35*a*). The length required for moment may be estimated by neglecting the horizontal welds and using the beam formula, $f = Mc/I = 6M/l^2t$, where t is the throat thickness. But since $ft = q$, we find from this equation $q = 6M/l^2$. The width of the angle is $2\frac{1}{2}$ in., so that $M = 23 \times 2.5 = 57.5$ in.-kips, and $l = \sqrt{6 \times 57.5/2.78} = 11.1$ in. for each weld. On the basis of these computations, we try the connection shown in Fig. 7-36*a*, whose length is about half the depth of the beam.

The distance x from the vertical weld to the centroid of the complete shop weld is $x = 2 \times 2 \times 1/14 = 0.29$ in. The polar moment of inertia, $J = I_x + I_y$, is

$$J = \frac{10^3}{12} + 2 \times 2 \times 5^2 + 10 \times 0.29^2 + 2\left(\frac{1.71^3}{3} + \frac{0.29^3}{3}\right) = 186 \text{ in.}^4$$

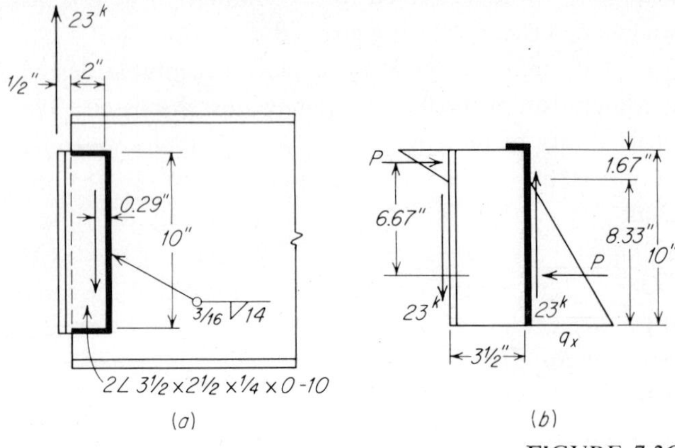

FIGURE 7-36

The first two terms of this equation represent I_x, and since they contribute 183 in.^{4}, there is not much point in bothering about I_y. The size of the weld does not appear in the calculations, since we are investigating the shear per inch, not the shearing stress.

The maximum stress occurs at the free end of the horizontal weld. The two components are

$$q_x = \frac{23 \times 2.21 \times 5}{186} = 1.37 \text{ kli}$$

$$q_y = \frac{23 \times 2.21 \times 1.71}{186} = 0.47 \text{ kli}$$

The component q_y must be combined with the vertical shear resulting from the 23-kip reaction, $q_y = 23/14 = 1.64$ kli. The resultant shear is therefore $q = \sqrt{2.11^2 + 1.37^2} = 2.52$ kli. This value checks closely with the allowable value, 2.78 kli.

The stress distribution for the field weld is shown in Fig. 7-36*b*. Assuming that the web of the beam and the one leg of the angle are in contact for a distance equal to one-sixth the length of the angle (Art. 7-21) and equating the couples, we have $(8.33q_x/2) \times 6.67 = 23 \times 3.5$, from which $q_x = 2.90$ kli. The vertical shear is $q_y = 23/10 = 2.30$ kli, and the resultant $q = \sqrt{2.90^2 + 2.30^2} = 3.70$ kli. The field weld must therefore be $\frac{1}{4}$ in., for which the allowable $q = 21 \times 0.707 \times \frac{1}{4} = 3.70$ kli. To assure flexibility of the connection, we use angles $\frac{1}{4}$ in. thick, although this will require that the $\frac{1}{4}$-in. shop welds be built out to ensure the full throat thickness.

PROBLEMS

7-32 A W24 × 100 beam spanning 40 ft and supporting a uniform load of 7.5 klf including its own weight connects at each end to the flange of a W12 × 72 column. Design a welded connection of the type shown in Fig. 7-33*c*. A36 steel and E70 electrodes. AISC specification.

7-33 Design a welded connection of the type shown in Fig. 7-33*c* for the stringers to the floor beams of Prob. 5-10. A36 steel, E70 electrodes. AASHO specifications.

7-34 A W18 × 70 beam spanning 15 ft and supporting a uniform load of 7 klf including its own weight connects at each end to the flange of a W12 × 79 column. Design a welded connection of the type shown in Fig. 7-33*c*. A441 steel, E70 electrodes. AISC specification.

7-35 For the data of Prob. 7-32 design a welded connection of the type shown in Fig. 7-33*d*.

7-36 For the data of Prob. 7-34 design a welded connection of the type shown in Fig. 7-33*e*.

7-22 UNSTIFFENED WELDED BEAM SEATS

Welded seat connections can be designed so as to have about the same flexibility as the welded framed connection, and where they can be used, they provide a very satisfactory and efficient support. As is the case in bolted or riveted construction, the seat may be either stiffened or unstiffened, and it must be used in conjunction with a top angle to assure stability of the beam. Typical unstiffened seats are shown in Fig. 7-37, one to a column flange, the other to the web. Two welds, one at each end of the angle, are used. These welds should be returned across the top of the seat for a distance of about $\frac{1}{2}$ in. in order to eliminate craters at the top of the vertical welds. Although a fillet weld across the top of the seat could also be used, it is likely to interfere with erection if the beam overruns in length. The top angle should be welded only on its toes, so that it is free to bend and thus contribute to the flexibility of the connection. If the welding clearance shown in section *A-A* is not available, the web seat may consist of a plate, set out far enough to clear the web fillets of the column, fitted and welded to the inside faces of the column flanges.

The design of the welded seat involves no principles different from those of the bolted or riveted seat. The reaction is usually assumed to be uniformly distributed over a length, measured from the end of the beam, just sufficient to satisfy the web-crippling requirement of the beam. Some designers assume the stress distribution in the vertical welds to be the same as for beams, with neutral axis at midheight, while others assume a distribution similar to that of the bolted bracket, so that the neutral axis is nearer the compression side. Those who hold the latter view usually take the zone of compression to be one-third the length of the weld. Either way, the design calculations are simple enough. After determining the dimensions of the horizontal leg of the seat, we assume the size of the other leg. This gives us the length of the vertical welds. It is then easy to compute the stress due to bending and combine it

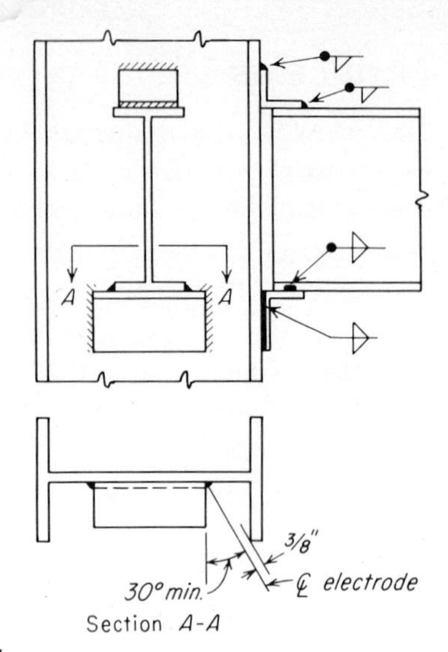

FIGURE 7-37

with the vertical shear due to the reaction, in the usual way for fillet welds. In these calculations we work with the force per inch of length of weld, as described in the example in the preceding article, so that the size of the weld need not be known in advance. The resultant force determines the required size of the weld. If the weld turns out to be too large or too small, we change the size of the angle and try again. If we prefer, we may assume the size of the weld and compute the length required to resist the shear alone, and the moment alone, as discussed in Example 7-3. This may lead to a better first guess for the size of the angle.

Since there is no computed force on the top angle, its proportions are a matter of judgment. A 4-in. vertical leg and a thickness of $\frac{1}{4}$ in. are suggested as minimums in the AISC Manual.

7-23 STIFFENED WELDED BEAM SEATS

The welded, stiffened beam seat is somewhat simpler than its riveted or bolted prototype. It may consist of a tee or of two plates welded together in the form of a tee. The two types are shown in DP7-4 and Fig. 7-41. The bracket may be cut as in DP7-4 when clearance is necessary, as for fireproofing; otherwise it may be cut square as in Fig. 7-41.

The eccentricity of the beam reaction is probably greater for the stiffened seat than for the unstiffened one, since the former is stiffer. The reaction is usually assumed to lie at the midpoint of the length of bearing which is demanded by the web-crippling

WELDED BRACKET DP7-4

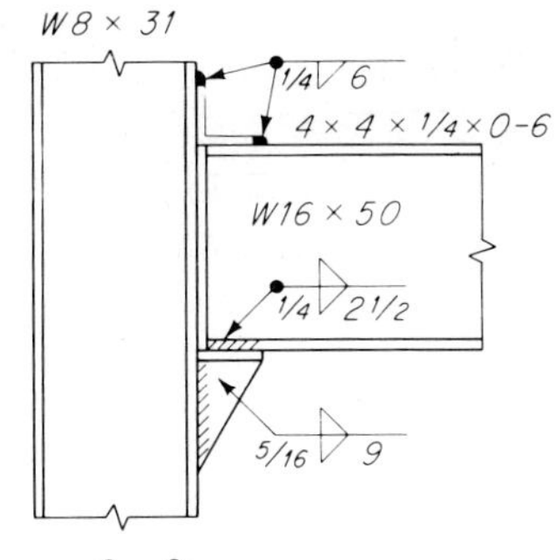

W16 × 50, A36 steel AISC Spec.
Span 12′-0″
Load 9 klf

$R = 9 \times 6 = 54^{k}$

$\text{Brg. lgth.} = \dfrac{54}{27 \times 0.38} - 1.25 = 4.01''$

Seat lgth. = 4.01 + 0.5 = 4.51″ Use 4 1/2″

$M = 54\,(4.5 - 2.0) = 135''^{k}$

Try vert. leg 8″
horiz. leg 2″

$\bar{y} = \dfrac{2 \times 8 \times 4}{16 + 4} = 3.2''$

$I_x = 2\left[\dfrac{3.2^3 + 4.8^3}{3} + 2\,(3.2)^2\right] = 137\ \text{in.}^3$

$q_z = \dfrac{135 \times 3.2}{137} = 3.15^{kli}$

$q_y = \dfrac{54}{20} = 2.70^{kli}$

$q = \sqrt{3.15^2 + 2.70^2} = 4.15^{kli}$

Use 5/16-in. fillet weld, $q = 4.64^{kli}$

Web of tee:

$0.40 \times 36 \times t = 2 \times 2.70$

$t = 0.375''$

W16 web $t = 0.380$

Use WT8 × 29

requirements of the beam, as in the unstiffened seat, but with this difference: the bearing length is measured from the outer end of the seat rather than from the end of the beam. The bracket is usually short enough to eliminate need for concern about local buckling of the stem.

Since the vertical welds along the stem are close together, it is advisable to weld along the underside of the flange so as to increase the torsional stiffness of the connection. This return also helps to avoid craters at the top of the vertical weld. This detail is shown on the figure in DP7-4. The length of each horizontal weld will usually be one-fifth to one-half that of the vertical weld, although it may be longer.

The design of the stiffened beam seat is illustrated in the following article.

7-24 DP7-4: WELDED BRACKET

The stress in the vertical welds is computed on the assumption that the moment is resisted entirely by the welds. Furthermore, the stress due to bending is determined at the horizontal weld. This is the usual practice, although theoretically the stress is larger at the lower end of the vertical weld. Tests have shown that stiffened brackets designed according to this procedure have ample factors of safety with respect to failure.

The required thickness of the web of the tee is based on the allowable shear $0.40F_y$; that is, the shear per inch of web must equal the sum of the vertical shears per inch of the two welds. The web should also be as thick as the web of the beam supported by the seat (if both are of the same steel) in order to have the same crippling strength. According to the AISC Manual, however, the web of stiffened seats should also be not less than twice the dimension of the weld. This is based on the idea that the web should be capable of reacting the full allowable shear of the welds:

$$2 \times 21 \times 0.707a = 0.40 \times 36t$$

which gives $t = 2a$. Since only the vertical component of the weld shear is in the direction to produce shear on the web of the seat, this would seem to be an unnecessarily severe requirement. It would require a $\frac{5}{8}$-in. web, instead of $\frac{3}{8}$ in., for the seat in this example.

PROBLEMS

7-37 Design a welded seat to support a W10 × 25 beam on the web of a W12 × 65 column. The beam reaction is 30 kips. A36 steel, E70 electrodes, AISC specification.

7-38 Design an A36 welded seat to support an A441 W12 × 50 beam on the web of an A441 W12 × 85 column. The beam reaction is 50 kips. E70 electrodes, AISC specification.

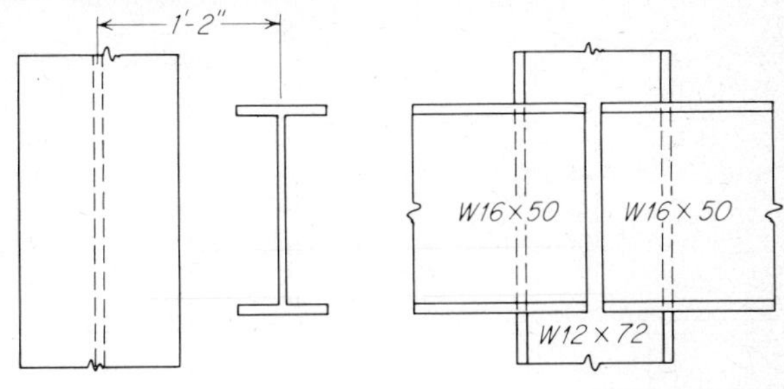

FIGURE 7-38
Problem 7-41.

7-39 Design a welded seat to support a W16 × 78 beam on the flange of a W14 × 87 column. The beam reaction is 70 kips. A36 steel, E70 electrodes, AISC specification.

7-40 Design a welded bracket for the crane girder of Prob. 7-15. AISC specification.

7-41 Design a welded connection to support the A441 W16 × 50 beams on the A441 W12 × 72 column shown in Fig. 7-38. Each beam reaction is 50 kips. E70 electrodes, AISC specification.

7-42 Design a welded connection to support the S12 × 31.8 beams on the W8 × 31 column shown in Fig. 7-39. Each beam reaction is 20 kips. A36 steel, E70 electrodes, AISC specification.

7-25 MOMENT-RESISTANT WELDED BEAM CONNECTIONS

Typical moment-resistant welded connections for beams are illustrated in Fig. 7-40. The most direct connection of a beam to a column flange is shown in *a*. Here the beam flanges are groove-welded and the web fillet-welded to the column flange. In *b* the web is field-welded to a plate which is shop-welded to the column flange. This allows a more lenient setback of the web and also eliminates the angle seat in the connection shown in *a*. In the connection shown in *c* bolts are used instead of field fillet welds in the web connection. A similar connection of a beam to a column web

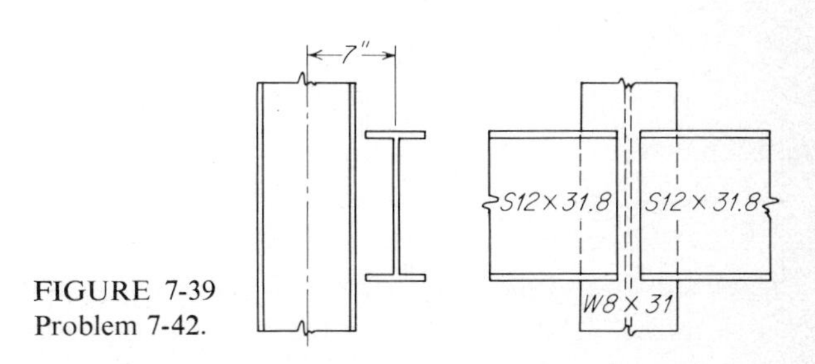

FIGURE 7-39
Problem 7-42.

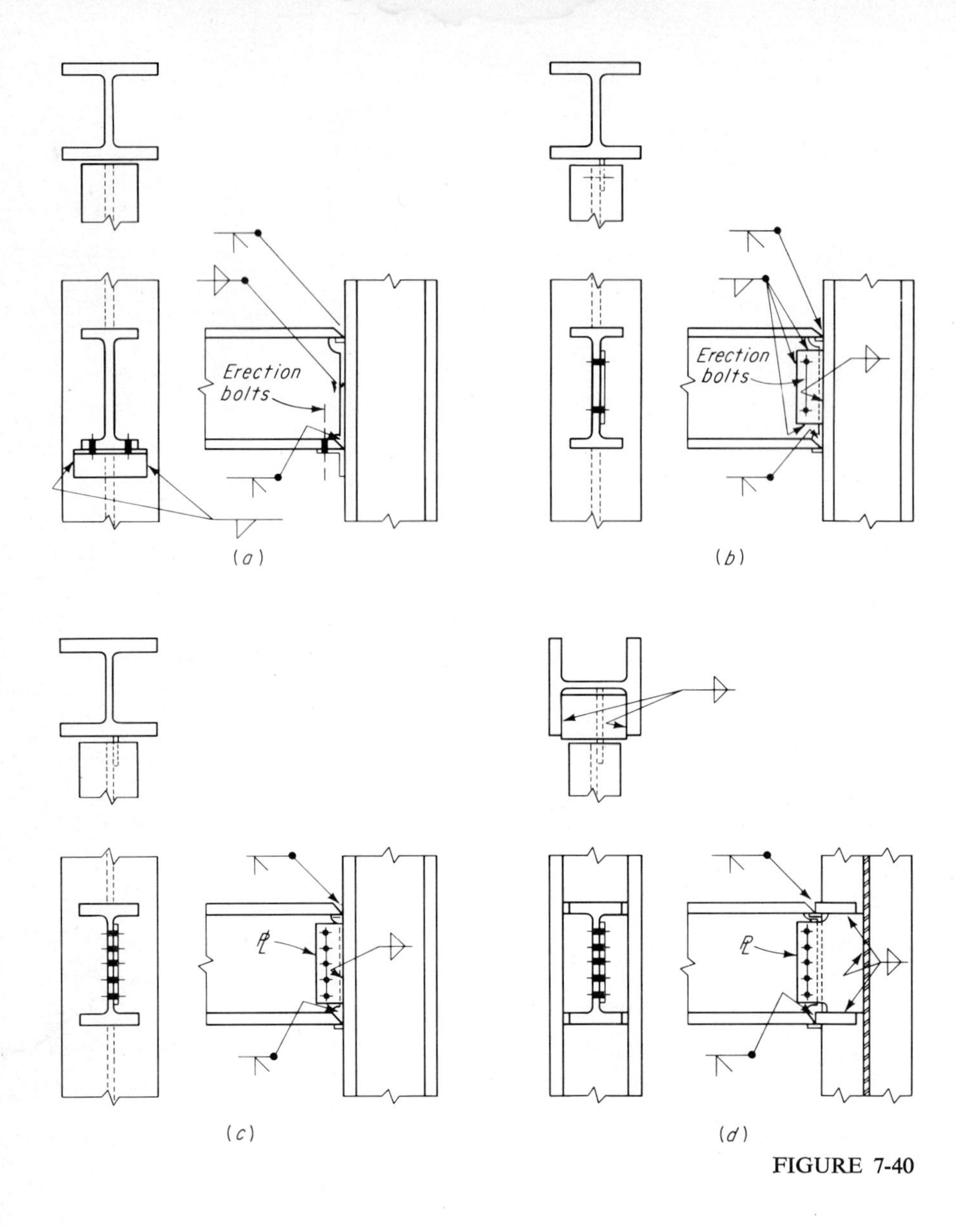

FIGURE 7-40

is shown in *d*. Tolerances must be closely controlled in these connections; otherwise considerable difficulty may be encountered in the field.

Less restrictive tolerances in beam length can be allowed with a connection of the type shown in Fig. 7-41. The top plate is shipped loose and positioned after the beam is in place. The top plate for the beam at the left is welded full length. This is con-

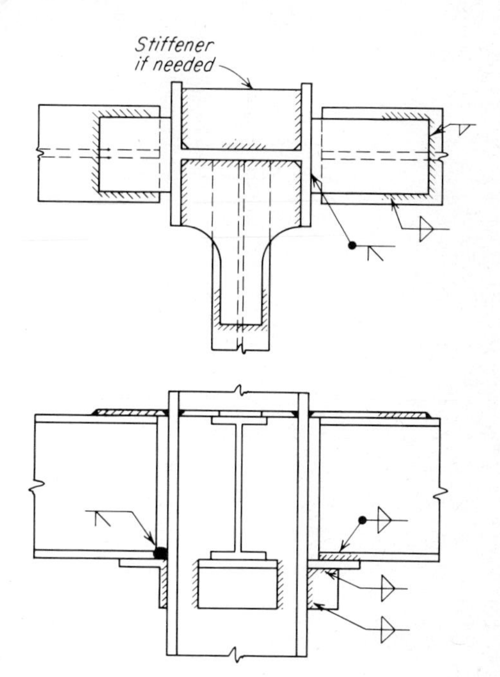

FIGURE 7-41

sidered to be a connection suitable for Type 1 construction (Art. 7-19). The top plate for the beam at the right is welded over only a part of its length; this connection is considered to be semirigid. In case the beam frames to the web of a column, the top plate may be widened and fillet-welded to the column web and flanges as shown. A variety of procedures, depending on the assumptions as to behavior of the connection and of the frame itself, are used for the design of top-plate connections.[11, 12]

Depending on the magnitude of the reaction, the seat may be unstiffened, as at the left in Fig. 7-41, or stiffened, as at the right. The horizontal plate of the stiffened seat may extend beyond the vertical plate if this is required for a sufficient length of weld between the beam flange and its seat. If the end moment is negative, the horizontal plate may be butted against the column flange without being welded to it. But if the moment may be either positive or negative, as with wind loads, the horizontal plate should be welded to the column flange. Alternatively, the beam flange may be welded directly to the column flange, as shown in the connection at the left. If this weld is required, clearance at the end of the beam must be watched, since a minimum of $\frac{3}{16}$ in. is required. The weld between the top plate and the column flange requires the same $\frac{3}{16}$-in. minimum gap.

Moment-resistant beam-to-girder connections are illustrated in Fig. 7-42. If the top flanges are at the same elevation, the top plate crosses the flange of the girder, as shown in Fig. 7-42*a*. If the flanges are at different elevations, the top plates may be

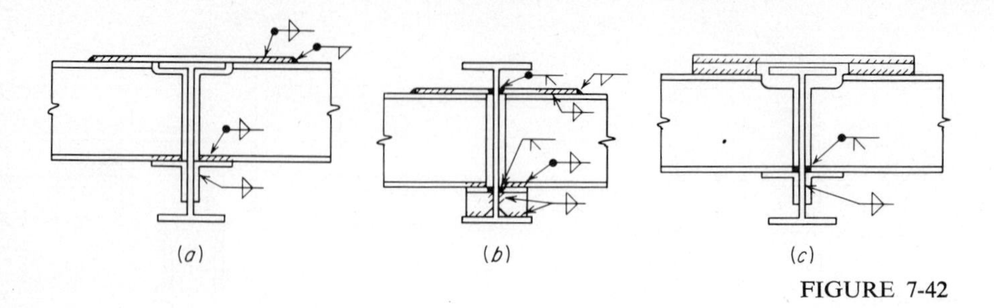

FIGURE 7-42

groove-welded to the web of the girder, as in Fig. 7-42*b*, or a single plate passing through a slot in the web may be used. If the difference in elevation of the flanges is small, the detail in *c* may be used. The seat may be an angle, as in Fig. 7-42*a*. If the distance between the bottom flanges of the two beams is too small for a sufficient length of weld on the seat angle, a tee may be used, as in Fig. 7-42*b*. The bottom flange of the beam may be groove-welded to the web of the girder, as in Fig. 7-42*c*, using the seat for backup, or the heel of the seat may be groove-welded in the shop to permit fillet welds in the field, as in Fig. 7-42*b*.

7-26 STIFFENERS IN BEAM-TO-COLUMN CONNECTIONS

The forces in the beam flanges of beam-to-column connections of the type shown in Fig. 7-40 may cause crippling and/or buckling of the column web opposite the compression flange of the beam. This phenomenon is analogous to the crippling and buckling of beam webs at concentrated loads, which was discussed in Art. 5-19. So far as buckling is concerned, Eq. (5-38*b*) gives a conservative estimate of the limiting slenderness which allows the web to reach the compressive yield stress without buckling. With $E = 29{,}000$ ksi, $\mu = 0.3$, and $F_{cr} = F_y$, the equation gives

$$\frac{b}{t} = \frac{162}{\sqrt{F_y}} \tag{a}$$

The AISC specification has adopted the value

$$\frac{d_c}{t} \ngtr \frac{180}{\sqrt{F_y}} \tag{7-14}$$

where d_c is the column-web depth between fillets.

According to tests reported in Ref. 13, the flange force can be assumed to distribute itself into the column web as in Fig. 7-43*a* for an evaluation of web crippling.

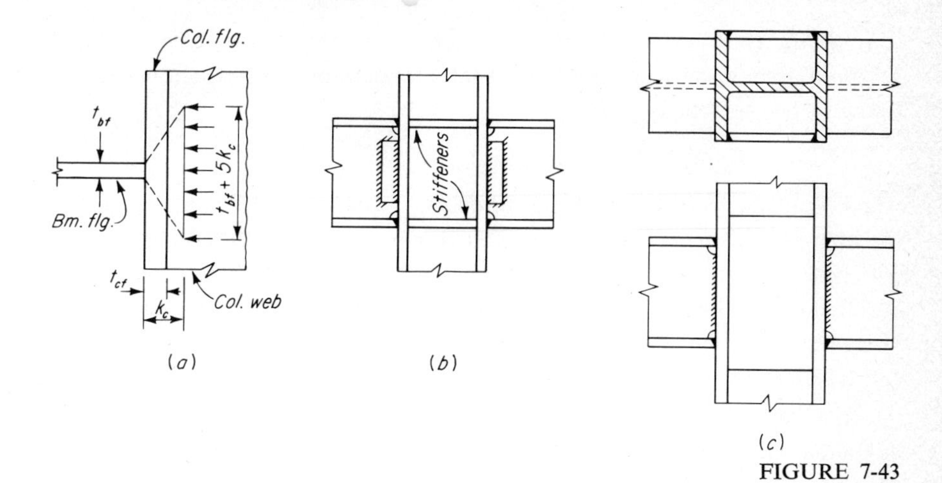

FIGURE 7-43

Therefore, if the column web yields over this length when the beam flange yields, we have

$$F_{yf}A_f = F_{yc}t_w(t_{bf} + 5k_c) \tag{b}$$

from which

$$t_w = \frac{A_f}{t_{bf} + 5k_c}\frac{F_{yf}}{F_{yc}} \tag{7-15}$$

where F_{yf} = yield stress for beam flange
F_{yc} = yield stress for column
A_f = area of beam flange
t_w = column web thickness

This formula is used in the AISC specification.

In tests to determine the strength of the connection at the beam's tension flange, some members failed by fracture of the flange weld in the vicinity of the column web, some by fracture at the fillet of the column web, and some by a tearing out of material in the column flange. A yield-line analysis of a portion of the column flange, together with observations from the tests, gave the following equation for the required thickness of the column flange

$$t_{cf} = 0.4\sqrt{A_f} \tag{c}$$

This equation has been extended for the AISC specification to the case of differing yield stresses in beam and column by writing it in the form

$$t_{cf} \geq 0.4\sqrt{\frac{F_{yf}A_f}{F_{yc}}}. \tag{7-16}$$

If the thickness of the column web is less than that given by Eq. (7-15), the required area A_s of a pair of horizontal stiffeners of the type shown in Fig. 7-43*b* is taken to be the deficiency in area of the web:

$$F_{ys}A_s = F_{yf}A_f - F_{yc}t_w(t_{bf} + 5k_c) \tag{7-17}$$

where F_{ys} is the yield stress of the stiffeners. Equation (7-17) is also used to determine the area of stiffeners if the column flange is too thin by Eq. (7-16). The ends of the stiffeners opposite the tension flange must be fully welded to the column flanges. Stiffeners opposite the compression flange may be fitted against the column flanges. However, if a beam connects to only one flange, as at an exterior column, the stiffeners can terminate at mid-depth of the column web. Stiffeners must also be proportioned for local buckling. The same slenderness limit as for compression elements free on one unloaded edge may be used, which is $95/\sqrt{F_y}$ in the AISC specification (Table 4-4). This limit should probably be reduced somewhat for plastically designed structures, although the limit established for I-beam flanges to guarantee rotation capacity is probably too severe.

Vertical stiffeners may be used instead of horizontal stiffeners (Fig. 7-43*c*). However, they are probably less effective than the column web since the column flange acts somewhat as a beam continuous over three supports. Therefore, vertical stiffeners are assumed to be only 50 percent efficient, so that

$$F_{ys}t_s(t_{bf} + 5k_c) = F_{yf}A_f - F_{yc}t_w(t_{bf} + 5k_c)$$

where t_s is the thickness of one stiffener. The stiffener slenderness limit can be determined as for the column web.

Beam-to-column moment connections sometimes require reinforcement of the column web for shear. Consider the welded corner connection shown in Fig. 7-44*a*. The tensile force $T = M/d_b$ in the top flange of the beam is assumed to be resisted by a shear force in the corner web (Fig. 7-44*b*). If the shear stress is uniform over the length d_c, we get

$$f_v d_c t_w = \frac{M}{d_b} \tag{d}$$

where t_w is the thickness of the corner web. For elastic design, the stress f_v in Eq. (*d*) will be the allowable shear stress F_v, which gives for the thickness of the web

$$t_w \geq \frac{M}{F_v d_b d_c} \tag{7-18}$$

The AISC specification uses the formula

$$t_w \geq \frac{32M}{F_y d_b d_c} \tag{7-19}$$

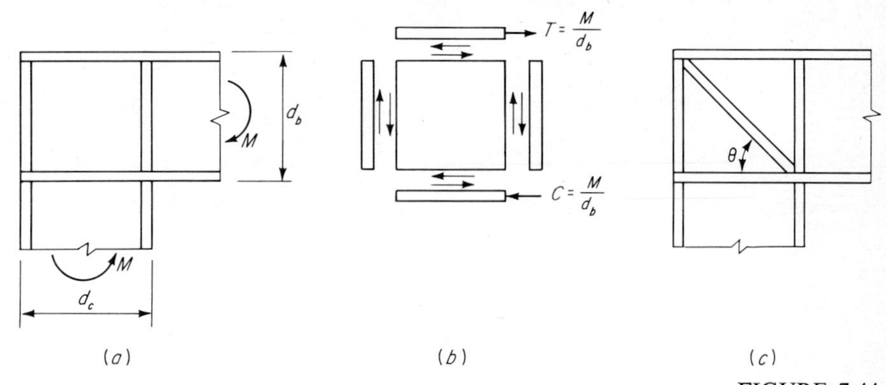

FIGURE 7-44

where M is in foot-kips, rather than inch-kips as in Eq. (7-18), while d_b and d_c are in inches and F_y in kips per square inch. This formula is derived from Eq. (7-18) by substituting $0.4F_y$ for F_v and taking the moment arm of the couple in Fig. 7-44*a* at $0.95d_b$ instead of d_b.

For plastic design, the moment M in Eq. (*d*) is assumed to reach the plastic value M_p when the column web yields, which gives

$$t_w \geq \frac{\sqrt{3}M_p}{F_y d_b d_c} \tag{7-20}$$

where, by the Mises yield criterion, $F_{vy} = F_y/\sqrt{3}$. This equation is used in the following form in the AISC specification:

$$t_w \geq \frac{23M_p}{F_y d_b d_c} \tag{7-21}$$

where the units are the same as in Eq. (7-19). The differing coefficients in Eqs. (7-19) and (7-21) result from the fact that M is the service-load moment while M_p is the service-load moment multiplied by the load factor.

If the web is not thick enough, the connection can be reinforced by welding a doubler plate to the web or by using a stiffener as shown in Fig. 7-44*c*. The area A_s of the stiffener must be sufficient for the horizontal component of the stiffener force to make up the deficiency in shear resistance. Therefore,

$$A_s f \cos\theta = T - f_v d_c t_w \tag{e}$$

For elastic design, $f = F_t$, the allowable tension, and $f_v = F_v$, the allowable shear, so that

$$A_s = \frac{1}{F_t \cos\theta}\left(\frac{M}{d_b} - F_v t_w d_c\right) \tag{7-22}$$

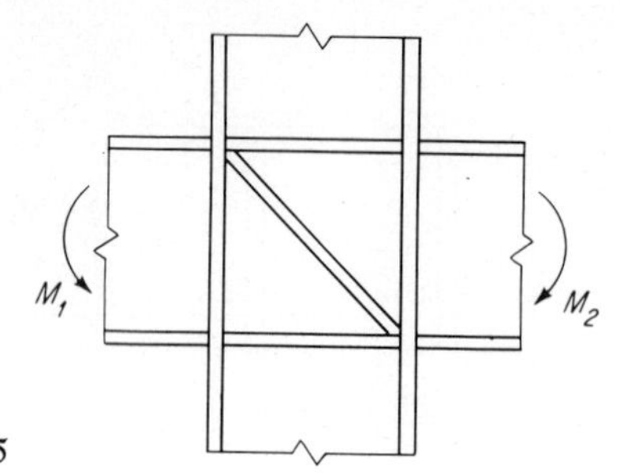

FIGURE 7-45

For plastic design, $f = F_{ys}$ and $f_v = F_{yc}/\sqrt{3}$, which gives

$$A_s = \frac{1}{F_{ys}\cos\theta}\left(\frac{M_p}{d_b} - \frac{F_{yc}t_w d_c}{\sqrt{3}}\right) \tag{7-23}$$

If the moments at beams connecting to opposite flanges of a column differ significantly in magnitude (Fig. 7-45), they may produce large shear stresses in the column web, as in the corner connection. An analysis similar to that in Fig. 7-44*b* shows that Eq. (7-18) can be used for this case by taking $M = M_1 - M_2$. This neglects the effect of the end shears in the column. Also, it can be shown that the area of stiffeners, when they are needed, can be determined from Eqs. (7-22) and (7-23), again using $M = M_1 - M_2$.

7-27 DP7-5: BEAM-TO-COLUMN MOMENT CONNECTION

This example illustrates the design of the connection of a W24 × 76 beam to a W14 × 150 column. The following are clarifications of the correspondingly lettered computations on the design sheet.

a The *DL* + *LL* shear and moment are both less than 75 percent of the total including wind, so that the latter values control the design.

b The width of the plate is 1½ in. less than the width of the beam flange to allow adequate space for the welds. In general, the difference in widths should be not less than four times the size of the weld. The required fillet weld is distributed 7½ in. across the end of the plate and 11 in. on each side. The connection plate is field-welded to the column with a full-penetration groove weld.

c The bottom flange connection plate is chosen 1½ in. wider than the beam flange to accommodate the ⅜-in. fillet welds. This plate is also field-welded to the column flange with a full-penetration groove weld.

d The shear plate is shop-welded to the column flange. It is not welded full length. Some designers would weld it full length even though the 10-in. welds shown give more than is required.

BEAM-TO-COLUMN MOMENT CONNECTION — DP7-5

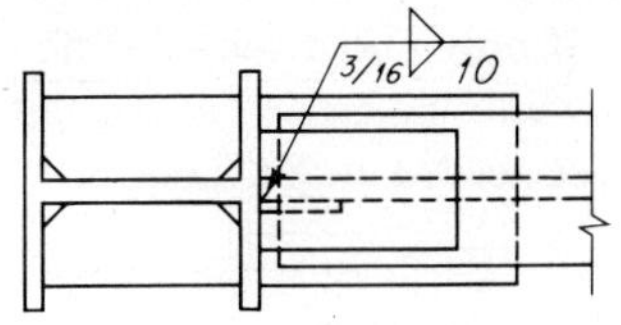

W24 × 76	W14 × 150	AISC Spec.
$d_b = 23.91$	$d_c = 14.88$	A36 steel
$b = 8.99$	$t_{cf} = 1.13$	E70 electrodes
	$t_w = 0.695$	A325 bolts (friction-type connection)
	$k_c = 1.81$	

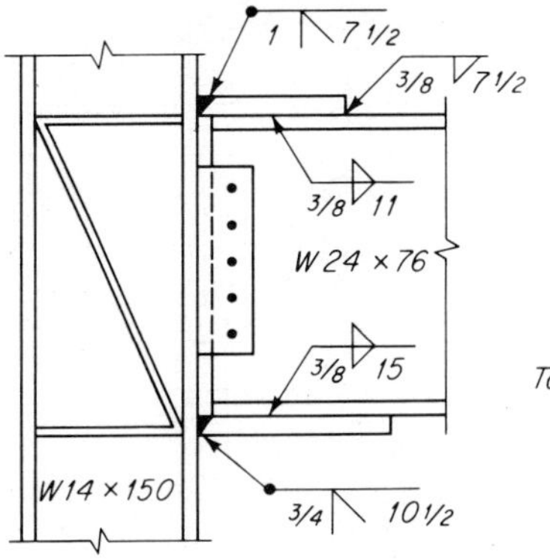

Load	Shear	Moment	
DL	22.1	100	
LL	8.3	50	
Total	30.4^k	$150'^k$	
WL	24.7	280	
Total	55.1	430	
× 3/4	41.3^k	$322'^k$	(a)

Top (tension-flange) plate

$T = 322 \times 12/23.91 = 162^k$

$A_s = 162/22 = 7.35 \text{ in}^2$ $1 \times 7\tfrac{1}{2}$ pl. $= 7.5 \text{ in}^2$ (b)

3/8" fillet weld $q = 21 \times 0.707 \times 3/8 = 5.56$ kli

Length of weld $= 162/5.56 = 29.1"$

Use $7\tfrac{1}{2}"$ across end and 11" each side

Bottom (compression-flange) plate

Use $3/4 \times 10\tfrac{1}{2}$ pl., $A = 7.87 \text{ in}^2$ Use 15" of 3/8" fillet weld each side (c)

Shear plate

A325 7/8" bolts s.s. $= 15 \times 0.6 = 9.0^k$, $n = 41.3/9 = 4.6$ use 5

Length of shear plate $= 4 \times 3 + 2 \times 1\tfrac{1}{2} = 15"$

$t = \dfrac{41.3}{0.4 \times 36 \times 15} = 0.19"$ Use 1/4" × 5" plate.

Length of 3/16" fillet weld $= 41.3/2.78 = 14.9"$. Weld 10" each side (d)

Column-flange stiffener at beam tension (top) flange

Eq. (7-16) $t_{cf} = 0.4\sqrt{A_f} = 0.4\sqrt{7.5} = 1.10 < 1.13$, stiffener not req'd.

Column-web stiffener at beam compression (bottom) flange

Eq. (7-14) $d_c/t \geq 180/\sqrt{F_y} = 30$, $d_c/t = 21.4$ O.K.

Eq. (7-15) $t_w \geq \dfrac{A_f}{t_{bf} + 5k_c} = \dfrac{7.87}{0.75 + 5 \times 1.81} = 0.802"$, $t_w = 0.695"$ stiffener req'd.

Eq. (7-17) $A_s = 7.87 - 0.695(0.75 + 5 \times 1.81) = 1.06 \text{ in}^2 = 0.53 \text{ in}^2$ each stiffener

Use two 5/16 × 5 stiffeners, $A_s = 1.57 \text{ in}^2$ each, $b/t = 16$ O.K. (e)

Column-web shear

Eq. (7-19) $t_w \geq \dfrac{32M}{F_y d_b d_c} = \dfrac{32 \times 322}{36 \times 23.91 \times 14.88} = 0.805$, $t_w = 0.695$

Eq. (7-22) $A_s = \dfrac{1}{22 \times 0.528}\left(\dfrac{322 \times 12}{23.91} - 0.4 \times 36 \times 0.695 \times 14.88\right) = 1.16 \text{ in}^2$ (f)

Use one 1/4 × 5 stiffener.

e The width of the stiffener is determined by the width of the connection plate. Once the width is established, the minimum thickness follows from the b/t limitation.

f Since a diagonal stiffener is required, a horizontal stiffener is used at the tension flange as well as at the compression flange. If only one diagonal stiffener is used, as suggested here, only one horizontal stiffener would be needed at the top flange of the beam, since it was shown that stiffeners are not needed at this flange at all so far as the column flange itself is concerned.

PROBLEMS

7-43 Design a welded connection for the beam of Prob. 7-30. E70 electrodes, AISC specification.

7-44 Design a welded connection for the beam of Prob. 7-31. E70 electrodes, AISC specification.

7-45 Design a welded connection for two W12 × 40 beams which frame opposite the web of a W18 × 50 girder. The end shear and end moment of each 12-in. beam are 37.5 kips and 80 ft-kips, respectively. The top flanges of the beams and girder are at the same elevation. A36 steel, E70 electrodes, AISC specification.

7-46 Design a welded connection for two W14 × 53 beams which frame opposite the web of a W21 × 62 girder. The end shear and moment of each beam are 48 kips and 128 ft-kips, respectively. The top flanges of the beams are 1 in. below the top flange of the girder. A441 steel, E70 electrodes, AISC specification.

7-28 HAUNCHED CONNECTIONS

Corner connections are sometimes tapered, as in Fig. 7-46*a*, or curved, as in Fig. 7-46*b*. This is usually done for architectural reasons. An example is shown in Fig. 2-3. Methods of analysis have been developed for both elastic design[14, 15] and plastic design[16] of tapered haunches and curved haunches. Tests of elastically designed haunches show that they have adequate strength but, because of premature local buckling, may lack the rotation capacity needed for plastically designed structures.

The following procedure is recommended for plastic design of the tapered haunch.[16, 17] The plastic section modulus must be adequate at all cross sections, of course, but it has been shown that it is sufficient to check it at sections 1, 2, and 3 (Fig. 7-46*a*). Figure 7-46*c* shows that the stress distribution on a cross section normal to the axis of the member is not symmetrical if both flanges are yielded and $A_c = A_t$. For symmetry, $F_y A_c \cos \beta$ must equal $F_y A_t$, from which $A_c = A_t/\cos \beta$. This condition could be fulfilled by making $t_c = t_t/\cos \beta$. However, even if $\beta = 20°$, A_c would be only 6 percent larger than A_t, which is small enough to be neglected. Therefore, for

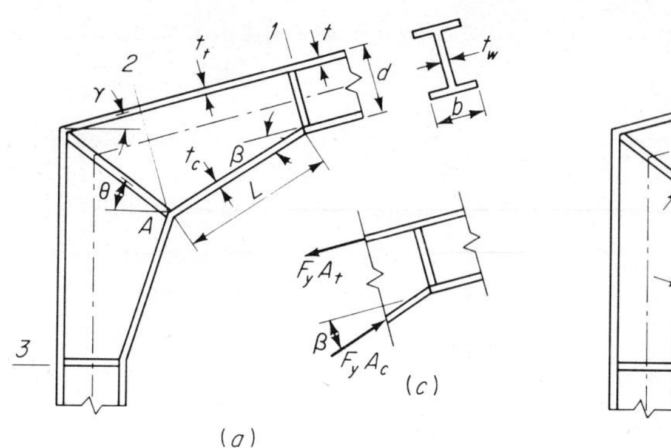

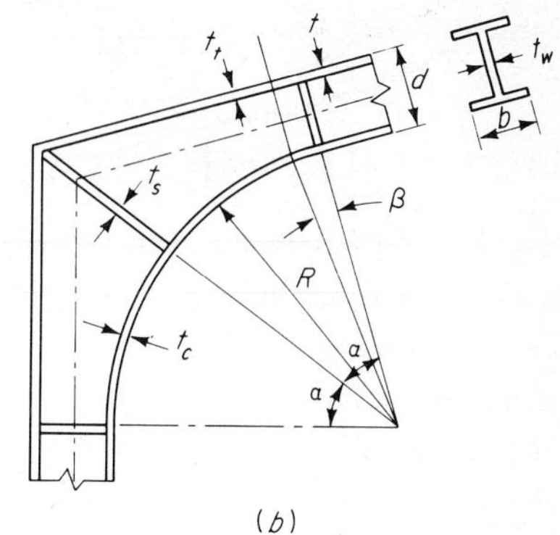

FIGURE 7-46

haunches of usual proportions the flange areas can be made equal and the cross section analyzed as a symmetrical one.

Connections of this type are generally used in one-story frames, so that axial compression and shear are not likely to influence proportioning for bending. Since interaction of bending and axial compression is given by Eqs. (4-38), it will be noted that the connection can be sized for bending alone if $P/P_y \lesssim 0.15$.

Lateral support must be provided at sections 1, 2, and 3, and the dimensions of the flange checked for stability. The required ratio b/t is the same as for beams of uniform depth (Art. 8-12). Thus, for A36 steel, the projecting-element slenderness is 8.5, so that, in terms of the full width b in Fig. 7-46, $b/t = 17$. An approximate solution for lateral-torsional buckling can be obtained by treating the flange as a column. For $\beta \approx 12°$, haunches of usual proportions are plastic throughout most of the length. To assure the necessary rotation capacity for plastic behavior, it is assumed that the flange is just beginning to strain harden, so that $\varepsilon = \varepsilon_s$. Thus,

$$F_{cr} = F_y = \frac{\pi^2 E_s}{(KL/r)^2} \tag{a}$$

where E_s is the strain-hardening modulus. With $E_s = 800$ ksi, $F_y = 36$ ksi, and $r = b/\sqrt{12}$, and taking $K = 0.8$ to allow for some rotational restraint, Eq. (a) gives $L/b = 5.3$ as the critical slenderness for an A36-steel haunch.[17] Tests show that this limit is not unreasonable and cannot be increased very much without a compensating increase in A_c. (For haunches in elastically designed structures, the flange need only reach ε_y. In this case Young's modulus applies, and $L/b = 5.3\sqrt{30{,}000/800} = 32$ for A36 steel.)

If the limit L/b exceeds 5.3, stability can be maintained by using a thicker flange. According to Ref. 16, the required increase in thickness is given by $\Delta t_c = 0.1(L/b - 6)$.

However, this was based on a limiting value $L/b = 6$ and should be changed for $L/b =$ 5.3 to

$$t = 0.1\left(\frac{L}{b} - 5.3\right) \tag{7-24}$$

The critical length can also be increased by proportioning the haunch with $\beta > 12°$. This results in a flange which yields over only part of its length and remains elastic over the remainder. In tests reported in Ref. 16, a haunch with $L/b = 10$ and $\beta = 21°$ developed a moment about 10 percent larger than, and rotation capacity equal to, that of a similar haunch with $L/b = 5$ and $\beta = 12°$. The flange of the haunch with $\beta = 21°$ yielded over about 20 percent of its length. Therefore, it would seem reasonable to allow values of L/b varying linearly with β within the range tested.

Since the critical length $L = 5.3b$ determined above is based on the assumption that the flange is yielded to the point of beginning of strain hardening for its full length, it may be conservative in some cases. This is because the range of available sizes of standard shapes, plates, etc., is such that there is usually excess moment capacity. There is no simple way to evaluate this effect. The curve in Fig. 7-47 shows the variation of L/b with the extent of the yielded zone according to Ref. 16. By determining the distance from section 1 in Fig. 7-46*a* to the cross section where the haunch is at the yield moment $M_y = F_y S$, the extent of the yielded zone and the corresponding value of L/b can be determined. However, it will be noted that the increase in L/b over the value for a flange yielded the full length of the haunch is not significant unless the taper is such as to produce M_y at a section less than halfway from section 1 to section 2. This will usually require a value of β considerably in excess of 12°.

Stiffeners can be proportioned by assuming that they equilibrate the forces in the flanges which meet at their ends. Thus, for the stiffener at section 1,

$$F_y A_s = F_y A_c \sin \beta \tag{7-25}$$

where A_s is the area of the two stiffeners, one on each side of the web. Equilibrium of forces at A in Fig. 7-46*a* leads to the following formula for the area of the pair of stiffeners at this point:[17]

$$A_s = \frac{A_{c1} \cos(\beta_1 + \gamma) - A_{c2} \sin \beta_2}{\cos \theta} \tag{7-26}$$

where A_{c1} = area of inner flange of rafter haunch
A_{c2} = area of inner flange of column haunch
β_1 = angle of taper of rafter haunch
β_2 = angle of taper of column haunch
θ = angle of inclination of stiffener
γ = angle of inclination of rafter

An example of the design of a tapered haunch is given in DP11-1, Art. 11-11.

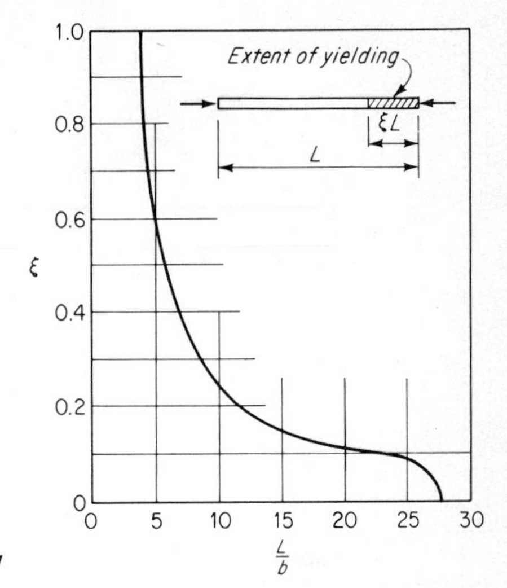

FIGURE 7-47

A typical curved knee is shown in Fig. 7-46*b*. It is shown in Ref. 16 that the critical section for bending in haunches of practical proportions will be at the cross section for which $\beta = 12°$. Therefore, the thickness t_t of the outer flange can be determined from $M_p = F_y Z$ at this cross section. For a simpler solution, it may be taken conservatively as $\frac{4}{3}t$. As is the case for the tapered haunch, $A_c = A_t/\cos\beta$ if the cross section is computed as a symmetrical one. However, since $\beta \approx 12°$ at the critical section, t_c can be made equal to t_t.

The curved compression flange imposes an additional restriction on the thickness t_c. Because the radial resultant of the flange stresses at two adjacent cross sections is directed inward (Fig. 7-48*a*), the result is a cross bending of the flange as shown in Fig. 7-48*b*. This has two effects: (1) it reduces the longitudinal bending stress at a point such as *a* in Fig. 7-48*b*, because of the reduced distance to the neutral axis, and (2) it produces transverse bending stresses in the flange, which are maximum at the juncture of the flange and web. The radial resultant P_R of the flange forces on a unit width of flange, taken at cross sections a unit distance apart (Fig. 7-48*a*), is

$$P_R = F_y \times 1 \times t_c \times \phi = \frac{F_y t_c}{R} \tag{b}$$

Therefore, the bending moment at the web (Fig. 7-48*b*) is $(F_y t_c/R) \times b^2/8$. Equating this to the plastic resisting moment of the unit strip, $F_y t^2/4$, gives

$$t_c \gtrless \frac{b^2}{2R} \tag{7-27a}$$

which can also be written

$$\frac{b}{t_c} \lessgtr \frac{2R}{b} \tag{7-27b}$$

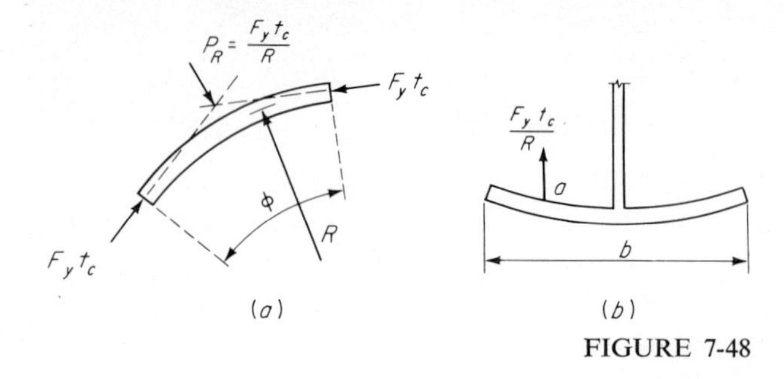

FIGURE 7-48

In this form, we have a flange slenderness limit which is in addition to the limit for local buckling.

Lateral-torsional buckling of the compression flange of the curved knee can be prevented by using the limiting length $L = 5.3b$ (for A36 steel) as for the tapered haunch, where L is the arc length of the curved flange between points of lateral support.[17]

Although radial components of the force in the curved flange can be equilibrated by the web of the knee, stiffeners at the points of tangency and at the corner should be provided to guard against buckling of the web. It is conservative to proportion these stiffeners for the resultant of the radial components on either side of the stiffener, midway to the adjacent stiffeners. This is easily determined as follows. The radial component for a unit length of flange is bP_R, where P_R is the component for a unit width given by Eq. (*b*). The component of this force parallel to the stiffener is $bP_R \cos \phi$, where ϕ is the angle between the radius through the stiffener and the radius to any element. Therefore, the force on the stiffener at the corner is given by

$$P = 2\int_0^{\alpha/2} \frac{bt_c F_y}{R} R \, d\phi \cos \phi = 2F_y A_f \int_0^{\alpha/2} \cos \phi \, d\phi = 2F_y A_f \sin \frac{\alpha}{2} \tag{7-28}$$

where α is half the central angle (Fig. 7-46). Of course, the force for the stiffeners at a point of tangency is one-half the value from Eq. (7-28).

EXAMPLE 7-4 Proportion the curved knee shown in Fig. 7-49. The adjoining rafter is a W24 × 76 and the column a W24 × 84, both A36 steel.

SOLUTION The dimensions of the adjoining sections are:

For the W24 × 76:	$d = 23.91$	$b = 8.99$	$t_f = 0.682$	$t_w = 0.440$
For the W24 × 84:	$d = 24.09$	$b = 9.01$	$t_f = 0.772$	$t_w = 0.470$

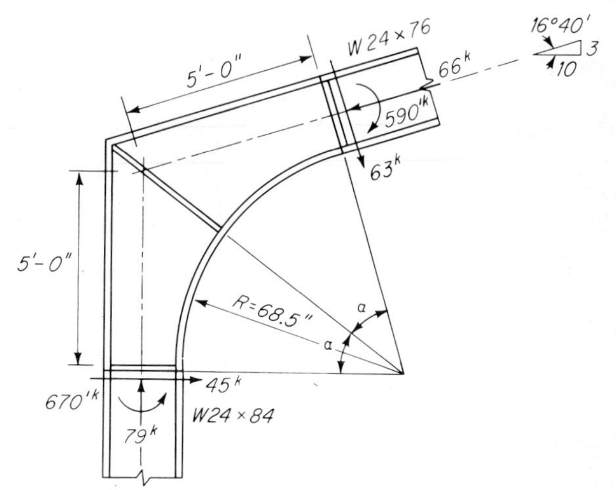

FIGURE 7-49

The central half angle α is $(90° - 16°40')/2 = 36°40' = 0.64$ radian. The critical length of flange is $L = \alpha R = 5.3b$. The flange width of the knee should be the same as that of the W24's, $b = 9$ in., so that

$$0.64R = 5.3 \times 9 \qquad R = 75 \text{ in.}$$

Therefore, the 68.5-in. radius of the knee is satisfactory.

A $\frac{7}{16}$-in. plate ($t = 0.438$ in.) will be used for the web of the knee to match the webs of the W24's. The required flange area will be determined by the moment at the critical section for bending, which is at $\beta = 12°$ from the point of tangency.

The depth of the knee at the critical section adjacent to the column is $24.09 + R(1 - \cos 12°) = 24.09 + 68.5 \times 0.022 = 24.09 + 1.51 = 25.60$ in. The moment at this section is

$$M = 670 \times 12 + 45\, R \sin 12° - 79\left(\frac{d_{12°}}{2} - \frac{d_{0°}}{2}\right)$$

$$= 8{,}040 + 45 \times 68.5 \times 0.208 - 39.5(25.60 - 24.09) = 8{,}620 \text{ in.-kips}$$

The depth of the knee and the moment at the critical section adjacent to the rafter are found similarly to be 25.4 in. and 7,930 in.-kips. The required area of the flange is found from $ZF_y = M$, which gives

$$Z = A_f d_f + \frac{t_w d_w^{\,2}}{4} = \frac{M}{F_y}$$

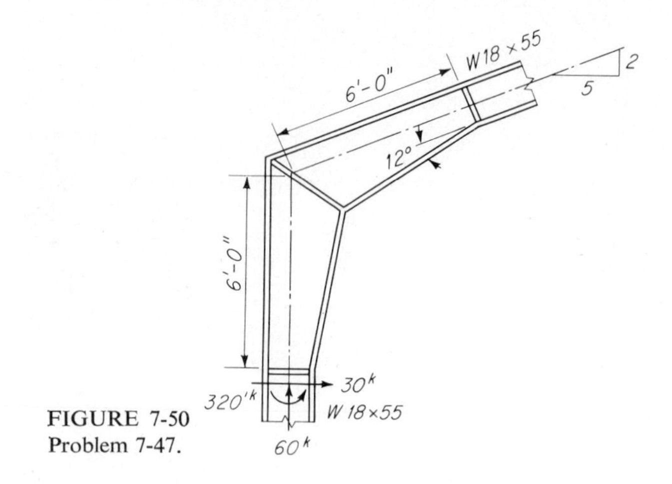

FIGURE 7-50
Problem 7-47.

Assuming the flange to be $\frac{7}{8}$ in. thick, $d_f = 25.60 - 0.88 = 24.72$ in., and $d_w = 24.72 - 0.88 = 23.84$. Therefore

$$24.72A_f + \frac{0.438 \times 23.84^2}{4} = \frac{8{,}620}{36}$$

from which $A_f = 7.2$ in.^{2} A $\frac{13}{16} \times 9$-in. plate gives 7.31 in.^{2} The radio b/t for this flange is $9/0.812 = 11.1$. The permissible limits are 17 for local buckling and, from Eq. (7-27*b*), $2R/b = 2 \times 68.5/9 = 15.2$ for cross bending of the flange. Therefore, the $\frac{13}{16} \times 9$-in. flange is satisfactory.

The stiffener force is given by Eq. (7-28). If the stiffeners are made of the same steel as the knee, the stiffener area is $A_s = 2A_f \sin(\alpha/2) = 2 \times 7.31 \times \sin 18°20' = 4.6$ in.^{2} for the stiffeners at the corner, and half this amount for the stiffeners at the points of tangency. The corner stiffeners can be two $\frac{9}{16} \times 4$ in. plates ($\frac{1}{2} \times 4$ should be ample) and those at the tangent points two $\frac{1}{4} \times 4$ in. plates.

PROBLEMS

7-47 Design the tapered knee shown in Fig. 7-50. A36 steel, AISC specification.

7-48 Same as Prob. 7-47 except design a curved knee.

7-29 COLUMN SPLICES

Columns for multistory buildings are usually fabricated in two-story lengths. Although the reduction in load at successively higher stories ordinarily would permit a different size for each story, with a consequent saving in weight, the extra cost of splices and

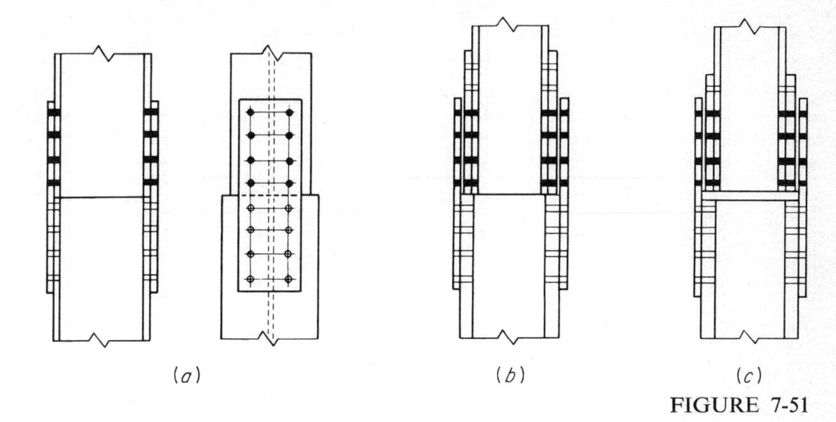

FIGURE 7-51

erection would more than offset any saving in cost of material. For an odd number of stories, the top section may be either one or three stories long. Column sections are usually spliced 2 to 3 ft above the finished floor line to avoid interference with beam and girder connections. In columns with end moments this also puts the splice at a section with a smaller moment.

The ends of each section usually are finished to a plane surface for good contact. This means that, as far as the axial load is concerned, no splice at all is needed, since the load is resisted by bearing over the area in contact. However, splice plates are required because of shears and moments due to wind or other loads, and even if no shears and moments existed, it is obvious that some sort of connection is called for. The AISC specification requires columns which are finished to bear at splices or which bear on bearing plates to have enough fasteners to "hold all parts securely in place." However, splice material for other compression members which are finished to bear must be proportioned for 50 percent of the stress at the splice. Furthermore, in both these cases the connection must also be proportioned to resist the tension, if any, that results from moments due to lateral forces acting together with only 75 percent of the dead load and no other gravity load.

Figure 7-51*a* shows a simple bolted or riveted splice which is sufficient if two column sections are of the same depth. If the depths are not the same, the difference must be taken up by fill plates connected to the flanges of the smaller section. However, if the two sections are of the same nominal depth, the fill plates need not provide any of the bearing area. This is because W shapes of the same nominal depth are rolled with a constant depth inside the flanges, so that the flanges of two abutting sections will be flush on the inside, with the difference in depth accounted for by the difference in thickness of the flanges. However, if one column section connects to another of different nominal depth, the flanges of the upper section will not bear on those of the

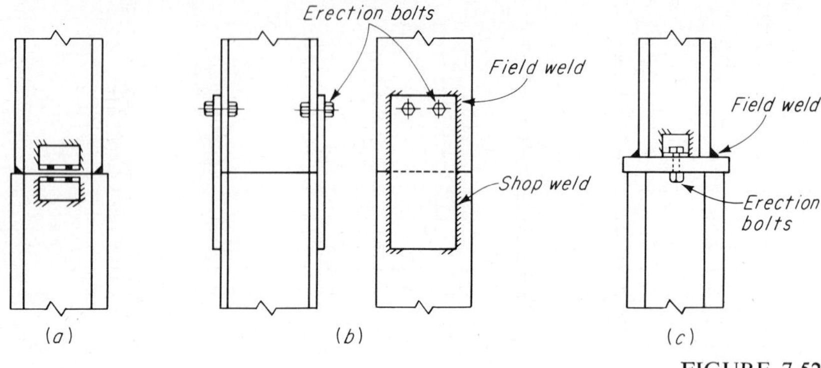

FIGURE 7-52

lower section. To provide for this condition, one or more fill plates may be fastened to the upper shaft with sufficient connection to develop their share of the load. In this case, the shaft must be finished with the fills attached (Fig. 7-51*b*). Alternatively, a bearing (butt) plate may be used between two sections, in which case the fills need not bear (Fig. 7-51*c*).

Typical welded splices are shown in Fig. 7-52. Columns of the same nominal depth may be spliced as shown in Fig. 7-52*a*, usually with partial-penetration groove welds, or by a splice similar to the corresponding riveted or bolted splice (Fig. 7-52*b*). If fill plates transfer a portion of the load, they must be welded to the shaft before the ends are finished for bearing. Figure 7-52*c* shows a butt-plate splice for two columns of different nominal depths.

7-30 DP7-6: DESIGN OF A TIER-BUILDING COLUMN SPLICE

The comments which follow are identified by letters corresponding to those on the design sheet.

a In this computation, the forces are assumed to be resisted entirely by the splice plates; this is equivalent to the assumption that the two shafts are not in contact. The wind moment is assumed to be counteracted by the moment due to only 75 percent of the dead load, as prescribed by the AISC specification. The splice-plate force can also be evaluated by assuming that the plate must supply the resultant tension that would exist on the cross section if there were no splice. This is computed as follows. The stresses at the outside faces of the flanges of the W10 × 49 are

$$f = -\frac{P}{A} \pm \frac{M}{S} = -\frac{\frac{3}{4} \times 42}{14.4} \pm \frac{39.5 \times 12}{54.6} = -2.2 \pm 8.7 = -10.9,\ +6.5 \text{ ksi}$$

From these stresses we find the distance from the outside face of the tension flange to the neutral axis to be $10 \times 6.5/(10.9 + 6.5) = 3.74$ in. Since the flange is 0.56 in. thick,

COLUMN SPLICE DP7-6

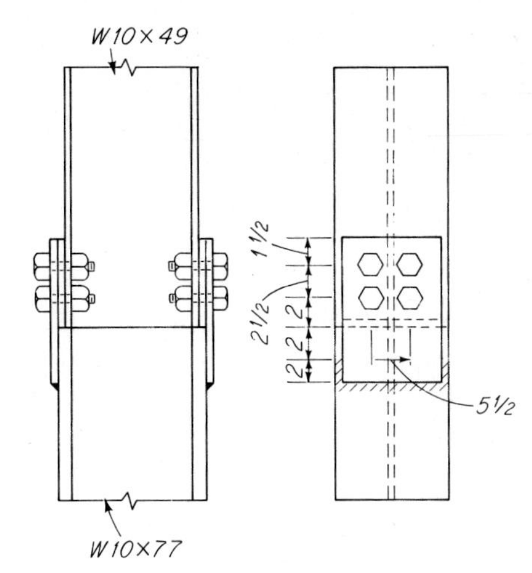

A 36 steel, AISC Spec.

Column Load at Point of Splice

	Axial	Moment*	Shear
DL	42^k	–	–
LL	75^k	–	–
WL	2^k	$39.5'^k$	3.3^k
Total	119^k	$39.5'^k$	3.3^k

* in plane of web

$\frac{3}{4}$-in. A325 bolts, E70 electrodes

Flange splice plates

$$\text{Tension} = \frac{39.5 \times 12}{10.62} - \frac{1}{2} \times \frac{3}{4} \times 42 = 44.6 - 15.7 = 28.9^k \qquad (a)$$

$$t = \frac{28.9}{4/3 \times 22\,(8 - 2 \times 7/8)} = 0.16'', \text{ use } 1/4 \times 8 \text{ plate} \qquad (b)$$

s.s. $22 \times 0.44 = 9.7^k$

brg. $48.6 \times \frac{3}{4} \times \frac{1}{4} = 9.1^k$

$$\frac{28.9}{4/3 \times 9.1} = 2.4$$

Use 4 bolts

Weld for shop connection

$$q = 21 \times 0.707 \times \frac{3}{16} = 2.78 \text{ kli}$$

$$\frac{28.9}{4/3 \times 2.78} = 7.8''$$

Use 12" of 3/16" weld

Splice plate 1/4 × 8 × 10

Fill plate 1/4 × 8 × 5 1/2

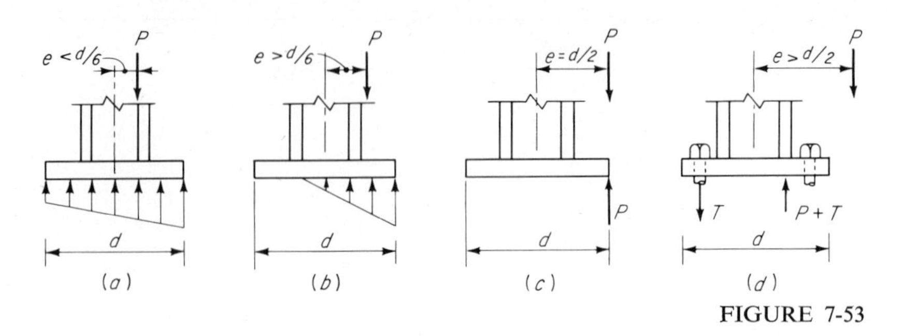

FIGURE 7-53

its inside face is 3.74 − 0.56 = 3.18 in. from the neutral axis, and the tensile stress here is 6.5 × 3.18/3.74 = 5.5 ksi. Therefore the average stress on the tension flange is 6.0 ksi, and the resultant tension $T_1 = 6.0 \times 10 \times 0.56 = 33.6$ kips. Adding to this the tension in the web, $T_2 = \frac{1}{2} \times 5.5 \times 3.18 \times 0.34 = 3.0$ kips, we have $T = T_1 + T_2 = 36.6$ kips, which is 25 percent larger than the value computed by the simpler calculation. It is impossible to say which is the better result, and since the prescribed loading condition is somewhat artificial, it makes little difference which is used.

b The allowable stress is increased by one-third for wind forces, and the diameters of two bolt holes are deducted to obtain the net width of the 8-in. splice plate. Since the shear is only 10 kips, it is safe to assume it to be resisted by friction. The coefficient of friction of steel on steel ranges from about 0.15 to 0.6 or more. Even with only dead load and wind load acting, the required value is only 3.3/44 = 0.08.

PROBLEMS

7-49 Design a splice between a W8 × 48 column and a W8 × 31 column which supports a concentric load of 50 kips, a shear of 8 kips, and a wind moment of 30 ft-kips in the plane of the web at the point of splice. AISC specification, A36 steel, A325 bolts.

7-50 Design a welded splice for the column shafts of Prob. 7-49. E70 electrodes.

7-51 Design a splice between a W12 × 85 column and a W10 × 60 column which supports a concentric load of 100 kips, a shear of 15 kips, and a wind moment of 70 ft-kips in the plane of the web at the point of splice. AISC specification, A36 steel, A325 bolts.

7-52 Design a welded splice for the column shafts of Prob. 7-51. E70 electrodes.

7-53 Design a splice to connect a W12 × 65 to a W12 × 99. The forces and moments at the splice are:

	DL	*LL*	*WL*
P, kips	133	70	10
M, ft-kips	25	15	75

A36 steel, A325 bolts, AISC specification.

7-54 Design a welded splice for the column of Prob. 7-53. E70 electrodes.

7-31 MOMENT-RESISTANT BASES FOR COLUMNS

In the design of base plates for axially loaded columns, we assume the bearing pressure between the plate and the footing to be uniformly distributed (Art. 4-16). Anchor bolts in such a base plate are needed only to hold the column in position. Even if the column must resist a moment, however, such resistance can be developed under certain conditions without the aid of anchor bolts. If the moment is small, the bearing pressures can be assumed to be distributed as shown in Fig. 7-53*a*. If b is the width of the plate, the pressures f_p at the edges are given by

$$f_p = \frac{P}{A} \pm \frac{Mc}{I} = \frac{P}{bd} \pm \frac{6M}{bd^2} \tag{7-29}$$

From this equation, we see that if $M/P = d/6$, the pressures are zero at one edge and $2P/bd$ at the other. For eccentricities greater than $d/6$ a line of zero pressure lies between the edges of the plate, and the maximum pressure exceeds $2P/bd$ (Fig. 7-53*b*). Finally, if the eccentricity is $d/2$, the bearing pressure is concentrated at the edge of the plate, as in Fig. 7-53*c*. Of course, this condition can never be realized, since the reactive force must be distributed over some area. However, it is an upper bound on the eccentricity of load which can exist without anchor bolts. Thus, if $e = M/P$ exceeds $d/2$, it is clear that equilibrium requires the system of forces shown in Fig. 7-53*d*, where T is the anchor-bolt tension and $P + T$ the resultant bearing pressure.

The stress distributions shown in Fig. 7-53 can be used for an allowable-stress design procedure for base plates. The size of the base plate is assumed. If the eccentricity M/P is more than half the assumed depth d, the plate is checked for the case shown in Fig. 7-53*d*, with the resultant $P + T$ distributed triangularly as in Fig. 7-53*b*. If the resulting anchor-bolt tension T yields anchor bolts of reasonable size, the assumed dimensions are adequate. However, if the eccentricity is less than half the assumed depth d, the bearing pressure according to the applicable Fig. 7-53*a* or *b* is computed first. If this value is less than the allowable, the assumed dimensions are adequate and only nominal anchorage is needed. If it is more than the allowable, the plate is checked for the stress distribution of Fig. 7-53*d*, as for the case where M/P exceeds $d/2$.

The analysis for the case in Fig. 7-53*d* can be made by assuming that both the concrete and the steel are linearly elastic and the plate rigid, from which an equation relating bolt tension and bearing pressure follows. However, it is more realistic to assume that the allowable anchor-bolt tension and the allowable bearing pressure on the concrete are independent, since this is consistent with conditions at ultimate load. Therefore, this assumption will be adopted in the following examples.

EXAMPLE 7-5 Check the adequacy of a 15 × 20 in. base plate for the A36 W10 × 33 column of Fig. 7-54*a*, according to the AISC specification. The footing is of 3,000-psi concrete, and the plate covers less than one-third the area of the footing.

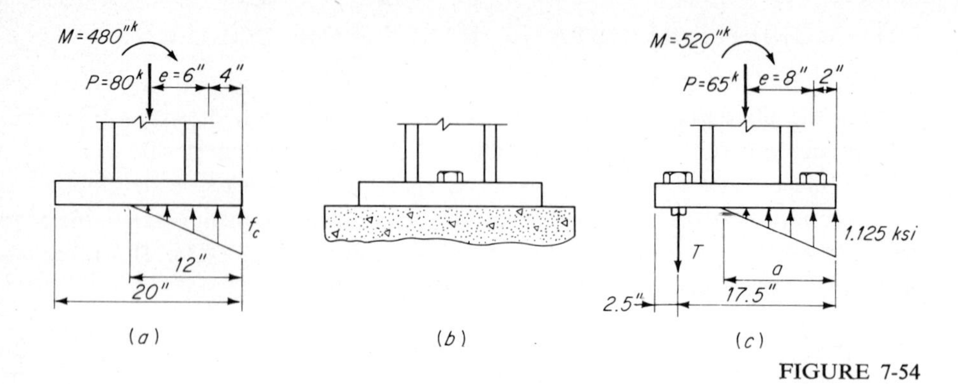

FIGURE 7-54

SOLUTION The eccentricity $e = 480/80 = 6$ in. is less than half the assumed length of the plate. For equilibrium with no anchor-bolt tension, the base of the triangular bearing-pressure diagram must be 12 in., as shown in the figure. Therefore,

$$\tfrac{1}{2}f_c \times 12 \times 15 = 80 \text{ kips} \qquad f_c = 0.89 \text{ ksi}$$

The allowable bearing pressure is $0.375f'_c = 0.375 \times 3 = 1.125$ ksi. Therefore, the assumed dimensions are adequate, and only nominal anchorage is required. Two $\frac{1}{2}$-in. A36 bolts,* one on each side of the web as in Fig. 7-54*b*, could be used, although most designers would probably use a larger bolt for a base plate the size of this one.

EXAMPLE 7-6 Same as Example 7-5, except that $P = 65$ kips and $M = 520$ in.-kips.

SOLUTION The eccentricity $e = 520/65 = 8$ in. The base of the triangular bearing-pressure diagram is $3(10 - 8) = 6$ in., so that

$$\tfrac{1}{2}f_c \times 6 \times 15 = 65 \text{ kips} \qquad f_c = 1.44 \text{ ksi}$$

Since this exceeds the allowable bearing pressure, we investigate the stress distribution shown in Fig. 7-54*c*. Taking moments about *T*, we get

$$\tfrac{1}{2} \times 1.125 \times 15a\left(17.5 - \frac{a}{3}\right) = 65 \times 15.5$$

$$a^2 - 52.6a + 358 = 0$$

This gives $a = 8.0$ in. The tension T is found from

$$T = \tfrac{1}{2} \times 1.125 \times 8 \times 15 - 65 = 67.5 - 65 = 2.5 \text{ kips}$$

* Nonheaded anchor bolts, straight or bent, are usually made of A36 steel.

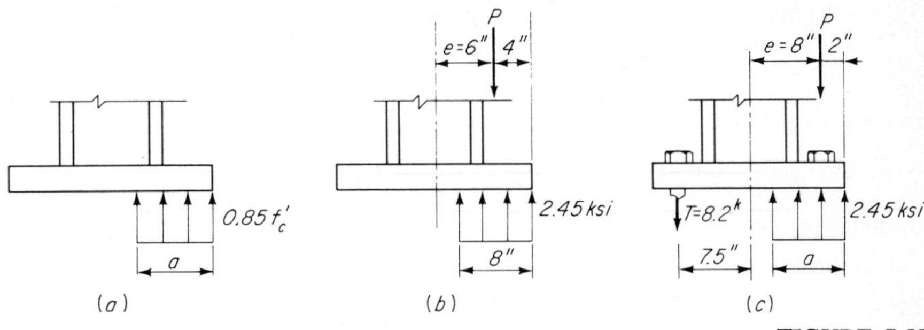

FIGURE 7-55

The AISC allowable load on a $\frac{1}{2}$-in. A36 anchor bolt is $0.6 \times 36 \times 0.142 = 3.07$ kips, where 0.142 is the tensile-stress area defined in Table 7-1. Therefore, two $\frac{1}{2}$-in. bolts, placed as shown in Fig. 7-54*c*, are adequate. Most designers would probably use larger bolts.

EXAMPLE 7-7 Compute the factor of safety (load factor) for the base plate of Example 7-5.

SOLUTION The conventional rectangular stress block of Fig. 7-55*a* can be used for an ultimate-load analysis. For equilibrium with no anchor-bolt tension, $a = 8$ in. (Fig. 7-55*b*). Then, with $0.85f_c' = 0.85 \times 3 = 2.45$ ksi, we get

$$P = 2.45 \times 15 \times 8 = 294 \text{ kips}$$

The service load is 80 kips. Therefore, $n = 294/80 = 3.7$.

EXAMPLE 7-8 Compute the factor of safety for the base plate of Example 7-6.

SOLUTION The ASTM minimum tensile strength of A36 steel is 58 ksi (Table 2-2). The tensile strength of the $\frac{1}{2}$-in. bolt is $58 \times 0.142 = 8.2$ kips. Summing vertical forces in Fig. 7-55*c* gives

$$8.2 + P = 2.45 \times 15a$$

$$P = 36.75a - 8.2 \qquad (a)$$

Taking moments at T gives

$$15.5P = 2.45 \times 15a\left(17.5 - \frac{a}{2}\right) \qquad (b)$$

Substituting from Eq. (*a*) into Eq. (*b*) gives

$$a^2 - 4a - 6.92 = 0$$

from which $a = 5.31$ in. Therefore

$$P = 36.75 \times 5.31 - 8.2 = 195 - 8.2 = 187 \text{ kips}$$

and the factor of safety is $n = 187/65 = 2.88$.

Of course, the ultimate-load analysis can also be used to design a base plate. The procedure is identical with that in Examples 7-5 and 7-6, except that the service load is multiplied by the required load factor and the rectangular stress block replaces the triangular one.

7-32 BASE-PLATE DETAILS

The connection of the column shaft to its base plate may take various forms. If the moment is relatively small, so that the combined stress $f = P/A + My/I$ is compressive at all points of the column cross section, the entire cross section will bear on the base plate. In this case the shaft need not be attached to the base plate to meet stress requirements, although practical considerations dictate at least a nominal connection, as in the case of columns subjected only to axial load. On the other hand, it is obvious that full-penetration groove welds (or the equivalent in fillet welds) for the entire cross section of the column would be adequate no matter how large the moment compared with the axial load. However, such a connection ordinarily is not required.

Except for the smaller columns it is usual to ship base plates unattached so that they can be leveled and grouted in place on the footings. The column can be fitted with angles attached in the shop (Fig. 7-56*a*). Of course, these angles are attached to the web if the anchor bolts are located as in Fig. 7-54*b*. Alternatively, the angles can be omitted and the column shaft welded to the base plate in the field. A direct connection between anchor bolt and column shaft, through the use of sleeves or boots attached in the shop, is usually needed for the larger moments (Fig. 7-56*b*). On the left in this figure is shown an angle welded to the column flange and capped by a plate welded both to the angle and to the column flange. Instead of the angle two bars may be fillet-welded both sides to the column flange, as shown at the right. This detail can be expanded to accommodate two anchor bolts.

EXAMPLE 7-9 Design the base plate for the A36 W12 × 72 column shown in Fig. 7-57. The axial load P is 80 kips, and the moment 1,600 in.-kips. The footing concrete is 3,000 psi. AISC specification.

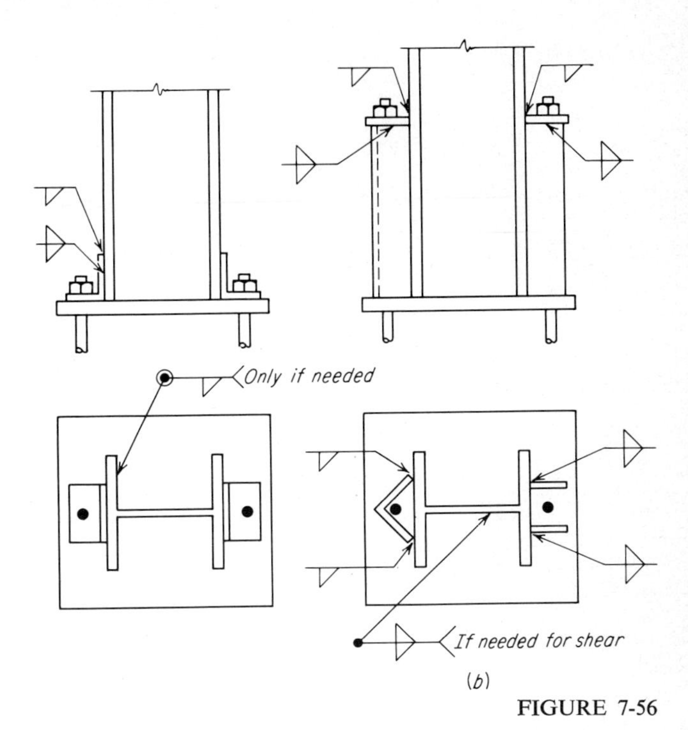

FIGURE 7-56

SOLUTION Since there is no way to make a direct determination of the size of the base plate, its dimensions must be assumed and then checked, as in Examples 7-5 and 7-6. Assume the plate to be 20 × 24 in. The eccentricity of load M/P is 20 in., which exceeds $d/2$, so that anchor-bolt tension is required. With the bolt lines $2\frac{1}{2}$ in. from the edges of the plate (Fig. 7-57), summation of moments at the line of action of T gives

$$\tfrac{1}{2} \times 1.125 \times 20a\left(21.5 - \frac{a}{3}\right) = 80 \times 9\tfrac{1}{2} + 1{,}600$$

$$a^2 - 64.5a + 629 = 0$$

from which $a = 12.0$ in. Summation of vertical forces gives

$$T + 80 = \tfrac{1}{2} \times 1.125 \times 12 \times 20 = 135 \text{ kips} \qquad T = 55 \text{ kips}$$

The required tensile-stress area for A36 anchor bolts is $55/22 = 2.50$ in.^{2} From the AISC Manual's table of standard screw threads we find the required area to be furnished by one 2-in. bolt.

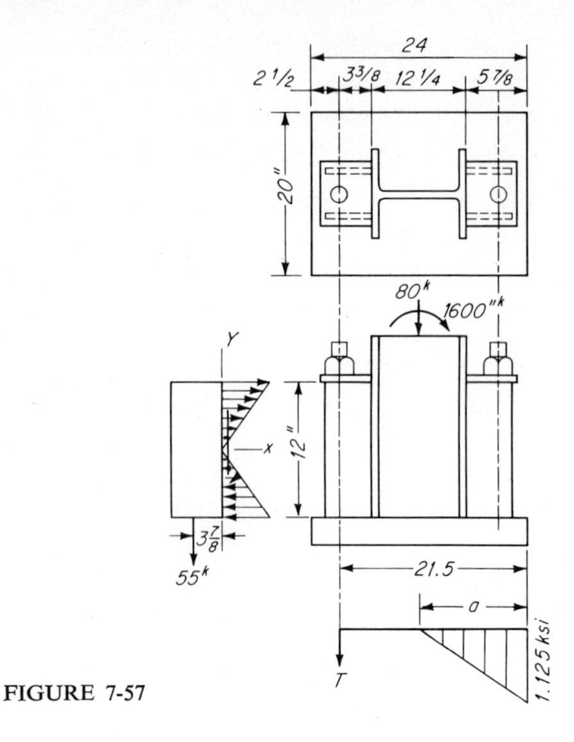

FIGURE 7-57

A detail of the type shown at the right in Fig. 7-56*b* will be used. Assume the boot to be 12 in. long, with each vertical plate fillet-welded both sides to the column flange. The force per weld is 55/4 = 13.8 kips, and the stresses on one weld are (Fig. 7-57)

$$q_y = \frac{13.8}{12} = 1.15 \text{ kli}$$

$$q_x = \frac{13.8 \times 3.375}{12^2/6} = 1.94 \text{ kli}$$

$$q = \sqrt{1.15^2 + 1.94^2} = 2.26 \text{ kli}$$

The required size of E70 weld is 2.26/(21 × 0.707) = 0.152 or $\frac{3}{16}$ in. A $\frac{3}{8} \times 5\frac{1}{2}$ in. plate will be used. Its thickness is ample for the size of the welds, and its b/t ratio is 5.5/0.375 = 14.7 < 16. The top plate will be $\frac{3}{8} \times 5\frac{1}{2} \times 4$.

The thickness of the base plate must be enough to assure the stiffness needed to distribute the load. A procedure for determining the thickness of base plates for axially loaded columns was discussed in Art. 4-16. Bending of the plate in Fig. 7-57 would be critical at or near the flange on the side where the plate bears on the footing. In this case, the stiffeners complicate evaluation of such a moment, but a conservative estimate of thickness can be made by ignoring this effect. Since a in Fig. 7-57 was found to be 12 in., we see that the projecting 6 in. of base plate is acted upon by a trapezoidal stress block varying in intensity from 0.56 ksi at the column flange to 1.125 ksi at the edge of the plate. The moment at the column flange for a 1-in. strip of plate is

$$M = \tfrac{1}{2} \times 1.125 \times 6 \times 4 + \tfrac{1}{2} \times 0.56 \times 6 \times 2 = 16.9 \text{ in.-kips}$$

The allowable bending stress for an A36 plate is $F_b = 0.75F_y = 27$ ksi. Therefore, from $M = F_y S$, we get

$$16.9 = \frac{27t^2}{6} \qquad t = 1.94 \text{ in.}$$

This is a reasonable thickness for a plate with the lateral dimensions of this one. Therefore, a 2 × 20 × 24 in. plate is satisfactory.

PROBLEMS

7-55 Design the base plates of Examples 7-5 and 7-6 on an ultimate-load basis using a load factor of 2.5.

7-56 A W8 × 31 column supports a concentric load of 50 kips and a moment of 30 ft-kips in the plane of the web. Design a base plate and its connection to the column shaft. AISC specification, A36 steel, E70 electrodes.

7-57 A W12 × 65 column supports a concentric load of 90 kips and a moment of 100 ft-kips in the plane of the web. Design a base plate and its connection to the column shaft. AISC specifications, A36 steel, E70 electrodes.

7-58 A W14 × 158 column supports a concentric load of 150 kips and a wind moment of 275 ft-kips in the plane of the web. Design a base plate and its connection to the column shaft. AISC specification, A36 steel, E70 electrodes.

7-33 PINNED CONNECTIONS

Pinned connections are used to permit relatively free end rotation of the connected members. Pinned joints in bascule bridges, crane booms, etc., must allow relatively large rotation. Pinned joints where the rotations are relatively small are found in hinged arches (Fig. 7-58), in the links by which the suspended span of cantilever systems may be fastened to the cantilever span, in main supports of heavy trusses and girders, and in light bracing systems (clevis pins). Steel pins, which may be cast, forged, or cold-rolled, and then machined, generally range in size from $1\tfrac{1}{4}$ to 10 in. in diameter, but sizes up to 24 in. are available.

FIGURE 7-58

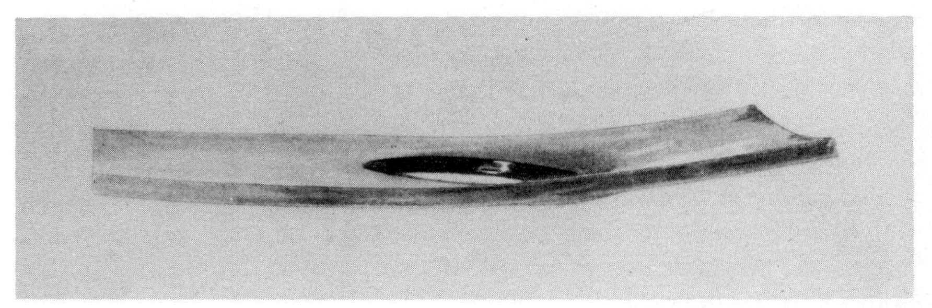

FIGURE 7-59
Buckling failure of pin plate.

Since the pin does not have parts bearing on its full length, its size is likely to be determined by bending. However, failure may occur by shearing of the pin, crushing of the plate, and, in tension members, shearing of the plate between the pinhole and the end of the plate or fracture of the plate on the net section through the pinhole. A pinned connection may also fail by dishing of the plate (Fig. 7-59).

The pin should fit snugly in the pinhole. The AISC specification requires the diameter of the hole to be not more than $\frac{1}{32}$ in. larger than the diameter of the pin. The allowable bearing stress is somewhat less than for bolts and rivets. In effect, this allows freer rotation. Where large rotations are expected, bearing stresses should be reduced. The AASHO specifications allow only 50 percent of the basic bearing stress if large rotations are expected. Since the required thickness of the parts is inversely proportional to the pin diameter, a relatively thin bearing may require an excessively large pin. To avoid this situation, members may be reinforced with pin plates at the pinhole.

The allowable bending stress for pins is usually about 50 percent larger than that for beams. The larger stress is primarily justified because of the shape of the cross section. Allowable shears are $0.40F_y$ in most specifications. Allowable bearing stress is $0.80F_y$ in the AASHO specifications and $0.90F_y$ in the AISC specification.

In a series of 106 tests on pin-connected plate links, fracture on the net section of the plate at the pinhole did not occur until the average stress on the net section reached the ultimate tensile strength of the steel.[18] Therefore, stress concentrations at the edge of the hole had no apparent effect on the strength of the plate under static loads. Both the AREA and AASHO specifications require that the net section through the pinhole of pin-connected tension members be at least 40 percent greater than the required net section of the body of the member. In the AISC specification, the allowable tension on the net section is only $0.45F_y$, compared with $0.60F_y$ on the body of the member. This gives a net section at the pinhole one-third larger than the net section of the body.

The tests also showed that the net area beyond the pinhole, on a longitudinal section of the member, must range from about 60 to 75 percent of the transverse net area at the pinhole if failure beyond the pinhole and failure on the net section at the pinhole are to be about equally probable. The smaller figure is for a member about twice as wide as its pin, while the larger figure is for one about four times as wide as its pin (this is the range of ratios of plate width to pin diameter covered by the tests). For pin-connected tension members (except eyebars), both the AREA and AASHO specifications require that the net section beyond the pinhole, parallel with the axis of the member, be not less than the required net section of the body of the member. This clause, together with the one relative to the transverse net section at the pinhole, results in a member for which the ratio of longitudinal net section beyond the pinhole to transverse net section at the pinhole is $1/1.40 = 0.715$. This compares favorably with the ratio 0.60 to 0.75 disclosed by Johnston's tests. According to the AISC specification the net section beyond the pinhole, parallel to the axis of the member, need be no more than two-thirds the required net section of the body of the member. Therefore, for these specifications the ratio of net section beyond the pinhole to that at the pinhole is $\frac{2}{3}/\frac{4}{3} = 0.5$.

Buckling of the plate beyond the pin (dishing) is a function of stability rather than strength. Therefore, dishing depends upon the ratios D/t, b/t, and a/t, where D is the diameter of the pinhole, b and a the edge and end distances, respectively, at the pinhole, and t the thickness of the plate. However, for the range of ratios of plate width to pin diameter covered by his tests, Johnston's results show that if a member is designed to be about equally strong on the net section through the pinhole and on the net section beyond the pin, it will be at least as strong with respect to dishing if the ratio of net width at the pinhole to thickness of the plate ranges from about 6 to 10. These ratios correspond, respectively, to the plates of Johnston's tests which were four times and two times as wide as the diameter of their pins. The AISC, AREA, and AASHO specifications all agree on a limit of 8 for the ratio of net width through the pinhole to thickness of the member at the pin.

REFERENCES

1 Higgins, T. R.: New Formulas for Fasteners Loaded off Center, *Eng. News-Rec.*, May 21, 1964.

2 Douty, R. T., and W. McGuire: High Strength Bolted Moment Connections, *J. Struct. Div. ASCE*, vol. 91, no. ST2, April 1965.

3 Grinter, Linton E.: "Design of Modern Steel Structures," p. 49, Macmillan, New York, 1941.

4 AISC: "Plastic Design in Steel," p. 38, New York, 1959.

5 Higgins, T. R., and W. H. Munse: How Much Combined Stress Can a Rivet Take, *Eng. News-Rec.*, Dec. 4, 1952, p. 40.

6 Chesson, E., Jr., N. L. Faustino, and W. H. Munse: High Strength Bolts Subjected to Tension and Shear, *J. Struct. Div. ASCE*, vol. 91, no. ST5, October 1965, pp. 155–180.

7 Nair, R. S., P. C. Birkemoe, and W. H. Munse: Behavior of Bolts in Tee-connections Subjected to Prying Action, *Univ. Ill. Civ. Eng. Stud., Struct. Res. Ser.* 1969.

8 Batho, Cyril: First, Second, and Final Reports, Steel Structures Committee, Department of Scientific and Industrial Research, Great Britain, 1931–1936.

9 Rathbun, J. C.: Elastic Properties of Riveted Connections, *Trans. ASCE*, vol. 101, p. 539, 1936.

10 Johnston, Bruce, and Lloyd Greene: Flexible Welded Angle Connections, *J. AWS*, October 1940.

11 LeMessurier, W. J., and H. W. Hagen: Elastic Design of Steel Structures, sec. 6 in "Structural Engineering Handbook," E. H. Gaylord and C. N. Gaylord (eds.), McGraw-Hill, New York, 1968.

12 Blodgett, O. W.: "Design of Welded Structures," The James F. Lincoln Arc Welding Foundation, Cleveland, Ohio, 1966.

13 Graham, J. D., A. N. Sherbourne, R. N. Khabbaz, and C. D. Jensen: "Welded Interior Beam-to-column Connections," AISC, New York, 1959.

14 Bleich F.: "Design of Rigid Frame Knees," AISC, July 1943 (reprinted March 1956).

15 Griffiths, J. D.: "Single Span Rigid Frames in Steel," AISC, 1948 (reprinted March 1956).

16 Fisher, J. W., G. C. Lee, J. A. Yura, and G. C. Driscoll, Jr.: Plastic Analysis and Tests of Haunched Corner Connections, *Weld. Res. Counc. Bull.* 91, October 1963.

17 "Plastic Design in Steel," 2d ed., ASCE Manuals and Reports on Engineering Practice no. 41, New York, 1971.

18 Johnston, Bruce G.: Pin-connected Plate Links, *Trans. ASCE*, vol. 104, p. 314, 1939.

19 Munse, W. H., K. S. Petersen, and E. Chesson, Jr.: Strength of Rivets and Bolts in Tension, *J. Struct. Div. ASCE*, March 1959.

8

PLASTIC ANALYSIS AND DESIGN

8-1 INTRODUCTION

It has been pointed out on several occasions in the preceding chapters that factors of safety based on yield stress give no necessary or consistent indication of the factor of safety with respect to the ultimate capacity of a structural member or the structure of which it is a part. Thus, it was shown in Art. 5-2 that, depending upon the shape of its cross section, a structural member may develop bending resistance ranging from 10 to 70 percent more than the moment at first yield, provided the metal has sufficient ductility. Similarly, in some cases, the load corresponding to the peak of the load-deflection curve of a beam-column may be only slightly larger than the load corresponding to first yield, while in other cases it may be much larger. Furthermore, a continuous structure can usually carry a load considerably in excess of the load at which the first plastic hinge develops, as discussed in Art. 5-13.

A rational procedure for predicting the strength of continuous structures is developed in this chapter. This method of design is usually called *plastic design* but is also called *limit design*, *collapse design*, and, occasionally, *ultimate-strength design*. However, since the latter term has a different meaning in the design of reinforced-concrete structures, it is not commonly used as a synonym for plastic design.

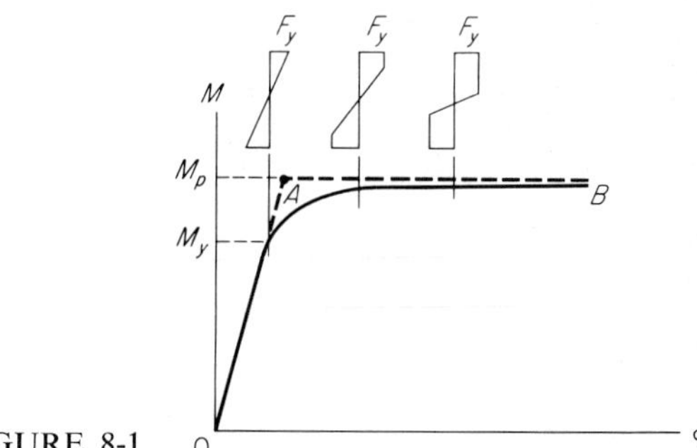

FIGURE 8-1

8-2 DEVELOPMENT OF COLLAPSE MECHANISM

The variation of bending moment with rotation of the cross section was discussed in Art. 5-2. For beams of materials for which stress is proportional to strain, this variation is linear until the yield moment M_y is reached (Fig. 8-1). The rate of increase of moment drops rapidly after M_y is exceeded and the moment is very nearly equal to the plastic moment M_p at strains only several times ε_y, even for cross sections with large shape factors (Fig. 5-4). This variation of moment with rotation is given to good approximation by OAB in Fig. 8-1. According to this approximation, we assume that moment is proportional to rotation until M_p is reached (OA), after which rotation continues indefinitely while the moment remains constant at the value M_p (AB). A cross section in this condition is said to have developed a *plastic hinge*.

The sequence of formation of plastic hinges and the resulting redistribution of moment which leads to the peak load on a uniformly loaded, fixed-ended beam was discussed in detail in Art. 5-13. It was shown that yielding develops first at the supports, following which the bending moments at these cross sections soon reach the plastic value M_p. At this stage, the bending moment at the center is $M_p/2$ if we assume the behavior OAB of Fig. 8-1. However, continued rotation at the supports under the constant moment M_p allows the moment at midspan to increase until it also reaches M_p. At this stage, rotation at midspan and at the supports can continue indefinitely with no increase in moment, provided strain hardening is neglected. Therefore, the load can increase no further. Also, with three plastic hinges, this beam is unstable and behaves like a linkage. Because of this analogy the beam is said to be a mechanism.

8-3 DETERMINATION OF COLLAPSE MECHANISM

In order for a beam or structure to reach its capacity load it must develop a sufficient number of plastic hinges to reduce it to a mechanism. It is evident that these hinges form at points of maximum moment, and it is immaterial whether they develop

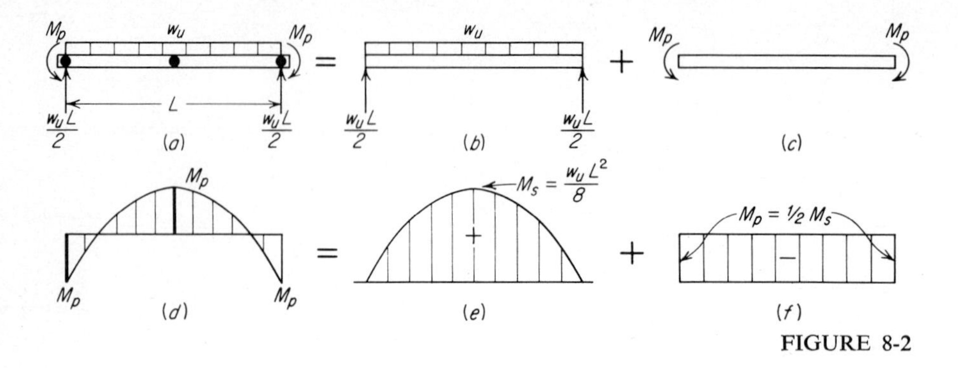

FIGURE 8-2

simultaneously or successively. The problem is to determine the moment diagram at collapse, which requires a knowledge of the number of plastic hinges the structure must develop to become a mechanism. The plastic-moment diagram for beams can readily be determined. Thus, the uniformly loaded beam with fixed ends (Fig. 8-2) is equivalent to a simply supported beam acted upon simultaneously by the forces shown separately in Fig. 8-2*b* and *c*. Therefore, the plastic-moment diagram of Fig. 8-2*d* can be found by superimposing the simple-beam moment diagrams of Fig. 8-2*e* and *f*.

The above procedure is illustrated in Fig. 8-3. Since the beam shown has a real hinge at the left support, two plastic hinges are enough to produce a mechanism (Fig. 8-3*b*). These hinges form at the point of application of the load and at the right support, where there are peak moments. The corresponding plastic-moment diagram is shown in Fig. 8-3*c*. The diagram is constructed by superimposing the simple-beam diagram ABC for the concentrated load and the simple-beam diagram ADC for the end moment. In this figure, $BE = DC = M_p$. Then since $BE = BF - EF$ and $EF = 0.6DC$, we get

$$M_p = M_s - 0.6M_p \qquad 1.6M_p = M_s$$

$$M_s = \frac{60 \times 8 \times 12}{20} = 288 \text{ ft-kips}$$

$$M_p = \frac{288}{1.6} = 180 \text{ ft-kips}$$

The location of the plastic hinges was self-evident in the preceding example. This is not always the case. For example, consider the beam shown in Fig. 8-4*a*. We construct the simple-beam moment diagram for the concentrated loads and superimpose the simple-beam moment diagram for the unknown plastic moment at the right support (Fig. 8-4*b*). Since the beam has a real hinge at the left end, it will become a mechanism upon the formation of two plastic hinges. One of these hinges

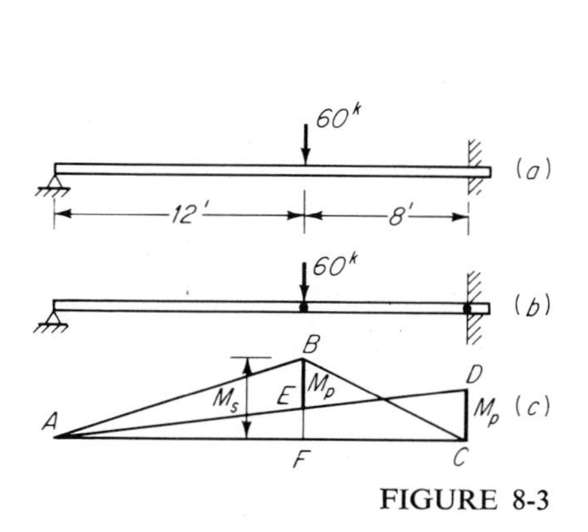

FIGURE 8-3

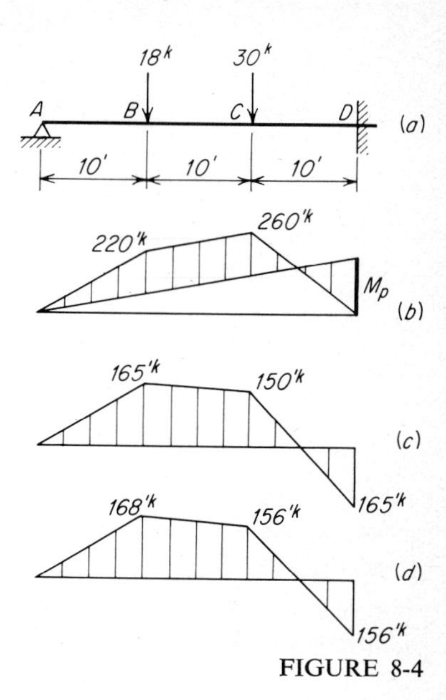

FIGURE 8-4

develops at the right support, but whether the other forms at B or at C of Fig. 8-4a is not immediately apparent. Therefore, we must try both possibilities. Assuming the second hinge to be at B, we find from the geometry of the moment diagram of Fig. 8-4b

$$220 - \frac{M_p}{3} = M_p \qquad M_p = 165 \text{ ft-kips}$$

Next, assuming the second hinge to be at C, we have

$$260 - \frac{2M_p}{3} = M_p \qquad M_p = 156 \text{ ft-kips}$$

To determine which of these values is correct, consider the corresponding moment diagrams in Fig. 8-4c and d. In Fig. 8-4c there is no moment larger than 165 ft-kips, and there are two points at which the moment is 165 ft-kips, namely, at points B and D where hinges were assumed to be located in obtaining $M_p = 165$ ft-kips. On the other hand, the moment at B in Fig. 8-4d exceeds the value 156 ft-kips based on plastic hinges at C and D. Therefore, this solution cannot be correct if the beam is to be of the same cross section throughout.

A uniformly loaded beam with only one end built-in is somewhat more difficult. Since the solution is useful in dealing with continuous beams, we shall develop it as a final example of the single-span beam. The beam and its plastic-moment diagram

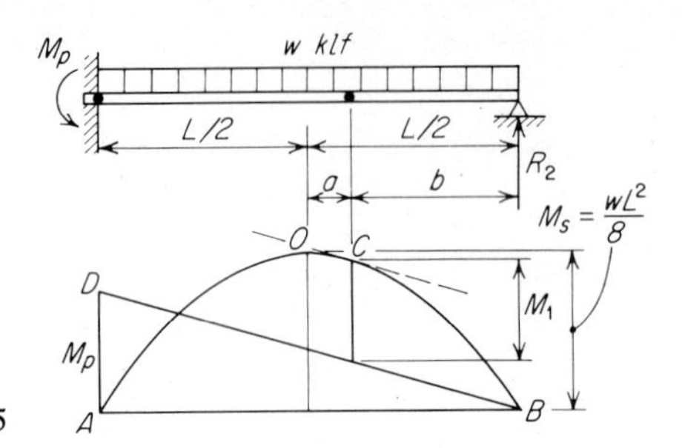

FIGURE 8-5

are shown in Fig. 8-5. One plastic hinge forms at the built-in end, the other at the point of maximum positive moment. The latter is not at midspan, however, but at the section of the beam corresponding to the point C, where the tangent to the parabola AOB is parallel to the line DB. Since C is a point of maximum moment, it is also a point of zero shear in the beam. Using the notation shown in Fig. 8-5, the position of point C is found as follows:

$$R_2 = \frac{wL}{2} - \frac{M_p}{L} \tag{a}$$

$$V = R_2 - wb = \frac{wL}{2} - \frac{M_p}{L} - wb = 0$$

Therefore,

$$b = \frac{L}{2} - \frac{M_p}{wL} \tag{b}$$

The bending moment M_1 at point C is

$$M_1 = R_2 b - \frac{wb^2}{2} \tag{c}$$

Substituting into Eq. (c) the values of R_2 and b from Eqs. (a) and (b), respectively, we get

$$M_1 = \frac{wL^2}{8} - \frac{M_p}{2} + \frac{M_p^2}{2wL^2} \tag{d}$$

Since the simple-beam moment $M_s = wL^2/8$, Eq. (d) may be written

$$M_1 = M_s - \frac{M_p}{2} + \frac{M_p^2}{16M_s} \tag{8-1}$$

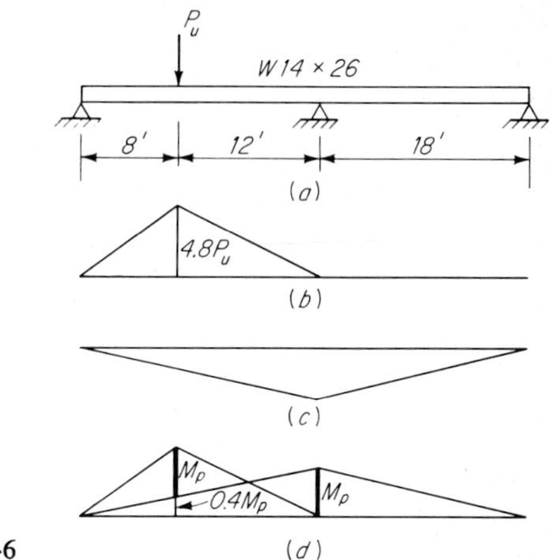

FIGURE 8-6

If there is a plastic hinge at the point of maximum positive moment as well as at the built-in end, $M_1 = M_p$. Making this substitution, we get from Eq. (8-1)

$$M_p^2 - 24M_s M_p + 16M_s^2 = 0 \qquad (e)$$

From Eq. (*e*)

$$M_p = 0.686M_s \qquad (8\text{-}2)$$

The distance a from midspan to the point of maximum positive moment (Fig. 8-5) is found from Eq. (*b*) to be

$$a = \frac{M_p}{wL} = \frac{L}{8}\frac{M_p}{M_s} \qquad (8\text{-}3)$$

In particular, if $M_p = 0.686M_s$,

$$a = \frac{L}{8}\frac{0.686M_s}{M_s} = 0.086L \qquad (8\text{-}4)$$

8-4 CONTINUOUS BEAMS

The procedure described in the preceding article can be used to advantage in plastic analysis of continuous beams. For example, to determine the ultimate load P_u for the A36 W14 × 26 continuous beam of Fig. 8-6*a* we superimpose the simple-beam moment diagrams for the load P_u (Fig. 8-6*b*) and the moment at the support (Fig. 8-6*c*).

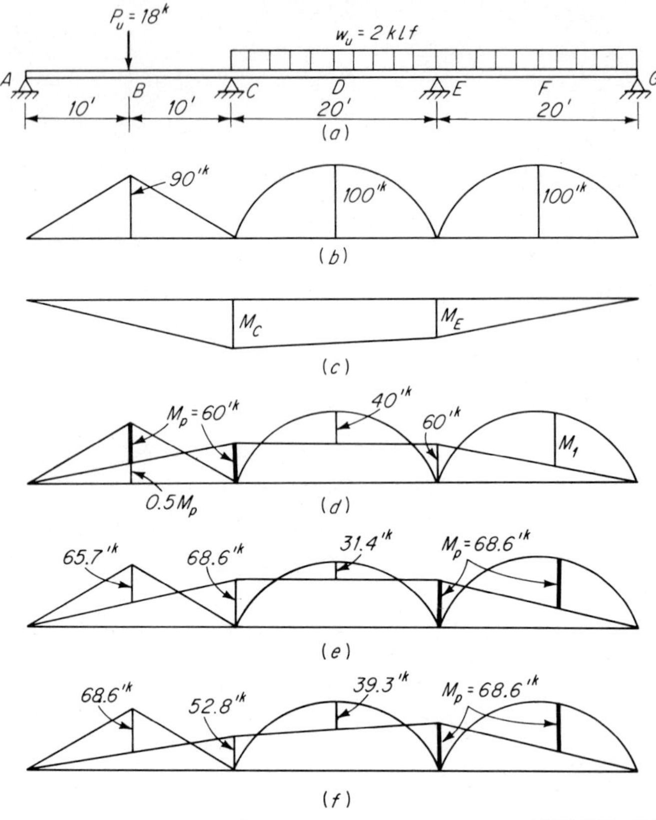

FIGURE 8-7

The result (Fig. 8-6*d*) shows that there are two points of peak moments at which plastic hinges can develop. Since these moments are equal, we have

$$4.8P_u - 0.4M_p = M_p$$

$$M_p = F_y Z = \frac{36 \times 40}{12} = 120 \text{ ft-kips}$$

$$P_u = \frac{1.4M_p}{4.8} = \frac{1.4 \times 120}{4.8} = 35 \text{ kips}$$

This value of P_u is the true ultimate load if there are enough plastic hinges so located as to form a mechanism and if the moment does not exceed M_p at any cross section of the beam. These conditions are fulfilled. (It is assumed that the beam has lateral support sufficient to prevent lateral-torsional buckling.)

As a second example a plastic-moment diagram for the laterally supported continuous beam of uniform cross section shown in Fig. 8-7*a* will be determined. Considering the structure to consist of three simple spans, we get the three moment diagrams shown in Fig. 8-7*b*. The moment diagram for the unknown moments at the interior supports is shown in Fig. 8-7*c*. The two figures must now be superimposed to produce a moment diagram that will satisfy the conditions for development of a mechanism. There are three possible mechanisms: one in span *AC* with plastic hinges at *B* and *C*, one in span *CE* with plastic hinges at *C*, *D*, and *E*, and one in span *EG* with plastic hinges at *E* and *F*. Assuming the first of these possible solutions, we get

$$90 - 0.5M_p = M_p \qquad M_p = 60 \text{ ft-kips}$$

The corresponding moment diagram for span *AC* is shown in Fig. 8-7*d*. If $M_p = 60$ ft-kips is the correct solution, it must be possible to complete the moment diagram without exceeding 60 ft-kips. This condition can be satisfied in span *CE* by taking $M_E = 60$ ft-kips, which gives $M_D = 40$ ft-kips. The corresponding maximum moment in span *EG* is determined from Eq. (8-1):

$$M_1 = 100 - \frac{60}{2} + \frac{60^2}{16 \times 100} = 72.3 \text{ ft-kips}$$

Since this moment exceeds 60 ft-kips and cannot be reduced without increasing M_E, the assumed mechanism is not correct. This suggests that the mechanism will form in span *EG*. Using Eq. (8-2),

$$M_p = 0.686 \times 100 = 68.6 \text{ ft-kips}$$

Figure 8-7*e* shows a moment diagram corresponding to this value of M_p, for which the moment nowhere exceeds M_p. Therefore, the assumed mechanism is the correct one. However, the moment diagram of Fig. 8-7*e* is not the only one for which *M* is everywhere less than M_p. We could also assume $M_B = 68.6$ ft-kips, in which case $M_C = 2(90 - 68.6) = 52.8$ ft-kips, and $M_D = 100 - 0.5(68.6 + 52.8) = 39.3$ ft-kips (Fig. 8-7*f*). Thus, of the two moments M_C and M_E which were originally statically indeterminate, only M_E has been determined. For the other moment, we know only

$$52.8 \leq M_C \leq 68.6$$

Thus, the structure is still statically indeterminate, but this is immaterial so far as the determination of the required value of M_p is concerned.

PROBLEMS

8-1 The loads shown on the laterally supported continuous beam of Fig. 8-8*a* are service loads. Determine the required A36 beam for a load factor of 1.5.

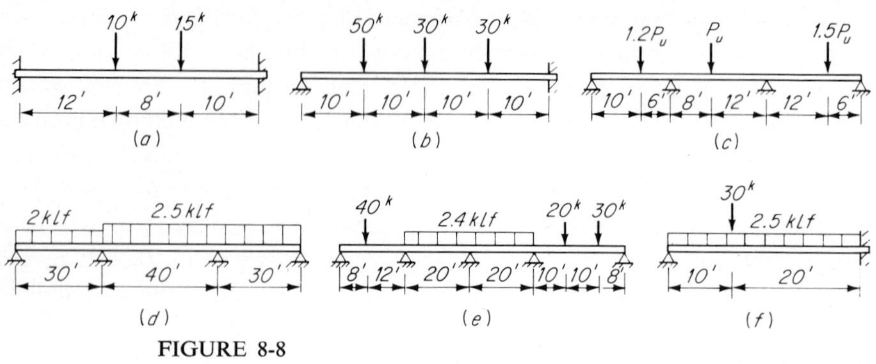

FIGURE 8-8
(*a*) Problem 8-1, (*b*) Problem 8-2, (*c*) Problem 8-3, (*d*) Problem 8-4, (*e*) Problem 8-5, (*f*) Problem 8-6.

8-2 Same as Prob. 8-1 for the beam of Fig. 8-8*b*.

8-3 Determine P_u for the A36 laterally supported W16 × 36 continuous beam of Fig. 8-8*c*.

8-4 The loads shown on the laterally supported continuous beam of Fig. 8-8*d* are service loads. Determine the required A36 beam for a load factor of 1.6.

8-5 Same as Prob. 8-4 for the beam of Fig. 8-8*e*.

8-6 Same as Prob. 8-4 for the beam of Fig. 8-8*f*.

8-5 MECHANISM ANALYSIS BY VIRTUAL DISPLACEMENTS

The collapse mechanism of a beam or structure can be determined by using the principle of virtual displacements. According to this principle, if a structure which is in equilibrium is given an arbitrary displacement, the work done by the external forces will equal the work done by the internal forces. In using the principle to determine the moment at collapse, the arbitrary displacement must be one for which only the internal moments at the plastic hinges contribute to the internal work. This is accomplished by allowing rotations of the structure only at points of simple support and at the points where plastic hinges are expected to occur in producing the mechanism.

Figure 8-9*a* shows the beam of Fig. 8-3*a*, for which the value of M_p is to be determined. Two plastic hinges are sufficient to form a mechanism. They will develop at the peak-moment points *B* and *C*. Therefore, we give the beam the virtual displacement shown in Fig. 8-9*b*, where the only rotations permitted are those at *A*, *B*, and *C*. Assigning a value to any one of the hinge rotations enables the other two to be determined. Thus, with $\theta_C = \theta$, $\theta_A = 2\theta/3$ and $\theta_B = 5\theta/3$ (Fig. 8-9*c*). These rotations produce a displacement $\delta = 8\theta$ at the point of application of the external 60-kip load.

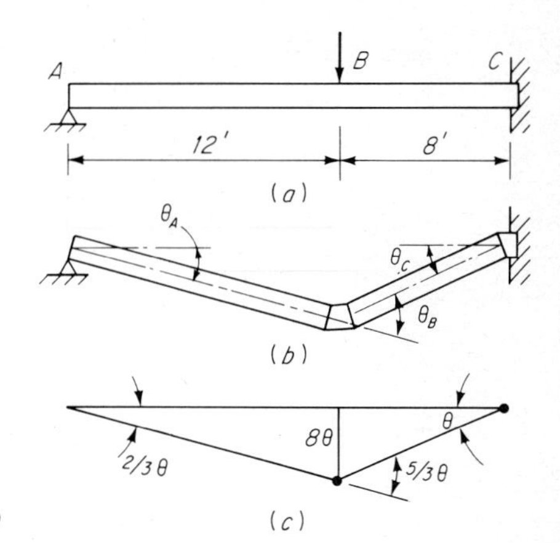

FIGURE 8-9

Equating the internal work of the moments M_p to the external work of the load, we get

$$(1 + \tfrac{5}{3})\theta M_p = 60 \times 8\theta$$
$$M_p = 180 \text{ ft-kips}$$

As a second example, the value of P_u will be determined for the beam of Fig. 8-10*a*, for which $M_p = 136$ ft-kips. Since two plastic hinges are enough to form a mechanism, there are two possible solutions. Assuming plastic hinges at *B* and *D* and giving the beam the virtual displacement shown in Fig. 8-10*b* gives

$$20 \times P_u + 10 \times 1.2P_u = (3 + 1)M_p$$

$$P_u = \frac{4}{32} M_p = \frac{4 \times 136}{32} = 17 \text{ kips}$$

This result is shown in Fig. 8-10*c*.

Assuming plastic hinges at *C* and *D* and giving the beam the virtual displacement shown in Fig. 8-10*e* gives

$$10 \times P_u + 20 \times 1.2P_u = (3 + 2)M_p$$

$$P_u = \frac{5}{34} M_p = \frac{5 \times 136}{34} = 20 \text{ kips}$$

This result is shown in Fig. 8-10*f*.

The moment diagrams for the two solutions are shown in Figs. 8-10*d* and *g*. It will be noted that the moment in Fig. 8-10*d* is nowhere larger than the specified plastic moment, 136 ft-kips. On the other hand, there are moments in excess of 136 ft-kips in the diagram for $P_u = 20$ kips (Fig. 8-10*g*). Therefore, P_u cannot be 20 kips.

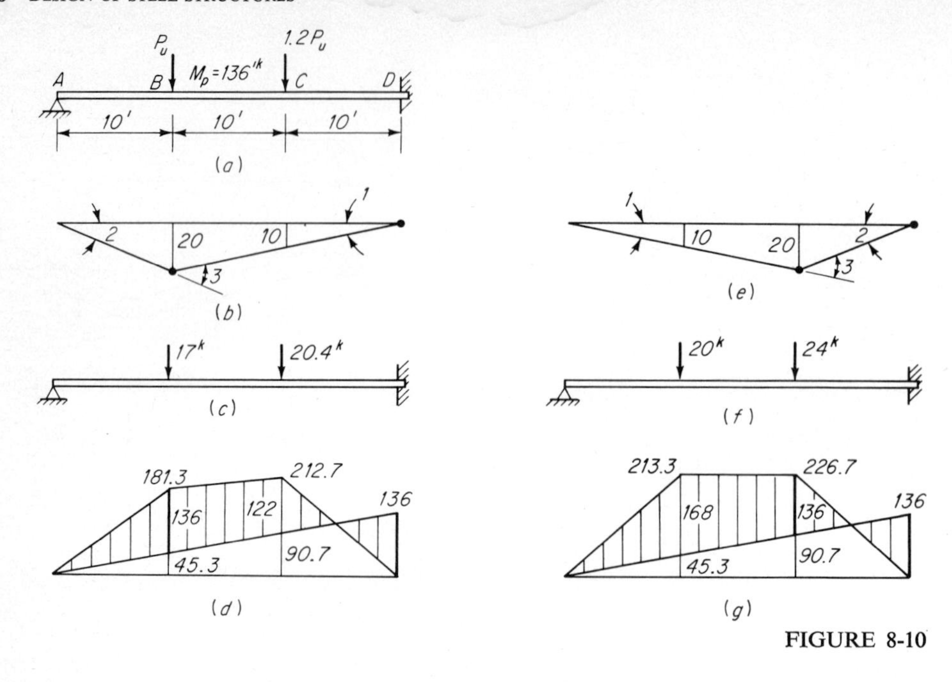

FIGURE 8-10

Had the moment diagram for $P_u = 17$ kips been constructed, it would have been unnecessary to investigate the second mechanism. Therefore, one has a choice of two procedures in using this method of analysis: (1) compute P_u for all possible mechanisms, in which case the true ultimate load is the smallest value thus found, or (2) construct the moment diagram for each P_u as it is determined, and continue until the true ultimate load is identified by $M \lesseqgtr M_p$ at all cross sections.

8-6 THE BOUND THEOREMS

In the analysis of the beam of the second example in Art. 8-5 it was found that only one of the two possible mechanisms gave the true ultimate load. This was the mechanism of Fig. 8-10*b*. Furthermore, the incorrect mechanism, Fig. 8-10*e*, gave a load greater than the true ultimate load. It can be proved that this is always true; i.e., of all the mechanisms that can be formed, all but the correct mechanism of collapse correspond to loads larger than the ultimate load the structure can support. This conclusion can be stated as follows.

Upper-bound theorem The load corresponding to an assumed mechanism is greater than or equal to the true ultimate load.

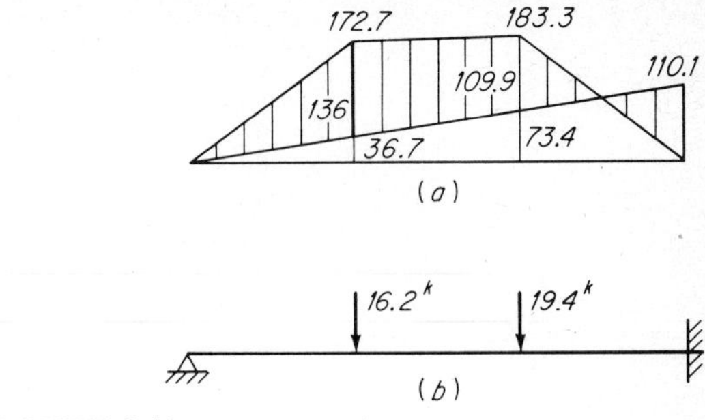

FIGURE 8-11

A second theorem which is useful in plastic analysis can be demonstrated by using the second example of Art. 8-5, as follows. If the moments in the diagram of Fig. 8-10*g* are reduced in the ratio 136/168, the moment diagram in Fig. 8-11*a* is obtained. The corresponding reduced loads, shown in Fig. 8-11*b*, are less than the true ultimate loads (Fig. 8-10*c*). Although the moment in this diagram is nowhere larger than the prescribed value of M_p (136 ft-kips), it equals M_p at only one point. Thus, there is only one plastic hinge, which is one short of the number required to produce a mechanism. This leads to the following statement.

Lower-bound theorem The load corresponding to an assumed moment distribution which is in equilibrium with the load, and for which the moment nowhere exceeds M_p, is less than or equal to the true ultimate load.

8-7 ANALYSIS OF RECTANGULAR FRAMES

EXAMPLE 8-1 The rectangular frame of Fig. 8-12*a* is statically indeterminate. Since it is once redundant, it would be statically determinate if there were a hinge at one other point. Therefore, if there were still another hinge, the frame would be unstable. Thus, two plastic hinges are required to form a mechanism.

Figure 8-12*b* shows the frame with the horizontal restraint at *A* removed. The resulting moment diagram is shown in Fig. 8-12*e*. (Moments are taken positive if the interior face of the member is in tension and are plotted positive on the tension face.) The redundant force *H* (whose direction is not yet known) is shown in *c* and the corresponding moment diagram in *f*. This moment diagram must be superimposed on the moment diagram of Fig. 8-12*e* in such a way as to produce a moment diagram for which the frame will develop the required two plastic hinges. If the three members of

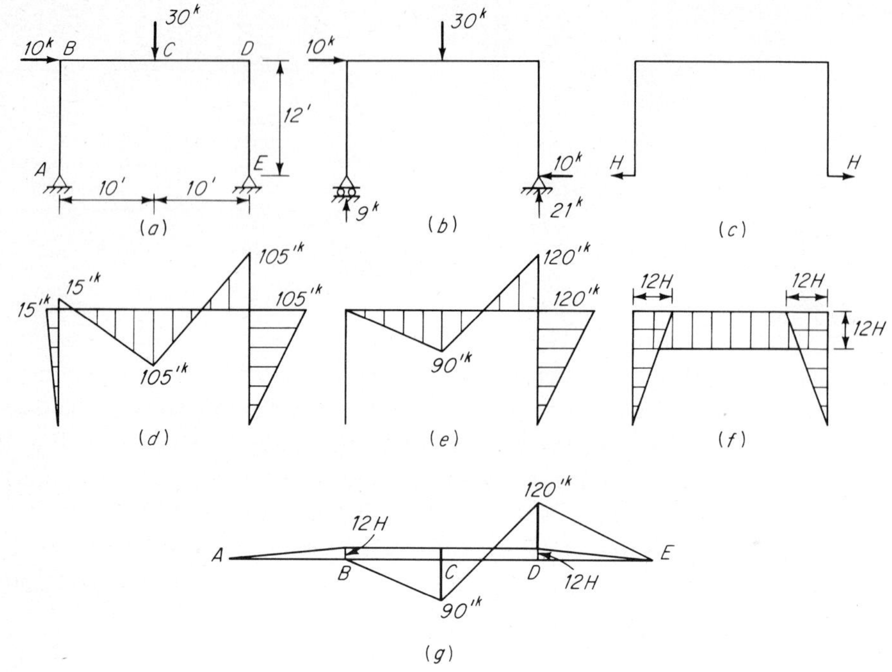

FIGURE 8-12

the frame have equal plastic resisting moments, these hinges will develop at C and D, as shown in Fig. 8-12g. Therefore,

$$120 - 12H = 90 + 12H \qquad H = 30/24 \text{ kips}$$
$$M_p = 90 + 12H = 90 + 15 = 105 \text{ ft-kips}$$

The resulting moment diagram is plotted on the outline of the frame in Fig. 8-12d.

If the frame of Fig. 8-12a is to be proportioned so that the plastic moment M_{pc} of the columns is only one-half the plastic moment M_{pb} of the beam, the required values of M_p are readily determined from Fig. 8-12g. Thus,

$$120 - 12H = \tfrac{1}{2}(90 + 12H) \qquad 18H = 75$$
$$M_{pc} = 120 - 12H = 120 - 50 = 70 \text{ ft-kips}$$
$$M_{pb} = 2M_{pc} = 140 \text{ ft-kips}$$

It will be noted that the final moment diagram of Fig. 8-12d can be determined once the relative moment-resisting capacities of members BD and DE have been decided. The members of the frame can then be proportioned to have resisting

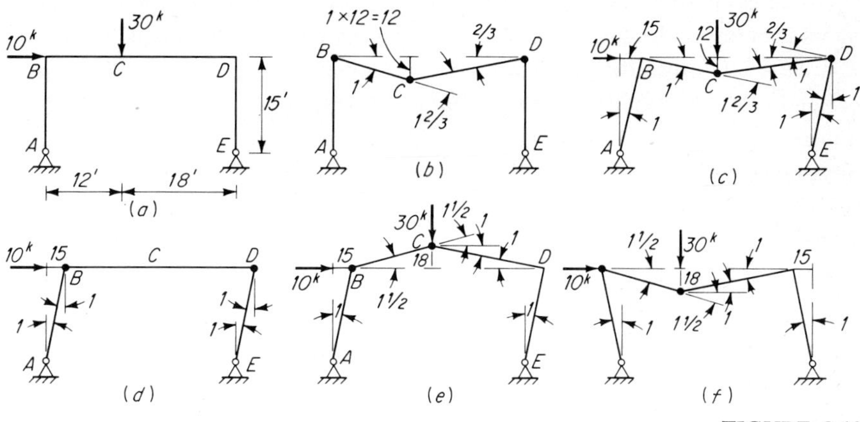

FIGURE 8-13

moments according to the requirements of the moment diagram. On the other hand, in an elastic analysis of this frame, the relative stiffnesses of the members must be assumed to effect a solution. Then, when the members are proportioned, they may have relative stiffnesses different from those assumed in advance, in which case a reanalysis is required. Thus, an advantage of plastic design is that members of a structure can be proportioned to have their resisting moments in the ratios that were assumed in advance.

EXAMPLE 8-2 Determine the A36 W shapes required for the frame of Fig. 8-13*a*. The three members are to have the same bending resistance.

SOLUTION Plastic hinges can form at *B*, *C*, and *D*. Should the frame collapse by forming hinges at all three of these points, the virtual displacement corresponding to the collapse mechanism would be as shown in Fig. 8-13*b*. This gives

$$(1 + 1\tfrac{2}{3} + \tfrac{2}{3})M_p = 12 \times 30 \qquad M_p = 108 \text{ ft-kips}$$

If the frame collapses by developing hinges at *C* and *D*, the virtual displacement corresponding to this collapse mechanism is as shown in Fig. 8-13*c*. This gives

$$(1\tfrac{2}{3} + 1\tfrac{2}{3})M_p = 15 \times 10 + 12 \times 30 \qquad M_p = 156 \text{ ft-kips}$$

The frame might also collapse by developing hinges at *B* and *D*. The mechanism displacement is shown in Fig. 8-13*d*, for which

$$(1 + 1)M_p = 15 \times 10 \qquad M_p = 75 \text{ ft-kips}$$

In evaluating the collapse moment in the preceding cases, the virtual displacements corresponding to the collapse mechanism were taken in such a way as to make the external forces do positive work. In other words, the virtual displacement was taken in the direction in which the frame collapses. If hinges were to form at B and C, the direction in which the frame would collapse is not apparent. If the virtual displacement is taken as in Fig. 8-13*e*, the external work is

$$W_e = 15 \times 10 - 30 \times 18 = -390 \text{ ft-kips}$$

Therefore, if the frame were to collapse with hinges at C and D, it would collapse in the direction shown in Fig. 8-13*f*, for which $W_e = +390$ ft-kips:

$$(2\tfrac{1}{2} + 2\tfrac{1}{2})M_p = 390 \qquad M_p = 78 \text{ ft-kips}$$

(It is shown in the next article that it is not necessary to consider mechanisms *e* and *f*.)

Since all possible mechanisms have been investigated, the largest value of M_p found is known to be the correct one. Thus, the frame must be designed for $M_p = 156$ ft-kips, for which $Z = 156 \times 12/36 = 52$ in.^{3} The W16 × 31 is the lightest section which can be used ($Z = 54$ in.^{3}).

8-8 GENERAL PROCEDURE FOR MECHANISM ANALYSIS OF FRAMES

Extension of the procedure discussed in Art. 8-7 to more complex frames becomes extremely involved because of the large number of possible modes of collapse. A systematic procedure to help resolve this difficulty was suggested by Symonds and Neal.[3,4] The fundamentals of this procedure can be explained in terms of Example 8-2.

Three moments, M_B, M_C, and M_D, must be known to define the moment diagram for the frame of Fig. 8-13. Since the structure is once redundant, one of these moments can be prescribed independently, after which the remaining two can be determined by equations of static equilibrium. Each of these equations can be obtained by a virtual displacement corresponding to a collapse mechanism. Therefore, of all the mechanisms shown in Fig. 8-13, only two can be considered to be independent. All the others are combinations of these two. The mechanisms which are considered to be independent are called *elementary mechanisms*, and their combinations are called *combined mechanisms*. Although the elementary mechanisms can be chosen arbitrarily, there is a logical classification for rectangular frames. According to this classification, mechanisms *b* and *d* are considered to be elementary. The former is called a *beam mechanism*, the latter a *sidesway* or *panel mechanism*. It is clear that

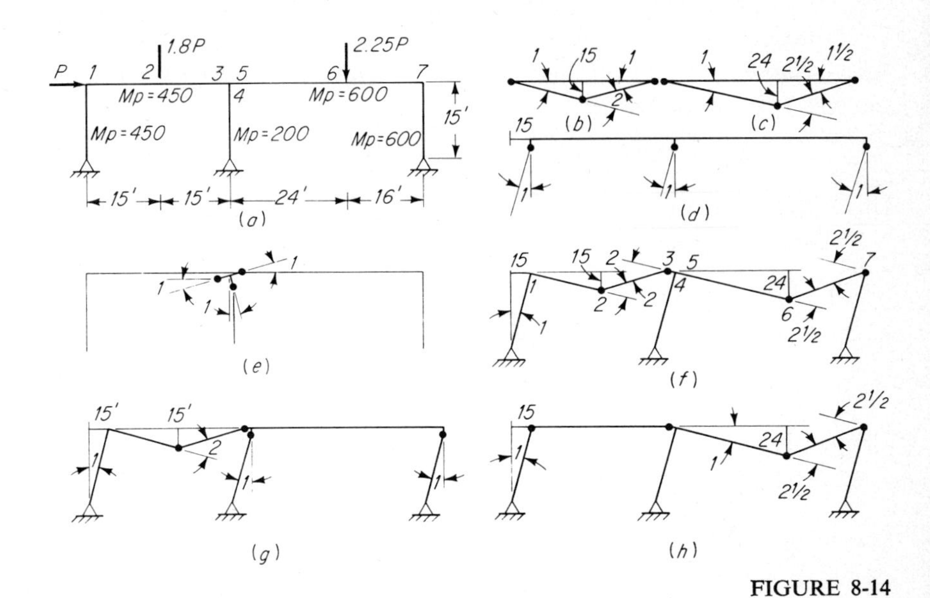

FIGURE 8-14

mechanism c can be obtained by combining these two mechanisms. Mechanisms e and f are eliminated because each is a combination of a positive mechanism with a negative one.

Since the number of elementary mechanisms is equal to the number of independent equations involving the moments at points where plastic hinges can develop, a rule for the number of independent mechanisms is established. Thus

$$N = M - R \tag{8-5}$$

where N = number of independent mechanisms
M = number of possible plastic hinges
R = number of redundants

8-9 TWO-BAY FRAME—CONCENTRATED LOADS

An example illustrating analysis by combining mechanisms is given in this article, where the collapse load P for the frame of Fig. 8-14a is determined. Plastic hinges can develop at any of the seven numbered locations in the frame. The frame has three redundants. Then, from Eq. (8-5), the number of elementary mechanisms is $N = M - R = 7 - 3 = 4$. There are only two beam mechanisms and one panel mechanism, as shown in b, c, and d, so that a third type of elementary mechanism must be invented.

This is taken as shown in Fig. 8-14*e* and is called a *joint mechanism*. However, this is not a collapse mode of the frame. Since there is no external work for a virtual rotation θ of this mechanism, we get

$$(M_3 + M_4 + M_5)\theta = 0$$

This is nothing more than a statement of equilibrium of the joint.

Using the beam mechanism in Fig. 8-14*b*, we get

$$(1 + 2 + 1)450 = 15 \times 1.8P \qquad P = 67 \text{ kips} \tag{a}$$

The beam mechanism of Fig. 8-14*c* gives

$$(1 + 2\tfrac{1}{2} + 1\tfrac{1}{2})600 = 24 \times 2.25P \qquad P = 56 \text{ kips} \tag{b}$$

The panel mechanism of Fig. 8-14*d* gives

$$1 \times 450 + 1 \times 200 + 1 \times 600 = 15P \qquad P = 83 \text{ kips} \tag{c}$$

According to the upper-bound theorem, we know that if any of these three values is correct, it is the smallest one. At this stage of the analysis, we could construct the moment diagram for $P = 56$ kips to determine whether the moment at any cross section in the frame exceeds the moment capacity of that cross section. However, it is usually advisable to investigate at least one combined mechanism.

Since we are seeking the smallest value of P, it is clear from an examination of the work equations (*a*), (*b*), and (*c*) that the elementary mechanisms must be combined in such a way as to try to decrease the internal work and/or increase the external work. This can be done by combining the work equations themselves, and for a systematic investigation of large frames this is the best procedure. However, in this example it is equally convenient to work with the combined mechanism. A combination of mechanisms *b*, *c*, and *d* eliminates the hinge at joint 1 but produces hinges at all the remaining points. However, it will be seen that the hinges at 4 and 5 can be eliminated by a combination with the joint mechanism of Fig. 8-14*e*. The result is shown in *f*. The work equation for this combined mechanism is

$$(2 + 2)450 + (2\tfrac{1}{2} + 2\tfrac{1}{2})600 = 15P + 15 \times 1.8P + 24 \times 2.25P$$

$$P = 50 \text{ kips}$$

Two other combined mechanisms are shown in Fig. 8-14*g* and *h*. However, instead of determining the collapse value for these, we shall check the value just determined by constructing the moment diagram.

Figure 8-15*a* shows free-body diagrams for segments 1-2 and 2-3 of beam 1-3. Since there are plastic hinges at 2 and 3, M_2 and M_3 are known to be 450 ft-kips. This enables the shear for segment 2-3 to be determined, which then gives the shear for segment 1-2. With this shear known, M_1 can be determined and is found to be zero.

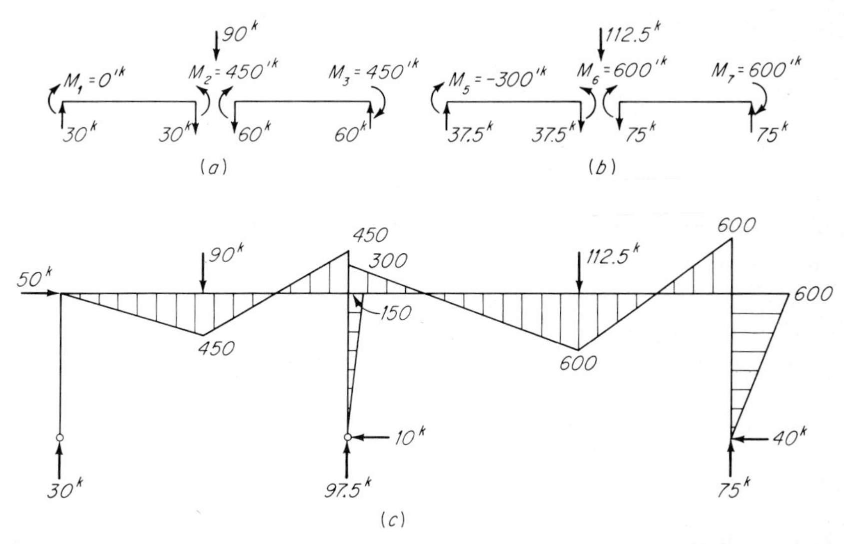

FIGURE 8-15

Similarly, for segments 5-6 and 6-7 of beam 5-7, shown in Fig. 8-15*b*, we find $M_5 = -300$ ft-kips. The resulting moment diagram and column-base reactions are shown in Fig. 8-15*c*. Since the moment is nowhere larger than the prescribed values for the members, the solution $P = 50$ kips is now known to be correct.

The unknown moments at points 1 and 5 can also be determined by using the principle of virtual work. To do this, a sign convention for moment and rotation must be adopted. We take moments and rotations positive when they produce tension at the bottom of the beam. Thus, with the virtual displacement shown in Fig. 8-16, rotations at 5 and 7 are negative while that at 6 is positive. Similarly, for the mechanism in Fig. 8-14*f* which is being investigated, M_6 is positive while M_7 is negative. Therefore, the work equation for the virtual displacement in Fig. 8-16 is

$$-1 \times M_5 + 600 \times 2.5 + (-600)(-1.5) = 112.5 \times 24$$

from which $M_5 = -300$ ft-kips.

It will be noted that the geometry of each combined mechanism in Fig. 8-14 is determined as soon as the rotation of one of the hinges is specified; i.e., each mechanism has one degree of freedom. This is an essential characteristic of a mechanism which is to be used to compute an ultimate load or a collapse moment. Since we are restricted to mechanisms with one degree of freedom, all possible combinations of the four elementary mechanisms in Fig. 8-14 are shown in *f*, *g*, and *h*. Therefore, we could have computed the value of *P* for mechanisms *g* and *h*, in which case the true ultimate

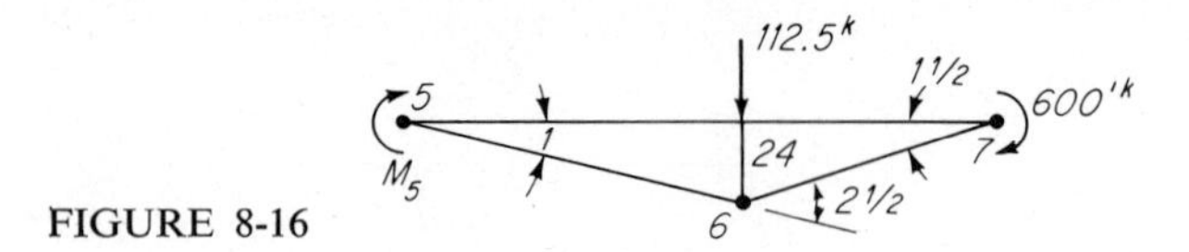

FIGURE 8-16

load would be the smallest value of P for the two beam mechanisms, the panel mechanism, and the three combined mechanisms. However, this method of determining the ultimate load is not generally satisfactory because the number of possible mechanisms becomes extremely large for complex frames and because it is always possible that some combined mechanism has been overlooked. Furthermore, it is usually necessary to know the peak moments at points which were not involved in the critical mechanism, so that the moment diagram must be determined in any case.

8-10 TWO-BAY FRAME—DISTRIBUTED LOADS

In frames which support distributed loads, the locations of the hinges under the distributed loads are not always self-evident. If the load is uniformly distributed over the entire span, it is convenient to assume the hinge to be at midspan for a first approxima-

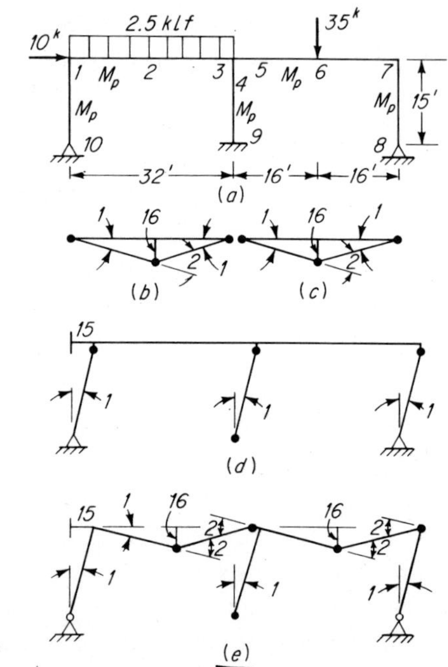

FIGURE 8-17

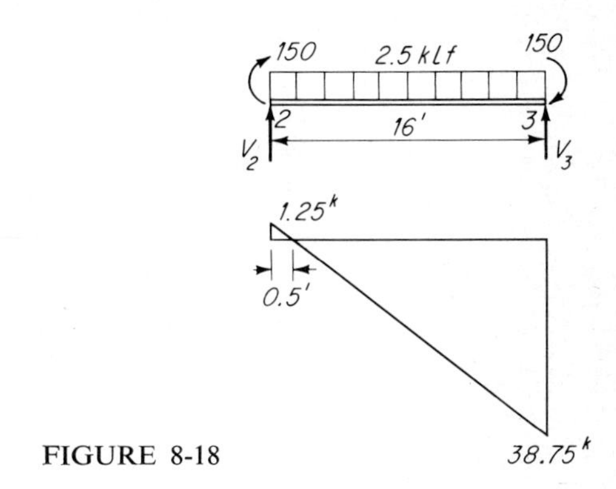

FIGURE 8-18

tion. In many cases this first approximation will be sufficiently accurate. This technique will be illustrated in the following example.

The frame shown in Fig. 8-17 has four redundants, and there are eight points at which plastic hinges can form. Therefore, from Eq. (8-5), the number of elementary mechanisms is $N = M - R = 8 - 4 = 4$. These are the beam mechanisms in *b* and *c*, the joint mechanism (not shown), and the panel mechanism in *d*. If all members of the frame have the same values of M_p, we get:

Mechanism *b*:	$(1 + 2 + 1)M_p = \frac{1}{2} \times 16 \times 32 \times 2.5$	$M_p = 160$ ft-kips
Mechanism *c*:	$(1 + 2 + 1)M_p = 16 \times 35$	$M_p = 140$ ft-kips
Mechanism *d*:	$(1 + 1 + 1 + 1)M_p = 10 \times 15$	$M_p = 37.5$ ft-kips

The combined mechanism shown in *e* gives

$$(2 + 2 + 1 + 2 + 2)M_p = \tfrac{1}{2} \times 16 \times 32 \times 2.5 + 16 \times 35 + 10 \times 15$$
$$M_p = 150 \text{ ft-kips}$$

Beam mechanism *b* gives the largest value of M_p. However, the value of M_p for the combined mechanism is an approximation based on formation of a plastic hinge at midspan of beam 1-3. To determine whether the combined mechanism might give a value larger than 160 ft-kips, we calculate the maximum moment for the approximate solution. A free-body diagram of segment 2-3 is shown in Fig. 8-18. With the known moments $M_p = 150$ ft-kips, the shear at section 2 is found to be $V_2 = 1.25$ kips. Since the load intensity is 2.5 klf, the point of zero shear is $1.25/2.5 = 0.5$ ft to the right of midspan. Therefore, the maximum moment is greater than M_2 by the amount

$$\Delta M = \tfrac{1}{2} \times 0.5 \times 1.25 = 0.3 \text{ ft-kips}$$

Thus, the second approximation to the moment for the combined mechanism is $M_p = 150.3$ ft-kips. A closer approximation to the theoretically correct value can be

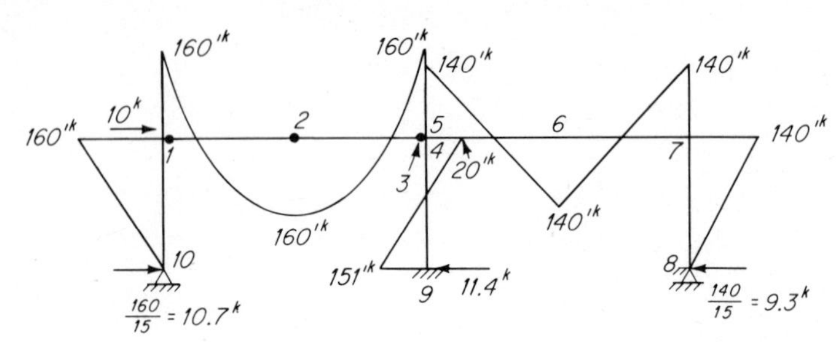

FIGURE 8-19

obtained by repeating the analysis, with the plastic hinge located at the point of maximum moment for the first approximation. However, it is clear that such a refinement is unnecessary in this example. This is usually the case for frames of practical proportions.

Instead of investigating additional mechanisms, a check is made to determine whether M is everywhere less than $M_p = 160$ ft-kips for mechanism b. M_1, M_2, M_3 are known, leaving M_4, M_5, M_6, M_7, and M_9 to be determined. Since only three equations of equilibrium remain, the structure is twice redundant when mechanism b forms. However, we need not calculate these redundants because, according to the lower-bound theorem, if we can construct a statically admissible moment diagram for which M is everywhere less than M_p, we know that the assumed collapse mechanism is the correct one.

Although values of any two of the remaining moments can be assumed, it would seem reasonable to assume M_5, M_6, and M_7 to be equal to their mechanism values, 140 ft-kips, determined above. In this way, we get the moment diagrams for the two beams, as shown in Fig. 8-19. Moments at the column tops can then be found by joint equilibrium, which leaves only M_9 undetermined. However, with M_1 and M_7 known, the horizontal reactions at 8 and 10 are found to be 9.3 and 10.7 kips, respectively. Therefore, the horizontal reaction at 9 is 11.4 kips. The required moment at 9 is then found by moment equilibrium of column 4-9. Then, since the moment diagram for the frame satisfies $M \lesseqgtr M_p$ at all points, we know that $M_p = 160$ ft-kips is correct.

Although the procedure for handling distributed loads that was used in this example is usually adequate, there are two other procedures that may be used. In one of these, the plastic hinges whose positions are unknown are located by coordinates x_1, x_2, etc. The work equation for the mechanism is then written and the correct locations of the hinges found by maximizing M_p by $\partial M_p/\partial x_1 = 0$, $\partial M_p/\partial x_2 = 0$, etc. If the structure is being analyzed to determine P, instead of M_p, the procedure is the same except that P is minimized by $\partial P/\partial x_1 = 0$, $\partial P/\partial x_2 = 0$, etc.

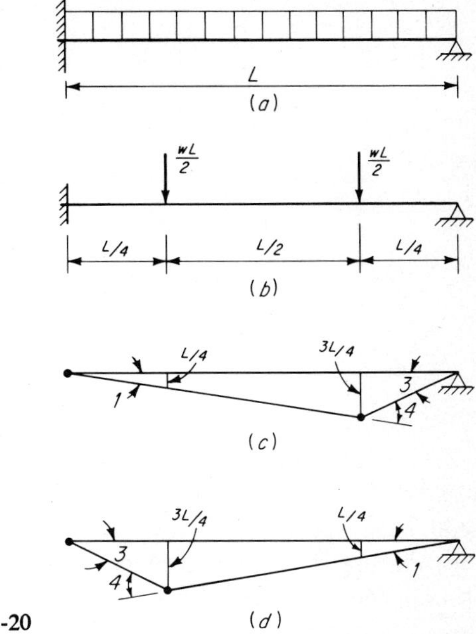

FIGURE 8-20

In the other alternative procedure the distributed load is replaced by one or more concentrated loads. The simplest replacement is a single concentrated load at mid-span, but this results in considerable error. The next simplest replacement is two equal concentrated loads, one at each quarter point. This will be illustrated for the beam of Fig. 8-20*a*, for which the load replacements are shown in Fig. 8-20*b*. The two possible mechanisms for the substitute beam are shown in Figs. 8-20*c* and *d*. The work equations for the two mechanisms are:

Mechanism *c*: $$(4+1)M_p = \frac{wL}{2}\left(\frac{3L}{4} + \frac{L}{4}\right) \qquad M_p = \frac{wL^2}{10}$$

Mechanism *d*: $$(4+3)M_p = \frac{wL}{2}\left(\frac{L}{4} + \frac{3L}{4}\right) \qquad M_p = \frac{wL^2}{14}$$

Therefore $M_p = wL^2/10$. This beam was analyzed in Art. 8-3 by finding the exact location of the interior plastic hinge, which turned out to be $0.414L$ from the simply supported end of the beam. The corresponding value of M_p is given by Eq. (8-2)

$$M_p = 0.686M_s = 0.686\frac{wL^2}{8} = \frac{wL^2}{11.7}$$

Thus, load replacement at the quarter points overestimates the required plastic moment by 17 percent.

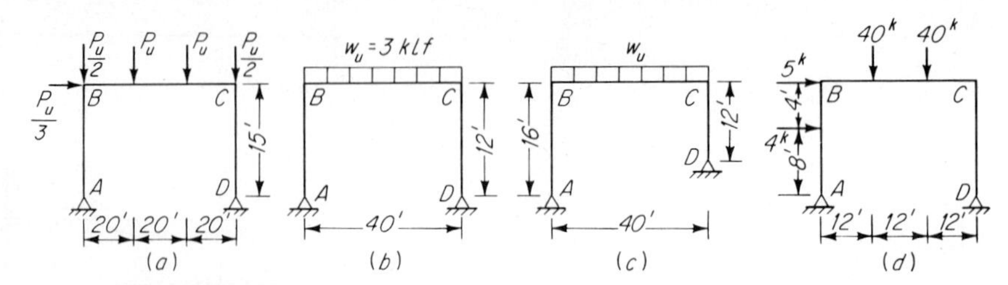

FIGURE 8-21
(*a*) Problem 8-8, (*b*) Problem 8-11, (*c*) Problem 8-13, (*d*) Problem 8-14.

Replacement of the uniform load in Fig. 8-20 by four equal concentrated loads $wL/4$ at the odd eighth points gives $M_p = wL^2/11$. This is 6 percent more than the correct value. It will be noted that a disadvantage of the load-replacement method is that each additional concentrated load increases the number of mechanisms by introducing an additional hinge. This can lead to an impractical number of mechanisms in large frames. The two procedures discussed above are covered in greater detail in Refs. 1, 2, and 5.

PROBLEMS

8-7 Compute the required plastic moment for the frame of Fig. 8-12 supporting the 30-kip vertical load and a horizontal load of 5 kips at *B*. The three members are to have equal resisting moments M_p.

8-8. Compute the required plastic moment for the frame of Fig. 8-21*a* with $P_u = 60$ kips. The three members have equal values of M_p.

8-9 Compute the value of P_u for the frame of Fig. 8-21*a* with $M_p = 240$ ft-kips for the beam and 180 ft-kips for each column.

8-10 Same as Prob. 8-8 except that the column bases *A* and *D* are fixed.

8-11 Compute the required plastic moment for the frame of Fig. 8-21*b*. The three members have equal values of M_p.

8-12 Same as Prob. 8-11 with a horizontal force $P_u = 8$ kips at *B*, in addition to the uniform load.

8-13 Compute w_u for the frame of Fig. 8-21*c*. Given $M_{pAB} = M_{pBC} = M_{pCD} = 160$ ft-kips.

8-14 Compute the required value of M_p for the frame of Fig. 8-21*d*. The loads shown are ultimate loads. Given $M_{pBC} = M_{pCD}$ and $M_{pAB} = 1.2M_{pBC}$.

8-15 Compute the value of w_u for the frame of Fig. 8-22*a*, with $M_p = 200$ ft-kips for *AB*, *BCD*, and *DE* and 50 ft-kips for *CF*.

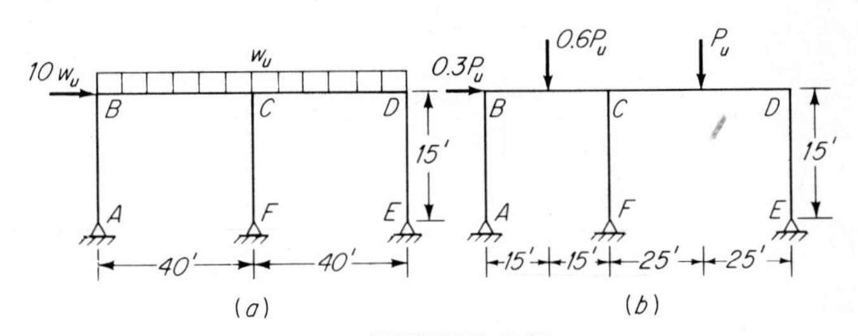

FIGURE 8-22
(*a*) Problem 8-15, (*b*) Problem 8-16.

8-16 Compute the value of P_u for the frame of Fig. 8-22*b*, with $M_{pAB} = M_{pBC} = 100$ ft-kips, $M_{pCD} = M_{pDE} = 300$ ft-kips, $M_{pCF} = 200$ ft-kips.

8-17 Same as Prob. 8-16 except that the horizontal load is $0.1P_u$.

8-11 GABLE FRAMES

One-story buildings are often built with sloping roofs, with the structure supported by nonrectangular frames called *gable frames.* Plastic analysis of gable frames is slightly more difficult than plastic analysis of rectangular frames because the mechanism displacements are not as readily determined. The concept of the instantaneous center of rotation facilitates their determination in gable frames and other frames with sloping members. This will be shown in the following example.

The frame of Fig. 8-23*a* has 1 redundant and 11 points at which plastic hinges can form. Therefore, the number of elementary mechanisms is $N = M - R = 11 - 1 = 10$. Eight of these are beam mechanisms, two of which are shown in Fig. 8-23*b*. A third elementary mechanism is the panel mechanism of Fig. 8-23*c*. Since there are no joint mechanisms (joint mechanisms need be considered only when three or more members frame in at the joint), we must invent a fourth elementary mechanism. This is usually taken in the form shown in Fig. 8-23*d*. It is essentially a variation of the panel mechanism but is sometimes called a gable mechanism.

Assuming the interior hinge to form under the third purlin, the beam mechanism of Fig. 8-23*b* gives

$$(1\tfrac{1}{2} + 2\tfrac{1}{2} + 1)M_p = 15(12 + 24 + 16 + 8) \qquad M_p = 180 \text{ ft-kips}$$

The same result is obtained if the hinge is assumed to act one panel farther to the right.

The panel mechanism of Fig. 8-23*c* gives

$$(1 + 1)M_p = 24 \times 20 \qquad M_p = 240 \text{ ft-kips}$$

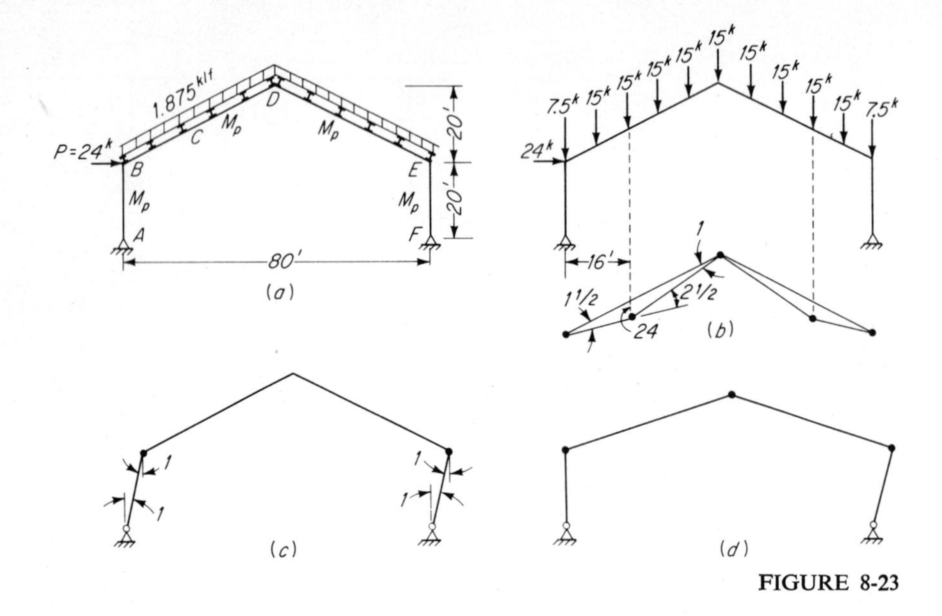

FIGURE 8-23

The displacements for the gable mechanism will be determined by using the instantaneous center. Figure 8-24 shows the gable mechanism, where column AB rotates about A to the position AB' and rafter GC rotates about G to the position GC'. End C of rafter BC also rotates about G, but end B rotates about A. Therefore, rafter BC rotates about a point found by extending GC and AB to their intersection at O. This point is called the *instantaneous center*. It can always be found as the intersection of two straight lines, each originating at a known nontranslating point (A and G) and passing through the adjacent hinge.

To determine the external work of the purlin loads, we need the vertical component of the displacement of each purlin. The displacement of any point on a rafter, such as D in Fig. 8-24, may be found in the following manner. If the angle of rotation about O is θ, $DD' = OD \times \theta$. The vertical component of this displacement is found from the similar triangles $DD'E$ and ODF. Hence

$$\frac{D'E}{DD'} = \frac{DF}{OD}$$

so that $D'E = DF \times \theta$. Therefore, the vertical displacement of any point on rafter BC is the product of the rotation about O and the horizontal distance from ABO to that point. Similarly, the vertical displacement of any point on rafter GC is the product of the rotation about G and the horizontal distance from GH to that point.

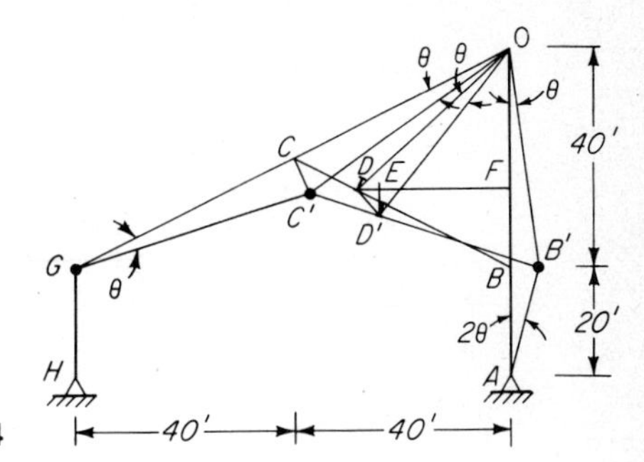

FIGURE 8-24

The external work for the frame of this example is, for rafter BC rotating about O,

$$W_{e,\,BC} = (8 + 16 + 24 + 32 + 40)15 \times \theta = 1{,}800\theta$$

and, for rafter GC rotating about G,

$$W_{e,\,GC} = (8 + 16 + 24 + 32)15 \times \theta = 1{,}200\theta$$

The hinge rotation at G in Fig. 8-24 is θ. Since CB also rotates (about O) through the angle θ, the hinge rotation at C is 2θ. Similarly, the hinge rotation at B is $\theta + 2\theta = 3\theta$. Therefore, the internal work is

$$W_i = (\theta + 2\theta + 3\theta)M_p = 6\theta M_p$$

Equating W_e and W_i, we get $M_p = 500$ ft-kips.

Although M_p is larger for the gable mechanism than for either the beam mechanism or the panel mechanism, there is a combined mechanism which, because it has only two hinges, could give a larger value. This mechanism is shown in Fig. 8-25. The hinge at H usually forms close to the ridge. In this case, we assume it to be at the first purlin to the left. After locating the instantaneous center at the intersection of GH and AB, we assume a rotation θ, about O, of the segment BDH. The corresponding rotation of AB is 3.5θ and of GCH 1.5θ. The hinge rotation at H is $1.5\theta + \theta = 2.5\theta$, and that at B is $3.5\theta + \theta = 4.5\theta$. Therefore, the internal work is

$$W_i = (2.5 + 4.5)\theta \times M_p = 7\theta M_p \qquad (a)$$

The external work is found as before by determining the vertical displacements of the purlins. Thus, for the segment BD

$$W_{e,\,BD} = (8 + 16 + 24 + 32 + 40)\theta \times 15 = 1{,}800\theta \qquad (b)$$

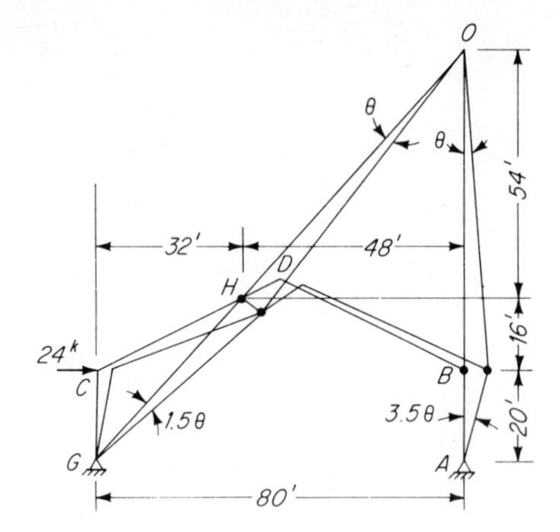

FIGURE 8-25

For the segment CH

$$W_{e,CH} = (8 + 16 + 24 + 32)1.5\theta \times 15 = 1{,}800\theta \tag{c}$$

Finally, the external work for the wind load at C is

$$W_{e,C} = 20 \times 1.5\theta \times 24 = 720\theta \tag{d}$$

Equating the internal work from Eq. (*a*) to the sum of the external work from Eqs. (*b*), (*c*), and (*d*), we get

$$7\theta M_p = 4{,}320\theta \qquad M_p = 617 \text{ ft-kips}$$

Since this value is the largest so far, we check to see if it satisfies the criterion $M \lesssim M_p$. With the moment at the two plastic hinges known, the frame is statically determinate and the reactions are readily computed (Fig. 8-26). A check of the moments at the purlins at J and D shows both to be less than M_p. Finally, the moment diagram shows $M \lesssim 617$ for all points, so that $M_p = 617$ ft-kips is the required plastic moment of resistance.

A detailed example of plastic design of a gable frame is given in Chap. 11, DP11-1.

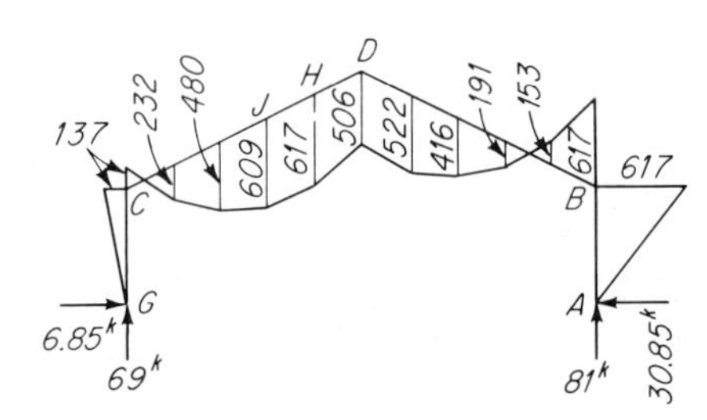

FIGURE 8-26

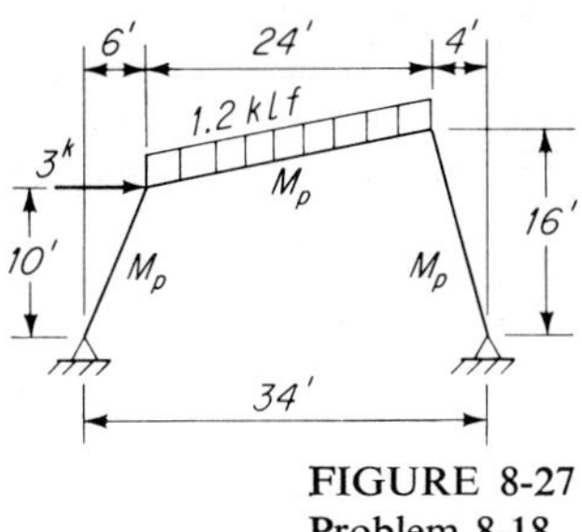

FIGURE 8-27
Problem 8-18.

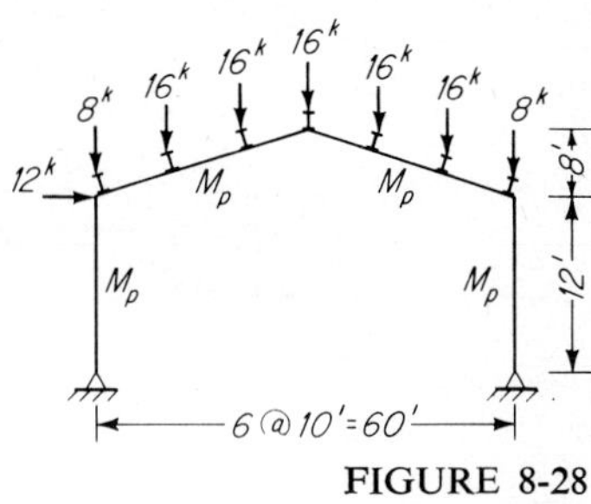

FIGURE 8-28
Problem 8-19.

PROBLEMS

8-18 The loads shown on the frame in Fig. 8-27 are service loads. Determine the required value of M_p for load factors of 1.7 on gravity load alone and 1.3 on gravity load combined with wind load. The wind can act from either direction.

8-19 The loads shown in the frame of Fig. 8-28 are service loads. Determine the required value of M_p for load factors of 1.5 on gravity load and 1.2 on gravity load combined with wind load.

8-20 Determine the value of P_u for the frame of Fig. 8-29. All the members are the same size, with $M_p = 180$ ft-kips.

8-12 LOCAL AND LATERAL-TORSIONAL BUCKLING

The manner in which collapse mechanisms develop was discussed in Arts. 5-13, 8-2, and 8-3. It was shown that the plastic hinges required to produce a mechanism do not, in general, form simultaneously and that rotation at the hinges which form in the early stages is assumed to continue at an essentially constant moment until the last hinge is formed. *OABC* in Fig. 8-30 shows this necessary performance, which is usually simplified in the form *OADC*. However, the last hinge to form need only reach M_p, for which the rotation ϕ_p is sufficient.

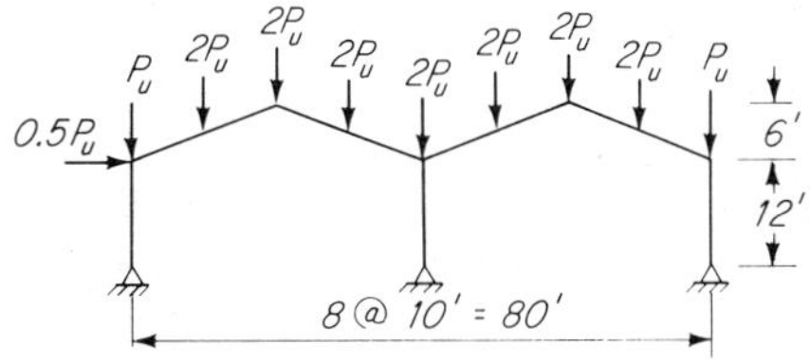

FIGURE 8-29
Problem 8-20.

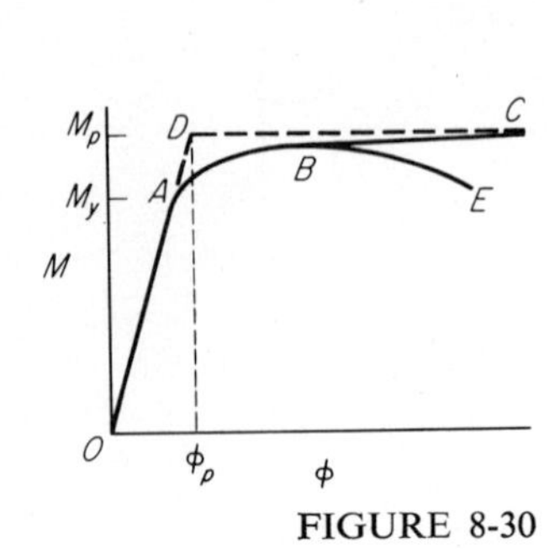

FIGURE 8-30

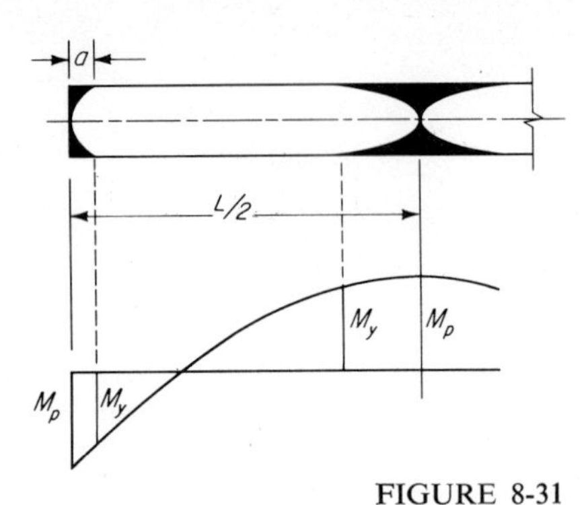

FIGURE 8-31

Hinges do not develop at a point, as is assumed in the mechanism analysis. Instead, they are distributed over finite lengths of the beam. Thus, for a fixed-end beam, the strain on the extreme fiber in the vicinity of the support varies from ε_y at the cross section where $M = M_y$ to $\varepsilon \gg \varepsilon_y$ at the end, where $M = M_p$. For a uniformly loaded fixed-end beam with a shape factor $M_p/M_y = 1.12$, this transition occurs in the length $a = 0.015L$ (Fig. 8-31).

Rotation capacity has been investigated both theoretically and experimentally. For example, it can be shown that the average rotation in the yielded length a of Fig. 8-31 is about 11 times the rotation ϕ_p in Fig. 8-30. However, rotations of this magnitude can be developed in beams of I cross section only if adequate precaution is taken against local buckling and against lateral-torsional buckling of the member itself. Without such precaution, a reduction in moment, as in *BE* of Fig. 8-30, may occur.

An investigation of inelastic buckling of compression flanges[8] is discussed in Art. 5-3. Results of this investigation indicate that flanges supported on both unloaded edges, like the flanges of box sections, can reach strain hardening if $b/t \lessapprox 192/\sqrt{F_y}$ [Eq. (5-3)]. In the AISC specification, this limit is set at $190/\sqrt{F_y}$. The same limit is also specified for elastic design of compact sections (Art. 5-5). The investigation also indicates that compression flanges supported on only one unloaded edge, as in the I cross section, can reach strain hardening if $b/t \lessapprox 49/\sqrt{F_y}$ [Eq. (5-2)]. In the AISC specification, this limit is set at $52.2/\sqrt{F_y}$ for elastic design of compact sections. However, a specific limit for each yield stress is set for plastic design. Thus, $b/t = 8.5$ for $F_y = 36$ ksi, 7 for $F_y = 50$ ksi, and 6 for $F_y = 65$ ksi. The latter is the highest yield stress of the steels which are presently recommended for plastic design.

An investigation of inelastic buckling of beam-column webs in Ref. 8 is discussed in Art. 5-17. Provisions of the AISC specification are based on this investigation.

Thus, for plastic design, slenderness of beam-column webs must not exceed the values given by

$$\frac{d}{t} = \begin{cases} \dfrac{412}{\sqrt{F_y}}\left(1 - 1.4\dfrac{P}{P_y}\right) & \dfrac{P}{P_y} \lesssim 0.27 \qquad (8\text{-}6a) \\[2ex] \dfrac{257}{\sqrt{F_y}} & \dfrac{P}{P_y} \gtrsim 0.27 \qquad (8\text{-}6b) \end{cases}$$

These are the same formulas as Eqs. (5-35) for elastic design of compact sections, except that they are given in terms of the load P instead of the stress F_a.

Lateral-torsional buckling was discussed in Arts. 5-7 and 5-8, and specification provisions for elastic design in Art. 5-11. However, the objective was to determine the reduction in in-plane bending strength as a function of the unbraced length, and the limiting unbraced length which will allow the beam to reach its yield moment (point A of Fig. 8-30). It was shown that lateral-buckling resistance depends on the moment gradient, which is accounted for by the coefficient C_b in the lateral-buckling formulas.

There have been a number of investigations of postelastic lateral-buckling behavior to determine rotation capacity. In general, it is assumed that a beam under moment gradient has elastic properties in the portions where moments are less than about $0.90M_p$ to $0.95M_p$ and strain-hardening properties in the remainder. This assumption is in recognition of the discontinuous nature of yielding, which was discussed in Art. 2-4. Provisions of the AISC specification are based on work reported in Ref. 9. St. Venant torsional resistance is negelcted in this investigation, and lateral-torsional buckling of the beam is assumed to be equivalent to lateral buckling of the compression half of the I cross section as a column. Since the moment is M_p, this compression half of the beam is subjected to a uniform compressive stress F_y. Furthermore, with St. Venant torsion neglected, the results are expressed in terms of L/r_y, as in the simplified formulas in Art. 5-10 [Eqs. (5-15)]. According to the AISC specification, the limiting values of L/r_y which assure the rotation capacity of a member which must develop a plastic hinge are given by

$$\frac{L}{r_y} = \begin{cases} \dfrac{1{,}375}{F_y} + 25 & 1 \geq \dfrac{M}{M_p} > -0.5 \qquad (8\text{-}7a) \\[2ex] \dfrac{1{,}375}{F_y} & -0.5 \geq \dfrac{M}{M_p} \geq -1 \qquad (8\text{-}7b) \end{cases}$$

In these formulas, M is the moment at the end opposite the plastic hinge, and M/M_p is taken positive for a member bent in reversed curvature. The formulas are plotted in Fig. 8-32. It will be noted that moment gradient is not taken into account by a coefficient such as C_b. Instead, a constant value of L/r_y is specified for each of two

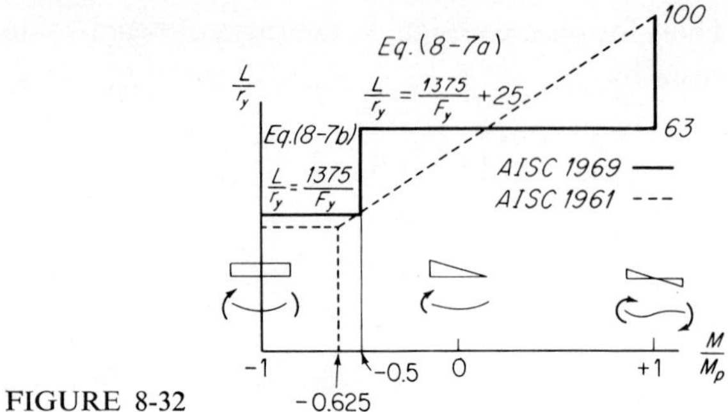

FIGURE 8-32

ranges of moment gradient, which results in a discontinuity at $M/M_p = -0.5$. The corresponding formulas in the 1961 edition of the specification did not give this discontinuity. A comparison of the provisions of the two versions is shown for A36 steel in the figure.

The L/r_y limits in Fig. 8-32 need not be applied in the region of the last hinge to form in a mechanism, since rotation capacity is not required here. Instead, this segment of the frame can be analyzed as an elastic member. However, since the segment is required to develop a plastic moment, the distance from the last plastic hinge to the adjacent lateral supports must satisfy the requirements for a compact section [Eqs. (5.19)]. On the other hand, the distance between lateral supports that bound a segment in which there is no plastic hinge may be checked instead for compliance with lateral-buckling requirements for noncompact sections [Eqs. (5-16) and (5-17)]. In the latter case, of course, the analysis must be based on service loads rather than ultimate loads. Studies of rectangular frames and gable frames show that the last hinge develops in the beam or rafter except for frames with unusually large ratios of column height to frame span (greater than about 0.6).[5,7] Therefore, the less severe lateral-buckling requirements will usually govern the spacing of lateral support at the hinges in beams and rafters except where they join the columns. Examples are given in the next article and in Chap. 11, DP11-1.

8-13 BEAM-COLUMNS

Beam-column strength was discussed in Arts. 4-12 to 4-14, where it was shown that values of axial load P and bending moment M which can be resisted simultaneously are given by Eqs. (4-36) and (4-38). These formulas were derived for the case of bending restricted to the plane of the end moments. However, it was shown in Arts 5-11 and 5-28 that they can be extended to the case where lateral-torsional buckling is not prevented by using the critical moment M_m, rather than M_p, as the limiting value for bending alone in Eq. (4-36). The result is

$$\frac{P}{P_{cr}} + \frac{M}{M_m} \frac{C_m}{1 - P/P_E} \lesssim 1 \tag{8-8}$$

$$\frac{P}{P_y} + 0.85 \frac{M}{M_p} \lesssim 1 \qquad \text{if } \frac{P}{P_y} \gtrsim 0.15 \tag{8-9a}$$

$$M = M_p \qquad \text{if } \frac{P}{P_y} \lesssim 0.15 \tag{8-9b}$$

P_{cr} in these equations is defined as the smaller of the critical loads for in-plane buckling and out-of-plane buckling. Lateral-torsional buckling due to bending moment was discussed in Arts. 5-7 and 5-8. Simplified formulas for elastic design, such as the AISC formulas [Eqs. (5-16) and (5-17)], were discussed in Art. 5-10. The corresponding AISC formula for plastic design is

$$M_m = \left(1.07 - \frac{\sqrt{F_y}}{3{,}160} \frac{L}{r_y}\right) M_p \lesssim M_p \tag{8-10}$$

This formula is based on the results of tests on W shapes of the proportions ordinarily used as columns.

It should be noted that the limiting value of L/r_y which was discussed in Art. 8-12 and which is shown in Fig. 8-32*a* enables a beam-column to develop the fully plastic moment and adequate rotation capacity, provided $P = 0$. Therefore, where the spacing of lateral support satisfies this limitation, $M_m = M_p$.

EXAMPLE 8-3 The W16 × 31 columns of the frame of Example 8-2 are checked in this example for compliance with the AISC specification. The frame is shown in Fig. 8-33*a* and the moment diagram at collapse in Fig. 8-33*b*.

Both flange and web must be checked for local buckling. The permissible flange slenderness is $b/t = 8.5$, while the actual value is $2.76/0.442 = 6.25$. The permissible web slenderness is given by Eqs. (8-6):

$$P_y = 36 \times 9.13 = 329 \text{ kips} \qquad \frac{P}{P_y} = \frac{17}{329} = 0.05 < 0.27$$

Therefore, Eq. (8-6*a*) applies, and the allowable slenderness is

$$\frac{d}{t} = \frac{412}{\sqrt{36}}(1 - 1.4 \times 0.05) = 64$$

For the W16 × 31, $d/t = 15.84/0.275 = 57.6$.

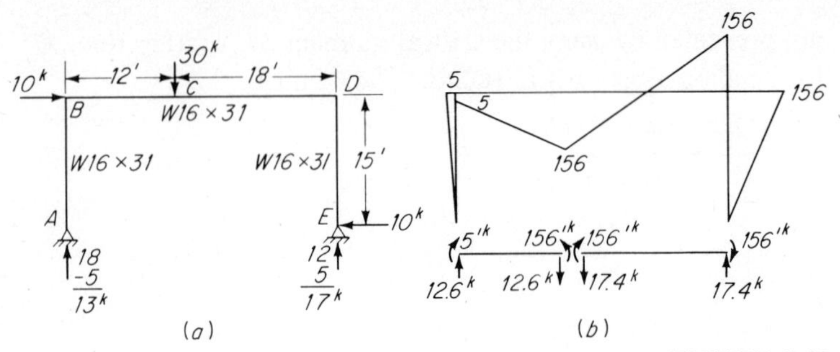

FIGURE 8-33

The lateral-bracing requirements will now be checked. Figure 8-33*b* shows that column *DE* controls. Since $M/M_p = 0$, Eq. (8-7*a*) applies and the permissible slenderness ratio is

$$\frac{L}{r_y} = \frac{1{,}375}{36} + 25 = 38 + 25 = 63$$

The actual slenderness ratio is $15 \times 12/1.17 = 154$. Therefore, intermediate support is required. In order for the limiting value 63 to apply, the moment at the intermediate braced point must satisfy $M > -0.5M_p$ (Fig. 8-32). This would require the braced point to be $L/2 = 7.5$ ft (or farther) from the column top. The slenderness ratio for a brace at this point is $7.5 \times 12/1.17 = 77$, which also exceeds the permissible value 63. Therefore, closer spacing is needed, for which Eq. (8-7*b*) applies, and the permissible slenderness is

$$\frac{L}{r_y} = \frac{1{,}375}{36} = 38$$

The distance from the plastic hinge to this point is $L = 38 \times 1.17/12 = 3.7$ ft.

Assuming a brace at 3.5 ft, the remaining 11.5-ft segment must be checked to determine whether additional lateral support is needed. Since there is no plastic hinge in this segment, formulas based on beginning of yield apply, as was noted in Art. 8-12. Therefore, the loads shown in Fig. 8-33*a* must be reduced to service values for this investigation. Assume the 10-kip load to be a wind load. The AISC load factor for a combination of wind and gravity load is 1.3, so that the column axial force and moment and the corresponding stresses are

$$P = \frac{17}{1.3} = 13.1 \text{ kips} \qquad M = \frac{11.5}{15} \times \frac{156}{1.3} = 92 \text{ ft-kips}$$

$$f_a = \frac{13.1}{7.27} = 1.8 \text{ ksi} \qquad f_b = \frac{92 \times 12}{47.2} = 23.5 \text{ ksi}$$

According to the AISC specification, allowable stresses may be increased by one-third for combinations of wind and gravity load. The value of F_a must be checked for both x-axis and y-axis buckling:

$$G_D = \frac{(I/L)_C}{(I/L)_B} = \frac{30}{15} = 2 \qquad G_E = 10 \qquad K = 2.1 \qquad \text{(Fig. 4-39}b\text{)}$$

$$\frac{KL}{r_x} = \frac{2.1 \times 15 \times 12}{6.40} = 59 \qquad \frac{L}{r_y} = \frac{11.5 \times 12}{1.17} = 118$$

$$F_a = 10.6 \times 1.33 = 14.1 \text{ ksi}$$

In determining the allowable bending stress, C_b must be taken equal to unity because C_m in the interaction formula provides for the moment gradient.

$$\frac{L}{r_T} = \frac{11.5 \times 12}{1.41} = 98 > \frac{L}{r} = \sqrt{\frac{102{,}000}{36}} = 53$$

Therefore, Eq. (5-16b), which is AISC formula 1.5-6a, applies. For A36 steel, this gives

$$F_b = 1.33\left[24 - \frac{(L/r_T)^2}{1{,}181}\right] = 1.33\left(24 - \frac{98^2}{1{,}181}\right) = 21.2 \text{ ksi}$$

According to Eq. (5-17), which is AISC formula 1.5-7,

$$\frac{Ld}{A_f} = 11.5 \times 12 \times 6.49 = 895$$

$$F_b = 1.33\,\frac{12{,}000}{Ld/A_f} = 1.33\,\frac{12{,}000}{895} = 17.8 \text{ ksi}$$

The larger of these two values of F_b is the correct one. However, it is less than f_b, so that an additional brace must be provided at, say, 5.75 ft from the base of the column. Had F_b been larger than f_b, the values of f_a, F_a, f_b, and F_b would have been checked for compliance with the interaction formulas Eqs. (4-39), which are AISC formulas 1.6-1.

It remains to check the columns according to Eqs. (8-8) and (8-9). Since $P/P_y = 0.06$, as determined above, $M = M_p$ [Eq.(8-9b)]. According to the AISC specification, P_{cr} and P_E in Eq. (8-8) are determined by multiplying the allowable loads according to the formulas for elastic design by the factors of safety, which are 1.7 for P_{cr} and 1.92 for P_E. The slenderness ratio for in-plane buckling was determined above, $KL/r_x = 59$. With lateral support at 5.75 ft and 11.5 ft above the base, $L/r_y = 5.75 \times 12/1.17 = 59$. Therefore, $F_a = 17.5$ ksi, $F'_E = 42.9$ ksi, and

$$P_{cr} = 1.7F_aA = 1.7 \times 17.5 \times 7.67 = 229 \text{ kips}$$

$$P_E = 1.92 \times 42.9 \times 7.67 = 633 \text{ kips}$$

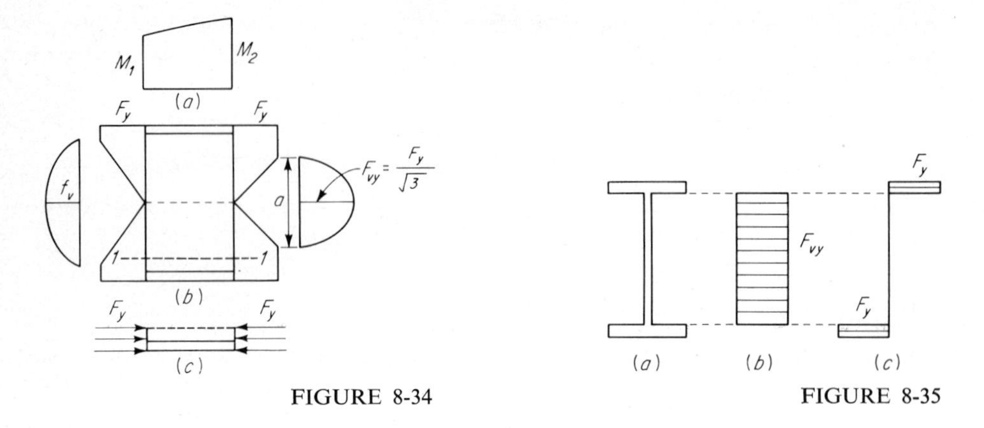

FIGURE 8-34

FIGURE 8-35

The specification prescribes $C_m = 0.85$ for columns in frames not braced against sidesway. Furthermore, since the column is braced in the weak direction, $M_m = M_p = 54 \times 36/12 = 162$ ft-kips. Then from Eq. (8-8)

$$\frac{17}{229} + \frac{156}{162} \frac{0.85}{1 - 17/633} = 0.074 + 0.840 = 0.914 < 1$$

8-14 SHEAR RESISTANCE

The shear resistance of webs of beams for which the limiting load is based on first yield was discussed in Arts. 5-15 and 5-16. The shear strength of beams at cross sections where there has been some penetration of yield stress is investigated in this article.

A portion of the bending-moment diagram for a beam and the corresponding bending stresses are shown in Fig. 8-34*a* and *b*. Consider the segment below a horizontal section 1-1 which lies entirely within the yielded zone (Fig. 8-34*b*). Since the horizontal forces acting on this segment are in equilibrium (Fig. 8-34*c*), there can be no shear on section 1-1. Therefore, neither can there be shear on the vertical faces of this segment. Thus, the shear resistance of the beam must be developed entirely in the portion of the web which is still elastic, i.e., in the portion of depth a shown in Fig. 8-34*b*. According to this analysis, no shear resistance can exist at a cross section where there is a plastic hinge.

Consider next a cross section at which the web has yielded uniformly in shear throughout its depth (Fig. 8-35*b*). In this case, the yield criterion states that there can be no bending stresses in the web, so that only the flanges can resist moment. The shear resistance is given by

$$V_u = F_{vy}\, t(d - 2t_f) \tag{a}$$

where t = thickness of web
t_f = thickness of one flange
d = depth of beam

For rolled I cross sections which are used as beams, $d - 2t_f$ averages about $0.95d$. Furthermore, $F_{vy} = F_y/\sqrt{3}$ by the Mises yield criterion. Therefore, Eq. (*a*) can be written

$$V_u = 0.95\, dt \frac{F_y}{\sqrt{3}} = 0.55 F_y\, dt \tag{8-11}$$

It would appear from this analysis that a fully plastic moment cannot develop at a cross section where $V = V_u$ since, according to the yield criterion, only the flanges can develop moment if the web is fully yielded in shear (Fig. 8-36*c*). However, tests show that this is not the case. Instead, M_p is reduced only slightly, if at all, so long as V does not exceed the value from Eq. (8-11). This is due to the effects of strain hardening. Therefore, Eq. (8-11) is prescribed in the AISC specification for plastic design.

8-15 DEFLECTIONS

As a rule, it is the service-load deflections of a structure which are of concern. However, plastic design is based on loads at collapse, so that there is a question how service-load deflections may be determined, other than by an elastic analysis at service-load intensity. Of course, this is no problem if computer facilities are available. However, sometimes it may be expedient to determine collapse-load deflections, from which a conservative estimate of service-load deflection can always be made. This will be shown for the uniformly loaded fixed-end beam, for which the load-deflection plot was determined in Art. 5-13 (Fig. 5-25). If only the deflection at ultimate load were known, the service-load deflection for this beam could be estimated by assuming linear load-deflection behavior from O to B. Then, if the load factor were 1.7, the service-load deflection would be $0.82/1.7 = 0.48$ in. However, with this load factor the service load is $153/1.7 = 90$ kips, and the correct service-load deflection is given by a point on OA in Fig. 5-25, that is, $\Delta = 0.31 \times 90/115 = 0.24$ in. Therefore, the approximate procedure overestimates the deflection by 100 percent in this case. While this is indeed a large error, the estimated value may still be useful, since if it is in itself acceptable, no further calculations are needed.

Although the service-load deflection can be computed closely by determining the load-deflection history, as in Fig. 5-25, this procedure is limited in application because a deflection computation is required at the formation of each additional plastic hinge until the mechanism is formed. However, collapse-load deflections can be determined without a hinge-by-hinge analysis if the location of the last hinge to form is known. The slope-deflection equations can be used for this analysis,[5] although other methods, such as the virtual-displacement method,[2] can also be used. Elastic continuity of the frame at the last hinge to form is essential to the solution, since it is the basis for one of the necessary equations. If the location of the last hinge to form is not known, each hinge of the mechanism can be assumed to be last to form, and the largest deflection so calculated is the true deflection.

As pointed out in Art. 5-22, deflection of beams in floors can be easily evaluated in terms of the ratio L/d of span to depth. Therefore, in most practical applications, deflection of floors can be approximated closely enough without the deflection analyses discussed above. An upper bound for the deflection of a beam in a frame can be obtained by assuming the beam to be simply supported, in which case Δ/L is readily determined.

8-16 MOMENT BALANCING

A scheme for plastic analysis of structures, called *moment balancing*, is described in this article. It is based on the two bound theorems discussed in Art. 8-6.

If one constructs for a frame any moment diagram that satisfies requirements for equilibrium and then proportions the frame to this moment diagram so that there are enough plastic hinges to form a mechanism, and so that M nowhere exceeds M_p, the lower-bound and upper-bound theorems are both satisfied. The lower-bound theorem is satisfied because the assumed moment diagram is in equilibrium with the loads and the moment nowhere exceeds M_p, and it follows that the corresponding load is less than, or at best equal to, the ultimate load the structure can support. The upper-bound theorem is satisfied because there are enough plastic hinges to form a mechanism, and it follows that the corresponding load is also greater than, or at best equal to, the ultimate load the structure can support. Thus, such a structure will just collapse at the load for which it is proportioned.

Figure 8-36 shows simple adjustments which can be made in a trial moment diagram for the purpose of developing a moment diagram which yields a collapse mechanism. Interior moments are taken positive if they produce tension on the bottom fiber; end moments are positive when they are clockwise on the member. Each of these distributions can be thought of as a shift of an initial base line. In Fig. 8-36*a* base *a-a* of the beam-mechanism moment diagram for a concentrated load is shifted

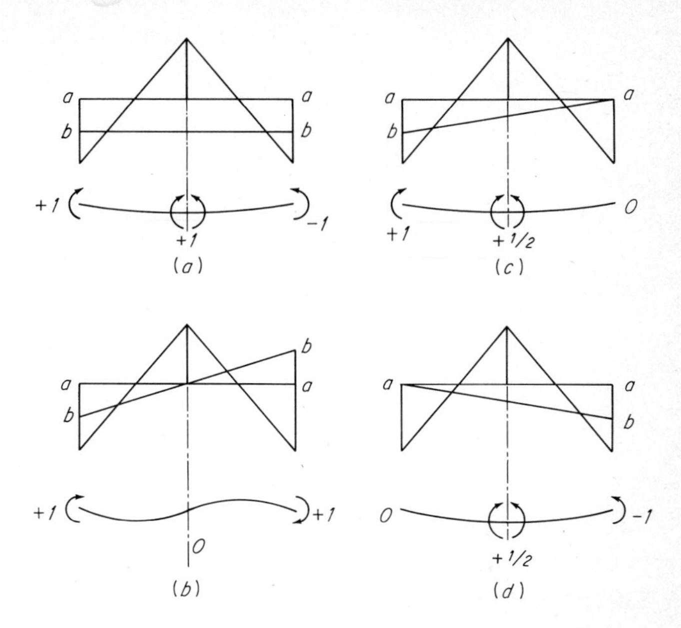

FIGURE 8-36

to *b-b* by adding uniform moment, i.e., moments in the ratio +1, +1, −1. Figure 8-36*b* shows the base line *a-a* rotated about its midpoint to *b-b*. In this case, the distribution of moment that is added to the initial distribution is in the ratio +1, 0, +1. The base may also be rotated about either end, as in Fig. 8-36*c* and *d*.

A simple example is shown in Fig. 8-37*a*.[10] The frame is to be designed in A36 steel for a gravity load of 1.3 klf and a wind load of 0.6 klf. The beam-mechanism moment for gravity load alone at a load factor of 1.7 is

$$M_p = \frac{1.7 \times 1.3 \times 96^2}{16} = 1{,}275 \text{ ft-kips}$$

The required section modulus is $Z = 1{,}275 \times 12/36 = 425$ in.^{3} The lightest W shape is the W33 × 130, for which $Z = 467$ in.^{3} and $M_p = 1{,}401$ ft-kips. However, if this shape is used for both beam and columns, the frame will not form a mechanism at the specified ultimate load. Instead, it will support a service load of $1.30 \times 1{,}401/1{,}275 = 1.43$ klf at a load factor of 1.7.

To see whether a lighter frame can be designed, we revise the assumed distribution of moments, −1,275, +1,275, +1,275, which is shown in Fig. 8-37*b*, so as to produce a plastic hinge at midspan. This is accomplished by superimposing the set of moments +126, +126, −126, which corresponds to Fig. 8-36*a*, on the initial values. The resulting moments at the ends of the beam are −1,149 and +1,149, which require

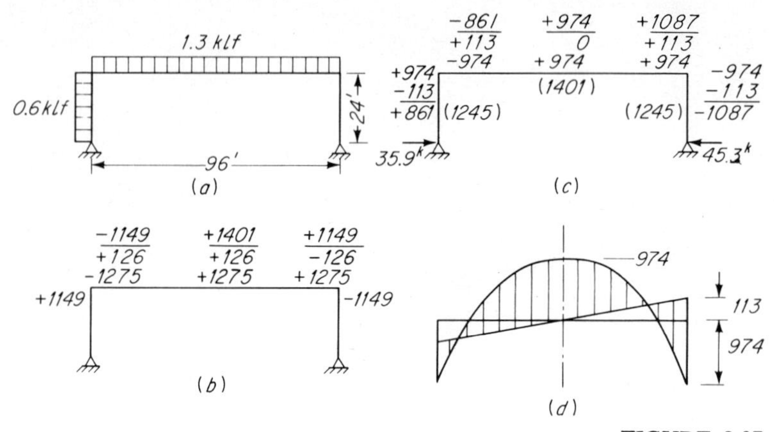

FIGURE 8-37

moments at the column tops of +1,149 and −1,149, as shown on the figure. Therefore, a mechanism will form if the columns can be sized to a plastic moment of 1,149 ft-kips. However, the lightest column which furnishes a plastic moment of at least 1,149 ft-kips is the W33 × 118, for which $M_p = 1{,}245$ ft-kips. Although no mechanism will form, the resulting frame is the lightest that can be built in A36 steel with standard rolled shapes and is 500 lb lighter than the frame with W33 × 130 members for both beam and columns.

The proposed frame is checked in Fig. 8-37*c* for adequacy against wind and gravity load. The load factor for this combination is 1.3. The gravity-load moments and sway moment due to wind are:

$$\text{Gravity:} \qquad 1.3 \times 1.3 \times \frac{96^2}{16} = 974 \text{ ft-kips}$$

$$\text{Wind:} \qquad 1.3 \times 0.6 \times 12 \times 24 = 226 \text{ ft-kips}$$

The analysis begins with the gravity-load moments, as shown in the figure. Since gravity and wind load together tend to produce a combination of beam and panel mechanism, the frame will tend to develop two plastic hinges, one near midspan of the beam and the other at the top of the leeward column. The wind moments will be counterclockwise on the column tops. If these moments are assumed to be equal, equilibrium is achieved with the beam-moment distribution shown in Fig. 8-36*b*. Therefore, the moments +113, 0, +113 are added to the beam-mechanism moments, with moments of −113 at the column tops for equilibrium, as shown in Fig. 8-37*c*. The resulting horizontal reactions on the frame are shown in the figure. Although it is not needed in the solution, the starting moment and the subsequent adjustment are

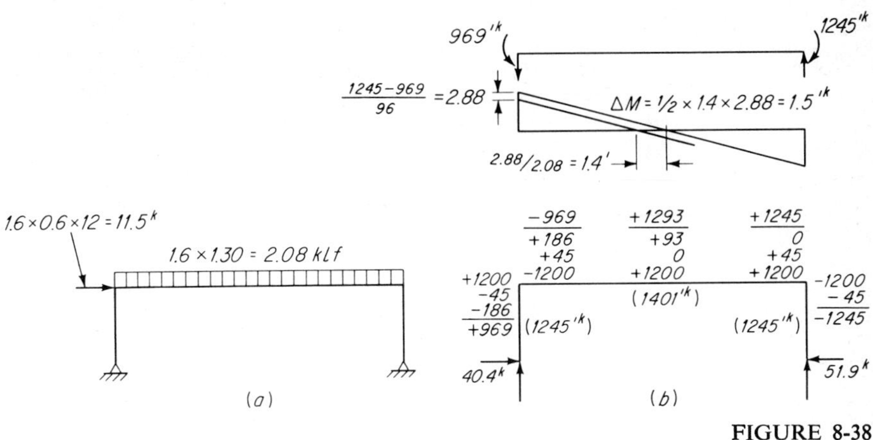

FIGURE 8-38

shown in Fig. 8-37*d*. Since the moments nowhere exceed the plastic-moment capacity of the members, the frame is adequate, provided local buckling, lateral-torsional buckling, and beam-column requirements are satisfied. Neither shape need be checked for local buckling of the flange, because the AISC Manual's plastic-design selection table excludes shapes for which the flange b/t exceeds 8.5. The Manual denotes with an asterisk those shapes for which local buckling of the web must be checked [by Eqs. (8-6)] if there is axial load in addition to bending. The W33 × 118 column does require this check. It must also be checked against Eqs. (8-7) to (8-9).

It should be noted that the final moments in Fig. 8-37*c* are not those which would exist under the assumed loading. In fact, one could find any number of moment distributions which would be in equilibrium with these loads and for which the moment would nowhere exceed M_p. However, we are interested only in determining whether the frame is adequate for the specified loads, and the precise moment distribution is not of interest.

If the required load factor for wind and gravity load were 1.6, the gravity-load beam moments for the frame of Fig. 8-37*a* would be 1,200 ft-kips, as shown in Fig. 8-38. The required wind-load moment is 276 ft-kips. However, only 45 ft-kips of this moment can be developed at the leeward column, since $M_p = 1{,}245$ ft-kips for that member. [If P/P_y exceeds 0.15, this column-top moment must be reduced to satisfy Eqs. (8-8) and (8-9).] Adding the distribution +45, 0, +45 to the starting moments gives 90 ft-kips of wind moment, which is 186 ft-kips short of the 276 ft-kips needed. This moment must be carried by the windward column. The joint-balancing beam moment is +186 ft-kips, which produces the set of moments +186, +93, 0 in the beam. The final moment distribution is one in which the moment is nowhere larger than M_p, except possibly for the beam, for which the maximum moment occurs to

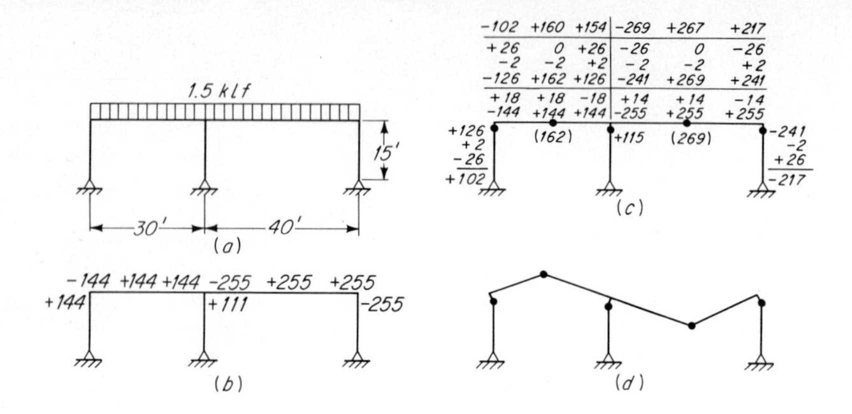

FIGURE 8-39

the left of midspan. The location of this point is easily found, as shown in the shear diagram, and the maximum moment determined by computing the increment $\Delta M =$ 1.5 ft-kips. Thus, the maximum moment is only 1,295 ft-kips. Therefore, $M \lessapprox M_p$ at all points of the frame. Furthermore, there is only one plastic hinge, instead of the two required for a mechanism. Therefore, the frame is more than adequate for these loads at a load factor of 1.6.

8-17 TWO-BAY FRAME BY MOMENT BALANCING

The two-bay frame in Fig. 8-39*a* is to be designed in A36 steel for a roof load of 1.5 klf at a load factor of 1.7. The initial beam moments, assuming beam mechanisms in each span, and the balancing column moments are shown in Fig. 8-39*b*.[10] The required section moduli are given in Table 8-1, for which the sections shown are chosen. The columns are restricted to not more than 12 in. wide in the plane of the frame. All the members satisfy local-buckling requirements. Each column is checked against the beam-column interaction formulas, as in Example 8-3, to determine the reduction in M_p, if any, because of the axial force P. The reduced moments are shown in the table.

Figure 8-39*c* shows how the initial distribution of moments of Fig. 8-39*b* can be improved to take advantage of the excess capacities of the members shown in Table 8-1. A midspan hinge is developed in each beam by adding the distributions +18, +18, −18, and +14, +14, −14. These do not upset the sway balance. Were it possible to choose columns with precisely the required moments, the resulting frame would develop plastic hinges at the points indicated in Fig. 8-39*c*. Although these are enough hinges to produce a mechanism, the possible displacements, one of which is

TABLE 8-1

Member	M	Z required	Section	M_p	Reduced M	Revised M_p	Revised Section
1	144	48	W16 × 31	162			
2	255	85	W18 × 45	269			
3	144	48	W12 × 36	155	153	102	W12 × 27
4	111	37	W12 × 40	173	138	115	
5	255	85	W12 × 58	260	260	217	W12 × 50

shown in Fig. 8-39*d*, involve rotations at certain hinges which are opposed in sense to the moments at these points. Therefore, this system of hinges does not produce a collapse mechanism, and it may be possible to reduce the column moments still further.

The collapse configuration in Fig. 8-39*d* suggests that a hinge is needed in at least one of the beams at the interior column. This is easily accomplished by using the distribution of Fig. 8-36*b*. However, the off-center location of the interior hinge of each beam must be anticipated. Therefore, the adjustment -2, -2, $+2$ is made in each span. This does not alter the sway balance. Furthermore, the +, 0, + distribution which is added to one beam must be accompanied by a −, 0, − distribution, with the same absolute values, in the other span in order to maintain equilibrium of sway moments. It turns out that the right span controls. That is, there cannot be a midspan hinge and one at the interior column in the left span without exceeding the moment capacity of the W18 × 45 in the right span. Therefore, the moment set $+26$, 0, $+26$ is added to the left span and -26, 0, -26 to the right span to produce the final set of moments shown in Fig. 8-39*c*. This gives a beam mechanism in the right span. A shear-diagram check for maximum moment in the right span, as made in Fig. 8-38, shows that M_p is not exceeded. The revised column moments are entered in Table 8-1. It will be noted that columns 3 and 5 can be reduced in size to save 9 and 8 lb per ft, respectively.

Moment balancing is also called moment distribution. However, it is significant that it bears little or no resemblance to moment distribution based on elastic behavior. In elastic design one must fit the moment diagram to the structure because the distribution of moments depends on relative stiffnesses and cannot be determined until the frame has been proportioned. On the other hand, plastic design allows one to construct a moment diagram and then proportion a frame to fit it; this can be done because the distribution of moments is according to plastic-moment capacities rather than relative stiffnesses. Thus, whether computations are by hand or by digital computer, plastic design offers greater freedom in the design of frameworks.

Additional examples of moment balancing are given in Refs. 1, 10, and 11.

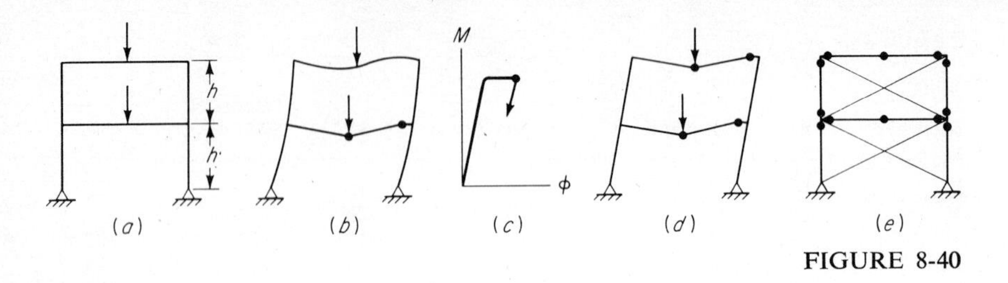

FIGURE 8-40

PROBLEMS

8-21 Design a one-bay, one-story A36-steel frame which spans 60 ft and is 18 ft high. The column bases are hinged. The frame carries concentrated vertical loads of 30 kips at midspan and each quarter point, 15 kips vertical at each column top, and a wind load of 12 kips at the roof line. Use a load factor of 1.7 for gravity load alone and 1.3 for gravity load plus wind load.

8-22 Determine whether the frame which was designed in Art. 8-17 can support the specified gravity loads together with a concentrated wind load of 8 kips at the roof line. The load factor is 1.5 for both wind and gravity load.

8-18 FRAME BUCKLING

Buckling of frames was discussed in Art. 4-20. It was shown that frames which are braced against sidesway are not subject to overall buckling and that only the stability of the individual members need be considered. Unbraced frames, however, even if they are symmetrical and symmetrically loaded, may become unstable and buckle in a sidewise mode at loads much less than they can support in the braced condition (Fig. 4-47). A procedure for predicting the critical load for this case, based on column effective-length coefficients K as functions of the relative stiffnesses $G = \sum (I/L)_c / \sum (I/L)_b$, is discussed in that article. Although the same procedure can be used for plastically designed frames, it is much more difficult because buckling will occur after plastic hinges are partially or fully developed in some of the beams. As a result, stiffnesses of the beam cross section vary along the beam, and its overall stiffness is difficult to evaluate.[12]

The resistance of a frame to sidesway buckling depends on its stiffness in sidewise bending, which is successively reduced with the formation of each additional plastic hinge. For example, a plastic hinge at one end of a beam reduces its effective stiffness at the opposite end from I/L to $0.75I/L$, as is known from moment-distribution theory. Formation of a second hinge at or near midspan reduces the effective stiffness to zero. Thus, there is a progressive deterioration of stiffness of the beams in a frame.

A symmetrically loaded, symmetrical frame may become unstable prior to the development of a collapse mechanism. If a mechanism develops in the first-floor beam of the two-story frame shown in Fig. 8-40*a*, the sidesway buckling shape will be as shown in Fig. 8-40*b*. The plastic hinge at the left end of the beam is not shown in Fig. 8-40*b* because the moment at this point at the beginning of buckling is opposite to the moment due to the load. Therefore, the beam unloads elastically at this end, as shown in Fig. 8-40*c*. It will be seen that, because of the beam mechanism, the frame buckles as a one-story frame $2h$ in height. If both beams develop mechanisms, the sidesway stiffness vanishes, as shown in Fig. 8-40*d*. A detailed investigation of an unbraced four-story one-bay frame with both gravity load and wind load is given in Ref. 2.

There is a further complication in the analysis of unbraced high-rise frames because sidesway causes additional gravity-load moments. Thus, if there is a horizontal displacement Δ at the roof line of a one-story frame, there is a moment $P\Delta$ which is not taken into account in the examples in this chapter. This effect is negligible for frames of a few stories, but it becomes significant in high-rise frames. Procedures for taking this effect into account have been developed.[13,14]

It will be noted that the braced frame shown in Fig. 8-40*e* remains stable even if hinges develop in the columns as well as in the beams. This and the absence of any significant $P\Delta$ effect in such frames is the reason the 1969 AISC specification permits plastic design of braced frames of any number of stories but restricts it for unbraced structures to frames not more than two stories high.

REFERENCES

1 Hodge, P. G., Jr.: "Plastic Analysis of Structures," McGraw-Hill, New York, 1959.
2 Massonnet, C. E., and M. A. Save: "Plastic Analysis and Design," Blaisdell, New York, 1965.
3 Symonds, P. S., and B. G. Neal: Recent Progress in the Plastic Method of Structural Analysis, *J. Franklin Inst.*, vol. 252, 1951.
4 Neal, B. G., and P. S. Symonds: The Rapid Calculation of the Plastic Collapse Load of a Framed Structure, *Proc. Inst. Civ. Eng. (London)*, vol. 1, 1952.
5 Beedle, L. S.: "Plastic Design of Steel Frames," Wiley, New York, 1958.
6 Neal, B. G.: "The Plastic Methods of Structural Analysis," Wiley, New York, 1956.
7 "Plastic Design in Steel," 2d ed., ASCE Manuals and Reports on Practice no. 41, New York, 1971.
8 Haaijer, G.: Plate Buckling in the Strain-hardening Range, *Trans. ASCE*, vol. 124, 1959.

9 Lay, M. G., and T. V. Galambos: Inelastic Beams under Moment Gradient, *J. Struct. Div. ASCE*, February 1967.
10 Gaylord, E. H.: Plastic Design by Moment Balancing, *AISC Eng. J.*, October 1967.
11 Horne, M. R.: A Moment Distribution Method for the Analysis and Design of Structures by the Plastic Theory, *Proc. Inst. Civ. Eng.*, vol. 3, pt. 3, April 1954.
12 Lu, Le-Wu: Inelastic Buckling of Steel Frames, *J. Struct. Div. ASCE*, December 1965.
13 Plastic Design of Multistory Frames, Lecture Notes, *Lehigh Univ. Fritz Eng. Lab. Rep.* 273.20, 1965.
14 Wright, E. W., and E. H. Gaylord: Analysis of Unbraced Multistory Steel Rigid Frames, *J. Struct. Div. ASCE*, May 1968.

9

STABILITY AND STRENGTH OF FLAT PLATES

9-1 STABILITY OF FLAT PLATES

Buckling of flat plates with various in-plane force systems acting on the edges has been discussed at several points in preceding chapters. The solution to all such problems derives from a single equation of equilibrium relating the edge forces to the displacement w normal to the plate.

Figure 9-1 shows a rectangular flat plate with stresses f_x and f_y (tension positive) and shear stresses f_v distributed along the middle line at the edges. The equation of equilibrium is

$$\frac{EI}{1-\mu^2}\left(\frac{\partial^4 w}{\partial x^4} + 2\frac{\partial^4 w}{\partial x^2\,\partial y^2} + \frac{\partial^4 w}{\partial y^4}\right) = t\left(f_x \frac{\partial^2 w}{\partial x^2} + 2f_v \frac{\partial^2 w}{\partial x\,\partial y} + f_y \frac{\partial^2 w}{\partial y^2}\right) \tag{9-1}$$

where $I = t^3/12$ = moment of inertia of cross-sectional area of a unit strip of plate

t = thickness of plate

μ = Poisson's ratio

w = deflection of a point in the middle plane of plate

The bending moments are given by

$$M_x = -\frac{EI}{1-\mu^2}\left(\frac{\partial^2 w}{\partial x^2} + \mu \frac{\partial^2 w}{\partial y^2}\right) \tag{9-2a}$$

$$M_y = -\frac{EI}{1-\mu^2}\left(\frac{\partial^2 w}{\partial y^2} + \mu \frac{\partial^2 w}{\partial x^2}\right) \tag{9-2b}$$

where M_x and M_y are the moments per unit of width of sections parallel to the y and x axes, respectively (Fig. 9-1). The shearing resultants Q_x and Q_y shown in Fig. 9-1 are

$$Q_x = -\frac{EI}{1-\mu^2}\frac{\partial}{\partial x}\left(\frac{\partial^2 w}{\partial x^2} + \frac{\partial^2 w}{\partial y^2}\right) \tag{9-3a}$$

$$Q_y = -\frac{EI}{1-\mu^2}\frac{\partial}{\partial y}\left(\frac{\partial^2 w}{\partial x^2} + \frac{\partial^2 w}{\partial y^2}\right) \tag{9-3b}$$

Details of the derivation can be found elsewhere.[1,2]

If f_x is compressive (negative) and w is independent of y, Eq. (9-1) reduces to

$$\frac{EI}{1-\mu^2}\frac{d^4 w}{dx^4} + f_x t \frac{d^2 w}{dx^2} = 0 \tag{a}$$

This is the differential equation for bending of a bar of unit width and thickness t acted upon by a compressive force $f_x t$. Except for the term $1 - \mu^2$, Eq. (*a*) is the fourth-order form of Eq. (*g*) of Art. 4-2. The solution for hinged ends is

$$f_x t = \frac{\pi^2 EI}{(1-\mu^2)a^2} \tag{9-4}$$

where a is the length of the strip in the direction of x (Fig. 9-1). This corresponds to Eq. (4-1) for the centrally loaded column. Similarly, with w independent of x, the third and sixth terms of Eq. (9-1) give the critical load for a strip of length b in the direction of y. The second term in Eq. (9-1) results from distortion of an element of the plate by twisting moments acting on the element.

Figure 9-2 shows a flat plate, hinged on all four edges, which has buckled under the uniform edge compression f_x. For this case, Eq. (9-1) reduces to

$$\frac{EI}{1-\mu^2}\left(\frac{\partial^4 w}{\partial x^4} + 2\frac{\partial^4 w}{\partial x^2\,\partial y^2} + \frac{\partial^4 w}{\partial y^4}\right) + f_x t \frac{\partial^2 w}{\partial x^2} = 0 \tag{9-5}$$

Equation (9-5) can be satisfied for certain values of f_x by

$$w = w_{mn} \sin\frac{m\pi x}{a} \sin\frac{n\pi y}{b} \qquad \begin{array}{l} m = 1, 2, 3, \ldots \\ n = 1, 2, 3, \ldots \end{array} \tag{9-6}$$

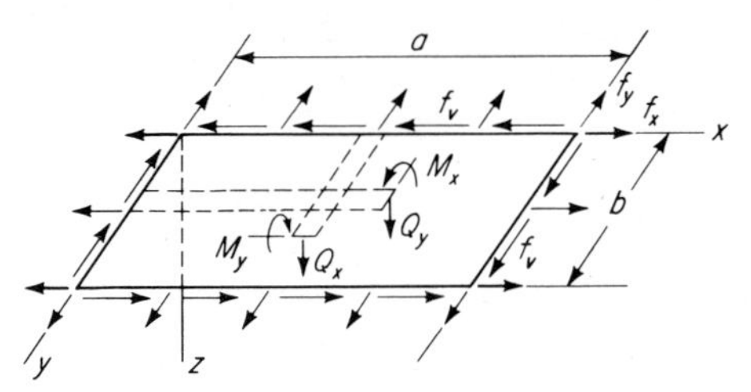

FIGURE 9-1

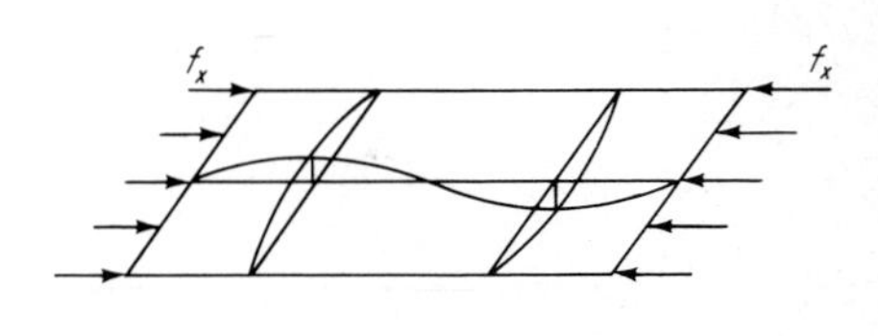

FIGURE 9-2

It is clear that this solution also satisfies the specified boundary conditions since $w = 0$ at the four edges and, from Eqs. (9-2), bending moments are zero at the four edges. Substituting w from Eq. (9-6) into Eq. (9-5) gives

$$\frac{EI}{1-\mu^2}\left(\frac{m^4\pi^4}{a^4} + 2\frac{m^2n^2\pi^4}{a^2b^2} + \frac{n^4\pi^4}{b^4}\right) - f_x t\,\frac{m^2\pi^2}{a^2} = 0$$

from which

$$f_x t = \frac{\pi^2 EI}{1-\mu^2}\,\frac{(m^2/a^2 + n^2/b^2)^2}{m^2/a^2} = \frac{\pi^2 EI}{1-\mu^2}\left(\frac{m}{a} + \frac{n^2}{m}\frac{a}{b^2}\right)^2$$

This equation gives an infinite number of values of the compressive force $f_x t$. However, we are interested in the smallest value at which a buckled configuration can exist. It is clear that $f_x t$ is smallest when $n = 1$, which means that the plate buckles in one half wave transverse to the direction of loading. Therefore,

$$f_x t = \frac{\pi^2 EI}{(1-\mu^2)a^2}\left(m + \frac{1}{m}\frac{a^2}{b^2}\right)^2 \tag{9-7}$$

where m is the number of half waves in the direction of f_x. If $m = 1$,

$$f_x t = \frac{\pi^2 EI}{(1-\mu^2)a^2}\left(1 + \frac{a^2}{b^2}\right)^2 \tag{9-8}$$

This equation is identical to Eq. (9-4) except for the factor in parentheses. Furthermore, $f_x t$ in Eq. (9-8) approaches the value given by Eq. (9-4) as a/b decreases. Thus, the second term in parentheses measures the stiffening of the plate which results from support of the unloaded edges.

It would appear from Eq. (9-8) that $f_x t$ increases without limit as the width b of the plate decreases. This would be true if the plate buckled in only one longitudinal half wave. However, the possibility of multiple-wave buckling modes must be investigated. For this purpose, it is convenient to rewrite Eq. (9-7) in the form

$$f_x t = \frac{\pi^2 EI}{(1-\mu^2)b^2}\left(m\frac{b}{a} + \frac{1}{m}\frac{a}{b}\right)^2 \tag{9-9}$$

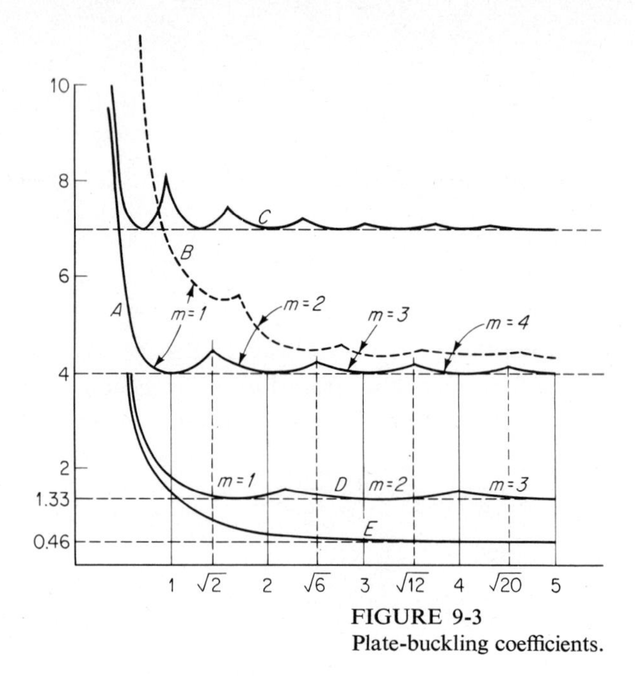

FIGURE 9-3
Plate-buckling coefficients.

Substituting $I = t^3/12$ into this equation and using the notation

$$k = \left(m\frac{b}{a} + \frac{1}{m}\frac{a}{b}\right)^2 \tag{9-10}$$

and denoting the critical value of f_x by F_{cr}, we get

$$F_{cr} = \frac{k\pi^2 E}{12(1-\mu^2)(b/t)^2} \tag{9-11}$$

The ratio of length a to width b of a plate is called its *aspect ratio*. Values of k from Eq. (9-10) are shown in curve A of Fig. 9-3. It will be noted that k has a minimum value of 4 for $a/b = 1, 2, 3, \ldots, n$. The plate buckles in one longitudinal half wave ($m = 1$) if $a/b \lessapprox \sqrt{2}$, two if $\sqrt{2} \lessapprox a/b \lessapprox \sqrt{6}$, etc. Except for the unlikely case of the extremely short plate (a/b less than about 0.5) the error in using $k = 4$ for all values of a/b is at most about 10 percent, and in the usual case, for which a/b is of the order of 10 or more, it is extremely small.

Values of k for the plate of Fig. 9-2 with the loaded edges clamped (unloaded edges hinged) are shown by the dashed line B in Fig. 9-3. This is also a family of intersecting curves. In this case, the minimum values of k for the various branches are not equal. However, it is clear that $k = 4$ is satisfactory for the usual case of large a/b.

Values of k for the plate of Fig. 9-2 with the unloaded edges clamped (loaded edges hinged) are shown in curve C of Fig. 9-3. Here there is a substantial increase in k; the minimum value for each curve of the system is 7. The effect of clamping the loaded edges in this case is similar to that of the preceding case, and can be ignored for plates of practical proportions.

Values of k for plates with one unloaded edge free are given by D and E of Fig. 9-3. For curve D the unloaded edge that is supported is clamped, while for curve E it is hinged. The minimum value of k for curve D is 1.33. The buckling behavior of the plate with the supported unloaded edge hinged differs from that of the other cases shown in this figure, in that such a plate buckles in only one longitudinal half wave regardless of the aspect ratio. The value of k decreases with increase in a/b and approaches the limiting value 0.456. However, even for a plate as short as $a = 5b$, k is only 9 percent larger than the minimum value. Therefore, except for very short plates, $k = 0.456$ is a very good approximation.

Equation (9-11) is valid only if stress is proportional to strain. Modification of the equation for the case where the critical stress exceeds the proportional limit is discussed in Art. 4-21.

Critical stresses for plates with two opposite edges subjected to bending stresses $f_x = My/I$ can also be determined from Eq. (9-1). This is discussed in Art. 5-17. The critical stress f_v for a plate subjected to shear stress on its edges is discussed in Art. 5-16.

Local buckling of plate elements in axially loaded compression members is shown in Fig. 4-49.

9-2 POSTBUCKLING STRENGTH OF FLAT PLATES

It was pointed out in Art. 4-2 that the difference between the buckling load and the postbuckling strength of the axially loaded column is so small that the buckling load is a practical measure of strength. In contrast, the postbuckling strength of a flat plate can be considerably larger than the buckling load. Figure 9-4*a* shows a uniformly compressed flat plate, simply supported on all edges, at the onset of buckling. At this stage, the stress is given by Eq. (9-11) with $k = 4$. If the critical stress is less than the yield stress, further shortening of the plate increases the stress at the edges because they must remain straight. On the other hand, this shortening produces little or no increase in strain on a vertical strip at the middle, because such shortening is more easily accommodated by an increase in the amplitude of the buckles. The result is a stress distribution such as that shown in Fig. 9-4*b*. The value of P corresponding

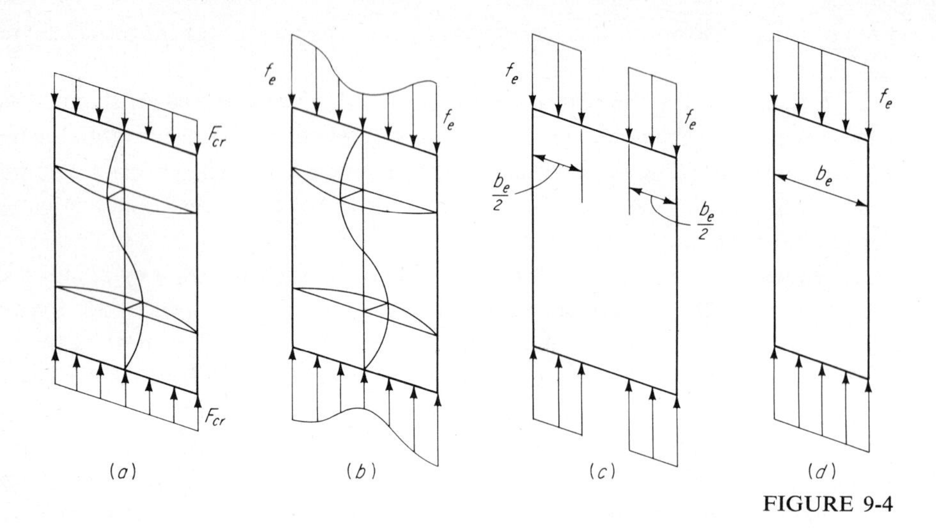

FIGURE 9-4

to this stress distribution is more difficult to determine than the buckling strength. Equation (9-1) is not applicable because it neglects the effect of membrane stresses in the middle plane of the plate; these become significant for large deflections. The following approximate procedure was given by Karman[3] in 1932. It is assumed that the nonuniform postbuckling stress distribution can be replaced by the two rectangular stress blocks of intensity f_e and width $b_e/2$, where f_e is the edge stress of Fig. 9-4*b* and b_e is the *effective width* of the plate (Fig. 9-4*c*). Therefore,

$$P = f_e b_e t \tag{a}$$

To determine the effective width, we assume that it is the same as the width b_e of a plate which buckles at the uniform stress f_e (Fig. 9-4*d*). Then, using Eq. (9-11),

$$f_e = \frac{4\pi^2 E}{12(1-\mu^2)(b_e/t)^2} \tag{b}$$

from which

$$\frac{b_e}{t} = 1.9\sqrt{\frac{E}{f_e}} \tag{9-12a}$$

The edge stress f_e continues to increase with increase in strain of the plate until a limiting value is reached. In the case of a flat-yield steel, this limit is F_y for practical purposes. For a gradually yielding steel the offset-yield stress is usually

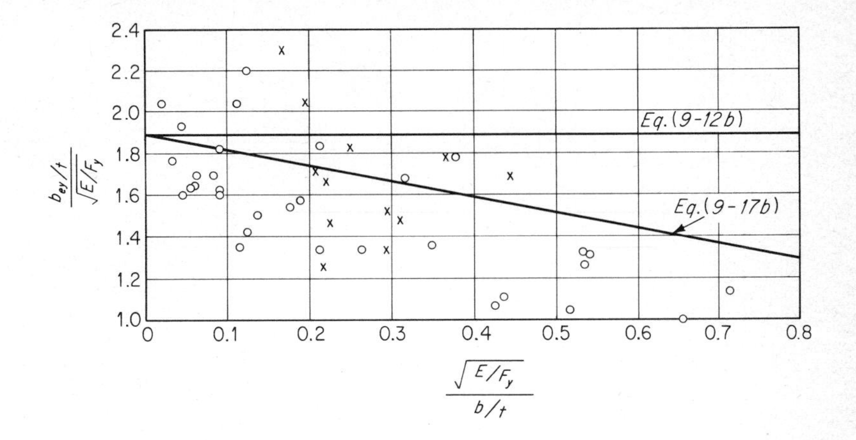

FIGURE 9-5
Postbuckling strength of uniformly compressed plates with both unloaded edges supported.

taken as the limiting stress. Therefore, the effective width at ultimate load is given by

$$\frac{b_{ey}}{t} = 1.9\sqrt{\frac{E}{F_y}} \tag{9-12b}$$

Figure 9-5 shows a comparison of Eq. (9-12*b*) with results of tests.[4] The ordinate in this figure is such that the equation plots as a straight line at 1.9. A test that checks the formula exactly plots on this line. It will be seen that most of the tests plot below, which means that Eq. (9-12*b*) overestimates the effective width, except for plates at the left in the figure, which are those with large b/t. Because of this, other formulas, one of which is shown in the figure, have been developed. These are discussed in Art. 9-3.

Assuming that the Karman formulation for effective width can be used for uniformly compressed plates with other boundary conditions, Eq. (*b*) becomes

$$f_e = \frac{k\pi^2 E}{12(1-\mu^2)(b_e/t)^2} \tag{c}$$

which gives the effective width

$$\frac{b_e}{t} = 0.95\sqrt{\frac{kE}{f_e}} \tag{9-13a}$$

Similarly, at ultimate load

$$\frac{b_{ey}}{t} = 0.95\sqrt{\frac{kE}{F_y}} \tag{9-13b}$$

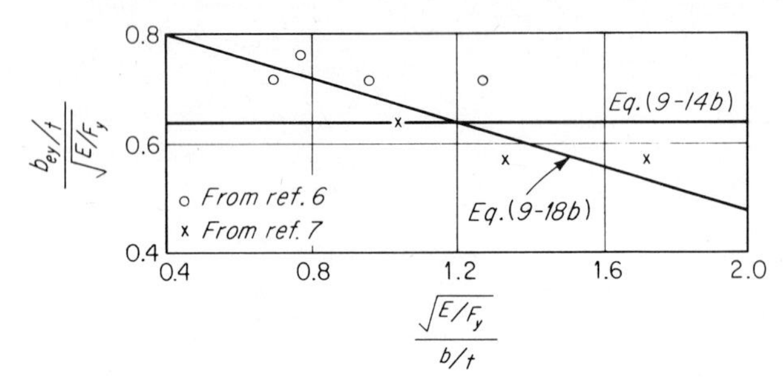

FIGURE 9-6
Postbuckling strength of uniformly compressed plates with one unloaded edge free.

For a plate with one unloaded edge free and the other hinged, $k = 0.456$ (Fig. 9-3), for which Eqs. (9-13) give

$$\frac{b_e}{t} = 0.64\sqrt{\frac{E}{f_e}} \tag{9-14a}$$

$$\frac{b_{ey}}{t} = 0.64\sqrt{\frac{E}{F_y}} \tag{9-14b}$$

Postbuckling strength according to the effective width from Eq. (9-14*b*) is compared in Fig. 9-6 with results of tests. The test results in this figure are from two sources. The specimens reported in Ref. 6 were of cruciform cross section, fabricated by welding three $\frac{1}{4}$-in. A514-steel plates for which $F_y \approx 100$ ksi. The four flanges were identical. Thus, they tended to buckle simultaneously, so that each acted as a single plate simply supported on one unloaded edge and free on the other. The b/t ratios of the six specimens were 4, 8, 12, 16, 20, and 22. The specimen with flanges of $b/t = 4$ failed by bend buckling, the other five by local buckling. However, only four of these five are shown in the figure because the specimen with $b/t = 8$ did not fail (by local buckling) until after it had reached yield stress. Each of the three specimens reported in Ref. 7 was composed of two cold-formed channels joined back to back by epoxy glue to form an I. The average yield stress was 78.6 ksi. The members were fabricated from 16-gage carbon steel ($t = 0.06$ in.), with b/t values of 11.40, 15.40, and 20.75. The L/r ratio was less than 20, and all three failed by local buckling of the flanges.

The effective width by Eqs. (9-13) can be expressed in different form by dividing Eq. (*c*) by Eq. (9-11). This gives

$$\frac{b_e}{b} = \sqrt{\frac{F_{cr}}{f_e}} \tag{9-15a}$$

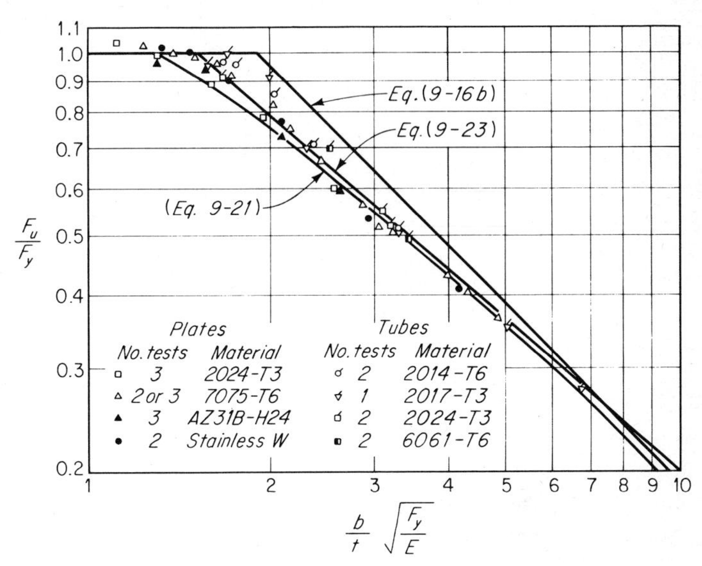

FIGURE 9-7
Postbuckling strength of uniformly compressed plates with both unloaded edges supported. (*Adapted from "Structural Stability Theory" by George Gerard, 1957. Used with permission of McGraw-Hill Book Company.*)

$$\frac{b_{ey}}{b} = \sqrt{\frac{F_{cr}}{F_y}} \tag{9-15b}$$

Using Eq. (9-15*a*), the average stress is given by

$$f_{\text{av}} = \frac{f_e b_e t}{bt} = \sqrt{F_{cr} f_e} \tag{9-16a}$$

Similarly, from Eq. (9-15*b*)

$$F_u = \sqrt{F_{cr} F_y} \tag{9-16b}$$

These are simple formulas for the postbuckling loads of a plate under edge compression, for all conditions of support at the boundaries.

Figure 9-7 shows a comparison of predicted strengths by Eq. (9-16*b*) with results of tests on square tubes and on plates whose edges were supported in V-shaped grooves.[2] These specimens were of steel and various aluminum and magnesium alloys. It will be noted that the equation tends to overestimate the postbuckling strength of plates with small b/t, which is consistent with the observation in Fig. 9-5 that the Karman formula overestimates the effective width of such plates. Nevertheless, it gives a surprisingly good fit, considering that it is based on such a simple

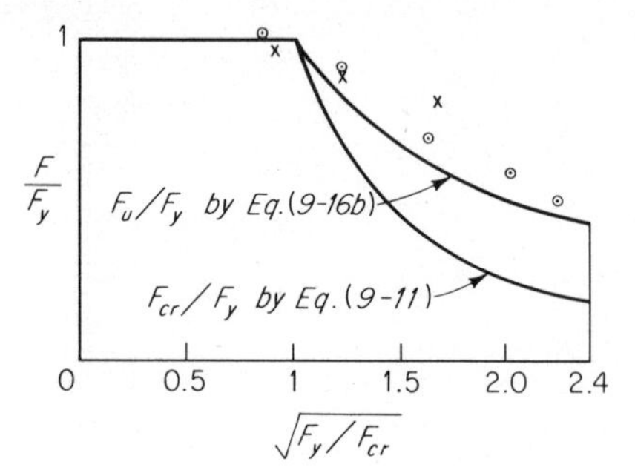

FIGURE 9-8
Postbuckling strength of uniformly compressed plates with one unloaded edge free.

concept. The other formulas shown in this figure are discussed in Art. 9-3. Figure 9-8 gives a comparison of predicted strengths by Eq. (9-16*b*) with the test results for plates with one unloaded edge free which are shown in Fig. 9-6. Again, the Karman effective width is seen to give good results.

9-3 ADDITIONAL FORMULAS FOR EFFECTIVE WIDTH

Equations (9-12) for the compressed plate supported on both unloaded edges can be brought into closer agreement with results of tests by a reduction factor, as follows:

$$\frac{b_e}{t} = 1.9\sqrt{\frac{E}{f_e}}\left(1 - 0.415\frac{\sqrt{E/f_e}}{b/t}\right) \tag{9-17a}$$

$$\frac{b_{ey}}{t} = 1.9\sqrt{\frac{E}{F_y}}\left(1 - 0.415\frac{\sqrt{E/F_y}}{b/t}\right) \tag{9-17b}$$

These formulas, except with a coefficient 0.475 instead of 0.415, were given in 1947 by Winter.[4] The smaller coefficient, which gives somewhat larger effective widths, was proposed in 1968. Equation (9-17*b*) is plotted in Fig. 9-5.

The following similar modifications of Eqs. (9-14) for the plate with one unloaded edge free were also presented in Ref. 4:

$$\frac{b_e}{t} = 0.8\sqrt{\frac{E}{f_e}}\left(1 - 0.202\frac{\sqrt{E/f_e}}{b/t}\right) \tag{9-18a}$$

$$\frac{b_{ey}}{t} = 0.8\sqrt{\frac{E}{F_y}}\left(1 - 0.202\frac{\sqrt{E/F_y}}{b/t}\right) \tag{9-18b}$$

Equation (9-18*b*) is plotted in Fig. 9-6. According to Ref. 4, the formula is valid to $\sqrt{E/F_y}/(b/t) = 1.75$.

Equations (9-17) and (9-18) can also be put in the form of Eqs. (9-15). Only the formulas for effective width b_{ey} at ultimate load are given here, since, as in the preceding formulas, the effective width b_e for a lesser edge stress is found by substituting f_e for F_y. Using Eq. (9-11), we get

$$\sqrt{E} = \frac{b/t}{0.95}\sqrt{\frac{F_{cr}}{k}} \tag{d}$$

Substituting $\sqrt{E}$ from this equation, with $k = 4$, into Eq. (9-17*b*) gives the effective width b_{ey} for the plate with both unloaded edges supported:

$$\frac{b_{ey}}{b} = \sqrt{\frac{F_{cr}}{F_y}}\left(1 - 0.218\sqrt{\frac{F_{cr}}{F_y}}\right) \tag{9-19}$$

Except for the factor in parentheses, this is the same as Eq. (9-15*b*). Similarly, using $k = 0.456$ in Eq. (*d*) and substituting into Eq. (9-18*b*), we get for the plate with one unloaded edge free

$$\frac{b_{ey}}{b} = \sqrt{\frac{F_{cr}}{F_y}}\left(1.25 - 0.395\sqrt{\frac{F_{cr}}{F_y}}\right) \tag{9-20}$$

Formulas for ultimate stress are found from Eqs. (9-19) and (9-20) by using $F_u = F_y b_e t/bt$, which gives for the plate with both unloaded edges supported

$$F_u = \sqrt{F_{cr}F_y}\left(1 - 0.218\sqrt{\frac{F_{cr}}{F_y}}\right) \tag{9-21}$$

and, for the plate with one unloaded edge free

$$F_u = \sqrt{F_{cr}F_y}\left(1.25 - 0.395\sqrt{\frac{F_{cr}}{F_y}}\right) \tag{9-22}$$

Equation (9-21) is plotted in Fig. 9-7, where the Karman value by Eq. (9-16*b*) was compared with results of tests. The following additional formula,[2] which is also based on curve fitting to test results, is shown in the figure:

$$F_u = 1.42F_y\left(\frac{\sqrt{E/F_y}}{b/t}\right)^{0.85} \tag{9-23}$$

Equations (9-16*b*), (9-21), and (9-23) for the compressed plate supported on both unloaded edges are compared in Fig. 9-9, where values of F_u for an A36 plate are plotted as functions of plate slenderness b/t. Also shown is the critical stress, curve *A*. It is of interest to note that Eq. (9-23) gives $b/t = 44$ as the limiting value to enable an A36 steel plate to just reach yield stress without buckling. This agrees very

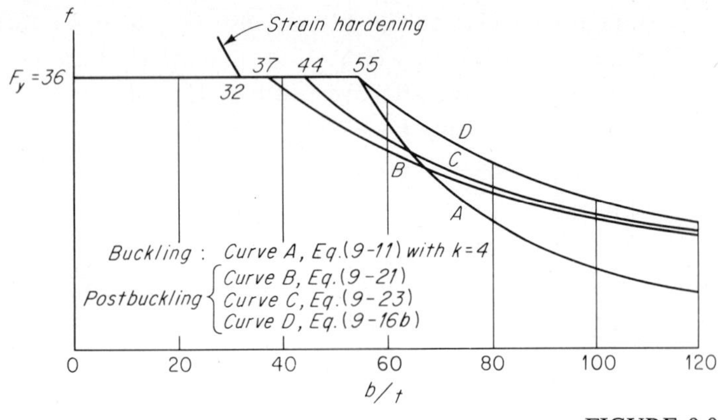

FIGURE 9-9

well with the AISC value 42 (Table 4-4). The limiting value $b/t = 32$, which enables strain hardening to begin, is also shown in the figure. This value was established in Ref. 5. It was adopted by AISC as a requirement for plastic design and for compact sections in elastic design, as shown in Table 5-1, that is, $190/\sqrt{F_y} = 190/\sqrt{36} = 32$.

An important aspect of the nature of the support at the unloaded edges of plates supported on all four edges remains to be discussed. If these unloaded edges are free to move in the plane of the plate, the plate contracts in width as it buckles and the unloaded edges remain free of stress in the transverse direction. On the other hand, if they resist contraction, reactive forces develop. These effects can be evaluated by using the large-deflection theory of plates (Ref. 1, p. 411). Most test results are on specimens in which lateral contraction is not restrained appreciably.

9-4 POSTBUCKLING STRENGTH OF SHORT COLUMNS

In this article the postbuckling strength of compression members with thin-plate elements is considered. The member is assumed to be short, so that there is no bend buckling.

EXAMPLE 9-1 Compute the critical load and the postbuckling strength of the A36 column of cruciform cross section shown in Fig. 9-10.

Critical load:

$$A = 4 \times 6 \times \tfrac{1}{4} = 6 \text{ in.}^2 \qquad \frac{b}{t} = \frac{6}{1/4} = 24$$

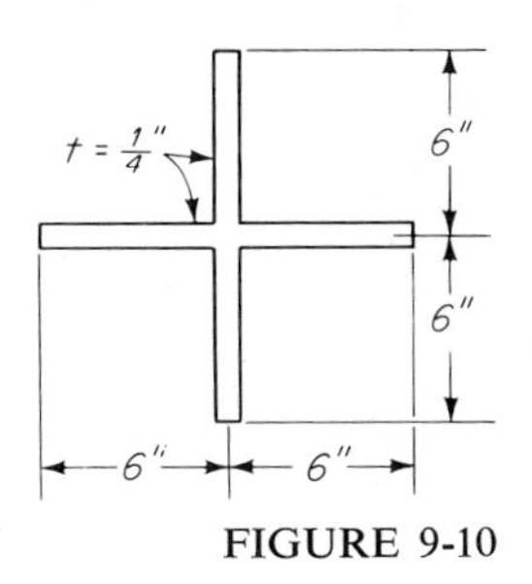

FIGURE 9-10

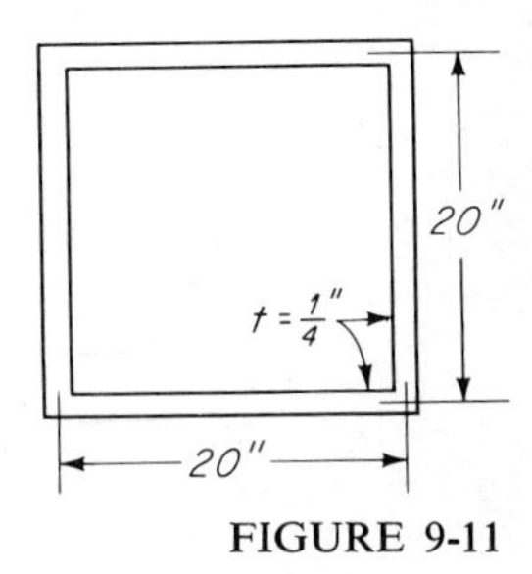

FIGURE 9-11

With $E = 30{,}000$ ksi and $\mu = 0.3$, Eq. (9-11) gives

$$F_{cr} = \frac{27{,}000k}{(b/t)^2} = \frac{27{,}000 \times 0.456}{24^2} = 21.4 \text{ ksi}$$

$$P_{cr} = 21.4 \times 6 = 128 \text{ kips}$$

Postbuckling strength:

$$F_u = \sqrt{F_{cr} F_y} = \sqrt{21.4 \times 36} = 27.8 \text{ ksi} \qquad [\text{Eq. } (9.16b)]$$

$$P_u = 27.8 \times 6 = 167 \text{ kips}$$

EXAMPLE 9-2 Compute the critical load and the postbuckling strength of the A36 box column of Fig. 9-11.

Critical load:

$$A = 4 \times 20 \times \tfrac{1}{4} = 20 \text{ in.}^2 \qquad \frac{b}{t} = \frac{20}{1/4} = 80$$

$$F_{cr} = \frac{27{,}000k}{(b/t)^2} = \frac{27{,}000 \times 4}{80^2} = 16.9 \text{ ksi} \qquad \text{Eq. (9-11)}$$

$$P_{cr} = 16.9 \times 20 = 338 \text{ kips}$$

The postbuckling strength according to Eq. (9-12b) is

$$\frac{b_{ey}}{t} = 1.9\sqrt{\frac{30{,}000}{36}} = 55 \qquad b_{ey} = 55 \times \tfrac{1}{4} = 13.8 \text{ in.}$$

$$A_{\text{eff}} = 4 \times 13.8 \times \tfrac{1}{4} = 13.8 \text{ in.}^2$$

$$P_u = 36 \times 13.8 = 497 \text{ kips}$$

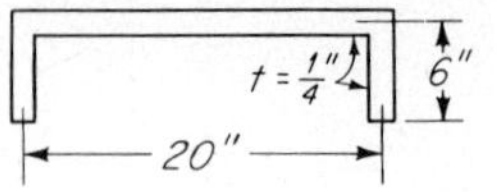

FIGURE 9-12

This result could also have been obtained from Eq. (9-16*b*) as in Example 9-1. The postbuckling strength according to Eq. (9-17) is

$$\sqrt{\frac{E}{F_y}} = \sqrt{\frac{30{,}000}{36}} = 28.9$$

$$\frac{b_{ey}}{t} = 1.9 \times 28.9\left(1 - 0.415 \times \frac{28.9}{80}\right) = 55 \times 0.85 = 46.8$$

$$b_{ey} = 46.8 \times \tfrac{1}{4} = 11.7 \text{ in.} \qquad A_{\text{eff}} = 4 \times 11.7 \times \tfrac{1}{4} = 11.7 \text{ in.}^2$$

$$P_u = 36 \times 11.7 = 421 \text{ kips}$$

This is 15 percent less than the value according to Karman. The postbuckling strength according to Eq. (9-23) is

$$\frac{F_u}{F_y} = 1.42\left(\frac{\sqrt{E/F_y}}{b/t}\right)^{0.85} = 1.42\left(\frac{28.9}{80}\right)^{0.85} = 0.597$$

$$F_u = 0.597 \times 36 = 21.5 \text{ ksi} \qquad P_u = 21.5 \times 20 = 430 \text{ kips}$$

This is 13 percent less than the value according to Karman.

EXAMPLE 9-3 Compute the critical load for the column with the cross section shown in Fig. 9-12. Compute the postbuckling strength.

Critical load:

Flange: $\quad \dfrac{b}{t} = \dfrac{6}{1/4} = 24 \qquad F_{cr} = 21.4 \text{ ksi} \qquad$ (Example 9-1)

Web: $\quad \dfrac{b}{t} = \dfrac{20}{1/4} = 80 \qquad F_{cr} = 16.9 \text{ ksi} \qquad$ (Example 9-2)

$$A = 2 \times 6 \times \tfrac{1}{4} + 20 \times \tfrac{1}{4} = 8 \text{ in.}^2$$

$$P_{cr} = 16.9 \times 8 = 135 \text{ kips}$$

The postbuckling strength according to Eq. (9-16*b*) is

Flange: $\quad F_u = \sqrt{21.4 \times 36} = 27.8 \text{ ksi}$

Web: $$F_u = \sqrt{16.9 \times 36} = 24.7 \text{ ksi}$$

$$P_u = 2 \times 27.8 \times 6 \times \tfrac{1}{4} + 24.7 \times 20 \times \tfrac{1}{4} = 83 + 123 = 206 \text{ kips}$$

The postbuckling strength according to Eqs. (9-21) and (9-22) is

Flange: $$F_u = \sqrt{21.4 \times 36}\left(1.25 - 0.395\sqrt{\frac{21.4}{36}}\right) = 26.3 \text{ ksi}$$

Web: $$F_u = \sqrt{16.9 \times 36}\left(1 - 0.218\sqrt{\frac{16.9}{36}}\right) = 21.0 \text{ ksi}$$

$$P_u = 2 \times 26.3 \times 6 \times \tfrac{1}{4} + 21.0 \times 20 \times \tfrac{1}{4} = 79 + 105 = 184 \text{ kips}$$

Although the ultimate strength of a short compression member is the sum of the strengths of its components, as noted above, tests on members composed of cold-formed elements show that the wave formation in elements supported on only one of the unloaded edges, as in the flange of the channel, may have large amplitudes at stresses only slightly larger than the local-buckling stress of the element. Information regarding postbuckling appearance of the thicker, hot-rolled elements is lacking. However, it is logical to expect it to be about the same. Therefore, if protection against unserviceability because of excessive waviness is required, the service-load stress must not be significantly larger than the local-buckling stress of the weakest element which has one of its unloaded edges free. Evaluation of the useful strength of a member in this case is illustrated in the next example.

EXAMPLE 9-4 Compute the axial load P for the member of Fig. 9-12 at which local buckling of the flanges occurs.

SOLUTION Critical stresses for the components were computed in Example 9-3. For the flange, $F_{cr} = 21.4$ ksi, while for the web $F_{cr} = 16.9$ ksi. Therefore, the web buckles first and will develop some postbuckling before the flange buckles. This will be the postbuckling strength corresponding to an edge stress of 21.4 ksi. Therefore, according to Eq. (9-16*a*),

$$f_{av} = \sqrt{16.9 \times 21.4} = 19.0 \text{ ksi}$$

$$P_u = 2 \times 21.4 \times 6 \times \tfrac{1}{4} + 19.0 \times 20 \times \tfrac{1}{4} = 64 + 95 = 159 \text{ kips}$$

It should be noted that if the proportions of the member are such that the critical stress for the flange is less than that for the web, there can be no postbuckling of the web if beginning of flange waviness is the criterion. In this case, the useful strength of the member is simply its area multiplied by the flange critical stress.

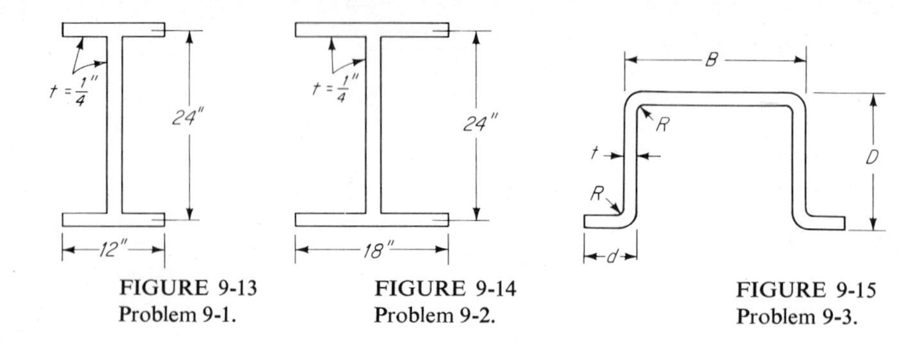

FIGURE 9-13
Problem 9-1.

FIGURE 9-14
Problem 9-2.

FIGURE 9-15
Problem 9-3.

PROBLEMS

9-1 Compute the short-column strength of the A36-steel welded member of Fig. 9-13 (*a*) with full postbuckling of both flange and web and (*b*) with the useful strength limited to that at which the flange buckles.

9-2 Same as Prob. 9-1 except that the cross section is that shown in Fig. 9-14.

9-3 Compute the short-column strength of the cold-formed A570 Grade D steel hat section of Fig. 9-15 (*a*) with full postbuckling of all components and (*b*) with the useful strength limited to that at which the projecting flange buckles. Given $B = 12$ in., $D = 10$ in., $d = 1.50$ in., $t = 0.06$ in., $R = \frac{3}{16}$ in.

9-5 INTERACTION OF LOCAL BUCKLING WITH BEND BUCKLING

The box column of Art. 9-4 will be used to illustrate the interaction of local buckling and bend buckling. Figure 9-16 shows the F-L/r plot for bend buckling, using the CRC parabola for the inelastic range. The local-buckling stress is 16.9 ksi (Example 9-2). Since this stress is in the elastic range of bend buckling, the corresponding slenderness ratio is found from the Euler formula:

$$\frac{L}{r} = \pi\sqrt{\frac{30{,}000}{16.9}} = 133$$

This gives point A in Fig. 9-16. Thus, bend buckling and local buckling occur simultaneously for this column if $L/r = 133$, while failure by bend buckling would be expected for $L/r > 133$. On the other hand, the postbuckling strength calculated in Example 9-2 can be counted on only if the column is short enough to preclude bend buckling. Of course, this condition is certain to be satisfied if $L/r = 0$. Using the value of P_u according to Karman's effective width, calculated in Example 9-2, $F_u = 497/20 = 24.9$ ksi, which gives point B in Fig. 9-16. Thus, it is clear that a

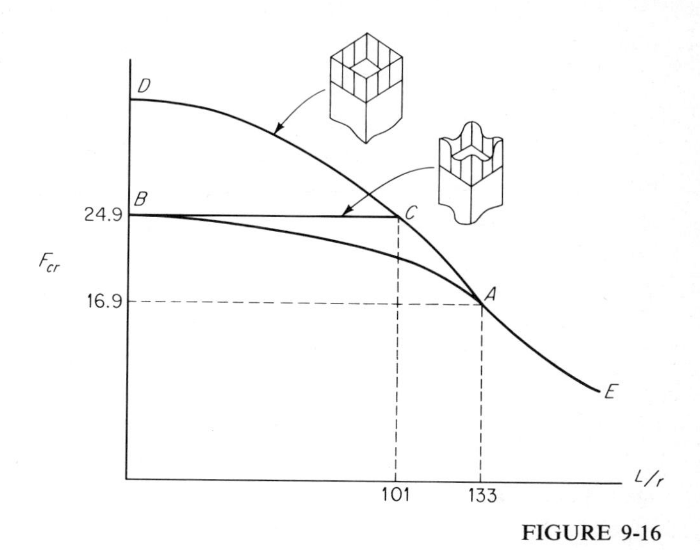

FIGURE 9-16

column for which $0 < L/r < 133$ will fail at $24.9 > F_u > 16.9$ ksi. Although the horizontal line *BC* might be taken as the ultimate stress for $0 < L/r < 101$, this leads to an inconsistency at *C* because the stress distribution corresponding to any point on *BC* is the nonuniform postbuckling distribution shown in the figure while the distribution corresponding to any point on the bend-buckling curve *DCAE* is uniform, as shown. Thus, point *C* represents two incompatible stress distributions. Therefore, the strength of this particular column must be represented by a line *BA* which is a function of both L/r and b/t. Except for whatever information is available from tests, the shape of *BA* is unknown.

9-6 SPECIFICATION PROVISIONS FOR POSTBUCKLING IN COLUMNS

Forces in members of steel frames in which hot-rolled shapes are commonly used are usually large enough to require plate elements thick enough to preclude local buckling at stresses less than yield stress. Thus, there is no provision in the AASHO specifications for taking postbuckling strength of plates into account; instead, upper limits of slenderness b/t are specified, as discussed in Art. 4-23. AREA also specifies upper limits for b/t. However, it allows the limit for plates supported on one edge to be exceeded if the area corresponding to the excess width is excluded in computations having to do with the stress on the member. Of course, this is an application of the simple Karman effective width. On the other hand, this option is not permitted for plates supported on both unloaded edges. However, the limiting b/t for such plates may be increased for columns in which the actual stress f_a is less than the allowable stress F_a by multiplying the limiting value of b/t by $\sqrt{F_a/f_a}$.

For many years the AISC specification also allowed limiting values in both types of plates to be exceeded, provided that stress computations were based on specified effective widths according to the Karman concept. The specification was revised in 1969, and still allows the b/t limits of plates to be exceeded. However, it requires the effective width for plates supported on both unloaded edges to be determined according to formulas of the type of Eqs. (9-17), while for plates supported on only one unloaded edge it uses an allowable stress which has a small factor of safety with respect to the local-buckling stress. The same procedure is used in the AISI specification for cold-formed steel members.

In both the AISC and the AISI specifications plates supported on both unloaded edges are called *stiffened elements* while those supported on only one unloaded edge are called *unstiffened elements*. The effective width of stiffened elements is given in terms of the maximum calculated stress f at service load. However, to allow for the fact that the load on which the factor of safety is based depends on the effective width at that load, rather than on the effective width at the service load, the service-load stress must be multiplied by the factor of safety before substituting into Eq. (9-17a). Then, with $f_e = 1.65f$ and $E = 29{,}000$ ksi, we get

$$\frac{b_e}{t} = \frac{253}{\sqrt{f}}\left(1 - \frac{55.3/\sqrt{f}}{b/t}\right) \tag{9-24}$$

This is the AISI formula for uniformly compressed stiffened elements in cross sections other than square and rectangular tubes. For the latter, the AISI formula is

$$\frac{b_e}{t} = \frac{253}{\sqrt{f}}\left(1 - \frac{50.3\sqrt{f}}{b/t}\right) \tag{9-25}$$

Similar formulas are used in the AISC specification. For example, the AISC formula for square and rectangular tubes is the same as that of the AISI [Eq. (9-25)].

Full advantage is not taken of the postbuckling strength of unstiffened elements in the AISC and AISI specifications because of the serviceability question regarding waviness, which was mentioned in Art. 9-4. Instead, the limiting stress is taken at a value between the critical stress and the postbuckling strength. Thus, for the single angle, which has little or no rotational restraint on the supported edge, the AISC specification is based on a limiting stress F_L which, in the elastic-buckling range, is given by

$$F_L = \frac{15{,}500}{(b/t)^2} \tag{a}$$

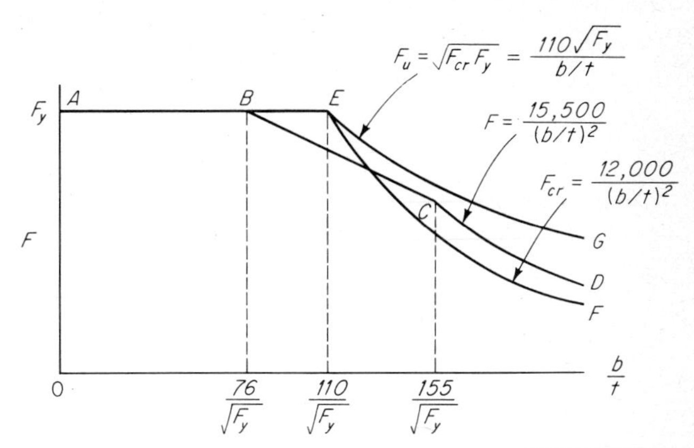

FIGURE 9-17

This is curve CD in Fig. 9-17. Also shown in the figure are EF, the critical stress for $k = 0.456$, and EG, the postbuckling strength, which intersect at E, corresponding to $b/t = 110/\sqrt{F_y}$, the upper limit of slenderness for which yield stress can be reached in a perfect plate. Point B at $b/t = 76/\sqrt{F_y}$ is the corresponding limiting value with allowance for out-of-flatness, residual stress, etc. These limiting values were discussed in Art. 4-23 and summarized in Table 4-4. Inelastic buckling is represented by BC, which, although somewhat arbitrary, is based on results of tests, most of which were on cold-formed members. This inelastic limiting stress is given by

$$F_L = F_y\left(1.340 - 0.00447\sqrt{F_y}\,\frac{b}{t}\right) \tag{b}$$

which intersects CD at $b/t = 155/\sqrt{F_y}$.

Similar formulas are used for other unstiffened elements, where some rotational restraint at the supported edge can be developed. The formulas for projecting compression flanges of beams and columns are

$$F_L = \begin{cases} F_y\left(1.415 - 0.00437\sqrt{F_y}\,\dfrac{b}{t}\right) & \dfrac{95}{\sqrt{F_y}} \lessapprox \dfrac{b}{t} \lessapprox \dfrac{176}{\sqrt{F_y}} \qquad (c) \\[2ex] \dfrac{20,000}{(b/t)^2} & \dfrac{176}{\sqrt{F_y}} \lessapprox \dfrac{b}{t} \qquad (d) \end{cases}$$

The limiting $b/t = 95/\sqrt{F_y}$ for Eq. (c) is the upper limit of slenderness for projecting compression flanges of beams and columns (Table 4-4).

It should be noted that the limiting stresses F_L discussed above are not allowable stresses. How they are used in the specification will be described next.

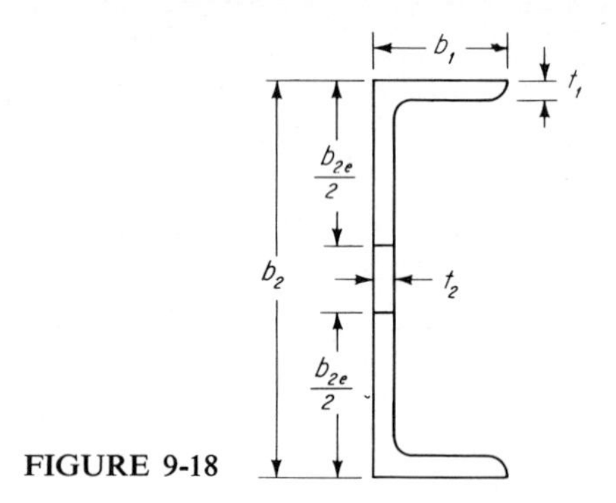

FIGURE 9-18

For the case of the column short enough not to fail by bend buckling, the useful strength for a member containing both stiffened and unstiffened components is determined as follows. For the channel of Fig. 9-18 as an example,

$$P = F_L(2b_1t_1 + b_{2e}t_2) \tag{e}$$

where F_L is the limiting average stress on the flanges given by Eqs. (*c*) and (*d*). From Eq. (*e*), the average stress on the member is

$$f_{\text{av}} = \frac{P}{A} = F_L\frac{2b_1t_1 + b_{2e}t_2}{2b_1t_1 + b_2t_2} \tag{f}$$

The ratio of the effective area of the member to its gross area is denoted by Q_a,

$$Q_a = \frac{\text{effective area}}{\text{gross area}} \tag{g}$$

With this notation and using $F_L = Q_sF_y$, Eq. (*f*) gives

$$f_{\text{av}} = Q_aQ_sF_y = QF_y \tag{9-26}$$

where $Q = Q_aQ_s$. Values of Q_s, obtained by dividing Eqs. (*a*) to (*d*) by F_y, are given in appendix C of the AISC specification.

Since QF_y in Eq. (9-26) is the average stress on the column short enough not to bend-buckle, it gives a point at $L/r = 0$ on the f-L/r plot (Fig. 9-19). Bend buckling would then be accounted for by a curve from this point to an intersection with the column curve *ABC* at the point corresponding to the critical stress for the plate component which buckles first, as shown in Fig. 9-16. However, it is more convenient to use the parabola whose vertex is at QF_y and which intersects and is tangent to the Euler curve at $QF_y/2$. The equation for this column curve is the same as the CRC column curve [Eq. (4-16)] with QF_y substituted for F_y. The corresponding allowable-stress formula is obtained by substituting QF_y for F_y in Eq. (4-17). Of course, C_c

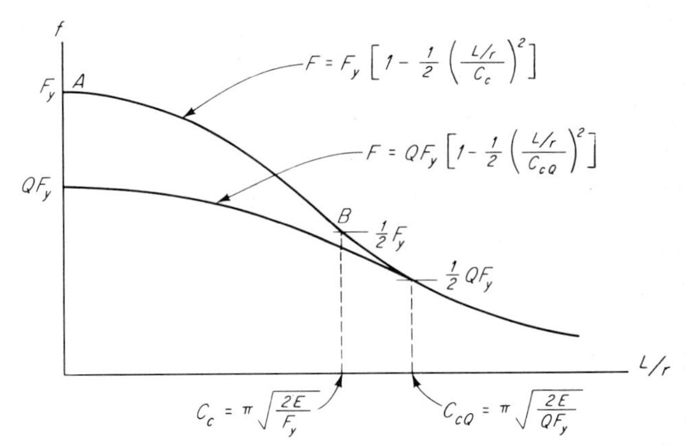

FIGURE 9-19

must be similarly redefined, so that $C_{cQ} = \pi\sqrt{2E/QF_y}$. The radius of gyration must depend on the stress distribution on the column cross section, so that it is different for the very short column, which fails at an average stress nearly equal to QF_y, than for the longer column, which fails at a smaller average stress. However, the specification defines the radius of gyration as that of the gross cross section.

Similar formulas and procedures are prescribed for cold-formed steel structural members in the AISI specification. The two specifications give essentially the same results.

EXAMPLE 9-5 Compute the AISC allowable load for a 20-ft pin-ended A36-steel column with the cross section shown in Fig. 9-20.

SOLUTION According to the specification, the effective width of plate components of columns is to be determined for a stress $f = 0.6F_y$ if the member has no unstiffened components and for $f = 0.6Q_sF_y$ if it does. For this member, $f = 0.6 \times 36 = 21.6 \approx 22$ ksi. Then, with $b/t = 20/0.25 = 80$, Eq. (9-25) gives

$$\frac{b_e}{t} = \frac{253}{\sqrt{22}}\left(1 - \frac{50.3/\sqrt{22}}{80}\right) = 46.5$$

$$b_e = 46.5 \times \tfrac{1}{4} = 11.6 \text{ in.}$$

$$A_{\text{eff}} = 4 \times 11.6 \times \tfrac{1}{4} = 11.6 \text{ in.}^2 \qquad A = 4 \times 20 \times \tfrac{1}{4} = 20 \text{ in.}^2$$

$$Q = \frac{11.6}{20} = 0.58 \qquad QF_y = 0.58 \times 36 = 20.9 \text{ ksi}$$

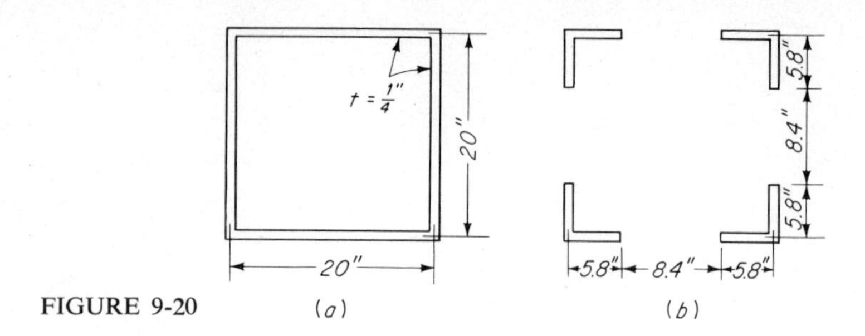

FIGURE 9-20

$$C_{cQ} = \pi\sqrt{\frac{2E}{QF_y}} = \pi\sqrt{\frac{60{,}000}{20.9}} = 168$$

$$I = 2 \times 20 \times \tfrac{1}{4} \times 10^2 + 2 \times \tfrac{1}{4} \times \frac{20^3}{12} = 1{,}000 + 333 = 1{,}333 \text{ in.}^4$$

$$r = \sqrt{\frac{1{,}333}{20}} = 8.20 \qquad \frac{L}{r} = \frac{20 \times 12}{8.20} = 29.3$$

$$F.S. = \frac{5}{3} + \frac{3}{8}\,\frac{29.3}{168} - \frac{1}{8}\left(\frac{29.3}{168}\right)^3 = 1.73$$

$$F_a = \frac{20.9}{1.73}\left[1 - \frac{1}{2}\left(\frac{29.3}{168}\right)^2\right] = 11.9 \text{ ksi}$$

$$P = 11.9 \times 20 = 238 \text{ kips}$$

PROBLEMS

9-4 Compute the AISC allowable load for a 16-ft A36 steel column with the cross section shown in Fig. 9-14.

9-5 Compute the allowable load for a 10-ft A570 Grade D steel column with the cross section shown in Fig. 9-15 with $B = 4$, $D = 4$, $d = 0.915$, $t = 0.075$, $R = \frac{3}{32}$, all in inches, and with a 5.68×0.06 in. plate spot-welded to the hat to form a closed section. Use the AISI specification, for which the allowable stress for axially loaded compression members is, in kips per square inch,

$$F_a = \begin{cases} 0.522QF_y - \left(\dfrac{QF_y}{1{,}494}\right)^2\left(\dfrac{KL}{r}\right)^2 & \text{for } \dfrac{KL}{r} \leq \dfrac{C_c}{\sqrt{Q}} \\[2ex] \dfrac{151{,}900}{(KL/r)^2} & \text{for } \dfrac{KL}{r} \geq \dfrac{C_c}{\sqrt{Q}} \end{cases}$$

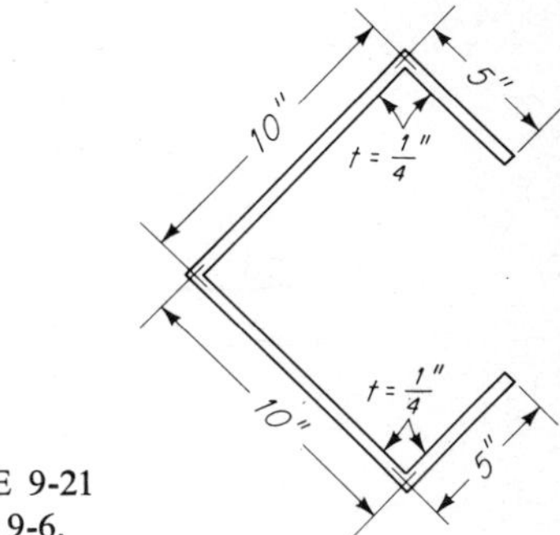

FIGURE 9-21
Problem 9-6.

9-6 Compute the allowable load for a 10-ft pin-ended A36-steel column with the cross section shown in Fig. 9-21 for a factor of safety of 1.8 based on accounting for the full post-buckling strengths of the components.

9-7 POSTBUCKLING STRENGTH OF BEAMS

The bending strength of beam cross sections with compression-flange components which buckle locally is discussed in this article. If the compression flange of the box shown in Fig. 9-22 buckles at a stress less than F_y, the bending strength of the cross section is attained with the stress distribution shown in Fig. 9-22*b*. The equivalent stress distribution on the effective cross section of *c* is shown in *d*. Since the neutral axis lies below mid-depth, the compression flange yields before the tension flange. Therefore, the edge stress f_e is known, and the effective width b_{ey} can be determined. On the other hand, if the cross section is one in which the tension flange yields first, the edge stress f_e on the compression flange is not known, so that either it or the effective width must be assumed and the resisting moment determined by successive approximations.

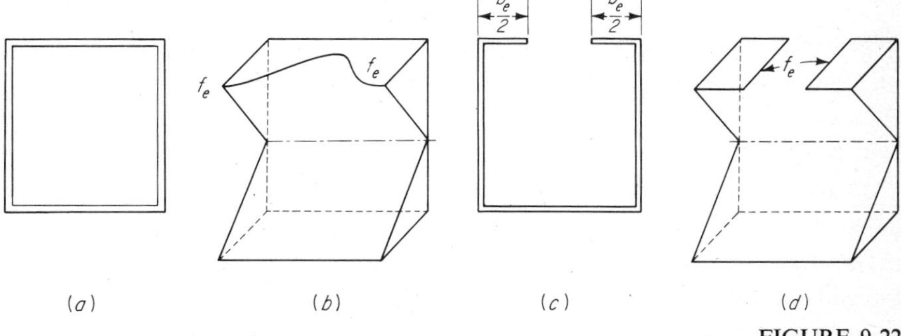

FIGURE 9-22

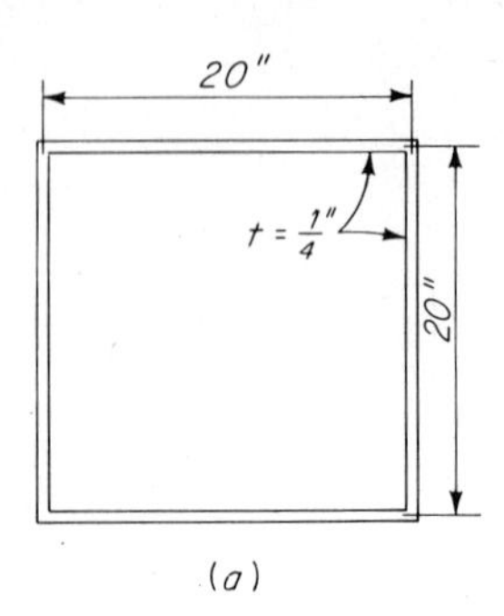

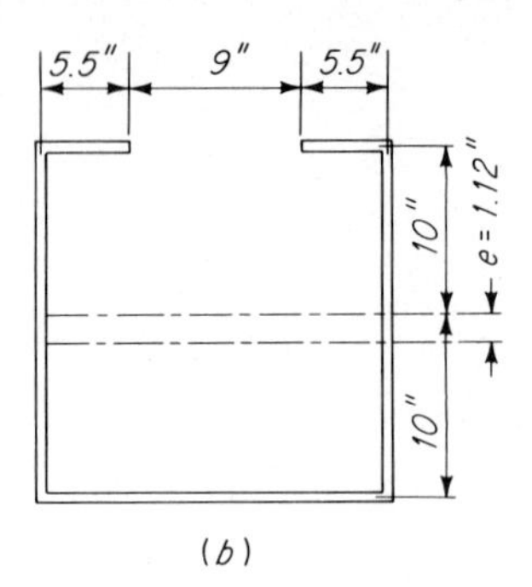

FIGURE 9-23

EXAMPLE 9-6 Compute the resisting moment at extreme-fiber yield of the $20 \times 20 \times \frac{1}{4}$ in. steel box shown in Fig. 9-23*a*. $F_y = 42$ ksi.

SOLUTION The compression flange yields first. Therefore, using Eq. (9-17*b*),

$$\sqrt{\frac{E}{F_y}} = \sqrt{\frac{30{,}000}{42}} = 26.8 \qquad \frac{b}{t} = \frac{20}{1/4} = 80$$

$$\frac{b_{ey}}{t} = 1.9 \times 26.8\left(1 - 0.415\,\frac{26.8}{80}\right) = 43.9 \qquad b_{ey} = 43.9 \times \tfrac{1}{4} = 11$$

$$A = 0.25(4 \times 20 - 9) = 17.75 \text{ in.}^2 \qquad e = \frac{0.25 \times 9 \times 10}{17.75} = 1.27 \text{ in.}$$

$$I = 0.25 \times 11 \times 11.27^2 = 365$$

$$0.25 \times 20 \times 8.73^2 = 382$$

$$2 \times 0.25 \times \frac{11.27^3}{3} = 238$$

$$2 \times 0.25 \times \frac{8.73^3}{3} = 111$$

$$\overline{1{,}096 \text{ in.}^4}$$

$$M = \frac{F_y I}{c} = \frac{42 \times 1{,}096}{11.27} = 4{,}080 \text{ in.-kips}$$

If the compression flange has only unstiffened elements, the resisting moment depends on the decision one makes relative to the postbuckling of such elements. If the postbuckling strength is not taken into account, the critical stress can be determined from Eq. (9-11). With this stress known, it is a simple matter to determine for an unsymmetrical cross section whether the compression flange buckles before the

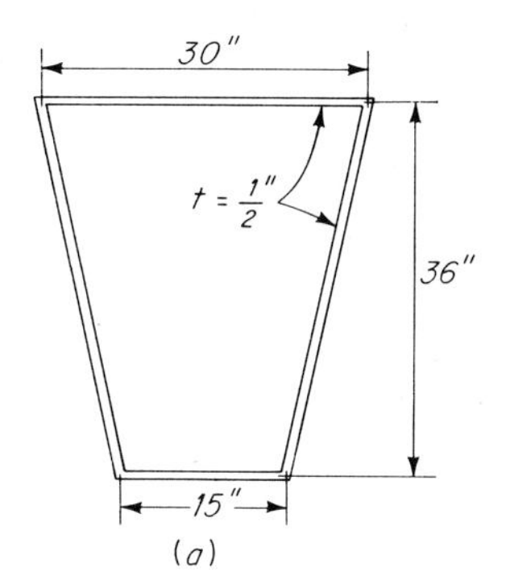

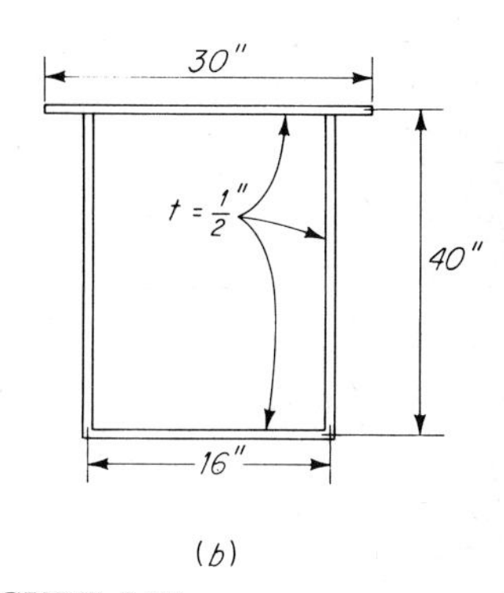

FIGURE 9-24
(*a*) Problem 9-9, (*b*) Problem 9-10.

tension flange yields. If postbuckling of the compression flange is considered, the effective width can be determined by one of the formulas from Art. 9-2, which requires a solution by successive approximation if the tension flange yields first. Of course, in the usual case, where design according to a standard specification is required, an allowable stress will be prescribed for the compression-flange element.

The effective widths which determine the bending strength of a beam cannot be used to calculate service-load deflections because they are the effective widths at extreme-fiber yield while at service load extreme-fiber stresses are less than yield. Therefore, effective widths are larger for deflection calculations. In the case of effective widths prescribed by specification formula, as in Eq. (9-24), the corresponding formula for deflection calculations is obtained by multiplying the two numerical coefficients by the square root of the factor of safety. This is explained by the fact that b_e/t is inversely proportional to the square root of the stress.

PROBLEMS

9-7 Compute the allowable bending moment for the A36-steel cross section of Fig. 9-13 according to the AISC specification. What factor of safety does this allowable moment have with respect to the bending resistance taking into account full postbuckling of the flange?

9-8 Compute the allowable bending moment for the A570 Grade D cross section of Fig. 9-15, with the top flange in compression, according to the AISI specification.

9-9 Compute for a factor of safety of 1.75 the allowable bending moment of the A441-steel cross section shown in Fig. 9-24*a*. The top flange is in compression.

9-10 Same as Prob. 9-9 for the cross section of Fig. 9-24*b*.

9-11 Compute for a factor of safety of 1.8 the allowable uniformly distributed load for a beam of the cross section of the example in Art. 9-7 simply supported on a span of 24 ft. Determine the service-load deflection.

REFERENCES

1 Timoshenko, S., and J. M. Gere: "Theory of Elastic Stability," 3d ed., McGraw-Hill, New York, 1969.

2 Gerard, G.: "Introduction to Structural Stability Theory," McGraw-Hill, New York, 1962.

3 Karman, T. von, E. E. Sechler, and L. H. Donnell: The Strength of Thin Plates in Compression, *Trans. ASME, APM*, vol. 54, no. 2, January 1932.

4 Winter, G.: Strength of Light-gage Steel Compression Flanges, *Trans. ASCE*, vol. 112, 1947.

5 Haaijer, G.: Plate Buckling in the Strain-hardening Range, *Trans. ASCE*, vol. 124, 1959.

6 Haaijer, G.: Selection and Application of Constructional Steels, in E. D. Verink, Jr. (ed.), *Methods of Materials Selection, Metall. Soc. Conf.*, vol. 40 (Gainesville, Fla., May 1966), Gordon and Breach, New York.

7 Dhalla, A. K.: Influence of Ductility on Structural Behavior of Cold-formed Steel Members, *Cornell Univ. Dept. Struct. Eng. Rept.* 336, 1971.

10

STEEL BRIDGES

10-1 INTRODUCTION

The factors that must be considered in the design of a bridge may be grouped into three interdependent categories: economic, functional, and physical. In addition, there are governmental regulations to be met if the bridge crosses navigable waters. There are often conflicts in requirements that can be resolved only by compromise, and judgment, skill, and broad experience of a high order are essential to the development of a sound plan. Economics plays an important role. Not only are there questions of economy with respect to the type of structure to be built, the materials to be used, the number of spans, the method of erection, and the like, but in addition there are questions that have to do with the economic feasibility of the project itself. The functional planning involves considerations of grade and alignment, number of tracks or traffic lanes, approach facilities for collecting and dispersing traffic of highway bridges, and appearance. Conditions at the site have a considerable influence on grade and alignment. Although beauty is not prerequisite to efficient performance, an ugly bridge is not functional in the broad sense of the word. The bridge should be in harmony with its setting, and in many cases this in itself can well determine the type that should be built. The physical planning involves consideration of the foundation materials, the character of the stream if one is involved, the type of bridge, the number of spans, and questions of similar nature. Borings to determine the character and depth of foundation materials are an obvious necessity. Information on the stream's flood stages and low-water stages, on its scouring propensities, and on the nature and amount of drift carried by floods is also vital. Both foundation and stream characteristics have a direct influence on the layout of the bridge. For example, the more expensive the pier, the longer will be the economic span, and it is evident that foundation materials at considerable depths and high flood stages both call for taller and therefore more expensive piers. Furthermore, as the distance between piers increases, the limit of the simple-span bridge is eventually passed, so that a structure suitable for long spans, say the cantilever or the suspension bridge, becomes a necessity.

The federal government has authority to regulate the bridging of navigable waters; permits for the construction of such bridges (except those over streams which have been placed in the "advanced approval" category by the Commandant, U.S. Coast Guard) must be obtained from the U.S. Coast Guard and other appropriate agencies.

The discussion in this article is intended only to highlight the problems involved in planning a bridge. Bridge design is treated in detail in Ref. 1. Except for the fact that certain elements such as floors are more or less independent of the type of bridge, discussion in the articles to follow is limited to simple-span beam bridges and truss bridges.

10-2 ECONOMICS OF SIMPLE-SPAN BRIDGES

Simple-span bridges were described and discussed in Art. 2-2. The rolled-beam bridge is likely to be an economical choice for railroad bridges of spans up to 50 ft, while the highway beam bridge may be economical for spans up to about 60 ft, and the composite beam bridge for spans up to about 100 ft. The plate-girder bridge becomes economical for spans of about 60 ft and is commonly used for spans to 300 ft or more. Welded girders made up of three plates compete with rolled beams in all but the shortest spans. The plate-girder bridge is likely to be cheaper than truss bridges up to at least the limit for shipment in one piece. The economical limiting span for both the beam bridge and the plate-girder bridge is greater for continuous structures than it is for simple spans. Continuous plate-girder bridges with spans exceeding 950 ft have been built.

Although trusses are usually used in highway bridges only for very long spans, they may be economical for shorter spans where aesthetics and (in the case of through bridges) safety with high-speed traffic are not critical. Deck truss bridges are preferable. The use of welding, in conjunction with H sections and box sections for truss members, produces a clean, light, cheap, and easy-to-maintain structure.[1]

The depth of simple-span bridge trusses usually ranges from about one-fifth to one-eighth the span, shorter spans being relatively deeper. The depth-span ratio is also somewhat dependent upon the live load, so that highway-bridge trusses are usually shallower than railroad-bridge trusses of comparable span. Trusses for deck bridges are often relatively shallower than trusses for through bridges. Both the AASHO and the AREA specifications prescribe one-tenth the span as the preferable minimum depth of truss. Trusses of economical proportions usually result if the angle between diagonals and verticals is between 35 and 40°. Thus, panel lengths increase with the span of the truss and eventually lead to excessively heavy floor systems unless subdivided trusses are adopted. Panel lengths of 16 to 32 ft are usually economical for the highway-bridge truss. For spans greater than about 320 ft, the K truss is advisable to reduce floor weight and inclination of the diagonals.

The relative economy of the parallel-chord truss and the curved-chord truss is not easy to evaluate. In the curved-chord truss, chord stresses are essentially uniform, and their vertical components relieve the web members. Thus, the curved-chord truss is lighter than the parallel-chord truss, but costs of fabrication and erection may be slightly higher. It should be noted, however, that for through bridges the curved-chord truss is more graceful. On the other hand, the lines of the parallel-chord truss are more in harmony with the deck bridge.

10-3 BRACING

Lateral bracing for the deck truss bridge usually consists of horizontal trusses in the planes of each of the two chords of the vertical trusses. Since lateral forces resulting from live load are resisted almost entirely by the lateral truss in the plane of the floor, the top lateral truss is more important than the bottom lateral truss. However, the top lateral truss is not as important for the deck highway bridge as for the deck railroad bridge, since the former usually has a concrete or steel floor, which, with the stringers and floor beams, can take over the function of the lateral truss. Nevertheless, a top lateral truss must be provided in any case to help in the erection of the bridge and to furnish wind resistance until the floor is in place. Except for the half-through (pony-truss) bridge, which is virtually obsolete, lateral bracing for through bridges also consists of two trusses, but the relative importance of the two systems is reversed. Two diagonals in each panel are used ordinarily, but single diagonals are common in single-track plate-girder railroad bridges.

Members of the lateral truss take various forms. Single angles are quite common for the relatively unimportant bottom lateral truss of the deck bridge, and it is usual to assume that the shear is resisted entirely by the diagonal that is in tension. Of course, this means that the member connecting opposite panel points must be designed as a

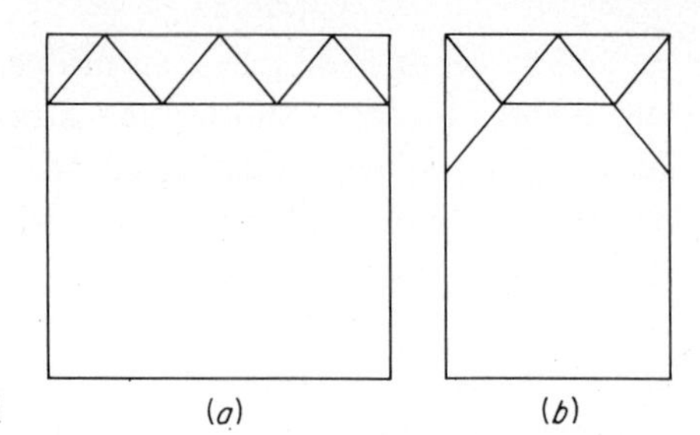

FIGURE 10-1

strut, so that it may need to be a double-angle member or perhaps an H. Diagonals of the bottom lateral truss in through bridges are usually single or double angles connected to gusset plates on the bottom flanges of the chords. This truss is often designed on the assumption that one diagonal in a panel resists half the shear in tension while the other resists the other half in compression. The top laterals of both deck and through bridges are usually the same depth as the chord so as to connect to both its flanges. A separate system of lateral bracing is usually employed to connect the stringers of an open-floor railroad bridge to relieve them of bending due to lateral forces from the train. A single system in the plane of the top flanges is sufficient.

Bracing in the transverse vertical planes is called *sway bracing*. In the deck truss bridge, two diagonals in the transverse planes at each panel point are used. The sway frames at the ends of the bridge transmit the end reactions of the top lateral truss to the abutment, and must be designed accordingly. Lateral forces on the intermediate frames are not usually calculated, since the lateral trusses are designed to transmit these forces to the abutments. Therefore, sizes of members of the intermediate frames are a matter of judgment, except for limitations as to slenderness ratio.

The end posts of the through bridge must be tied together to form a rigid frame capable of transferring the end reaction of the top lateral truss to the abutments. This combination of end posts and bracing is called a *portal frame*. In order to minimize bending stresses in the end posts and provide a maximum of rigidity, portal bracing should be as deep as headroom allows. The portal frame shown in Fig. 10-1*a*, or some modification of it, is quite common for highway bridges and double-track railroad bridges. The frame shown in Fig. 10-1*b*, or some modification of it, is common for single-track railroad bridges. The frame is almost invariably a two-plane structure connecting to both flanges of the end posts. Its members are usually of the same type as the top lateral diagonals.

10-4 WEIGHT OF BRIDGES

The dead-load force is often a large percentage of the total force in a bridge-truss member. For this reason, it is helpful to have a good estimate of the weight of a bridge before its members are designed. The weight is dependent on the span, depth, width of roadway, number of panels, and loading, as well as on the design specifications and the individuality of the designer. Many rules and formulas for estimating weights of bridges have been devised. Probably the most complete data were published by Waddell.[2] A simple and quite accurate method of estimating the weight per foot of two trusses and their bracing was developed by Hudson and is given by the equation

$$w = \frac{100A_n}{3} \tag{10-1}$$

where w is the weight per foot of bridge of two trusses and their bracing and A_n is the required net area, in square inches, of the largest member of the tension chord. The net area A_n is calculated for the total force due to live load, impact, and dead load, including a guessed weight of trusses and their bracing.

10-5 BRIDGE FLOORS

The most common floor system for the steel highway bridge consists of a reinforced-concrete slab supported on steel stringers parallel to the direction of traffic. The stringers are supported by floor beams connecting to girders or trusses. In the beam bridge, the beams themselves perform the function of the stringers; and since the beams rest on abutments or piers, no floor beams are needed. The concrete slab may be bonded mechanically to the top flanges of the stringers by using connectors to develop horizontal shear between the slab and the stringer. This is called *composite construction*, and is discussed in Chap. 5 and Art. 10-11.

Since the floor constitutes a major part of the dead load of a bridge, several forms of lightweight floor construction have been developed. The battle-deck floor consists of steel plate welded to the top flanges of the stringers. The plate is usually less than 1 in. thick, and the stringers are usually spaced 18 to 32 in. apart. The plate not only distributes the wheel load over several stringers but also serves as part of the compression flange of each stringer. Unless it is covered with a bituminous surfacing, battle-deck flooring is likely to be slippery when wet. Another type of steel flooring is made by riveting together alternate straight and crimped bars to form an open grid. The straight bars are the main load-carrying members and are braced laterally by the crimped bars. The top edges of the bars are flush. This type of open-grid flooring is welded to a secondary system of structural members (called *sills*) placed transverse to, and supported by, stringers. In another type of open-grid floor the sill is eliminated by special carrying beams that are an integral part of the grid. Several other types of grid flooring are manufactured. In some cases the grid is filled with concrete. Orthotropic steel-deck-plate construction is described in Art. 2-2 (Fig. 2-11) and discussed in detail in Ref. 1.

The most common type of floor for the railroad bridge consists of timber crossties resting on stringers which frame into floor beams. Since this is an open floor, it can be used only for stream and secondary highway crossings; over city streets and main highways a ballasted roadway on a solid floor is necessary. The solid floor may consist of an all-metal deck or of a concrete or timber trough supported on stringers and floor beams or on floor beams alone. In addition to the protection it offers, this type of floor provides a continuous ballasted roadbed.

10-6 DESIGN OF FLOOR SLABS

The principal moment in the bridge-floor slab is the positive moment in the transverse direction midway between stringers. The design procedure prescribed by the AASHO specifications is based on theoretical studies by Westergaard.[3] For a slab whose span is perpendicular to the flow of traffic, the live-load moment for simple spans in foot-kips per foot width of slab is given by

$$M = \frac{S + 2}{32} P \tag{10-2}$$

where S is the effective span length (for continuous slabs the center-to-center distance between stringers minus half the width of the flange) and P is the load on the rear wheel. The positive moment for slabs continuous over three or more supports is assumed to be 80 percent of the simple-span moment. According to the specifications, Eq. (10-2) applies to both composite and noncomposite construction.

In addition to the positive bending moments, there are, of course, negative transverse moments at the supports (stringers). Furthermore, certain positions of the truck wheels produce negative transverse moments in the midsection of the slab between stringers, while other positions produce positive transverse moments over the supports. All these moments have been found to be less than the positive transverse moments at the midsection of the slab panel. The ratios of these moments to the positive transverse moment vary along the span of the bridge, and experimental results show that they are smaller than theory would indicate. The AASHO specifications prescribe a negative moment at the support equal in magnitude to the positive moment at midpanel.

A wheel load in the central portion of a slab panel produces not only transverse bending moments but longitudinal bending moments as well. The existence of these moments in the longitudinal direction is easy to see if one visualizes the saucerlike depression surrounding the wheel. For this reason, and also to control cracking of the slab due to shrinkage and temperature change, longitudinal reinforcement is needed at both the top and the bottom surfaces of the slab. AASHO specifications prescribe a percentage of the positive transverse reinforcement of $220/\sqrt{S}$, with a maximum of 67 percent. This amount is placed in the bottom in the middle half of the slab span; in the outer quarter of the span an amount of not less than 50 percent of that required in the middle half must be provided.

Shear and bond stresses in reinforced-concrete floor slabs that meet the requirements for moment discussed in this article can be depended upon to be within allowable limits, so that they need not be investigated.

10-7 DESIGN OF STRINGERS AND FLOOR BEAMS

A wheel load applied directly over an interior stringer of a bridge floor is shared by the several stringers in the panel because of the stiffness of the slab in the direction transverse to the stringers. Of course, the load is not distributed equally, and the distribution depends upon the relative stiffnesses of the slab and the stringers. However, since at least two wheels (on the same axle) occupy a given position on the bridge, and since the maximum moment in a given stringer results when both lanes of a two-lane bridge are loaded, the maximum moments for the various stringers do not differ appreciably. The maximum moment for one stringer can be determined with sufficient accuracy by assuming that it supports the portion kP of a single wheel load P, where k is given by

$$k = \frac{b}{s} \qquad (a)$$

where b is the center-to-center spacing of stringers and s is a constant that depends upon the span of the stringer and the relative stiffnesses of the stringer and the slab. Within practical limits for bridge floors of the type under consideration, values of s range from 5 to 6 ft.

For bridges of two or more traffic lanes, the AASHO specifications require that the live-load moment for one interior stringer be computed for the portion kP of a single wheel load P given by

$$k = \frac{S}{5.5} \qquad S < 14 \text{ ft} \qquad (10\text{-}3)$$

where S is the average stringer spacing in feet. No distinction is made between composite and noncomposite construction. If S exceeds 14 ft, the load on one stringer is to be determined by assuming that the floor slab acts as a series of simple beams supported by the stringers. The live load on the outside stringer is to be determined by assuming the slab to act as a simple beam between stringers unless the floor is supported on four or more stringers, in which case the fraction of the wheel load must not be less than

$$\frac{S}{5.5} \qquad S \lesssim 6 \text{ ft}$$
$$\frac{S}{4.0 + 0.25S} \qquad 6 < S < 14 \text{ ft} \tag{10-4}$$

where S is the distance in feet between the outside stringer and the adjacent interior stringer. When $S > 14$ ft, the slab is assumed to act as a simple beam between stringers.

In determining the dead-load moment for one stringer, the stringer can be assumed to support a strip of floor extending halfway to the adjacent stringer on each side. Curbs, railings, and wearing surface, if placed after the slab has cured, may be considered equally distributed to all stringers.

The live-load moment in a floor beam is determined according to the type of load (lane or truck) that produces the larger moment. According to the AASHO specifications, no distribution of wheel loads on floor beams is to be assumed; i.e., wheel loads over the floor beam are considered to be concentrated loads. The dead-load moment can be determined by treating the weight of the slab, stringers, and floor beam as a load distributed uniformly along the floor beam or as a series of concentrated loads acting at the points of connection of the stringers.

10-8 END BEARINGS

Bridge bearings are designed to transmit the loads to the foundation and to provide for expansion of the superstructure. They are of two general types, fixed and expansion. Fixed bearings act as hinges in that they permit rotation but not expansion. Expansion bearings permit rotation as well as movements of the superstructure resulting from temperature change, deflection, etc.

The span beyond which the simple bearing plate is unsatisfactory is largely a question of judgment and experience. According to the AASHO specifications, spans of less than 50 ft may be arranged to slide upon metal plates with smooth surfaces, and no provisions for deflection of the spans need be made. The AREA specifications set the limit at 70 ft. AASHO permits the use of elastomeric pads for both fixed and expansion bearings for all types of bridges. It is the least expensive bearing for light and intermediate reactions. Figure 10-2 shows a bearing that makes use of a rocker between the bearing plate and the beam or girder. A similar detail in which the anchor bolts do not pass through the rocker is shown in Fig. 10-3. In this case, the beam is held in position by means of pintles shaped like gear teeth. This type of support

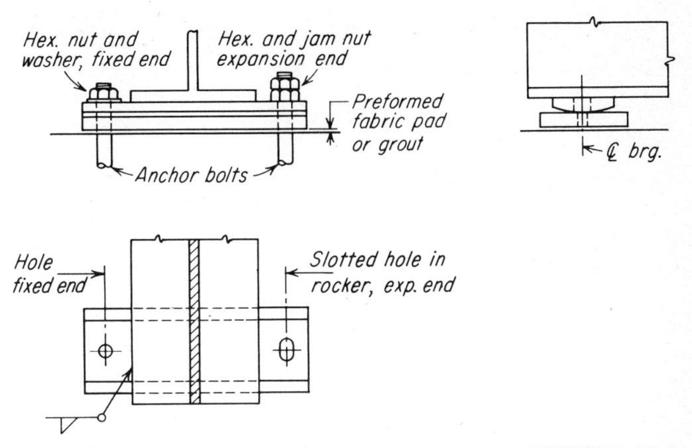

FIGURE 10-2

may be used where resistance to uplift need not be provided. For example, it may be used for the inside beams of the beam bridge, with the outside beams supported by bearings of the type shown in Fig. 10-2.

Figure 10-4*a* shows an expansion bearing for larger bridges. Several variations are shown in the view at the right. The soleplate may be bolted to the girder, as at the left of the centerline, or welded as shown at the right. Resistance to uplift may be provided by using a hinge plate, as at the left; if such resistance is not needed, lateral movement is prevented by a plate such as that shown at the right. A corresponding fixed end bearing is shown in Fig. 10-4*b*.

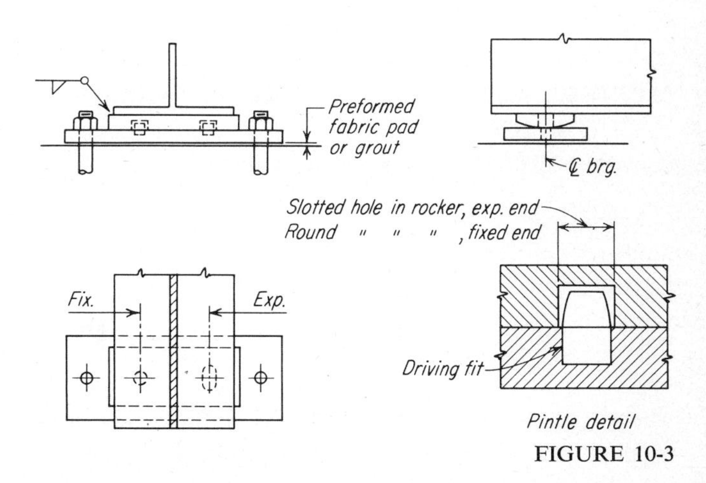

FIGURE 10-3

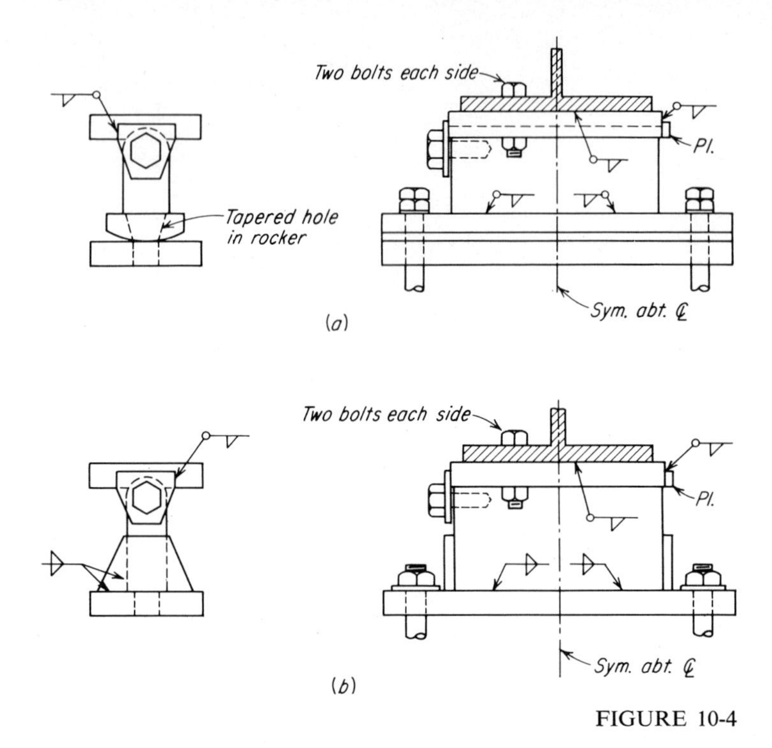

FIGURE 10-4

Although there is only a line of contact between an unloaded rocker and its bearing plate, deformation under load distributes the reaction over a finite area. Evidently, at a given load this area increases with increase in the radius of the rocker, since a rocker of infinitely large radius would have a plane surface to begin with. The allowable load must be evaluated in terms of a limiting permanent deformation. Thus the yield point of the material is also a factor. Empirical formulas based on the results of tests are used for purposes of design. According to both the AREA and the AASHO specifications, the allowable bearing pressure p in pounds per lineal inch between rocker and bearing plate is given by

$$p = \begin{cases} \dfrac{F_y - 13{,}000}{20{,}000} 600d \text{ psi} & d \le 25 \text{ in.} \\[2ex] \dfrac{F_y - 13{,}000}{20{,}000} 3{,}000\sqrt{d} \text{ psi} & 25 \le d \le 125 \text{ in.} \end{cases} \tag{10-5}$$

where F_y is the yield point of the metal in the rocker or the bearing plate, whichever is smaller, and d is the diameter of the rocker in inches.

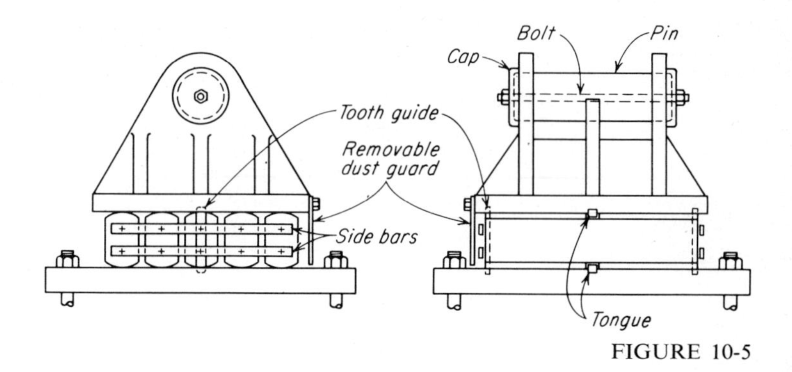

FIGURE 10-5

Since the necessary radius and length of a given rocker depend upon the load to be supported, it is obvious that for a very large reaction the required radius may be too large for satisfactory results, or else the rocker may be too long if the radius is held within reasonable bounds. This dilemma is resolved by using a series of rollers. Segmental rollers (Fig. 10-5) are ordinarily used since they occupy less space than cylindrical rollers. The rollers may be coupled with the sidebars shown and the entire nest held in position by tooth guides which engage slots in the shoe and in the bearing plate. Sidebars may be omitted if each roller is held by teeth. Lateral movement is prevented by the tongues shown in the view at the right. Resistance to uplift may be provided by lugs that have projections extending over the upper surface of the base of the shoe or by enlarging the base of the shoe and providing slotted holes for the anchor bolts. The roller assembly may be enclosed with removable dust guards; they are shown on only two sides in Fig. 10-5 to indicate that they are optional.

A detailed example of the design of bearings is given in the next article.

10-9 DP10-1: ELDERSBURG–LOUISVILLE ROAD BRIDGE

The Eldersburg–Louisville Road Bridge (Figs. 10-6 and 10-7), which crosses Morgan Run in Carroll County, Md., consists of two 216-ft truss spans and two 70-ft beam spans. It was designed for the city of Baltimore by J. E. Greiner Company, Consulting Engineers, of Baltimore. Except for some of the joints and splices, the complete calculations for the design of the truss-span superstructure are presented here through the courtesy and cooperation of the consulting engineers and the Baltimore Department of Public Works. The authors have made some changes in notation and have revised the design in A36 steel to conform to the 1969 edition of the AASHO specifications, but otherwise the calculations are given substantially in the same form as

FIGURE 10-6
Eldersburg–Louisville Road Bridge. (*J. E. Greiner Company, Inc., Consulting Engineers.*)

FIGURE 10-7
Eldersburg–Louisville Road Bridge. (*J. E. Greiner Company, Inc., Consulting Engineers.*)

presented by the Greiner Company. The following comments are identified by letters corresponding to those alongside the computations on the design sheets. Pertinent clauses in the AASHO specifications are noted in parentheses.

Sheet 2

a These calculations determine the location of the center of gravity and the weight of the structure that overhangs the outside stringer. Moment arms are measured (in feet) from the center line of the stringer.

b The sidewalk is designed as a continuous slab, using the clear span between supports. The dead load consists of the weights of the parapet and sidewalk. The live load is 85 psf (1.2.11). The minimum concrete cover measured from the top of the slab is $1\frac{1}{2}$ in. (1.5.6).

Sheet 3

a Spans of roadway slab were discussed in Art. 10-6. A 9-in. flange width for the stringer was assumed.

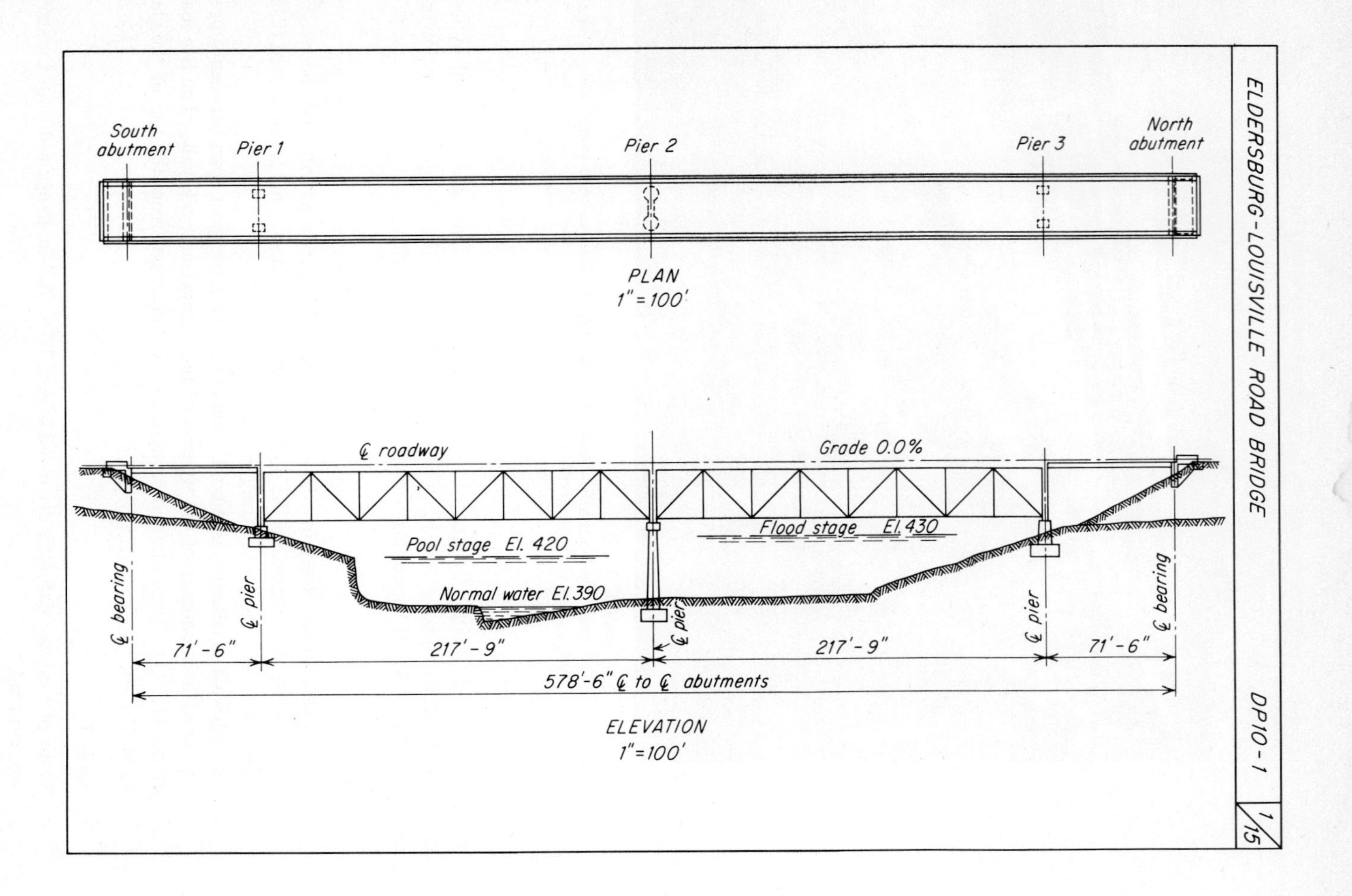
ELDERSBURG-LOUISVILLE ROAD BRIDGE
DP10-1
1/15
South abutment
Pier 1
Pier 2
Pier 3
North abutment
PLAN
1"=100'
℄ roadway
Grade 0.0%
Pool stage El. 420
Flood stage El. 430
Normal water El. 390
℄ bearing
℄ pier
℄ pier
℄ pier
℄ bearing
71'-6"
217'-9"
217'-9"
71'-6"
578'-6" ℄ to ℄ abutments
ELEVATION
1"=100'

ELDERSBURG-LOUISVILLE ROAD BRIDGE DP10-1 

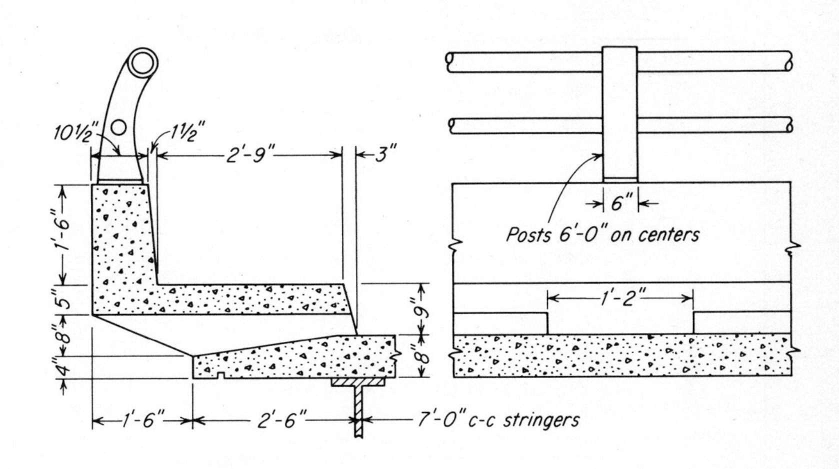

Section		Weight	Arm	Moment	(a)
Top rail	(3½ std. steel pipe)	9.2	3.4	31	
Bott. rail	(2½ std. steel pipe)	5.8	3.7	22	
Post	65 × 1/6 (cast iron)	10.8	3.6	39	
Parapet	0.875 × 1.5 × 150	= 197.5	3.56	703	
	½ × 0.125 × 1.5 × 150	= 14.0	3.08	43	
Sidewalk	0.42 × 3.82 × 150	= 241.0	2.09	504	
Sidewalk support	½ × 0.67 × 1.5 × 1.17 × 150 × 1/6	= 14.6	3.00	44	
	0.33 × 2.5 × 1.17 × 150 × 1/6	= 24.3	1.25	30	
	½ × 0.33 × 2.17 × 1.17 × 150 × 1/6	= 10.5	1.78	19	
		527.7	2.72	1435	

SIDEWALK Span = 6.0 − 1.17 = 4.83′ (b)

DL 197.5 + 14.0 + 241 = 452.5 #/′ × $4.83^2/10$ = 1055

LL 85 × 3 = 255.0 #/′ × $4.83^2/10$ = 595

1650′#

$d = 5 - 1.5 - 0.31 = 3.19''$

$$A_s = \frac{1650 \times 12}{20{,}000 \times 0.88 \times 3.19} = 0.35 \text{ in.}^2$$

#5 @ 8″ $A_s = 0.46$ in.²

ELDERSBURG-LOUISVILLE ROAD BRIDGE DP10-1 3/15

ROADWAY SLAB

Stringers 7'-0" c.c.

$S = 7.0 - 0.37 = 6.63'$ (a)

$M_D = 1/10 \times 0.1 \times 6.63^2 = 0.44$ (b)

$M_L = 0.8 \times 16 \times (6.63+2)/32 = 3.46$ (b)

$M_I = 0.30 \times 3.46 = 1.04$ (c)

$4.94'^k$

8 in. slab $d = 8.0 - 0.75 - 1.50 - 0.31 = 5.44"$ (d)

#5 @ 6" $A_s = 0.62 \text{ in.}^2 \quad p = \dfrac{0.62}{12 \times 5.44} = 0.0095$

$k = 0.351; \quad j = 0.882$

$$f_s = \frac{M}{A_s jd} = \frac{4.94 \times 12}{0.62 \times 0.882 \times 5.44} = 19.9 \text{ ksi}$$

$$f_c = \frac{2pf_s}{k} = \frac{2 \times 0.0095 \times 19.9}{0.351} = 1.08 \text{ ksi}$$

Section over ℄ of stringer at face of curb
(carried by 4 ft. section of slab) (e)

		Shear	Arm	Moment	
DL	0.528 × 6	3.17	2.72	8.62'k	(f)
	0.1 × 2.5 × 4	1.00	1.25	1.25	(g)
LL sdwk	85 × 3 × 6	1.53	1.50	2.31	
LL curb	0.5 × 6	3.00	0.75	2.25	(h)
				14.43'k	

$d = 8 - 2 = 6"$ #5 @ 6" $A_s = 8 \times 0.31 = 2.48 \text{ in.}^2$ (i)

$p = \dfrac{2.48}{48 \times 6} = 0.00861 \quad k = 0.338 \quad j = 0.887$

$$f_s = \frac{14.43 \times 12}{2.48 \times 0.887 \times 6.0} = 13.1 \text{ ksi}$$

$$f_c = \frac{2 \times 0.00861 \times 13.1}{0.35} = 0.67 \text{ ksi}$$

ELDERSBURG-LOUISVILLE ROAD BRIDGE — DP10-1 — 4/15

EXTERIOR STRINGER

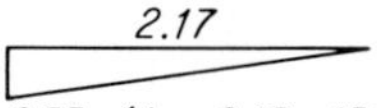

$0.33 \times 1/2 \times 2.17 \times 150 \times 4.83/6.0 = 44.5^{\#/'}$

Sdwk & parapet	$528 \times 9.72/7$	$= 734$	(a)
Fl. slab	$(100 \times 9.5 \times 4.75 - 44.5 \times 8.78)/7$	$= 587$	
Stringer		$= 82$	
		1403	

Sidewalk LL $= 85 \times 3 \times 8.5/7 = 310^{\#/'}$

Fraction of wheel load to stringer $= 7/(4 + 0.25 \times 7) = 1.22$ (b)

		Shear		Moment	
DL	1.403×13.5	$= 19.0$	$1/8 \times 1.403 \times 27^2$	$= 128$	(c)
Sdwk LL	310×13.5	$= 4.2$	$1/8 \times 0.310 \times 27^2$	$= 28$	
LL	$1.22 \times 47.4 \times 1/2$	$= 28.9$ (d)	$1.22 \times 237 \times 1/2$	$= 145$	(d)
I	0.30×28.9	$= 8.7$	0.30×145	$= 43$	
		V 60.8^k		M $344'^k$	

$$\frac{I}{c} = \frac{344 \times 12}{20 \times 1.25} = 165 \text{ in.}^3 \qquad \underline{W24 \times 76} \qquad (e)$$

<u>7/8 A325 bolts</u> Friction-type connection: $ss = 8.12^k$
$60.8/8.12 = 7.5$ req'd.

$$\text{DL defl.} = \frac{5ML^2}{48EI} = \frac{5 \times 128 \times 27^2 \times 1728}{48 \times 29{,}000 \times 2100} = 0.276''$$

no camber req'd. (f)

INTERIOR STRINGER

Fl. slab	100×7	$= 700$	
Stringer		90	
		790	

Fraction of wheel load to stringer $= 7/5.5 = 1.27$ (g)

		Shear		Moment
DL	0.79×13.5	$= 10.7$	$1/8 \times 0.79 \times 27^2$	$= 72$
LL	$1.27 \times 47.4 \times 1/2$	$= 30.1$	$1.27 \times 237 \times 1/2$	$= 151$
I	0.30×30.1	$= 9.0$	0.30×151	$= 45$
		V 49.8^k		M $268'^k$

$$\frac{I}{c} = \frac{268 \times 12}{20} = 161 \text{ in.}^3 \qquad \underline{W24 \times 76}$$

<u>7/8 A325 bolts</u> $49.8/8.12 = 6.2$ req'd.

Effect of cantilevered sidewalk on first interior stringer
Use 90% DL

$0.9 \times 528 \times 2.72/7 = 185^{\#/'}$ uplift (h)

$M = 0.185 \times 27^2/8 = 17'^k$

$M = 268 - 17 = 251'^k$

$$\frac{I}{c} = \frac{251 \times 12}{20} = 151 \text{ in.}^3 \qquad \underline{W24 \times 68}$$

ELDERSBURG-LOUISVILLE ROAD BRIDGE DP10-1 5/15

INTERIOR FLOOR BEAM

7' — 7' — 7' — 7'
4' — 20' c-c trusses — 4'

LL per lane
$32 + (8 + 32)\ 13/27 = 51.3^k$ (a)

DL ext. stringer	$1.40 \times 27 + 0.11 \times 2$	$= 38.0^k$	(b)
DL int. stringer	$(0.79 - 0.20)\ 27 + 0.11 \times 5$	$= 16.5^k$	(c)
DL ctr. stringer	$0.79 \times 27 + 0.11 \times 7$	$= 22.1^k$	
Sdwk LL = $0.085 \times 3 \times 27 \times 8.5/7.0$		$= 8.4^k$	

Max. neg. moment

DL	38.0×4	= 152
Sdwk	8.4×4	= 34
LL	$51.3 \times 1/2 \times 1.22 \times 4.0$	= 125
I	0.30×74	= 22
		$333'^k$

LL to ext. stringer	$51.3 \times 1/2 \times 1/7$	$= 3.7^k$	(d)
LL to int. stringer	$51.3 \times 1/2 \times 8/7$	$= 29.3^k$	
LL to ctr. stringer	$51.3 \times 5/7$	$= 36.7^k$	

Max. pos. moment

DL	$38.0\ (10-14) + 16.5\ (10-7) + 11.05 \times 10$	= 8
LL	$3.7\ (10-14) + 29.3\ (10-7) + 18.35 \times 10$	= 257
I	0.30×257	= 77
		$342'^k$

$$\frac{I}{c} = \frac{342 \times 12}{20} = 205 \text{ in.}^3 \quad \underline{W27 \times 84}$$

Net $I = 2830 - 0.490\ (2^2 + 6^2)\ 2 = 2790$ in.^{4}

Net $I/c = 2790/13.35 = 208$ in.^{3}

END FLOOR BEAM

LL per lane	$1/27\ (32 \times 28.25 + 32 \times 14.25 + 8 \times 0.25)$	$= 50.6^k$	(e)
DL ext. stringer	$1.40 \times 28.25 \times 14.12/27 + 0.11 \times 2$	$= 20.9^k$	
DL int. stringer	$(0.79 - 0.20) \times 28.25 \times 14.12/27 + 0.11 \times 5$	$= 9.2^k$	
DL ctr. stringer	$0.79 \times 28.25 \times 14.12/27 + 0.11 \times 7$	$= 12.5^k$	
Sdwk LL	$\dfrac{0.085 \times 3 \times 28.25 \times 14.12}{27} \times \dfrac{8.5}{7.0}$	$= 4.6^k$	

Max. neg. moment

DL	20.9×4	= 83.6
Sdwk	4.6×4	= 18.4
LL	$50.6 \times 1/2 \times 1.22 \times 4.0$	= 123.5
I	0.30×72.5	= 21.8
		$247.3'^k$

ELDERSBURG-LOUISVILLE ROAD BRIDGE DP10-1 6/15

END FLOOR BEAM (continued)

LL to ext. stringer $50.6 \times 1/2 \times 1/7 = 3.6^k$
LL to int. stringer $50.6 \times 1/2 \times 8/7 = 28.9^k$
LL to ctr. stringer $50.6 \times 5/7 = 36.2^k$

Max. pos. moment

DL	$20.9 \times (-4) + 9.2 \times 3 + 6.25 \times 10$	= 6.4
LL	$3.5 \times (-4) + 28.9 \times 3 + 18.1 \times 10$	= 253.7
I	0.30×253.7	= 76.1
		336.2'k

$\frac{I}{c} = \frac{336 \times 12}{20} = 202 \text{ in.}^3$ W27×84

TRUSS

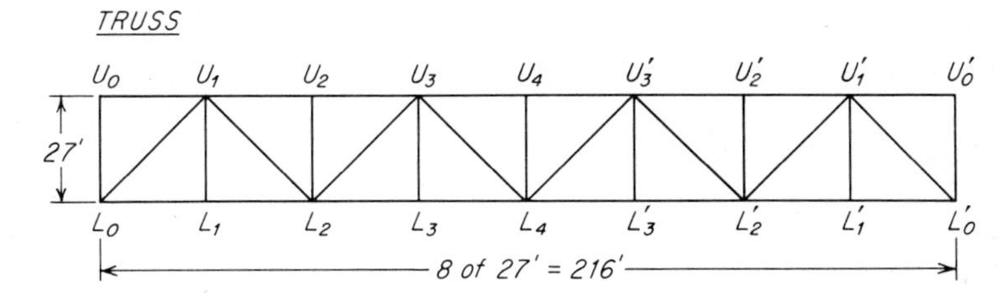

DL per truss	plf	Panel load (kips)
Sdwk., parapet, railing	528	14.3
Fl. slab (100 × 16.5) − 44	1606	43.4
Stringers (76 + 76 + 38) 1.02 (a)	190	5.1
Fl. bms. 84 × 1.20 × 14.5/27	54	1.5
Truss (assumed)	580	15.8
Bracing (assumed)	150	4.1
	3108	84.2

Dead load stresses (b)

Panel pt.	Panel load	V	ΣV	Diagonal	Top chd.	Bott. chd.
0	42.0	336				
		294		−416	0	+294
1	84.0		294			
		210		+298	−504	+294
2	84.0		504			
		126		−178	−504	+630
3	84.0		630			
		42		+ 60	−672	+630
4	84.0		672			

ELDERSBURG-LOUISVILLE ROAD BRIDGE DP10-1 7/15

TRUSS (continued)

Sidewalk live load

Load per ft. on truss $P \times 3 \times 25.5/20 = 3.83P$ (a)

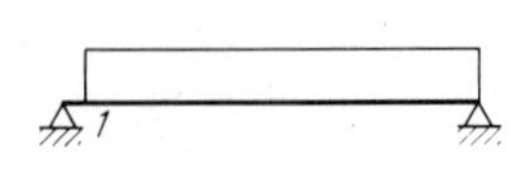

$$P = \left(30 + \frac{3000}{216}\right)\left(\frac{55-3}{50}\right) = 45.6\text{psf} \times 3.83 = 175\text{plf}$$

Panel load $= 0.175 \times 27 = 4.72^k$

$V = 4.72 \times 3.5 = 16.5^k$

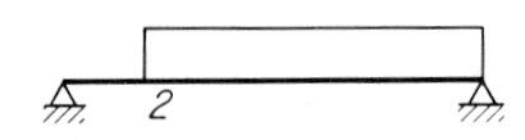

$$P = \left(30 + \frac{3000}{189}\right)(1.04) = 47.6\text{psf} \times 3.83 \times 27 = 4.93^k$$

$V = 4.93 \times 6 \times 7/16 = 13.0^k$

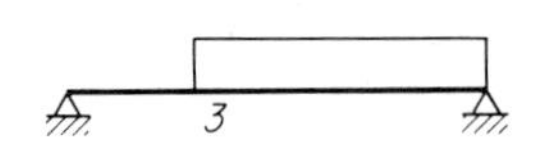

$$P = \left(30 + \frac{3000}{162}\right)(1.04) = 50.5\text{psf} \times 3.83 \times 27 = 5.22^k$$

$V = 5.22 \times 5 \times 3/8 = 9.8^k$

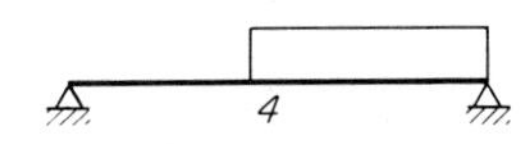

$$P = \left(30 + \frac{3000}{135}\right)(1.04) = 54.3\text{psf} \times 3.83 \times 27 = 5.60^k$$

$V = 5.60 \times 4 \times 5/16 = 7.0^k$

Top chord

$U_0 U_1 = 0$

$U_1 U_3 = -16.5 \times 2 - 4.72 \times 1 = -28.3^k$

$U_3 U_3' = -16.5 \times 4 - 4.72 \times 6 = -37.7^k$

Bottom chord

$L_0 L_2 = 16.5 \times 1 = +16.5^k$

$L_2 L_4 = 16.5 \times 3 - 4.72 \times 3 = +35.4^k$

Diagonals

$L_0 U_1 = 16.5\sqrt{2} = -23.4^k$

$U_1 L_2 = 13.0\sqrt{2} = +18.4^k$

$L_2 U_3 = 9.8\sqrt{2} = -13.9^k$

$U_3 L_4 = 7.0\sqrt{2} = +9.9^k$

Verticals

$U_2 L_2$ & $U_4 L_4 = 60 \times 3.83 \times 27 = -6.2^k$

$U_0 L_0 = 1/2 \times 6.2 = -3.1^k$

ELDERSBURG-LOUISVILLE ROAD BRIDGE	DP10-1	8/15

TRUSS (continued)

Live loads

Fraction of lane load to truss = $(19+5)/20 = 1.2$ (a)

$$\left.\begin{aligned} 0.64 \times 1.2 &= 0.768 \times 27 = 20.7^k \\ 26 \times 1.2 &= 31.2^k \\ 18 \times 1.2 &= 21.6^k \end{aligned}\right\} \quad (b)$$

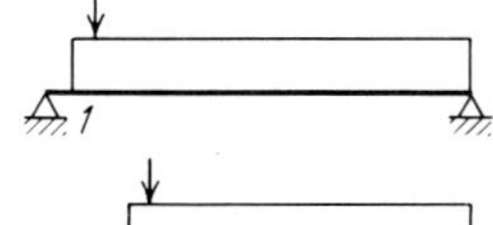

$V = 20.7 \times 3.5 + 31.2 \times 7/8 = 72.5 + 27.3 = 99.8^k$

$I = 50/(189+125) = 0.159$

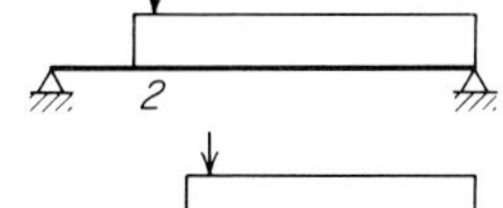

$V = 20.7 \times 6 \times 7/16 + 31.2 \times 3/4 = 54.3 + 23.4 = 77.7^k$

$I = 50/(162+125) = 0.174$

$V = 20.7 \times 5 \times 3/8 + 31.2 \times 5/8 = 38.8 + 19.5 = 58.3^k$

$I = 50/(135+125) = 0.192$

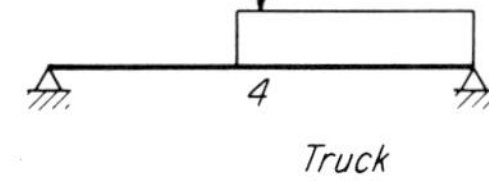

$V = 20.7 \times 4 \times 5/16 + 31.2 \times 1/2 = 25.9 + 15.6 = 41.5^k$

$I = 50/(108+125) = 0.215$

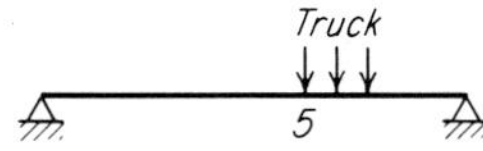

$V = 72 \times 1.2 \times 71.67/216 = 28.7^k$ (c)

$I = 50/(81+125) = 0.243$

Uniform load over entire span: $I = 50/(216+125) = 0.147$

Top chords

$U_0 U_1 = 0$

$U_1 U_3 = 72.5 \times 2 - 20.7 \times 1 + 21.6 \times 3/4 \times 2 = -157^k$

$U_3 U_3' = 72.5 \times 4 - 20.7 \times 6 + 21.6 \times 1/2 \times 4 = -209^k$

Bottom chords

$L_0 L_2 = 72.5 \times 1 + 21.6 \times 7/8 \times 1 = +92^k$

$L_2 L_4 = 72.5 \times 3 - 20.7 \times 3 + 21.6 \times 5/8 \times 3 = +196^k$

Diagonals

$L_0 U_1 = 99.8\sqrt{2} = -142^k$

$U_1 L_2 = 77.7\sqrt{2} = +110^k$

$L_2 U_3 = 58.3\sqrt{2} = -83^k$

$U_3 L_4 = 41.5\sqrt{2} = +59^k$

Verticals

$U_0 L_0 = 32 + 32 \times 13/27 = 47.4 \times 1.2 = -57^k$

$U_2 L_2 = 32 + 40 \times 13/27 = 51.3 \times 1.2 = -62^k$

ELDERSBURG-LOUISVILLE ROAD BRIDGE — DP10-1 — 9/15

TRUSS (continued)

Mem.	Stresses in kips					L	r	L/r	Allow. stress	Area req'd.	Area furnished			Section
	DL	LL	I	Sdwk LL	Total						Gross	Net*	Eff.†	
U_0U_1					Nom.	27.0								W14×61
U_1U_3	−504	−157	−23	−28	−712	27.0	4.02	81	14.0	50.9	51.7			W14×176
U_3U_3'	−672	−209	−31	−38	−950	27.0	4.11	79	14.1	67.3	69.7			W14×237
L_0L_2	+294	+92	+14	+17	+417	27.0		88	20	20.9	25.6	22.0		W14×87
L_2L_4	+630	+196	+29	+36	+891	27.0		81	20	44.6	51.7	44.8		W14×176
L_0U_1	−416	−142	−23	−24	−605	38.18	4.02	114	12.1	50.0	51.7			W14×176
U_1L_2	+298	+110	+19	+19	+446	38.18		124	20	22.3	25.6	22.8		W14×87
							{3.81	120	11.7	24.9	27.9		26.5	W14×95
L_2U_3	−178	−83	−16	−14	−291	38.18	3.71	124	11.4	25.5	27.9			
U_3L_4	+60	+59	+13	+10	+142	38.18	2.45	187	20	7.1	17.9	15.3		W14×61
							{2.70	120	11.7	10.5	17.9		14.8	W14×61
U_0L_0	−46	−57	−17	−3	−119	27.0	2.45	132	10.7	11.1	17.9			
U_1L_1					Nom.	27.0								W14×30
U_2L_2	−84	−62	−19	−6	−171	27.0	{2.70 2.45	120 132	11.7 10.7	14.6 16.0	17.9 17.9		14.8	W14×61
U_3L_3					Nom.	27.0								W14×30
U_4L_4	−84	−62	−19	−6	−171	27.0	{2.70 2.45	120 132	11.7 10.7	14.6 16.0	17.9 17.9		14.8	W14×61

*Deductions for two 1-in. holes in each flange and two 1-in. holes in web. No holes in webs of diagonals.

†The actual radius of gyration of members corresponding to the entries in this column is less than that required to give maximum allowable L/r. AASHO specifications allow this provided the required area is not more than an effective area based on the actual value of r (1.7.12). Thus for L_2U_3, $L/r = 120$ if $r = 3.81$. But for the W14×95 $r = 3.71$. Therefore, the effective area = $27.94\ (3.71/3.81)^2 = 26.5$ in.^{2} This applies only to compression members, furthermore, it is also required that the allowable load based on properties of the actual area be adequate.

ELDERSBURG-LOUISVILLE ROAD BRIDGE DP10-1 10/15

LATERAL BRACING

Wind loads (a)

Railing, parapet & floor $0.29 + 0.21 + 1.92 + 0.67 = 3.09\ ft^2/ft$
Railing posts $(0.5 \times 2.0 + 1.17 \times 0.33)/6 = 0.23$
Stringers $= 2.0$
1/2 truss $= 3.0$ (b)
8.32

Wind on top chord $= W = 8.32 \times 75 = 624\ plf$
" " bott. " $= W = 3.0 \times 75 = 225\ plf$
" " live load $= WL = 100\ plf$ (c)

Chords

$M = 1/8 \times 216^2 (0.30 \times 624 + 100) = 1670'^k$ (d)
$U_3U_3' = 1670/20 = 84^k$
$DL + LL + I + 84 = 950 + 84 = 1034^k < 1.25 \times 14.1 \times 69.7 = 1230^k$ (e)
$M = 1/8 \times 216^2 \times 624 = 3630'^k$
$U_3U_3' = 3630/20 = 182^k$
$DL + 182 = 672 + 182 = 854^k < 1230^k$ (e)

Top laterals

D1 D2 D3 D4 20'

D1 Panel shear $= 624 \times 27 \times 3.5 = 59.2^k$ $l = 33.6'$
Diag. stress $= 59.2 \times 33.6/20 = 100^k$ @ $20 = 5.00$ in.² net (f)
or 50^k @ $11.4 = 4.90$ in.² gross
Use W8×28 : $8.23 - 1.85 = 6.38$ in.² net 12 bolts
For struts $r = 20 \times 12/140 = 1.72$ Use W8×31 12 bolts

Bottom laterals

D1 Stress $= 100 \times 225/624 = 36^k$ @ $20 = 1.80$ in.² net
$r = 33.6 \times 0.5 \times 12/240 = 0.84$ req'd
$5 \times 5 \times 5/16$ ∠ $3.03 - 0.31 - 0.78 = 1.94$ in.² net 3 bolts (g)
Bottom struts W8×31

Sway frames (at panel points 2, 4, 2')

$l = 33.6'$ $r = 33.6 \times 0.5 \times 12/240 = 0.84$ req'd
Use $5 \times 5 \times 5/16$ ∠

End cross frames

Diag. stress $= 0.624 \times 108 \times 33.6/20 = 113^k$ @ $20 = 5.65$ in.² net
Use W8×28 12 bolts

ELDERSBURG-LOUISVILLE ROAD BRIDGE DP10-1 11/15

TRUSS SHOES

Shoe reaction

1. DL + LL + I

DL 3.11 × 108	= 336	(sheet 6)
Sdwk LL = 0.175 × 108	= 19	(sheet 7)
LL (0.64 × 108 + 18) × 1.2	= 105	(sheet 8)
I 0.147 × 105	= 15	
	475^k/shoe at 100% all. stress	

2. DL + LL + I + 30% W+WL

Wind top chord and WL 0.624 × 0.3 + 100 = 0.287 × 29.75 = 8.5 (a)

Wind bott. chord 0.225×0.3 = 0.068 × 2.75 = 0.2

0.355^k/ft 8.7$'^k$/ft

Lat. shear = 0.355 × 108 = 38.4^k/shoe

Mom. = 8.7 × 108 = 940$'^k$

Shoe reaction = 475 ± 940/20 = 522^k/shoe at 125% all. stress (b)

= 428^k/shoe (no uplift)

3. DL + W

Wind top chord 0.624 × 29.75 = 18.6$'^k$/ft.

Wind bott. chord 0.225 × 2.75 = 0.6

0.849 19.2$'^k$/ft.

Lat. shear = 0.849 × 108 = 91.8^k/shoe

Mom. = 19.2 × 108 = 2070$'^k$

Shoe reaction = 336 ± 2070/20 = 440^k/shoe at 125% all. stress

= 232^k/shoe (no uplift)

4. Long. forces (applied to fixed shoes)

Traction 0.05 (18 + 0.64 × 216) = 7.8^k/shoe (c)

Roller friction 0.03 × 336 = 10.1^k

$$45° \text{ wind } (DL + LL + I + 30\% W + WL) = \frac{0.355}{2} \times \frac{216}{2} = 19.2^k \quad (d)$$

$$45° \text{ wind } (DL + W) = \frac{0.849}{2} \times \frac{216}{2} = 45.9^k$$

Expansion shoes (see dwg. on Sheet 12)

Assume rocker radius = 1'-5" Dia. = 34"

All. brg. stress = $3000\sqrt{34}\,(36-13)/20 = 20.1^{k/''}$ (e)

Req'd. lgth. brg. = 475/20.1 = 23.6"

Shear area req'd. = 91.8/(14 × 1.25) = 5.25 in.^{2} (f)

Use 2-2"ϕ dowels Area = 6.28 in.^{2}

Req'd. pin brg. area = 475/14 = 34.0 in.^{2}

Req'd. pin dia. = 34.0/(18-3) = 2.26 in.^{2} (g)

Use 4"ϕ bearing surface (see figure Sheet 12)

Req'd. allowance for exp. and contr. = 1.25 × 2.16 = 2.7" (h)

Use 7" billet for shoe

ELDERSBURG-LOUISVILLE ROAD BRIDGE — DP10-1 — 12/15

Expansion shoes (continued)

Bearing plate

Req'd. area = 475/1 = 475 in.^{2} Req'd. width = 23.6 + 4.25 + 8 = 35.9" (a)

Try 36 × 20 pl. I = 24,000 in.^{4}

Assume max. eccentricity due to contraction = 1.6" (b)

$$\frac{475}{720} \pm \frac{475 \times 1.6 \times 10}{24{,}000} = 0.660 \pm 0.317 = 0.977 \text{ ksi} = 0.343 \text{ ''}$$

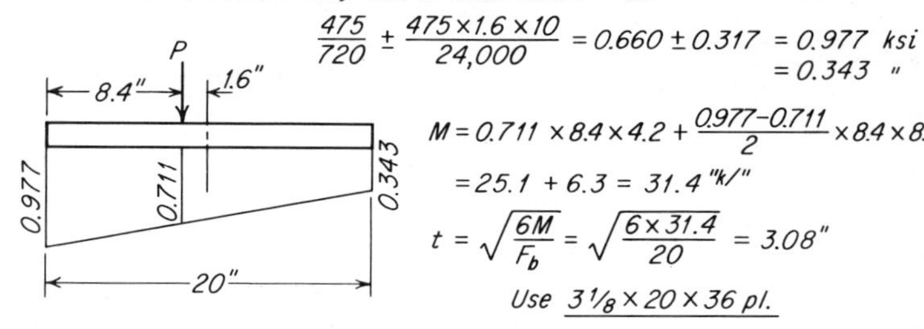

$$M = 0.711 \times 8.4 \times 4.2 + \frac{0.977 - 0.711}{2} \times 8.4 \times 8.4 \times 2/3$$

$$= 25.1 + 6.3 = 31.4 \text{ "k/"}$$

$$t = \sqrt{\frac{6M}{F_b}} = \sqrt{\frac{6 \times 31.4}{20}} = 3.08"$$

Use 3 1/8 × 20 × 36 pl.

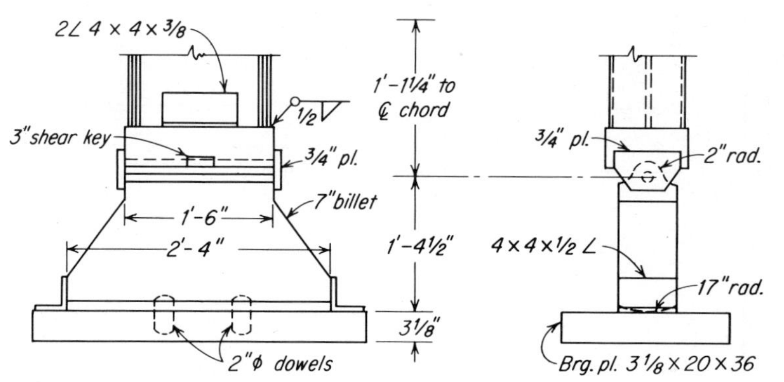

Fixed shoes

Bearing plate

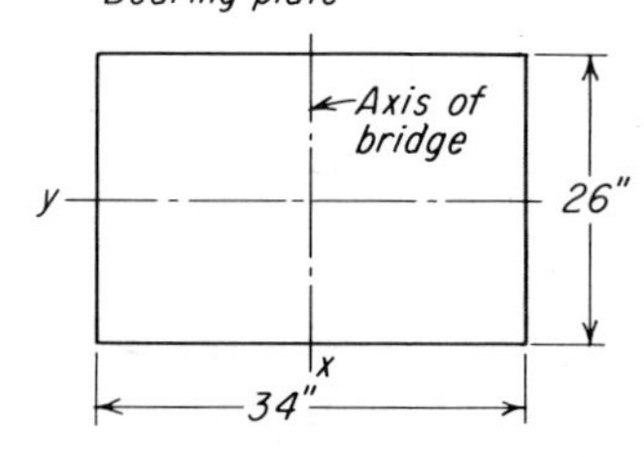

Trial area 475/0.6 = 792 in.^{2} (c)

Try 26 × 34 pl. A = 884 in.^{2}

$$S_x = \frac{26 \times 34^2}{6} = 5009 \text{ in.}^3$$

$$S_y = \frac{34 \times 26^2}{6} = 3831 \text{ in.}^3$$

1. DL + LL + I + 30%W + WL + Tr. + Fr. @ 125% all. = 1.25 ksi brg. (d)

$$\frac{522}{884} \pm \frac{38.4 \times 19.6}{5009} \pm \frac{17.9 \times 19.6}{3831} = 0.591 \pm 0.151 \pm 0.092 = 0.842 \text{ ksi} \quad (e)$$

$$\frac{428}{884} \pm \text{ " } \pm \text{ " } = 0.486 \pm 0.151 \pm 0.092 = 0.243 \text{ ksi} \quad (f)$$

ELDERSBURG-LOUISVILLE ROAD BRIDGE DP10-1 13/15

Fixed shoes (continued)

2. DL + W

$$\frac{440}{884} \pm \frac{91.8 \times 19.6}{5009} \pm \frac{10.1 \times 19.6}{3831} = 0.498 + 0.360 + 0.052 = +0.910 \text{ ksi}$$

$$\frac{232}{884} \pm \quad '' \quad \pm \quad '' \quad = 0.262 \pm 0.360 \pm 0.052 = +0.674 \text{ ksi}$$

$$= -0.150 \text{ ksi} \quad (a)$$

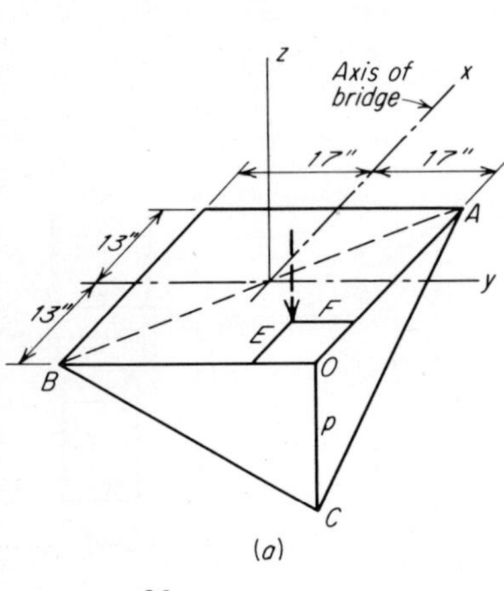

(a)

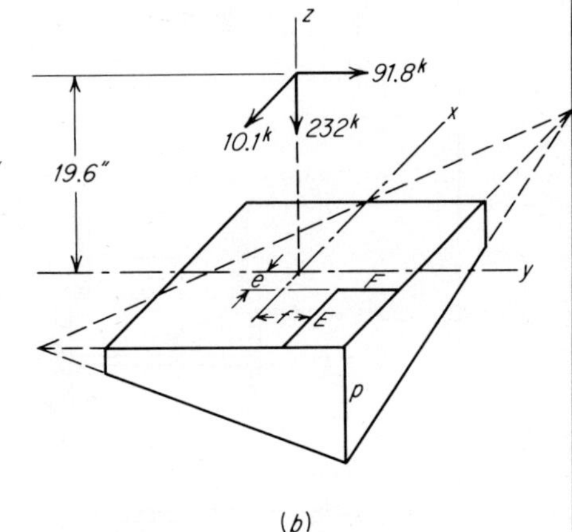

(b)

$$E = \frac{26}{4} = 6.5''$$

$$F = \frac{34}{4} = 8.5''$$

$$p = \frac{3V}{8EF} = \frac{3 \times 232}{8 \times 6.5 \times 8.5} = 1.57 \text{ ksi}$$

$$e = \frac{10.1 \times 19.6}{232} = 0.9, \quad f = \frac{91.8 \times 19.6}{232} = 7.8$$

$$E = 13 - 0.9 = 12.1, \quad F = 17 - 7.8 = 9.2$$

$$p = \frac{3 \times 232}{8 \times 12.1 \times 9.2} = 0.780 \text{ ksi}$$

3. 45° wind (DL + LL + I + 30% W + WL + Tr + Fr)

$$\text{Shoe react.} = 475 \pm \frac{470}{20} = \begin{matrix} +499 \\ +451 \end{matrix}$$

W = 19.2
Tr. = 7.8
Fr. = 10.1
Long. force = 37.1^k/shoe

$$\frac{499}{884} \pm \frac{19.2 \times 19.6}{5009} \pm \frac{37.1 \times 19.6}{3831} = 0.565 + 0.075 + 0.190 = +0.830 \text{ ksi}$$

$$\frac{451}{884} \pm \quad '' \quad \pm \quad '' \quad = 0.510 - 0.075 - 0.190 = +0.245 \text{ ksi}$$

ELDERSBURG - LOUISVILLE ROAD BRIDGE DP10-1 14/15

Fixed shoes (continued)

4. 45° wind (DL + W + Fr.)

$$\text{Shoe react.} = 336 \pm \frac{1035}{20} = \begin{matrix} +388 \\ +284 \end{matrix}$$

W = 45.9
Fr. = 10.1
Long. force = 56 k/shoe

$$\frac{388}{884} \pm \frac{45.9 \times 19.6}{5009} \pm \frac{56 \times 19.6}{3831} = 0.439 + 0.180 + 0.287 = +0.906 \text{ ksi}$$

$$\frac{284}{884} \pm \quad '' \quad \pm \quad '' \quad = 0.322 \pm 0.180 \pm 0.287 = +0.789 \text{ ksi}$$

$$= -0.145 \text{ ksi}$$

Since uplift occurs on only one corner, maximum pressure will not exceed +0.789 ksi by enough to approach the allowable pressure of 1.25 ksi (a)

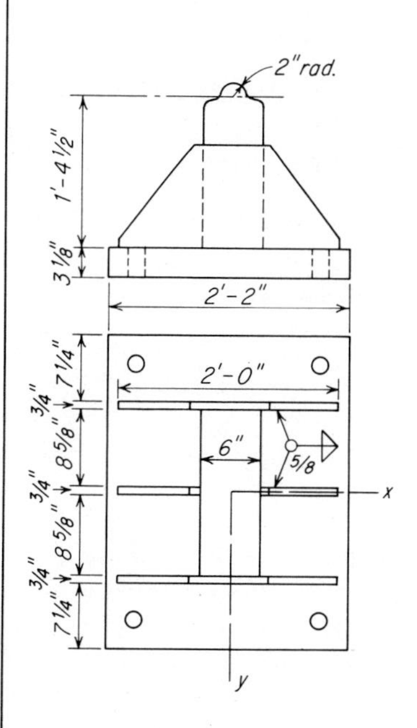

Billet stresses on section at bottom

$A = 6 \times 18 = 108 \text{ in.}^2$

$$S_x = \frac{6 \times 18^2}{6} = 324 \text{ in.}^3$$

$$S_y = \frac{18 \times 6^2}{6} = 108 \text{ in.}^3$$

All. stress = 1.25 × 20 = 25 ksi

1. DL + LL + I + 30% W + WL + Tr. + Fr.

$$\frac{522}{108} \pm \frac{38.4 \times 16.5}{324} \pm \frac{17.9 \times 16.5}{108}$$

$$4.83 \pm 1.96 \pm 2.74 = +9.53, +0.13 \text{ ksi}$$

3. 45° wind (DL + LL + I + 30%W + WL)

$$\frac{499}{108} \pm \frac{19.2 \times 16.5}{324} \pm \frac{37.1 \times 16.5}{108}$$

$$4.62 \pm 0.98 \pm 5.66 = +11.26, -2.02 \text{ ksi}$$

Overturning moment due to longitudinal force taken by diaphragm.

$$I \text{ billet} = 18 \times 6^3/12 = 324$$

$$I \text{ diaphragm} = \frac{3 \times 3/4 \times 24^3}{12} - \frac{3/4 \times 6^3}{12} = \underline{2578}$$

$$2902$$

$$\pm \frac{37.1 \times 16.5 \times 12}{2902} = \pm 2.53 \text{ ksi}$$

The −2.53 ksi represents uplift on welds at base of diaphragm.

All. stress on welds 2 × 12.4 × 1.25 × 0.707 × 0.625 = 13.7 kli

Force per inch of diaphragm = 2.53 × 1 × 3/4 = 1.90 kli < 13.7

ELDERSBURG-LOUISVILLE ROAD BRIDGE — DP10-1 — 15/15

TRUSS DETAILS

Joint L_0 (Fig. 10-15) 7/8 A325 bolts ss = 8.12^k (friction-type connection)

L_0U_1 W14×176, $A = 51.7$ in.², $F_a = 12.1$ ksi, $d = 15.25''$

$P = -605^k$, $P_{all} = -626^k$, 616/8.12 = 76 bolts

L_0L_2 W14×87, $A_g = 25.6$ in.², $A_n = 22.0$ in.², $F_t = 20$ ksi, $d = 14.0''$

$P = 417^k$, $P_{all} = 440^k$, $P_{av} = 428^k$, 428/8.12 = 53 bolts

1-11/16 fill and 1-5/8 fill

$2 \times 0.688 \times 20 = 27.5^k$ @ 8.12 = 4 bolts required before 4 holes can be deducted from flange (a)

U_0L_0 W14×61, $A = 17.9$ in.², $A_{eff} = 14.8$ in.², $F_a = 11.7$ ksi, $d = 13.91''$

$P = -119^k$, $P_{all} = -173^k$, $P_{av} = 149^k$, 149/8.12 = 19 bolts

2-11/16 fills

Bolts to W14×61 stub (b)

Vert. comp. L_0U_1 — $605 \times 0.707 = 430^k$

Approx. load from U_0L_0 — 62^k (c)

492^k

Gusset plates in brg. = $2 \times 10 \times 5/8 \times 20$ = 250

$242^k/8.12$ = 30 bolts

Stress in 5/8 gusset plate

Shear on vertical section $430/(0.625 \times 2 \times 43) = 8.0$ ksi (d)

Ecc. load on horizontal section at end of diagonal (e)

$P = 430 + 62 - 12 \times 8.12 = 395^k$

$A = 40 \times 0.625 \times 2 = 50.0$ in.²

$I = 2 \times 0.625 \times 40^3/12 = 6667$ in.⁴ $e = 4''$

$$\frac{395}{50} \pm \frac{395 \times 4 \times 20}{6667} = 7.9 \pm 4.8 = 12.7 \text{ ksi}$$

Joint U_1 (Fig. 10-16)

U_1U_3 W14×176, $A = 51.7$ in.², $F_a = 14.0$ ksi, $d = 15.25''$

$P = -712^k$, $P_{all} = -726^k$, $P_{av} = 719^k$, 719/8.12 = 89 bolts

L_0U_1 W14×176, $d = 15.25''$, 76 bolts

U_1L_2 W14×95, $A_g = 27.9$ in.², $A_n = 24.9$ in.², $F_t = 20$ ksi, $d = 14.12''$

$P = 446^k$, $P_{all} = 498^k$, 472/8.12 = 58 bolts, 2-9/16 fills

U_0U_1 W14×61, $A_g = 17.9$ in.², $A_{eff} = 14.8$ in.², $F_a = 11.7$ ksi, $d = 13.91''$

$P = 0$, $P_{all} = 17.9 \times 11.7 \times 0.75 = 155^k$, 155/8.12 = 20 bolts (f)

1-5/8 fill, 1-11/16 fill

U_1L_1 W14×61, $A_g = 17.9$ in.², $A_n = 15.3$ in.², $F_t = 20$ ksi, $d = 13.91''$

$P = 0$, $P_{all} = 15.3 \times 20 \times 0.75 = 230^k$, 230/8.12 = 28 bolts (g)

1-5/8 fill, 1-11/16 fill

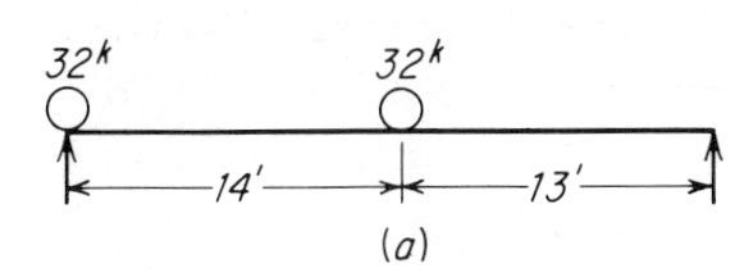

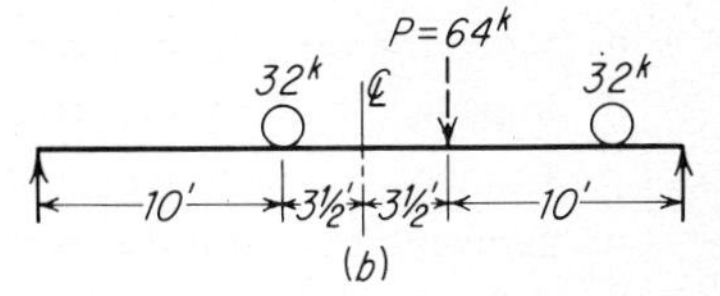

FIGURE 10-8

b The dead-load and live-load moments are taken to be 80 percent of the corresponding moments for a simply supported slab. The formula for the live-load moment [Eq. (10-2)] is given by the specifications (1.3.2C, case A).

c The allowance for impact is the fraction of the live load given by $I = 50/(L + 125)$, in which L is the length in feet of the portion of the span that is loaded to produce the maximum stress in the member. The maximum value of I is 0.3 (1.2.12C).

d The roadway slab is 8 in. thick and is reinforced for positive moment with no. 5 bars 6 in. on centers. Allowances of $\frac{3}{4}$ in. for wearing surface and at least $1\frac{1}{2}$ in. of protective covering for the reinforcement are provided. The calculated stresses compare favorably with the allowable values, 20 ksi for the reinforcement and 1.2 ksi for the concrete.

e The sidewalk and parapet are supported on concrete pedestals 14 in. wide on 6-ft centers, as shown in the sketch on Sheet 2. The pedestals are supported in turn on the cantilevered portion of the roadway slab. The width of slab over which the stresses can be considered to be distributed uniformly is not known; however, it is certainly more than the width of the pedestal but less than the distance between pedestals. The effective width is assumed to be 4 ft.

f This is the dead load computed on Sheet 2.

g This is the weight of the 4-ft section of slab which is assumed to support the pedestal.

h The live load on the curb is a lateral force representing the effect of a vehicle striking the curb and is assumed to act at the top of the curb (1.2.11B). It should be noted that since the curb load is lateral, the shear on the section of slab under consideration is only $3.17 + 1.00 + 1.53 = 5.70$ kips. This is so small that the shearing stress need not be computed.

i The slab is reinforced for negative moment by bending up alternate positive-moment bars and supplementing them with no. 5 bars 12 in. on centers.

Sheet 4

a To calculate the dead load supported by the exterior stringer, the slab is assumed to act as a simple beam supported by the exterior stringer and the first interior stringer. The sketch and other information on Sheet 2 will help to clarify these computations.

b This is the wheel-load distribution formula of 1.3.1B2(*a*).

c The stringers span 27 ft (see sketch on Sheet 6).

d The quantities 47.4 and 237 are taken from the table of maximum moments, shears, and reactions per lane of load (AASHO specifications, appendix A). These values are easy to compute. For shear (Fig. 10-8*a*) $R = 32 + 32 \times 13/27 = 47.4$ kips. For moment (Fig. 10-8*b*) $M = 64 \times 10 \times 10/27 = 237$ ft-kips. Since the wheels on only one side of the truck are effective for the exterior stringer, the shear and moment for this stringer are one-half the values for the lane load.

e Since the outside stringer supports sidewalk live load as well as traffic live load and impact, the allowable stress may be increased 25 percent (1.3.1B).

f Camber is secured by cold-gagging the beam in the fabricating shop. Since small cambers may not be permanent, camber is usually not specified if the dead-load deflection is less than the minimum camber that is likely to remain permanent. According to the AISC Manual, camber less than about $\frac{3}{4}$ in. is not likely to be permanent for this beam.

g The distribution of wheel loads to stringers was discussed in Art. 10-7. In this case, the fraction of one wheel load supported by one stringer is given by Eq. (10-3) (1.3.1B).

h Since the sidewalk produces an uplift on the first interior stringer, a check is made to determine whether the reduction in bending moment on this stringer is enough to justify a lighter shape. The design of the stringers assuming the weight of the sidewalk, curb, and railing to be equally distributed to all four stringers (1.3.1B) is as follows:

Interior stringer:

Floor slab $100 \times 7 = 700$

Sidewalk and parapet $528/4 = 132$

Stringer $= 80$

912

$$\text{Fraction of wheel load to stringer} = \frac{7}{5.5} = 1.27$$

	Shear	Moment
DL:	$0.9 \times 13.5 = 12.3$	$\frac{1}{8} \times 0.91 \times 27^2 = 83$
LL:	$1.27 \times 47.4 \times \frac{1}{2} = 30.1$	$1.27 \times 237 \times \frac{1}{2} = 151$
I:	$0.30 \times 30.1 = 9.0$	$0.30 \times 151 = 45$
	$V = 51.4$ kips	$M = 279$ ft-kips

$$\frac{I}{c} = \frac{279 \times 12}{20} = 167 \text{ in.}^3 \qquad \text{W24} \times 76$$

Exterior stringer:

Sidewalk and parapet $\dfrac{528}{4} = 132$

Floor slab $\dfrac{100 \times 9.5 \times 4.75 - 44.5 \times 8.87}{7} = 587$

Stringer 82

801

Sidewalk $LL = 310$ lb/ft

Fraction of wheel load to stringer $= 1.22$

	Shear	Moment
DL	$0.801 \times 13.5 = 10.8$	$\frac{1}{8} \times 0.801 \times 27^2 = 73$
Sidewalk *LL*	$0.310 \times 13.5 = 4.2$	$\frac{1}{8} \times 0.310 \times 27^2 = 28$
LL	$1.22 \times 47.4 \times \frac{1}{2} = 28.9$	$1.22 \times 237 \times \frac{1}{2} = 145$
I	$0.30 \times 28.9 = 8.7$	$0.30 \times 145 = 43$
	$V = 52.6$ kips	$M = 289$ ft-kips

$$\frac{I}{c} = \frac{289 \times 12}{20 \times 1.25} = 138.7 \text{ in.}^3$$

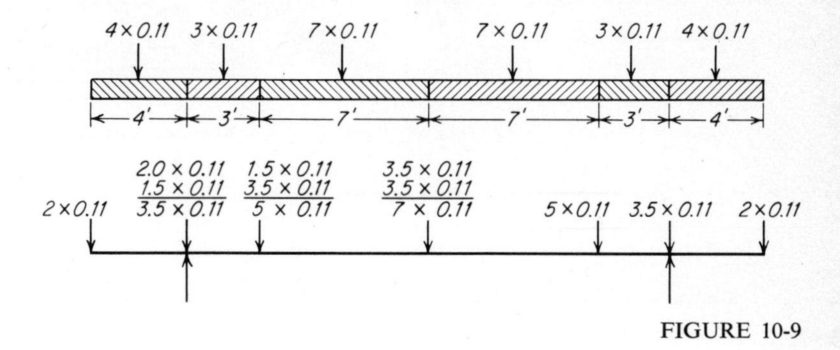

FIGURE 10-9

A W21 × 68 provides sufficient strength. However, AASHO specifications require that the outside stringer provide a carrying capacity not less than that of the interior stringer. Therefore a W24 × 76 is required.

Sheet 5

a The part of the live load in each lane that is supported by the floor beam consists of the 32-kip rear axle, directly over the floor beam, and an 8-kip front axle and a 32-kip trailer axle each 14 ft from the floor beam.

b In this calculation the number 1.40 is the combined weight, in kips, of the sidewalk, parapet, floor slab, and stringer, taken from Sheet 4. The floor beam is assumed to weigh 0.11 klf. Figure 10-9 shows how the weight of the floor beam is converted to a statically equivalent system of forces acting at the points of attachment of the stringers.

c In the computation for this stringer the uplift from the cantilevered sidewalk is taken into account.

d Figure 10-10 explains the calculations for the live-load reactions of the stringers on the floor beam. The floor slab is assumed to act as a series of 7-ft simple beams.

e The slab overhangs the end floor beam a distance of 1.25 ft. Therefore, the maximum live load on the end floor beam will result with the truck placed so that the trailer axle is at the tip of the overhang.

Sheet 6

a In this calculation the factor 1.02 provides an allowance of 2 percent for the weight of bolt heads, nuts, details, paint, etc. In the next line, the factor 1.20 provides an

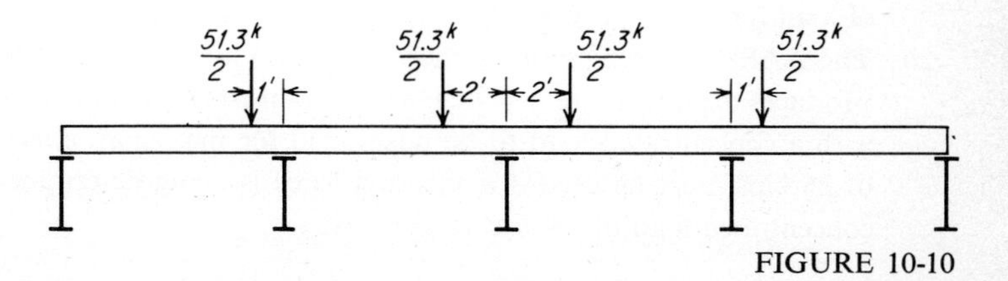

FIGURE 10-10

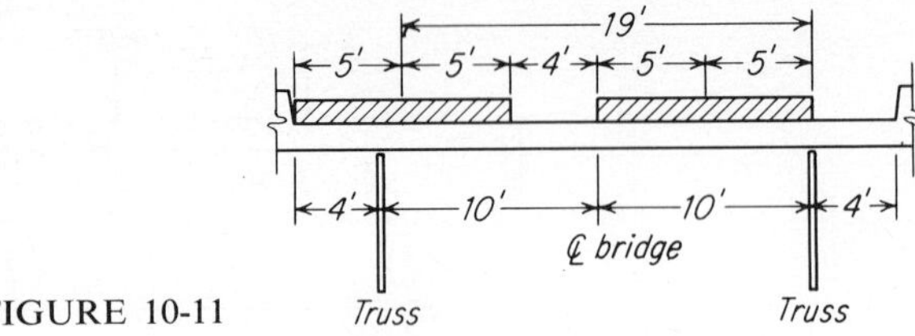

FIGURE 10-11

allowance of 20 percent for the weight of stringer connection angles and floor-beam details, paint, etc.

b The forces in the bars of the truss are computed by the method of index stresses.

Sheet 7

a Since the structure supports cantilevered sidewalks, maximum stresses in the truss members will occur with only one sidewalk loaded. The specifications allow a reduction in the intensity of live load if a long stretch of sidewalk is to be loaded (1.2.11A). This is because of the improbability that it will be fully loaded. For loaded lengths ranging from 26 to 100 ft the live load is 60 psf, while for loaded lengths in excess of 100 ft it is given by the equation

$$P = \left(30 + \frac{3{,}000}{L}\right)\frac{55 - W}{50}$$

where P = load, psf (not to exceed 60 psf)
L = loaded length, ft
W = width of sidewalk, ft

The various lengths of load considered here are chosen so as to give maximum stresses. The approximate method, based on the assumption that full panel loads exist at all panel points to the right of a given panel, with none to the left, is used. For example, the load extending from the right end of the bridge leftward to panel point 1 gives full panel loads at joints 2 to 1′ inclusive and half panel loads at 1 and 0′. The half panel load at joint 1 is neglected. This is the position of load for maximum stress in U_1L_2.

Sheet 8

a For the purpose of determining the live load on the truss, the roadway is assumed to be divided into two traffic lanes of equal width (1.2.6.). The truck or lane loads are assumed to occupy a width of 10 ft and are placed each within its own traffic lane but in such a position as to produce the maximum load on the truss. The resulting position of load for the 28-ft roadway is shown in Fig. 10-11.

b The AASHO specifies that either the truck load or the lane load be used, whichever produces the larger stress. The lane load consists of a uniform load of 0.64 klf together with a concentrated load to be positioned for maximum effect. A concentrated load of 26 kips must be used if a shearing force is being determined, while for moment a concentrated load of 18 kips is to be used.

c The truck load is shown here in position to produce a maximum compressive force in $L_4 U_3'$. The object is to determine whether there can be a reversal of stress in this member. Since the live-load shear including impact is less than the dead-load shear for this panel, which was computed on Sheet 6, no reversal occurs.

Sheet 9 The W14's chosen for this truss range in depth from 13.86 to 16.12 in. Differences are taken care of at the joints by fillers (Sheet 15). Three of the members have only nominal loads; i.e., their calculated loads are zero. The lightest available W is used for all these members except in the case of $U_0 U_1$. The flange of the W14 × 30 is only 6.73 in. wide; the difference between it and the adjacent 15.65-in. flange of $U_1 U_3$ would have been considerable. Hence the lightest shape in the 14 × 10 series, the W14 × 61, is used for $U_0 U_1$.

Sheet 10

a AASHO specifications require that bridge trusses be designed for a wind pressure of 75 psf. The specifications further require total forces of not less than 300 plf and 150 plf in the plane of the loaded chord and the plane of the unloaded chord, respectively. Since gusts can be relatively concentrated, it is assumed that only part of the structure may be subject to wind pressure. Therefore, the wind is assumed to act on whatever area will result in the maximum force in a particular member. For this reason, wind pressure is usually referred to as a "moving" load (1.2.14).

b The area of the truss is estimated by considering the exposed area of its members plus an allowance for gusset plates.

c Wind pressure on the live load is 100 psf (1.2.14).

d Only 30 percent of the wind force on the structure need be taken in combination with wind on the live load.

e Allowable stresses may be increased 25 percent for combinations of dead load, live load, impact, and wind. Thus, if the sum of $DL + LL + I +$ wind or of $DL +$ wind is less than 1.25 times the allowable force in the member at the nominal allowable stress, the member is acceptable.

f Here the lateral truss is assumed to act as a single-diagonal truss, with only the diagonal member that is subjected to tension considered to be active. In the next line, the truss is assumed to be a double-diagonal system, with each diagonal resisting half the shear in the panel. The allowable compressive stress is taken to be that for $L/r = 120$. Although the wind loads for *D*2, *D*3, and *D*4 will be successively smaller than that in *D*1, the W8 × 28 is used throughout.

g To determine the net area, deduction is made for one bolt hole in the connected leg and for half the area of the unconnected leg (1.7.15).

Sheet 11

a Resultant wind forces are assumed to act at the centerlines of the chords. The base of the bearing plate is 2 ft $8\frac{7}{8}$ in., or 2.75 ft, below the centerline of the bottom chord (see sketch on Sheet 12).

b The trusses are spaced 20 ft apart.

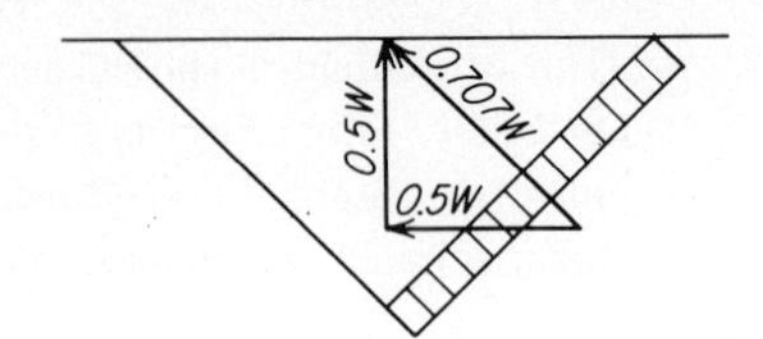

FIGURE 10-12

c Fixed shoes are designed for a longitudinal force of 5 percent of the live load in all lanes, using the uniform lane loads plus the 18-kip concentrated load, with no impact. Provision for a frictional force at expansion bearings is also required (1.2.13). The frictional force is taken here at 3 percent of the dead load.

d Although it is not required by the specifications, a quarterly wind is assumed to act as shown in Fig. 10-12. The component parallel to the bridge is resisted by the fixed shoes.

e The allowable bearing pressure on rollers and rockers is given by Eqs. (10-5).

f The dowels which hold the rocker in place on the base plate must transmit the lateral shear of 91.8 kips.

g The length of the bearing surface is 18 in., as shown in the sketch on Sheet 12. Deduction is made for the 3-in. shear key shown in the figure. Note that a pin is not used; instead, the billet is machined to a semicylindrical surface.

h The specifications require an allowance for expansion and contraction of 1.25 in. per 100 ft of span (1.7.17).

Sheet 12

a The required bearing length of the rocker is 23.6 in. (Sheet 11). The slots in the rocker for the 2-in. dowels are $2\frac{1}{8}$ in., and 4×4 angles are used to fasten the rocker to the base plate, as shown on the sketch.

b Since it is assumed that the reaction of the truss on the shoe is centered on the base plate at an approximate mean temperature, the 2.7-in. allowance for expansion and contraction determined on Sheet 11 is assumed to be about equally divided between expansion and contraction.

c A trial area of the bearing plate is found by using the reaction for the first load combination on Sheet 11. Since this is not the worst condition of loading, the bearing pressure on the concrete is assumed at only 0.6 ksi instead of the allowable value 1 ksi. The guessed size of the plate is then checked for various combinations of load.

d The load combination here is that of the second group on Sheet 11 together with traction and roller friction from the fourth group. The 38.4-kip force from the second group is a lateral force; i.e., it is parallel to the y axis of the bearing plate. This lateral force and the longitudinal force from traction and roller friction are assumed to act at the centerline of the pin, which is $19\frac{5}{8}$ in. above the base. Note that these forces have the same value for the shoe on the leeward side of the bridge as for the windward shoe,

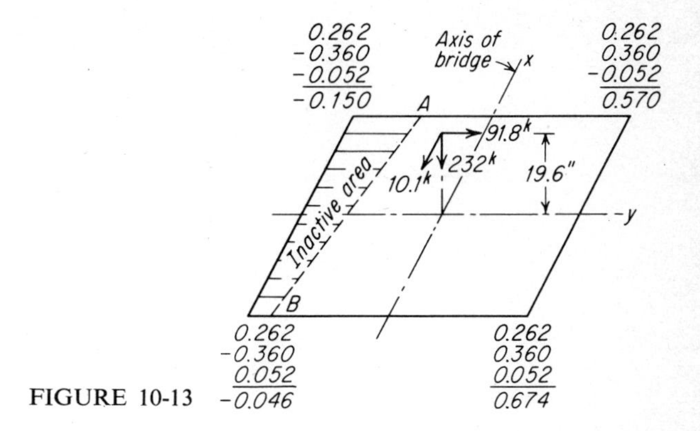

FIGURE 10-13

while the vertical reaction from group 2 is 522 kips on the leeward shoe but only 428 kips on the windward shoe.

e This is the largest bearing pressure under the leeward shoe. However, we must check to see that the combination of the plus and the two minus signs does not result in a minus value, since this would mean that only part of the area of the plate bears on the masonry and this is contrary to the assumption on which the calculation is based. Inspection of the figures shows that there is no uplift.

f The only reason for investigating the windward shoe is that uplift is more likely here, so that even though the vertical reaction is smaller, the bearing pressure might be larger than for the leeward shoe. Using both minus signs, we get +0.243 ksi. Therefore no uplift occurs, and there is no need to compute the maximum bearing pressure on this shoe.

Sheet 13

a The load combination considered here produces uplift. Computing the bearing pressures at the four corners of the plate (Fig. 10-13), we find uplift indicated to the left of *AB*. Since this represents a sizable inactive area, the maximum bearing pressure may exceed the value 0.674 ksi by a considerable margin. Assume that a line of zero pressure extends diagonally across the plate, as in Fig. *a* of Sheet 13. The resultant pressure passes through the centroid of the pressure pyramid *OABC*. But for a pyramid whose edges *OA*, *OB*, and *OC* are mutually prependicular, the coordinates *E* and *F* of the centroid are one-fourth of the distances *OA* and *OB*, respectively. In terms of these coordinates, the volume of the pyramid is given by $V = 8pEF/3$. For the reaction of 232 kips, we find from this equation $p = 1.57$ ksi, as calculated on Sheet 13. The required location of the 232-kip resultant bearing pressure is found by computing the eccentricities *e* and *f* in Fig. *b* of Sheet 13. Then $E = 12.1$ in., and $F = 9.2$ in. Since both these values are larger than those computed previously, we know that the resultant force in Fig. *a* is too far from the center of the plate. Therefore, the actual

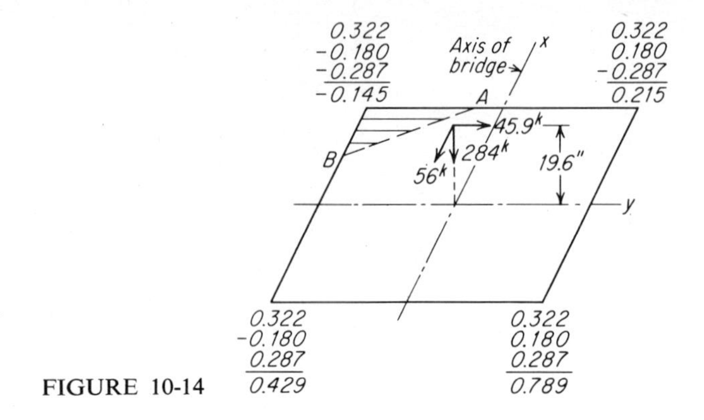

FIGURE 10-14

maximum bearing pressure must be less than 1.57 ksi. Since a pressure pyramid corresponding to $E = 12.1$ and $F = 9.2$ extends beyond the corners of the plate, as shown in Fig. *b*, the actual maximum bearing pressure must be more than the value of *p* corresponding to this pyramid. We assume then that the maximum bearing pressure lies between 0.780 and 1.57 ksi, and since the allowable value is 1.25 ksi, we decide that the bearing plate is adequate for this combination of forces.

Sheet 14

a The pressures at the four corners of the plate for this combination of forces are shown in Fig. 10-14.

Sheet 15

a There are four rows of bolts in each flange (Fig. 10-15). Two holes were deducted from each flange in determining the net section of the tension members (Sheet 9).

b The calculation here is to determine the number of bolts needed to connect the gusset plates at L_0 to the short piece of W14 × 61 that bears on the soleplate (Fig. 10-15).

c Since the positions of live load for maximum effect are not the same for U_0L_0 and L_0U_1, the 119-kip force in U_0L_0 does not occur simultaneously with the 605-kip force in L_0U_1. The value here is estimated, although it could be determined by calculating the force in U_0L_0 with the live load in position for maximum stress in L_0U_1.

d This is the calculated shearing stress on a vertical section through the gusset plates just to the left of L_0L_1 (Fig. 10-15). A discussion of the assumptions that are usually made in analyzing gusset plates is given in Art. 3-17.

e The section here is taken horizontally through the end of L_0U_1 at its centerline (Fig. 10-15). The load on this section from the upper portion of the gusset plates is the vertical component of L_0U_1 plus the 62-kip load in U_0L_0.

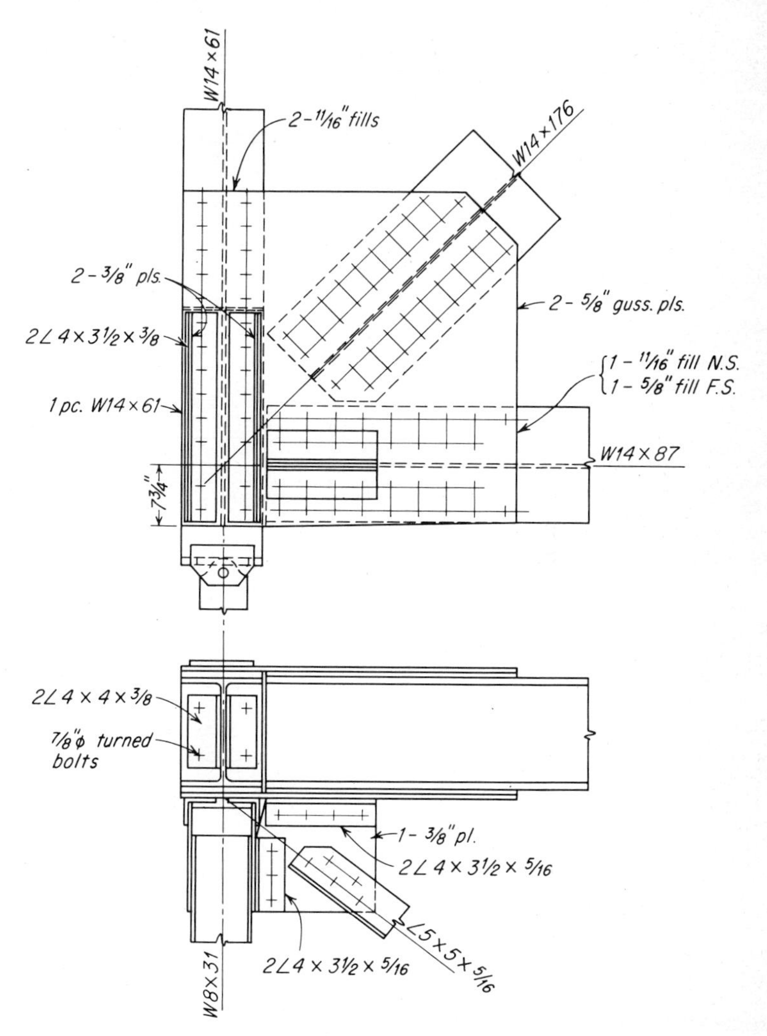

FIGURE 10-15
Detail at L_0, Eldersburg–Louisville Road Bridge.

f Since there is no calculated load in this member, the number of bolts must satisfy the requirement that no connection be designed for less than 75 percent of the allowable load (1.7.21).

Joint U_1 is shown in Fig. 10-16. The splice of $L_0 L_2$ to $L_2 L_4$ is designed in DP3-3 and shown in Fig. 3-29.

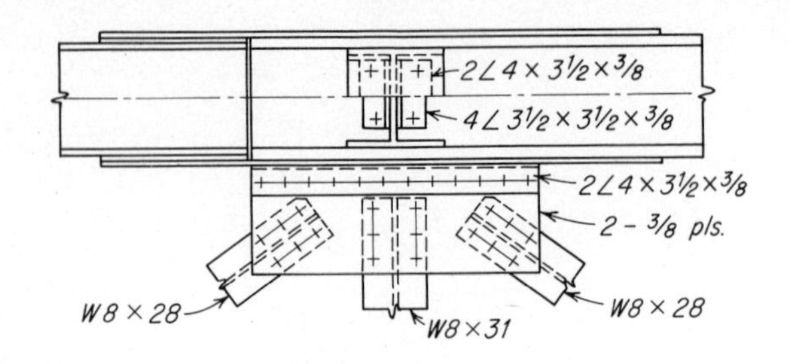

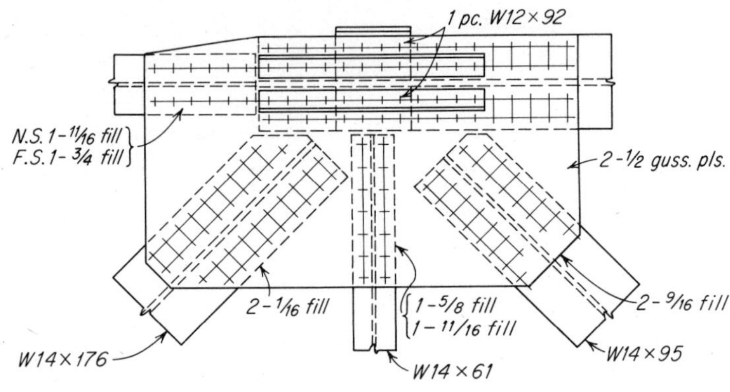

FIGURE 10-16
Detail at U_1, Eldersburg–Louisville Road Bridge.

10-10 BEAM BRIDGES

The highway beam bridge is widely used because of its simplicity of design and construction. The bridge consists of a series of parallel steel beams (stringers) supporting a continuous reinforced-concrete slab. The construction may be either composite or noncomposite. The stringers may be W's, W's reinforced with cover plates, or plate girders. They are usually spaced 5 to 8 ft on centers. The multiple-span beam bridge may be continuous over its piers or may consist of a series of simply supported spans. The two-lane bridge usually has at least four stringers. Since the beam bridge is essentially a bridge floor in which abutments and piers replace the floor beams of the truss bridge, the procedures discussed in Art. 10-7 apply also to it. Thus, unless the construction is composite, all the information needed for the design of the beam bridge has already been covered. The remainder of this chapter will be devoted to a discussion of design procedures for the composite beam bridge and, as an illustrative example, the design of the beams for the 70-ft approach spans of the Eldersburg–Louisville Road Bridge.

10-11 COMPOSITE-BEAM BRIDGES

The behavior and design of composite beams is discussed in Chap. 5, where it is shown that a reinforced-concrete floor supported on steel beams can be designed as a system of parallel T beams, provided the slab is mechanically bonded to the beam. This bond is provided by stud or channel shear connectors, which are discussed in Art. 5-31. With the beam spacing and slab thickness common to highway bridges, considerable savings in weight of steel may be realized by reinforcing the bottom flange with a cover plate or by using a welded unsymmetrical I.

Since the spacing of the beams of a beam bridge is a function of the width of the roadway and the number of beams, and since the thickness of the slab depends in turn upon the spacing of the beams, this much is known when the proportioning of the beam itself is begun. Except for tables of cross-sectional properties and allowable loads, however, there is no direct procedure for determining the size of a composite beam. Procedures for preliminary design are given in Ref. 4.

In the usual case of the beam bridge built without shoring, the dead load of the beams and slab is supported by the steel beam alone. Therefore, there is no stress in the concrete due to this load, and the section modulus of the beam itself determines the stresses. Since the live load is supported by composite action, the stresses from this source are determined from the modulus of the transformed section. Properties of the transformed section are determined in the usual way, assuming elastic behavior. According to the AASHO specifications, the effective width of the slab as a T-beam flange is as follows.

For an interior beam, the smallest of:

1 One-fourth of the span of the beam
2 The center-to-center distance of beams
3 Twelve times the least thickness of the slab

For a beam having a flange on only one side, the smallest of:

1 One-twelfth of the span of the beam
2 Half the center-to-center distance of the adjacent beam
3 Six times the thickness of the slab

Sidewalks, curbs, railings, and wearing surface are usually added after the slab is built, and the weight of this additional construction produces a sustained compressive stress that may result in plastic flow of the slab. An allowance for this effect can be made by using a section modulus based on an increased value of the modular ratio n; the value $3n$ is usually recommended.

10-12 LENGTH OF COVER PLATES

A formula for the length of cover plates for a uniformly loaded composite beam can be determined as follows. The bending stresses f at midspan and f' at the cross section where the cover plate ends are

$$f = \frac{wL^2}{8S_c} \qquad f' = \frac{w(L^2 - L'^2)}{8S'_c}$$

where L = span
L' = length of cover plate
S_c = section modulus of composite beam at midspan
S'_c = section modulus of composite beam without cover plate

Assuming both cross sections stressed to their allowable values, $f = f'$, and dividing the second equation by the first and simplifying the result, we get

$$\frac{L'}{L} = \sqrt{1 - \frac{S'_c}{S_c}} \tag{10-6}$$

Equation (10-6) is on the unsafe side for beams that support moving concentrated loads. For this case, the equation may be modified as explained in Art. 6-7. However, the following procedure is more conservative. Figure 10-17*a* shows the wheels of an H truck of the AASHO specifications placed in position for maximum moment. The corresponding moment diagram, including the effect of a uniform dead load, is *ABCD* in Fig. 10-17*b*. If we assume that the segment *AB* is a parabola with vertex at *B*, Eq. (10-6) may be used to determine the distance from *B* to the left end of the cover plate. To this must be added the distance $a/2$ from *B* to the midpoint of the span. Thus, to get the half length $L'/2$ of the cover plate, we add $a/2$ to the value obtained by substituting $L/2 - a/2$ for L in Eq. (10-6). The resulting value of L' is given by

$$L' = a + (L - a)\sqrt{1 - \frac{S'_c}{S_c}} \tag{10-7}$$

Values of a for the truck loads of the AASHO specifications are given in Table 10-1. For spans shorter than 26.5 ft, maximum moment occurs under the rear

Table 10-1

H load		HS load	
Span, ft	a, ft	Span, ft	a, ft
Less than 26.5	0	Less than 23.9	0
Greater than 26.5	2.8	Between 23.9 and 33.8	7
		Greater than 33.8	4.7

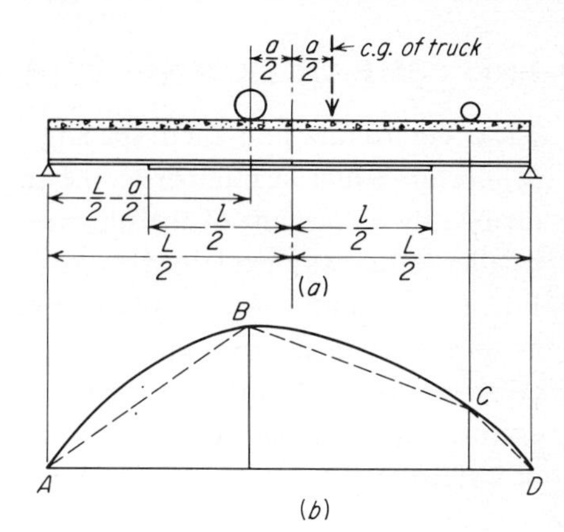

FIGURE 10-17

wheel of the H truck, so that the wheel is at midspan and $a = 0$. For spans longer than 26.5 ft, maximum moment occurs with the loads in the position shown in Fig. 10-17*a*, and $a = 2.8$ ft. The three axles of the HS truck result in three possibilities. For spans less than 23.9 ft, maximum moment occurs with one wheel at midspan and $a = 0$. For spans ranging from 23.9 to 33.8 ft, maximum moment occurs with the two equally heavy trailer axles on the span, and $a = 7$ ft. For spans longer than 33.8 ft, maximum moment occurs when three axles are on the span, and $a = 4.7$ ft.

The cover plate must be extended beyond the theoretical cutoff point and attached adequately. According to AASHO, this extension must be at least twice the width of the plate if it is not welded across the end and $1\frac{1}{2}$ times the width if it is so welded. The weld connecting the plate within the length of this extension must be sufficient to develop the force in the plate corresponding to the computed stress at the theoretical cutoff point, where the stress is computed for the cross section including the cover plate. Thus, if M is the bending moment at the theoretical cutoff point, we have

$$M = F_b S_c' \tag{a}$$

where F_b is the allowable bending stress and S_c' is the section modulus without the cover plate. The stress f at this cross section, with the cover plate included, is

$$f = \frac{M}{S_c} \tag{b}$$

where S_c is the modulus of the cross section with the cover plate. Therefore, the force P on which the required weld is based is

$$P = fA = \frac{M}{S_c} A = F_b A \frac{S_c'}{S_c} \tag{10-8}$$

10-13 SHEAR-CONNECTOR SPACING

Shear connectors and their spacing were discussed in Art. 5-31. AASHO requires connectors equal in number to the smaller of T by Eq. (5-59) and C_1 by Eq. (5-60*a*) divided by 85 percent of the ultimate strength of one connector. The reason for the smaller value was explained in Art. 5-31. Shear-connector strengths are given by Eqs. (5-61*a*) and (5-61*b*). AASHO also requires that shear connectors be spaced so as to protect them from fatigue failure. The range of shear to be considered at any cross section is the difference between the maximum shear and the minimum shear, exclusive of dead load, at the section. The fatigue strengths of channel connectors, and of welded studs, in pounds, are:

Number of cycles	Channel	Welded stud
100,000	$4{,}000w$	$13{,}000d^2$
500,000	$3{,}000w$	$10{,}600d^2$
2,000,000	$2{,}400w$	$7{,}850d^2$

An example is given in DP10-2.

10-14 DP10-2: ELDERSBURG–LOUISVILLE ROAD BRIDGE

The two 70-ft approach spans of the Eldersburg–Louisville Road Bridge (DP10-1, Sheet 1) are composite-beam bridges. Each span has five beams spaced 7 ft on centers and supporting a concrete slab $7\frac{1}{4}$ in. thick topped with a 2-in. bituminous wearing surface. Computations for the design of the slab are omitted, since the procedure is a repetition of that for the slab of the truss spans.

Calculations for the bearing stiffeners and the bearing plates shown on the elevation on Sheet 4 are omitted in this example.

REFERENCES

1. Elliott, A. L.: Steel and Concrete Bridges, and Wolchuk, R., Steel-plate-deck Bridges, sec. 18 in E. H. Gaylord and C. N. Gaylord (eds.), "Structural Engineering Handbook," McGraw-Hill, New York, 1968.
2. Waddell, J. A. L.: Weights of Metal in Steel Trusses, *Trans. ASCE*, vol. 101, 1936.
3. Westergaard, H. M.: Computation of Stresses in Bridge Slabs Due to Wheel Loads, *Public Roads*, vol. 11, no. 1, March 1930.
4. Viest, I. M.: Composite Construction, sec. 14 in E. H. Gaylord and C. N. Gaylord (eds.), "Structural Engineering Handbook," McGraw-Hill, New York, 1968.

ELDERSBURG-LOUISVILLE ROAD BRIDGE DP10-2 1/4

Span 70'-0". Beams spaced 7'-0" on centers

BEAM B (See plan Sheet 4)

Dead load

Slab (7¼") = 7 × 150 × 7.25/12 = 0.634 klf

Beam 0.226 klf

0.860 klf

Wearing surface (2" bituminous) = 0.175 klf

Live load

Fraction wheel load to beam = 7/5.5 = 1.27

$I = \frac{50}{125 + 70} = 0.256$

	Shear	Moment
LL	1.27 × 62.4 × ½ = 39.7ᵏ	1.27 × 985.6 × ½ = 626'ᵏ
I	0.256 × 39.7 = 10.2	0.256 × 626 = 160'ᵏ
	49.9ᵏ	786 × 12 = 9,450"ᵏ
DL	0.86 × 35 = 30.1ᵏ	⅛ × 0.86 × 70² = 527 × 12 = 6,320"ᵏ
Wear. sur.	0.175 × 35 = 6.1ᵏ	⅛ × 0.175 × 70² = 107 × 12 = 1280"ᵏ

Properties of section

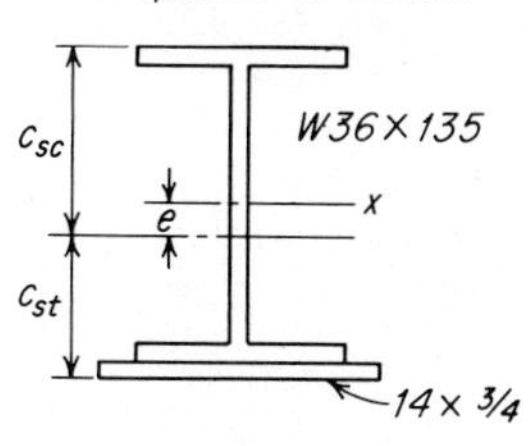

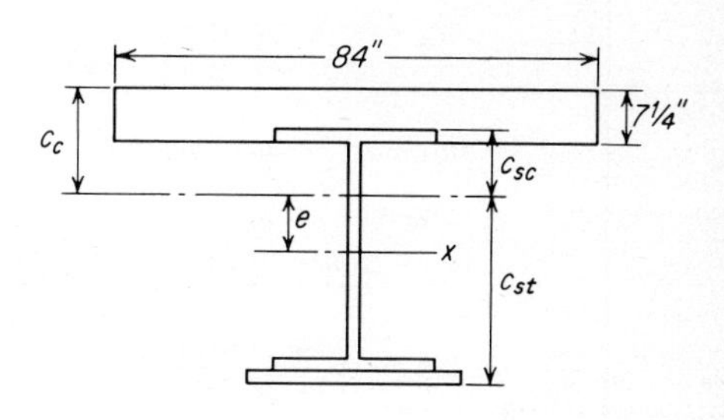

Beam and flange plate

Section	Area	Mom. abt. x	I_x
W36×135	39.8	0	× 3.80² = 574
14 × ¾	10.5	×(−18.15) = −191	×14.35² = 2160
	50.3	−191	7820
		e = −3.80 in.	10,554 in.⁴

c_{sc} = 3.80 + 35.55/2 = 21.58 in. c_{st} = 17.78 − 3.80 + 0.75 = 14.73 in.

Beam, flange plate and slab (n = 10)

Section	Area	Mom. abt. x	I_x
W36 × 135	39.8		× 9.58² = 3650
14 × ¾	10.5	×(−18.15) = − 191	× 27.73² = 8060
84 × 7.25/10	60.9	×(+20.60) = +1254	× 11.03² = 7400
	111.2	+1063	7820
		e = + 9.58 in.	268
			27,198 in.⁴

c_{sc} = 17.78 − 9.58 = 8.20 in. c_{st} = 17.78 + 9.58 + 0.75 = 28.11 in.

c_c = 8.20 + 7.25 − 0.79 = 14.66 in.

ELDERSBURG-LOUISVILLE ROAD BRIDGE — DP10-2 — 2/4

Properties of section (continued)

Beam, flange plate and slab $(n = 30)$

Section	Area	Mom. abt. x	I_x
W36 × 135	39.8		× 3.22^2 = 411
14 × 3/4	10.5	× (−18.15) = −191	× 21.37^2 = 4780
84 × 7.25/30	20.3	× (+20.60) = +418	× 17.39^2 = 6120
	70.6	+227	7820
		e = +3.22 in.	89
			19,220 in.^{4}

$c_{sc} = 17.78 - 3.22 = 14.56$ in. $\quad c_{st} = 17.78 + 3.22 + 0.75 = 21.75$ in.

$c_c = 14.56 + 7.25 - 0.79 = 21.02$ in.

Beam and slab $(n = 10)$

Section	Area	Mom. abt. x	I_x
W36 × 135	39.8		× 12.48^2 = 6183
84 × 7.25/10	60.9	× (+20.60) = +1254	× 8.13^2 = 4025
	100.7	+1254	7820
		e = 12.48 in.	268
			18,296 in.^{4}

$c_{sc} = 17.78 - 12.48 = 5.30$ in.

$c_c = 5.30 + 7.25 - 0.79 = 11.76$ in. $\quad c_{st} = 17.78 + 12.48 = 30.26$ in.

Maximum bending stresses

$$f_{sc} = \frac{9450 \times 8.20}{27{,}198} + \frac{6320 \times 21.58}{10{,}554} + \frac{1280 \times 14.56}{19{,}220} = 16.72 \text{ ksi}$$

$$f_{st} = \frac{9450 \times 28.11}{27{,}198} + \frac{6320 \times 14.73}{10{,}554} + \frac{1280 \times 21.75}{19{,}220} = 20.03 \text{ ksi}$$

$$f_c = \frac{9450 \times 14.66}{10 \times 27{,}198} + \frac{1280 \times 21.02}{30 \times 19{,}220} = 0.56 \text{ ksi}$$

Length of cover plate

$$l = a + (L - a)\sqrt{1 - \frac{S_c'}{S_c}}$$

$a = 4.67$ ft. $\quad S_c' = 18{,}296/30.26 = 607$ in.^{3} $\quad S_c = 27{,}198/28.11 = 967$ in.^{3}

$$l = 4.67 + (70 - 4.67)\sqrt{1 - \frac{607}{967}} = 44.7 \text{ ft.}$$

Force in cover plate at theoretical end

$$P = F_b A \frac{S_c'}{S_c} = 20 \times 10.5 \times \frac{607}{967} = 132^k$$

Length 5/16-in. fillet weld

$$l = \frac{132}{12.4 \times 0.707 \times 5/16} = 49''$$

Use cover plate 3/4 × 14 × 47′-3″

ELDERSBURG-LOUISVILLE ROAD BRIDGE DP10-2 3/4

Shear connectors

Vertical shear

End: $I = \frac{50}{70+125} = 0.256$

$V_{max} = \left[16 + \frac{7.0}{5.5}\left(16 + \frac{70-14}{70} + 4 \times \frac{70-28}{70}\right)\right] \times 1.256 = 44.3^k$

$V_{min} = Q$

10 ft from end: $I = \frac{50}{60+125} = 0.27$

$V_{max} = \frac{7.0}{5.5}\left(16 \times \frac{60}{70} + 16 \times \frac{60-14}{70} + 4 \times \frac{60-28}{70}\right) \times 1.27 = 42.1^k$

$V_{min} = -\frac{7.0}{5.5} \times 4 \times \frac{10}{70} \times 1.30 = -0.90^k$

20 ft from end: $I = \frac{50}{50+125} = 0.286$

$V_{max} = \frac{7.0}{5.5}\left(16 \times \frac{50}{70} + 16\,\frac{50-14}{70} + 4 \times \frac{50-28}{70}\right) \times 1.286 = 34.2^k$

$V_{min} = -\frac{7.0}{5.5}\left(4 \times \frac{20}{70} + 16 \times \frac{6}{70}\right) \times 1.30 = -4.2^k$

30 ft from end: $I = \frac{50}{40+125} = 0.30$

$V_{max} = \frac{7.0}{5.5}\left(16 \times \frac{40}{70} + 16 \times \frac{40-14}{70} + 4 \times \frac{40-28}{70}\right) 1.30 = 26.1^k$

$V_{min} = -\frac{7.0}{5.5}\left(4 \times \frac{30}{70} + 16 \times \frac{16}{70} + 16 \times \frac{2}{70}\right) 1.30 = -9.6^k$

Use groups of two 7/8-in. diameter studs

$Z_r = 2 \times 10.6 \times 0.875^2 = 16.2^k$

For beam and slab: $\frac{Z_r I}{Q} = \frac{16.2 \times 18{,}296 \times 10}{7.25 \times 84\,(11.76-3.62)} = 550$

For beam, flange plate and slab: $\frac{Z_r I}{Q} = \frac{16.2 \times 27{,}198 \times 10}{7.25 \times 84\,(14.66-3.62)} = 654$

Spacing

End = $S = \frac{550}{44.3} = 12.4''$

10 ft from end: $S = \frac{598}{42.1+0.9} = 13.9''$

20 ft from end: $S = \frac{654}{34.2+4.2} = 17.1''$

30 ft from end: $S = \frac{654}{26.1+9.6} = 18.3''$

Number connectors for ultimate strength

$P = A_s F_y = 50.20 \times 36 = 1810^k \quad P = 0.85 f_c' bc = 0.85 \times 3 \times 84 \times 7.25 = 1550^k$

$S_u = 930 d_s^2 \sqrt{f_c'} = 930 \times 0.875^2 \sqrt{3000} = 39^k$

$N = \frac{P}{0.85 S_u} = \frac{1550}{0.85 \times 39} = 47$ studs

Fatigue strength governs

Spacing

20 @ 12 + 10 @ 16 + 2 @ 18 + 10 @ 16 + 20 @ 12

Place first row of studs 2in. inside span

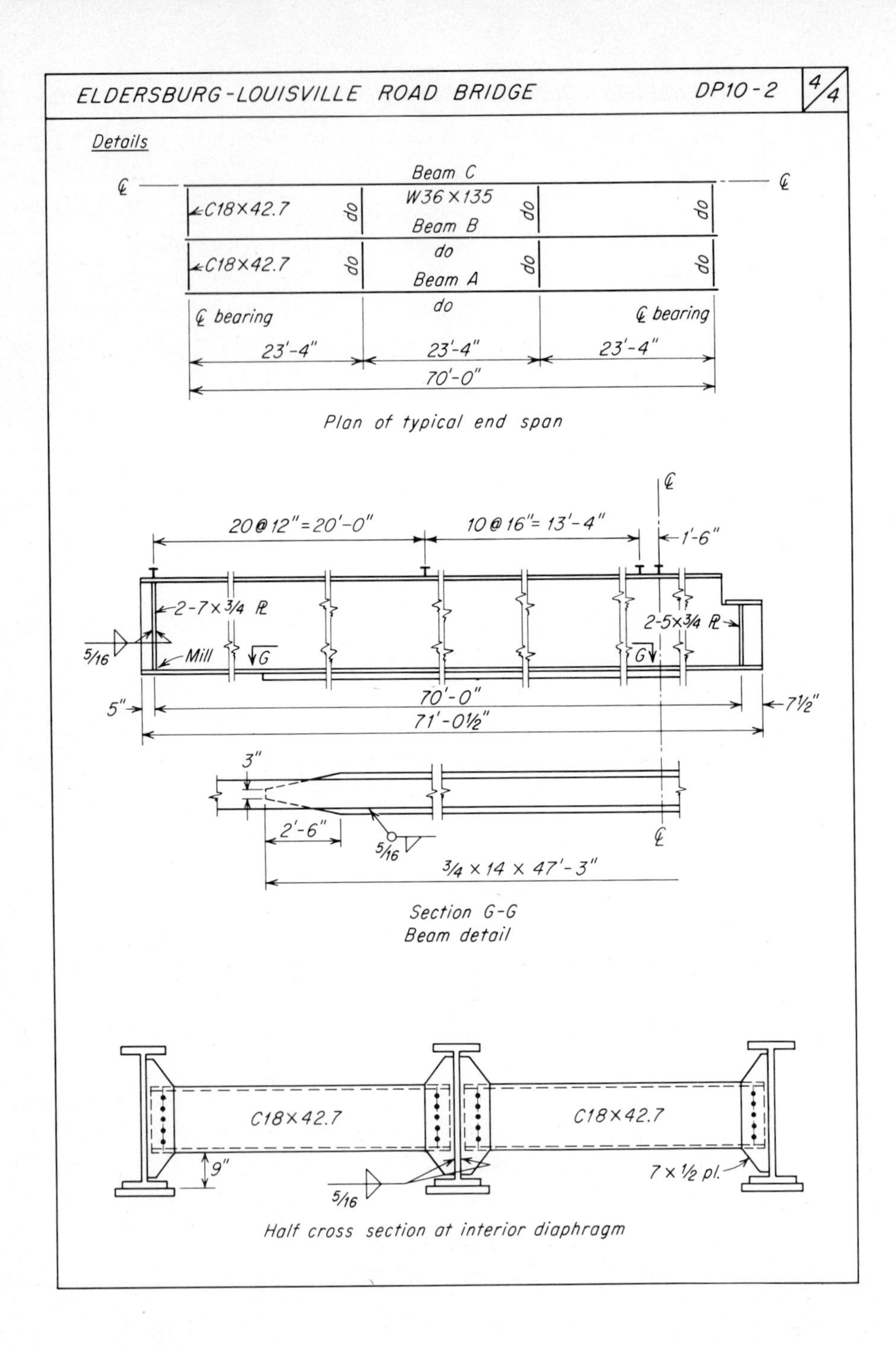

ELDERSBURG-LOUISVILLE ROAD BRIDGE
DP10-2
4/4
Details
Beam C
W36×135
Beam B
do
Beam A
do
C18×42.7
C18×42.7
do
℄ bearing
℄ bearing
23'-4"
23'-4"
23'-4"
70'-0"
Plan of typical end span
20 @ 12" = 20'-0"
10 @ 16" = 13'-4"
1'-6"
2-7×3/4 PL
2-5×3/4 PL
5/16
Mill
G
70'-0"
71'-0½"
5"
7½"
3"
2'-6"
5/16
3/4 × 14 × 47'-3"
Section G-G
Beam detail
C18×42.7
C18×42.7
9"
5/16
7 × ½ pl.
Half cross section at interior diaphragm

11

INDUSTRIAL BUILDINGS

11-1 INTRODUCTION

The function of a modern industrial building is to support and house a manufacturing process or to store the raw materials for (or products of) a manufacturing process. Many industrial plants are being located in small population centers because of cheaper land, adequate space for parking and future expansion, and the opportunity for providing a pleasing environment away from congested city areas. Other factors which must be considered in the site selection are topography, subsoil conditions, transportation, and utilities.

With the site selected, the planning of the production process and the structure to house it can proceed. Using the total area and volume requirements established for the preliminary planning, the exterior dimensions can be developed. Square or nearly square areas are usually more economical because they require less exterior wall length. Dimensions of individual bays within the structure are often dictated by the manufacturing process. Large, clear areas unobstructed by columns and partitions are desirable so as to provide sufficient flexibility and to facilitate later changes in the production layout without major building alterations. The plan must also allow for future expansion, preferably by expanding the plant on any or all of its four sides without shifting existing production layouts. The single-story industrial building,

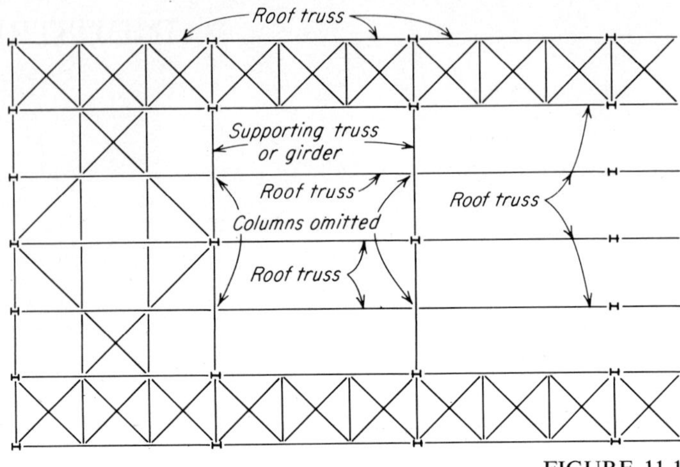

FIGURE 11-1

consisting of a floor slab on grade and a flat-roof superstructure, is the most economical from the standpoint of first cost and maintenance. In most industrial manufacturing processes and warehousing, it is also the most economical from the standpoint of operating cost.

11-2 FRAMING

Upon completion of the production layout from which the general plan of the structure derives, the structural framing scheme may be developed somewhat as described in the following paragraphs. These steps are not dissociated but closely interrelated.

Column rows are located to provide adequate clearance for the production layout. The height of the building will partly determine the economical spacing of column rows, since several tall, closely spaced columns with short-span trusses may require more steel than fewer columns with trusses of longer span. For a structure 180 ft wide, for example, two 90-ft aisles may be more economical than three 60-ft aisles. Wider aisles also allow greater flexibility in future modification of the arrangement of the plant. With the column rows tentatively located, cross-sectional sketches of the building must be studied to develop a design that will be not only structurally efficient but pleasing in proportion as well. The columns in each row must be distributed so as to minimize interference with the mechanical layout. Although 20 ft may prove to be more economical, a spacing of 25 ft or more provides greater flexibility. In some areas of the building, spatial requirements may necessitate a spacing of more than 20 to 25 ft in the interior rows. This may be managed by enlarging a bay for the full

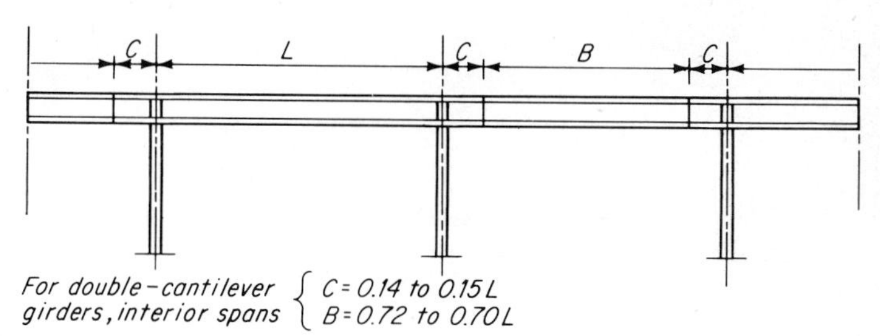

FIGURE 11-2
Double-cantilever construction. [*From E. A. Picardi, sec. 19, pt. 2, in E. H. Gaylord and C. N. Gaylord (eds.), "Structural Engineering Handbook," McGraw-Hill, New York, 1968.*]

width of the building or by an arrangement such as the one shown in Fig. 11-1, in which the roof trusses in one or more aisles are supported by transverse girders or trusses instead of by columns. Trusses are not likely to be economical for spans less than 70 ft. For short spans, the framing system may consist of simple-span, continuous, or double-cantilevered girders (Fig. 11-2) over the columns in one direction with simple-span bar joists in the other direction. If heavy loads are to be suspended from the roof, rolled sections may be substituted for bar joists. Where spans exceed 70 ft in both dimensions of a bay, the designer should investigate the use of primary trusses in one direction with secondary trusses framing between them. The secondary trusses should be spaced 20 to 30 ft on centers to afford an economical purlin span.

After the general outline of the trusses has been determined, their panel lengths and the location of the purlins can be decided upon. Finally, lateral bracing may be laid out. The lateral bracing may be in the plane of the top chord, in the plane of the bottom chord, or in both planes, depending largely on the designer's judgment in the matter.

The selection of materials for floors, roofs, walls, and partitions, and the framing of roofs, trusses, and walls are discussed briefly in subsequent articles.

11-3 RIGID-FRAME BUILDINGS

The rigid-frame bent may be used for the main structural framing of industrial buildings, auditoriums, gymnasiums, churches, and other structures which require large areas unobstructed by columns. The framing is simple to erect and pleasing in appearance. Several types of rigid-frame bents are shown in Fig. 11-3. The moments developed at the connections in these frames reduce the moment at midspan of the transverse members, resulting in lighter and shallower transverse members but

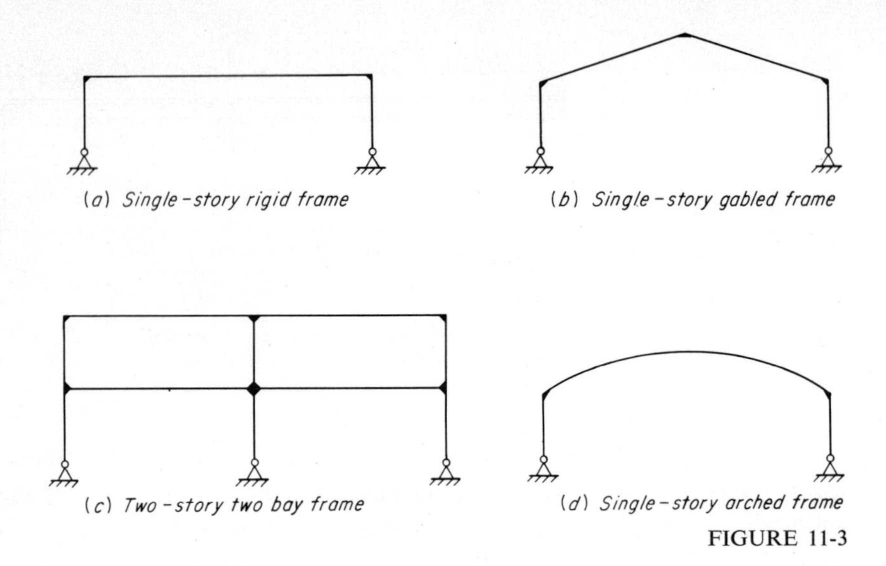
(a) Single-story rigid frame
(b) Single-story gabled frame
(c) Two-story two bay frame
(d) Single-story arched frame

FIGURE 11-3

heavier column sections. Therefore a rigid frame may require a greater amount of steel than a braced bent of equal rise and span.

Economical spans for single-story single-span rigid frames may range from 30 to about 200 ft. For buildings with average roof load the following spacing of frames will usually prove economical:

Span, ft	Frame spacing, ft
30–40	16
40–60	18
60–100	20
Over 100	$\frac{1}{5}$–$\frac{1}{6}$ span

Columns of rigid frames may be designed as rotationally restrained (fixed) or rotationally free (hinged). Pin-connected bases are rarely required; instead, a flat base plate with a single line of anchor bolts on the neutral axis of the column is common, and is usually considered to act as a hinge. The horizontal reaction in smaller frames can be developed by the footing, but for spans greater than 60 to 80 ft a tie between the column bases or the footings may be necessary. Such a tie is usually located under the floor slab.

The design of a rigid frame is illustrated in Art. 11-11.

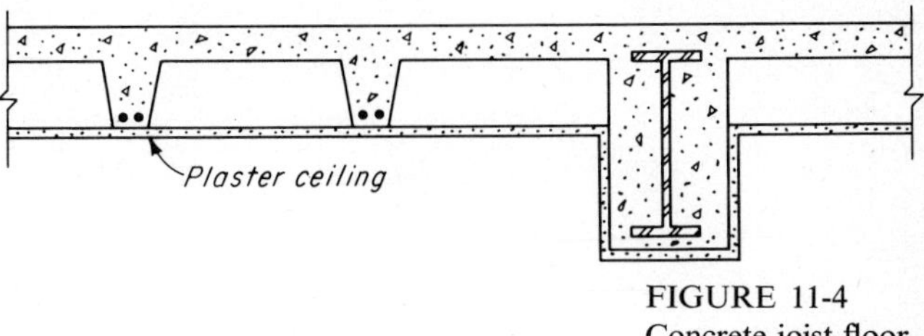

FIGURE 11-4
Concrete joist floor.

11-4 FLOOR CONSTRUCTION

Although most industrial buildings are one-story structures, the part devoted to offices may consist of two or more stories. For this portion of the building, the most economical arrangement usually results if the area is divided into equal rectangles, with columns spaced not less than 20 ft. Various types of floor deck can be used with this system. The span, weight, fire rating, insulating value, acoustical properties, maintenance, etc., must be considered in choosing the floor construction. A comprehensive comparison of floor systems is given in Ref. 1.

The one-way reinforced-concrete slab is one of the heaviest floor decks, but it makes a stiff floor and has a good fire rating. The minimum thickness is $3\frac{1}{2}$ to 4 in. Lightweight aggregates may be used to reduce the weight. Composite construction, in which the steel beam is encased in concrete or else bonded to the slab by means of shear connectors, may be advantageous for heavily loaded floors. The design of composite beams is discussed in Chap. 5.

A floor system consisting of a reinforced-concrete slab $2\frac{1}{2}$ to 3 in. thick, supported by, and cast with, reinforced-concrete joists spaced 24 to 36 in., is economical for spans up to 30 ft (Fig. 11-4). The joists are formed by using removable metal pans. A similar two-way system is made by using small, square pans. Both systems may also be built using lightweight filler blocks of clay tile, gypsum tile, etc., instead of metal pans. This system provides an excellent surface for plastering, and no metal lath is required except on the bottom flanges of the steel beams if they are to be fireproofed with concrete or plaster. Most floors of this type require substantial forms and shoring during construction.

Cellular floor decking of light-gage steel, which may be used for spans up to about 25 ft, is available from a number of manufacturers. In one form, the unit for this type of floor consists of a flat sheet of steel spot-welded to a second sheet bent into a series of alternating troughs. Almost any kind of floor covering, e.g., asphalt tile, may be applied to the flat surface. Other forms consist of two sheets of steel both bent into a series of alternating troughs and spot-welded together (Fig. 11-5). These units must be topped with concrete or other material to produce the floor surface.

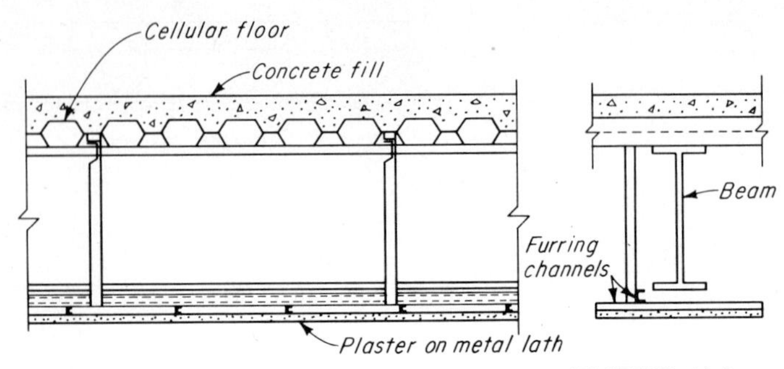

FIGURE 11-5
Cellular steel floor.

Either form gives parallel, lengthwise cells which are useful as electrical raceways. If a fire-rated floor is required, at least 2 in. of concrete topping must be used in conjunction with a suspended, plastered ceiling.

Open-web steel joists produce the lightest floor system (Fig. 2-1). These joists may be welded or bolted to the top flange of the supporting beam or attached by anchor rods fastened over the beam flange. A concrete slab 2 to $2\frac{1}{2}$ in. thick and cast in place over steel mesh or welded wire fabric backed by heavy paper is ordinarily used. Standard open-web joists are available in depths ranging in 2-in. increments from 8 to 24 in. The 24-in. joist may be used on spans up to 48 ft. Longspan steel joists, with depths of 18 and 20 in. and 24 to 48 in. in 4-in. increments are available for clear spans of 25 to 96 ft. Deep longspan joists with depths of 52, 56, and 60 in. can span up to 120 ft.

There are many variations and adaptations of the floor systems described above, some of which are proprietary. Precast planks of gypsum or concrete are also used.

11-5 ROOF SYSTEMS

The structural systems described previously for floors can be used for flat roofs. The longspan open-web joist is available with the top chord sloped to give either a single or a double pitch, but not exceeding $\frac{1}{8}$ in. per ft, where roof drainage is required. In addition to these systems, corrugated cement-asbestos board, corrugated metal, and patented steel roof-deck materials should be considered. Corrugated cement-asbestos board is surfaced with a rigid insulating material and a built-up covering consisting of several layers of roofing felt cemented together with coal-tar pitch or asphalt and topped with slag or gravel. Corrugated metal may also be insulated and surfaced with a built-up covering.

Metal plates formed with interlocking ribs which increase strength and stiffness are manufactured in many different styles. These decks are usually covered with a vapor seal, a rigid insulating board, and a built-up roofing. They are installed either with ribs up or with ribs down. With ribs up there results a smooth ceiling which may be shop-painted with a baked-on enamel. If the appearance is objectionable with ribs down, an acoustical material may be applied to the lower surface.

A variety of precast units in the form of flat slabs, channel sections, and interlocking tiles, with or without glass insert panels, is used for pitched roofs as well as for flat roofs. These units may be manufactured from portland cement and lightweight aggregate, chemically expanded concrete, gypsum, and portland cement and asbestos fiber. Corrugated sheets of metal, cement-asbestos board, and wire glass are particularly adaptable to industrial buildings.

11-6 EXTERIOR WALLS

Many forms of exterior wall construction are available. Walls must be structurally sound, have adequate fire-resistance and insulating properties, offer protection against condensation of water vapor, be durable and easily maintained and pleasing in appearance.

Bearing walls must be strong enough to support, in addition to their own weight, any loads from floors and the roof which frame into them. According to the Uniform Building Code, bearing walls of plain masonry must be not less than 12 in. thick, except that walls of one-story business buildings and residential buildings not over three stories high may be 8 in. thick. Each successive 35-ft height of wall below the topmost 35 ft must be increased 4 in. in thickness. The minimum thickness of unreinforced, grouted, brick masonry walls may be 2 in. less than that required for plain masonry walls. Similar, but less severe, limitations are prescribed for reinforced-concrete bearing walls. Panel walls (also called curtain walls) in skeleton frame construction are usually specified in terms of their fire-resistance rating. For example, the National Building Code requires that the fire-resistance rating of panel walls be not less than 4 hr for both fireproof and semifireproof construction. Except for reinforced-concrete walls, this requirement necessitates masonry walls ranging in thickness from 8 to 12 in.

Corrugated cement-asbestos board and corrugated-metal siding are used extensively for walls of industrial buildings. Wall sections consisting of two steel sheets with a layer of insulation between have also been widely accepted. Very attractive architectural effects are produced with fluted metal wall panels, which may be aluminum, stainless steel, galvanized copper-bearing steel, or porcelain-enameled iron or steel.

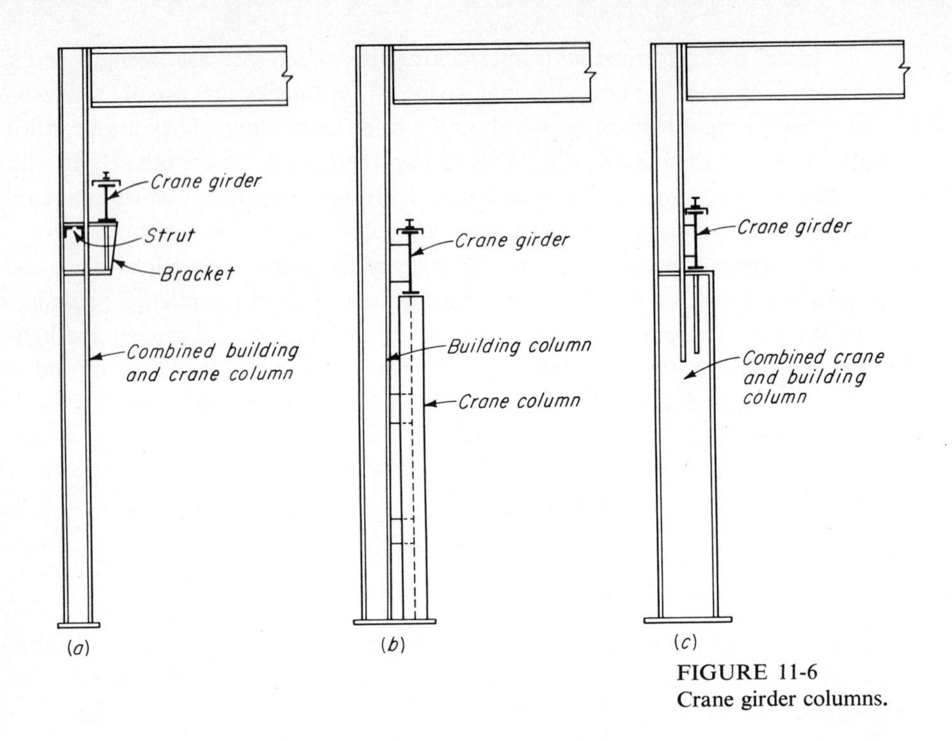

FIGURE 11-6
Crane girder columns.

Structural glass blocks, which are hollow, colorless, and translucent, are appropriate for exterior walls when ventilating sash are not required or when clear glass is not needed for visibility. Since their coefficient of heat transfer is less than one-half that of single-thickness common glass, heating costs are reduced and air conditioning improved. Glass blocks should not be subjected to vertical loads other than their own weight.

11-7 PARTITIONS

Some of the factors to be considered in selecting partitions are appearance, weight, acoustical properties, and ease of erection. Although they usually carry no load other than their own weight, partitions for a building of only a few stories may be designed as bearing walls.

Clay tile, cinder block, and gypsum tile are used extensively, but for certain types of occupancy, such as general office areas, hollow partitions and other lightweight assemblies are more popular. Prefabricated hollow partitions may consist of metal lath and plaster supported on steel channel studs. Solid metal lath and plaster parti-

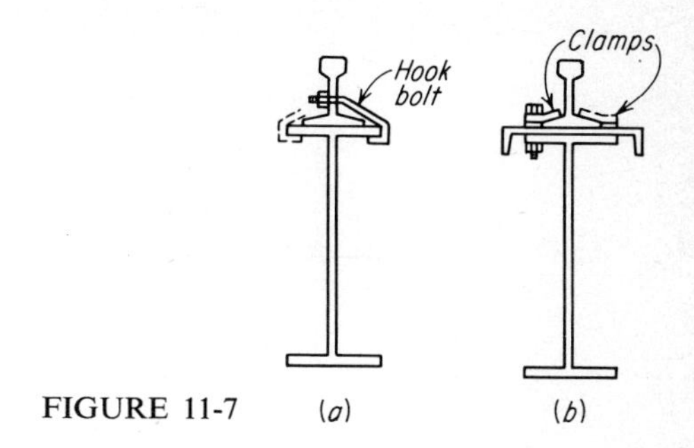

FIGURE 11-7

tions offer the advantage of increased floor area because of their 2-in. thickness. Metal and wire glass are adaptable to conditions requiring lightweight partitions which can be removed, altered, and relocated easily.

11-8 CRANEWAYS

Crane girders for industrial buildings may be supported on column brackets, as in Fig. 11-6*a*, but heavy cranes are supported as in Fig. 11-6*b* or *c*. The column shown in Fig. 11-6*b* has the advantage of being fabricated from rolled sections and has good stiffness in both principal directions. The crane girder may be a rolled shape or plate girder, depending upon the span and the load. The section shown in Fig. 11-7*b*, which consists of a channel attached to the top flange of a W shape, is also used. Crane rails are attached to the girder with bolted clamps, but hook bolts are used with the lighter rails if the girder flanges are too narrow for clamps (Fig. 11-7).

Large industrial buildings sometimes require craneways located at two or three levels in the same aisle and one or more runways in adjacent aisles. Crane girders in adjacent aisles can be supported on brackets connected to a single line of columns for the lighter cranes, but the heavier ones require separate columns.

The design of crane girders is illustrated in Arts. 5-12 and 5-24.

11-9 BRACING

The bent of Fig. 11-8*a* can resist vertical loads and lateral loads in its plane without dependence upon adjacent bents. However, it has little resistance to forces normal to its plane, and bracing such as *ABCD* in the plane of the top chord (Fig. 11-8*b*) and *BCEF* in the plane of the columns (Fig. 11-8*c*) is required. Two adjacent bents and their bracing form a braced bay, which, through the purlins and eave struts, supports adjacent bents. However, supplementary bracing such as that shown in Fig. 11-8*b*

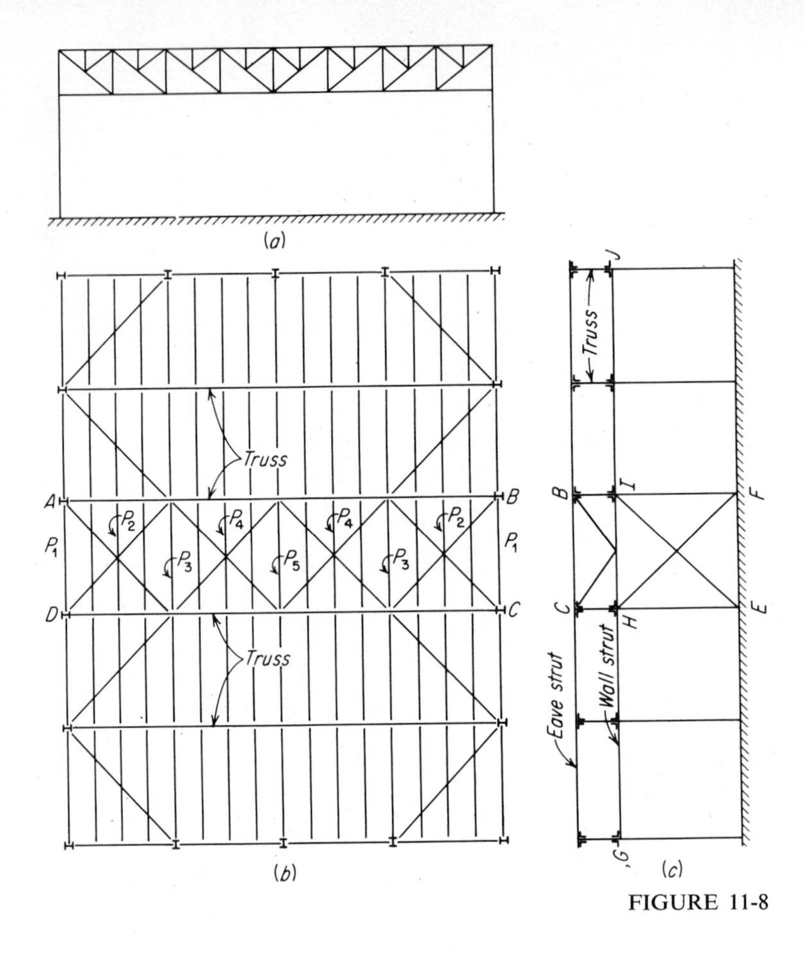

FIGURE 11-8

is usually provided for additional stiffness and to facilitate erection. The number and spacing of the braced bents depends to some extent on the magnitude of the horizontal forces to be resisted. There should be at least one braced bay between adjacent expansion joints in the building.

Different arrangements of bracing may be used in the wall panel *BCEF*. For example, the wall strut *GHIJ* is usually omitted in a building with girts, in which case the bay may be braced with diagonals from *C* to *F* and from *B* to *E* or, depending on the height of the building, with one or more intermediate lines of struts dividing *BCEF* into equal panels, with two diagonals in each panel. In some cases knee braces from *H* and *I* to the eave strut *CB* are used instead of diagonal bracing.

In addition to the bracing just described, the bottom chords of the trusses should be tied together at occasional panel points by lines of struts parallel to the wall strut *GHIJ*. As a general rule, these lines of struts should be spaced so that the slenderness ratio of the bottom chord, with respect to bending in the horizontal plane, does not exceed the specification limit. If the spacing of the trusses is too large for a single strut, cross frames similar to *BCHI* are used instead, with the purlins and the verticals of the main trusses forming part of the system.

A system of bracing like that described above will, in general, be sufficient to produce a stable structure. Bracing is expensive, and should not be overdone. Just what is necessary in a given situation is, to a considerable extent, a question of judgment that depends on the engineer's qualitative evaluation. Forces in the members of the bracing system are rarely large enough to control their design. Therefore, except for whatever guidance is offered by slenderness-ratio limitations, their size is largely a question of judgment.

Trusses at the ends of a building are not necessary unless future extensions are to be provided for. Purlins in the end bays can be supported on rafters and other gable framing attached to the columns in the end walls. End walls without trusses must also have X- or knee-braced panels.

Bracing for simpler steel buildings which consist of beams continuous over columns, with rolled-section purlins or open-web joists framing between, usually consists of bridging between joists to stiffen the roof system and knee braces or moment-resistance beam-to-column connections.

Craneways must be braced to resist the longitudinal forces from starting and stopping the crane. If the girder is supported as in Fig. 11-6*b* or *c*, knee braces at the crane column will be adequate for all but the heaviest cranes. Runways for the latter may require occasional X-braced panels. If the girder is supported on brackets (Fig. 11-6*a*), a strut consisting of two angles, laced or battened, may be used, as shown. This strut may be a part of the bracing system in the braced bays.

11-10 TRUSSES

It is impossible to formulate rules which will automatically determine the most economical truss for a given set of conditions. Certain principles which are helpful are discussed in this article.

The slope of a roof is determined principally by considerations of climate, drainage, and the type of roofing. Slopes must be 20 percent or more for lapped shingles, such as wood, asphalt, clay tile, etc. Flat roofs should be built with a slope of at least $\frac{1}{8}$ in. per ft to facilitate drainage.

The depth of the Fink truss and similar types is determined largely by the slope requirements of the roofing. Since they are not ordinarily suitable for slopes less than about 5 in. per ft, their depths will be at least one-fifth the span, which is ample for stiffness. Trusses for roofs of lesser slope and for flat roofs will usually be of the Pratt or Warren triangulation. The economical depth for these trusses is likely to be one-twelfth to one-eighth the span. The former figure is probably ample for continuous trusses, but simply supported trusses ordinarily should be somewhat deeper, perhaps about one-tenth the span as a minimum. However, clearance for shipping must be taken into consideration. It may be impossible to ship a truss deeper than about 10 ft, depending of course upon clearances on the route it must travel. Thus, if a truss with a span of, say, 120 ft and a depth of 11 or 12 ft cannot be shipped shop-fabricated in one or two sections, a depth of 10 ft—even though it is only one-twelfth the span—will almost certainly prove to be more economical.

The depth of a truss must also be considered in relation to the length of the panels. The panel length depends in turn upon the spacing of the purlins and upon the necessity of getting a favorable slope for the diagonal web members. Spacing of the purlins is governed, of course, by the type of roof deck. In general, the slope of the diagonals should be 40 to 50° with the horizontal. A slope of more than 50° may require an excessive number of diagonals, while a slope of less than 40° is likely to result in needlessly large stresses in the web system. Steep diagonals can be avoided in deep trusses with relatively close-spaced purlins by using a subdivided truss, such as that of Fig. 11-8*a*.

The spacing of longspan trusses (longer than about 150 ft) should be carefully considered. Such trusses may be more economical on 40- to 50-ft centers, rather than 20- to 25-ft, with the purlins supported on longitudinal trusses framing into the longspan trusses at 20- to 25-ft intervals.

Before the stresses in the members can be determined, the weight of the truss must be estimated and added to the dead load of the purlins, roof deck, and roofing. The weight of a truss is influenced by its pitch and span, by the spacing of the trusses, by the magnitude and distribution of loading, by differences in specifications, and by the ingenuity of the designer. The weight of a truss of ordinary span, together with the lateral bracing, may be estimated at about 10 percent of the load it supports. Longspan trusses, especially those supporting other trusses, are likely to be heavier. Many formulas for estimating the weight of a truss have been published, but because of the factors just mentioned, they cannot be depended upon for accurate estimates. The weight of a truss should be computed after it has been designed to make sure that it is close enough to the assumed weight.

11-11 DP11-1: DESIGN OF A STEEL-BUILDING RIGID FRAME FOR A SINGLE-STORY BUILDING

The design will be governed by the Basic Building Code (BOCA 1965) and the AISC specification. The dimensions of the building and the purlin spacing are shown in the figure. The bases are assumed to be hinged. The following comments are identified by letters corresponding to those alongside the computations on the design sheets.

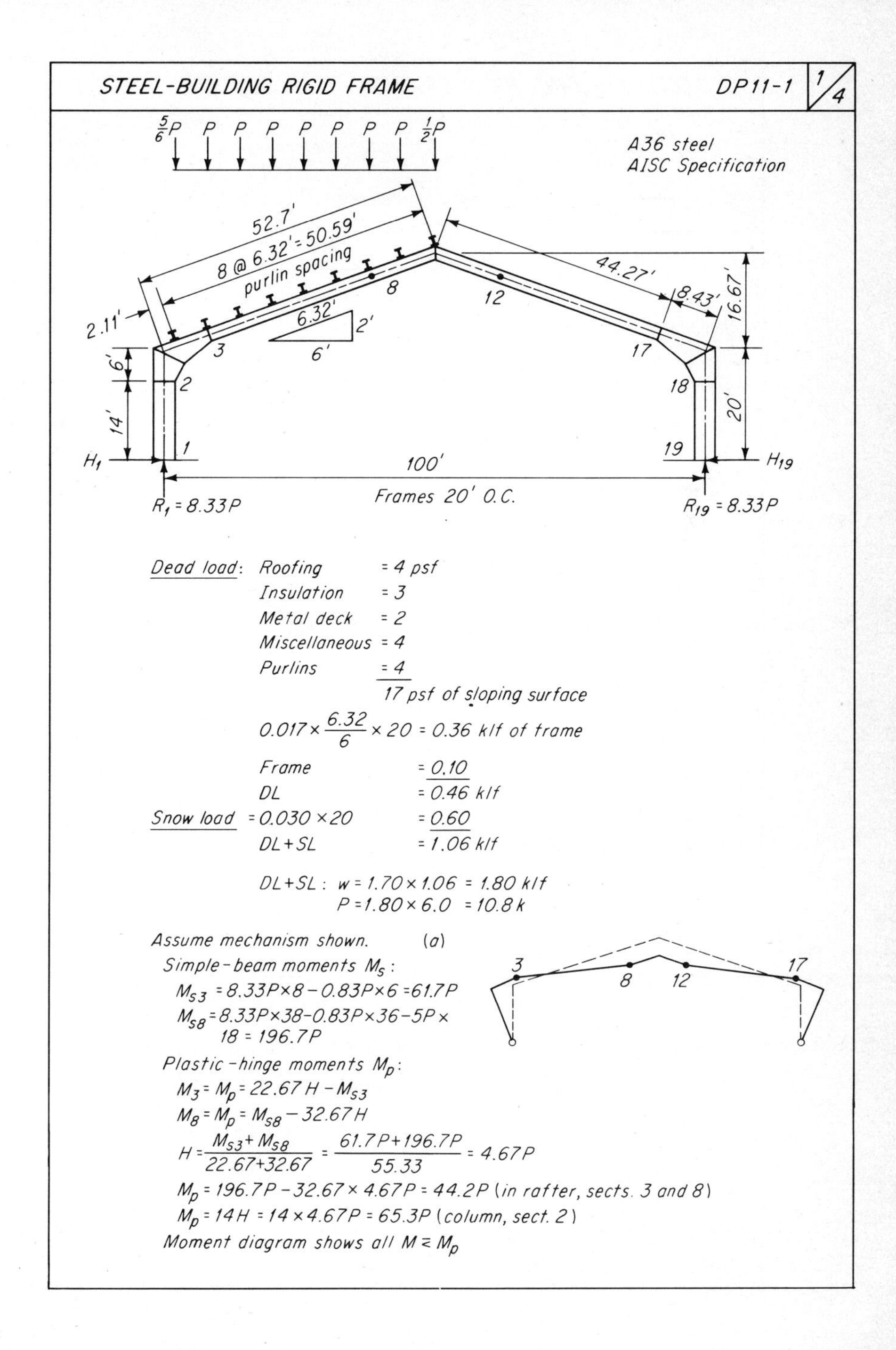

Dead load: Roofing = 4 psf
Insulation = 3
Metal deck = 2
Miscellaneous = 4
Purlins = 4
17 psf of sloping surface

$$0.017 \times \frac{6.32}{6} \times 20 = 0.36 \text{ klf of frame}$$

Frame = 0.10
DL = 0.46 klf
Snow load = 0.030 × 20 = 0.60
DL + SL = 1.06 klf

DL + SL: $w = 1.70 \times 1.06 = 1.80$ klf
$P = 1.80 \times 6.0 = 10.8$ k

Assume mechanism shown. (a)

Simple-beam moments M_s:

$M_{s3} = 8.33P \times 8 - 0.83P \times 6 = 61.7P$

$M_{s8} = 8.33P \times 38 - 0.83P \times 36 - 5P \times 18 = 196.7P$

Plastic-hinge moments M_p:

$M_3 = M_p = 22.67H - M_{s3}$

$M_8 = M_p = M_{s8} - 32.67H$

$$H = \frac{M_{s3} + M_{s8}}{22.67 + 32.67} = \frac{61.7P + 196.7P}{55.33} = 4.67P$$

$M_p = 196.7P - 32.67 \times 4.67P = 44.2P$ (in rafter, sects. 3 and 8)

$M_p = 14H = 14 \times 4.67P = 65.3P$ (column, sect. 2)

Moment diagram shows all $M \lessapprox M_p$

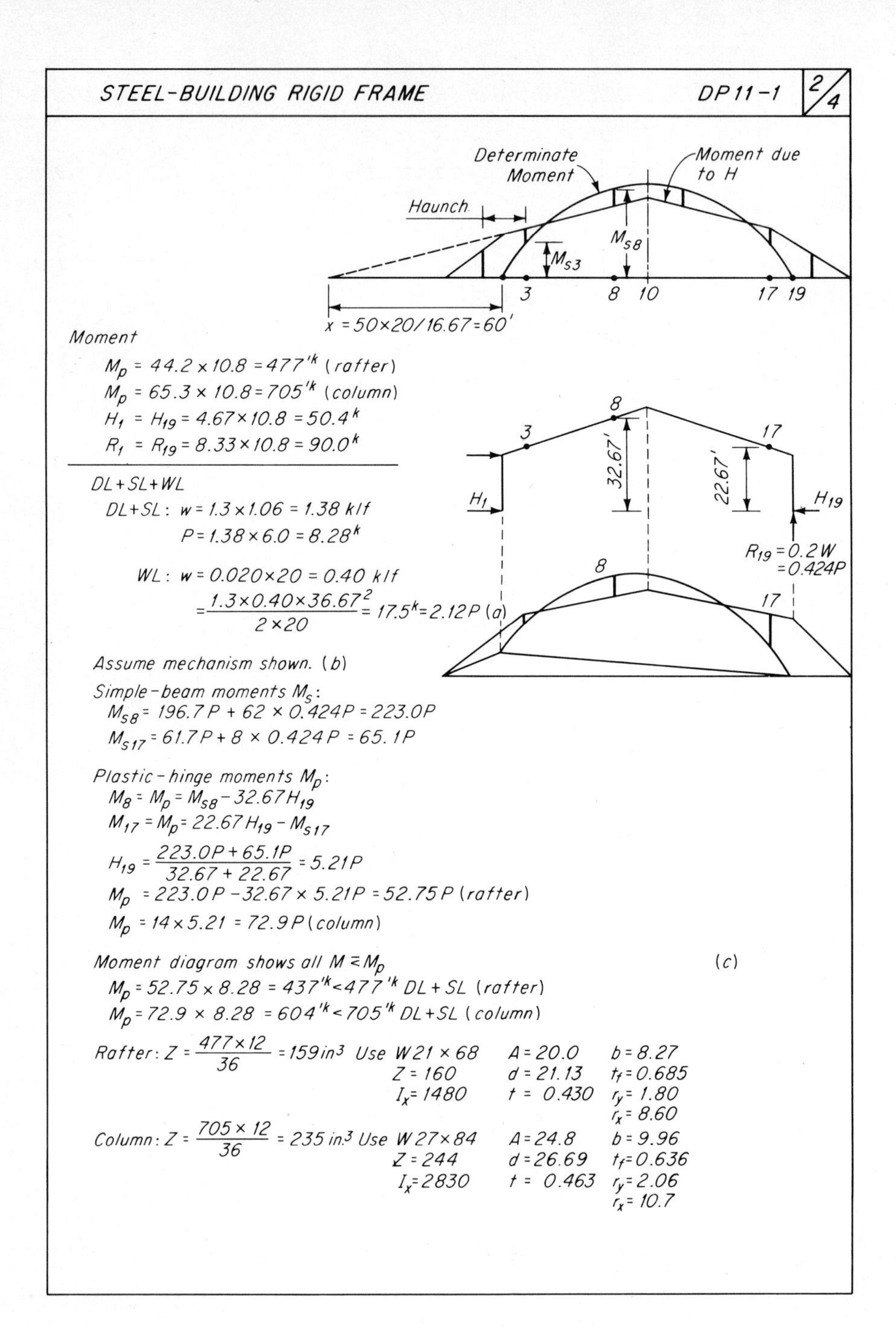

STEEL-BUILDING RIGID FRAME — DP 11-1 — 2/4

Moment

$M_p = 44.2 \times 10.8 = 477'^k$ (rafter)

$M_p = 65.3 \times 10.8 = 705'^k$ (column)

$H_1 = H_{19} = 4.67 \times 10.8 = 50.4^k$

$R_1 = R_{19} = 8.33 \times 10.8 = 90.0^k$

DL + SL + WL

DL + SL: $w = 1.3 \times 1.06 = 1.38$ klf

$P = 1.38 \times 6.0 = 8.28^k$

WL: $w = 0.020 \times 20 = 0.40$ klf

$$= \frac{1.3 \times 0.40 \times 36.67^2}{2 \times 20} = 17.5^k = 2.12P \quad (a)$$

Assume mechanism shown. (b)

Simple-beam moments M_s:

$M_{s8} = 196.7P + 62 \times 0.424P = 223.0P$

$M_{s17} = 61.7P + 8 \times 0.424P = 65.1P$

Plastic-hinge moments M_p:

$M_8 = M_p = M_{s8} - 32.67H_{19}$

$M_{17} = M_p = 22.67H_{19} - M_{s17}$

$$H_{19} = \frac{223.0P + 65.1P}{32.67 + 22.67} = 5.21P$$

$M_p = 223.0P - 32.67 \times 5.21P = 52.75P$ (rafter)

$M_p = 14 \times 5.21 = 72.9P$ (column)

Moment diagram shows all $M \leqq M_p$ (c)

$M_p = 52.75 \times 8.28 = 437'^k < 477'^k$ DL + SL (rafter)

$M_p = 72.9 \times 8.28 = 604'^k < 705'^k$ DL + SL (column)

Rafter: $Z = \dfrac{477 \times 12}{36} = 159\ \text{in}^3$ Use W21 × 68

	$A = 20.0$	$b = 8.27$
$Z = 160$	$d = 21.13$	$t_f = 0.685$
$I_x = 1480$	$t = 0.430$	$r_y = 1.80$
		$r_x = 8.60$

Column: $Z = \dfrac{705 \times 12}{36} = 235\ \text{in}^3$ Use W27 × 84

	$A = 24.8$	$b = 9.96$
$Z = 244$	$d = 26.69$	$t_f = 0.636$
$I_x = 2830$	$t = 0.463$	$r_y = 2.06$
		$r_x = 10.7$

STEEL-BUILDING RIGID FRAME — DP11-1 — 3/4

Combined axial force and bending (a)

Column: At top $G = \frac{2830}{20} \times \frac{2 \times 52.7}{1480} = 10.1$

At bottom $G = 10$ (hinged base)

$K = 3$ (Fig. 4-39b)

$KL/r_x = 3 \times 14 \times 12/10.69 = 47$

$P_{cr} = 1.70 \times 24.8 \times 18.61 = 785^k$

$P_y = 24.8 \times 36 = 893^k$

$P_E = 1.92 \times 24.8 \times 67.51 = 3215^k$

$P = 90^k$

$M = 705'^k$

$M_p = 244 \times 36/12 = 732'^k$

$$\frac{P}{P_y} + \frac{M}{1.18M_p} = \frac{90}{893} + \frac{705}{1.18 \times 732} = 0.101 + 0.819 = 0.920 < 1.0$$

$$\frac{P}{P_{cr}} + \frac{C_m M}{(1-P/P_E)M_m} = \frac{90}{785} + \frac{0.85 \times 705}{(1-90/3215)732} = 0.115 + 0.845 = 0.960 < 1.0$$

Rafter: At column $G = \frac{1480}{52.7} \times \frac{20}{1.5 \times 2830} = 0.13$

At ridge $G = 1$

$K = 0.66$ (Fig. 4-39a)

$KL/r_x = 0.66 \times 44.27 \times 12/8.60 = 41$

$P_{cr} = 1.70 \times 20.0 \times 19.11 = 650^k$

$P_y = 20.0 \times 36 = 720^k$

$P_E = 1.92 \times 20.0 \times 88.71 = 3410^k$

$$P = \frac{6}{6.32}H_1 + \frac{2}{6.32}(R_1 - 0.83P)$$

$$= \frac{6}{6.32} \times 50.41 + \frac{2}{6.32}(90-9) = 47.8 + 25.6 = 73.4^k$$

$M = 477'^k$

$M_p = 160 \times 36/12 = 480'^k$

$$\frac{P}{P_y} + \frac{M}{1.18M_p} = \frac{73.4}{720} + \frac{477}{1.18 \times 480} = 0.102 + 0.844 = 0.946 < 1.0$$

$$\frac{P}{P_{cr}} + \frac{C_m M}{(1-P/P_E)M_m} = \frac{73.4}{650} + \frac{0.85 \times 477}{(1-70/3410)480} = 0.113 + 0.865 = 0.978 < 1.0$$

Shear

Column: $V_{all} = 0.55 \times 36 \times 0.463 \times 26.69 = 245^k > 50.4^k$ (b)

Rafter: $V_{all} = 0.55 \times 36 \times 0.430 \times 21.13 = 180^k$

$$V_{max} = \frac{6}{6.32}(90 - 0.83P) - \frac{2}{6.32}H_1$$

$$= \frac{6}{6.32}(90-9) - \frac{2}{6.32} \times 50.4 = 77.0 - 16.0 = 61.0^k < 180^k$$

Local buckling

Column: flange, $b/t_f = 9.96/0.636 = 15.7 < 17$ (c)

web, $P/P_y = 90/890 = 0.101 < 0.27$

$$d/t = \frac{412}{\sqrt{36}}(1 - 1.4 \times 0.101) = 58.8 > 26.69/0.463 = 57.7$$

STEEL-BUILDING RIGID FRAME — DP11-1 — 4/4

Rafter: flange, $b/t_f = 8.27/0.685 = 12.1 < 17$

web, $P/P_y = 69.6/720 = 0.097 < 0.27$

$$d/t = \frac{412}{\sqrt{36}}(1-1.4\times0.097) = 59.1 > 21.13/0.430 = 49 \qquad (a)$$

Lateral bracing

Rafter: purlins 6.32′ o.c. $\frac{L}{r_y} = \frac{6.32\times12}{1.80} = 42.1$

$M_4 = M_{S4} - 24H_1 = 100.7P - 24.7\times4.67P$
$= -14.6P = -14.6\times10.8 = -158'^{k}$

Segment 3-4 in reverse curvature, so end moment ratio is positive

$$\frac{M_4}{M_3} = \frac{158}{477} = 0.334$$

$$\frac{L}{r_y} = \frac{1375}{F_y} + 25 = \frac{1375}{36} + 25 = 38.2 + 25 = 63.2 > 42.1$$

Therefore lateral support OK at 3-4

Hinge at 8 forms last (Art. 8-12), so elastic-design formulas apply

$$f_b = \frac{M}{1.7S_x} = \frac{477\times12}{1.7\times140} = 24 \text{ ksi}$$

$$\frac{76b_f}{\sqrt{F_y}} = \frac{76\times8.27}{\sqrt{36}} = 105'' = 8.8' > 6.32'$$

$$\frac{20{,}000}{(d/A_f)F_y} = \frac{20{,}000}{3.73\times36} = 149'' = 12.4' > 6.32'$$

Therefore lateral support OK at 7-8

Column: $\frac{M_1}{M_2} = 0$

$$\frac{L}{r} = \frac{1375}{36} + 25 = 63.2 < \frac{14\times12}{2.06} = 81.6$$

Provide brace midway between points 1 and 2

Haunch design (b)

$d = 40''$, $t_f = 1\tfrac{1}{16}''$, $b = 8\tfrac{1}{4}''$, $t_w = \tfrac{7}{16}''$, $l = 7.17'$

$Z = 10\times0.687(40-0.687) + \tfrac{1}{4}(40-1.38)^2\times0.438$
$= 270 + 163 = 433 \text{ in.}^3$

$M_p = 433\times36/12 = 1299'^{k} > 20\times50.4 = 1008'^{k}$

Check for lateral buckling (c)

$l/b = 86/8.25 = 10.4$

$\Delta t = 0.1(10.4-5.3) = 0.51$ in.

$t_f = t_c = 1\tfrac{1}{16} + \tfrac{1}{2} = 1\tfrac{3}{16}$ in.

Web stiffeners (d)

at sections "a" and "c"

$A_{st} = A_f \sin 14°41' = 9.96\times1.19\times0.250 = 2.96 \text{ in.}^2$

Use 2 stiffeners $4\times\tfrac{1}{2} = 4.0 \text{ in.}^2$

at section "b": $$A_{st} = \frac{A_{c1}\cos 121°18' - A_{c2}\sin 14°41'}{\cos 35°47'}$$

$$= \frac{8.25\times1.19(0.5195-0.2535)}{0.8112} = 3.22 \text{ in.}^2$$

Use 2 stiffeners $4\times\tfrac{1}{2} = 4.0 \text{ in.}^2$

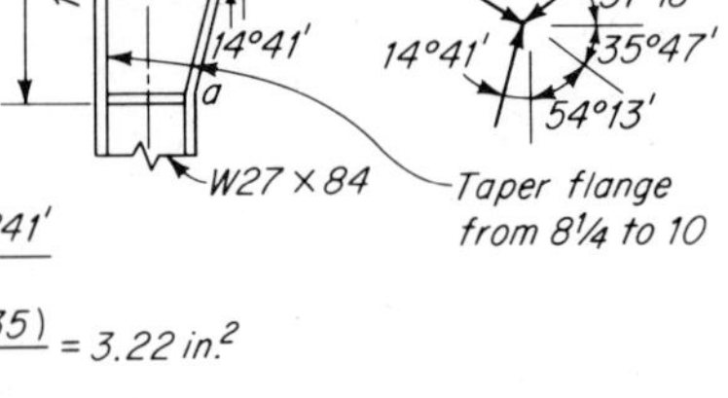

Sheet 1

a Plastic moments are assumed to form in the rafters at the haunches and at purlin locations 8 and 12. The simple-beam moments and plastic moments are computed at sections 3 and 8. The moment diagram (Sheet 2) shows that all $M \lesssim M_p$.

Sheet 2

a The distributed wind load is replaced with single concentrated loads at the eaves, which produce the same moment about the base.

b The assumed mechanism has hinges at 8 and 17. The simple-beam moments of 196.7P and 61.7P for vertical loads were computed on Sheet 1.

c The moment diagram shows all $M \lesssim M_p$. The moments due to $DL + SL$ at the load factor 1.7 are larger than those due to $DL + SL + WL$ at the load factor 1.3.

Sheet 3

a The column and rafter are checked by Eqs. (8-8) and (8-9).

b The shear resistance of the web is given by Eq. (8-11).

c The slenderness of the column and rafter webs is checked by Eq. (8-6).

Sheet 4

a Lateral bracing requirements are discussed in Art. 8-12; Eqs. (8-7) are applicable.

b The design procedure for tapered haunches is discussed in Art. 7-28. The depth of 40 in. was chosen so that β would be greater than 12°. The width of the flange along the rafter portion of the haunch is made the same as the W21 × 68 rafter. The haunch is tapered to 10 in. at the juncture with the W27 × 84.

c The l/b ratio of 10.4 requires that the flange thickness be increased to $1\frac{3}{16}$ in. [Eq. (7-24)].

d Stiffener areas are determined from Eqs. (7-24) and (7-25). An elastic design of the frame of this example is given in Ref. 3.

REFERENCES

1 Tang, S. J. Y., and J. E. Ambrose: Buildings, General Design Considerations, sec. 19, pt. 1 in E. H. Gaylord and C. N. Gaylord (eds.), "Structural Engineering Handbook," McGraw-Hill, New York, 1968.

2 Picardi, E. A.: Industrial Buildings, sec. 19, pt. 2 in E. H. Gaylord and C. N. Gaylord (eds.), "Structural Engineering Handbook," McGraw-Hill, New York, 1968.

3 Kavanagh, T. C., and R. C. Y. Young: Arches and Rigid Frames, sec. 17 in E. H. Gaylord and C. N. Gaylord (eds.), "Structural Engineering Handbook," McGraw-Hill, New York, 1968.

12

MULTISTORY BUILDINGS

12-1 INTRODUCTION

The steel skeleton of a multistory building is a simple affair in outward appearance. However, because it must be built within the limitations imposed by architectural considerations and must be adequately stiff and at the same time practicable and economical, it presents a number of difficult and challenging problems. Judgment, experience, and a good sense of proportion play an indispensable role in designing such a framework.

The three types of framework—simple, semirigid, and rigid—were discussed in Art. 7-19. Design of the various members for gravity loads was considered in preceding chapters, while the loads that must be provided for were discussed in Chap. 1. Design for wind forces is discussed in following articles.

12-2 WIND BRACING

Wind forces are of no consequence in the design of the structural frame for multistory buildings that have a small ratio of height to width. The inherent stiffness of the standard framing connection produces a skeleton which has the strength to resist lateral forces of considerable magnitude without the help of special wind bracing.

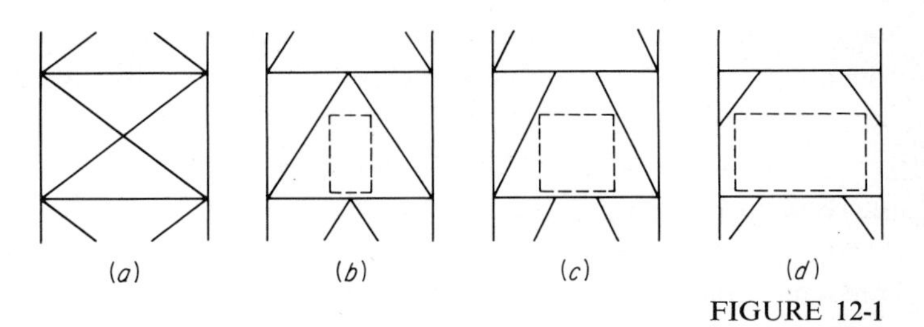

FIGURE 12-1

When a building must be braced for wind or other lateral force, the obvious solution is a system of diagonal bracing that forms with the columns and beams a series of vertical trusses (Fig. 12-1*a*). It is one of the early types of wind bracing, but modern architectural requirements usually relegate it to walls enclosing shaft or service areas where there are few or no openings to interfere. The K-brace system (Fig. 12-1*b*), in which the horizontal member (the floor beam) is supported at midspan, is more efficient. The K brace also allows greater freedom in the use of aisle space, since it is possible to fit doors beneath its apex. Larger openings may necessitate a degenerate form of the K frame (Fig. 12-1*c*). Still larger openings can be provided by using knee braces (Fig. 12-1*d*). If the aisle must be free of bracing, moment-resistant beam-to-column connections must be used. A bent that is provided with any form of bracing is called a *braced bent*.

A bent which is designed to resist wind forces is called a *wind bent*. It may be a braced bent, a rigid frame, or a combination of the two. Since the floors are usually of a type that serve as horizontal diaphragms, not every bent in a tier building need be a wind bent. In some cases wind bents in the outside walls may be sufficient. If the plan provides a sufficiently large, centrally located service area, this area may be surrounded with braced bents to form a core through which wind forces are delivered to the ground. Service areas may sometimes be enclosed by shear walls, so as to eliminate braced bents altogether. The shear wall is nothing more than a substantial wall, usually of reinforced concrete or masonry. It need not be a monolithic wall—it may consist of securely anchored panels built into the spaces framed by the columns and girders of the frame.

Where full usable aisle space is required in wind bents, moment-resistant beam-to-column connections must be used instead of bracing. However, rigid-frame bents cannot always be used in the modern multistory building. Ducts for mechanical systems occupy a considerable portion of the depth allowed for floor construction, so that beams and girders must be as shallow as possible to minimize story height. Furthermore, where large column-free space is required, beam spans are long. Under these conditions, it is difficult and sometimes impossible to design efficient rigid-frame bents. Although this problem can be solved by using deep beams or girders with web openings for mechanical ducts, such openings generally require reinforcement, which also increases the cost of the framing system.

The braced bent is generally stiffer than the rigid frame and requires little or no increase in the sizes of beams and columns (over that which is required for gravity loads) to achieve acceptable horizontal deflection. On the other hand, moment-resistant frames usually require larger member sizes, at the lower levels of the building, to limit these deflections. Every high-rise building requires a bracing system which is compatible with the architectural scheme and which provides adequate strength and stiffness. Braced frames of different configurations may be required in the same system, and in many cases braced frames must be combined with rigid frames.

According to Ref. 1, frames with simply connected beams are likely to be economical for buildings up to about six stories high, with masonry in-filled exterior walls. Semirigid frames will normally be efficient for buildings up to about 15 stories high, with some help from in-filled walls. Rigid frames may be used up to about 40 stories, but their economy reduces gradually after about 20 stories. Braced frames are efficient up to about 40 stories. A combination of braced frame and rigid frame is efficient for buildings of about 40 to 60 stories. For buildings over 60 stories, the optimum bracing system is the box-type frame, in which all the structural elements of the exterior walls act like walls of a tube. This can be accomplished by using closely spaced exterior columns with relatively stiff spandrels (Fig. 12-2), by closely spaced diagonals on the exterior wall (Fig. 12-3), or by making the exterior framing a braced bent the full width of the building (Fig. 12-4).

12-3 WIND BENT ANALYSIS

A braced bent of a high-rise building frame is highly statically indeterminate. However, the analysis is usually based on the assumption that the wind shear in any panel is equally divided among the diagonals of the panel. Thus, the analysis is relatively simple; examples are given in Ref. 2.

Rigid-frame bents for high-rise buildings are also highly statically indeterminate. For example, a 20-story frame three bays wide, with moment-resistant beam-to-column connections throughout, is 180 times indeterminate. The analysis of such frames can be easily and economically performed on a computer. Such solutions are generally based on the slope-deflection equations. However, the elastic properties of the members of the frame must be known, or estimated, in advance. Therefore, approximate methods, which may be used as a guide to the final design or to furnish properties for a more refined analysis, are useful. The two most commonly used approximate methods are discussed in the following articles.

FIGURE 12-2
World Trade Center, New York City. (*Port of New York Authority*.)

FIGURE 12-3
IBM Building, Pittsburgh, Pa. (*United States Steel Corporation.*)

FIGURE 12-4
Alcoa Building, San Francisco. (*American Institute of Steel Construction.*)

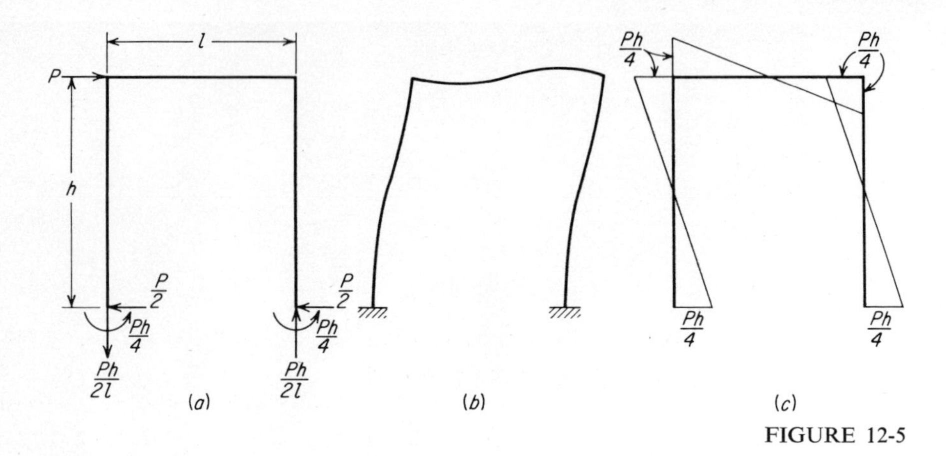

FIGURE 12-5

12-4 THE PORTAL METHOD

The simple portal of Fig. 12-5*a* is statically indeterminate to the third degree, so that three assumptions are needed to make an analysis by statics alone. If we assume (1) the magnitude of one of the horizontal reactions, (2) the location of the point of inflection of the left-hand column, and (3) the location of the point of inflection of the right-hand column, we can solve for the remaining horizontal reaction, the vertical reactions, and the end moments. Figure 12-5*a* shows the reactions, Fig. 12-5*b* the deflected shape of the portal, and Fig. 12-5*c* the bending-moment diagrams, assuming points of inflection to lie at the midpoint of each column and the horizontal reactions to be equal.

Consider next the single-story, three-bay bent shown in Fig. 12-6*a*. If we had instead three independent portals (Fig. 12-6*b*) and were to make for each the same assumptions as were described for Fig. 12-5, each bent would be statically determinate. The portal method of analysis is based on the assumption that the lateral force P acting on the multibay bent of Fig. 12-6*a* is divided equally among the corresponding number of independent portals (Fig. 12-6*b*) and that the reactive forces are those which are obtained by simple superposition of the forces acting on the independent portals. This leads to the conclusion that the interior columns of a multibay bent carry twice the shear of the exterior columns and that each column and each girder has a point of inflection at midpoint. By a simple extension of this idea, we assume for a multistory, multibay bent that the shear in each story is distributed in the same manner as for a single-story, multibay bent and that every column and every girder has a point of inflection at midpoint.

The portal method is easy to use. Once the shears in the columns of each story are known, moments at the top and bottom of each column in a story are found by

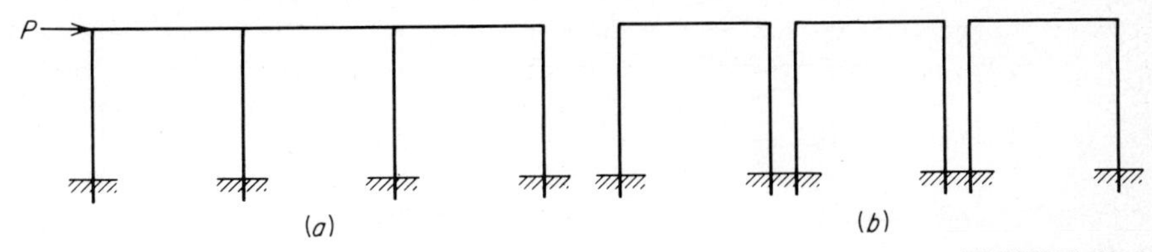

FIGURE 12-6

multiplying the shear by the half length of the column. This follows from the assumption that the point of inflection is at midheight. Next, the end moment in a girder connecting to an outside column is equal to the sum of the column moments at that joint, as is evident from Fig. 12-7*a*. But if there is a point of inflection at the midpoint of the girder, the end moments on the girder are numerically equal, so that the moment at the interior end of the girder is also known (Fig. 12-7*b*). Next, the end moment for the first interior girder is found from the condition that the sum of the girder moments must equal the sum of the column moments at an interior joint (Fig. 12-7*c*). The shear in any girder is found by dividing the sum of the end moments on the girder by the span (Fig. 12-7*b*). Evidently, the axial force in any column of any story is equal to the sum of the shears on all the girders connecting to it above the story in question.

The portal method is considered to be generally satisfactory for buildings of moderate height-width ratio and not over about 25 stories high. Story heights and girder spans should be approximately equal and the configuration reasonably symmetrical.

12-5 THE CANTILEVER METHOD

In the cantilever method of analysis, points of inflection of all the columns and girders of a building frame are assumed to lie at midpoint, just as in the portal method. However, instead of an assumption about the distribution of the shear among the

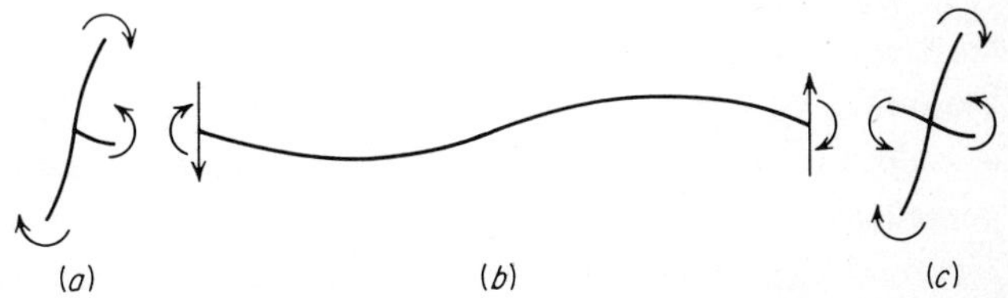

(a) (b) (c)

FIGURE 12-7

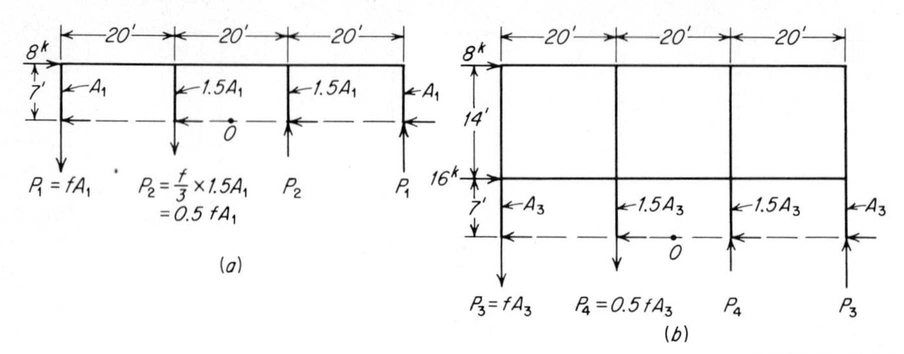

FIGURE 12-8

columns of a given story, in the cantilever method an assumption is made about the distribution of the axial stresses in the columns of a given story. It is assumed that the bent acts as a cantilever beam and that in resisting the bending moments produced by the lateral forces it develops in its columns axial stresses distributed linearly about a neutral axis, as in beams. Since it is the stresses rather than the forces that are distributed linearly, the axial forces in the columns cannot be determined without an estimate of the relative cross-sectional areas. However, some users of this method assume that the axial forces, rather than the axial stresses, are distributed linearly, which is another way of saying that they assume the columns to have equal cross-sectional areas.

As an example, assume that the columns of a three-bay framework are spaced 20 ft on centers and that the story height, measured between centerlines of floor beams, is 14 ft. Isolating the portion that lies above the points of inflection of the columns in the top story, we get Fig. 12-8*a*. The wind force at the roof is 8 kips. Assume that the cross-sectional area of each interior column is 1.5 times the cross-sectional area A_1 of the exterior column. Let the axial stress on the exterior column be f. Then on each interior column the axial stress is $f/3$. The corresponding forces P_1 and P_2 are shown on the figure. From the equation of moments about the neutral axis O, we get

$$60P_1 + 20P_2 = 60fA_1 + 20 \times 0.5fA_1 = 8 \times 7$$

$$fA_1 = 0.8$$

Therefore, $P_1 = fA_1 = 0.8$ kip, and $P_2 = 0.5fA_1 = 0.4$ kip.

If Fig. 12-8*a* represents the top story of a multistory frame, we can find by proportion the axial forces in the columns of any other story for which the cross-sectional areas of the columns are in the same ratio as are those in the top story. Thus if we write the equation for moments at point O of Fig. 12-8*b*, we get the same equation as before except that A_3 replaces A_1 and the right member, $8 \times 7 = 56$ ft-kips, becomes $8 \times 21 + 16 \times 7 = 280$ ft-kips. Then, since $280/56 = 5$, we have $P_3 = 5P_1 = 4$ kips and $P_4 = 5P_2 = 2$ kips.

Once the axial forces in the columns are known, the shears in the girders are easy to determine. Thus in Fig. 12-8*a* it is evident that the shear in the outside girder at the roof is equal to the axial force P_1 in the outside column while that in the interior girder is equal to the sum of the forces P_1 and P_2. Similarly, the shear in the outside girder at the top floor (Fig. 12-8*b*) is equal to the difference between the axial forces P_3 and P_1 in the adjacent outside columns. With the shears in the girders known, end moments in the girders are determined by multiplying each shearing force by the half span of the corresponding girder. This follows from the assumption that the points of inflection are at midspan. Moments in the columns may be found next, starting at the roof and taking advantage of the assumed equality of the end moments in each column of each story. Following this, determination of the shears in the columns and the axial forces in the girders completes the analysis.

The cantilever method is used less often than the portal method. In the opinion of a committee of the ASCE, it is suitable for buildings of moderate height-width ratio and not more than 25 to 35 stories high. It is considered to be superior to the portal method for high, narrow buildings.[3] Story heights and girder spans should be approximately equal and the configuration reasonably symmetrical. A more detailed explanation of the portal and the cantilever methods may be found in Ref. 4. Other methods are described elsewhere.[5–9]

12-6 LIMITATIONS OF PORTAL AND CANTILEVER METHODS

The deflection of a frame is a combination of flexural and shear distortions. Since the portal method is based on a distribution of lateral shear to the columns of a bent, it tends to produce a result which is primarily the effect of shear distortion. The distortion of frames with low height-to-width ratio is primarily that of shear. Therefore, the portal analysis is appropriate for such frames. On the other hand, the cantilever method is based on a distribution of column axial forces similar to the distribution of bending stress in a flexural member, so that it tends to produce a result which is primarily a result of flexural distortion. The primary distortion of building frames with a high height-to-width ratio is due to bending as a cantilever. Therefore, the cantilever method is appropriate for such frames.

FIGURE 12-9
South Central Bell Telephone Building. (*Weiskopf and Pickworth.*)

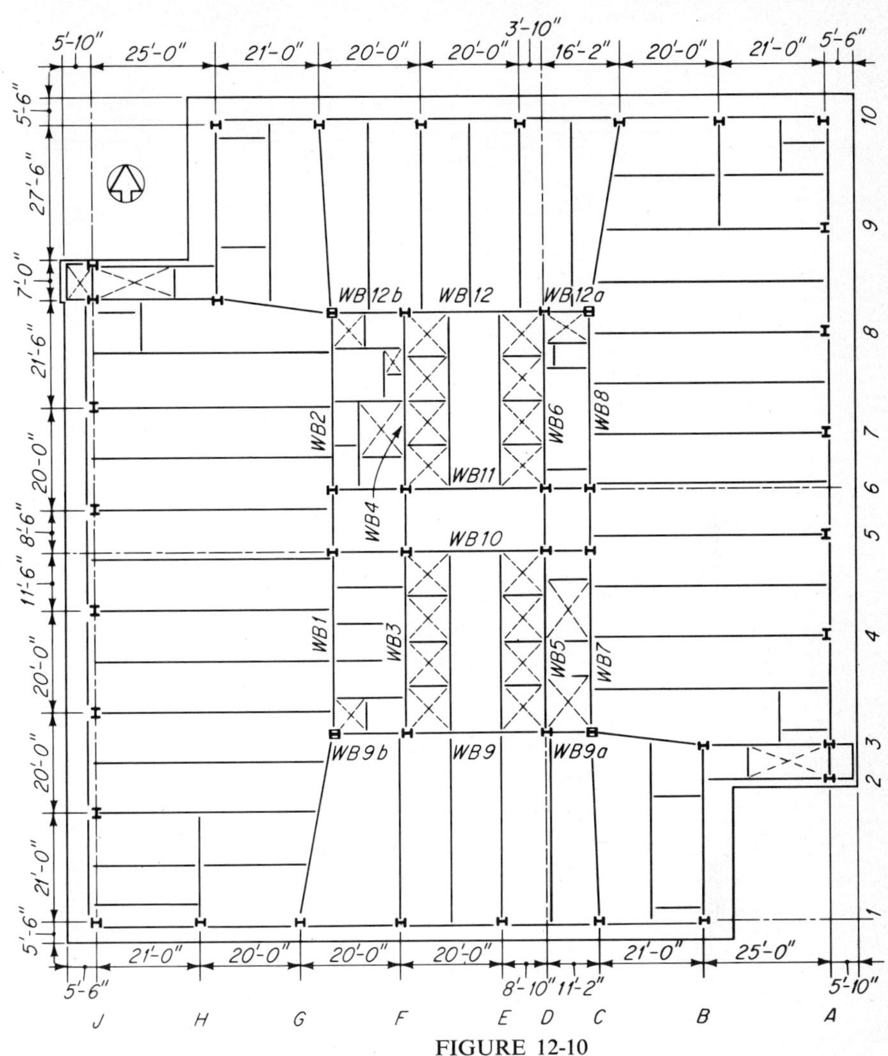

FIGURE 12-10
Seventh through twentieth floor framing plan.

12-7 SOUTH CENTRAL BELL TELEPHONE COMPANY BUILDING

Wind bracing for the headquarters building of the South Central Bell Telephone Company at Birmingham, Ala. (Fig. 12-9) is described in this article. This 30-story building is 390 ft high and has 723,000 gross square feet of floor space. There are 8,350 tons of structural-steel framing. The floor framing plan for the seventh to twentieth floors, inclusive, is shown in Fig. 12-10. Unobstructed space from the

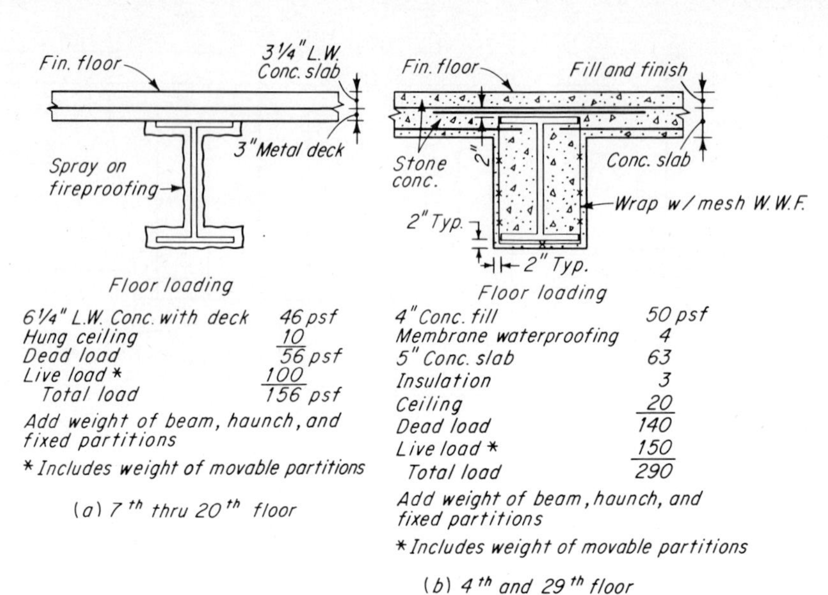

FIGURE 12-11
(*a*) Typical floor section. (*b*) Mechanical-equipment floors.

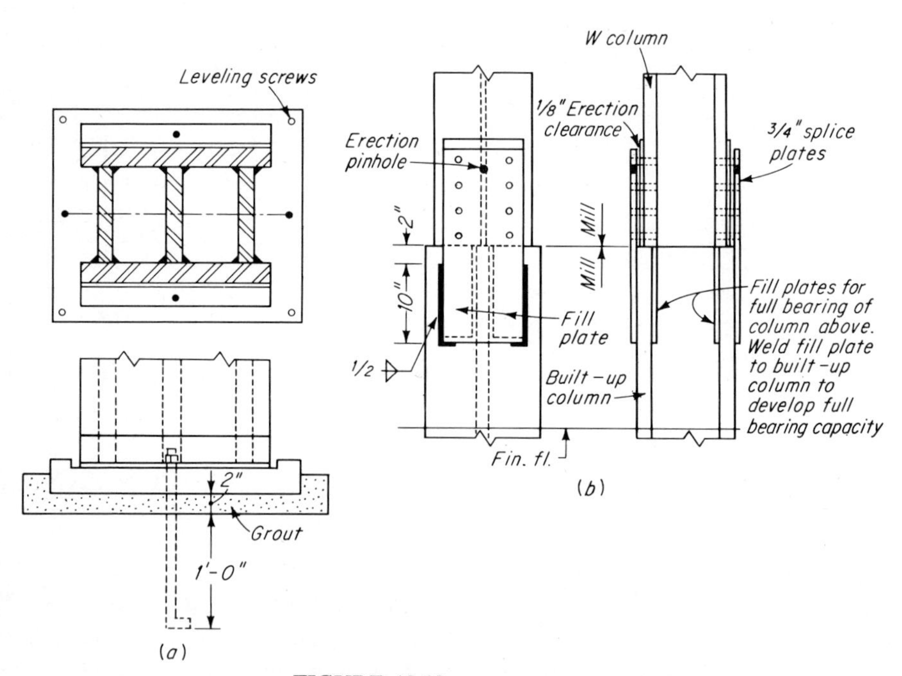

FIGURE 12-12
(*a*) Built-up column and base. (*b*) Splice at built-up column.

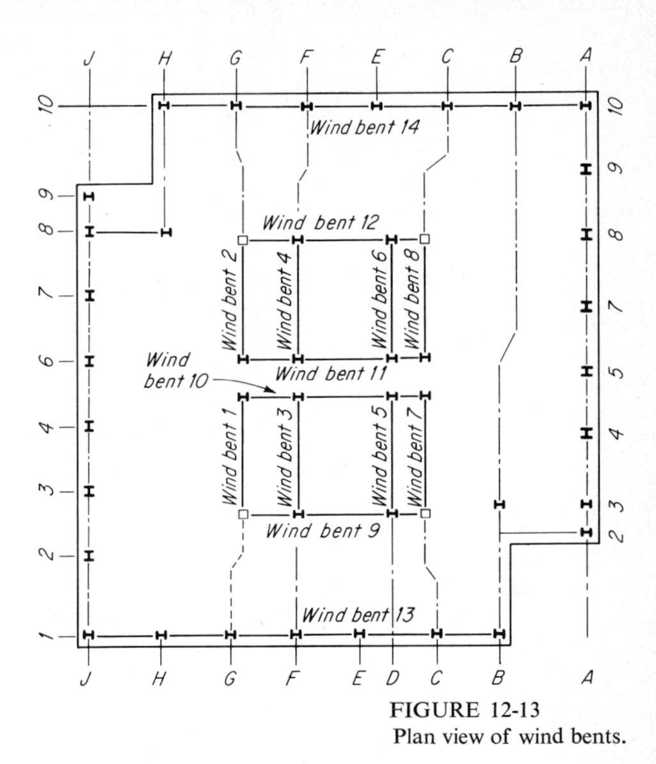

FIGURE 12-13
Plan view of wind bents.

center core to the perimeter is 45 ft on the east and west sides and 35 ft on the north and south. The typical floor is a 3-in. metal deck with a $3\frac{1}{4}$-in. lightweight-concrete slab (Fig. 12-11*a*). Mechanical-equipment floors (fourth and twenty-ninth) are 5-in. concrete slabs composite with the floor beams (Fig. 12-11*b*).

The frame is A36 steel throughout, and was designed according to the AISC specification. Columns range from W14 × 43 at the two-story top tier to the largest size rolled (W14 × 730) at the two-story bottom tier, except for the built-up columns in the core. A typical built-up column and its base are shown in Fig. 12-12*a*. The heaviest of the built-up columns has 4 × 32 flange plates and 4 × 14 web plates; it weighs 1,443 plf and carries a load of 6,416 kips. Its base plate is $10\frac{1}{2}$ × 78 × 83 and weighs 21,100 lbs. A typical splice of a W14 column to the built-up section is shown in Fig. 12-12*b*. Except for the moment-resistant connections in wind bents, beam-to-column connections are AISC standard framed connections sufficient to support the beams carrying uniform loads producing a bending stress $f_b = 24$ ksi. They were shop-welded and field-bolted.

The building required a variety of braced bents, as well as rigid frames at the upper levels, for wind bracing. The system is located in the service core, and was made

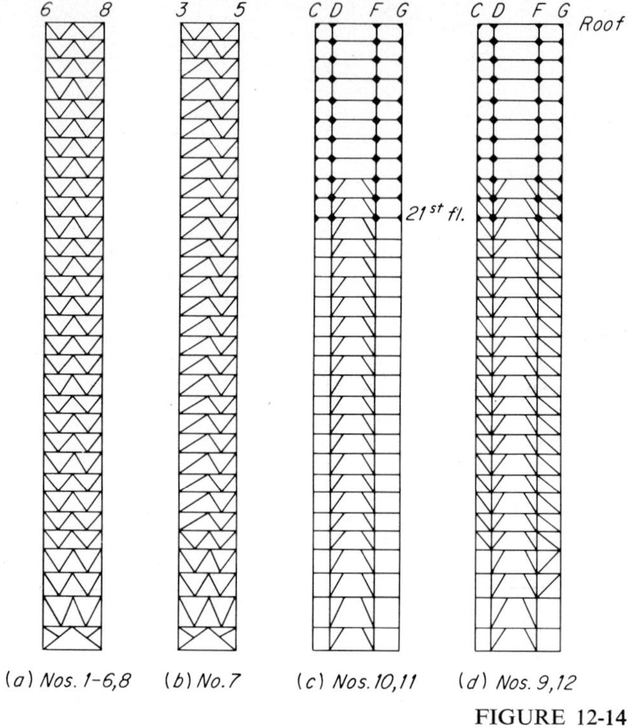

FIGURE 12-14
Wind bents.

as symmetrical as possible to minimize secondary effects of torsion. Figure 12-13 shows the location of the wind bents. Eight braced bents in the north-south direction (numbers 1 to 8) extend from basement to roof. These bents are shown in Fig. 12-14*a* and *b*. Each is 35 ft 6 in. between columns. Bents 3 to 6 are located in the rear walls of the four elevator banks. There are no penetrations of these walls, which allows freedom in determining the diagonal configuration (Fig. 12-15). Bents 1, 2, and 8 are located in walls with minor penetrations for mechanical services and a door (Fig. 12-16). However, the door could be located to clear the diagonals, so that the same configuration as for bents 3 to 6 was used. Bent 7 is in a partition where the door to a main exit stair and access openings to a main mechanical shaft influenced the configuration (Fig. 12-17). Because of beams framing at the interior panel points, these trusses support gravity load as well as wind load.

Four bents (9 to 12) located in the core walls brace the building in the east-west direction (Fig. 12-13). A main elevator corridor located between column lines *D* and *F* and extending from lines 3 to 8 imposes different and more severe limitations on this bracing compared to the north-south direction. These bents are braced from the

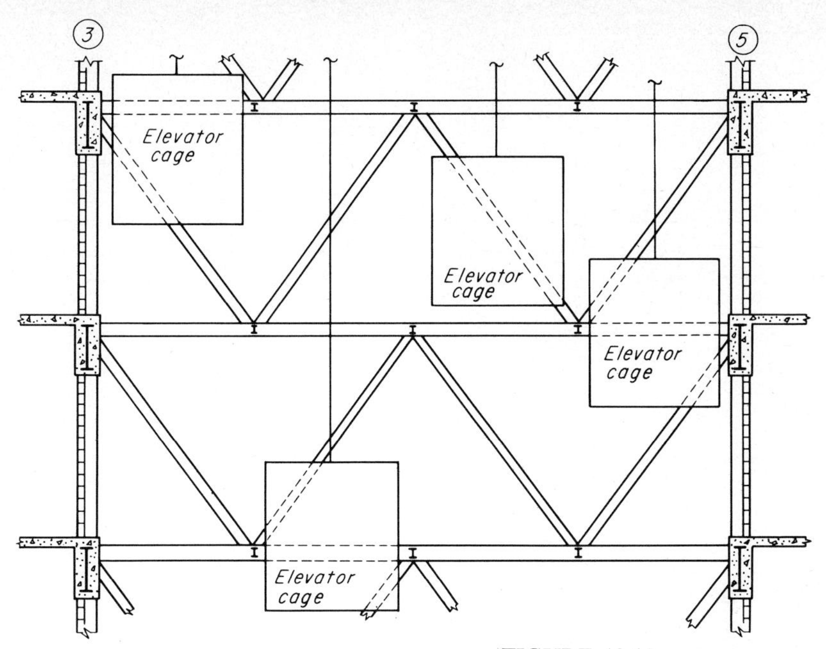

FIGURE 12-15
Wind bent 3. Bents 4, 5, 6 same.

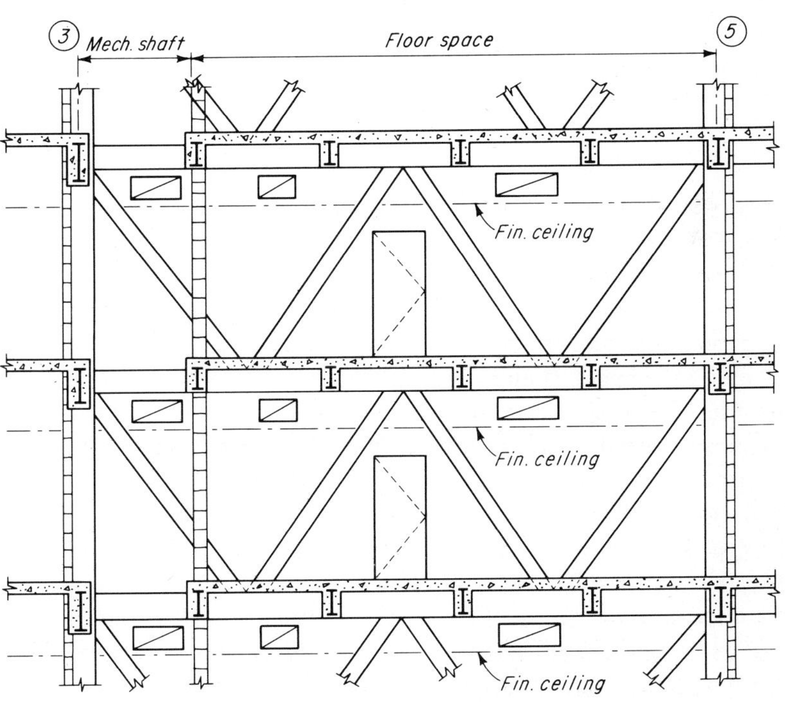

FIGURE 12-16
Wind bent 1. Bents 2 and 8 same.

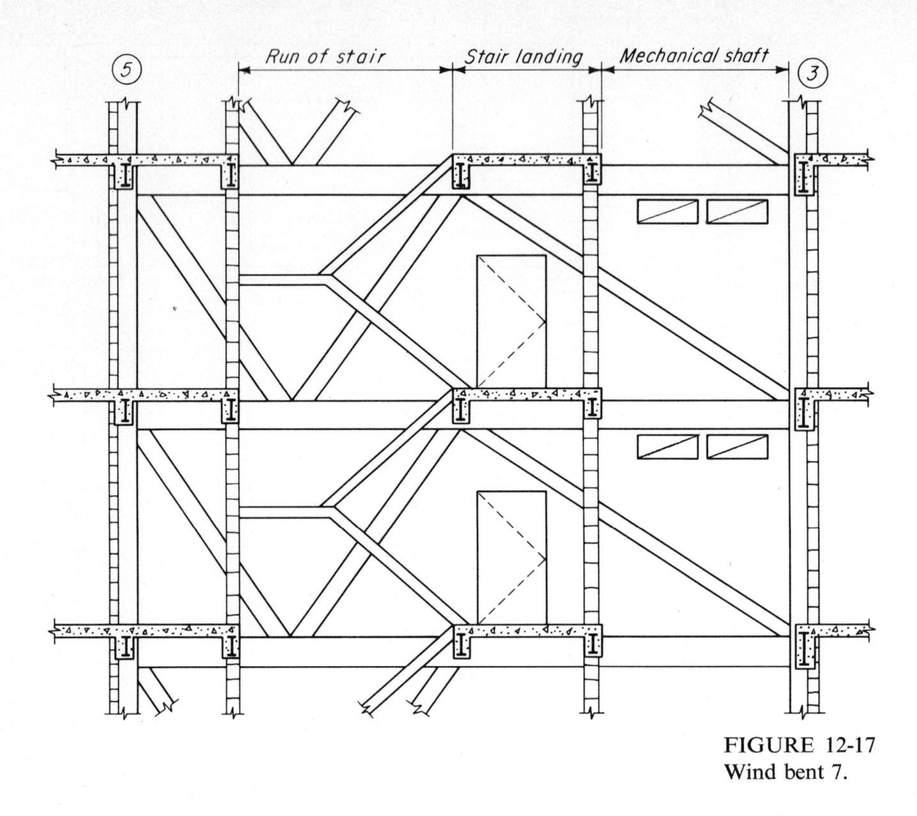

FIGURE 12-17
Wind bent 7.

basement to the twenty-third floor, and are 27 ft 8 in. wide between column lines *D* and *F* (Fig. 12-14*c*, *d*). The system changes at the twentieth floor, for reasons which are discussed later. Because of the elevator corridor, the configurations shown in Figs. 12-18 and 12-19 are used. Although this results in bending in the floor members, it is a stiffer, more efficient frame than an all moment-resisting frame. No diagonals could be used in panels *CD* and *FG* on either side of bents 10 and 11 due to major openings (Fig. 12-18). However, the core walls in these bays on either side of bents 9 and 12 had only small openings, which allowed single diagonals. These additional braced bents are called 9*a*, 9*b*, 12*a*, and 12*b* (Fig. 12-19).

Two elevator banks (eight elevators) are omitted above the twentieth floor because of the reduced traffic. Since the space thus released should be relatively open to provide flexibility in its use, vertical trusses become less desirable in this area. However, they also lose some of their advantage compared to rigid frames because wind shears are considerably smaller. The floor girders which serve as members of a rigid-frame wind bent at the upper levels of a structure often require little or no increase in size compared to gravity-load requirements because of the one-third

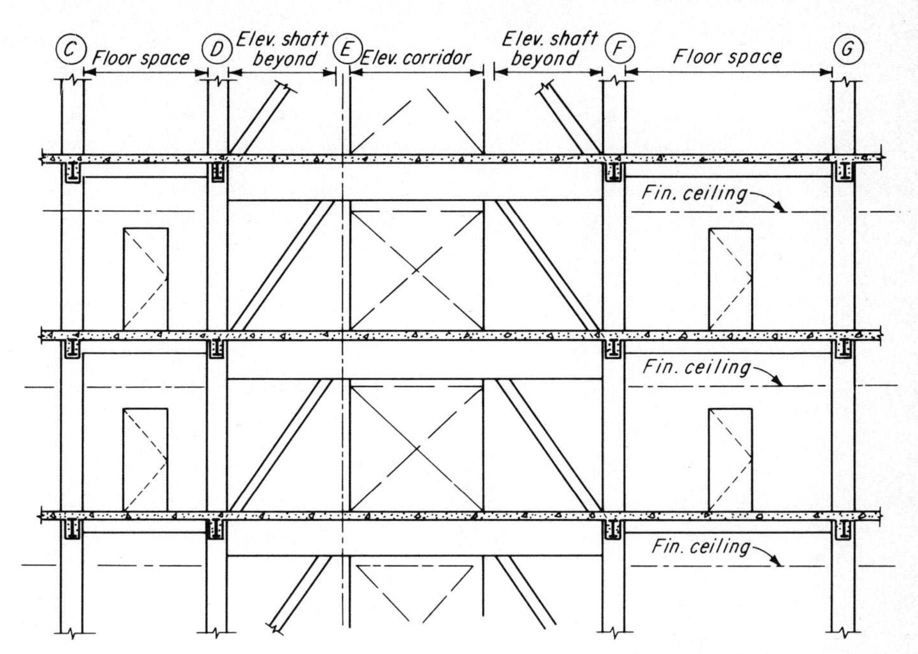

FIGURE 12-18
Wind bent 10. Bent 11 same.

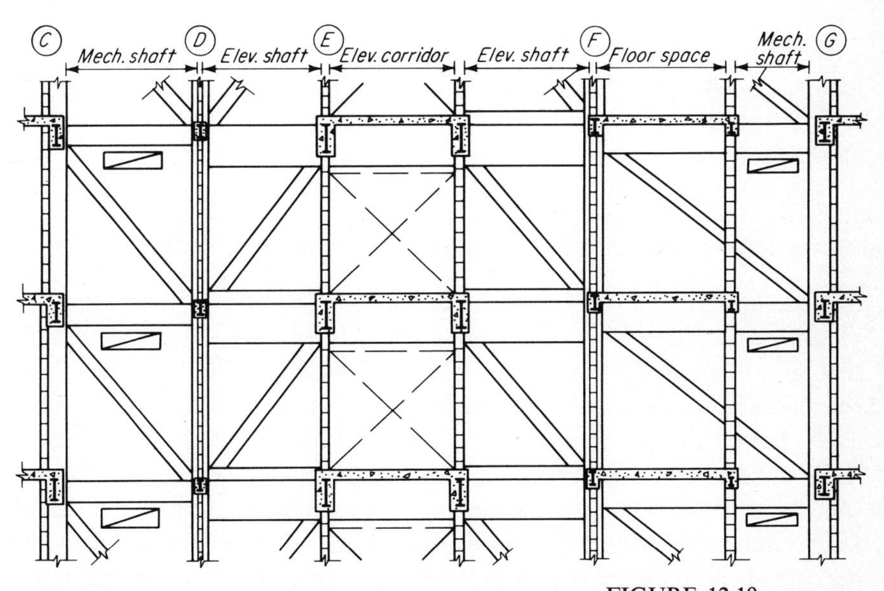

FIGURE 12-19
Wind bent 9. Bent 12 same.

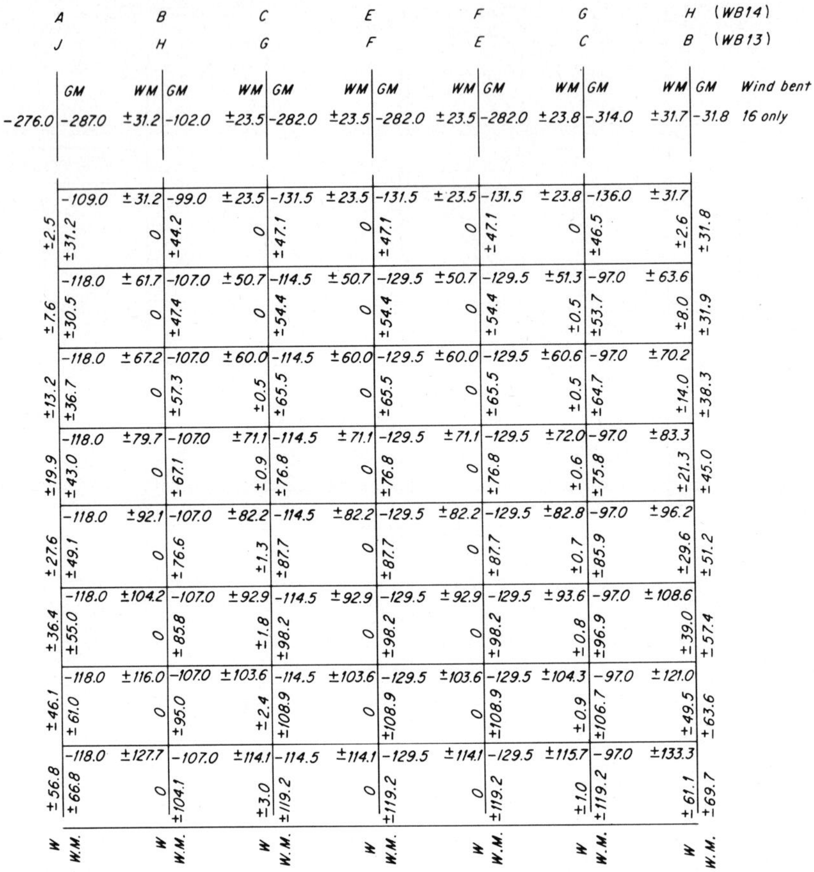

FIGURE 12-20
Rigid-frame bents 13 and 14.

increase in allowable stress for combinations of wind and gravity load. For these reasons, rigid-frame wind bents were used in the east-west direction from the twenty-first floor to the roof. Two additional bents, 13 and 14 (Fig. 12-13) were used in order to incorporate as many columns and floor members as possible in the east-west system. These frames are shown in Fig. 12-20. Lateral transfer of the wind shear from these bents to those in the core is through the floor framing. Because of the floor system (fluted metal deck with concrete topping), it was decided that the floor slab at only one level should not be relied upon to transfer all the wind forces from bents 13 and 14 to the core. Instead, distribution was divided among the twenty-first, twenty-second, and twenty-third floors by overlapping the two systems (Fig. 12-14).

12-8 DP12-1: WIND-BRACING ANALYSIS OF BELL BUILDING

Analysis of the north-south wind-bracing system of the headquarters building of the South Central Bell Telephone Company at Birmingham, Ala., which was described in Art. 12-7, is given in this example. It is presented through the courtesy of the owner, the architects, Kahn and Jacobs and Warren, Knight and Davis, and the structural engineers, Weiskopf and Pickworth and Hudson Wright and Associates.

The analysis is based on the assumption that the horizontal deflections of all bents at a particular level of the building are equal, from which it follows that the story shear is distributed among the bents in proportion to their stiffnesses.

Sheet 1 The stiffness of one panel of bents 1 to 6, and 8, is calculated on this sheet. The wind shear P is assumed to be divided equally among the four diagonals of the panel, since they have the same slope. The deflection Δ due to P is determined by the equation $\Delta = \sum Ffl/AE$. The diagonal-member forces for the unit load corresponding to Δ are also assumed to be equal. The stiffness of the panel, which is the shear P' for a unit value of Δ, is then computed.

Sheet 2 The stiffness of one panel of bent 7 is computed on this sheet. The wind shear P is assumed to be divided among the three diagonals in proportion to their slopes.

Sheet 3 Story deflection at the tenth floor is calculated on this sheet. Sizes of the members of the bents were determined first for the gravity loads from the beams which connect at the panel points, as shown in Fig. 12-15. If the calculated wind deflection for these members was excessive, member sizes were increased. Examination of the formulas for P' shows that the diagonals are the most efficient members of the bent; sizes of these can be increased in larger proportion to achieve a stiffer structure at least cost.

Since the columns in these bents are not generally of equal cross-sectional area at a story, the average of the areas was used. Since the columns are large, the deflection is relatively insensitive to variations in their areas. The stiffness P' at this floor is 10,700 kips per in.

Sheet 4

a The wind loading diagram is that specified by the Southern Building Code. The total wind force is 1,240 kips, and the corresponding deflection is found by proportion.

b Distribution of the north-south wind shear among bents 1 to 8 is determined on this sheet. It will be noted that bent 7 is somewhat stiffer than the others. However, the shear is distributed quite uniformly among the eight bents. This means that the stiffnesses of the bents are about equal. Twisting or excessive deflection in parts of a framework can result if the various bents are not adjusted to have approximately equal stiffnesses. Adjusting the stiffnesses of the wind bents is called *tuning*.

c The total deflection of the building above a given level (in this example, the tenth floor) is the sum of the shear deflection and the flexural deflection. To calculate the flexural deflection (column shortening) the building is assumed to act as a uniformly loaded cantilever beam for which the deflection $\Delta = wl^4/8EI$. Since $M = wl^2/2$, $\Delta = Ml^2 4/EI$. The moment is calculated for the wind pressure shown above.

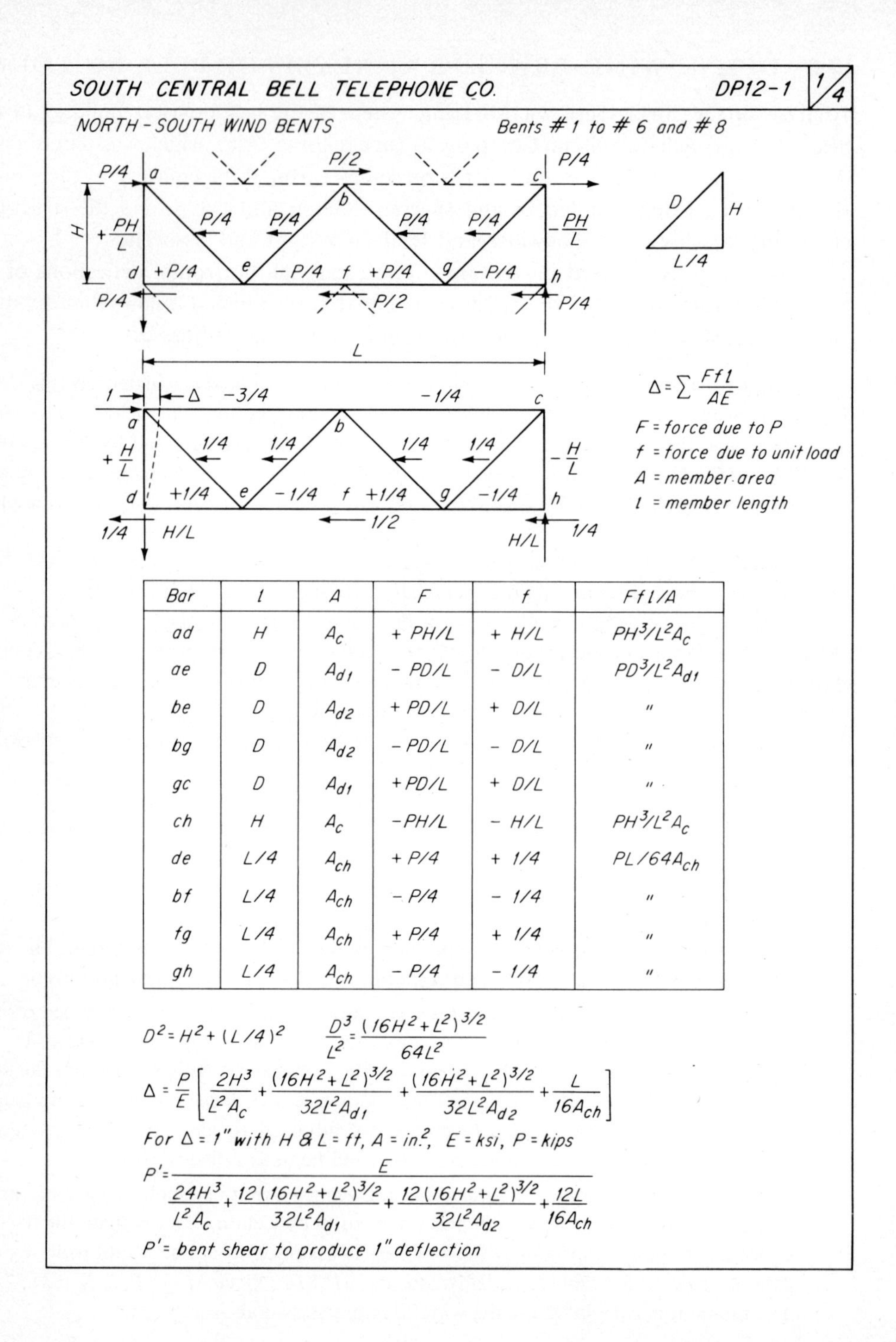

SOUTH CENTRAL BELL TELEPHONE CO. DP12-1 1/4

NORTH-SOUTH WIND BENTS Bents # 1 to # 6 and # 8

$$\Delta = \sum \frac{Ffl}{AE}$$

F = force due to P
f = force due to unit load
A = member area
l = member length

Bar	l	A	F	f	Ffl/A
ad	H	A_c	+ PH/L	+ H/L	PH^3/L^2A_c
ae	D	A_{d1}	− PD/L	− D/L	PD^3/L^2A_{d1}
be	D	A_{d2}	+ PD/L	+ D/L	"
bg	D	A_{d2}	− PD/L	− D/L	"
gc	D	A_{d1}	+ PD/L	+ D/L	"
ch	H	A_c	− PH/L	− H/L	PH^3/L^2A_c
de	L/4	A_{ch}	+ P/4	+ 1/4	$PL/64A_{ch}$
bf	L/4	A_{ch}	− P/4	− 1/4	"
fg	L/4	A_{ch}	+ P/4	+ 1/4	"
gh	L/4	A_{ch}	− P/4	− 1/4	"

$$D^2 = H^2 + (L/4)^2 \qquad \frac{D^3}{L^2} = \frac{(16H^2 + L^2)^{3/2}}{64L^2}$$

$$\Delta = \frac{P}{E}\left[\frac{2H^3}{L^2A_c} + \frac{(16H^2+L^2)^{3/2}}{32L^2A_{d1}} + \frac{(16H^2+L^2)^{3/2}}{32L^2A_{d2}} + \frac{L}{16A_{ch}}\right]$$

For $\Delta = 1''$ with H & L = ft, A = in.^{2}, E = ksi, P = kips

$$P' = \frac{E}{\dfrac{24H^3}{L^2A_c} + \dfrac{12(16H^2+L^2)^{3/2}}{32L^2A_{d1}} + \dfrac{12(16H^2+L^2)^{3/2}}{32L^2A_{d2}} + \dfrac{12L}{16A_{ch}}}$$

P′ = bent shear to produce 1″ deflection

SOUTH CENTRAL BELL TELEPHONE CO. DP12-1 2/4

NORTH-SOUTH WIND BENTS

Bent # 7

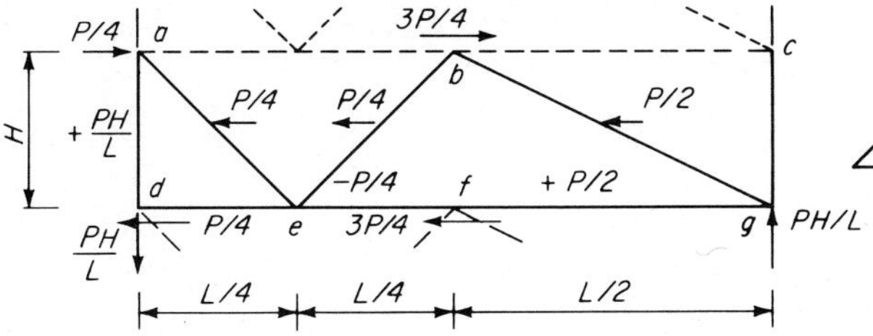

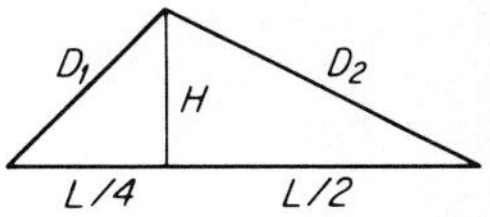

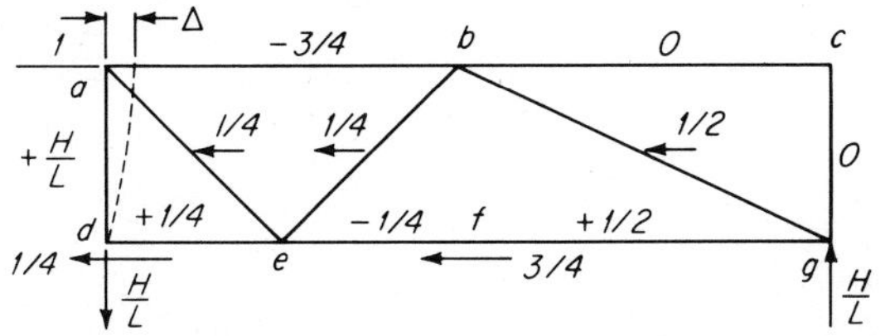

Bar	1	A	F	f	Ff1/A
ad	H	A_c	PH/L	H/L	$PH^3/L^2 A_c$
ae	D_1	A_{d1}	$-PD_1/L$	$-D_1/L$	$PD_1^3/L^2 A_{d1}$
eb	D_1	A_{d1}	$+PD_1/L$	$+D_1/L$	"
bg	D_2	A_{d2}	$-PD_2/L$	$-D_2/L$	$PD_2^3/L^2 A_{d2}$
de	$L/4$	A_{ch}	$+P/4$	$+1/4$	$PL/64 A_{ch}$
ef	$L/4$	A_{ch}	$-P/4$	$-1/4$	"
fg	$L/2$	A_{ch}	$+P/2$	$+1/2$	"

$$D_1^2 = H^2 + (L/4)^2 \qquad \frac{D_1^3}{L^2} = \frac{(16H^2 + L^2)^{3/2}}{64L^2}$$

$$D_2^2 = H^2 + (L/2)^2 \qquad \frac{D_2^3}{L^2} = \frac{(4H^2 + L^2)^{3/2}}{8L^2}$$

$$\Delta = \frac{P}{E}\left[\frac{H^3}{L^2 A_c} + \frac{(16H^2 + L^2)^{3/2}}{32L^2 A_{d1}} + \frac{(4H^2 + L^2)^{3/2}}{8L^2 A_{d2}} + \frac{5L}{32A_{ch}}\right]$$

For $\Delta = 1''$ with H & L = ft, A = in.^{2}, E = ksi, P = kips

$$P' = \frac{E}{\dfrac{12H^3}{L^2 A_c} + \dfrac{12(16H^2 + L^2)^{3/2}}{32L^2 A_{d1}} + \dfrac{12(4H^2 + L^2)^{3/2}}{8L^2 A_{d2}} + \dfrac{60L}{32A_{ch}}}$$

P' = bent shear to produce 1" deflection

SOUTH CENTRAL BELL TELEPHONE CO. DP12-1 3/4

North-south wind bents

Bents #1 thru #6 and #8: $H = 12.33'$, $L = 35.5'$

$$P' = \frac{E}{35.7/A_c + 66.7/A_{d1} + 66.7A_{d2} + 26.6/A_{ch}}$$

Bents #1 and #2

Bar	Section	A
1	2-C9×13.44 []	7.88
2	2-C7×9.8 []	5.74
3	W16×36	10.60
4	Column	188

$$P_1' = P_2' = \frac{29{,}000}{\frac{35.7}{188} + \frac{66.7}{7.88} + \frac{66.7}{5.74} + \frac{26.6}{10.60}} = \frac{29{,}000}{0.19 + 8.46 + 11.61 + 2.54} = 1{,}275 \text{ kips}$$

Bents #3, #4, #5, #6

Bar	Section	A
1	2-C6×8.2 []	4.80
2	2-C6×8.2 []	4.80
3	2-C8×11.5][	6.76
4	W14×455	134

$$P_3' = P_4' = P_5' = P_6' = \frac{29{,}000}{\frac{35.7}{134} + \frac{66.7}{4.80} + \frac{66.7}{4.80} + \frac{26.6}{6.76}} = \frac{29{,}000}{0.28 + 13.90 + 13.90 + 3.96} = 1{,}310 \text{ kips}$$

Bent #8

Bar	Section	A
1	2-C10×15.3 []	8.98
2	2-C7×9.8 []	5.74
3	W16×36	10.60
4	Column	158

$$P_8' = \frac{29{,}000}{\frac{35.7}{158} + \frac{66.7}{8.98} + \frac{66.7}{5.74} + \frac{26.6}{10.60}} = \frac{29{,}000}{0.23 + 7.44 + 11.61 + 2.50} = 1{,}360 \text{ kips}$$

Bent #7: $H = 12.33'$, $L = 35.5'$

$$P' = \frac{E}{17.8/A_c + 66.7/A_{d1} + 94.5/A_{d2} + 66.5/A_{ch}}$$

Bar	Section	A
1	2-C9×13.4 []	7.98
2	2-C12×20.7 []	12.18
3	W18×77	22.70
4	Column	158

$$P_7' = \frac{29{,}000}{\frac{17.8}{158} + \frac{66.7}{7.98} + \frac{94.5}{12.18} + \frac{66.5}{22.7}} = \frac{29{,}000}{0.11 + 8.36 + 7.76 + 2.83} = 1{,}550 \text{ kips}$$

$\Sigma P' = 2 \times 1275 + 4 \times 1{,}310 + 1{,}360 + 1{,}550 = 10{,}700$ kips (for $\Delta = 1''$)

SOUTH CENTRAL BELL TELEPHONE CO. DP12-1 4/4

Total wind force above 10th level

$P = 158.33\ (32 \times 87 + 30 \times 100 + 28 \times 73)$

$P = 1{,}240$ kips (a)

Horiz. defl. due to shear: 11th to 10th fl.

$$\Delta = \frac{1{,}240}{10{,}700} = 0.116''$$

Deflection index

$$\frac{\Delta}{H} = \frac{0.116}{148.0} = 0.0008 < 0.0025$$

Distribution of story shear among bents 1 to 8 (b)

Bent	Distribution		Wind shear at 10th level
1	1,275 / 10,700	= 11.9 %	× 1,240 = 148 kip
2	"	= 11.9 %	× 1,240 = 148
3	1,310 / 10,700	= 12.3 %	× 1,240 = 152
4	"	= 12.3 %	× 1,240 = 152
5	"	= 12.3 %	× 1,240 = 152
6	"	= 12.3 %	× 1,240 = 152
7	1,550 / 10,700	= 14.4 %	× 1,240 = 179
8	1,360 / 10,700	= 12.6 %	× 1,240 = 157
		100 %	1,240 kip

Flexural deflection

$$M = 158.33\ (32 \times 87 \times 216.5 + 30 \times 100 \times 123 + 28 \times 73^2 \times 1/2) = 166{,}000'^k \quad (c)$$

Bent No	A_1 ft²	A_2 ft²	d_1 ft	d_2 ft	I ft⁴
1 & 2	1.58	1.02	13.9	21.6	760
3 & 4	1.13	0.93	16.0	19.5	643
5	0.93	0.81	16.5	19.0	545
6	0.81	1.02	19.8	15.7	570
7 & 8	1.39	0.81	13.1	22.4	644

(d)

$$d_1 = \frac{35.5\,A_2}{A_1 + A_2}$$

$$I = 2 \times 760 + 2 \times 643 + 545 + 570 + 2 \times 644 = 5209 \text{ ft}^4$$

$$\Delta = \frac{ML^2}{4EI} = \frac{166{,}000 \times 260^2}{4 \times 29{,}000 \times 144 \times 5209} = 0.129' = 1.55''$$

$\Delta / h = 1.55/3120 = 0.0005$ (flexural)

0.001 (shear)

Total $= 0.0015 < 0.0025$ (e)

d The moment of inertia I of the structure is the sum of the $\sum Ad^2$ for each bent.

e The shear-deflection index computed above is 0.0008. It is assumed that all floors above the tenth floor have a cumulative index of 0.001. Since the total index is less than 0.0025, it is acceptable.

The wind-force analysis that is described and illustrated in this example was programmed for solution by computer.

REFERENCES

1 Khan, F. R.: Design of High-rise Buildings, paper presented at A Symposium on Steel sponsored by ASCE, AISC Chicago Fabricators, and University of Illinois at Chicago Circle, October 1965.

2 Eligator, M. H., and A. F. Nassetta: Multistory Buildings, sec. 19, pt. 3, in E. H. Gaylord and C. N. Gaylord (eds.), "Structural Engineering Handbook," McGraw-Hill, New York, 1968.

3 Wind Bracing in Steel Buildings, Final Report of Subcommittee 31, *Trans. ASCE*, vol. 105, p. 1713, 1940.

4 Norris, C. H., and J. B. Wilbur: "Elementary Structural Analysis," p. 303, McGraw-Hill, New York, 1960.

5 Spurr, Henry V.: "Wind Bracing," McGraw-Hill, New York, 1930.

6 Grinter, Linton E.: Wind Stress Analysis Simplified, *Trans. ASCE*, vol. 99, p. 610, 1934.

7 Goldberg, John E.: Wind Stresses by Slope Deflections and Converging Approximations, *Trans. ASCE*, vol. 99, p. 962, 1934.

8 Gottschalk, Otto: Simplified Wind-stress Analysis of Tall Buildings, *Trans. ASCE*, vol. 105, p. 1019, 1940.

9 Witmer, F. P.: Wind Stress Analysis by the K-percentage Method, *Trans. ASCE*, vol. 107, p. 925, 1942.

INDEXES

NAME INDEX

Page numbers in *italics* indicate references.

SUBJECT INDEX

APPENDIX

Table A-1 APPROPRIATE RADII OF GYRATION*

$r_x = 0.29h$ $r_y = 0.29b$	$r_x = 0.42h$ $r_y = 0.42b$	$r_x = 0.31h$ $r_y = 0.48b$
$r_x = 0.40h$ h = mean h	r_y = same as for 2 ∠	$r_x = 0.37h$ $r_y = 0.28b$
$r_x = 0.25h$	$r_x = 0.42h$ r_y = same as for 2 ∠	$r_x = 0.31h$
$r = \sqrt{\frac{H^2+h^2}{16}}$ $r = 0.35H_m$	$r_x = 0.39h$ $r_y = 0.21b$	$r_x = 0.31h$
$r_x = 0.31h$ $r_y = 0.31h$ $r_z = 0.197h$	$r_x = 0.45h$ $r_y = 0.235b$	$r_x = 0.40h$ $r_y = 0.21b$
$r_x = 0.29h$ $r_y = 0.32b$ $r_z = 0.18\frac{h+b}{2}$	$r_x = 0.36h$ $r_y = 0.45b$	$r_x = 0.38h$ $r_y = 0.22b$
$r_x = 0.31h$ $r_y = 0.215b$ $= b(0.21+0.02s)$	$r_x = 0.36h$ $r_y = 0.60b$	$r_x = 0.39h$
$r_x = 0.32h$ $r_y = 0.21b$ $= b(0.19+0.02s)$	$r_x = 0.36h$ $r_y = 0.53b$	$r_x = 0.35h$
$r_x = 0.29h$ $r_y = 0.24b$ $= b(0.23+0.02s)$	$r_x = 0.39h$ $r_y = 0.55b$	$r_x = 0.435h$ $r_y = 0.25b$
$r_x = 0.30h$ $r_y = 0.17b$	$r_x = 0.42h$ $r_y = 0.32b$	$r_x = 0.42h$
$r_x = 0.25h$ $r_y = 0.21b$	$r_x = 0.44h$ $r_y = 0.28b$	$r_x = 0.42h$
$r_x = 0.21h$ $r_y = 0.21b$ $r_z = 0.19h$	$r_x = 0.50h$ $r_y = 0.28b$	$r_x = 0.285h$ $r_y = 0.37b$
$r_x = 0.38h$ $r_y = 0.19b$	$r_x = 0.39h$ $r_y = 0.21b$	$r_x = 0.42h$ $r_y = 0.23b$

* J. A. L. Waddell, "Bridge Engineering," John Wiley & Sons Inc., New York, 1925.